Timber Bridges
Design, Construction, Inspection, and Maintenance

Michael A. Ritter, Structural Engineer
United States Department of Agriculture
Forest Service

Ritter, Michael A.   1990. Timber Bridges: Design, Construction, Inspection, and Maintenance.
Washington, DC: 944 p.

ACKNOWLEDGMENTS

The author acknowledges the following individuals, Agencies, and Associations for the substantial contributions they made to this **publication:**

For contributions to Chapter 1, Fong Ou, Ph.D., Civil Engineer, USDA Forest Service, Engineering Staff, Washington Office.

For contributions to Chapter 3, Jerry Winandy, Research Forest Products Technologist, USDA Forest Service, Forest Products Laboratory.

For contributions to Chapter 8, Terry Wipf, P.E., Ph.D., Associate Professor of Structural Engineering, Iowa State University, Ames, Iowa.

For administrative overview and support, Clyde Weller, Civil Engineer, USDA Forest Service, Engineering Staff, Washington Office. For consultation and assistance during preparation and review, USDA Forest Service Bridge Engineers, Steve Bunnell, Frank Muchmore, Sakee Poulakidas, Ron Schmidt, Merv Eriksson, and David Summy; Russ Moody and Alan Freas (retired) of the USDA Forest Service, Forest Products Laboratory; Dave Pollock of the National Forest Products Association; and Lorraine Krahn and James Wacker, former students at the University of Wisconsin at Madison. In addition, special thanks to Mary Jane Baggett and Jim Anderson for editorial consultation, JoAnn Benisch for graphics preparation and layout, and Stephen Schmieding and James Vargo for photographic support.

CONTENTS

CHAPTER 1
TIMBER AS A BRIDGE MATERIAL

CHAPTER 2
TYPES OF TIMBER BRIDGES

v

CHAPTER 3
PROPERTIES OF WOOD AND STRUCTURAL WOOD PRODUCTS

CHAPTER 4
PRESERVATION AND PROTECTION OF TIMBER BRIDGES

CHAPTER 5
BASIC TIMBER DESIGN CONCEPTS FOR BRIDGES

CHAPTER 6
LOADS AND FORCES ON TIMBER BRIDGES

CHAPTER 7
DESIGN OF BEAM SUPERSTRUCTURES

Part I: Glued-Laminated Timber (Glulam) Systems

Part II: Sawn Lumber Systems

CHAPTER 8
DESIGN OF LONGITUDINAL DECK SUPERSTRUCTURES

CHAPTER 9
DESIGN OF LONGITUDINAL STRESS-LAMINATED DECK SUPERSTRUCTURES

CHAPTER 10

RAIL SYSTEMS FOR TIMBER DECKS

CHAPTER 11

WEARING SURFACES FOR TIMBER DECKS

CHAPTER 12
TIMBER BRIDGE FABRICATION AND CONSTRUCTION

CHAPTER 13
BRIDGE INSPECTION FOR DECAY AND OTHER DETERIORATION

CHAPTER 14
MAINTENANCE AND REHABILITATION OF TIMBER BRIDGES

CHAPTER 15
BRIDGE MAINTENANCE, REHABILITATION, AND REPLACEMENT: CASE HISTORIES

CHAPTER 16
REFERENCE INFORMATION

CHAPTER 17
GLOSSARY OF TERMS

CHAPTER 18
TIMBER BRIDGE BIBLIOGRAPHY

CHAPTER 1

TIMBER AS A BRIDGE MATERIAL

1.1 INTRODUCTION

The age of wood spans human history. The stone, iron, and bronze ages were dramatic interims in human progress, but wood-a renewable resource-has always been at hand. As a building material, wood is abundant, versatile, and easily obtainable. Without it, civilization as we know it would have been impossible. One-third of the area of the United States is forest land. If scientifically managed and protected from natural disasters caused by fire, insects, and disease, forests will last forever. As older trees are harvested, they are replaced by young trees to replenish the wood supply for future generations. The cycle of regeneration, or sustained yield, can equal or surpass the volume being harvested.

Wood was probably the first material used by humans to construct a bridge. Although in the 20th century concrete and steel replaced wood as the major materials for bridge construction, wood is still widely used for short- and medium-span bridges. Of the bridges in the United States with spans longer than 20 feet, approximately 12 percent of them, or 71,200 bridges, are made of timber. In the USDA Forest Service alone, approximately 7,500 timber bridges are in use, and more are built each year. The railroads have more than 1,500 miles of timber bridges and trestles in service. In addition, timber bridges recently have attracted the attention of international organizations and foreign countries, including the United Nations, Canada, England, Japan, and Australia.

Timber's strength, light weight, and energy-absorbing properties furnish features desirable for bridge construction. Timber is capable of supporting short-term overloads without adverse effects. Contrary to popular belief, large wood members provide good fire resistance qualities that meet or exceed those of other materials in severe fire exposures. From an economic standpoint, wood is competitive with other materials on a first-cost basis and shows advantages when life cycle costs are compared. Timber bridges can be constructed in virtually any weather conditions, without detriment to the material. Wood is not damaged by continuous freezing and thawing and resists harmful effects of de-icing agents, which cause deterioration in other bridge materials. Timber bridges do not require special equipment for installation and can normally be constructed without highly skilled labor. They also present a natural and aesthetically pleasing appearance, particularly in natural surroundings.

The misconception that wood provides a short service life has plagued timber as a construction material. Although wood is susceptible to decay or insect attack under specific conditions, it is inherently a very durable material when protected from moisture. Many covered bridges built during the 19th century have lasted over 100 years because they were protected from direct exposure to the elements. In modern applications, it is seldom practical or economical to cover bridges; however, the use of wood preservatives has extended the life of wood used in exposed bridge applications. Using modern application techniques and preservative chemicals, wood can now be effectively protected from deterioration for periods of 50 years or longer. In addition, wood treated with preservatives requires little maintenance and no painting.

Another misconception about wood as a bridge material is that its use is limited to minor structures of no appreciable size. This belief is probably based on the fact that trees for commercial timber are limited in size and are normally harvested before they reach maximum size. Although tree diameter limits the size of sawn lumber, the advent of glued-laminated timber (glulam) some 40 years ago provided designers with several compensating alternatives. Glulam, which is the most widely used modern timber bridge material, is manufactured by bonding sawn lumber laminations together with waterproof structural adhesives. Thus, glulam members are virtually unlimited in depth, width, and length and can be manufactured in a wide range of shapes. Glulam provides higher design strengths than sawn lumber and provides better utilization of the available timber resource by permitting the manufacture of large wood structural elements from smaller lumber sizes. Technological advances in laminating over the past four decades have further increased the suitability and performance of wood for modern highway bridge applications.

1.2 HISTORICAL DEVELOPMENT OF TIMBER BRIDGES

The history and development of timber bridges can be divided into four periods: (1) prehistory through the Middle Ages (to 1000 A.D.), (2) the Middle Ages through the 18th century (1000-1800), (3) the 19th century (1800-1900), and (4) the 20th century (1900 to present). The definition of these periods is based on the sophistication of timber bridge design and construction, and the periods closely parallel human cultural and industrial evolution. From prehistoric times through the Middle Ages, our ancestors adapted available materials, such as logs and vines, to span crossings. From the end of the Middle Ages through the 18th century, scientific knowledge developed and influenced the design and construction of timber bridges. In the 19th century, the sophistication and use of timber bridges increased in response to the growing need for public works and transportation systems associated with the industrial revolution. With the

20th century came major technological advances in wood design, laminating, and preservative treatments.

PREHISTORY THROUGH THE MIDDLE AGES

In prehistorical times, bridges were built using adaptable materials within the environment. Where trees abounded, the first timber bridge was probably a tree that fell across a waterway. The first humanmade timber bridge is assumed to have been built by a Neolithic human who felled a tree across a chasm with a hand-fashioned stone axe circa 15,000 B.C.[10] Ideas for prototype suspension bridges probably came from hanging vines or stems. In subtropical parts of central Asia, palms with lengthy stems were used for constructing suspension bridges. In areas where plants with woody stems grew, native residents could build rope bridges constructed of twisted vines. Bridges of this type ranged in complexity from two or three stretched ropes to more sophisticated configurations employing several ropes to support a floor of tree limbs and branches (Figure 1-1).

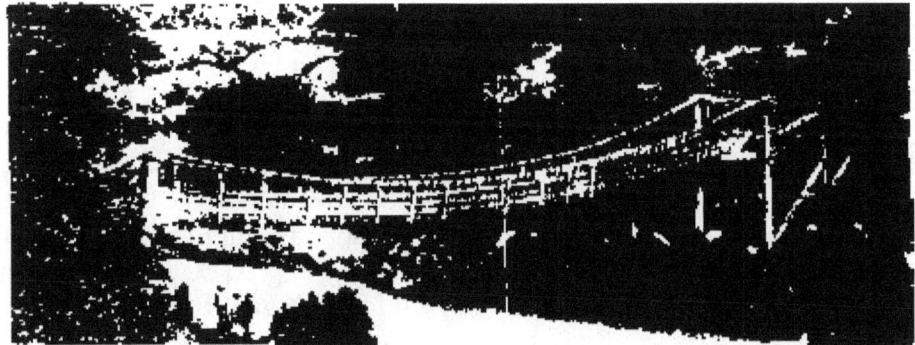

Figure 1-1.--Early highway type of rope bridge. This example is from the island of Java and has an apparent span of approximately 100 feet (photo courtesy of the American Society of Civil Engineers: © 1976. Used by permission).

Many timber bridges were probably built in the last 800 years B.C. by the Persians, Babylonians, Greeks, Romans, and Chinese, although there is little available literature describing specific designs. One of the oldest bridges on record was 35 feet wide and 600 feet long, built in 783 B.C. over the Euphrates River in Babylon.[11] It is theorized that most prehistoric timber bridges in remote areas remained virtually unchanged in design at least to the period of Julius Caesar (100-44 B.C.). One such prehistoric bridge, used by the Gauls in the hills of Savor in Italy, was viewed by Julius Caesar, who described it as follows.[12]

> It is a timber bridge or empilage, piled together rudely, not con-
> structed by art. It needs no carpentry.... On each bank of the
> stream a rough foundation of water-worn boulders was laid, about
> fifteen feet square; upon this a criss-cross of the tree trunks was
> built so that the logs in the direction of travel, in the alternative

layers, were made to jut out farther and farther over the water, narrowing the gap to be bridged later by a few logs serving as beams.

A particular Roman bridge, known as Caesar's Bridge, was built about 2,000 years ago to carry the Roman army into Germany. This bridge was documented by the Venetian architect Palladio (1518-1580), who made an exhaustive study of the remains of the Roman empire. In his treatise *Architecture,* Palladio describes the bridge and renders a drawing of his interpretation of its configuration (Figure 1-2). The structure consisted of a series of beams and inclined struts that fit together in notches so that the bridge could be erected and removed quickly. The imposed weight of the structure and of passing loads served to make the joints tighter. It is rather doubtful, however, that the actual structure utilized timbers as square and smooth as Palladio's drawing indicates.

Approximately one century after Caesar's Bridge (104 A.D.), Roman history mentions one of the most noteworthy works ever undertaken by the Romans. Trajan's Bridge across the Danube River reportedly rested on 20 timber piers, 150 feet high and 170 feet apart. The bridge spans between the piers were circular timber arches. During the same period, evidence shows that builders were concerned with extending the life of wood in structures. A book by a Roman architect covered various means of preserving trees after they were cut, gave remedies to protect against disorders, and included recommendations that (1) fresh cut timber be covered with ox dung to protect it from rapid drying, (2) wood be anointed with Lees of Oil to preserve it from all manner of worms, and (3) pitch was the best defense against deterioration caused by water.[10]

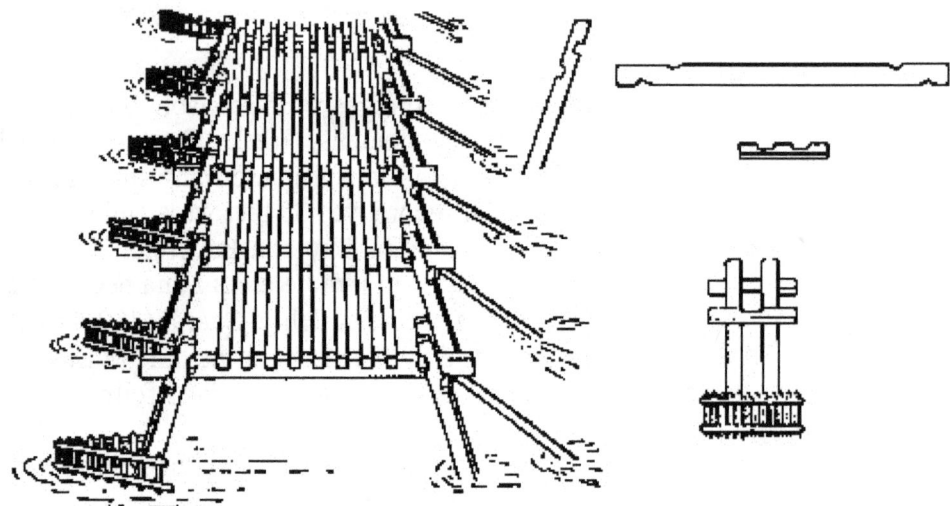

Figure 1-2.–Caesar's Bridge according to Palladio (photo courtesy of the American Society of Civil Engineers: © 1976. Used by permission.)

During the period from the Middle Ages to the end of the 15th century, literature documenting timber bridges is limited and incomplete. No significant developments are found until the 16th century, when Palladio composed *Architecture* around 1550. In his work, Palladio provides several timber bridge designs, or inventions as he called them, including a timber arch and the first illustration of a framed truss (Figure 1-3). The arches were apparently capable of spans of approximately 100 feet, while the framed truss was used for spans in the range of 50 to 60 feet. Although they were meaningful contributions to timber bridge evolution, Palladio's bridges attracted little attention, and there was no further development of timber bridges in Europe until the middle of the 18th century.

The 18th century was a period of rapid progress in which attention focused on the development of public works projects, including bridges. It was the period when civil engineering became recognized as a profession. In Europe, the French excelled in engineering developments and constructed numerous timber bridges in spans ranging from 65 to 150 feet. Most French designs were characterized by level floors and flat arches and were built from layers of planks that were clamped together. Covered or roofed bridges were not a common feature in European construction, although several such bridges were constructed by the Grubenmann brothers in Switzerland. The most notable of these bridges was the Schaffhausen Bridge constructed across the Rhine River in 1758 (Figure 1-4). This bridge was built in two spans (171 feet and 193 feet) and was top heavy with a needless amount of timber in the roof system.[12] It was destroyed by the French in 1799. Several other notable timber bridges were constructed in Europe during the 18th century, including a single-span crossing of 390 feet at Wittingen, Germany. However, the most significant timber bridge progress in the latter part of the century was made in the United States and Russia.[17]

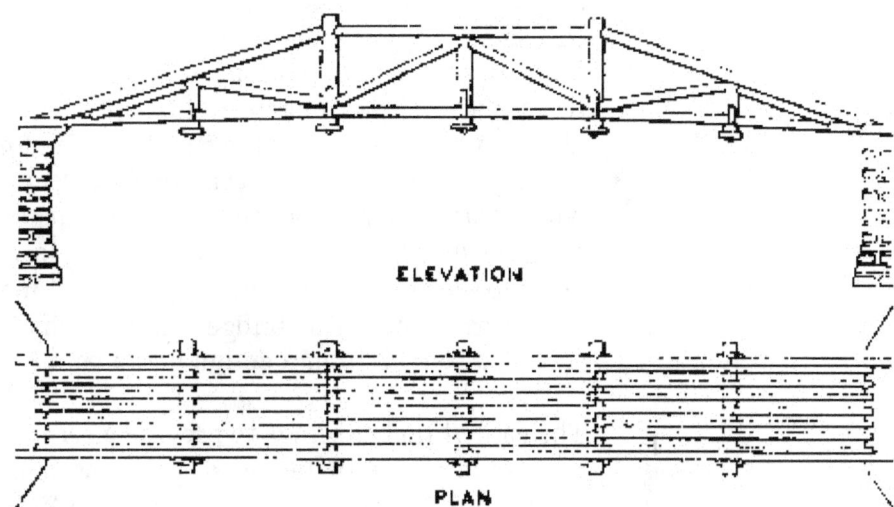

ELEVATION

PLAN

Figure 1-3.— Palladio's design for a framed truss, dated about 1550 (photo courtesy of the American Society of Civil Engineers; © 1976. Used by permission).

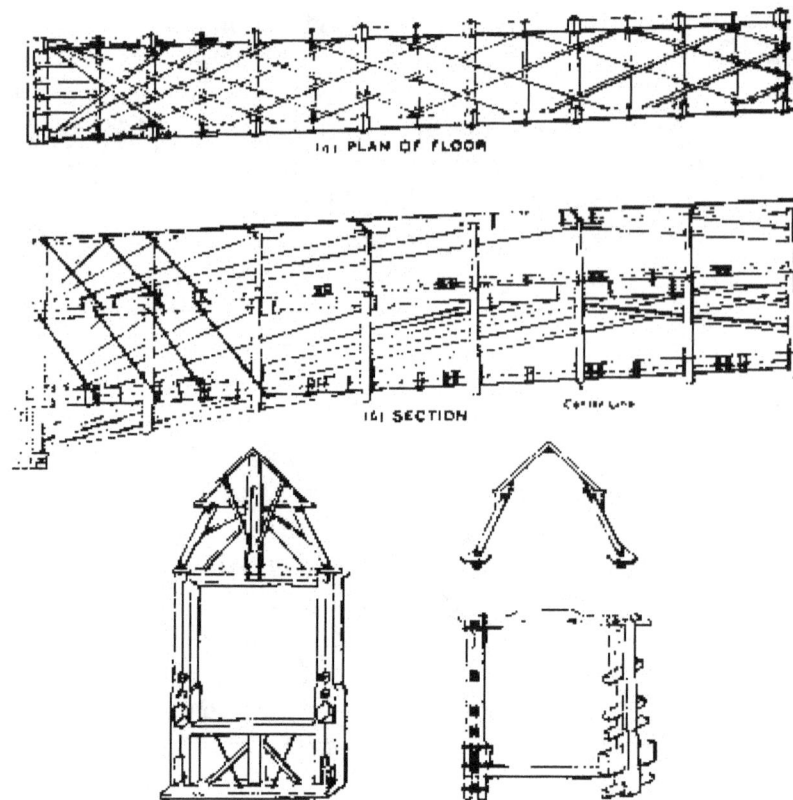

(a) PLAN OF FLOOR

(b) SECTION

Figure 1-4.-The Schaffhausen Bridge constructed in 1758 over the Rhine River in Switzerland (photo courtesy of the American Society of Civil Engineers; © 1976. Used by permission).

In the United States, most timber bridges built before the 18th century were pioneer bridges with short spans. During the mid-18th century, longer spans were made with trestle bridges consisting of timber beams placed between closely spaced pile piers. The first may have been constructed in 1761 over the York River at York, Maine, by Samuel Sewall. This bridge was 270 feet long, 25 feet wide, and supported on four-pile bents spaced approximately 19 feet apart. It also included a draw span to allow boat passage under the structure. The timber bents, including the pile cap and bracing, were completely assembled and driven as a unit, which was quite an engineering achievement in itself. Pile driving was accomplished by hoisting the butt ends of large logs (with their tips fastened to the previously driven bent) and letting them fall with considerable impact on the cap. This bridge is noteworthy because it is the first on record to be built from a design based on a survey of the site.

The earliest timber bridge to provide clear spans greater than could be negotiated with a single log or beam was completed by Colonel Enoch Hale in 1785, 2 years after the end of the Revolutionary War. It was constructed over the Connecticut River at Bellows Falls, Vermont, and was a 365-foot-long, two-span structure with center support provided by a

1-6

natural rock pier (Figure 1-5). This was the first bridge over the Connecticut River at any point, and residents reportedly looked on it as a foolhardy experiment.[12] After it was constructed, the bridge was widely noted and considered a remarkable feat of construction. It stood until about 1840.[10]

One of the most ingenious and famous bridge builders of the late 18th century and early 19th century was Timothy Palmer (1751-1821), a distinguished civil engineer from Newburyport, Massachusetts, In 1794, Palmer built the Piscataqua Bridge, 7 miles north of Portsmouth, New Hampshire. The bridge was 2,362 feet long and 38 feet wide. Approach spans were pile trestles that led to three arched trusses, the largest of which had a span of 244 feet. This bridge was considered a wonder of its time and became known as the Great Arch. The arch ribs were made from crooked timbers so that the grain was nearly in the direction of the curves.[7] In 1794 Palmer built a similar bridge at Haverhill, Massachusetts. It consisted of three arches, each 180 feet long, and included a short 30-foot draw span on one end (Figure 1-6). Ten years later, from 1804 to 1806, Palmer built the first American covered bridge over the Schuylkill River in Philadelphia.[28] This was a continuous three-span arch truss consisting of two 150-foot spans and one 195-foot span. It is recorded that the city bridge committee insisted that the heavy timbers be covered with a roof and siding to preserve and protect the structure from weathering. The bridge thus became known as the Permanent Bridge.

19TH CENTURY

With the 19th century came a tremendous demand for bridges in the United States both for highway use and, beginning in about 1830, to meet

Figure 1-5. Hale's bridge at Bellows Falls, Vermont, built about 1785. This is a sketch from an oil painting of the locality, showing the original structure, or a successor to it. The date of the painting is unknown (photo courtesy of the American Society of Civil Engineers; © 1976. Used by permission).

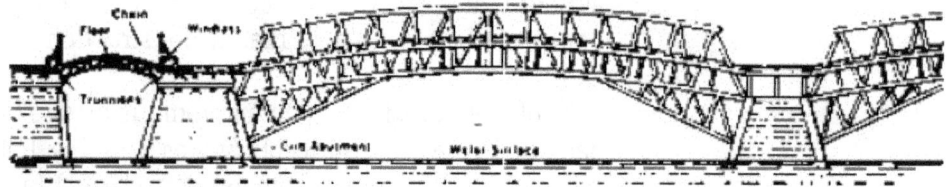

Figure 1-6.- Palmer's arch bridge built at Haverhill, Massachusetts, in 1794 (photo courtesy of the American Society of Civil Engineers; © 1976. Used by permission).

the demands of the railroad boom. During this period, truss and arch bridges became predominant in timber bridge design. Although both arches and trusses were adapted by Palmer in the late 18th century, large-scale application of these structures did not take place until the turn of the 19th century. In the early 1800's, bridge builders strived not only to fulfill design requirements, but also to make their designs bolder and superior to any before. The U.S. Patent Office issued 51 patents for timber bridges between 1797 and 1860.[28]Insistence on careful protection from weather for most of these bridges inaugurated the distinctly American covered bridge (Figure 1-7). An estimated 10,000 covered bridges were built in the United States between 1805 and 1885. Wernwag, Burr, Town, and Long were the four men who led the pioneering efforts during the first four decades of this period. A brief summary of some of the major American bridge accomplishments from 1785 to 1868 is shown in Table 1-1.

Figure 1-7.- Typical example of an American covered bridge. An estimated 10,000 covered bridges were built in the United States between 1805 and 1885.

Table 1-1.-Some major American timber bridges built between 1785 and 1868.

Enoch Hale's braced-stringer bridge, at Bellows Falls, VT 1785

Timothy Palmer's Essex Merrimac Bridge 1792

Timothy Palmer's Piscataqua and Haverhill Bridges 1794

Timothy Palmer's Georgetown, Washington, DC, Bridge 1796

Timothy Palmer's Permanent Bridge, at Philadelphia, PA 1804-06

Timothy Palmer's Easton, PA, Bridge 1805-06

Graves' (second) Connecticut River Bridge at Hanover, NH 1796

Windsor, VT, Bridge, contemporary with the Graves' Bridge 1796

Theodore Burr's Waterford, NY, Bridge 1804

Theodore Burr's Trenton, NJ, Bridge 1806

Theodore Burr's Mohawk River Bridge 1808

Theodore Burr's Harrisburg, PA, Bridge 1816

Lewis Wernwag's Colossus Bridge at Philadelphia, PA 1812

Lewis Wernwag's New Hope, PA, Bridge 1813-14

Lewis Wernwag's Economy Bridge 1810

Earliest lattice-truss bridge of which there is a record 1813

Ithiel Town's plank-lattice truss, patented 1820

Truss of Stephen H. Long, patented 1830

Ithiel Town's timber-lattice truss, patented 1839

Wernwag's Cheat River Bridge, WV 1834

Wernwag's Camp Nelson Bridge, near Lexington, KY (Standing in 1933 after 95 years.
    In both these bridges the arch is on the center line of the truss.) 1838

The Ramp Creek Bridge, IN, Burr trusses (Renovated and in service, 1933, after 96 years.)...1837

The Raccoon Creek Bridge, IN, Burr trusses (Still in use, 1933, after 95 years.).....................1838

Brunel's experiments with preservatives in England 1835

Wooden lattice bridges on British railways after Ithiel Town's visit about 1840 (before 1846) 1846

William Howe's patent for the Howe truss 1840

William Howe's Connecticut River Bridge, at Springfield, MA 1840

The Tucker Bridge, at Bellows Falls, VT, plank lattice. 1840

The trusses of Thomas W. and Caleb Pratt, patented 1844

Typical Burr truss railroad bridge (framed with white pine),
    at White River Junction, VT 1848

Howe truss bridge with double arches, at Bellows Falls, VT 1850

The unclassified truss of Nicholas Powers, North Blenheim, NY 1855

First bridge across the Mississippi River, five spans, Howe trusses with double arches,
    at Rock Island, IL 1853-56

Second historic bridge at Rock Island, IL, Howe trusses with curved upper chords and
    no arches, some time before 1868

The Ledyard Bridge, at Hanover, NH, timber lattice 1859

Howe truss bridge with double arches (12 spans) at Havre de Grace, MD 1862-66

Adapted from Fletcher and Snow.[12] © 1976. Used by permission.

In his bridge building career of 27 years, Lewis Wernwag (1770-1843) built a total of 29 timber bridges in the States of Pennsylvania, Maryland, Virginia, Kentucky, Ohio, and Delaware. His most noteworthy accomplishment was the Colossus Bridge built in 1812 over the Schuylkill River in Pennsylvania (Figure 1-8). This bridge was composed of five parallel arched trusses, each with a rise of 20 feet, that spanned a clear distance of 340 feet. The design, which was not patented until 1829, used iron tension rods, which also served as points of adjustment for joints in each panel. Other major bridges built by Wernwag include the Economy Bridge and the New Hope Bridge. The Economy Bridge was a timber cantilever structure built in 1810 across the Nashammony River in Pennsylvania. It incorporated provisions for tipping the center panel to allow passage of masted vessels and, according to Wernwag, could be used to advantage for spans up to 150 feet. The New Hope Bridge was built during 1813-14 over the Delaware River, at New Hope, Pennsylvania (Figure 1-9). It consisted of a parallel-chord truss arrangement with six arch spans of 175 feet. It was Wernwag's practice to saw all timbers through the heart in order to detect unsound wood and allow seasoning. He used no timbers greater than 6 inches thick and separated all arch timbers with cast iron washers to allow free air circulation.7

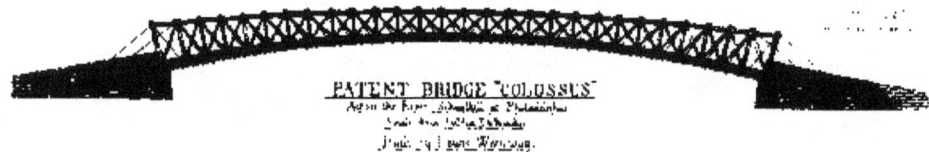

Figure 1-8.- Wernwag's Colossus bridge built over the Schuylkill River at Upper Darby, Pennsylvania, in 1812 (photo courtesy of the American Society of Civil Engineers; © 1976. Used by permission).

Figure 1-9.- Wernwag's New Hope bridge built over the Delaware River at New Hope, Pennsylvania, in 1814 (photo courtesy of the American Society of Civil Engineers; © 1976. Used by permission).

Theodore Burr is credited with building many famous timber bridges in the first two decades of the 19th century. His designs were based primarily on the combination of parallel-chord trusses with one or more reinforcing arches projecting from the supports, below the point of truss bearing. The first was a 176-foot span crossing the Hudson River between Waterford and Lansingburgh, New York, in 1804 (Figure 1-10). In 1817, Burr was granted a patent based on this Waterford design, which became widely used and known as the Burr truss. Another good example of a Burr bridge was built over the White River at White River Junction, Vermont, in 1848 (Figure 1-11). Constructed as a railroad bridge, it was as strong and serviceable after 54 years of service as when it was built.[12] Although it was capable of much longer service, it was removed in 1890 and replaced with an iron bridge. Hundreds of highway bridges, based to some degree on the Burr principle, were built in various parts of the East, Midwest, and New England States. Most were over 50 feet in span and were constructed as covered bridges of naturally durable white pine. Their longevity has been remarkable, with many providing service in excess of 100 years.

Ithiel Town (1784-1884) was a New Haven architect who recognized the need for a covered bridge truss that could be built at a low cost by good carpenters. In 1820, he was granted a patent on a plank-lattice bridge truss design that represented a first step toward modern truss form. Town's bridge included a web of light planks, 2 to 4 inches thick and 8 to 10 inches wide, that were criss-crossed at a 45 to 60-degree angle (Figure 1-12). The webs were fastened together at their intersections with wooden pins (trunnels). Town lattice trusses could be built for spans up to 220 feet, were lightweight and inexpensive, and could be assembled in a few days. They generally used sawn lumber with uniform sections throughout. Although this feature is often criticized as being wasteful of material, such waste was more than offset by the simplicity of framing and construction. A great number of covered Town lattice trusses were built for highway and railroad traffic in many parts of the United States where wood was abundant (Figure 1-13). Town was a promoter and salesman

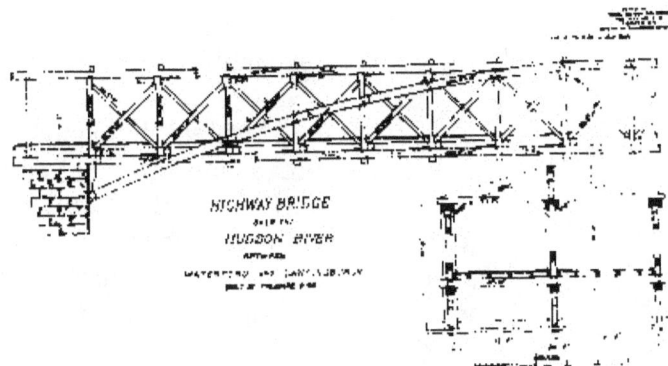

Figure 1-10.- Burr bridge built in 1804 over the Hudson River between Waterford and Lansingburgh, New York (photo courtesy of the American Society of Civil Engineers; © 1976. Used by permission).

Figure 1-11.- Burr bridge built in 1848 over the White River at White River Junction, Vermont. This photo was taken as the bridge was being removed in 1890, to be replaced by an iron bridge (photo courtesy of the American Society of Civil Engineers; © 1976. Used by permission).

rather than a builder.[12] He sold rights to build his design and published advertising pamphlets.

Many bridges built during the 19th century were designed using a trial-and-fail method by local carpenters. In 1910, more than 100 bridges of this type were in existence on the Boston and Maine Railroad system. Although built without any knowledge of stresses and strains, many of these bridges provided satisfactory service for the trains using them. Notwithstanding several common defects resulting from a lack of scientific design, it is remarkable how well the trial-and-fail method served.

In 1830, Brevet-Lieutenant Colonel Stephen H. Long patented a parallel-chord truss bridge that was modified in 1836 and again in 1839. The truss was of the panel type with crossed timbers between wooden posts. His 1830 patent drawing also included braces extending to the first and second panel points for an assisted truss arrangement (Figure 1-14). Connections were made by framing parts together or by using wooden keys or treenails (treenails are wooden pins, pegs, or spikes driven in holes to fasten lumber together). Although Long's bridges did not become widely popular, many highway and railroad bridges that were hybrids of his design were built by local carpenters. Most of them were for clear spans well over 150 feet.

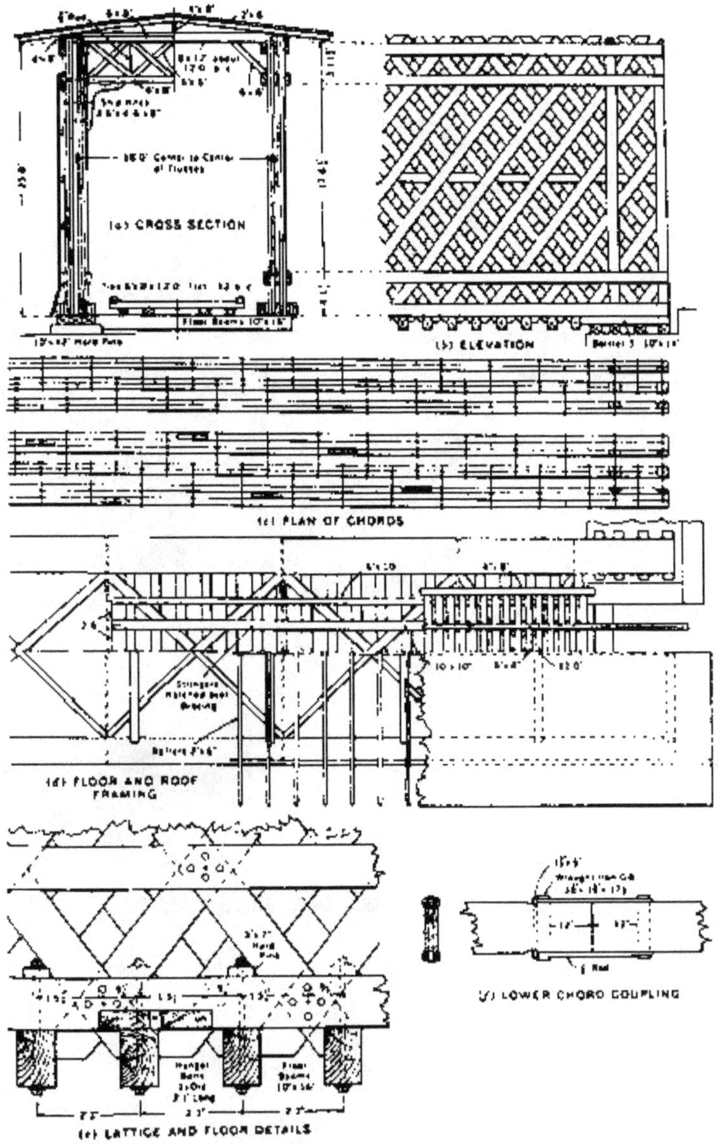

Figure 1-12.-Town's lattice truss patented in 1820 (photo courtesy of the American Society of Civil Engineers; © 1976. Used by permission).

The 1840's marked a turning point for timber bridge development. Until this time, most timber bridges, including those of Wernwag, Burr, Town, and Long, were built almost totally from wood. Iron components, when used, were limited to small fasteners or other hardware that could be forged by blacksmiths. From 1830, rapid railroad expansion provided great motivation for bridge development, and cast iron bridges were introduced. Although wood continued to be used as a primary bridge material, iron became a structural component for timber bridges, and the so-called combination bridges were born. It is obvious that until 1840, the development of timber bridges was empirical. The concepts of earlier designs were often used as a basis for developing newer bridge types.

Figure 1-13.-Typical Town lattice truss covered bridge.

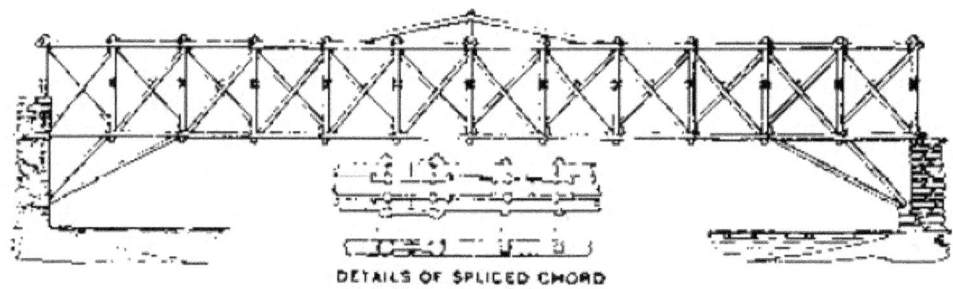

DETAILS OF SPLICED CHORD

Figure 1-14.--Drawing of Long's truss bridge as patented in 1830 (photo courtesy of the American Society of Civil Engineers; © 1976. Used by permission).

Although many pioneer builders may have considered the use of mathematical rules when determining structural elements for their bridges, no substantiating records of this exist.

After the Long trusses, no significant timber bridge developments occurred until William Howe of Massachusetts patented his bridge in 1840. The Howe truss was a parallel-chord truss design that used two systems of web members (Figure 1-15). The chords and diagonal braces were made of timber and the vertical web-tension members were made of round cast-iron rods. This was the first design to use iron as an essential structural element of a timber truss system. Howe's patent was also the first to include a complete stress analysis of the design by mathematical practices then in use. In 1840, Howe, in company with Amasa Stone (who bought the Howe patent in 1841), built the great bridge over the Connecticut River at Springfield, Massachusetts. This bridge was constructed to carry the new Western Railroad and consisted of seven spans, each measuring 190 feet, measured from the center of one pier to the center of the other pier (Figure 1-16). After a number of years, several modifications were made to the original Howe design to more accurately reflect the actual stresses the members sustained. The design continued to be widely used for railroads and highways and became the most popular truss for the last half of the 19th century.

In 1844, shortly after the Howe truss became popular, Thomas W. Pratt and Caleb Pratt patented their truss design. The Pratt truss was a panel type parallel-chord truss that used vertical timber posts in compression and crossed iron diagonals in tension, just the reverse of the Howe design (Figure 1-17). The advantage of the Pratt truss was that it used timber web members in the simplest and most efficient manner, by confining them to the verticals. The disadvantages were that the truss required a large quantity of expensive material and needed awkward angle blocks for the diagonals. Although numerous timber Pratt trusses were built, the design

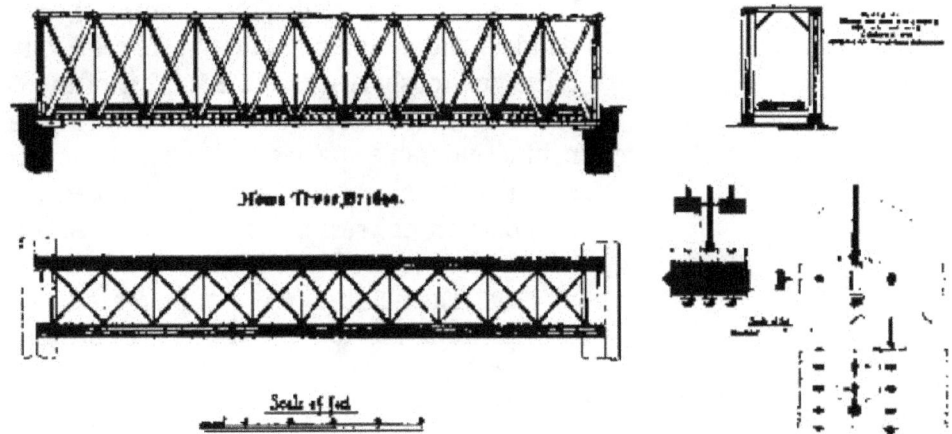

Figure 1-15.- Howe truss bridge patented in 1840 (photo courtesy of the American Society of Civil Engineers; © 1976. Used by permission).

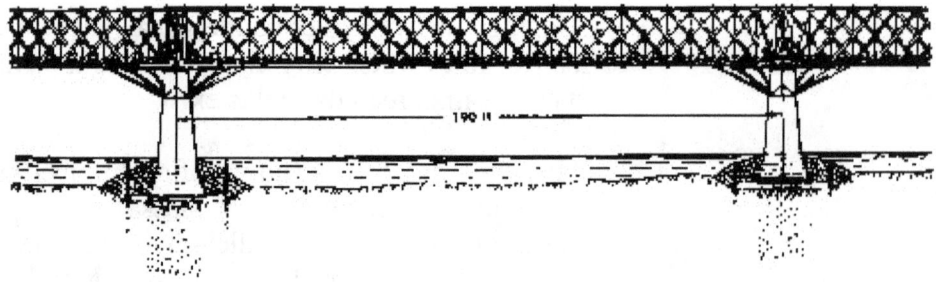

Figure 1-16.- Howe truss built over the Connecticut River at Springfield, Massachusetts, in 1840 (photo courtesy of the American Society of Civil Engineers; © 1976. Used by permission).

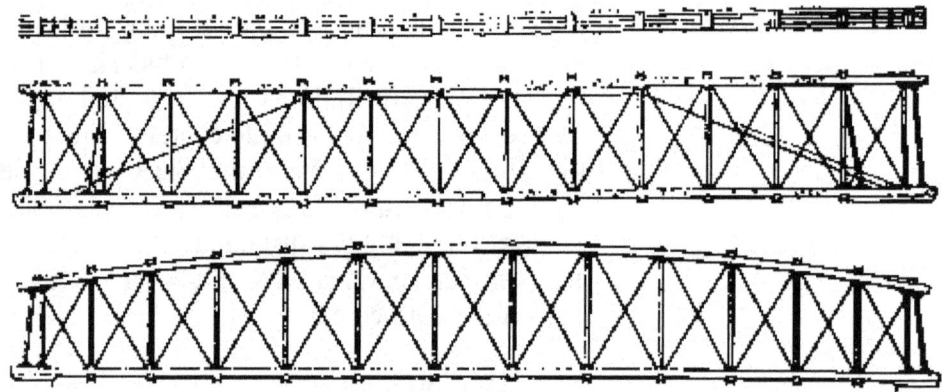

Figure 1-17.- Pratt truss as patented in 1844 (photo courtesy of the American Society of Civil Engineers; © 1976. Used by permission).

was not well suited for the joint use of wood and iron, and it never achieved the popularity of the Howe truss. However, it did become a favored form for constructing totally iron bridges, and thus was a major step in the development of American bridges.

For the remainder of the 19th century, there were other timber bridge builders and designs, but they were relatively minor in comparison to those previously discussed. For most of the century, bridges were constructed of untreated wood, and builders relied mainly on the use of naturally durable species and covers to provide long service lives. The first major development that improved timber bridge performance was the introduction of pressure preservative treatments. The fast pressure creosoting plant in the United States was built in Somerset, Massachusetts, in 1865. The number of plants increased steadily to 70 by 1910.[18] Thus, by the end of the 19th century timber bridges could be built with preservative-treated wood without the covers that had been traditionally used for protection.

In the latter half of the 1800's, iron bridges became increasingly popular and began to compete strongly with timber. In 1859, Howard Carroll built the first all-wrought-iron railroad bridge. In the last decade of the 19th century, steel took the place of iron as the most popular bridge material. Although timber continued to be used for bridges, its use began to decline as new materials were introduced.

20TH CENTURY

Technology in the steel industry developed rapidly in the early part of the 20th century, leading to a more expanded and economical use of steel as a bridge material. Until about 1890, timber lattice bridges could be built with spruce lumber (then costing about $18 per thousand board feet) for one-half the cost of iron bridges.[12] Twenty years later (1910), steel bridges could be built as economically as those of wood. By the mid-1930's, steel was less expensive than wood on a first-cost basis and took the lead as the primary bridge material. Also during the early 20th century, the popularity of reinforced concrete increased and became a primary material for bridge decks.

During this rapid technological development of other bridge materials, progress in timber bridge development slowed. Although there was substantial progress in the areas of wood fasteners and preservative treatments, it was not until the mid-1940's that the biggest single advancement in timber bridges occurred with the introduction of glulam as a bridge material. In the 1960's and 1970's, glulam continued to develop and became the primary material for timber bridge construction. In the 1980's, new glulam bridge designs have evolved, and the innovative concept of stress-laminated lumber has been introduced. As a result, there is a renewed interest in timber as a bridge material and a corresponding increase in the number of timber bridges constructed each year. A more complete description of the types of timber bridges currently in use in the United States is given in Chapter 2.

1.3 THE FUTURE OF TIMBER AS A BRIDGE MATERIAL

Deterioration of the Nation's infrastructure has been well publicized in recent years. Despite this recognition, bridge deterioration continues at an alarming rate. Over the next two decades, the role of timber in bridge applications has the potential to increase significantly, not only in the construction of new timber bridges, but also in the rehabilitation of existing structures constructed of timber, steel, and concrete. According to the 1987 Federal Highway Administration's national bridge inventory,[26] there are 575,607 bridges in the United States with spans of 20 feet or more. Among them, 304,307 are off the Federal aid system on city, county, and township roads. Of these bridges, 95,241 or 33.4 percent are classified as structurally deficient, and 71,542 or 27.4 percent are classified as functionally obsolete. A 1987 summary of substandard bridges by State is shown in Table 1-2.

Table 1-2.– 1987 Summary of substandard bridges by State.

State	Total Interstate & State Bridges	Total Substandard	Total City/County/ Township Bridges	Total Substandard	Total All Bridges	Combined Total Substandard
Alabama	5.373	2.014(37 5%)	10.090	6.372(63.2%)	15,463	8.386(54.2%)
Alaska	704	75(10.6%)	80	25(31.2%)	784	100(12.7%)
Arizona	3,936	117(2.9%)	1.691	172(10.3%)	5,627	289(5 1%)
Arkansas	6.644	1.786(26.9%)	6.307	4.175(66.2%)	12,951	5.961(46 0%)
California	11,848	931(8.0%)	11,661	4.238(36 0%)	23,509	5.169(22 0%)
Colorado	3,577	470(13.1%)	4,040	2.194(54.3%)	7,617	2.664(35 0%)
Connecticut	2.565	548(21.0%)	1,208	422(35.0%)	3,773	970(26 0%)
Washington, D C	202	60(30.0%)	13	5(38.5%)	220	65(29 5%)
Delaware	716	180(25.1%)	7	2(28.6%)	723	182(25 2%)
Florida	5.618	669(11 9%)	4.462	1,477(33.1%)	10.080	2,146(21 3%)
Georgia	6.308	1.275(20.2%)	8.242	3,262(39.6%)	14,550	4,537(31 2%)
Hawaii	691	148(21.4%)	415	131(31.5%)	1,106	279(25 2%)
Idaho	1,347	308(22.9%)	2,459	1,044(42 5%)	3,806	1,352(35 5%)
Illinois	8.110	1,679(20.7%)	17,255	5,846(33 9%)	25,365	7.525(29 7%)
Indiana	5.279	2,819(53 4%)	12,408	7,392(59.6%)	17,687	10.211(57 7%)
Iowa	3.859	912(23 6%)	22,462	12,207(54.3%)	26,321	13.119(49 8%)
Kansas	5,096	1,356(26 6%)	20,781	12,187(58.6%)	25,877	13.543(52 3%)
Kentucky	8,273	4,250(51.4%)	4,464	3,773(84.6%)	12,737	8,028(63 0%)
Louisiana	7,720	2,323(30.1%)	6,972	4,487(64.3%)	14,692	6,810(46.3%)
Maine	1,900	400(21.1%)	488	373(76.4%)	2,388	773(32 4%)
Maryland	2.529	1,036(40.9%)	2,509	1,087(43.3%)	5,038	2,123(42 1%)
Massachusetts	3,320	560(16.9%)	1,685	675(40.0%)	5,005	1,235(24 7%)
Michigan	4.133	320(7.7%)	6,452	2.813(43.9%)	10,585	3.133(29 6%)
Minnesota	2,860	420(14.7%)	10,012	3,086(30.8%)	12,872	3,506(27 2%)
Mississippi	4,715	2,036(43.2%)	12,420	8,498(68.4%)	17,135	10,534(61.5%)
Missouri	9.351	3,731(39.9%)	14,308	12,269(85.7%)	23,659	16,000(67.6%)
Montana	2,660	1,232(46.3%)	2,063	1,349(65.3%)	4,723	2,581(54.6%)
Nebraska	3.085	743(24 1%)	12,918	8,838(68 4%)	16,003	9,581(59 9%)
Nevada	898	13(1 4%)	230	31(13.4%)	1,128	44(3 9%)
New Hampshire	1,453	372(25 6%)	1,062	821(77.3%)	2,515	1,193(47 4%)
New Jersey	2,291	492(21 5%)	3,725	1,260(33.8%)	6,016	1.752(29 1%)
New Mexico	3,046	413(13.6%)	479	220(45.9%)	3,525	633(17 9%)
New York	7,264	2,687(36.9%)	12,290	6,330(51.5%)	19,554	9,017(46.1%)
North Carolina	16,831	8,992(53.4%)	556	272(48.9%)	17,387	9,264(53 3%)
North Dakota	1,432	287(20.0%)	4,129	2,670(64.7%)	5,561	2,892(52 0%)
Ohio	11,364	1,918(16.9%)	18,780	4,462(23.8%)	30,144	6,380(21.2%)
Oklahoma	6,729	2,554(37.9%)	15,936	10,517(65.9%)	22,665	13,071(57 7%)
Oregon	2,550	240(9.4%)	4,030	782(19.4%)	6,580	1,022(15 5%)
Pennsylvania	15,812	5,315(33.6%)	6,618	2,903(43.9%)	22,430	8,218(36.6%)
Rhode Island	530	68(12 8%)	*195	*72(36.9%)	*725	*140(19.3%)
South Carolina	7,969	1,197(15.0%)	989	630(63.7%)	8,958	1,827(20.4%)
South Dakota	1,759	179(9 7%)	5,094	3,035(59.6%)	6,853	3,214(46.9%)
Tennessee	6,907	2,695(39.0%)	11,456	6,263(55.0%)	18,363	8,958(49.0%)
Texas	31,243	5,841(18.7%)	15,069	10,486(69.6%)	46,312	16,327(35 3%)
Utah	1,603	61(3.8%)	960	251(26 1%)	2,563	312(12 2%)
Vermont	1,336	391(29.2%)	1,401	858(61.2%)	2,737	1,249(45 6%)
Virginia	11,461	3,703(32.3%)	893	192(21.5%)	12,354	3,895(31.5%)
Washington	3,034	1,107(36.5%)	4,264	857(20.1%)	7,298	1,964(26 9%)
West Virginia	6,869	4,292(62.5%)	177	103(58.2%)	7,046	4,395(62.4%)
Wisconsin	4,431	1,862(42.0%)	8,477	4,213(49.7%)	12,908	6,075(47.1%)
Wyoming	1,894	102(5.4%)	933	574(61.5%)	2,827	676(23.9%)
Totals	271.125	77,179(28.4%)	315,555	166,201(52.7%)	586,680	243,380(41.5%)

* Includes local railroad bridges.

Numbers vary slightly from those published by the Federal Highway Administration[26] due to differences in survey techniques.

From an exclusive survey conducted by Better Roads Magazine[2]; © 1987. Used by permission.

Over the past four decades, properly designed and preservative-treated timber bridges have demonstrated good performance with long service lives, given proper maintenance. Over the same period, timber has continued to be economically competitive with other bridge materials, both on a first-cost basis and a life-cycle basis. Despite these beneficial attributes, there has been a marked hesitation on the part of bridge designers to use timber, although this has been changing since the 1970's. Perhaps the biggest obstacle to the acceptance and use of timber has been a persistent lack of understanding related to design and performance of the material. Although well educated about other materials, such as steel and concrete, most bridge designers lack the same level of knowledge about wood. The following perspective on why wood has not received the same recognition as other materials was presented by Ken Johnson.[27]

> The practice of engineering, as it evolved over the years, has been shaped by the persuasive efforts of the steel and cement industries. This persuasion has been beneficial, in some ways, in that it produced and distributed good technical information about the design and the use of their respective products. In fact, many engineering schools use industry produced textbooks in their curriculum. That advantage has led to an increase in the reliance, use, prestige and position of those materials and to a corresponding decline, in the same factors, for other construction materials from those industries that have not provided the same level of technical information.

> The timber industry is one of those industries that has not made a substantial unified effort to generate and distribute technical information. This has been interpreted by some engineers as a reflection on the suitability of the material itself, and not as an indictment of the industry for failing to provide the information. The reason the timber industry has not met the challenge is quite obvious once one looks at the respective industries.

> The methods by which basic materials are produced provide the answers as to why steel and cement provide technical information and why timber has not. The basic difference between steel/cement and timber is the ability of steel/cement to form single industry-wide institutions to do the necessary research and to publish the results. This is possible because of the relatively small number of companies actually producing the product. The production of only three steel companies account for about ninety percent of the steel produced in the United States. The number of companies producing cement is somewhat larger, but still relatively small when compared to the timber industry.

> The timber industry, by contrast, consists of a multiplicity of sawmills, both large and small, resource based companies and many other independent operations such as treating plants. The

production is then further diversified by different species. Each of these entities is fiercely independent. The task to organize all of these independent operations is something akin to trying to organize all the farmers. However, the fact that the farmers do not have a single voice does not make their choice beef and Durham wheat less acceptable as steak and bread.

Given the potential market and the economic and performance advantages of wood, the future success of timber in bridge applications depends primarily on (1) the education of engineers on the basic design and performance characteristics of timber, (2) continued coordinated research to develop new bridge systems and improve existing ones, and (3) development of an effective technology transfer system to disseminate current design, construction, and maintenance information to users. Over the past several years, the Forest Service, in cooperation with the timber industry and other public and private agencies, established an Industry-Federal Government Cooperative Program on timber bridge technology to meet needs in these three areas. [25] One of the efforts of this program is to prepare and distribute information that provides engineers and educators with state-of-the-art information on timber bridges. This manual is one step in providing such information.

1.4 SELECTED REFERENCES

1. Archibald, R. 1952. A survey of timber highway bridges in the United States. Civil Engineering. September: 171-176.
2. Better Roads. 1987. Exclusive bridge inventory update. Better Roads 57(11): 30.
3. Bohannan, B. 1972. Glued-laminated timber bridges-reality or fantasy. Paper presented at the annual meeting of the American Institute of Timber Construction; 1972 March 13-16; Scottsdale, AZ. Madison, WI: U.S. Department of Agriculture, Forest Service, Forest Products Laboratory. 12 p.
4. Bruesch, L.D. Timber bridge systems. 1977. Paper presented at the 1977 FCP review conference on new bridge design concepts; 1977 October 3-7; Atlanta, GA. 7 p.
5. Civil Engineering. 1971. Who says wooden bridges are dead? Civil Engineering. June: 53.
6. Congdon, H.W. 1941. The covered bridge. Brattleboro, VT: Stephen Dayle Press. 150 p.
7. Cooper, T. 1976. American railroad bridges. In: American Society of Civil Engineers. American wooden bridges. ASCE Hist. Pub. No. 4. New York: American Society of Civil Engineers: 7-27.
8. Culmann, K. 1966. Brown's timber railroad bridges. Translated by M. Steinhaus, Civil Engineering 36(11): 72-74.

9. Culmann, K. 1968. Remington's wood bridges. Translated by M. Steinhaus. Civil Engineering 38(3): 60-61. 1968.
10. Eby, R.E. 1986. Timber & glulam as structural materials, general history. Paper presented at the Engineered Timber Workshop; March 17; Portland, OR. 15 p.
11. Edwards, L.N. 1976. The evolution of early American bridges. In: American Society of Civil Engineers. American wooden bridges. ASCE Hist. Pub. No. 4. New York: American Society of Civil Engineers: 143-168.
12. Fletcher, R.; Snow, J.P. 1976. A history of the development of wooden bridges. In: American Society of Civil Engineers. American wooden bridges. ASCE Hist. Pub. No. 4. New York: American Society of Civil Engineers: 29-123.
13. Freas, A.D. 1952. Laminated timber permits flexibility of design. Civil Engineering 22(9): 173-175.
14. Hardwood Record. 1913. The wooden bridge. Hardwood Record 36(8): 28.
15. Jakeman, A.M. 1935. Old covered bridges. Brattleboro, VT: Stephen Dayle Press. 107 p.
16. Jelly, I.A. 1941. Anatomy of an old covered bridge. Civil Engineering 2(1): 12-14.
17. Kuzmanovic, B.O. 1976. History of the theory of bridge structures. Preprint 2738; 1976 American Society of Civil Engineers Annual Convention and Exposition; 1976 September 27- October 1; Philadelphia, PA. New York: American Society of Civil Engineers. 29 p.
18. Neilson, G. 1971. Rubbing shoulders with the past. DuPont Magazine 65(6): 10-13.
19. Quimby, A.W. 1974. The Cornish-Windsor covered bridge. The Plain Facts 2(1): 1-2.
20. Sackowski, A.S. 1963. Reconstructing a covered timber bridge. Civil Engineering. October: 36-39.
21. Schneider, C.C. 1976. The evolution of the practice of American bridge building. In: American Society of Civil Engineers. American wooden bridges. ASCE Hist. Pub. No. 4. New York: American Society of Civil Engineers: 1-5.
22. Schuessler, R. 1972. America's antique bridges. Passages, Northwest Orient's Inflight Magazine 3(1): 16-19.
23. Tuomi, R.L. 1972. Advancements in timber bridges through research and engineering. In: Proceedings, 13th annual Colorado State University bridge engineering conference; 1972; Ft. Collins, CO. Colorado. State University: 34-61.
24. Tuomi, R.L.; McCutcheon, W.J. 1973. Design procedure for glued laminated bridge decks. Forest Products Journal 23(6): 36-42.
25. U.S. Department of Agriculture, Forest Service. 1988. Build better and save with modem timber bridges, a technology transfer plan for timber bridges. Washington, DC: U.S. Department of Agriculture, Forest Service. 28 p.

26. U.S. Senate. 1987. Highway bridge replacement and rehabilitation program. Eighth annual report of the Secretary of Transportation. Document 100-22. Washington, DC: U.S. Senate. 55 p.

27. Wheeler Consolidated, Inc. 1986. Timber bridge design. St. Louis Park, MN: Wheeler Consolidated, Inc. 42 p.

28. Wilson, R.E. 1976. Twenty different ways to build a covered bridge. In: American Society of Civil Engineers. American wooden bridges. ASCE Hist. Pub. No. 4. New York: American Society of Civil Engineers: 125-141.

CHAPTER 2

TYPES OF TIMBER BRIDGES

2.1 INTRODUCTION

Timber bridges are seen today in many types and configurations. Some of these bridges evolved from designs developed many years ago, while others have developed as a result of modem technological advances in timber design and fabrication. Regardless of the specific configuration, all timber bridges consist of two basic components, the superstructure and the substructure (Figure 2-1). The superstructure is the framework of the bridge span and includes the deck, floor system, main supporting members, railings, and other incidental components. The five basic types are the beam, deck (slab), truss, arch, and suspension superstructures. The substructure is the portion of the bridge that transmits loads from the superstructure to the supporting rock or soil. Timber substructures include abutments and bents. Abutments support the two bridge ends, while bents provide intermediate support for multiple-span crossings.

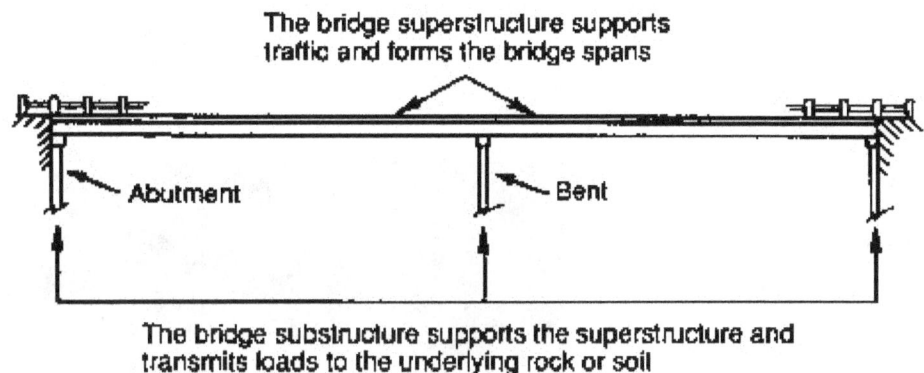

The bridge superstructure supports
traffic and forms the bridge spans

Abutment   Bent

The bridge substructure supports the superstructure and
transmits loads to the underlying rock or soil

Figure 2-1.- Basic components of a timber bridge.

This chapter provides an introduction to the many types of timber bridges currently used in the United States. Superstructures are discussed first, followed by decks and substructures. Although decks are technically part of the superstructure, they are addressed separately because of their varied application on many superstructure types.

Longitudinal beam superstructures are the simplest and most common timber bridge type (in bridge design, the longitudinal direction is measured in the direction of the traffic flow). Longitudinal beam superstructures consist of a deck system supported by a series of timber beams between two or more supports. Bridge beams are constructed from logs, sawn lumber, glued-laminated timber, or laminated veneer lumber (LVL). Individual beams may be termed stringers or girders, depending on the relative size of the member. Girders are larger than stringers; however, there is no clear-cut definition for either. For clarity, the word *beam* is used here to collectively define all longitudinal beam elements, including stringers and girders.

LOG BEAMS

The simplest type of timber bridge is the log beam or native timber bridge. It is constructed by placing round logs alternately tip to butt and binding them together with steel cables. A transverse (perpendicular to traffic flow) distributor log or needlebeam is normally attached to the bridge underside at centerspan to aid in load distribution. The deck for log beam bridges is formed by spiking sawn lumber planks across the log tops (Figure 2-2), or by placing soil and rocks on the logs (Figure 2-3).

Figure 2-2.-Log beam bridge with a transverse plank deck.

Figure 2-3.- Log beam bridge with a gravel deck. The two large "brow" logs along each side serve to delineate the roadway and function as a type of railing.

The span of log beam bridges is limited by the available species and the diameter and length of trees. Clear spans of 20 to 60 feet are most common; however, spans approaching 100 feet have been built that support off-highway trucks weighing in excess of 100 tons. Log bridges are generally not treated with preservatives and are primarily used as temporary structures. Service life typically ranges from 10 to 20 years, depending on log species and local conditions of use. Although log beam bridges may appear to be rather crude, they have proven to be very functional. Hundreds of these bridges are currently in use in the United States and Canada, primarily on logging and other low-volume roads. The basic concept has been adapted into many configurations, some of which are quite sophisticated.

SAWN LUMBER BEAMS

Sawn lumber beam bridges are constructed of closely spaced lumber beams that are commonly 4 to 8 inches wide and 12 to 18 inches deep (Figure 2-4). Solid timber blocking or lumber bridging is placed between beams for alignment and lateral beam support. Sawn lumber beam bridges are limited in span by the availability of lumber beams in the required sizes. They are most commonly used for clear spans of 15 to 25 feet with a practical maximum for highway loads of approximately 30 feet (Figure 2-5). Longer crossings are achieved by using a series of simple spans with intermediate supports.

Figure 2-4.- Underside of a sawn lumber beam bridge showing the characteristic close beam spacing. This photo is of the center bent of a two-span crossing, where beams from the two spans overlap at the support.

Figure 2-5.- Typical sawn lumber beam bridge. Most lumber beam bridges of this type span 25 feet or less, but longer spans have been built where large beams are available.

Sawn lumber beam bridges have been built in the United States for generations. They are economical, easy to construct, and well suited to secondary and local roads where long clear spans are not required. The service life of lumber bridges treated with preservatives averages about 40 years. Although their use has declined significantly since the introduction of glulam, many of the sawn lumber beam bridges built in the 1930's and 1940's are still in service.

GLUED-LAMINATED TIMBER BEAMS

Glulam bridges are constructed of glulam beams manufactured from 1-1/2- or 1-3/8-inch-thick lumber laminations that are bonded together on their wide faces with waterproof structural adhesive. The beams are available in standard widths ranging from 3 inches to 14-1/4 inches, with beam depth limited only by transportation and pressure-treating size considerations. Because of the large size of glulam beams, glulam beam bridges require fewer beams and are capable of much longer clear spans than conventional sawn lumber beam bridges (Figure 2-6). They are most commonly used for spans of 20 to 80 feet, but have been used for clear spans over 140 feet (Figure 2-7). The length of the beams, and thus the bridge, is normally limited only by transportation restrictions for moving the beams to the construction site.

Figure 2-6.- Underside of a glulam beam bridge. Because glulam beams are manufactured in a wide range of sizes, glulam bridges typically have larger beams and a greater beam spacing compared to conventional sawn lumber beam bridges (photo courtesy of Weyerhaeuser Co.).

Figure 2-7.- Glulam beam bridge over Dangerous River, near Yukatat, Alaska. This bridge consists of three 143-foot spans, each of which is supported by four glulam beams that are 91-1/2 inches deep (photo courtesy of the Alaska Department of Transportation and Public Facilities).

The first glulam beam bridges were built in the mid-1940's. Since that time, they have become the most common type of timber bridge in both single- and multiple-span configurations. Glulam beam bridges are completely prefabricated in modular components and are treated with preservatives after fabrication. When properly designed and fabricated, no field cutting or boring is required, resulting in a service life of 50 years or more.

LAMINATED VENEER LUMBER BEAMS

Laminated veneer lumber, a subcategory of new wood products called structural composite lumber, is a relatively new material for use in bridge construction. It is made from sheets of thin veneer that are glued together to form structural members. The veneer laminations are approximately 1/10 inch to 1/2 inch thick and are oriented vertically, instead of horizontally, as in glulam beams (Figure 2-8). Although LVL is made from veneer, it is more like glulam than like plywood because the grain directions of adjacent plies are parallel rather than at right angles. The advantages of LVL are its high strength, stiffness, and excellent treatability with wood preservatives.

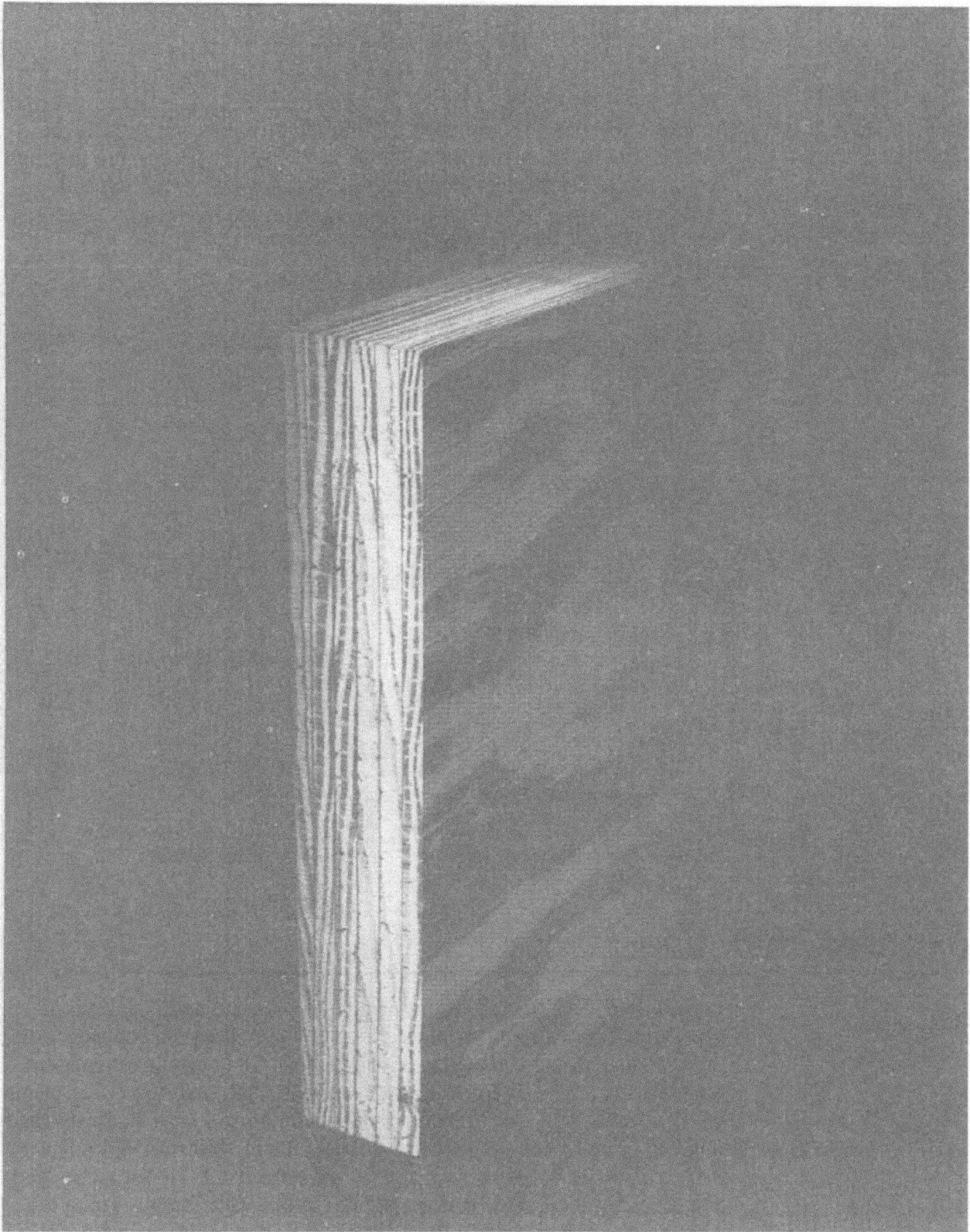

Figure 2-8.- End section of an LVL beam; LVL beams are manufactured by gluing together sheets of veneer. The grain direction of the veneer layers is oriented in the same direction, parallel to the direction of the beam span.

The only LVL beam bridge constructed to date is made of press-lam, a type of LVL developed at the USDA Forest Service, Forest Products Laboratory (FPL). This prototype structure, jointly sponsored by the Forest Service and the Virginia State Highway Department, consists of a 3-1/8-inch deck supported by 4-1/2- by 20-inch press-lam beams, spaced 30 inches on center (Figure 2-9). Design requirements and stresses for LVL are not included in current bridge design specifications, but they may be adopted in the future. Additional information on construction and performance of the press-lam demonstration bridge is given in references listed at the end of this chapter. [8,18,19,32]

Figure 2-9.- Press-lam LVL bridge built in 1977 on the George Washington National Forest in Virginia. The bridge spans 20 feet and carries a 26-foot-wide roadway.

2.3 LONGITUDINAL DECK SUPERSTRUCTURES

Longitudinal deck or slab superstructures are constructed of glulam or nail-laminated sawn lumber placed longitudinally between supports, with the wide dimension of the laminations vertical. The deck is designed to resist all applied loads and deflection without additional supporting members or beams; however, transverse distributor beams are usually attached to the deck underside to assist in load distribution. Glulam longitudinal deck bridges are constructed of panels that are 6-3/4 to 14-1/4 inches deep and 42 to 54 inches wide (Figure 2-10). Sawn lumber bridges use 2- to 4-inch-wide lumber, 8 to 16 inches deep, that is nailed or spiked together to form a continuous surface (Figure 2-11). Longitudinal deck bridges are economical and practical for maximum clear spans up to approximately

36 feet. Longer crossings are achieved with multiple spans. The low profile of these bridges makes them desirable when vertical clearance below the bridge is limited.

Figure 2-10.- Longitudinal glulam deck bridge over Au Train Creek on the Hiawatha National Forest. This bridge is 58 feet long over three spans and supports a 26-foot roadway width.

Figure 2-11.- Sawn lumber longitudinal deck bridge. Note the transverse distributor beams attached to the deck underside between bents (photo courtesy of Wheeler Consolidated, Inc.).

Trusses are structural frames consisting of straight members connected to form a series of triangles. In bridge applications, a typical truss superstructure consists of two main trusses, a floor system, and bracing (Figure 2-12). These superstructures are classified as deck trusses or through trusses, depending on the location of the floor system or deck. For deck trusses, the deck is at or above the level of the top chord. For through trusses, the deck is near the bottom chord. When the height of a through truss is insufficient for overhead bracing, it is referred to as a half-through or pony truss.

Timber trusses are constructed in many geometric configurations (Figure 2-13). Two of the most popular are the bowstring truss and parallel-chord truss (top chord and bottom chord parallel). In the bowstring truss,

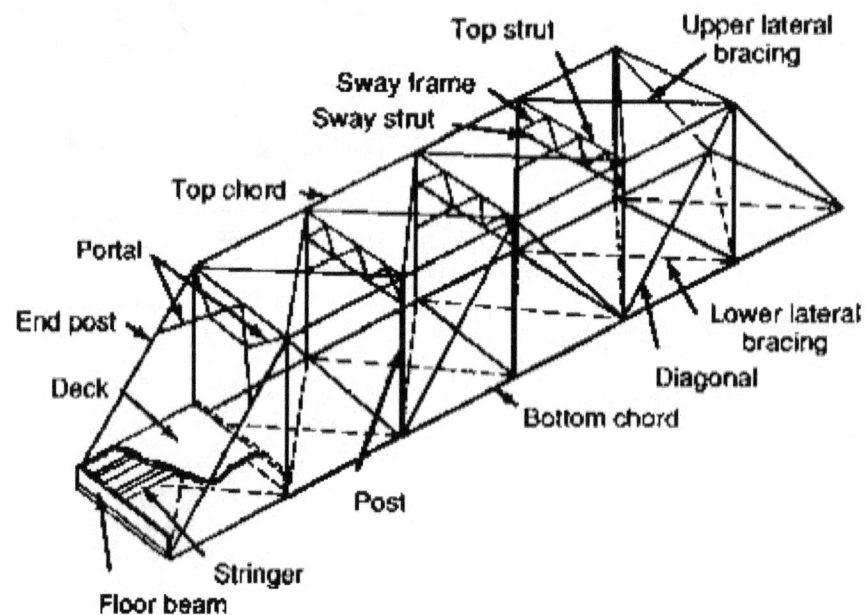

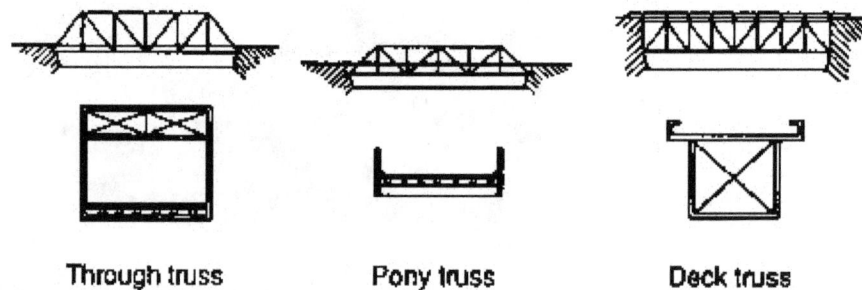

Figure 2-12.—Truss bridge nomenclature and classifications.

the top chord is constructed of curved glulam members or a series of straight sawn lumber members (Figure 2-14). As a pony truss, bowstrings are generally the most economical of all truss types for spans up to 100 feet. For longer spans, the bowstring is designed as a through truss. Parallel-chord trusses are constructed in various through-truss or deck-truss configurations for spans up to approximately 250 feet. As a deck truss, parallel-chord designs are practical when vertical clearance is sufficient for the truss depth and are especially economical for deep crossings where reduced bent height can result in substructure savings (Figure 2-15).

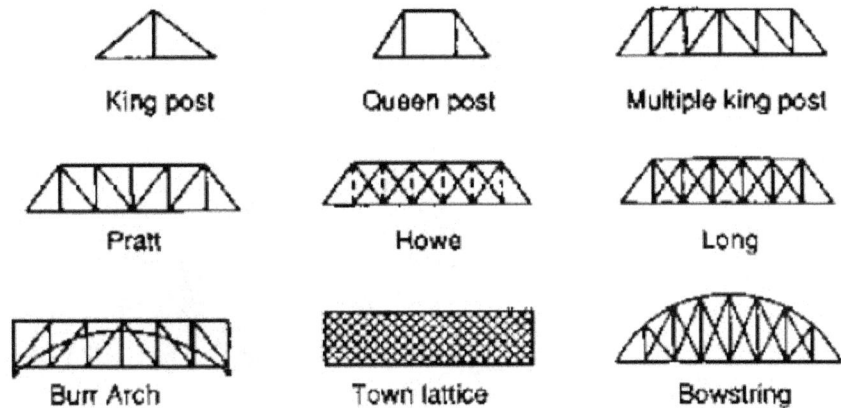

King post	Queen post	Multiple king post
Pratt	Howe	Long
Burr Arch	Town lattice	Bowstring

Figure 2-13—Typical truss configurations for timber bridges.

Figure 2-14.- Lumber bowstring truss over Dinkey Creek on the Sierra National Forest in Central California. This truss spans 90 feet and was built in 1934 (photo courtesy of Raul Gonzalez, USDA Forest Service).

2-11

Figure 2-15.- A multiple-span parallel-chord deck truss bridge.

Timber trusses were used extensively for vehicle bridges through the late 1950's, but their popularity has declined because of the high cost of truss fabrication and erection. Trusses are also more costly to maintain than many other bridge superstructures because of the large number of members and joints. Most timber trusses are built today for aesthetic reasons or when the light weight and relatively small individual members make them advantageous for transportation or erection.

2.5 TRESTLES

A trestle is a series of beam, deck, or truss superstructures supported on timber bents (Figure 2-16). Trestles are used for long crossings when lengthy clear spans are unnecessary, impractical, or not economical. Superstructure support for trestle bridges is provided by bents constructed of timber piles or sawn lumber frames (Section 2.10). The spacing between bents is controlled by the span capability of the superstructure. The most common trestle configuration is a series of simply supported sawn lumber beams spanning 20 to 30 feet. Longer spans can be achieved with trusses or glulam beams.

Figure 2-16.- Sewall's bridge is a timber trestle vehicle bridge in York, Maine. The bridge was built in 1933 using the same design features of the original bridge, built in 1761, that it replaced. This bridge became a designated landmark of the American Society of Civil Engineers in 1986 (photo courtesy of the American Society of Civil Engineers; Used by permission).

Trestle bridges have been used in the United States since the mid-1700's. Most were constructed as railroad bridges between 1900 and 1950 (Figure 2-17). In the mid-1950's, approximately 1,800 miles of timber trestles were in service on the Nation's railroads. Trestles were used for vehicle bridges through the 1950's, but their use has since declined because of the high cost of bent construction and the longer clear-span capabilities of glulam. With an average service life of 40 years or more, many treated-timber trestle bridges remain in service today.

2.6 GLULAM DECK ARCHES

The versatility of glulam in bridge construction is perhaps best demonstrated by glulam deck arch bridges. These structures are constructed of glulam arches manufactured in segmental circular or parabolic shapes and can be used for clear spans in excess of 200 feet. Two basic arch types are used, the two-hinge arch and the three-hinge arch (Figure 2-18). Two-hinge designs are practical for short spans of approximately 80 feet or less. Three-hinge designs are more appropriate for longer spans and are most common for vehicle bridges. The roadway for deck arch bridges is supported by glulam post bents connected to the arches with steel gusset plates.

Figure 2-17.- Early railroad trestle on the Verona, South Park, and Sunset Steam Railroad. Many long-span timber trestles of this type were built for railroad use, requiring large volumes of wood for the complex bent substructures (photo from the Forest Service Collection, National Agriculture Library).

Moment splice (optional) to facilitate
fabrication and transportation

Hinge

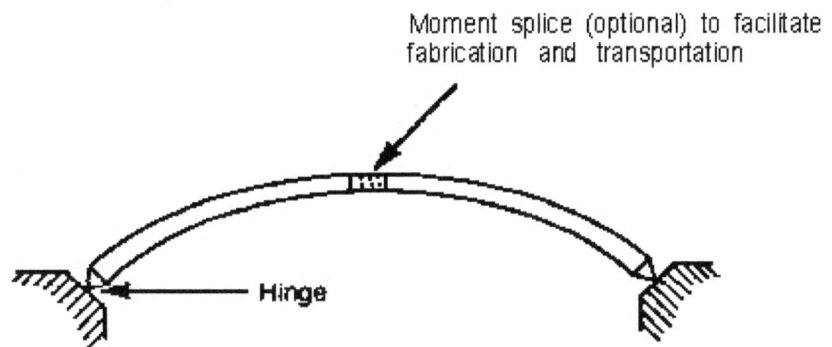

Two-hinge arch is hinged at reactions only.

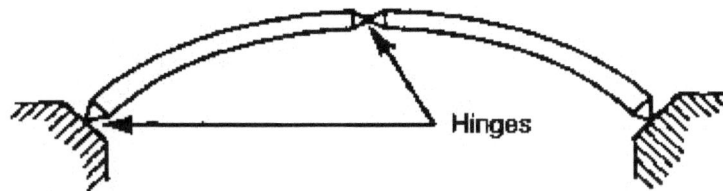

Hinges

Three-hinge arch is hinged at reactions and at the arch apex.

Figure 2-18.-Glulam arch configurations used for bridges.

The first glulam deck arches for vehicles were constructed in Oregon in
the late 1940's (Figure 2-19). They have since been used in many applications, including the highly publicized Keystone Wye interchange in South
Dakota (Figure 2-20). The design is most practical in applications where
considerable height is required and where foundations can be constructed
to resist horizontal end reactions. It is particularly suitable for deep
crossings where savings in substructure costs over other bridge types
make it economically competitive.

2.7 SUSPENSION BRIDGES

Timber suspension bridges consist of a timber deck structure suspended
from flexible steel cables (or chains) that are supported by timber towers
(Figure 2-21). They are capable of long clear spans (over 500 feet) and are
normally used only when other bridge types are impractical because of
span requirements or when the use of intermediate bents is not feasible.
Most timber suspension bridges in the United States have been constructed
for pedestrian or trail crossings. Although timber suspension bridges have
been built for vehicle traffic, their number is small in relation to other
timber bridge types.

Figure 2-19.-- The Loon Lake Bridge is a three-hinge glulam deck arch design, built near Roseburg, Oregon, in 1948. The bridge spans 104 feet and supports a 20-foot roadway.

Figure 2-20.- Three-hinge glulam deck arch bridge at the Keystone Wye interchange off U.S. Highway 16, near Mount Rushmore, South Dakota. The arch spans 155 feet and supports a 26-foot-wide roadway (photo courtesy of Wheeler Consolidated, Inc.).

Figure 2-21.- Typical timber suspension bridge designed for vehicle traffic.

2.8 DECKS

The deck is the portion of the bridge superstructure that forms the roadway and distributes vehicle loads to supporting elements of the structure. The type, thickness, and material of the deck are based on the weight and volume of traffic it must support. Timber decks are typically constructed of one of three materials: sawn lumber planks, nail-laminated lumber, and glulam. Composite timber-concrete decks are also used on timber superstructures in some applications.

SAWN LUMBER PLANKS

Sawn lumber plank decks are the oldest and simplest type of timber deck. They are constructed of lumber planks, 3 to 6 inches thick and 10 to 12 inches wide, that are placed flatwise and spiked to supporting beams. The planks are generally laid in the transverse direction and are attached directly to closely spaced timber beams with spikes (Figure 2-22). They are also used longitudinally on transverse floorbeams. Plank decks are most practical on low-volume or special-use bridges. They are not watertight and afford little protection to supporting members from the effects of weathering. Asphalt paving is not practical on plank decks because of large deck deflections that cause asphalt cracking and deterioration.

2-17

Figure 2-22.- Sawn lumber plank decks (A) in a transverse orientation and (B) in a longitudinal orientation.

NAIL-LAMINATED LUMBER

Nail-laminated lumber decks are constructed of sawn lumber laminations that are generally 2 inches thick and 4 to 12 inches deep. The laminations are placed with the wide dimension vertical and are nailed or spiked together to form a continuous surface (Figure 2-23). Nail-laminated decks are most commonly used in a transverse orientation on sawn lumber or steel beams spaced 2 to 6 feet apart. They are also used longitudinally over

Figure 2-23.- Nail-laminated lumber deck as viewed from (A) the deck top and (B) the deck edge.

transverse floorbeams in a manner discussed for longitudinal deck super-structures (Section 2.3).

Nail-laminated lumber decks were the most commonly used type of timber deck from the 1920's through the mid-1960's. Their use has declined significantly since the introduction of glulam. Although many nail-laminated decks have provided satisfactory performance for over

40 years, the design is generally not suitable unless supporting beams are closely spaced. As beam spacing increases, deflection of the deck and dimensional changes, from variations in moisture content, cause delamination or loosening of the deck, reducing structural integrity and service life.

GLUED-LAMINATED TIMBER

Glulam decks are constructed of glulam panels that are normally 5-1/8 to 8-3/4 inches thick and 3 to 5 feet wide. They are used in both transverse and longitudinal orientations on glulam or steel beams.

The design criteria for glulam deck panels were developed in the mid-1970's at the FPL. They are the most common type of timber deck and are used in two basic configurations, noninterconnected and doweled (Figure 2-24). Noninterconnected panels are placed edge to edge, with no connection between adjacent panels. Doweled panels are interconnected with steel dowels to improve load distribution and reduce differential displacements at the panel joints. Doweled panels are more costly to fabricate and construct but can result in thinner decks and better performance for asphalt wearing surfaces.

Glulam decks are stronger and stiffer than conventional plank or nail-laminated decks because of the homogeneous bond between laminations and the dispersion of strength-reducing characteristics of glulam. Glulam panels can be constructed to form a watertight surface and afford protection for supporting beams and other components. Because of their increased stiffness, glulam decks also provide a firm base for asphalt pavement, which is frequently used as the wearing surface. Panels are completely fabricated and drilled for deck attachment prior to preservative treatment, producing estimated service lives of 50 years or more.

COMPOSITE TIMBER-CONCRETE

A composite timber-concrete deck consists of a concrete slab rigidly interlocked to supporting timber components so that the combination functions as a unit. On single, simple spans, the concrete resists compression, while the timber carries tension. At intermediate supports of continuous spans, the opposite is true. There are two basic types of composite timber-concrete decks: T-beam decks and slab decks (Figure 2-25). Composite T-beam decks are constructed by casting a concrete deck, which forms the flange of the T, on a glulam beam, which forms the web of the T. Composite action between the timber and concrete is developed by shear connectors along the beam tops. Numerous T-beam composite decks have been constructed in recent years, but they are not widely used because of the high cost of beam fabrication and the cost of in-place casting of concrete (Figure 2-26).

Composite slab decks are constructed by casting a concrete layer on a continuous base of longitudinal nail-laminated sawn lumber. The lumber

Figure 2-24.- Glued-laminated timber deck in the (A) noninterconnected and (B) doweled configurations.

is placed edgewise in the direction of traffic flow, with alternate laminations raised 1-3/8 to 2 inches to form grooves in the base. Composite action between the timber and concrete is most commonly achieved through the use of triangular steel shear developers driven into the grooves. Composite slab decks were first built in 1932 and were used mostly during the 1930's and 1940's. They are not commonly used today.

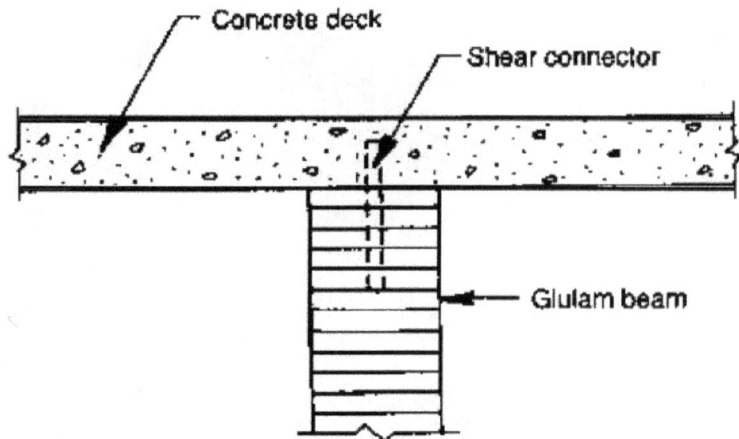

Composite timber-concrete T-beam

Note: Traffic direction is into page. Concrete-reinforcing steel is omitted for clarity.

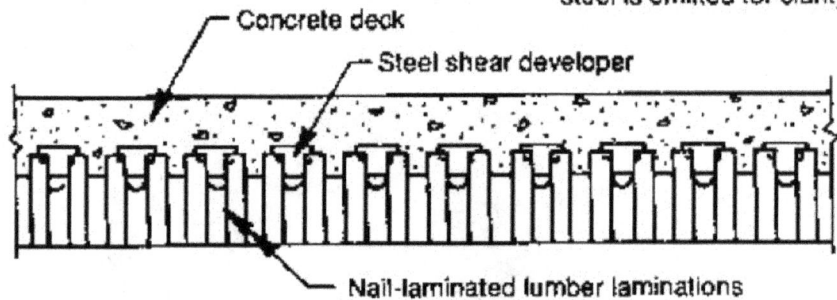

Composite timber-concrete slab

Figure 2-25.- Types of composite timber-concrete decks.

2.9 STRESS-LAMINATED TIMBER

Stress-laminated timber is a relatively new concept for timber bridge applications. Using this system, vertical sawn lumber laminations are clamped together on their wide faces by high-strength steel stressing rods. These stressing rods are placed on the outsides of the laminations (external) or through the laminations (internal), depending on the type of structure (Figure 2-27). For both configurations, the stressing pressure is transferred to the timber through bearing plates located along the outer laminations. This pressure develops sufficient friction between the laminations to cause them to perform structurally as a unit, in a manner similar to the performance of glulam.

Stress-laminated timber has been used successfully in bridge construction and rehabilitation. In new construction, it is used primarily for longitudinal decks (Figure 2-28), but it has also been applied to other superstructure types (Figure 2-29). Stressing is also practical for rehabilitating nail-

Figure 2-26.- Composite glulam-concrete T-beam bridge located in northern California. Although numerous bridges of this type have been built, they are not common.

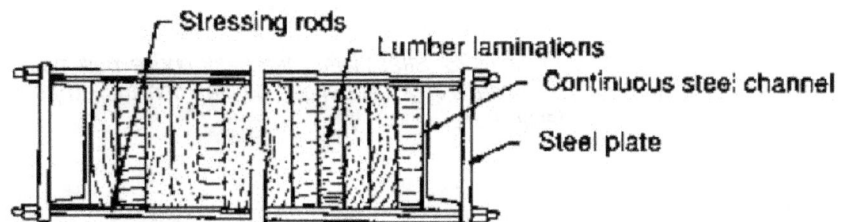

External rod configuration
(rods placed above and below the lumber laminations)

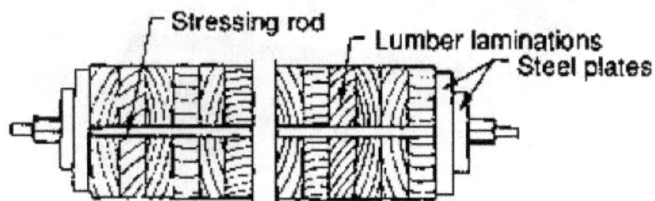

Internal rod configuration
(rods placed through the lumber laminations)

Figure 2-27.- Typical rod configurations for stress-laminated timber bridges.

laminated decks where load distribution characteristics of the deck have been reduced by delamination. The clamping action produced by the stressing rods restores deck integrity, increases load capacity, and substantially extends service life.

Figure 2-28- Stress-laminated deck bridge built near State College, Pennsylvania, in 1987. The bridge is 28 feet wide and was constructed from 4-inch-wide by 16-inch-deep lumber laminations.

Figure 2-29.- Stress-laminated deck bridge with stress-laminated slant-leg supports, built near Espanola, Ontario, Canada, in 1981. The bridge spans approximately 55 feet and supports two traffic lanes (photo courtesy of the Ontario Ministry of Transportation).

2-24

Stress-laminated timber for bridges was originally developed in Ontario, Canada, and adopted for use in the Ontario Highway Bridge Design Code in 1976. Although it has been successfully used in Canada, the system is relatively new in the United States and is not currently included in bridge design specifications. Research on stress-laminated timber, including the construction of several prototype structures, has been completed by the Forest Service in cooperation with the University of Wisconsin and West Virginia University. It is expected that the stress-laminated timber bridge system will be adopted in United States design specifications in the near future.

2.10 TIMBER SUBSTRUCTURES

The substructure is the portion of the bridge that supports the superstructure and transfers loads to the supporting soil or rock. The type of substructure used depends on the site conditions, quality of foundation material, and magnitude of the loads it must support. Timber bridges are adaptable to virtually any type of substructure constructed of timber, steel, or concrete. Discussions in this section will be limited to abutments or bents constructed of timber piles, sawn lumber, or glulam.

ABUTMENTS

Abutments support the bridge ends and contain roadway embankment material. The simplest timber abutment is a sawn lumber or glulam spread footing placed directly on the surface of the embankment (Figure 2-30). This type of abutment is used only when foundation material is of sufficient quality to support loads without excessive settlement, erosion, or scour. Another type of footing abutment is the post abutment (Figure 2-31). On post abutments, the superstructure is supported on sawn lumber or glulam posts connected to a spread footing located below the ground surface. Post abutments are used to elevate the superstructure and are provided with a backwall and wingwalls for retaining fill embankment.

When the quality of the foundation is not sufficient to support footings, pile abutments may be used (Figure 2-32). These abutments are constructed of timber piles driven to sufficient depth to develop the required load capacity by end bearing, or through friction between the pile surface and surrounding soil. The superstructure is connected to the piles by a continuous cap attached to the piles and to the superstructure at the bearings. Pile abutments are typically provided with backwalls and wingwalls to retain the embankment material.

BENTS

Bents are intermediate supports between abutments for multiple-span crossings. They are constructed of timber piles or sawn lumber frames, depending on required height and the suitability of foundation material.

Figure 2-30.- Surface bearing spread footing constructed of glulam (photo courtesy of Tim Chittenden, USDA Forest Service).

Figure 2-31.- Sawn lumber post abutment.

Pile bents are practical when foundation material is suitable and the required bent height, including pile penetration, is within the available length of timber piles (Figure 2-33). Frame bents are used for higher elevations or when rock or other foundation materials are not suitable for piles (Figure 2-34). Frames may be supported on footings or piles,

Figure 2-32.- Timber pile abutment.

Figure 2-33.- Timber pile bents.

Figure 2-34.- Sawn lumber frame bent.

depending on the quality of the foundation. For both pile and frame bents, bracing is provided between members to provide stability and resist lateral loads. Superstructure bearing is on heavy timber caps fastened to the tops of the piles or frame posts.

2.11 SELECTED REFERENCES

1. American Institute of Timber Construction. 1973. Modem timber highway bridges, a state of the art report. Englewood, CO: American Institute of Timber Construction. 79 p.
2. American Institute of Timber Construction. 1985. Timber construction manual. 3d ed. New York: John Wiley and Sons, Inc. 836 p.
3. American Society of Civil Engineers. 1975. Wood structures, a design guide and commentary. New York: American Society of Civil Engineers. 416 p.

4. American Wood-Preservers' Association. 1941. Timber-concrete composite decks. Chicago: American Wood Preservers' Association. 28 p.

5. Archibald, R. 1952. A survey of timber highway bridges in the United States. Civil Engineering. September: 171-176.

6. Bohannan, B. 1972. Glued-laminated timber bridges-reality or fantasy. Paper presented at the annual meeting of the American Institute of Timber Construction; 1972 March 13-16; Scottsdale, AZ. Madison, WI: U.S. Department of Agriculture, Forest Service, Forest Products Laboratory. 12 p.

7. Bruesch, L.D. 1977. Timber bridge systems. Paper presented at the 1977 FCP review conference on new bridge design concepts; 1977 October 3-7; Atlanta, GA. 7 p.

8. Gromala, D.S.; Moody, R.C.; Sprinkel, M.M. 1985. Performance of a press-lam bridge-a S-year load testing and monitoring program. Res. Note FPL-0251. Madison, WI: U.S. Department of Agriculture, Forest Service, Forest Products Laboratory. 7 p.

9. Gurfinkel, G. 1981. Wood engineering. 2d ed. Dubuque, IA: Kendall/Hunt Publishing Co. 552 p.

10. Gutkowski, R.M.; Williamson, T.G. 1983. Timber bridges: state-of-the-art. Journal of Structural Engineering. 109(9): 2175-2191.

11. Kirkwood, C.C. 1970. The use of timber for county bridges. Wood Preserving. 48(1): 14-24.

12. Kozak, J.J.; Leppmann, J.F. 1976. Bridge engineering. In: Merritt, F.S., ed. Standard handbook for civil engineers. New York: McGraw-Hill. Chapter 17.

13. Nagy, M.M.; Trebett, J.T.; Wellburn, G.V. 1980. Log bridge construction handbook. Vancouver, Can.: Forest Engineering Research Institute of Canada. 421 p.

14. Oliva, M.G.; Dimakis, A.G.; Tuomi, R.L. 1985. Interim report: behavior of stressed-wood deck bridges. Report 85-1/A. Madison, WI: University of Wisconsin, College of Engineering, Structures and Materials Test Laboratory. 40 p.

15. Oliva, M.G.; Tuomi, R.L.; Dimakis, A.G. 1986. New ideas for timber bridges. In: Trans. Res. Rec. 1053. Washington, DC: National Academy of Sciences, National Research Council, Transportation Research Board: 59-64.

16. Ou, Fong L. 1985. The state of the art of timber bridges: a review of the literature. Washington, DC: U.S. Department of Agriculture, Forest Service. [30 p.].

17. Scarisbrick, R.G. 1976. Laminated timber logging bridges in British Columbia. Journal of the Structural Division, American Society of Civil Engineers. 102(ST1). [10 p.].

18. Sprinkel, M.M. 1978. Evaluation of the performance of a press-lam timber highway bridge. Interim rep. 2. Charlottesville, VA: Virginia Highway and Transportation Research Council. 13 p.

19. Sprinkel, M.M. 1982. Final report of evaluation of the performance of a press-lam timber bridge. Bridge performance and load test after 5 years. VHTRC 82-R56. Charlottesville, VA: Virginia Highway and Transportation Research Council. 21 p.

20. Sprinkel, M.M. 1982. Prefabricated bridge elements and systems. National Cooperative Highway Research Program, Synthesis of Highway Practice 119. Washington, DC: National Academy of Sciences, National Research Council, Transportation Research Board. 75 p.

21. Taylor, R.J.; Batchelor, B.; Van Dalen, K. 1983. Prestressed wood bridges. SRR-83-01. Downsview, ON, Can.: Ministry of Transportation and Communications. 15 p.

22. Taylor, R.J.; Csagoly, P.F. 1979. Transverse post-tensioning of longitudinally laminated timber bridge decks. Downsview, ON, Can.: Ministry of Transportation and Communications. 16 p.

23. Taylor, R.J.; Walsh, H. 1984. A prototype prestressed wood bridge. SRR-83-07. Downsview, ON, Can.: Ministry of Transportation and Communications. 75 p.

24. Timber Structures, Inc. [1955]. Permanent timber bridges. Portland, OR: Timber Structures, Inc. 4 p.

25. Tuomi, R.L. 1972. Advancements in timber bridges through research and engineering. In: Proceedings, 13th annual Colorado State University bridge engineering conference; 1972; Ft. Collins, CO. Colorado State University: 34-61.

26. West Coast Lumbermen's Association. 1952. Highway structures of Douglas fir. Portland, OR: West Coast Lumbermen's Association. 55 p.

27. Weyerhaeuser Company. 1980. Weyerhaeuser glulam wood bridge systems. Tacoma, WA: Weyerhaeuser Co. 114 p.

28. White, K.R.; Minor, J.; Derocher, K.N.; Heins, C.P., Jr. 1981. Bridge maintenance inspection and evaluation. New York: Marcel Dekker, Inc. 257 p.

29. Wilson, T.R.C. 1939. The glued laminated wooden arch. Tech. Bull. 691. Washington, DC: U.S. Department of Agriculture. 123 p.

30. Wipf, T.J.; Klaiber, F.W.; Sanders, W.W. 1986. Load distribution criteria for glued-laminated longitudinal timber deck highway bridges. In: Trans. Res. Rec. 1053. Washington, DC: National Academy of Sciences, National Research Council, Transportation Research Board: 31-40.

31. Wood Preserving. 1969. Pressure-treated wood bridges win civil engineering award. Wood Preserving News 47(4): 12-22.

32. Youngquist, J.A.; Gromala, D.S.; Jokerst, R.W. [and others]. 1979. Design, fabrication, testing, and installation of a press-lam bridge. Res. Pap. FPL 332. Madison, WI: U.S. Department of Agriculture, Forest Service, Forest Products Laboratory. 19 p.

CHAPTER 3

PROPERTIES OF WOOD AND STRUCTURAL WOOD PRODUCTS

3.1 INTRODUCTION

Wood differs from other construction materials because it is produced in a living tree. As a result, wood possesses material properties that may be significantly different from other materials normally encountered in structural design. Although it is not necessary to have an in-depth knowledge of wood anatomy and properties, it is necessary for the engineer to have a general understanding of the properties and characteristics that affect the strength and performance of wood in bridge applications. This includes not only the anatomical, physical, and mechanical properties of wood as a material, but also the standards and practices related to the manufacture of structural wood products, such as sawn lumber and glulam.

In the broadest terms, trees and their respective lumber are classified into two general classes, hardwoods and softwoods. Hardwoods normally have broad leaves that are shed at the end of each growing season. Softwoods have needlelike leaves that normally remain green year round. The classification as hardwood or softwood has little to do with the comparative hardness of the wood. Several species of softwoods are harder than many low- to medium-density hardwoods. With few exceptions the structural wood products used in bridge applications throughout North America are manufactured primarily from softwoods. Although hardwoods are not widely used at this time, structural grading procedures for hardwoods have been developed recently, and their use is increasing in some regions of the country.

This chapter discusses the structure of wood, its physical and mechanical properties, and the manufacturing and grading processes for sawn lumber and glulam. The scope of coverage is limited to softwood species, although many of the general characteristics are applicable to hardwoods. Additional information on wood properties and characteristics is given in references listed at the end of this chapter.

3.2 STRUCTURE OF WOOD

To fully understand and appreciate wood as a structural material, one must first understand wood anatomy and structure. This can be considered at two levels: the microstructure, which can be examined only with the aid of a microscope, and the macrostructure, which is normally visible to the unaided eye.

MICROSTRUCTURE

The primary structural building block of wood is the wood cell, or tracheid. When closely packed, these wood cells form a strong composite system that is often compared to a bundle of drinking straws (Figure 3-1). As a unit, the straws (wood cells) weigh very little, but if restrained from lateral buckling, they will support a substantial load in compression parallel to their longitudinal axis. If the straws are loaded in compression perpendicular to their longitudinal axis, they will yield under relatively light loads. Using this analogy, it is easy to visualize the superior strength-to-weight ratio of a cellular composite such as wood. Yet, each individual wood cell is even more structurally advanced because it is actually a multilayered, filament-reinforced, closed-end tube rather than just a homogeneous, nonreinforced straw (Figure 3-2).

MACROSTRUCTURE

The cross section of a tree can be divided into three basic parts: bark, cambium, and wood (Figure 3-3). Bark is the exterior layer and is composed of an outer layer of corky material with a thin inner layer of living

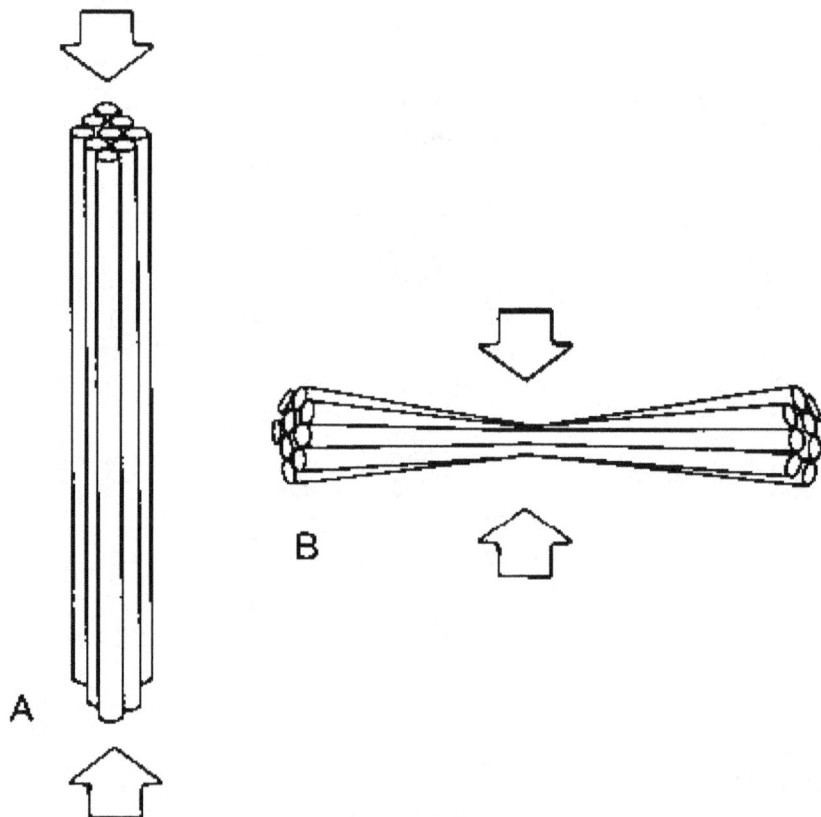

Figure 3-1.- Simplified depiction of the structure of wood, comparing it to a bundle of thin-walled drinking straws. (A) Parallel to their longitudinal axis, the straws (wood cells) can support loads substantially greater than their weight. (B) When loaded perpendicular to their longitudinal axis, the straws yield under much lower loads.

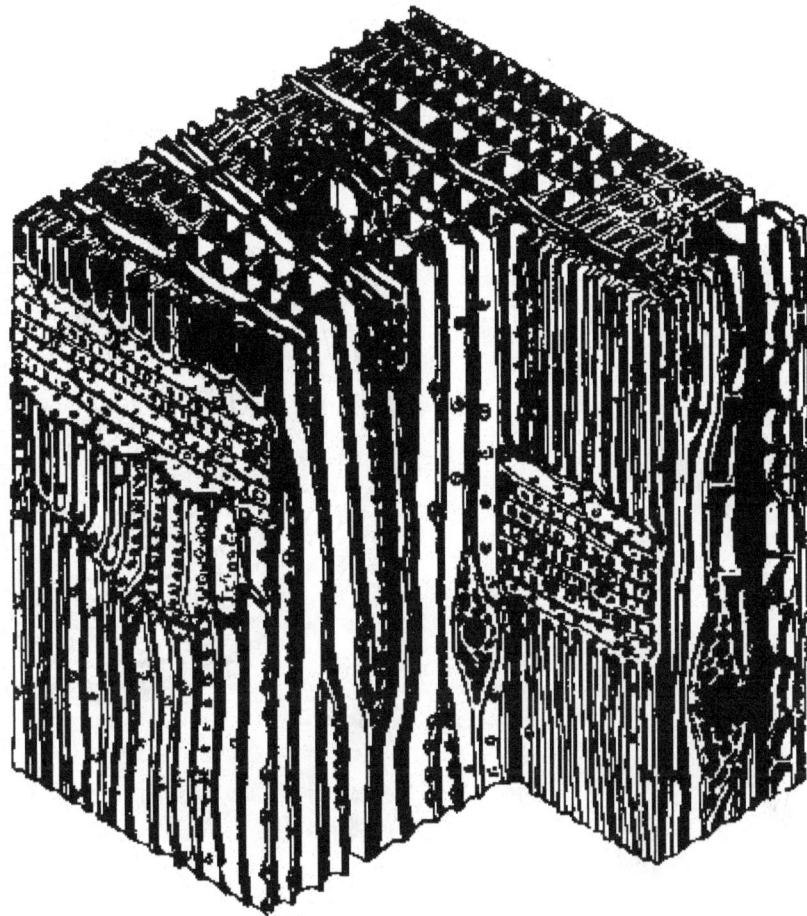

Figure 3-2.- Drawing of the magnified structure of a softwood.

cells. It functions to protect the tree and to conduct nutrients. The cambium is a thin, continuous ring of reproductive tissue located between the wood and the bark. It is the only portion of the tree where new wood and bark cells are formed and is usually only one to ten cells thick, depending on the season of the year. All material inside the cambium layer is wood, which conducts and stores nutrients and provides the tree with structural support. At the center of the wood, where tree growth began, is the pith or heart center.

Wood is divided into two general classes, sapwood and heartwood. The sapwood consists of both active and inactive cells and is located on the outside of the tree, next to the cambium. It functions primarily in food storage and the transport of sap. The radial thickness of sapwood is commonly 1-1/2 to 2 inches for most species, but it may be 3 to 6 inches thick for some species. Heartwood, which was once sapwood, is composed mostly of inactive cells that differ both chemically and physically from sapwood cells. The heartwood cells do not function in either food storage or sap transportation. In most species, the heartwood contains extractive

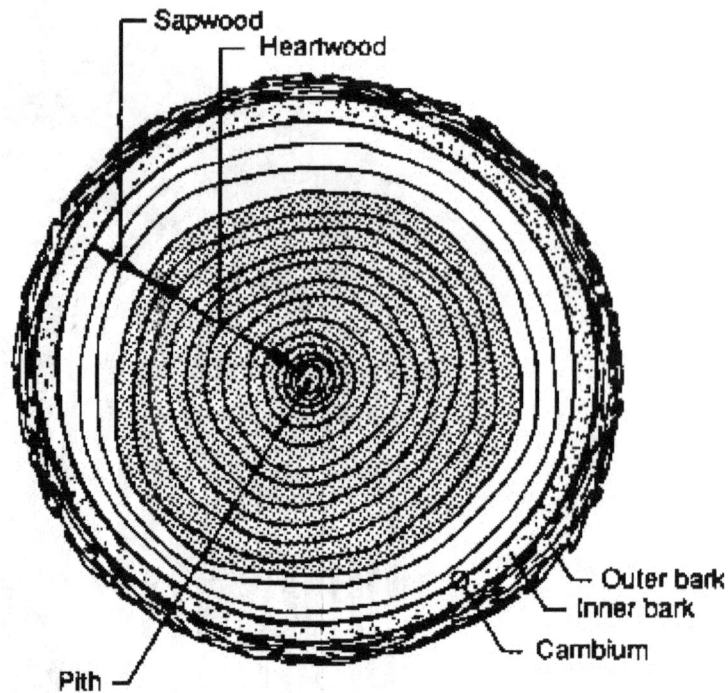

Figure 3-3.- Tree cross section showing elements of the macrostructure that are normally visible without magnification.

substances that are deposited in the cell during the conversion from sap-wood to heartwood. These deposits frequently give the heartwood a much darker color than sapwood; however, in several species the heartwood is not dark and looks virtually the same as sapwood. The extractives also serve to make the heartwood of several species more resistant to attack by decay fungi and insects. Because all heartwood was once sapwood, there is generally little difference in their dry weight or strength.

Growth in wood cells varies between cells that are formed early in the growing season, earlywood cells, and those formed late in the growing season, latewood cells. Earlywood cells are usually formed during the first or second month of the growing season and have relatively large cell cavities and thin walls. Latewood cells are formed later in the growing season and have smaller cell cavities and thicker walls. The contrast between the earlywood and latewood cells forms the characteristic growth rings common to most species (Figure 3-4). These growth rings vary in width, depending on species and site conditions. In many species of softwood, such as Douglas-fir and Southern Pine, there is a marked con-trast between the earlywood and latewood, and growth rings are plainly visible. In other species, such as spruces and true firs, the change from earlywood to latewood is less obvious, and rings are more difficult to see. Environmental conditions can also affect growth rings. Rings formed

3-4

Figure 3-4.- Cross section of a pine log showing growth rings. Light bands are earlywood, dark bands are latewood. A growth ring is composed of the earlywood ring and the latewood ring outside it.

during short or dry seasons are narrower than those formed under more favorable growing conditions.

3.3 PHYSICAL PROPERTIES OF WOOD

Physical properties describe the quantitative characteristics of wood and its behavior to external influences other than applied forces. Included are such properties as moisture content, density, dimensional stability, thermal and pyrolytic (fire) properties, natural durability, and chemical resistance. Familiarity with physical properties is important because those properties can significantly influence the performance and strength of wood used in structural applications.

DIRECTIONAL PROPERTIES Wood is an orthotropic material with unique and independent properties in different directions. Because of the orientation of the wood fibers, and the manner in which a tree increases in diameter as it grows, properties vary along three mutually perpendicular axes: longitudinal (L), radial (R), and tangential (T). The longitudinal axis is parallel to the grain direction, the radial axis is perpendicular to the grain direction and normal to the growth rings; and the tangential axis is perpendicular to the grain direction and tangent to the growth rings (Figure 3-5). Although wood properties differ in each of these three directions, differences between the radial and tangential directions are normally minor compared to their mutual differences with the longitudinal direction. As a result, most wood properties for structural applications are given only for directions parallel to grain (longitudinal) and perpendicular to grain (radial and tangential).

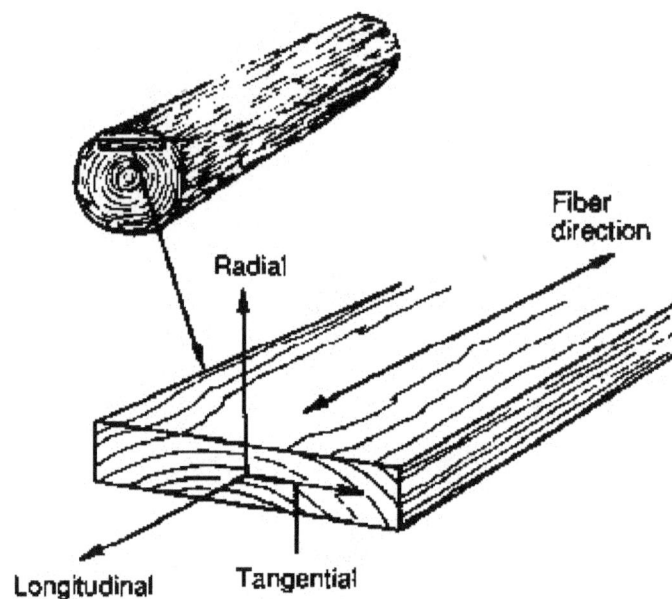

Figure 3-5- The three principal axes of wood with respect to grain direction and growth rings.

MOISTURE CONTENT The moisture content of wood (MC) is defined as the weight of water in wood given as a percentage of ovendry weight:

$$MC = \frac{\text{moist weight} - \text{ovendry weight}}{\text{ovendry weight}} \times 100 \text{ percent} \qquad (3\text{-}1)$$

In living trees, water is required for growth and development, and water constitutes a major portion of green wood. Depending on the species and type of wood, the moisture content of living wood ranges from approximately 30 percent to more than 250 percent (two-and-a-half times the weight of the solid wood material). In most species, the moisture content of the sapwood is higher than that of the heartwood (Table 3-1).

Table 3-1.-Average moisture content of green wood.

Species	Moisture content (percent)	
	Heartwood	Sapwood
Western redcedar	58	249
Douglas-fir (coast)	37	115
White fir	98	160
Western hemlock	85	170
Eastern hemlock	97	119
Larch (western)	54	110
Loblolly pine	33	110
Ponderosa pine	40	148
Sitka spruce	41	142
Average of 27 softwood species	55	149

From *Wood Handbook.* "

Water exists in wood as bound water, which is molecularly bonded within the cell walls, and as free water, which is present in the cell cavities (Figure 3-6). When moist wood dries, free water separates first and at a rate faster than bound water. The moisture content at which the cell walls are saturated with water, but at which virtually no free water exists in the cell cavities, is called the fiber saturation point. The fiber saturation point for most woods averages about 30 percent, but may vary by a few percentage points among different species.

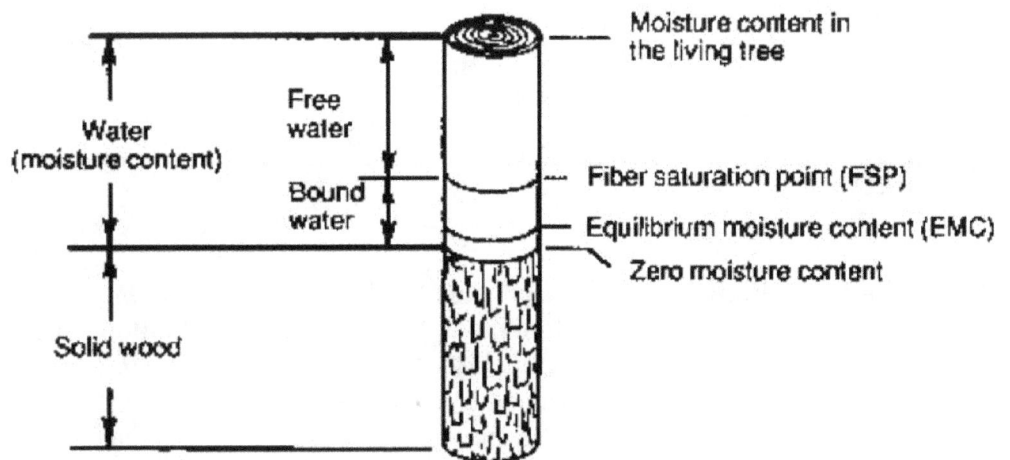

Figure 3-6.- Diagrammatic representation of wood moisture content.

Wood is a hygroscopic material that absorbs moisture in humid environments and loses moisture in dry environments. Therefore, the moisture content of wood is a result of atmospheric conditions and depends on the relative humidity and temperature of the surrounding air. Under constant temperature and humidity conditions, an equilibrium moisture content (EMC) is reached. The equilibrium moisture content represents a balance point where the wood is neither gaining nor losing moisture and is in equilibrium with the environment. In bridge applications, wood moisture content is almost always undergoing some changes as temperature and humidity conditions vary. These changes are usually gradual, short-term fluctuations that influence only the wood surface. Over a period of time, however, the wood will approach an equilibrium moisture content related to the environment. The time required to reach the equilibrium moisture content depends on the size and permeability of the member, the temperature, and the difference between the initial moisture content of the wood and the eventual equilibrium moisture content for the environment. The relationship between equilibrium moisture content, relative humidity, and temperature is generally independent of species and is shown in Table 3-2.

Table 3-2.- Moisture content of wood in equilibrium with stated dry-bulb temperature and relative humidity.

Temperature (dry-bulb)	Moisture content at various relative humidities (percent)										
	10	20	30	40	50	60	70	80	90	95	98
30	2.6	4.6	6.3	7.9	9.5	11.3	13.5	16.5	21.0	24.3	26.9
40	2.6	4.6	6.3	7.9	9.5	11.3	13.5	16.5	21.0	24.3	26.9
50	2.6	4.6	6.3	7.9	9.5	11.2	13.5	16.5	20.9	24.3	26.9
60	2.5	4.6	6.2	7.8	9.4	11.1	13.3	16.2	20.7	24.1	26.8
70	2.5	4.5	6.2	7.7	9.2	11.0	13.1	16.0	20.5	23.9	26.6
80	2.4	4.4	6.1	7.6	9.1	10.8	12.9	15.7	20.2	23.6	26.3
90	2.3	4.3	5.9	7.4	8.9	10.5	12.6	15.4	19.8	23.3	26.0
100	2.3	4.2	5.8	7.2	8.7	10.3	12.3	15.1	19.5	22.9	25.6
110	2.2	4.0	5.6	7.0	8.4	10.0	12.0	14.7	19.1	22.4	25.2
120	2.1	3.9	5.4	6.8	8.2	9.7	11.7	14.4	18.6	22.0	24.7

From *Wood Handbook*.[30]

DIMENSIONAL STABILITY

Wood is dimensionally stable when the moisture content is above the fiber saturation point. Below the fiber saturation point, wood shrinks when moisture is lost and swells when moisture is gained. This susceptibility to dimensional change is one of the few wood properties that exhibit significant differences for the three orthotropic axes. In the longitudinal direction, average shrinkage values from green to ovendry conditions are between 0.1 and 0.2 percent, which is generally of no practical concern. In

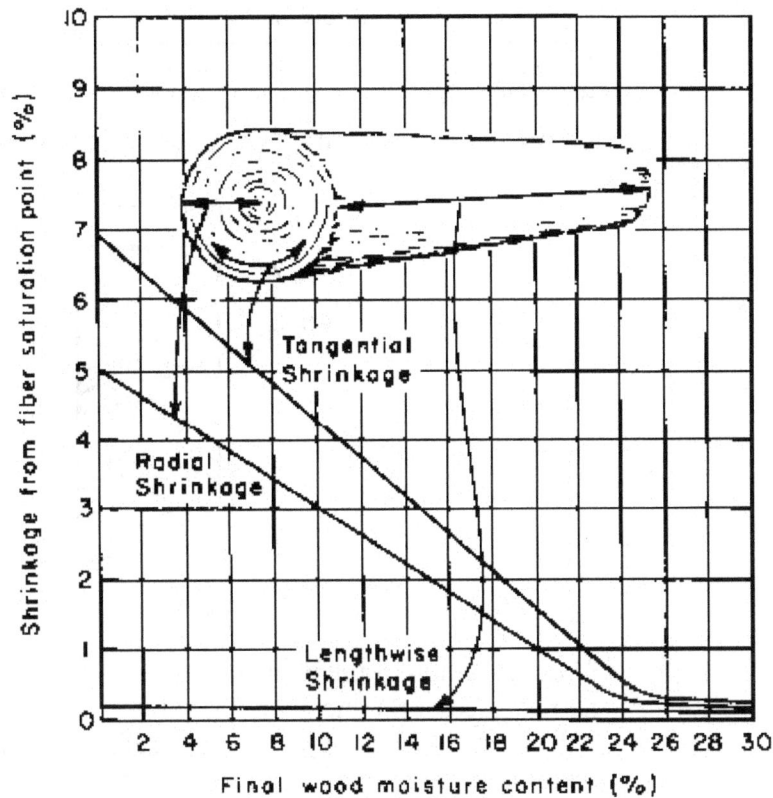

Figure 3-7.- Approximate wood shrinkage relationships below the fiber saturation point for the three orthotropic axes (adapted from the Canadian Wood Council [1]). Used by permission.

the tangential and radial directions, however, shrinkage is much more pronounced (Figure 3-7).

Wood shrinkage is approximately a linear function of moisture content, and dimensional changes below approximately 24 percent can be determined with reasonable accuracy. An example of shrinkage calculations based on the values given in Figure 3-7 is shown in Example 3-1. More accurate methods for computing shrinkage are given in the *Wood Handbook*.[30] Although formal shrinkage calculations are normally not required in structural design, the designer must be aware that wood is not a static material and that dimensional changes occur.

Example 3-1- Wood shrinkage from a decrease in moisture content

Determine the approximate changes in depth and width occurring when the wood member shown below dries from an initial moisture content of 28 percent to an equilibrium moisture content of 18 percent.

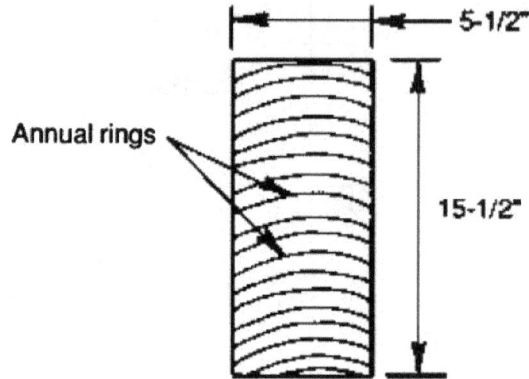

Dimensions at 28% MC

Solution

The orientation of the annual rings is approximately parallel with the narrow face of the member. Therefore, shrinkage in the tangential direction will affect member width, while shrinkage in the radial direction will affect member depth.

Approximate dimensional changes between two moisture contents are obtained from Figure 3-7. In the radial direction, shrinkage from fiber saturation to 28-percent moisture content is approximately 0.2 percent. At M-percent moisture content, the change is approximately 1.4 percent. The percent shrinkage in the radial direction between 28-percent and 18-percent moisture content is the difference between the two values:

Percent radial shrinkage = 1.4% - 0.2% = 1.2%

Applying the percentage shrinkage to the dimension at 28-percent moisture content gives the shrinkage in inches:

Radial shrinkage = 0.012(15.5 in.) = 0.2 in.

Shrinkage in the tangential direction is determined in the same manner. From fiber saturation, tangential shrinkage is approximately 0.3 percent at 28-percent moisture content and 2.1 percent at 18-percent moisture content:

Percent tangential shrinkage = 2.1% - 0.3% = 1.8%

Tangential shrinkage = 0.018(5.5 in.) = 0.1 in.

In summary, the member will shrink about 0.2 inch in depth and 0.1 inch in width.

The effects of uneven drying plus shrinkage differences in the tangential and radial direction can cause wood pieces to distort or warp (Figure 3-8). In addition, the uncontrolled drying or seasoning of wood frequently causes lengthwise separations of the wood across the annual rings, commonly known as checks (Figure 3-9). Most checks are not of structural significance; however, when checking extends from one surface to the opposite or adjoining surface (through-checks) the strength and other properties of the piece may be affected.

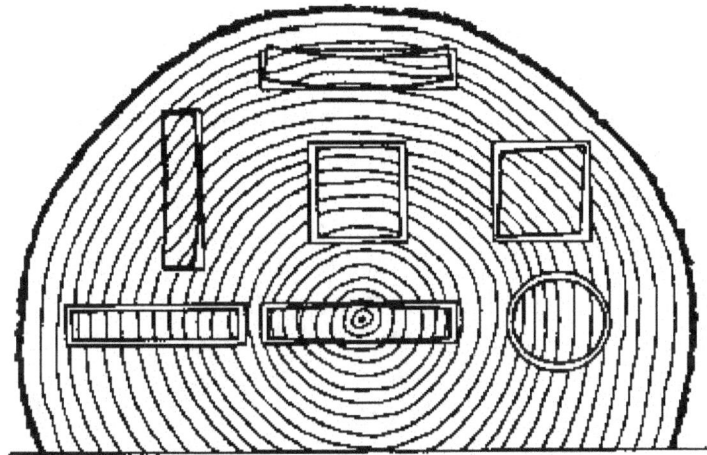

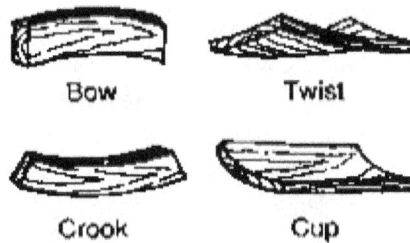

Bow   Twist

Crook   Cup

Figure 3-8.- Characteristic shrinkage and distortion of wood as affected by the direction of the growth rings. Such distortion can result in warp, which is generally classified as bow, twist, crook, and cup.

DENSITY

The density of a material is defined as the mass per unit volume at some specified condition. For a hygroscopic material such as wood, density depends on two factors, the weight of the basic wood substance and the weight of the moisture retained in the wood. Wood density varies with moisture content and must be given relative to a specific condition in order to have practical meaning. Values for density are generally based on the wood weight and volume at one of three moisture conditions: (1) ovendry, where the moisture content is zero; (2) green, where the moisture content is greater than 30 percent; or (3) in-use, where the moisture content is specified between ovendry and green.

3-11

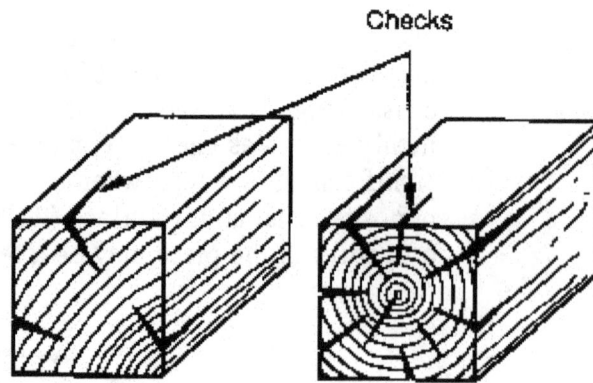

Figure 3-9.- Checks are lengthwise separations of the wood, perpendicular to the growth rings, caused by uncontrolled shrinkage in the tangential direction.

The density of ovendry wood varies within and among species. While the density of most species is between 20 and 45 lb/ft³, the range in densities extends from approximately 10 lb/ft³ for balsa to more that 65 lb/ft³ for some imported woods. Average densities for green wood and wood at different moisture contents are given in several reference publications.[4,30] For bridge applications, a density of 50 lb/ft³ is normally used as an average density for all species and moisture contents (Chapter 6).

SPECIFIC GRAVITY

Specific gravity provides a relative measure of the amount of wood substance contained in a sample of wood. It is a dimensionless ratio of the weight of a volume of wood at a specified moisture content to the weight of an identical volume of water at 62.4 lb/ft³. For example, a volume of wood with a specific gravity of 0.50 at some moisture content would have a density of 31.2 lb/ft³ (0.50 x 62.4 lb/ft³). In most applications, specific gravity is either reported on the basis of ovendry weight and green volume or ovendry weight and volume at 12 percent moisture content. For engineering purposes, specific gravity is normally based on the ovendry weight and the volume at 12 percent moisture content.

THERMAL EXPANSION

Thermal expansion of dry wood is positive in all directions; it expands when heated and contracts when cooled. The linear expansion coefficients of dry wood parallel to grain are generally independent of species and specific gravity and range from approximately 0.0000017 to 0.0000025 per degree Fahrenheit. The expansion coefficients perpendicular to grain are proportional to density and range from five to ten times greater than parallel to grain coefficients. Wood is a good insulator and does not respond rapidly to temperature changes in the environment. Therefore, its thermal expansion and contraction lag substantially behind temperature changes in the surrounding air.

Wood that contains moisture reacts to temperature changes in a manner different from that of dry wood. In most cases, thermal expansion and contraction are negligible compared to the expansion and contraction from moisture content changes. When moist wood is heated, it tends to expand because of normal thermal expansion and to shrink because of moisture loss from increased temperature. Unless the initial moisture content of the wood is very low (3 to 4 percent), the net dimensional change on heating is negative. Wood at intermediate moisture contents of approximately 8 to 20 percent will expand when first heated, then gradually shrink to a smaller volume as moisture is lost in the heated condition. In most bridge applications, the effects of thermal expansion and contraction in wood are negligible.

COEFFICIENTS OF FRICTION

The coefficients of friction for domestic softwoods vary little among species and depend on wood moisture content and roughness of the surface. On most materials, friction coefficients for dry wood increase as moisture increases to the fiber saturation point. Above the fiber saturation point, friction coefficients remain fairly constant until considerable free water is present. When the surface is flooded with water, the coefficients of friction decrease. The sliding coefficient of friction for wood is normally less than the static coefficient and depends on the speed of sliding. Sliding coefficients vary slightly with speed when the moisture content is less than approximately 20 percent. At higher moisture contents, sliding coefficients decrease substantially as speed increases. Coefficients of sliding friction for smooth, dry wood against a hard smooth surface average from 0.3 to 0.5. At intermediate wood moisture contents, values range from 0.5 to 0.7 and increase to 0.7 to 0.9 as the moisture content nears fiber saturation. Average coefficients of friction for several conditions are given in Table 3-3.

Table 3-3.- Average coefficients of friction for wood.

Wood condition	Friction against	Average coefficient of friction	
		Static	Sliding*
Dry	Unpolished steel	0.70	0.70
Green	Unpolished steel	0.40	0.15
Dry, smooth	Dry, smooth wood	0.60	—
Green, smooth	Green, smooth wood	0.83	—

* Based on a relative movement of 13 ft/sec.

ELECTRICAL CONDUCTIVITY

Dry wood is a good electrical insulator and exhibits only minor variations in conductivity relative to variations in species and density, but significant alterations in conductivity can be related to variations in grain orientation, temperature, and moisture content. The conductivity of wood is approximately twice that for parallel to grain than for perpendicular to grain, and generally doubles for each 18 °F increase in temperature. Although electrical properties generally have little effect on bridge design, the correlation between the electrical conductivity and moisture content is the basis for electrical resistance-type moisture meters that are commonly used in bridge inspection and other activities related to product manufacturing (Chapter 13).

PYROLYTIC PROPERTIES

The pyrolytic or fire properties of wood are perhaps the most misunderstood of all wood properties. Because wood burns, it is intuitively assumed that the performance of wood under fire conditions must be poor. In fact, the heavy wood members typically used in bridges provide a fire resistance comparable to, or greater than, that of other construction materials.

When wood is exposed to fire, the exterior portions of the member may ignite. If enough energy is focused on the member, sustained, self-propagating flaming will occur. The wood beneath the flame undergoes thermal decomposition and produces combustible volatiles that sustain the flame. However, as the wood burns, a char layer is formed that helps insulate the unburned wood from engrossing flames (Figure 3-10). As the surface char layer increases, the amount of combustible volatiles released from the uncharred wood decreases, and the rate of combustion slows. The depth of the char layer under constant fire exposure increases at a rate of approximately 1-1/2 inches per hour for Douglas-fir, but varies for other species and fire exposure conditions.

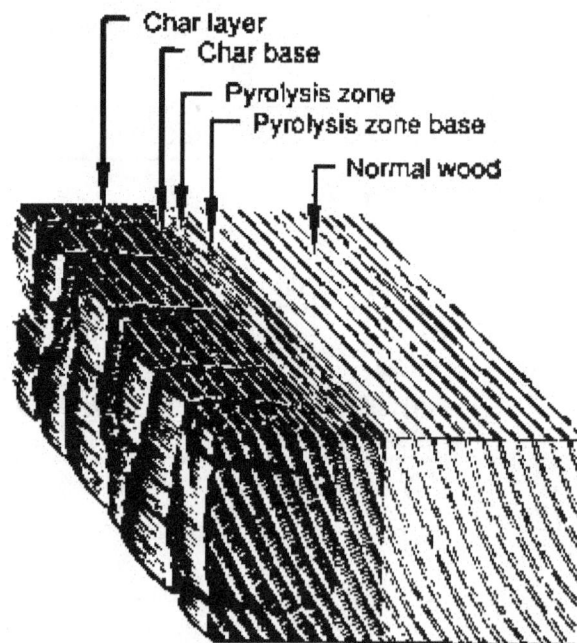

Figure 3-10.- Degradation zones in a wood section exposed to fire.

Although wood burns, its low specific gravity and thermal conductivity, combined with the insulating char layer, result in a slow rate of heat transmission into the solid, unburned wood. The surface chars, but the undamaged inner wood below the char remains at a relatively low temperature, thereby retaining its strength. As a result, the member will support loads equivalent to the capacity of the remaining uncharred section. It is this charring that allows wood to retain residual strength with surface temperatures of 1,500 °F or more. In addition, wood does not appreciably distort under high temperatures as most other materials do. When steel is subjected to elevated surface temperatures, its high mass density and thermal conductivity transport heat relatively quickly throughout the member. At temperatures of 1,500 °F, the yield strength of steel is less than 20 percent of that at room temperature.[21] Thus, under fire exposure, a steel member reaches its yield temperature and fails rapidly under structural load. A classic example of this scenario is shown in Figure 3-11.

Figure 3-11.- Damage resulting from a large building fire. Steel members yielded by the heat are supported by a charred wood beam.

NATURAL DURABILITY

The natural durability of wood, or its resistance to decay and insect attack, is related to species and anatomical characteristics. In general, the sapwood of all species has little resistance to deterioration and fails rapidly in adverse environments. When heartwood is considered, natural durability depends on species. Since the time of the Phoenicians, carpenters have known that the heartwood of some species exhibits greater durability in ground or marine environments. As discussed earlier, heartwood forms as the living sapwood cells gradually become inactive. In some species, sugars and other extraneous materials present in the cells are converted to highly toxic extractives that are deposited in the wood cell wall. In addition, some heartwoods contain internal crystalline deposits that inhibit attack by marine borers and insects. There are many species of wood in the world that provide durable heartwood, but few are found in North America. Baldcypress (old growth), cedars, and redwood are three North American commercial species that are recognized as naturally durable; however, durability varies within a tree and among species (Table 3-4). Because of this variability, it is unreliable to depend on natural durability for protection in structural applications, although many electric utilities continue to use untreated cedar poles installed in the 1930's. To ensure uniform performance, wood used in bridge applications is treated with wood preservatives that protect the structure from decay and deterioration for many years (Chapter 4).

Table 3-4.- Grouping of some domestic species according to approximate relative heartwood decay resistance.

Resistant or very resistant	Moderately resistant	Slightly or nonresistant
Baldcypress (old growth)	Baldcypress (young growth)	Hemlocks
Cedars	Douglas-fir	Pine (other than longleaf, slash, eastern white)
Redwood	Western larch	Spruces
	Eastern white pine	True firs (western and eastern)
	Longleaf pine	
	Slash pine	
	Tamarack	

From *Wood Handbook.*[30]

CHEMICAL RESISTANCE

Wood is resistant to many chemicals. In the chemical processing industry, it is the preferred material for processing and storing chemicals that are very corrosive to other materials. In isolated cases, the presence of strong acids or bases can cause wood damage. Strong bases attack the hemicellulose and lignin, leaving the wood a bleached white color. Strong acids attack the cellulose and hemicellulose, causing weight and strength losses. Chemical resistance is normally not a concern in bridge applications with the exception of de-icing chemicals that are used in some parts of the country. Because wood is resistant to these chemicals, it has a marked advantage over more vulnerable materials, such as steel and concrete.

Mechanical properties describe the characteristics of a material in response to externally applied forces. They include elastic properties, which measure resistance to deformation and distortion, and strength properties, which measure the ultimate resistance to applied loads. Mechanical properties are usually given in terms of stress (force per unit area) and strain (deformation per unit length).

The basic mechanical properties of wood are obtained from laboratory tests of small, straight-grained, clear wood samples free of natural growth characteristics that reduce strength. Although not representative of the wood typically used for construction, properties of these ideal samples are useful for two purposes. First, clear wood properties serve as a reference point for comparing the relative properties of different species. Second, they may serve as the source for deriving the allowable properties of visually graded sawn lumber used for design (Chapter 5).

ELASTIC PROPERTIES

Elastic properties relate a material's resistance to deformation under an applied stress to the ability of the material to regain its original dimensions when the stress is removed. For an ideally elastic material loaded below the proportional (elastic) limit, all deformation is recoverable, and the body returns to its original shape when the stress is removed. Wood is not ideally elastic, in that some deformation from loading is not immediately recovered when the load is removed; however, residual deformations are generally recoverable over a period of time. Although wood is technically considered a viscoelastic material, it is usually assumed to behave as an elastic material for most engineering applications, except for time-related deformations (creep), discussed later in this chapter and in Chapter 5.

For an isotropic material with equal properties in all directions, elastic properties are described by three elastic constants: modulus of elasticity (E), shear modulus (G), and Poisson's ratio (μ). Because wood is orthotropic, 12 constants are required to describe elastic behavior: 3 moduli of elasticity, 3 moduli of rigidity, and 6 Poisson's ratios. These elastic constants vary within and among species and with moisture content and specific gravity. The only constant that has been extensively derived from test data, or is required in most bridge applications, is the modulus of elasticity in the longitudinal direction. Other constants may be available from limited test data but are most frequently developed from material relationships or by regression equations that predict behavior as a function of density. General descriptions of wood elastic properties are given below with relative values for a limited number of species in Table 3-5. For additional information, refer to the references listed at the end of the chapter.[10,30]

Table 3.5. - Elastic ratios for selected species.

Species	Modulus of elasticity (E)		Shear modulus (G)			Poisson's ratios (μ)					
	E_T/E_L	E_R/E_L	G_{LR}/E_L	G_{LT}/E_L	G_{RT}/E_L	μ_{LR}	μ_{LT}	μ_{RT}	μ_{TR}	μ_{RL}	μ_{TL}
Coast Douglas-fir[a]	.050	.068	.064	.078	.007	.29	.45	.39	.37	.04	.03
Sitka Spruce[b]	.043	.078	.064	.061	.003	.37	.47	.44	.24	.04	.02
Loblolly Pine[c]	.079	.113	.081	.081	.013	.33	.29	.38	.36	—	—
Longleaf Pine[d]	.055	.102	.071	.061	.012	.33	.37	.38	.34	—	—

[a] Approximate specific gravity of 0.50 based on ovendry weight and volume at approximately 12 percent. From *Wood Handbook*.[30]

[b] Approximate specific gravity of 0.38 based on ovendry weight and volume at approximately 12 percent. From *Wood Handbook*.[30]

[c] Approximate specific gravity of 0.42 based on weight and volume at approximately 13 percent moisture content. From Bodig and Goodman.[10]

[d] Approximate specific gravity of 0.46 based on weight and volume at approximately 12 percent moisture content. From Bodig and Goodman.[10]

Modulus of Elasticity

Modulus of elasticity relates the stress applied along one axis to the strain occurring on the same axis. The three moduli of elasticity for wood are denoted E_L, E_R, and E_T to reflect the elastic moduli in the longitudinal, radial, and tangential directions, respectively. For example, E_L, which is typically denoted without the subscript L, relates the stress in the longitudinal direction to the strain in the longitudinal direction.

Shear Modulus

Shear modulus relates shear stress to shear strain. The three shear moduli for wood are denoted G_{LR}, G_{LT} and G_{RT} for the longitudinal-radial, longitudinal-tangential, and radial-tangential planes, respectively. For example, G_{LR} is the shear modulus based on the shear strain in the LR plane and the shear stress in the LT and RT planes.

Poisson's Ratio

Poisson's ratio relates the strain parallel to an applied stress to the accompanying strain occurring laterally. For wood, the six Poisson's ratios are denoted μ_{LR}, μ_{LT}, μ_{RT}, μ_{TL}, and μ_{TL}. The first letter of the subscript refers to the direction of applied stress, the second letter the direction of the accompanying lateral strain. For example, μ_{LR} is Poisson's ratio for stress along the longitudinal axis and strain along the radial axis.

STRENGTH PROPERTIES

Strength properties describe the ultimate resistance of a material to applied loads. They include material behavior related to compression, tension, shear, bending, torsion, and shock resistance. As with other wood properties, strength properties vary in the three primary directions, but differences between the tangential and radial directions are relatively minor and are randomized when a tree is cut into lumber. As a result, mechanical properties are collectively described only for directions parallel to grain and perpendicular to grain, as previously discussed.

Compression

Wood can be subjected to compression parallel to grain, perpendicular to grain, or at an angle to grain (Figure 3-12). When compression is applied parallel to grain, it produces stress that deforms (shortens) the wood cells along their longitudinal axis. Recalling the straw analogy discussed in Section 3.2, each cell acts as an individual hollow column that receives lateral support from adjacent cells and from its own internal structure. At failure, large deformations occur from the internal crushing of the complex cellular structure. The average strength of green, clear wood specimens of coast Douglas-fir and loblolly pine in compression parallel to grain is approximately 3,784 and 3,511 lb/in^2, respectively.[7]

Compression parallel to grain tends to shorten wood cells along their longitudinal axes.

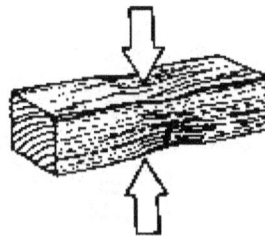

Compression perpendicular to grain compresses the wood cells perpendicular to their longitudinal axes.

Compression at an angle to grain results in compression acting both parallel and perpendicular to grain.

Figure 3-12.- Compression in wood members.

When compression is applied perpendicular to grain, it produces stress that deforms the wood cells perpendicular to their length. Again recalling the straw analogy, wood cells collapse at relatively low stress levels when loads are applied in this direction. However, once the hollow cell cavities are collapsed, wood is quite strong in this mode because no void space exists. Wood will actually deform to about half its initial thickness before complete cell collapse occurs, resulting in a loss in utility long before failure. For compression perpendicular to grain, failure is based on the accepted performance limit of 0.04 inch deformation. Using this convention, the average strength of green, clear wood specimens of coast Douglas-fir and loblolly pine in compression perpendicular to grain is approximately 700 and 661 lb/in², respectively.[7]

Compression applied at an angle to grain produces stress acting both parallel and perpendicular to the grain. The strength at an angle to grain is therefore intermediate to these values and is determined by a compound strength equation (the Hankinson formula) discussed in Chapter 5.

Tension

The mechanical properties for wood loaded in tension parallel to grain and for wood loaded in tension perpendicular to grain differ substantially (Figure 3-13). Parallel to its grain, wood is relatively strong in tension. Failure occurs by a complex combination of two modes, cell-to-cell slippage and cell wall failure. Slippage occurs when two adjacent cells slide past one another, while cell wall failure involves a rupture within the cell wall. In both modes, there is little or no visible deformation prior to complete failure. The average strength of green, clear wood specimens of interior-north Douglas-fir and loblolly pine in tension parallel to grain is approximately 15,600 and 11,600 lb/in², respectively.[20]

Tension parallel to grain stretches wood cells parallel to their longitudinal axes.

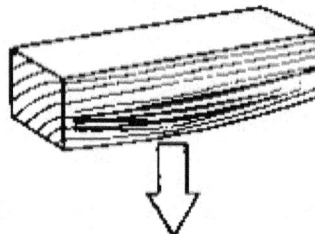

Tension perpendicular to grain tends to separate wood cells perpendicular to their axes, where resistance is low. Situations that induce this type of stress should be avoided in design.

Figure 3-13.- Tension in wood members.

In contrast to tension parallel to grain, wood is very weak in tension perpendicular to grain. Stress in this direction acts perpendicular to the cell lengths and produces splitting or cleavage along the grain that significantly affects structural integrity. Deformations are usually low prior to failure because of the geometry and structure of the cell wall cross section. Strength in tension perpendicular to grain for green, clear samples of coast Douglas-fir and loblolly pine averages 300 and 260 lb/in^2, respectively.[30] However, because of the excessive variability associated with tension perpendicular to grain, situations that induce stress in this direction must be recognized and avoided in design.

Shear

There are three types of shear that act on wood: vertical, horizontal, and rolling (Figure 3-14). Vertical shear is normally not considered because other failures, such as compression perpendicular to grain, almost always occur before cell walls break in vertical shear. In most cases, the most important shear in wood is horizontal shear, acting parallel to the grain. It produces a tendency for the upper portion of the specimen to slide in relation to the lower portion by breaking intercellular bonds and deforming the wood cell structure. Horizontal shear strength for green, small clear samples of coast Douglas-fir and loblolly pine averages 904 and 863 lb/in^2, respectively.[7]

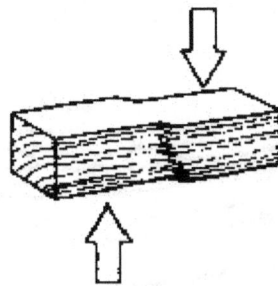

Vertical shear tends to deform wood cells perpendicular to their longitudinal axes. This type of shear in normally not considered for wood because other types of failures will occur before failure in vertical shear.

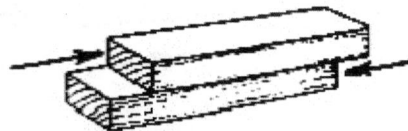

Horizontal shear produces a tendency for wood cells to separate and slide longitudinally. It is normally the controlling type of shear for wood members.

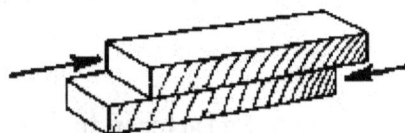

Rolling shear produces a tendency for the wood cells to roll over one another, transverse to their longitudinal axes. This type of shear is normally not a consideration for solid or laminated

Figure 3-14.- Shear in wood members.

In addition to vertical and horizontal shear, a less common type called rolling shear may also develop in wood. Rolling shear is caused by loads acting perpendicular to the cell length in a plane parallel with the grain. The stress produces a tendency for the wood cells to roll over one another. Wood has low resistance to rolling shear, and failure is usually preceded by large deformations in the cell cross sections. Test procedures for rolling shear in solid wood are of recent origin and few test values are available. In general, rolling shear strength for green, clear wood specimens average 18 to 28 percent of the shear strength parallel to grain.

Bending

When wood specimens are loaded in bending, the portion of the wood on one side of the neutral axis is stressed in tension parallel to grain, while the other side is stressed in compression parallel to grain (Figure 3-15). Bending also produces horizontal shear parallel to grain, and compression perpendicular to grain at the supports. A common failure sequence in simple bending is the formation of minute compression failures followed by the development of macroscopic compression wrinkles. This effectively results in a sectional increase in the compression zone and a section decrease in the tension zone, which is eventually followed by tensile failure. The ultimate bending strength of green, clear wood specimens of coast Douglas-fir and loblolly pine are reached at an average stress of 7,665 and 7,300 lb/in^2, respectively.[7]

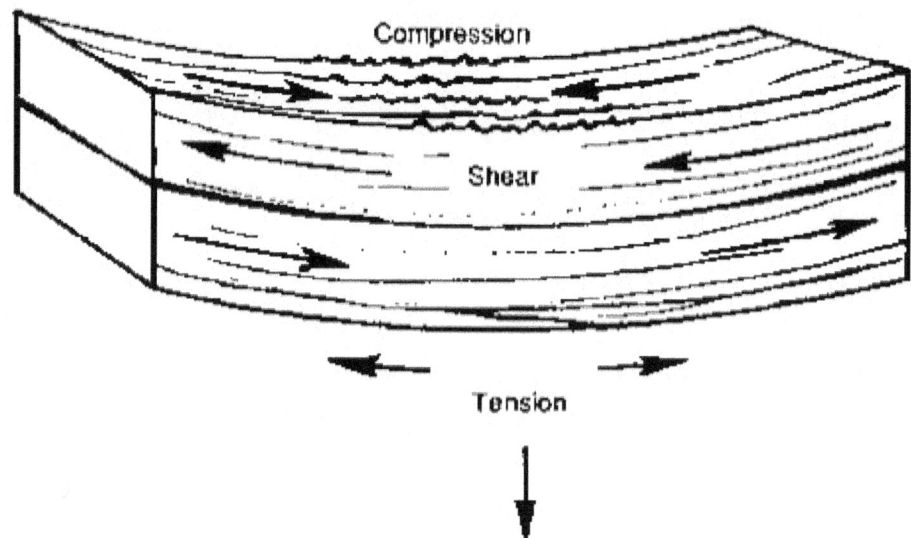

Figure 3-15.- Bending in wood members produces tension and compression in the extreme fibers, horizontal shear, and vertical deflection.

Torsion

Torsion is normally not a factor in timber bridge design, and little information is available on the mechanical properties of wood in torsion. Where needed, the torsional shear strength of solid wood is usually taken as the shear strength parallel to grain. Two-thirds of this value is assumed as the torsional strength at the proportional limit.[30]

Shock Resistance

Shock resistance is the ability of a material to quickly absorb, then dissipate, energy by deformation. Wood is remarkably resilient in this respect and is often a preferred material when shock loading is a consideration. Several parameters are used to describe energy absorption, depending on the eventual criteria of failure considered. Work to proportional limit, work to maximum load, and toughness (work to total failure) describe the energy absorption of wood materials at progressively more severe failure criteria.[30]

3.5 FACTORS AFFECTING THE STRENGTH AND OTHER PROPERTIES OF WOOD

Prior to this point, discussions of wood properties have been based on small, clear, straight-grain wood without strength-reducing characteristics. Clear wood properties are important, but they by no means represent the characteristics or performance of wood products used in structural applications. Because wood is a biological material, it is subject to variations in structure or properties or both resulting from (1) anatomical factors related to growth characteristics, (2) environmental factors related to the environmental conditions where wood is used, and (3) service factors related to applied loads or chemical treatments.

ANATOMICAL FACTORS

Anatomical factors involve variations in wood structure caused primarily by natural processes or growth influences. They include specific gravity, slope of grain, knots, abnormal wood, compression failures, and shake and pitch pockets.

Specific Gravity

The strength of clear wood is generally related to the relative weight of wood per unit volume, or specific gravity. The higher the specific gravity, the more wood material per unit volume and the higher the strength. However, because specific gravity depends on the amount of water in the wood, comparisons have no practical meaning unless measured at the same moisture content. In addition, specific gravity can be misleading in some specimens because gums, resins, and extractives increase specific gravity but contribute little to mechanical properties. In general, the specific gravity of wood is directly proportional to the amount of latewood. Therefore, the higher the percentage of latewood, the higher the specific gravity and strength of the specimen.

Slope of Grain

Slope of grain or cross grain are terms used to describe the deviation in wood fiber orientation from a line parallel to the edge of the specimen (Figure 3-16). It is expressed as a ratio such as 1 in 6 or 1 in 14 and is measured over sufficient distance along the piece to be representative of the general slope of the wood fibers. Slope of grain has a significant effect on wood mechanical properties, and strength decreases as the grain deviation increases. Specimens with severe cross grain are also more susceptible to warp and other dimensional deformations because of changes in moisture content. Two common types of cross grain are spiral grain and diagonal grain (Figure 3-17).

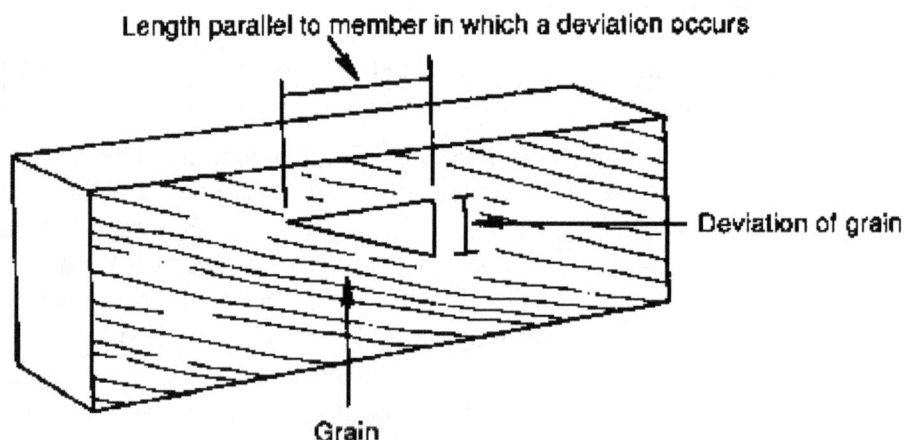

Length parallel to member in which a deviation occurs

Deviation of grain

Grain

Figure 3-16.- Slope of grain measurement in wood members.

Knots

As a tree grows, buds develop and branches grow laterally from the trunk. The branches produce deviations in the normal wood growth patterns that result in two types of knots when the wood is cut, intergrown knots and encased knots (Figure 3-18). Intergrown knots are formed by living branches, while encased knots result from branches that have died and subsequently have been surrounded by the wood of the growing trunk.

Knots reduce the strength of wood because they interrupt the continuity and direction of wood fibers. They can also cause localized stress concentrations where grain patterns are abruptly altered. The influence of a knot depends on its size, location, shape, soundness, and the type of stress considered. In general, knots have a greater effect in tension than in compression, whether stresses are applied axially or as a result of bending. Intergrown knots resist some kinds of stress but encased knots or knotholes resist little or no stress. At the same time, grain distortion is greater around an intergrown knot than around an encased knot of equivalent size. As a result, the overall effects of each are approximately the same.

3-24

Abnormal Wood

Several growth characteristics or influences can lead to the formation of abnormal wood, which differs in structure and properties from normal wood. The most important abnormal wood formations are associated with reaction wood and juvenile wood.

Reaction Wood

Reaction wood is abnormal wood produced by a tree in response to irregular environmental or physical stresses associated with a leaning trunk

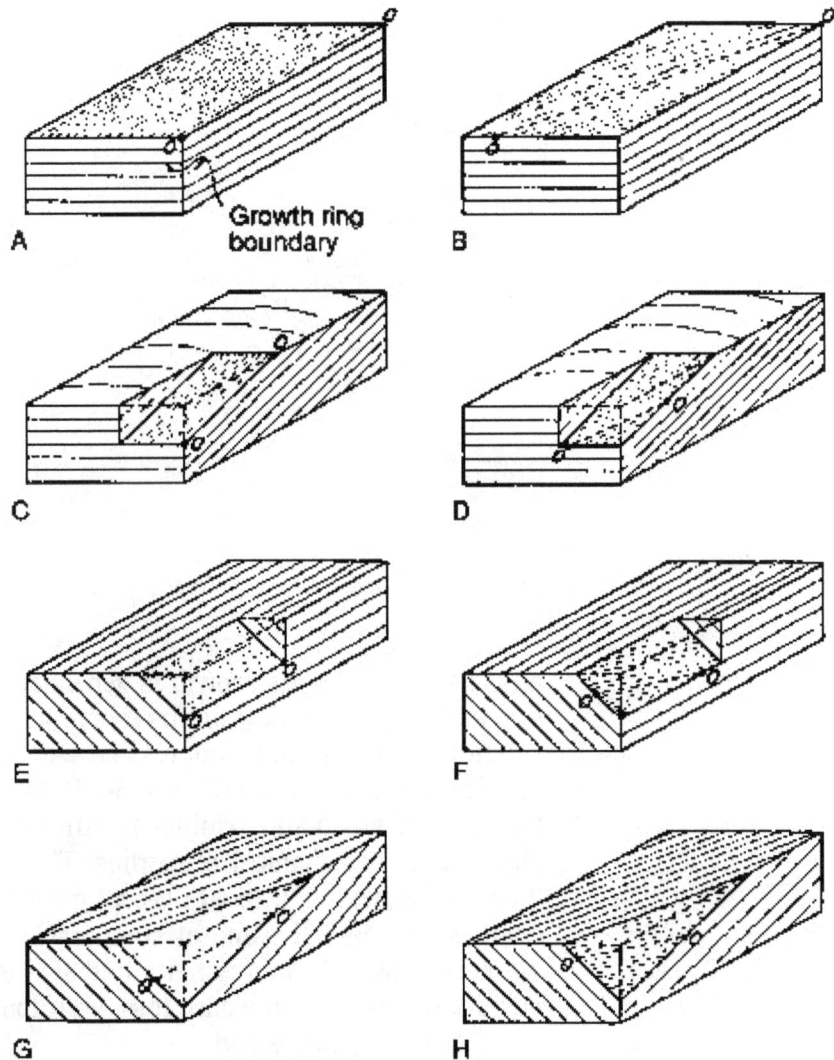

Figure 3-17- Schematic views of wood specimens containing straight grain and cross grain to illustrate the relationship of fiber orientation (O-O) to the axes of the piece. Specimens A through D have radial and tangential surfaces; E through H do not. A through E contain no cross grain. B, D, F, and H have spiral grain. C, D, G, and H have diagonal grain.

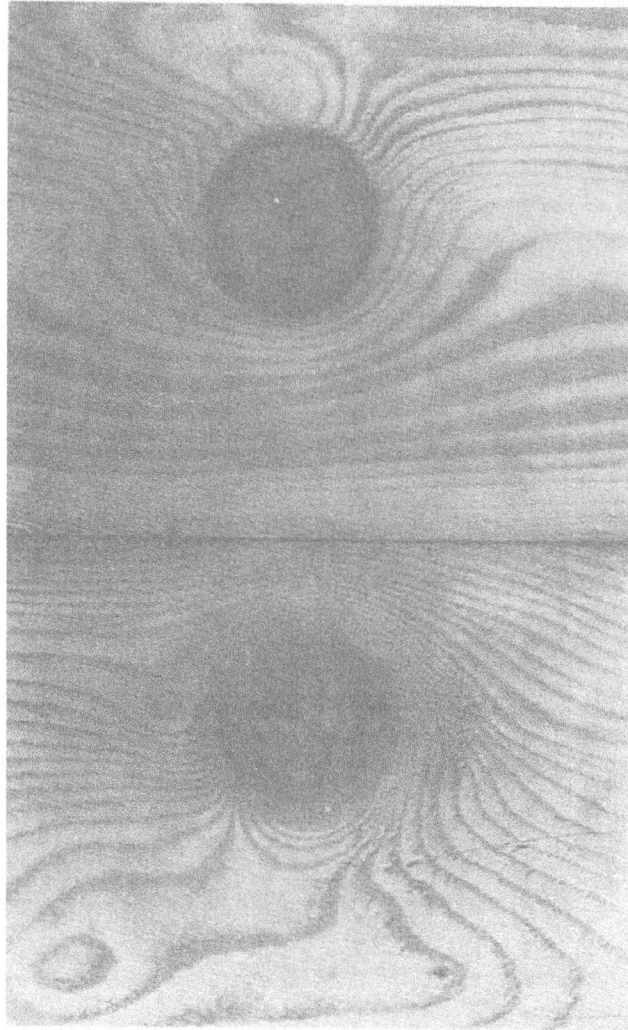

Figure 3-18.- Types of knots. (Top) encased knot and (bottom) intergrown knot.

and crooked limbs. Its growth is generally believed to be a response by the tree to return the trunk or limbs to a more natural position. In softwoods, reaction wood is called compression wood and is found on the lower side of a leaning tree or limb (Figure 3-19). It is denser and generally weaker than normal wood and exhibits significant differences in anatomical, physical, and mechanical properties. The specific gravity of compression wood is frequently 30 to 40 percent greater than normal wood, but when compared to normal wood of comparable specific gravity, compression wood is weaker. Compression wood also exhibits abnormal shrinkage characteristics from moisture loss, with longitudinal shrinkage up to 10 times that of normal wood.

Juvenile Wood

Wood cells produced by a tree in the first years of growth exhibit variations in wood cell structure distinct from cell structure in wood that develops in later years. This wood, known as juvenile wood, has lower

Figure 3-19.-(A) Eccentric growth about the pith in a cross section containing compression wood. The dark area in the lower third of the cross section is compression wood. (B) Axial tension break caused by excessive longitudinal shrinkage of compression wood. (C) Warp caused by excessive longitudinal shrinkage of compression wood.

strength properties and an increased susceptibility to warpage and longitudinal shrinkage. The duration of juvenile wood production varies for species and site conditions from approximately 5 to 20 years. In large-diameter, old-growth trees, the proportion of juvenile wood is small, and its effects in structural applications have been negligible. However, juvenile wood has recently become a more prevalent consideration within the wood industry because of the trend toward processing younger, smaller diameter trees as the large-diameter, old-growth trees become difficult to obtain.

Compression Failures

Extreme bending in trees from environmental conditions or mishandling during or after harvest can produce excessive compression stress parallel to grain that results in minute compression failures of the wood structure. In some cases, these failures are visible on the wood surface as minute lines or zones formed by the crumpling or buckling of the cells (Figure 3-20 A). They may also be indicated by fiber breakage on the end grain (Figure 3-20 B). Compression failures can result in low shock resistance and strength properties, especially in tension where strength may be less than one-third that of clear wood. Even slight compression failures, visible only with the aid of a microscope, can seriously reduce strength and cause brittle fractures.

3-27

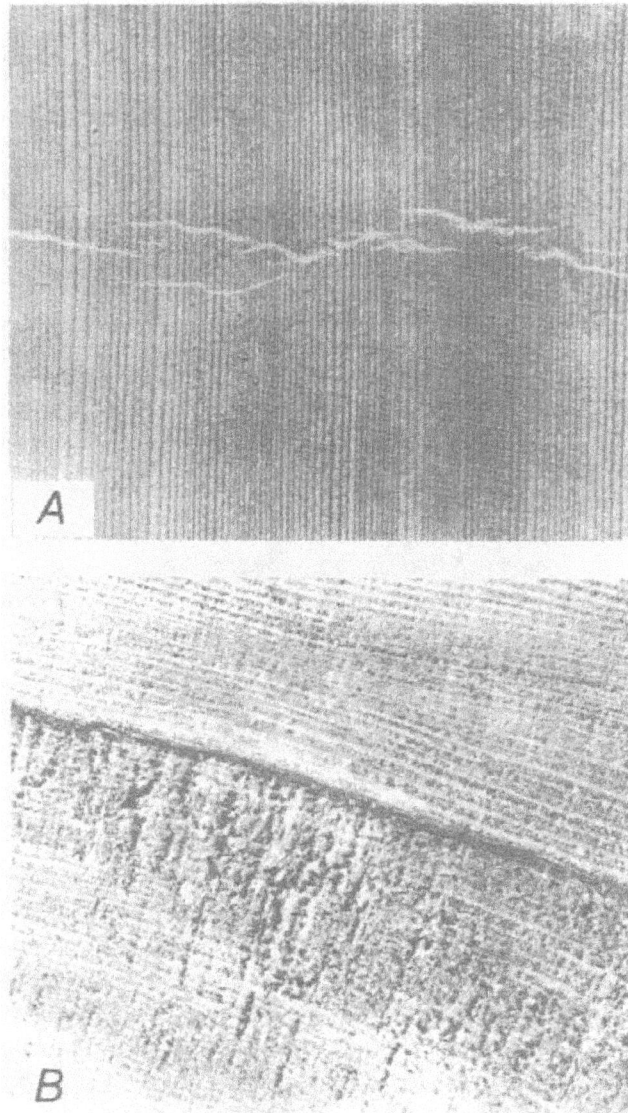

Figure 3-20.–(A) Compression failure is shown by irregular lines across the grain. (B) End-grain surface showing fiber breakage caused by compression failures below the dark line.

Shake and Pitch Pockets

Two natural characteristics in wood structure that can affect strength are shake and pitch pockets (Figure 3-21). A shake is a separation or plane of weakness between two adjacent growth increments. It is thought to occur because of excessive stresses imposed on the standing tree, or during harvest, and can extend a substantial distance in the longitudinal direction. Pitch pockets are well-defined openings that contain free resin. They extend parallel to the annual growth rings and are usually flat on the pith side and curved on the bark side. Pitch pockets are normally localized and do not extend far in the longitudinal direction. In bending specimens, shakes can severely reduce shear strength but usually have little effect on specimens subjected only to tension or compression. Pitch pockets generally have no significant effect on strength, but a large number of pitch pockets may indicate the presence of shake and a lack of bond between annual growth layers, which may result in some strength loss, particularly in shear.

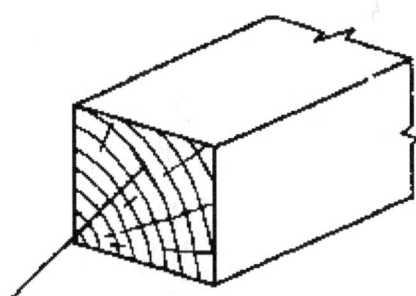

A pitch pocket is a well-defined opening that contains free resin. Pitch pockets are usually localized and do not extend far in the longitudinal direction.

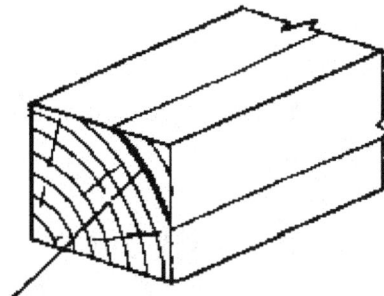

A shake is a separation or plane of weakness between growth increments that may extend a substantial distance in the longitudinal direction.

Figure 3-21.-Drawing of a pitch pocket and a shake.

ENVIRONMENTAL FACTORS

Environmental factors are related to the effects of the surroundings on the performance and properties of wood. They include moisture content, temperature, decay and insect damage, and ultraviolet degradation.

Moisture Content

The strength and stiffness of wood are related to moisture content between the ovendry condition and the fiber saturation point.[17] When clear wood is dried below the fiber saturation point, strength and stiffness increase. When clear wood absorbs moisture below the fiber saturation point, strength and stiffness decrease. Wood properties in both directions are recoverable to their original values when the moisture content is restored. The approximate middle-trend effects of moisture content on the mechanical properties of clear wood are shown in Table 3-6.

When wood contains strength-reducing characteristics (primarily knots), wood properties are currently assumed to be linearly related to moisture content for specimens up to 4 inches thick. However, recent research indicates that the effects of moisture content are not linear.[20] In wood with small strength-reducing characteristics, properties increase linearly with decreasing moisture content. In wood with large strength-reducing characteristics, however, there may be no increase in strength as the wood dries because the potential strength increases are offset by losses from shrinkage and seasoning defects. Although these effects have not yet been recog-

Table 3-6.- Approximate middle-trend effects of moisture content on the mechanical properties of clear wood at about 68 °F.

| Property | Relative change in property from 12% moisture content | |
	At 6% moisture content	At 20% moisture content
Modulus of elasticity parallel to grain	+9	-13
Modulus of elasticity perpendicular to grain	+20	-23
Shear modulus	+20	-20
Bending strength	+30	-25
Tensile strength parallel to grain	+8	-15
Compressive strength parallel to grain	+35	-35
Shear strength parallel to grain	+18	-18
Tensile strength perpendicular to grain	+12	-20
Compressive strength perpendicular to grain at proportional limit	+30	-30

From *Wood Handbook.* [30]

nized in existing codes and standards, it is likely that they will be incorporated in the near future.

Temperature

In general, the mechanical properties of wood decrease when it is heated and increase when it is cooled. This temperature effect is immediate and, for the most part, recoverable for short heating durations as long as wood is not exposed to temperatures higher than 150 °F for extended periods. A permanent reduction in strength results from degradation of the wood substance if exposure to temperatures higher than 150 °F occurs. The magnitude of these permanent effects depends on the moisture content, heating medium, temperature, exposure time, and, to a lesser extent, species and specimen size. [17,22] In most cases, temperature is not a factor in bridge design (Chapter 5).

Decay and Insect Damage

Under certain conditions, wood may be subject to deterioration from decay or insect damage. Decay effects on strength can be many times greater than visual observation indicates, with possible strength losses of 50 to 70 percent for a corresponding weight loss of only 3 percent. Insects that use wood as food or shelter can also remove a substantial portion of the wood structure and severely alter strength and other properties. Fortunately, wood preservatives have been developed that protect wood from decay and insect attack (Chapter 4). Additional discussions on the agents of wood deterioration and decay effects on strength are in Chapter 13.

Ultraviolet Degradation

Wood exposed to ultraviolet radiation in sunlight undergoes chemical reactions that cause photochemical degradation, primarily in the lignin component. This produces a characteristic grayish wood color in a process commonly known as weathering (Figure 3-22). As the wood surface degrades, cells erode and new wood cells are exposed, continuing the process. However, because this degradation is very slow, occurring at an estimated rate of only 1/4 inch per century, its impact is mainly one of aesthetics without serious effects on mechanical properties. Most wood preservative treatments (except waterborne preservatives discussed in Chapter 4) and opaque and semitransparent finishes inhibit weathering, which is normally not a concern in structural applications.

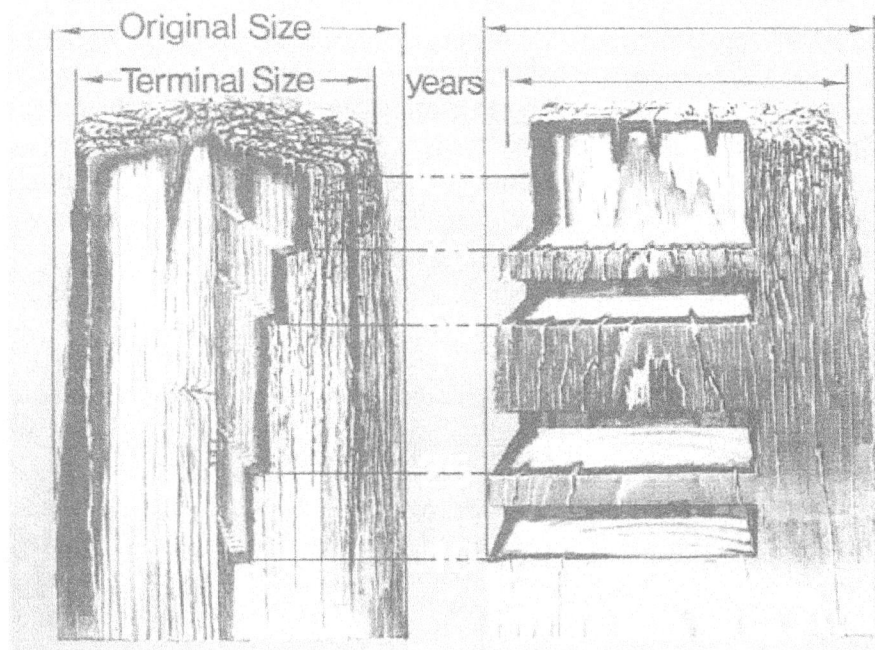

Figure 3-22.- Artist's rendition showing the weathering process of round and square timbers. Cutaway shows that interior wood below the surface is relatively unchanged.

SERVICE FACTORS

Service factors are related to the loading and chemical treatment of wood. They include duration of load, creep, fatigue, and treatment factors.

Duration of Load

Wood exhibits the unique property of carrying substantially greater maximum loads for short durations than for long periods. The shorter the duration of load, the higher the ultimate strength of the wood. Long-term tests have also shown that a series of intermittent loads produces the same cumulative effects on strength as a continuous load of equivalent duration.[18] For example, a load applied for alternating years over a 50-year period would have the same effect as the same load applied continuously for 25 years. For structural applications, wood strength values are based on an assumed normal load duration of 10 years (Chapter 5). Based on this assumption, the relationship of strength to duration of load is shown in Figure 3-23.

Creep

Duration of load affects the deformation of wood specimens subject to bending. For loads of relatively short duration, wood deflects elastically and essentially recovers its original position when the load is removed. Under sustained loading, however, wood exhibits an additional time-dependent deformation known as creep, which is not recoverable when the load is removed. Creep develops at a slow but persistent rate that increases with temperature and moisture content. Creep is discussed in more detail in Chapter 5.

Fatigue

Fatigue is the progressive damage that occurs in a material subjected to cyclic loading. The fibrous structure of wood is resistant to fatigue failure. At comparable stress levels relative to ultimate strength, the fatigue

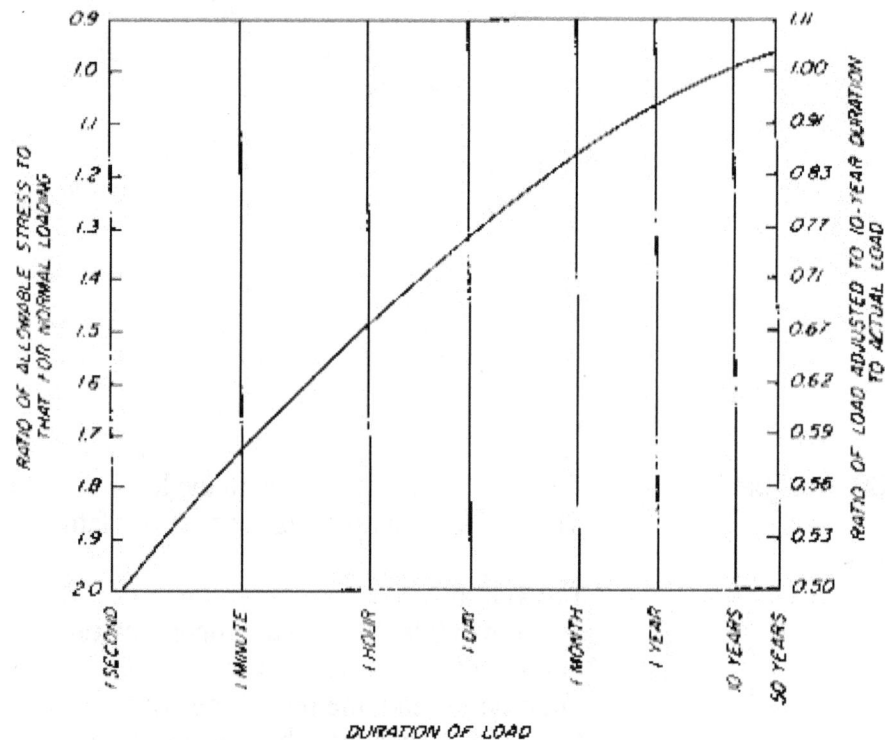

Figure 3-23.- Duration of load adjustment curve for wood.

3-32

strength of wood is often several times that of most metals.[21,30] The potential for fatigue-related failures in wood is generally considered to be minor, provided the stress cycles in bending do not exceed the proportional limit in bending. Fatigue is not normally a consideration in bridge design.

Treatment Factors

During the manufacturing process, wood may be treated with preservatives or fire-retardant chemicals to improve its performance and longevity in adverse environments (Chapter 4). Applied chemicals or treatment processes can affect the properties of wood in some situations. When wood preservatives are considered, oil-type preservatives do not react with the cell wall components, and no appreciable strength loss from the chemicals occurs.[9] When waterborne preservatives are used, the preservatives do react with cell wall components, and strength can be affected.[37] The only strength reduction currently recognized for waterborne preservatives is related to load duration increases for members treated with heavy retentions required for saltwater use (see Chapters 4 and 5); however, research is in progress to investigate additional effects of some waterborne chemicals on wood strength and ductility. For both oil-type and waterborne preservatives, significant reductions in strength and other wood properties can occur when treatment processes exceed the temperatures or pressures allowed by treating specifications. When proper preservative treatment procedures and limitations are followed, no significant alteration in wood properties is found.

In contrast to wood preservatives, treatment with fire-retardant chemicals can have a marked effect on wood strength and other properties. With fire retardants, the chemicals react with the cell wall components and cause substantial strength reductions.[37] As a general rule, fire retardants are not used in bridge applications. When they are, strength values must be reduced accordingly (Chapter 5).

3.6 PROPERTIES OF SAWN LUMBER

Square or rectangular lengths of wood that are cut from logs are called sawn lumber. Sawn lumber is the product of a sawmill and planing mill and is usually not manufactured beyond sawing, resawing, passing lengthwise through a standard planing machine, crosscutting to length, and matching. Sawn lumber is the most widely used of all timber products and is a primary material for timber bridge construction. Millions of board feet of lumber are produced each year from sawmills located in all parts of the United States.

As lumber is cut from a log, its quality and properties vary. To enable users to purchase the material that suits their particular purposes, sawn lumber is graded into categories of quality or appearance or both. Generally, the grade of a piece of lumber is based on the number and type of features that may lower the strength, durability, or utility of the lumber. Sawn lumber categories and grades are intended for a variety of purposes. For bridges and many other structural uses, sawn lumber categorized as stress-graded structural lumber is used almost exclusively. Structural lumber is graded primarily to provide design values in strength and stiffness. Further discussions in this section are limited to structural lumber only.

PRODUCT STANDARDS

Prior to the early 1900's, the manufacture and grading of sawn lumber was comparatively simple because most sawmills marketed their lumber locally, and grades had only local significance. As new timber sources were developed and lumber was transported to distant points, the need for some degree of standardization in lumber size, grade characteristics, and grade names became necessary. The U.S. Department of Commerce, in cooperation with lumber producers, distributors, and users, formulated a voluntary American Softwood Lumber Standard. The current version of that standard is the American Softwood Lumber Standard PS 20-70 (ALS).[31] The ALS serves as the basic product standard for structural lumber produced in the United States. When lumber conforms to the basic size, grading, labeling, and inspection provisions of the ALS, it may be designated as *American Standard Lumber*.

The objective of the ALS is to provide a reliable level of product standardization, yet allow enough flexibility for more specialized products on a regional basis. To accomplish this with structural lumber, the ALS provides for a National Grading Rule (NGR). The ALS and the NGR prescribe the ways in which stress-grading principles can be used to formulate grading rules said to be American Standard. Specifically, they contain information and standards related to lumber sizes, grade names, and grade descriptions. A grade description denotes the maximum number and location of strength-reducing characteristics that are allowed in a particular grade of lumber, and places limitations on other non-strength-reducing characteristics. All American Standard Lumber that is less than 5 inches thick must conform to the NGR and its requirements for standard sizes, grade names, and grade descriptions. For lumber that is 5 inches or more in thickness, the ALS specifies standard sizes, but grade names and grade descriptions are written, published, and certified by independent industry groups called grading rules agencies. Although grade names and descriptions for these lumber grades basically follow the NGR, there are minor differences among different grading rules agencies. Grade names and descriptions written by the grading rules agencies must be certified by the American Lumber Standards Committee before they can be considered as American Standard Lumber. A listing of the United States agencies that are currently certified to write grading rules is given in Table 3-7.

LUMBER MANUFACTURE

Lumber production starts when merchantable timber is felled, limbed, cut into logs, and transported to a sawmill for conversion into lumber. At the sawmill, the first step in lumber manufacture is generally log debarking, after which the logs are sawn into lumber (Figure 3-24). Softwood lumber can be cut from a log in two ways: tangent to the growth rings to produce flat-grain lumber or radially to the rings to produce edge-grain lumber. In commercial practice, most lumber falls somewhere in between, and lumber with rings at angles of 45 to 90 degrees to the wide face is considered edge-grain lumber, while lumber with rings at angles less than 45 degrees to the wide face is considered flat-grain lumber (Table 3-8). After cutting, lumber can be surfaced (planed) and shipped green, or dried and surfaced later. Most lumber 2 inches thick or less is either air-dried or kiln-dried before it is surfaced. For larger lumber sizes, it is impractical to dry the lumber, and it is generally shipped green.

Figure 3-24.-A log being sawn into lumber at a modern sawmill (photo courtesy of Frank Lumber Co.).

Lumber Species

Lumber is manufactured from a great variety of species. The commercial names of these species may vary from the official tree names adopted by the USDA Forest Service. In addition, some species with approximately the same mechanical properties are marketed together in species groups. The commercial designation Southern Pine, for example, is actually a species group comprised of loblolly pine, shortleaf pine, longleaf pine, slash pine, and others. Standard lumber names adopted by the ALS are shown in Table 3-9. Information regarding species and species groups not

Table 3-7.- U.S. grading rules agencies certified to write grading rules.

Agency	Lumber type
Northeastern Lumber Manufacturers Association (NELMA) 272 Tuttle Road P.O. Box 87A Cumberland Center, ME 04021	Aspen, balsam fir, beech, birch, eastern hemlock, eastern white pine, red pine, black spruce, white spruce, red spruce, pitch pine, tamarack, jack pine, northern white cedar, hickory, maple, red oak, white oak
Northern Hardwood and Pine Manufacturers Association (serviced by NELMA) 272 Tuttle Road P.O. Box 87A Cumberland Center, ME 04021	Aspen, cottonwood, balsam fir, eastern white pine, red pine, eastern hemlock, black spruce, white spruce, red spruce, pitch pine, tamarack, jack pine, yellow poplar
Redwood Inspection Service (RIS) 591 Redwood Highway, Suite 3100 Mill Valley, CA 94941	Redwood
Southern Pine Inspection Bureau (SPIB) 4709 Scenic Highway Pensacola, FL 32504	Longleaf pine, slash pine, shortleaf pine, loblolly pine, Virginia pine, pond pine, pitch pine
West Coast Lumber Inspection Bureau (WCLIB) Box 23145 6980 SW. Varns Road Portland, OR 97223	Douglas-fir, western hemlock, western redcedar, incense-cedar, Port-Orford-cedar, Alaska-cedar, western true firs, mountain hemlock, Sitka spruce
Western Wood Products Association (WWPA) 1500 Yeon Building Portland, OR 97204	Ponderosa pine, western white pine, Douglas-fir, sugar pine, western true firs, western larch, Engelmann spruce, incense-cedar, western hemlock, lodgepole pine, western redcedar, mountain hemlock, red alder, aspen

From Wood Handbook.[30]

listed in this table should be obtained from the appropriate grading rules organizations (Table 3-7).

Lumber Sizes
Structural lumber is manufactured in many sizes depending on use requirements. Lengthwise, it is normally produced in even, 2-foot increments. In width and thickness, common sizes vary from 2 to 16 inches,

Table 3-8.- Some relative advantages of flat-grain and edge-grain lumber.

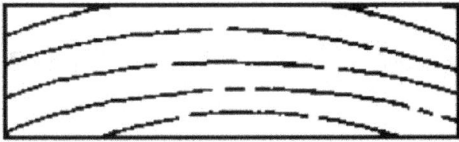

Flat-grain plank; rings form an angle less than 45° with the wide surface.

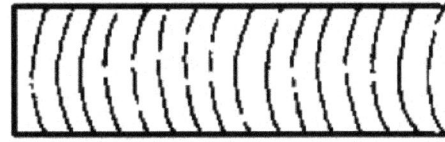

Edge-grain plank; rings form an angle of 45° to 90° with the wide surface.

Round or oval knots that may occur in flat-grain lumber affect the surface appearance less than spike knots that may occur in edge-grain lumber. Also, lumber with a round or oval knot is not as weak as lumber with a spike knot.

Shakes and pitch pockets, when present, extend through fewer pieces from the same log.

It is less susceptible to collapse in drying.

It shrinks and swells less in thickness.

It may cost less because it is generally easier to obtain.

Edge-grain lumber shrinks and swells less in width.

It twists and cups less.

It surface-checks and splits less in seasoning and in use.

It wears more evenly.

It does not allow liquids to pass into or through it so readily in some species.

The sapwood appearing in lumber is at the edges and its width is limited according to the width of the sapwood in the log.

From Wood Handbook.[30]

although larger sizes are obtainable for some species. Because available lumber sizes vary with species and locations, it is advisable to confirm size availability with local suppliers.

Lumber Size Classifications
During the evolution of stress grading in the United States, lumber size served as a guide in anticipating the final use of the piece. As a result, lumber came to be categorized into size classifications based on thickness and width. The three size classifications for structural lumber are Dimension Lumber, Beams and Stringers, and Posts and Timbers (Table 3-10).

Table 3-9.- Nomenclature of commercial softwood lumber.

Standard lumber names under ALS [31]	Official Forest Service tree name	Standard lumber names under ALS [31]	Official Forest Service tree name
Cedar		Pine	
Alaska	Alaska-cedar	Idaho white	Western white pine
Eastern red	Eastern redcedar	Jack	Jack pine
Incense	Incense-cedar	Lodgepole	Lodgepole pine
Northern white	Northern white-cedar	Longleaf yellow"	Longleaf pine
Port Orford	Port-Orford-cedar		Slash pine
Southern white	Atlantic white-cedar	Northern white	Eastern white pine
Western red	Western redcedar	Norway	Red pine
Cypress		Ponderosa	Ponderosa pine
Red (coast type), yellow (inland type), white (inland type)	Baldcypress	Southern (Major)	Longleaf pine
			Shortleaf pine
			Loblolly pine
Douglas-fir	Douglas-fir		Slash pine
Fir		Southern (Minor)	Pitch pine
Balsam	Balsam fir		Pond pine
	Fraser fir		Sand pine
Noble	Noble fir		Table mountain pine
White	California red fir		Virginia pine
	Grand fir	Sugar	Sugar pine
	Pacific silver fir	Redwood	Redwood
	Subalpine fir	Spruce	
	White fir	Eastern	Black spruce
Hemlock			Red spruce
Eastern	Eastern hemlock		White spruce
Mountain	Mountain hemlock	Engelmann	Blue spruce
West Coast	Western hemlock		Engelmann spruce
Juniper, western	Alligator juniper	Sitka	Sitka spruce
	Rocky Mountain juniper	Tamarack	Tamarack
	Utah juniper	Yew, Pacific	Pacific Yew
	Western juniper		
Larch, western	Western larch		

"The commercial requirements for longleaf pine are that it be produced from the species *Pinus elliottii* and *P. palustris* and that each piece must average either on one end or the other not less than 6 annual rings per inch and not less than 1/3 latewood. Longleaf pine lumber is sometimes designated as pitch pine in the export trade.

From Wood *Handbook*[20]

Table 3-10.- Lumber size classifications.

| Name | Symbol | Nominal dimensions | | Typical sizes |
		Thickness	Width	
Dimension Lumber				
Light Framing	LF	2 to 4 in.	2 to 4 in.	2x4, 4x4
Joist and Plank	J&P	2 to 4 in.	5 in. and wider	2x6, 2x12, 4x12, 4x16
Decking[a]	-	2 to 4 in.	4 in. and wider	2x6, 2x10 4x10, 4x12
Beams and Stringers	B&S	5 in. and thicker	More than 2 in. greater than thickness	6x10, 6x14 8x16, 10x18
Posts and Timbers	P&T	5 in. and thicker	Not more than 2 in. greater than thickness	6X6, 10X12 10x10, 12x14

[a] Decking sizes are the same as those designated for Light Framing and Joist and Plank. Decking is intended for flatwise use while LF and J&P are intended for edgewise use.

1.   **Dimension Lumber** is lumber that is 2 to 4 inches thick and 2 or more inches wide. This classification is further divided into a number of subcategories, the most common of which are Light Framing (LF), Joists and Planks (J&P), and Decking. LF and J&P are graded primarily for edgewise loading, while Decking is graded primarily for use in the flatwise orientation.

2.   **Beams and Stringers** (B&S) are rectangular pieces that are 5 or more inches thick, with a width more than 2 inches greater than the thickness. B&S are graded primarily for use as beams, with loads applied to the narrow face.

3.   **Posts and Timbers (P&T)** are pieces with a square or nearly square cross section, 5 by 5 inches and larger, with the width not more than 2 inches greater than the thickness. Lumber in the P&T size classification is graded primarily for resisting axial loads where strength in bending is not especially important.

An important point to understand about lumber size classifications is that they are based on the most efficient *anticipated* use of the member, rather than the actual use. The classifications are relevant to grading, which will be discussed later, but there are no restrictions on actual use for any size classification, provided the design stresses are within the stresses allowed for the grade.

Lumber Dimensions

Lumber dimensions for thickness and width, including those for defining size classifications, are traditionally recorded in nominal dimensions. These dimensions are usually in 2-inch increments such as 2 by 4 or 8 by 12. The nominal dimensions are normally greater than the actual net dimensions of the piece. In timber design, structural calculations must be based on the net lumber dimensions for the anticipated use conditions. These net dimensions depend on the type of surfacing, whether dressed, rough-sawn, or full-sawn (Figure 3-25).

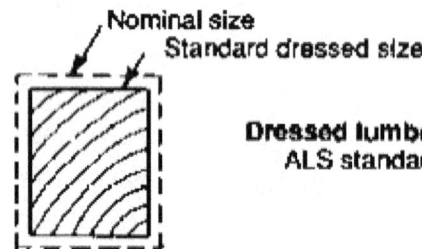

Nominal size
Standard dressed size

Dressed lumber is smaller than the stated nominal size. ALS standard dressed sizes are shown in Table 3-11.

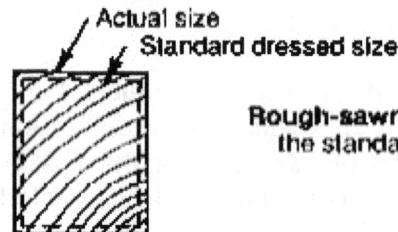

Actual size
Standard dressed size

Rough-sawn lumber is approximately 1/8" larger than the standard dressed size.

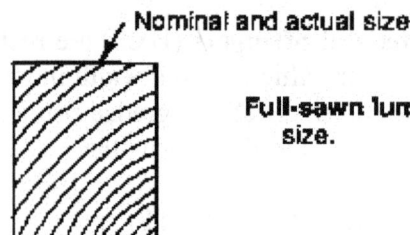

Nominal and actual size

Full-sawn lumber is the same size as the stated nominal size.

Figure 3-25.- Dressed, rough-sawn, and full-sawn lumber sizes.

Dressed lumber is surfaced on all four sides (S4S) at the time of manufacture to a minimum net dimension that is standardized by the ALS (Table 3-11). Lumber in the B&S and P&T size classifications, collectively referred to as Timbers, are normally surfaced green, and only green sizes are specified by the ALS. Dimension lumber may be surfaced green or dry at the prerogative of the manufacturer; therefore, both green and dry standard sizes are given. The green sizes are slightly larger in anticipation that the piece, as it dries, will shrink to the standard dry size. The ALS definition of dry lumber is lumber with a moisture content of 19 percent or less. For design purposes, the minimum dry dressed dimensions are used for structural calculations regardless of the moisture content at the time of manufacture or in use. For example, section properties for a standard dressed 2 by 6 are based on a section 1-1/2 inches by 5-1/2 inches, while for a standard 8 by 12, dimensions are 7-1/2 inches by 11-1/2 inches. Tables of section properties for standard dressed lumber are given in Chapter 16.

Table 3-11.- American Standard Lumber sizes for structural lumber.

Size classification	Thickness (in.)			Face width (in.)		
	Nominal dimension	Minimum dressed dimension		Nominal dimension	Minimum dressed dimension	
		dry	green		dry	green
Dimension Lumber	2	1-1/2	1-9/16	2	1-1/2	1-9/16
	2-1/2	2	2-1/16	3	2-1/2	2-9/16
	3	2-1/2	2-9/16	4	3-1/2	3-9/16
	3-1/2	3	3-1/16	5	4-1/2	4-5/8
	4	3-1/2	3-9/16	6	5-1/2	5-5/8
	4-1/2	4	4-1/16	8	7-1/4	7-1/2
				10	9-1/4	9-1/2
				12	11-1/4	11-1/2
				14	13-1/4	13-1/2
				16	15-1/4	15-1/2
Beams and Stringers and Posts and Timbers	5 and greater		1/2 less than nominal	5 and greater		1/2 less than nominal

From *Wood Handbook*.[30]

Rough-sawn lumber is normally about 1/8 inch larger than standard dry dressed sizes. Full-sawn lumber, which is not widely used, is cut to the same dimensions as the nominal size. In both cases, thickness and width dimensions are variable depending on the sawmill equipment. It is impractical to use rough-sawn or full-sawn lumber in a structure that requires close dimensional tolerances. For more accurate dimensions, surfacing can be specified on one side (S1S), two sides (S2S), one edge (S1E), two edges (S2E), combinations of sides and edges (S1S1E, S2S1E, S1S2E), or all sides (S4S). When designing with either rough-sawn or full-sawn lumber, the applicable moisture content and dimensions used in design must be clearly noted on the plans and material specifications.

LUMBER GRADING

When lumber is cut from a log, the properties of the individual pieces vary considerably in strength and stiffness. To have practical use in engineering applications, the lumber must be graded into categories for which reliable engineering properties can be assigned. Stress grades for lumber are characterized by one or more sorting criteria, a set of engineering properties, and a unique grade name. For each species or species group in each grade and size classification, the information related to sorting criteria and design values is contained in the grade description. Based on the information in the grade description, values for structural lumber are established for the following:

1. Modulus of elasticity
2. Tensile stress parallel to grain
3. Compressive stress parallel to grain
4. Compressive stress perpendicular to grain
5. Shear stress parallel to grain (horizontal shear)
6. Bending stress

Lumber grading is accomplished using visual grading criteria or non-destructive measurement using mechanical grading equipment. Grading for most lumber takes place at the sawmill. Generally, the grade of a piece of lumber is based on the number, character, and location of features that lower the strength, stiffness, or durability of the piece. Among the more common features that are evaluated during grading are knots and slope of grain, but many potential strength-reducing characteristics are considered.

Visual Stress Grading

Visual stress grading is the oldest and most widely used lumber grading method. It is based on the premise that the mechanical properties of lumber differ from those of clear wood because of growth characteristics that can be judged by the eye. After the lumber is sawn and surfaced (when required), each piece is examined by a lumber grader who is certified by one of the grading agencies (Figure 3-26). If the piece of lumber meets the grade description requirements for a particular grade, it is assigned that grade and the associated mechanical properties. If the piece does not meet requirements for one grade, it may qualify for a lower grade, or be rejected.

As previously discussed, all grades for dimension lumber are standardized by the NGR. However, grade names and descriptions for lumber in the B&S and P&T size classifications are not standardized and may vary among different grading rules agencies. There are many visual grades of structural lumber for the species groups and size classifications. Some of the typical grade names for Douglas Fir-Larch are shown in Table 3-12.

The engineer need not have an in-depth knowledge of all grade descriptions and the specifics of how they are derived. If a piece of lumber is graded, the tabulated values for that grade (discussed in Chapter 5) can be used for structural computations. It is beneficial, however, to have a basic understanding of how various grade requirements and tabulated design values are derived to better understand wood as an engineering material.

Figure 3-26.- Sawn lumber being visually graded and marked at a sawmill (photo courtesy of Funk Lumber Co.).

The process of establishing design properties for visually graded lumber is addressed in detail in ASTM D 245, Standard Methods for Establishing Structural Grades and Related Allowable Properties for Visually Graded Lumber.[6] A brief summary of the process is as follows:

1. The mechanical properties for each lumber grade may be established by adjusting test results conducted on small, clear green wood specimens of the species or species group (small, clear procedure), or by testing a representative sample of full-size members (in-grade procedure). The mechanical properties of virtually all lumber currently used in the United States have been derived using the small, clear procedure outlined in ASTM D 2555, Standard Methods for Establishing Clear Wood Strength Values 7. A comprehensive re-evaluation based on in-grade testing is currently in progress, and it is anticipated that mechanical properties will be based on test results from full-size specimens in the next few years.[8]

   Based on the small, clear procedures, large numbers of clear wood specimens are tested for each species or species group to determine ultimate stress and stiffness values. These values then serve as the starting point for deriving tabulated design values for lumber grades in that species. The ultimate stress is based on the 5-percent exclusion limit for the sample of small, clear specimens. This value is established from a statistical analysis and indicates that out of all clear wood samples tested, 95 percent would be

Table 3-12.- Typical grade names for visually graded structural lumber.

Dimension Lumber	Beams and Stringers	Posts and Timbers
Structural Light Framing (LF)		
Dense Select Structural	Dense Select Structural	Dense Select Structural
Select Structural	Select Structural Dense No. 1	Select Structural
Dense No. 1	No. 1	Dense No. 1
No. 1	Dense No. 2	No. 1
Dense No. 2	No. 2	Dense No. 2
No. 2		No. 2
No.3		
Appearance		
Stud		
Joists and Planks (J&P)		
Dense Select Structural		
Select Structural		
Dense No. 1		
No. 1		
Dense No. 2		
No. 2		
No. 3		
Appearance		
Stud		
Decking		
Select Decking		
Commercial Decking		

The grade names in this table are for Douglas Fir-Larch graded to Western Wood Products Association (WWPA) rules. Other species and grading rules agencies have similar names.

expected to fail at or above the 5-percent exclusion limit, while less than 5 percent would be expected to fail below the limit. For modulus of elasticity, the small, clear value is based on the average of all sample values, rather than the 5-percent exclusion limit.

2. Clear wood properties are next adjusted for strength-reducing characteristics by strength ratios for strength properties and quality factors for modulus of elasticity (Table 3-13). Strength ratios are factors that reduce clear wood strength properties to compensate for such growth characteristics as knots, slope of grain, shake, pitch pockets, and other defects. An individual piece of lumber will often have several characteristics that affect a particular strength property. The characteristic that gives the lowest strength ratio is used to derive the estimated strength.

Strength ratio values vary for lumber grades, depending on the maximum number and location of strength-reducing characteristics permitted for the grade. For example, high-strength grades have higher strength ratios because they have more restrictive requirements on the number, type, and location of defects. Lower grades are less restrictive on strength-reducing characteristics and have lower strength ratios. For modulus of elasticity, the clear wood average value is multiplied by empirically derived quality factors to represent the reduction that occurs in lower lumber grades. The value of the quality factor for each grade is based on the value of the strength ratio for bending for that grade.

Table 3-13.- Strength ratios and quality factors for some visual grades described in the National Grading Rule.

Lumber size classification	Grade name	Strength ratio	Quality factor
Light Framing	Select Structural	0.67	1.0
(2 to 4 in. thick,	No. 1	0.55	1.0
2 to 4 in. wide)	No. 2	0.45	0.90
	No. 3	0.26	0.80
Joists and Planks	Select Structural	0.65	1.0
(2 to 4 in. thick,	No. 1	0.55	1.0
6 in. and wider)	No. 2	0.45	0.90
	No. 3	0.26	0.80

Sizes shown are nominal.
From *Wood Handbook* [30] and *ASTM D 2555*.[7]

3. In addition to strength-reducing characteristics, engineering properties for small clear samples are adjusted to compensate for differences between sample test conditions and those for the actual lumber. Specific adjustments may be made for (1) moisture content, which equates strength and modulus of elasticity for the green samples to the actual moisture content of the lumber at the time of manufacture; (2) duration of load, which equates the short 5-minute load duration used for clear sample testing to an equivalent load duration of 10 years; and (3) size adjustments for bending strength, which equate the size of the sample to the actual lumber size. In addition, each strength property except compression perpendicular to grain is adjusted by a factor of safety. Because the modulus of elasticity is judged primarily by performance criteria rather than safety criteria, a factor of safety is not applied.

For visual lumber grades, two points deserve additional emphasis. With the exception of dimension lumber (less than 5 inches thick), grade descriptions are not standardized and may vary among grading rules agencies. It is therefore possible to have lumber with the same grade name and size classification, but different tabulated strength and stiffness values. Grading also differs among different size classifications, even though the grade name may be the same. For example, lumber graded No. 1 in the B&S size classification may have more restrictive requirements for edge knots, since that classification is graded with respect to bending. For a piece graded No. 1 in P&T, edge knot requirements may be less restrictive because this size classification is graded with respect to axial loading. As a result, the bending stress for a No. 1 P&T could be less than a No. 1 B&S for the same species.

Mechanical Stress Grading

Mechanical stress grading, commonly called machine stress rating (MSR), has been used as a method of lumber grading for more than 25 years. It is based on an observed relation between modulus of elasticity *(E)* and the bending strength of lumber. The sorting criterion, *E*, is measured for individual pieces of lumber by mechanical devices that operate at high rates of speed (Figure 3-27). The *E* used as a sorting criterion can be measured in a variety of ways, but is usually an apparent *E* based on deflection. Because lumber is heterogeneous, the apparent *E* depends on a number of factors, including the lumber span length, orientation, mode of test, and method of loading. Any apparent *E* can be used, so long as the grading machine is calibrated to assign the appropriate design property. Most grading machines in the United States are designed to detect the lowest flatwise bending stiffness that occurs in any approximate 4-foot span. Although the machine-measured *E* is the primary sorting criterion in this grading process, MSR lumber is also subject to visual override because the size of edge knots in combination with *E* is a better predictor of strength than *E* alone. Maximum edge knots are limited to a specified proportion of the cross section, depending on the grade level. Other visual restrictions also apply to checks, shakes, splits, and warp.

Bending strength is derived by correlations with modulus of elasticity determined by the machine rating process. Assigned properties in tension and compression parallel to grain are estimated from bending strength, although some procedures estimate tension directly from *E*. Strengths in shear parallel to grain and in compression perpendicular to grain are assigned the same values as the equivalent species of visual lumber grades.

In the United States, the number of machine stress grades available reflects specific market needs that have developed for MSR lumber. Grade designations for MSR lumber differ from visual grades and include the tabulated strength in bending and modulus of elasticity. For example, the MSR grade 2100 F-1.8E designates a tabulated bending strength of

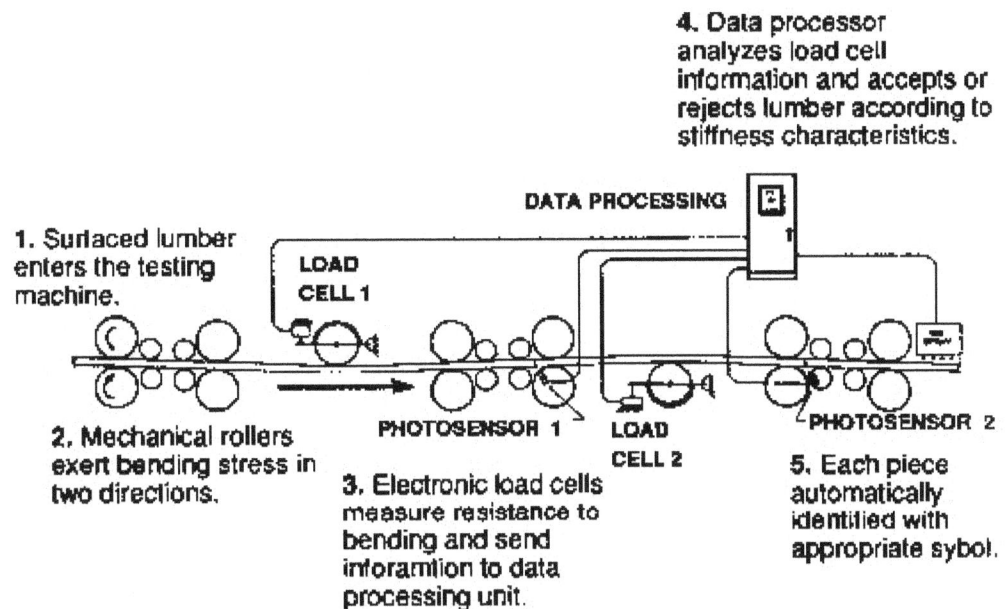

4. Data processor analyzes load cell information and accepts or rejects lumber according to stiffness characteristics.

DATA PROCESSING

1. Surfaced lumber enters the testing machine.

LOAD CELL 1

2. Mechanical rollers exert bending stress in two directions.

PHOTOSENSOR 1

LOAD CELL 2

PHOTOSENSOR 2

3. Electronic load cells measure resistance to bending and send inforamtion to data processing unit.

5. Each piece automatically identified with appropriate sybol.

Figure 3-27.- Schematic diagram of a continuous lumber tester (CLT) used for machine stress grading lumber (courtesy of the Western Wood Products Association; used by permission).

2,100 lb/in^2 and a modulus of elasticity of 1.8 million lb/in^2, both in an edgewise orientation. Additional MSR grades are discussed in Chapter 5.

QUALITY CONTROL AND MARKING

Quality control and marking requirements for American Standard Lumber are established in the ALS PS 20-70. The ALS also includes provisions for lumber inspection and reinspection in cases where a dispute arises between buyer and seller. Responsibility for grading accuracy and certification is with the grading rules and inspection agencies that are certified by the American Lumber Standards Committee. These agencies are required to conduct regular grading inspections and spot checks at the mill to ensure grading efficiency and conformity to all established rules. When the lumber is graded in accordance with the specific grade requirements, each piece is marked to certify that the lumber conforms to the size, grade, and seasoning provisions of the rules under which it is graded. The ALS requires that these grade marks include specific information, including:

1. The lumber grade description (e.g., No. 1 for visually graded lumber or 2100f-1.8E for MSR lumber)

2. The commercial name of the lumber species or species group (e.g., Douglas Fir-Larch or Southern Pine)

3. The identification of the grading rules agency under whose rules the lumber was graded (e.g., WCLIB or SPIB)

4. The registered symbol of the certified inspection agency, when the inspection agency is different from the grading rules agency

5. The lumber moisture content at the time of surfacing: S-DRY when lumber is surfaced at a moisture content of 19 percent or less; S-GRN when lumber is surfaced at a moisture content in excess of 19 percent; KD-15 or KD-19 when the lumber is kiln dried and surfaced at a moisture content of 15 or 19 percent or less, respectively; MC-15 when the lumber is either kiln dried or dried by other means and surfaced at a moisture content of 15 percent or less

6. The designated mill number or mill name

Examples of typical lumber grade marks are shown in Figure 3-28.

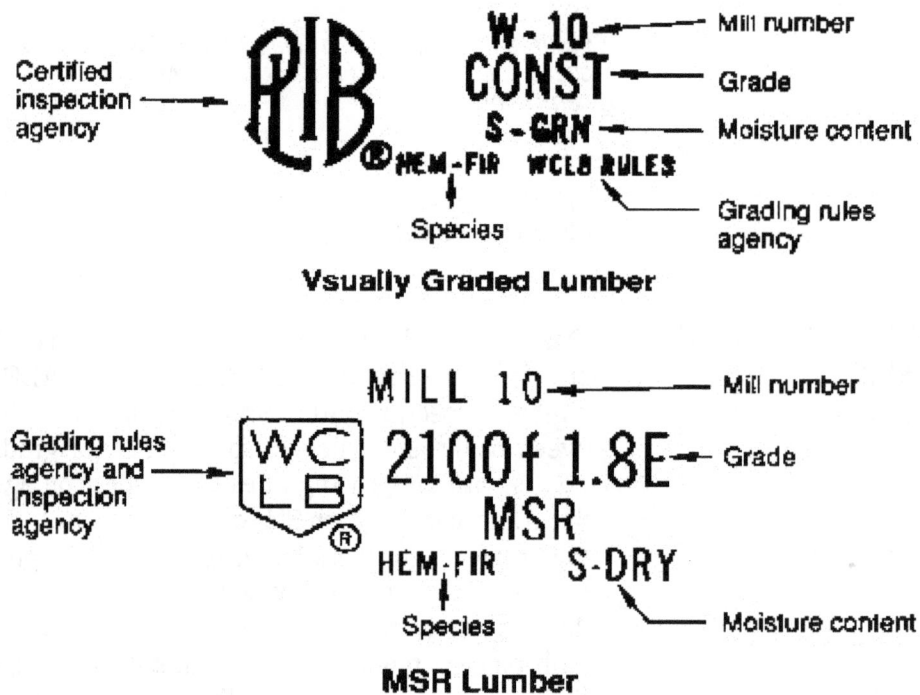

Figure 3-28.- Typical lumber grade marks.

MATERIAL SPECIFICATIONS

Accurate specifications are critical for obtaining the proper lumber for the required use. They must contain all applicable information related to the manufacture, grade, size, moisture content, and species for the required lumber. To some degree, this information must be obtained from tables of tabulated values, which are discussed in Chapter 5. The following is a summary of the some of the requirements and recommendations for specifying lumber:

1. **Manufacturing Standard.** All structural lumber shall be American Standard Lumber, manufactured and graded in accordance with the latest edition of the ALS PS 20-70. Each piece shall contain a grade mark conforming to the requirements of that standard and those of the applicable grading rules agency.

2. **Species.** Lumber species or species group must be clearly stated.

3. **Grade.** Grade designations must include the commercial grade name, size classification, and the grading rules agency under which the lumber is graded.

4. **Size.** Lumber thickness and width are specified in nominal dimensions. The length in feet may be specified on the basis of the nominal average length, limiting length, or single uniform length. For some lumber, random lengths can be specified when uniform lengths are not required. In this case, upper and lower limits should also be specified.

5. **Surfacing.** Lumber surfacing is specified as dressed (surfaced all sides), rough-sawn, or full-sawn. When partial surfacing is required on one or more sides or edges of rough-sawn or full-sawn lumber, the abbreviations previously discussed are used (lumber abbreviations are also given in Chapter 16).

6. **Moisture Content.** When moisture content at the time of surfacing is important, lumber may be specified as surfaced green, surfaced dry, or kiln-dried using the same definitions described for quality control and marking.

3.7 PROPERTIES OF GLUED-LAMINATED TIMBER (GLULAM)

Glulam is an engineered, stress-rated product of a timber-laminating plant. It consists of selected and prepared lumber laminations that are bonded together on their wide faces with structural adhesive (Figure 3-29). Glulam has been used successfully as a structural material in Europe since the early 1900's. In the United States, it has been used with excellent performance in buildings since approximately 1935 and in bridges since the mid-1940's. An important point about glulam is that it is an engineered timber product rather than simply wood glued together. Laminated beams made with pieces of lumber that are nailed and glued together should not be confused with glulam.

Figure 3-29.-A glulam bridge beam in the final stages of fabrication. The beam, which measures 12-1/4 inches wide, 71-1/2 inches deep and 103 feet long, required 12,800 board feet of structural grade Douglas-fir and 500 pounds of glue.

Glulam is a very versatile material that provides distinct advantages over sawn lumber. Because it is a manufactured product, glulam can be produced in a wide range of shapes and virtually any size. Most of the glulam used in bridges involves straight or slightly curved members, but curved or tapered members are used in some applications. Glulam also provides increased strength over sawn lumber because the laminating process disperses strength-reducing characteristics throughout the member (Figure 3-30). A knot in sawn lumber, for example, may substantially reduce the section and strength of a member. In glulam, the knots are distributed among the laminations so their effect at any section is minimized. Glulam also provides better dimensional stability because it is manufactured from dry lumber as compared to most large sawn lumber members, which are sawn green and seasoned after installation.

Glulam is available from a number of manufacturers strategically located across the United States. A partial listing of manufacturers is given in Chapter 16.

PRODUCT STANDARDS

The national product standard for glulam is the American National Standard for Wood Products-Structural Glued Laminated Timber, ANSI/AITC A190.1.[1] This standard, which was approved by the American National

Strength-reducing characteristics in sawn lumber can occupy much of the cross section and substantially reduce strength.

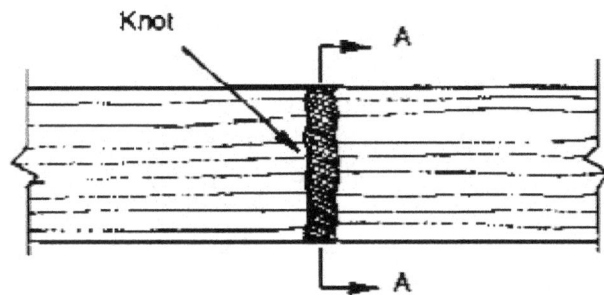

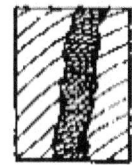

Laminating disperses strength-reducing characteristics, reducing their effects on strength.

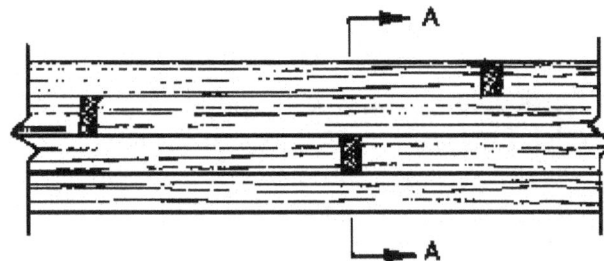

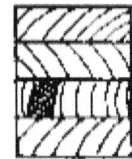

Figure 3-30.- Relative effects of strength-reducing characteristics on sawn lumber and glued-laminated timber.

Standards Institute (ANSI) in 1983, contains nationally recognized requirements for the production, inspection, testing, and certification of structural glulam. It also provides material producers, suppliers, and users with a basis for a common understanding of the characteristics of glulam. The requirements in ANSI/AITC A190.1 are intended to allow the use of any suitable method of manufacture that will produce a product equal or superior in quality to that specified, provided the methods of manufacture are approved in accordance with requirements of the standard.

The sponsor of ANSI/AITC A190.1 is the American Institute of Timber Construction (AITC), which is the national technical trade association of the structural glued laminating industry. Its members manufacture, fabricate, assemble, erect and/or design wood structural systems and related wood products for construction applications. AITC publishes standards related to the design, manufacture, and construction of glulam that are incorporated by reference in ANSI/AITC A190.1.

MANUFACTURING CRITERIA FOR GLULAM

Glulam can be manufactured from any softwood or hardwood lumber provided it meets necessary grading and stiffness requirements. In practice, most glulam is manufactured from two commercial species groups, western species (primarily Douglas Fir-Larch and Hem-Fir) and Southern Pine. The laminations are selected from stress-graded sawn lumber, but the lumber must be regraded using additional criteria before it can be laminating stock for glulam. Lamination regrading is accomplished using either visual grading or E-rating criteria (E-rated laminations are regraded

3-51

for stiffness and edge knots as previously discussed for MSR lumber). ANSI/AITC A190.1 specifies that the moisture content of the laminations at the time of gluing not exceed 16 percent; however, moisture contents lower than 16 percent or up to 20 percent can be specified depending on the anticipated moisture content of the components in service.

The maximum lamination thickness permitted for glulam is 2 inches. The actual thickness of the laminations depends on the species of the laminations and the shape of the member. Industry standard practice is to use 2-inch nominal lumber to produce straight or slightly curved members of the type normally used for bridge construction. This results in a dressed lamination thickness of 1-1/2 inches for western species and 1-3/8 inches for Southern Pine. When sharp radius curves are required, nominal 1-inch-thick lumber (3/4-inch dressed thickness) is generally used. Lamination thickness has a significant effect on glulam economics because manufacturing costs are related to the required number of glue lines in the member. Thus, the thinner the laminations the higher the relative cost of manufacture.

Standard Glulam Sizes

Glulam is most efficiently and most economically manufactured when standard dressed lumber is used for laminations. As a result, standard glulam sizes are related to the dressed sizes for sawn lumber, but are slightly less in width to account for surfacing after the material is glued. Standard glulam widths are available in increments from 2-1/8 inches to 14-1/4 inches based on nominal lamination widths of 3 to 16 inches (Table 3-14). Note that the net widths for nominal 4-, 6-, 10-, and 12-inch laminations are different for western species and Southern Pine. Glulam depth is equal to the lamination thickness times the number of laminations, or 1-1/2 inch multiples for western species and 1-3/8-inch multiples for Southern Pine. Unlike sawn lumber, the sizes specified for glulam are the actual dimensions of the member, rather than the nominal size.

Table 3-14.- Standard widths of glulam.

Nominal width (in.)	Western species net finished width (in.)	Southern Pine net finished width (in.)
3	2-1/8	-
4	3-1/8	3
6	5-1/8	5
8	6-3/4	6-3/4
10	8-3/4	8-1/2
12	10-3/4	10-1/2
14	12-1/4	-
16	14-1/4	-

Glulam members can hypothetically be manufactured to any depth or length by simply adding more laminations. From a practical standpoint, however, the size of the member must be limited because of handling and transportation considerations. In bridge applications, another primary consideration related to member size is the size capacity of the cylinders used for pressure treatment with preservatives. Because size capabilities vary among treaters, it is recommended that the designer verify treatment capabilities prior to requiring glulam depths in excess of 60 inches or lengths in excess of 80 feet.

Glulam dimensions may vary slightly because of minor variations in manufacturing processes. Dimensional tolerances permitted at the time of manufacture are as follows:[1]

1. **Width: ±1/16** inch

2. **Depth:** +1/8 inch. per foot of specified depth; -1/16 inch per foot of specified depth, or -1/8 inch, whichever is greater

3. **Length: ±1/16** inch for lengths up to 20 feet and ±1/16 inch per 20 feet of length for lengths over 20 feet, except where length dimensions are not specified or critical

4. **Squareness:** the cross section must be square within ±1/8 inch per foot of specified depth unless a specially shaped section is specified

5. **Straightness: ±1/4** inch for beams up to 20 feet in length and ±1/8 inch per 20 feet or fraction thereof for beams greater than 20 feet in length, but not greater than a total of ±3/4 inch; tolerances intended for straight or slightly cambered beams only, not for curved members such as arches

Adhesives

Laminations for glulam are joined together with structural adhesives that are capable of developing shear strength in excess of the wood capacity. Two types of adhesives are permitted, dry-use adhesives and wet-use adhesives. Dry-use adhesives (usually casein) are allowed only when the glulam moisture content in service will not exceed 16 percent. These adhesives set or cure by the dissipation of water in the adhesive to the surrounding air and wood. Wet-use adhesives, which cure by chemical polymerization, are required for exposed uses where moisture content may exceed 16 percent in service, as in bridge applications. Wet-use adhesives are also required when laminations are chemically treated with wood preservatives before or after gluing. In practice, the wet-use adhesive phenol-resorcinol is used almost exclusively for all glulam. This adhesive can withstand severe exposure conditions and offers a manufacturing advantage of rapid curing rates. All adhesives for glulam used in bridge applications must be wet-use adhesives.

Joints

In most cases, the size of glulam members substantially exceeds the size of available lumber, and laminations must be spliced with end joints or edge joints or both. End joints are used to splice laminations longitudinally before assembly into a glulam member. The two most common types of end joints are scarf joints and finger joints (Figure 3-31). In scarf joints, the ends of the laminations are cut at opposing slopes of 1:8 to 1:12 and are glued together on the sloping surfaces. For finger joints, ends are cut with horizontal or vertical fingers that are glued and mated together. In practice, finger joints are used almost exclusively because they require less material and are self-aligning during the gluing process.

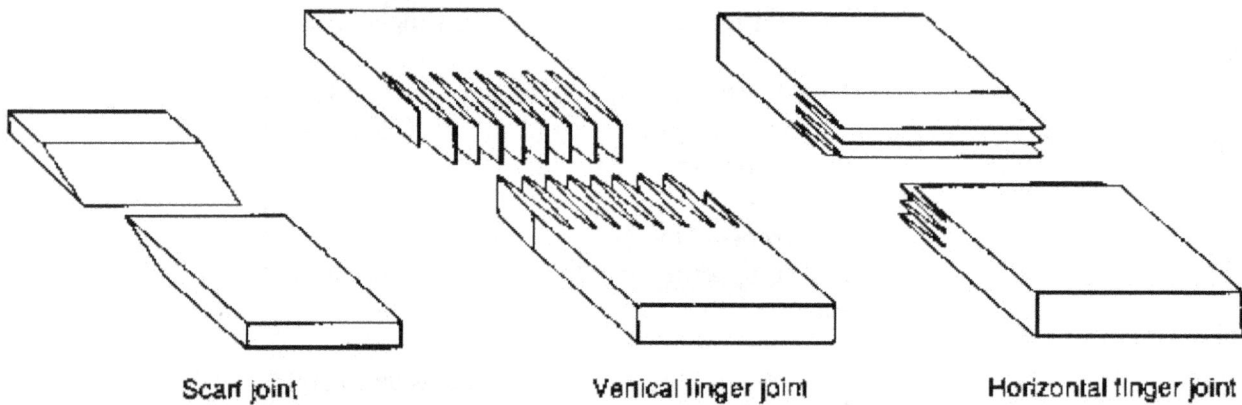

Scarf joint   Vertical finger joint   Horizontal finger joint

Figure 3-31.- Types of lamination end joints used in glulam.

Edge joints are required when the width of the glulam member is greater than available laminations. The most common edge joint configuration uses a staggered layup with edge joints offset between adjacent laminations (Figure 3-32). Edge joints may be glued or unglued depending on the type of member and applied stresses; however, because joint strength influences shear capacity, reduced design values for shear are required when edge joints are not glued (Chapter 5). When unglued edge joints are used for bridge members, it is recommended that edge joints in the top face be glued to prevent water and debris from becoming trapped in the joint.

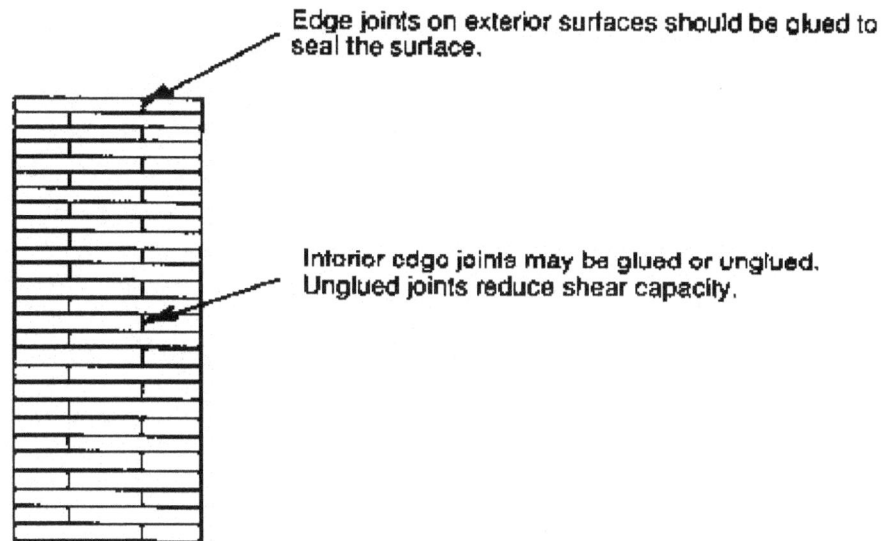

Edge joints on exterior surfaces should be glued to seal the surface.

Interior edge joints may be glued or unglued. Unglued joints reduce shear capacity.

Figure 3-32.- Edge joints in a glulam member manufactured with a staggered layup.

Appearance

Glulam is available in three appearance grades: industrial, architectural, and premium. Appearance grades apply to the glulam surface and include considerations related to growth characteristics, void filling, and surfacing operations. They do not address surface treatments, stains, or varnishes, and do not alter member strength or manufacturing controls. In bridge applications, glulam is normally finished to an industrial appearance grade. An architectural grade may be used in exceptional cases where an improved appearance is required. A more complete description of appearance grades is given in AITC 110, Standard Appearance Grades for Structural Glued-Laminated Timber, which can be found in the AITC *Timber Construction Manual.* [4]

GRADES OF GLULAM

Glulam is not graded in the same manner as sawn lumber. Rather, members are identified by a combination symbol that represents the combination of lamination grades used to manufacture the member. These combination symbols are divided into two general classifications consisting of bending combinations and axial combinations. The classifications are similar to lumber size classifications because they are based to some degree on the anticipated use of the member. Bending combinations anticipate that the member will be used as a beam, while the axial combinations anticipate the member will be used as a column or tension member. These anticipated uses are based on criteria for the most efficient and economical use of material rather than restrictions on actual use. For both types of combinations, members may be used in any loading situation or configuration provided the resulting stresses are within allowable limits for the specific combination symbol.

Bending Combinations

Glulam bending combinations were developed to provide the most efficient and economical section for resisting bending stress caused by loads applied perpendicular to the wide faces of the laminations. The quality and strength of the laminations are varied over the member depth to provide a wide range of strengths to accommodate different loading conditions (Figure 3-33). For example, a lower grade of lamination is used for the center portion of the member where bending stress is low, while a higher grade of material is placed on the outside faces where bending stress is relatively high. Bending combinations may also be used for axial loading or for bending applied parallel to the wide face of the laminations. In these cases, however, the strength of the member is controlled by the lower grade center laminations, and the higher strength outer laminations provide little benefit. The axial combinations normally provide the most economical member for these loading conditions.

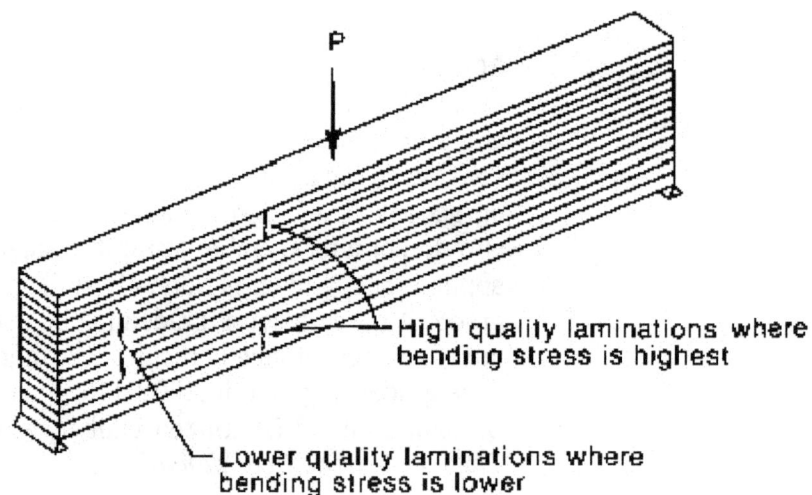

Figure 3-33.-Glulam bending combinations are intended primarily for applications where loads are applied perpendicular to the wide face of the laminations. The quality of the lumber laminations is varied over the member cross section to provide higher strength where bending stress is highest.

Combination symbols for bending combinations are specified by a series of numbers and letters that indicate the tabulated bending stress and the basis for lamination regrading. For a typical combination symbol such as 24F-V4, the 24F indicates a tabulated design stress in bending of 2,400 lb/in^2. Following the F, the letter V or E indicates if the combination is manufactured from visually graded or E-rated lumber, respectively. More detailed information on combination symbols and their associated design stresses is provided in Chapter 5.

Axial Combinations

Glulam axial combinations were developed to provide the most efficient and most economical section for resisting axial forces and bending stress applied parallel to the wide faces of the laminations (Figure 3-34). Unlike bending combinations, the same grade of lamination is used throughout the member. Axial combinations may also be loaded perpendicular to the wide face of the laminations, but the nonselective material placement often results in a less efficient and less economical member than the bending combination. Combination symbols for axial combinations are unrelated to strength or lamination grading and consist of a numerical value only. Examples include combination symbol 1 and combination symbol 47.

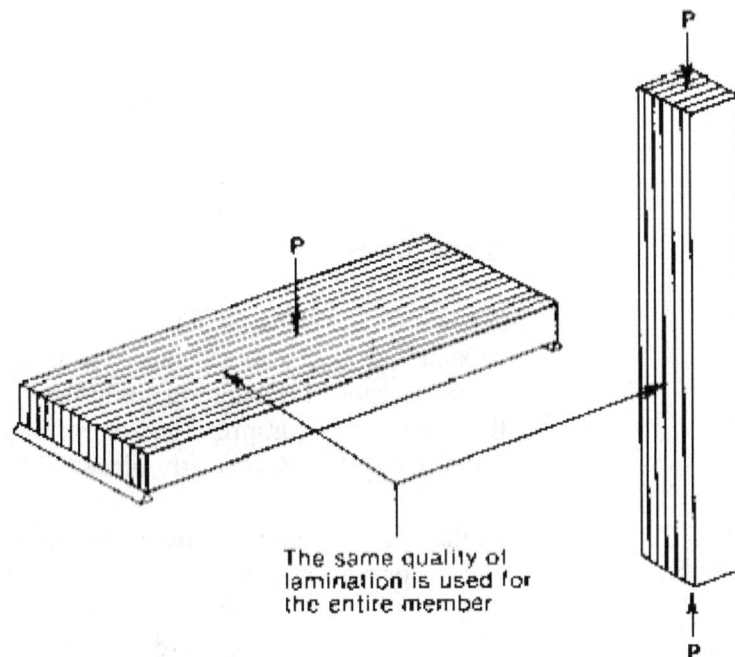

The same quality of lamination is used for the entire member

Figure 3-34.- Glulam axial combinations are intended primarily for applications where loads are applied parallel to the wide face of the laminations. The same quality of lumber lamination is used for the entire member.

QUALITY CONTROL AND MARKING

ANSI/AITC A190.1 requires that each glulam manufacturer maintain a strict quality control program for the production of glulam. This program must include continuing inspection and evaluation in areas related to manufacturing procedures, material testing, and quality control records. The inspections must be supervised by an independent third party to the manufacturer that meets specific qualification requirements outlined in the standard. The AITC operates a continuing quality program for its members; however, any independent inspection agency may be used, provided it meets the requirements of the ANSI/AITC standard.

To indicate compliance with quality control requirements, each glulam member must be distinctively marked. Marking requirements are given in ANSI/AITC A190.1 for two types of glulam products, custom products and noncustom products. Custom products, which are used almost exclusively in bridge applications, are manufactured to specific specifications for a known use. Noncustom products are manufactured in accordance with ANSI/AITC A190.1, but are not intended for a particular use (they are usually stock members maintained by a supplier for any applicable use). Marking for custom and noncustom products must include (1) identification that the glulam was manufactured in accordance with the requirements of ANSI/AITC A190.1, (2) identification of the qualified inspection and testing agency, and (3) identification of the laminating plant. Marking for noncustom products must also include additional information outlined in ANSI/AITC A190.1 (Figure 3-35). For all glulam, ANSI/AITC A190.1 also requires that a certificate of material conformance be issued when requested by the purchaser. It is recommended that such a certificate be required for all bridge members, because treatment of the member with preservatives often makes the quality mark difficult or impossible to read.

In addition to quality marks, straight or slightly curved glulam beams must be stamped TOP at both ends to indicate the proper orientation of the beam. Because the bending strength of glulam beams is often different for the tension and compression zones, this marking is important to ensure that the member is correctly placed.

MATERIAL SPECIFICATIONS

Glulam can be specified by combination symbol or by minimum required values for strength and stiffness (bending, shear, compression, and so forth). In both cases, familiarity with available combination symbols, tabulated design values, and modification factors is required. These items are discussed in detail in Chapter 5, as are glulam specification examples. At this point, however, it is important that the designer understand the basics of glulam specification summarized as follows:

1. **Manufacturing Standard.** Materials, manufacture, and quality control for glulam shall be in conformance with the latest edition of ANSI/AITC A190.1, Structural Glued Laminated Timber.

2. **Laminating Combinations.** It is recommended that combination symbols requiring E-rated laminations be specified only after availability is verified. Visually graded material should be specified with an E-rated substitution permitted when material is available.

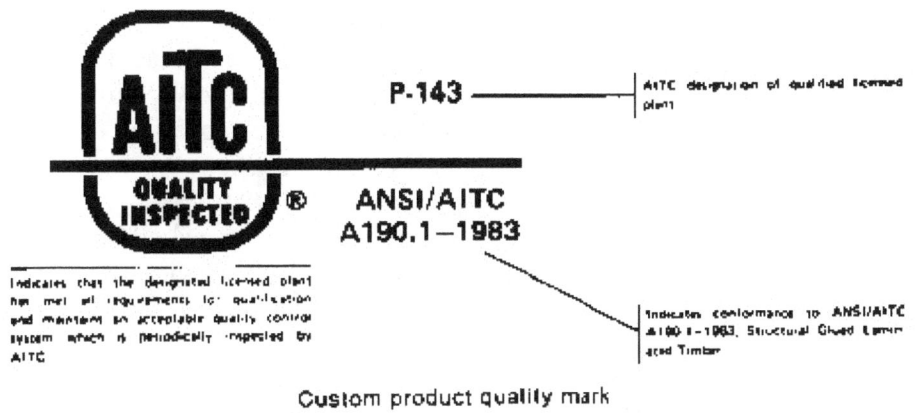

Custom product quality mark

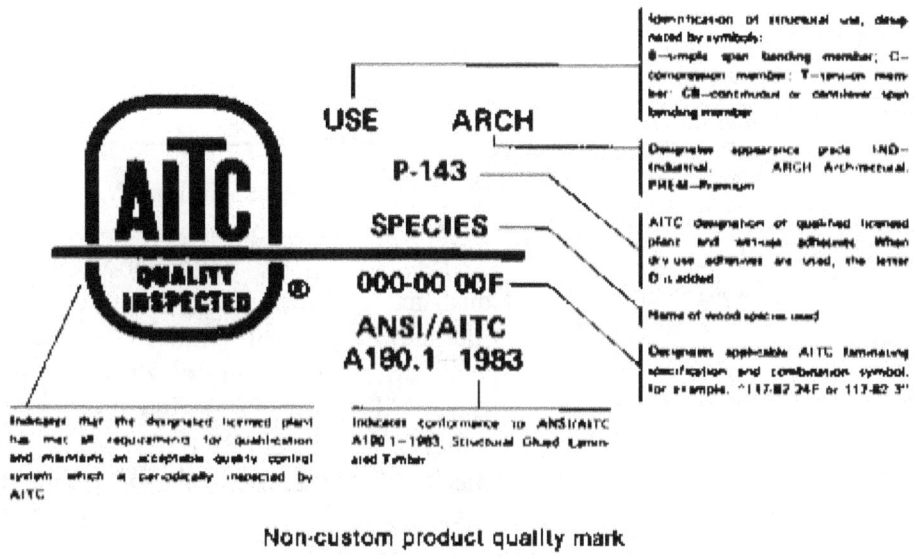

Non-custom product quality mark

Figure 3-35.- Typical glulam quality marks for custom and noncustom products. These marks are issued through the AITC quality program. Other types of marks with the same information may be used by other agencies (photo courtesy of the American Institute of Timber Construction; used by permission).

3. **Lamination Species.** The species or species group of lamination must be specified because the same combination symbol may be applicable to both western species and Southern Pine.

4. **Size.** Glulam members are specified by the actual member size rather than the nominal size commonly used for sawn lumber. Section properties for standard glulam sizes are given in Chapter 16.

5. **Adhesives.** All glulam manufactured for bridge applications shall use wet-use adhesives.

6. **Moisture Content.** The moisture content of glulam at manufacture is 16 percent or less. In arid regions where the equilibrium moisture content in service is expected to be significantly less, a lower glulam moisture content (as low as 10 percent) can be specified to minimize the potential for checking in service.

7. **Appearance.** Glulam shall be manufactured to industrial appearance unless aesthetic considerations warrant an improved surface condition.

8. **Quality Marks and Certificates.** Glulam members shall be marked with a quality mark and provided with a certificate of conformance to indicate conformance with ANSI/AITC A190.1, Structural Glued Laminated Timber.

3.8 SELECTED REFERENCES

1. American Institute of Timber Construction. 1983. American national standard for wood products-structural glued laminated timber. ANSI/AITC A190.1. Englewood, CO: American Institute of Timber Construction. 16 p.
2. American Institute of Timber Construction. 1987. Design standard specifications for structural glued laminated timber of softwood species. AITC 117-87-Design. Englewood, CO: American Institute of Timber Construction. 28 p.
3. American Institute of Timber Construction. 1973. Inspection manual. AITC 200-73. Englewood, CO: American Institute of Timber Construction. 33 p.
4. American Institute of Timber Construction. 1985. Timber construction manual. 3d ed. New York: John Wiley and Sons, Inc. 836 p.
5. American Lumber Standards Committee. 1982. Certified agencies and typical grade stamps. Germantown, MD: American Lumber Standards Committee. 4 p.
6. American Society for Testing and Materials. [See current edition.] Standard methods for estimating structural grades for visually graded lumber. ASTM D 245. Philadelphia, PA: ASTM.
7. American Society for Testing and Materials. [See current edition.] Standard methods for establishing clear wood strength values. ASTM D 2555. Philadelphia, PA: ASTM.
8. American Society of Civil Engineers. 1975. Wood structures, a design guide and commentary. New York: American Society of Civil Engineers. 416 p.

9. Barnes, H.M.; Winandy, J.E. 1986. Effects of seasoning and preservatives on the properties of treated wood. Proceedings of the American Wood Preservers' Association 82: 95-105.

10. Bodig, J.; Goodman, J.R. 1973. Prediction of elastic parameters for wood. Wood Science 5(4): 249-264.

11. Canadian Wood Council. 1985, Canadian wood construction. Glued laminated timber specifications. CWC datafile WS-2. Ottawa, Can.: Canadian Wood Council. 12 p.

12. Canadian Wood Council. 1985. Canadian wood construction. Lumber specifications. CWC datafile WS-1. Ottawa, Can.: Canadian Wood Council. 20 p.

13. Canadian Wood Council. 1986. Canadian wood construction. Structure and properties. CWC datafile SP-1. Ottawa, Can.: Canadian Wood Council. 16 p.

14. Ethington, R.L.; Galligan, W.L.; Montrey, H.M.; Freas, A.D. 1979. Evolution of allowable stresses in shear for lumber. Gen. Tech. Rep. FPL 23. Madison, WI: U.S. Department of Agriculture, Forest Service, Forest Products Laboratory. 16 p.

15. Galligan, W.L.; Snodgrass, D.V.; Crow, G.W. 1977. Machine stress rating: practical concerns for lumber producers. Gen. Tech. Rep. FPL 7. Madison, WI: U.S. Department of Agriculture, Forest Service, Forest Products Laboratory. 83 p.

16. Gerhards, C.C. 1977. Effect of duration and rate of loading on strength of wood and wood-based materials. Res. Pap. FPL-283. Madison, WI: U.S. Department of Agriculture, Forest Service, Forest Products Laboratory. 24 p.

17. Gerhards, CC. 1982. Effect of moisture content and temperature on the mechanical properties of wood: An analysis of immediate effects. Wood and Fiber 14(1): 4-36.

18. Gerhards, C.C. 1979. Time-related effects of loading on wood strength: a linear cumulative damage theory. Wood Science 11(3): 139-144.

19. Green, D.W.; Evans, J.W. 1987. Mechanical properties of visually graded lumber: Volume 1, a summary. Madison, WI: U.S. Department of Agriculture, Forest Service, Forest Products Laboratory. 131 p.

20. Green, D.W.; Evans, J.W. 1988. Moisture content-property relationships for dimension lumber: decisions for the future. For publication in the proceedings of the Workshop on In-Grade Testing of Structural Lumber; 1988; April 25-26, Madison, WI.

21. Gurfinkle, G. 1981. Wood engineering. 2d ed. Dubuque, IA: Kendall/Hunt. 552 p.

22. Millett, M.A.; Gerhards, C.C. 1972. Accelerated aging: residual weight and flexural properties of wood heated in air at 115 to 175 °C. Madison, WI: Wood Science 4(4).

23. National Forest Products Association. 1986. Design values for wood construction. A supplement to the national design specification for wood construction. Washington, DC: National Forest Products Association. 34 p.

24. National Forest Products Association. 1986. National design specification for wood construction. Washington, DC: National Forest Products Association. 87 p.

25. National Forest Products Association. 1980. Wood structural design data. Washington, DC: National Forest Products Association. 240 p.

26. Salmon, S.G.; Johnson, J.E. 1980. Steel structures, design and behavior. 2d ed. New York: Harper & Row. 1007 p.

27. Southern Pine Inspection Bureau. 1977. Grading Rules. Pensacola, FL: Southern Pine Inspection Bureau. 222 p.

28. U.S. Department of Agriculture, Forest Service, Forest Products Laboratory. 1981. Wood: Its structure and properties. Wangaard, F.F., ed. Clark C. Heritage Memorial Series on Wood. Compilation of educational modules; 1st Clark C. Heritage Memorial Workshop; 1980, August; Madison, WI. University Park, PA: The Pennsylvania State University. Vol. 1. 465 p.

29. U.S. Department of Agriculture, Forest Service, Forest Products Laboratory. 1982. Wood as a structural material. Dietz, A.G.H.; Schaffer, E.L.; Gromala, D.S., eds. Clark C. Heritage Memorial Series on Wood. Compilation of educational modules; 2d Clark C. Heritage Memorial Workshop; 1980, August; Madison, WI. University Park, PA: The Pennsylvania State University. Vol. 2. 282 p.

30. U.S. Department of Agriculture. 1987. Wood handbook: wood as an engineering material. Agric. Handb. No. 72. Madison, WI: U.S. Department of Agriculture, Forest Service, Forest Products Laboratory. 466 p.

31. U.S. Department of Commerce, National Bureau of Standards. 1970. American softwood lumber standard. Voluntary Prod. Stand. PS 20-70. Washington, DC: U.S. Department of Commerce, National Bureau of Standards. 26 p.

32. West Coast Lumber Inspection Bureau. 1984. West coast lumber standard grading rules. No. 16. Portland, OR: West Coast Lumber Inspection Bureau. 223 p.

33. Western Wood Products Association. 1984. Grade stamp guide. WWPA tech. guide TG-6. Portland, OR: Western Wood Products Association. 2 p.

34. Western Wood Products Association. 1981. Western lumber grading rules. Portland, OR: Western Wood Products Association. 222 p.

35. Western Wood Products Association. 1985. Western lumber product use manual. Portland, OR: Western Wood Products Association. 19 p.

36. Western Wood Products Association. 1983. Western woods use book. 3d ed. Portland, OR: Western Wood Products Association. [350 p.]

37. Winandy, J.E. 1987. Effects of treatment and redrying on the mechanical properties of wood. Proceedings of the Forest Products Research Society on Wood Protection Techniques and the use of Treated Wood in Construction; 1987 October 28-30; Memphis, TN. Madison, WI: Forest Products Research Society. [100 p.].

CHAPTER 4

PRESERVATION AND PROTECTION OF TIMBER BRIDGES

4.1 INTRODUCTION

Wood has been successfully used as a bridge material for thousands of years, but before the early 1900's most structures were built of untreated timber. Protection from decay and deterioration was afforded by using the heartwood of naturally durable species or by covering the structure to protect it from weathering. Although many bridges constructed of untreated timber performed well (some lasting longer than 100 years), the use of untreated timber declined as naturally resistant North American wood species became unavailable in the quantities and sizes necessary for bridge construction. Additionally, it became economically and functionally impractical to cover timber bridges for protection. In spite of the attractiveness of using naturally durable wood, modem timber bridges must be preservatively treated to obtain adequate performance.

Wood will last for centuries if kept dry. However, if it is used in an unprotected environment, it becomes susceptible to attack by living and nonliving agents capable of degrading the wood structure. Nonliving or physical agents, including heat, abrasion, ultraviolet light, and strong chemicals, generally act slowly to decrease wood strength. Although these physical agents may be significant in some applications, the greatest hazard to timber bridges results from living or biotic agents, such as decay fungi, bacteria, insects, and marine borers. These agents can cause serious damage to untreated wood in a relatively short period in a variety of environments (see Chapter 13 for more detailed discussions on the agents and processes of deterioration).

Most of the biotic agents that enter and decay untreated wood require four basic conditions for survival: (1) moisture levels in the wood above the fiber saturation point, (2) free oxygen, (3) temperature in the range of 50 to 90°F, and (4) food, namely the wood. Although most biotic agents can be controlled by limiting moisture, oxygen, or temperature, it is often difficult or impractical to control these conditions. As a result, the most common method for controlling deterioration in adverse environments involves removing the food source by introducing toxic preservative chemicals into the wood cells using a pressure treatment process.

This chapter was coauthored by Michael A. Ritter and Jeffrey J. Morrell, Ph.D., Associate Professor, Department of Forest Products, Oregon State University, Corvallis, Oregon.

Wood preservatives are toxic chemicals that penetrate and remain in the wood structure. They should not be confused with protective coatings, such as paints or stains, which do nothing to kill or prevent the spread of biotic agents. A wood preservative must have the ability to penetrate the wood and persist in sufficient quantities for long periods. The degree of protection depends on the type of preservative used, the treatment process, the species of wood, and the environment to which the structure will be exposed. Applied correctly, wood preservatives can increase the life of timber structures by as much as five times or more.

A complete approach to the preservation and protection of timber bridges involves many considerations related to materials, preservative treatments, design details, and construction practices. This chapter addresses design requirements and considerations related to preservative treatments, including types of preservatives, treatment processes, design specifications, and quality assurance. Additional information related to design details and construction practices is presented in subsequent chapters.

4.2 TYPES OF WOOD PRESERVATIVES

Wood preservatives are broadly classified as oil-type or waterborne preservatives. These classifications are based on the chemical composition of the preservative and the type of solvent or carrier employed in treating. Oil-type preservatives are generally used in petroleum solutions ranging from heavy oils to liquefied petroleum gas. Waterborne preservatives are water soluble and are applied in solutions with water. The advantages and disadvantages of each type of preservative/solvent system depend on the specific characteristics of the preservative and solvent and on the environmental conditions to which the treated wood will be exposed.

To adequately protect wood, conventional preservatives must be toxic to the intended targets, be they fungi, insects, or animals. Unfortunately, the same characteristics that make a preservative effective can, at higher levels, render it unsafe for humans. With the exception of one preservative, copper naphthenate, all the preservatives addressed in this section are restricted-use pesticides and can be obtained and used only by licensed applicators. Use of wood treated with these chemicals is not restricted, although it must be accompanied by a consumer information sheet that describes proper handling procedures and precautions (see Chapter 16). While current environmental concerns have stimulated the search for new, less toxic wood preservatives, most of these formulations are still in the evaluation process and are several years away from commercial service.

OIL-TYPE PRESERVATIVES The three oil-type preservatives used in bridge applications are coal-tar creosote (creosote), pentachlorophenol (penta), and copper naphthenate. The characteristics of these preservatives vary significantly depending on the specific type of preservative and the carrier or solvent in which they are mixed. With the exception of some solutions of penta, oil-type preservatives generally leave the surface of the wood with an oily, unpaintable surface that may exude or bleed preservative. This bleeding can be minimized or eliminated when appropriate precautions are observed.

For bridge applications, oil-type preservatives are used almost exclusively for treating such structural components as beams and decks. They provide good protection from decay and other deterioration, are noncorrosive, and generally afford good physical protection of the wood surface from the effects of weathering. Because most oil-type preservatives can cause skin irritations, they should not be used for applications that require repeated human or animal contact, such as handrails.

Creosote

Creosote, which was first patented in 1831, ushered in the age of effective wood protection. It is a black or brownish oil consisting of a complex mixture of polynuclear aromatic hydrocarbons. Creosote is derived either from the destructive distillation of coal to produce coke (a byproduct of steel production) or by distillation of oil shale. Although creosote can be manufactured from other materials, such as wood or oil, all creosote used for commercial wood treatment is derived from coal tar. Because it is not a primary product, the composition of creosote has varied widely over the years. However, more restrictive requirements now ensure the availability of relatively uniform creosote. Because it is a complex mixture of nearly 300 compounds, the toxic mechanisms and migration of creosote from wood are still poorly understood more than 150 years after the chemical was patented.

Creosote has a long record of satisfactory use as a wood preservative, with many case histories documenting more than 50 years of proven performance in both railroad and highway use. This chemical has performed well in almost every environment except in areas where marine borer hazards are high because of attack by *Limnoria tripunctata* (this species of borer is capable of attacking creosoted wood in warmer marine saltwaters). Creosote provides the added advantages of protecting the wood from the effects of weathering and retarding the checking and splitting associated with changes in moisture content.

At one time, creosote was the most commonly used wood preservative for timber products, but an increased desire for clean surfaces, coupled with complaints about handling creosoted wood, has led to a gradual decline in the percentage of wood treated with this chemical. Today, creosote is frequently used to treat bridge components, utility poles, marine piling, and railroad ties. All these applications involve minimal human contact with the treated wood. Recently, a clean creosote with reduced surface deposits has been developed that leaves the wood a light brown color and has a reduced risk of preservative exudation on the wood surface.

As a wood preservative, creosote is commonly available in both its undiluted or straight form, and also as a blend in solvents. The following paragraphs discuss the various creosote preservatives and their use in timber bridge applications.

Coal-tar creosote in its straight or undiluted form is the most commonly used creosote preservative for sawn lumber, glulam, piling, and poles. This form of creosote preservative is preferred for bridge applications.

Creosote/coal-tar solutions are a blend of creosote and coal tar. There are four creosote/coal-tar solutions: Types A, B, C, and D. The percentage (by volume) of coal-tar distillate (creosote) in each type of solution is 80, 70, 60, and 50 percent, respectively. Creosote/coal-tar solutions have been used with some success for treating poles and piling in marine exposures. They are not commonly used in bridge applications because the high level of insolubles in the solutions can produce excessive bleeding of the treatment from the timber surface, contributing to environmental concerns. The number of creosote/coal-tar solutions available in the future is expected to decline because of the expense required to meet Environmental Protection Agency (EPA) requirements.

Creosote/petroleum-oil solutions consist of a blend of not less than 50 percent creosote (by volume) in a solution of petroleum oil. Although this type of preservative performs well in bridge applications when a minimum 50-percent volume of creosote is in the solution, there is currently no method of determining the percentage of creosote in the mixture after the creosote and oil are blended. There have been cases where treatments of this type contained insufficient quantities of creosote to adequately protect wood from deterioration. Until analytical or other methods are developed that ensure the level of creosote in oil solutions, this treatment is not recommended for bridge applications unless blending of the creosote and oil is observed and verified by the purchaser or a designated representative.

In addition to the preservatives mentioned already, creosote has been blended with naphthalene, penta, copper naphthenate, and sulfur. While some of these chemicals were effective in preventing wood deterioration, technical problems or costs have precluded their use.

Pentachlorophenol

First patented in 1935, penta was among the first of many synthetic pesticides that revolutionized the way people dealt with pests. Because penta could be easily synthesized by chlorinating phenol, there were few variations in the product, and the supply could meet demand. As a result, oil-borne penta and the waterborne pentachlorophenate salt became two of our most important biocides. As a wood preservative, penta is a highly effective inhibitor of oxidative phosphorylation, which prevents the affected organism from obtaining energy. However, penta is not effective against marine borers and is not recommended for marine use.

Although penta is still widely used, the presence of trace contaminants known as dioxins has led to increased pressure to ban this preservative, and EPA has placed penta on its list of restricted-use chemicals (the dioxins present in penta are not the more highly toxic tetrachlorodioxins). Restricted-use chemicals can be used only by applicators who have passed a test on pesticide safety in their respective States; however, use of wood treated with this chemical is not restricted. In addition to these restrictions, EPA has placed limits on the permissible levels of dioxins present in penta. This combination of regulations should reduce the hazard of using penta. In spite of these restrictions, penta is used on approximately 30 percent of the wood treated each year, primarily for poles, posts, and timbers.

Penta is generally applied a solution of approximately 5 to 9 percent (by weight) in one of four hydrocarbon solvents, Type A, B, C, or D. The use of penta preservatives is characterized by the type of solvent.

Type A solvent is an oil solvent that is generally referred to as heavy oil. It is commonly used to treat sawn lumber, poles, and glulam after gluing. This is the preferred solvent for most bridge applications because the oil provides some protection from weathering, resulting in reduced checking and splitting in members. It is not paintable and should not be used in applications subject to human or animal contact.

Type B solvent is a liquefied petroleum gas (butane) that evaporates from the wood to leave a clean, paintable surface. It is used (with limited availability) to treat sawn lumber and lumber laminations for glulam, and may also be used to treat small glulam members after gluing. Penta in Type B solvent can be used in bridge applications for treating handrails and floors on pedestrian crossings. It is not recommended for main structural components or members subjected to ground contact because it provides no surface protection from weathering.

Type C solvent is a light petroleum solvent that gives the wood a light color that can be painted. For bridge applications, penta in Type C solvent is the preferred treatment for lumber laminations in glulam that must be treated before gluing. Although the light petroleum does provide some initial protection against weathering, its effectiveness diminishes with time.

Type D solvent is methylene chloride that provides a treatment similar to that produced by Type B solvent; however, the solvent recovery process for this treatment may result in raised grain and checking of the wood.

In addition to these oil-type solvents, efforts have been made to develop waterborne penta formulations (Type E solvents); however, these formulations are currently approved only for aboveground use. Stake tests are now underway to determine appropriate ground contact levels for waterborne penta formulations.

A considerable body of literature has accumulated to suggest that the solvent used to deliver penta to the wood has a significant impact on preservative performance. This effect is most notable with penta treatments using the gaseous solvents (Types B and D). Because penta must enter the target organism to be effective, the solvent must permit the preservative to come in contact with the target organism. Types B and D solvents apparently limit the ability of penta to move in this manner, and there are several reports of surface decay in poles treated with these formulations. Studies are now underway to better understand the nature of this effect.

Copper Naphthenate

In addition to creosote and penta, a third oil-type preservative, copper naphthenate, has received increased attention and use in the past few years. Originally developed in the 1940's, copper naphthenate is produced by complexing copper with napthenic acid derived from petroleum. As with penta, it can be blended with several types of oil solvents and has performed well in long-term stake tests. Its primary advantage is that it is considered an environmentally safe preservative and is not currently included on the EPA list of restricted-use pesticides. Although the use of copper naphthenate has been limited in the past because of its high cost relative to other preservatives, its future use will undoubtedly increase as environmental considerations become more restrictive for other oil-type preservatives.

WATERBORNE PRESERVATIVES

Waterborne preservatives include formulations of inorganic arsenical compounds in a water solution. These chemicals leave the wood surface relatively clean with a light green, gray-green, or brown color, depending on the type of chemical used. Unlike most oil-type preservatives, water-borne formulations usually do not cause skin irritations and are suitable for use where limited human or animal contact is likely. After drying, wood surfaces treated with these preservatives can also be painted or stained.

The first waterborne preservatives were developed in the late 1800's; however, most of those formulations were susceptible to leaching from the wood and performed poorly in service. In the late 1930's, several waterborne formulations were developed that employed chromium along with copper and arsenic. The chromium bonds strongly with the wood to prevent leaching of the preservative system. The first of these formulations, chromated copper arsenate (CCA), was approved for wood use in the late 1940's, but did not receive extensive usage until the 1960's, when demand for clean and paintable wood increased. As CCA was being approved for use on wood, a second formulation, ammoniacal copper arsenate (ACA), was developed and approved for use on wood in 1953. Ammoniacal copper arsenate is the preferred waterborne preservative for difficult-to-treat species, such as Douglas-fir, because it penetrates the wood more effectively. A number of other waterborne formulations have also been developed, including acid copper chromate (ACC), ammoniacal copper zinc arsenate (ACZA), and chromated zinc chloride (CZC).

Of the numerous waterborne preservatives, CCA, ACA, and ACZA are most commonly used in bridge applications. Each of these preservatives is strongly bound to the wood, thereby reducing the risk of chemical leaching. Chromated copper arsenate is generally used to treat Southern Pine, ponderosa pine, and red pine, while ACA and ACZA are for refractory (difficult to treat) wood species, such as Douglas Fir-Larch; however, large quantities of western wood species, such as Hem-Fir, are treated with CCA. There are reports of incomplete penetration of Douglas-fir treated with CCA, and this matter is under study by the American Wood Preservers' Association. There are also reports that CCA and ACA are corrosive to galvanized hardware. However, the tendency for corrosion seems to vary with the wood species, preservative formulation, treatment conditions, and the service conditions to which the wood is exposed. Such corrosion has not been reported to be a problem for hot-dipped galvanized hardware commonly used for bridges.

While the treatment processes for ACA and ACZA use combinations of steam in higher temperature solutions to sterilize wood during the treatment process, CCA treatments are ambient temperature processes that do not result in wood sterilization. While this poses little problem in dimension lumber, failure to sterilize larger material during treatment can permit fungi already established in the central core to continue decaying the wood. Where CCA treatments are used on larger wood members with a high percentage of heartwood, the use of high-temperature kiln cycles to heat the center of the wood to at least 155 °F for 75 minutes to eliminate established decay fungi is highly recommended.

Waterborne preservatives are used most frequently for railings and floors on pedestrian sidewalks or other areas that may receive human contact. In some situations, they are also used to treat laminations for glulam before gluing. Waterborne preservatives are also very effective in treating piling for marine exposures where borer hazards are high. Test results based on seawater exposure have shown that a dual treatment of waterborne preservatives followed by creosote is possibly the most effective method of protecting wood where marine borer hazards are extremely high. Water-borne preservatives are not recommended for large glulam members because the wetting and drying process associated with treatment can cause dimensional changes as well as warping, splitting, or cracking of members. Additionally, they provide little resistance to weathering, which may result in more pronounced checking and splitting from moisture changes than would occur with oil-type preservatives.

4.3 PRESERVATIVE TREATMENT

Preservative treatment of wood involves the introduction of chemical preservatives into the wood structure. To be effective, the treatment must provide sufficient preservative penetration (the depth to which the preservative enters the wood) and adequate retention (the amount of preservative chemicals remaining in the wood after treatment). In the direction parallel to grain, fluids flow relatively easily, and adequate penetration is usually not difficult to achieve. In the directions perpendicular to grain, however, movement is much more restrictive and pressure processes are normally required to force the preservatives into the wood structure. Even with effective wood preservatives, adequate performance cannot be achieved without sufficient preservative penetration and retention.

The degree of protection provided by preservative treatment depends not only on the protective value of the preservative chemicals but also on the material properties of the wood, the manner in which it is prepared, and the treating process used to apply the preservative. Each of these factors can have an effect on preservative penetration and retention, and thus on the service life of the component being treated.

MATERIAL FACTORS AFFECTING TREATMENT

There are several factors related to the material character of wood that can affect its ability to accept preservatives. The most significant of these factors are the wood species, geographic source, moisture content at the time of treatment, harvest-treatment interval, and storage conditions before treatment.

Wood Species and Source

Wood species vary considerably in their ability to accept preservative treatments. In general, the sapwood of any species is much more receptive to treatment than heartwood, which in many cases is nearly impenetrable (Figure 4-1). Unfortunately, not all commercial species have large quantities of sapwood. This poses a major challenge to treaters faced with treating species characterized by high percentages of difficult-to-treat heartwood (Table 4-1). Such species as Southern Pine, ponderosa pine, and red pine have a high percentage of sapwood and are relatively easy to

Figure 4-1.- Cross section of a coastal Douglas-fir pile treated with creosote. Note that the preservative treatment penetrates the outer sapwood ring but stops at the less permeable heartwood.

Table 4-1.- Relative treatability of selected domestic species.

Heartwood least difficult to penetrate	Heartwood moderately difficult to penetrate	Heartwood difficult to penetrate	Heartwood very difficult to penetrate
Bristlecone pine	Baldcypress	Eastern hemlock	Alpine fir
Pinyon pine	California red fir	Engelmann spruce	Corkbark fir
Redwood	Douglas-fir (coast)	Grand fir	Douglas-fir (Rocky Mtn.)
	Eastern white pine	Lodgepole pine	Northern white-cedar
	Jack pine	Noble fir	Tamarack
	Loblolly pine	Sitka spruce	Western redcedar
	Longleaf pine	Western larch	
	Ponderosa pine	White fir	
	Red pine	White spruce	
	Shortleaf pine		
	Sugar pine		
	Western hemlock		

From Gjovik and Baechler.[9]

treat. Other species, such as Douglas-fir, have a low percentage of sapwood and are more difficult to treat. The amount of sapwood can also affect the rate at which wood must be processed after harvesting. Southern Pine has a high percentage of decay-susceptible sapwood and must be rapidly processed to prevent decay in the warm, humid southeastern climate. Conversely, such species as Douglas-fir have a lower percentage of sapwood and can be air-seasoned for long periods with relatively little degradation.

The effects of wood species and sapwood percentage on treatability differs for round material, such as piles and poles, and for sawn lumber. For round material, the sapwood of many species is treatable, resulting in a well-treated sapwood shell surrounding an untreated heartwood core. When some of the same species are sawn into lumber, however, many pieces contain little or no sapwood and are untreatable. Lodgepole pine, for example, has a treatable sapwood ring when used for piles or poles, but as sawn lumber, it may be totally untreatable.

Another species-related factor affecting treatment involves the elevation at which the wood is grown. Wood grown at higher elevations appears to be more difficult to treat than that grown at or near sea level. While this poses few problems in the eastern half of the country, a large percentage of western species are harvested from high-elevation stands. In one particular study,[11] it was found that treatability of Douglas-fir was highest in wood from the Oregon coastal range and steadily declined until the wood from trees grown east of the Cascade Mountains was classified as refractory, or untreatable (Figure 4-2). Although studied to a lesser extent, there are also reports that lodgepole pine, ponderosa pine, and many of the true firs *(Abies* sp.) are also affected in this manner. This variation in treatability places added importance on the need to adequately select the species and origin of wood to be treated and is recognized in national treating standards, which differentiate treatments for coastal Douglas-fir and intermountain Douglas-fir.

Moisture Content
In addition to wood species and source, moisture content at the time of treatment has a significant impact on preservative penetration and retention. Excessive moisture can result in incomplete penetration or areas totally void of treatment. It is generally accepted that wood must be below the fiber saturation point before treatment. Methods for reducing the moisture content of wood or conditioning before treatment are discussed under mechanical preparation.

Harvest-Treatment Interval
In the interim between harvesting and preservative treatment, wood is susceptible to attack by a variety of stain and decay organisms. Stain fungi generally attack the sapwood of freshly cut wood and cause discol-

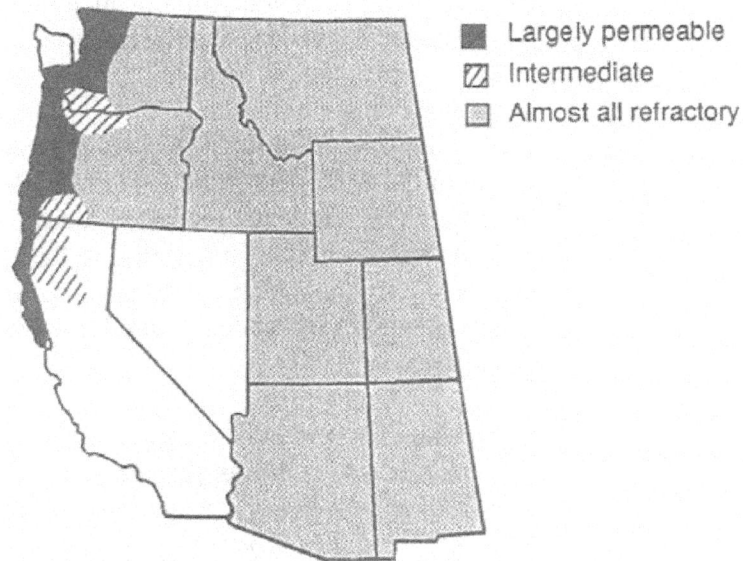

Figure 4-2.- Heartwood permeability of Douglas-fir varies with geographic source. Generally, coastal sources are permeable, Cascade Mountain sources are moderately impermeable, and intermountain sources are impermeable (refractory). From Morrell, Helsing, and Graham.''

oration, increased permeability to liquids, and reduced wood toughness (Chapter 13). Increased permeability can improve the treatability of difficult-to-treat species, but it can also result in bleeding of preservative from the wood after treatment. Stain prevention can be accomplished by drying the wood as quickly as possible or by dipping the freshly cut or peeled wood into fungicidal chemicals immediately after cutting. When wood is inspected before treatment, care should be taken to ensure the absence of stain, because this defect may indicate improper handling procedures. Where feasible, wood should be processed as soon as possible after cutting. Thick sapwood species should not be air seasoned for long periods, and care should be taken to ensure that all air-seasoned wood is sterilized during the treatment process. Species with thin sapwood are less susceptible to decay and stain fungi.

The length of time between harvest and treatment also seems to affect treatability. Although no detailed studies have been performed, treatability seems to decrease in Douglas-fir with increased length of air seasoning below the fiber saturation point.

MATERIAL PREPARATION

In addition to the need to choose treatable material that is free of defects, there are a number of mechanical processes that can substantially improve preservative treatment. These processes, which include debarking, prefabrication, incising, radial drilling, through-boring, kerfing, and pretreatment

4-11

conditioning, are intended to enhance the penetration and retention of preservatives to provide maximum protection.

Debarking

One of the first processing steps in preservative treatment involves removal of the bark. This zone contains cells that are extremely resistant to fluid flow and can leave untreated, decay-susceptible sapwood pockets near the wood surface. In addition to the effect on treatability, many insects require the presence of bark to infest the log. Removing the bark before the insect larvae hatch and burrow into the wood can limit this type of damage.

Debarking of round logs is usually accomplished by mechanically rotating shavers, wheels, or drums (Figure 4-3). These devices also remove some sapwood, and care must be taken to ensure that thin sapwood species are not overpeeled. For most sawn timber products, bark is removed in this manner before sawing or is removed during the sawing process. Sawn lumber should be inspected before treatment for the presence of bark on the edges. When this material is present, it should be removed before treatment.

Prefabrication

One of the most damaging, yet common, practices in the construction of timber bridges is field fabrication of treated wood (for example, attaching connectors or other wood members). Preservative treatment creates an envelope of protection around the wood. Any field fabrication involving cutting or drilling after treatment breaks this envelope, exposing untreated wood to attack by decay fungi and insects (Figure 4-4). Decay potential in field-drilled holes and sawn surfaces can be reduced by field treatment of the cut surfaces during construction; however, wood treated by superficial field methods (Chapter 12) is less resistant to decay than wood treated by pressure processes. A more effective prevention method involves complete fabrication (cutting and boring) prior to preservative treatment. This practice results in thoroughly protected wood, reducing the risk of decay, minimizing potential maintenance costs, and reducing the time required for field erection. The latter benefit can reduce the cost of construction and make timber more competitive as a material. All timber members should be fabricated before preservative treatment.

Incising

The sapwood of most species is easily penetrated by liquids, but adequate penetration of species containing mostly heartwood can pose much difficulty. Because fluids move more easily through end-grain, one approach to improving the preservative penetration of these species is to increase the amount of cross-sectional area exposed to the fluid. This can be accomplished by cutting or boring a series of slits or holes into the wood. This practice, called incising, is required for the adequate treatment of many wood species and results in a deeper, more uniform treatment.

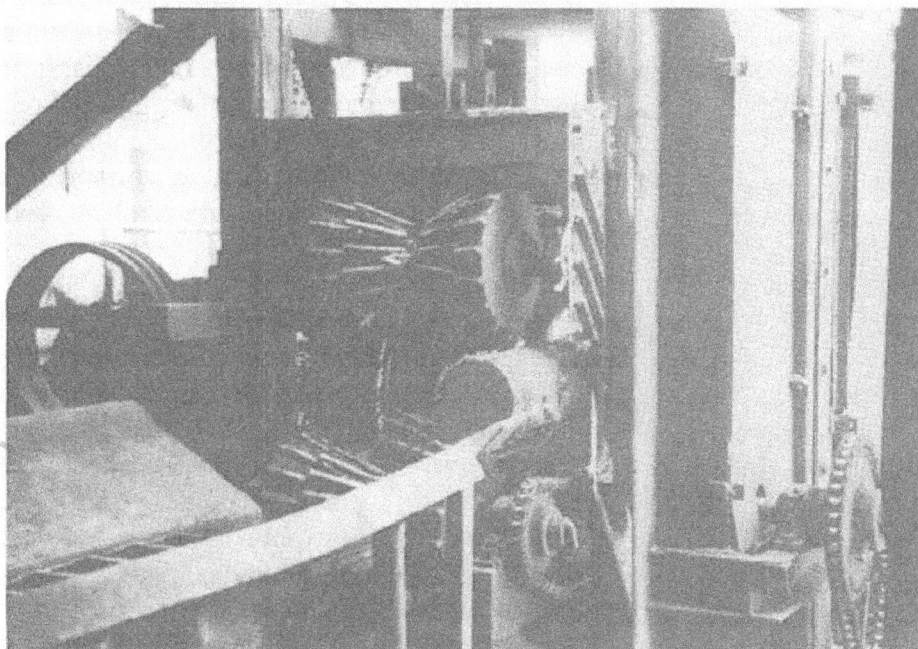

Figure 4-3.- Removing bark is an important part of the treating process. In this photo, logs are debarked by rotating wheels prior to being sawn (photo courtesy of Kevin Rockwell, Southern Pine Inspection Bureau).

Figure 4-4.- Decay can originate in field-bored holes that are made after the wood is treated with preservatives. In this member, a hole drilled after treatment exposed untreated wood and eventually led to decay at the center of the member (the horizontal split across the bolt hole was made as the member was removed). Preboring holes prior to pressure treatment can prevent this damage.

Incising practices vary considerably, depending on the commodity being treated and the wood species. Current standards for preservative treatment of incised wood are results oriented. That is, incised material must meet preservative requirements for penetration and retention, but there is no standard incising pattern. While this approach poses little problem in large timbers used for railroad ties and other nonstructural applications, the effect of incising on wood strength can be considerable for smaller dimension lumber. [13]

Incising is most commonly performed by pressing teeth into the wood surface to a predetermined depth, generally 1/4 to 3/4 inch (Figure 4-5). The teeth are spaced to give the desired distribution of preservative with the minimum number of incisions. Studies are under way to develop other, less destructive incising methods. To date, needle incising, water-jet incising, and laser incising have been or are being explored. Although none of these has yet proven commercially feasible, the results of preliminary work in these areas is promising.

Incising improves preservative penetration and retention and is required for treating most species. It is not normally required for Southern Pine, ponderosa pine, or red pine. With some species, such as coastal Douglas-fir, western hemlock, eastern white pine, and many of the true firs grown at lower altitudes, incising can greatly improve preservative penetration and retention. With other more refractory species, such as western larch, intermountain Douglas-fir, and lodgepole pine, the effects of incising are beneficial but less pronounced. With the exception of Southern Pine, ponderosa pine, and red pine, incising is an important part of the treating process and should not be waived for a lack of incising equipment at a treating plant. When large, glued-laminated members exceed the size capacity of incising equipment, individual laminations should be edge incised before gluing, or the entire member manually incised after gluing.

Radial Drilling

In some applications, incising can be replaced by radial drilling. In this process, a series of small-diameter holes are drilled into the sapwood to the desired depth of treatment. Radial drilling is required by many utilities for the treatment of electric transmission poles in high-decay-hazard areas. It also may be used for the treatment of piling but is not commonly used for sawn lumber or glued-laminated timber. As with incising, radial drilling decreases the cross-sectional area of the wood and may have some effect on strength.

Through-Boring

In addition to incising and radial drilling, preservative penetration and retention can be greatly improved by through-boring. This process, which is used by some utilities to reduce the decay hazard in poles at the groundline, involves drilling a series of angled holes through the wood approximately 4 feet above and below the theoretical groundline

Figure 4-5.-(A) The most common method of incising involves pressing small metal teeth into the wood surface. (B) The openings in the wood improve the penetration and retention of preservatives in many difficult- to-treat species.

(Figure 4-6). When performed properly, through-boring results in nearly complete preservative penetration of the heartwood. Although there is a reduction in strength associated with through-boring (approximately 5 percent in bending strength in utility poles), it is a feasible method for providing maximum protection for poles and piling in areas of severe decay hazard.

4-15

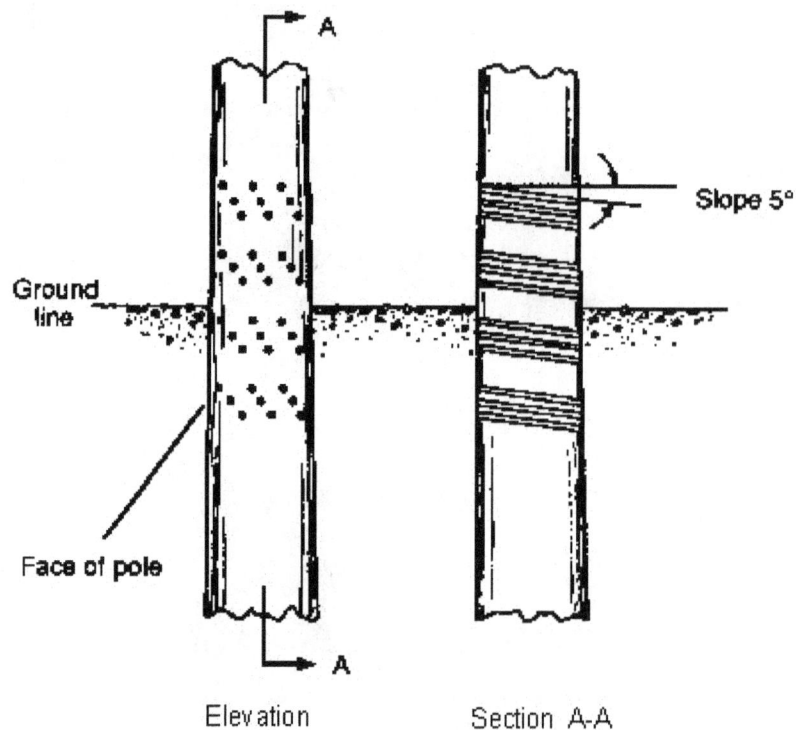

Figure 4-6.- Through-boring in areas of high decay hazard can result in nearly complete preservative penetration.

Kerfing

Most large wood members cannot be fully dried before preservative treatment. As a result, the wood continues to dry in service, resulting in splitting and checking from shrinkage. These checks penetrate beyond the preservative-treated shell of the wood member, providing avenues of entry for decay organisms. One method for limiting check development is to saw a narrow, longitudinal kerf to the center of the wood before preservative treatment (Figure 4-7). The kerf serves to allow some movement and relieve stresses from dimensional changes (shrinkage) that would otherwise cause the wood member to check. Although not commonly used in bridge applications, kerfing seems to work equally well in round or sawn timbers. While kerfing may reduce wood strength, the presence of a deep split has the same effect and, with kerfing, the location of the split can be controlled to minimize strength effects.

Conditioning

Conditioning is the process used to reduce the moisture content of wood before to preservative treatment. Although there are many methods of conditioning, the four most common methods are air drying, kiln drying, steaming, and Boulton drying. Air drying and kiln drying are often employed to process sawn lumber products for both treated and untreated uses. In contrast, steaming and Boulton drying are performed in a treating cylinder and are used exclusively as a method of conditioning wood before treatment. None of the conditioning methods completely dry large

Figure 4-7.—Kerfing large wood members, such as this pole, can reduce the potential for checking in service.

members, which would be uneconomical, but they do adequately dry the zone to be treated. As a result, large sawn lumber members may continue to dry and check after they are placed in service.

Air Drying

Air drying is the least intensive drying method and is extensively used for large western conifers and eastern hardwoods (Figure 4-8). Generally, the species must exhibit some resistance to decay to prevent fungi from causing damage during the conditioning period. Air-drying periods vary, ranging from as short as 6 months to as long as 3 years or more, and in most cases the wood is colonized by decay fungi during the process. While these fungi do not seem to cause damage if the seasoning period is limited, their presence places added importance on the need to adequately sterilize the wood during the treatment process.

Kiln Drying

In kiln drying, sawn lumber or timbers are placed in an enclosed structure and subjected to elevated temperatures and forced ventilation until the desired moisture content is achieved (Figure 4-9). The process increases

Figure 4-8.- Sawn lumber stacked for air drying. Note the thin wood strips or "stickers" placed between the lumber to permit free air circulation.

Figure 4-9.- Sawn lumber stacked for drying in a dry kiln (photo courtesy of Kevin Rockwell, Southern Pine Inspection Bureau).

the drying rate considerably over air drying and is commonly used for dimension lumber. The temperatures for conventional kiln drying typically range from 110 to 180°F, although high-temperature drying may reach temperatures in excess of 212°F. Drying time depends on the wood species, initial moisture content, lumber size, and the temperature maintained in the kiln. For 2-inch material dried to 19-percent moisture content at conventional temperatures, average times vary from approximately

4-18

41 hours for Southern Pine to approximately 72 hours for Douglas-fir. In the South and, at an increasing level, the West, kiln drying is the preferred method for reducing moisture content of dimension lumber before treating.

Steaming

In steam conditioning, green wood is placed in a treating cylinder and heated by steam to temperatures up to 245°F for several hours. After the steaming process is complete, a vacuum is applied to the cylinder, reducing the boiling point of water and causing the moisture in the outer zone of the wood to evaporate. The steaming and vacuum generally reduce the moisture content of the wood slightly, and the elevated temperature of the wood significantly facilitates preservative penetration. A sufficient steaming period also will sterilize the wood and exterminate decay fungi. Steaming is used primarily for conditioning wood that will be treated with waterborne preservatives, but steaming is not used when the planned treatment will be with CCA.

Boulton Drying

Boulton drying is a process developed in the 1870's that involves heating wood in oil under vacuum. Boulton drying is extensively used in western species, especially Douglas-fir, to condition green or partially air-seasoned timber before pressure treatment with oil-type preservatives. The Boulton drying period lasts from 24 to 48 hours and employs temperatures of 180 to 220°F. It permits seasoning of green, freshly cut, or peeled material to treatable moisture levels, with a minimal impact on wood strength. Although the Boulton process is still extensively used, it is under increasing scrutiny because the moisture removed from the wood is contaminated by trace amounts of wood preservative. Because of this, the wastewater, which can approach 5,000 gallons from a single charge, must be used to make up new solution or be disposed of. This adds to the expense of using this energy-intensive process and may ultimately preclude its use.

METHODS OF APPLYING PRESERVATIVES

There are two basic types of methods for applying preservative treatment to wood, nonpressure methods and pressure methods. Nonpressure methods include brushing, soaking, dipping, and the thermal process. With the exception of the thermal treatment of western redcedar and lodgepole pine, nonpressure processes are not used to any significant extent to initially treat wood used in bridge construction. Brushing and soaking are used to protect field cuts and bore holes made after pressure treatment (Chapter 12).

Wood used in bridges and other exposed environments is treated by using processes involving combinations of vacuum and pressure in a confined cylinder (retort) to deliver a specified amount of chemical into the wood (Figure 4-10). These pressure processes date back to 1836, and with few exceptions, the basic processes used today were patented before 1904.

Figure 4-10.-(A) Treatment cylinders (retorts) for pressure-treating wood. (B) As vie wed from the inside of a cylinder, wood ready for treating is loaded on carts that are rolled into the cylinder on steel tracks.

4-20

Although there have been many process variations to improve chemical penetration and fixation, or to reduce exudation of chemical from the wood, the overall treatment processes have remained fairly stable since the 1950's.

The objectives of the pressure processes are to kill any fungi that may be growing in the wood and ensure that a sufficient amount of preservative is delivered to the proper depth in the wood. The two types of pressure processes are the full-cell process and the empty-cell processes. The names *full-cell* and *empty-cell* are somewhat representative of the results of the respective treating processes. In the full-cell process, wood preservative coats the wood cell walls and, to various degrees, fills the empty-cell cavities. In the empty-cell processes, the cell walls also are penetrated, but the cell cavities are left relatively empty of preservative.

Full-Cell Process
The full-cell (or Bethell) process uses an initial vacuum in the treating cylinder for 30 minutes or longer to remove as much air as possible from the wood. Following this vacuum, preservative is added to the cylinder and pressure is applied up to 150 lb/in^2. Once a sufficient amount of chemical has been forced into the wood, the pressure is released and the preservative is withdrawn (Figure 4-11 A). At this point, a vacuum may be introduced in the cylinder, or the wood may be steamed to hasten recovery of excess preservative and to clean the wood surface.

The full-cell process produces the maximum solution retention for a given depth of penetration and is most often used for treatments with waterborne preservatives and for treating marine piling with creosote. For waterborne preservatives, solution strength can be varied to achieve the desired retention. With the exception of wood members in ground contact in areas of high decay hazard, the full-cell process is not recommended for wood bridge members treated with creosote or other preservatives in oil carriers (unless the required retention cannot be provided by empty-cell processes discussed below). High retentions of oil-type preservative in cell cavities can result in excessive bleeding of preservatives on the wood surface.

Empty-Cell Processes
The empty-cell processes, which include the Lowry and Rueping processes, do not use the initial vacuum treatment employed in the full-cell process (Figure 4-11 B). In the Lowry process, the preservative solution is admitted into the cylinder containing the wood, and the pressure on the solution is gradually increased. This pressure is held until a sufficient amount of solution is forced into the wood. As the pressure is released, air that was compressed into the wood forces out excessive preservative in a process termed *kickback,* resulting in a lower preservative retention for a given depth of penetration. At the end of the pressure period, the cylinder is drained, and a final vacuum is generally applied to remove any surplus preservative from the wood.

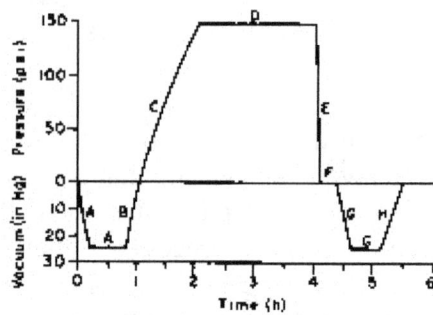

A. Preliminary vacuum
B. Filling cylinder with preservative
C. Pressure rising to maximum
D. Maximum pressure maintained
E. Pressure released
F. Preservative withdrawn
G. Final Vacuum
H. Vacuum released

A. Full-cell treatment cycle

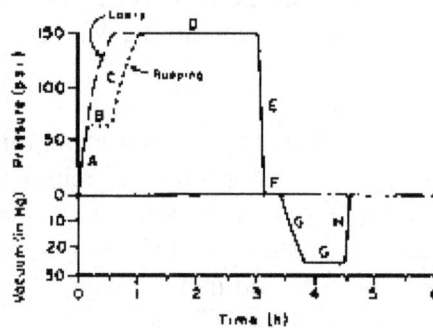

A. Preliminary air pressure applied
B. Filling cylinder with preservative
C. Pressure rising to maximum
D. Maximum pressure maintained
E. Pressure released
F. Preservative withdrawn
G. Final Vacuum
H. Vacuum released

B. Empty-cell treatment cycle

Figure 4-11- Diagrammatic representations of the full-cell and empty-cell processes for pressure-treating wood.

In the Rueping process, the cylinder containing the wood is initially pressurized at 25 to 100 lb/in² for 30 to 60 minutes before the preservative solution is added. After this period, preservative is forced into the cylinder, causing air in the cylinder to escape into an equalizing or Rueping tank at a rate that keeps the pressure constant in the cylinder. When the treating cylinder is filled with preservative, additional pressure is applied, and the treating process is completed in the same manner as the Lowry process.

Both the Lowry and Rueping processes are widely and successfully used in the treating industry. One advantage of the Lowry process is that it uses the same treating equipment used for the full-cell process. The Rueping process requires an equalizing tank and additional equipment to force the preservative into the pressurized cylinder.

Empty-cell processes are used for oil-type treatment of sawn lumber, glulam, piling, and poles. The objective of the processes is to achieve deep penetration with a relatively low net retention. As a result, the potential for substantial surface bleeding of preservative is less than with a full-cell process. It is recommended that empty-cell processes be used for all bridge treatments involving oil-type preservatives, provided retention requirements can be met.

Modified Pressure Processes

One variation in the pressure processes is the use of solvents that carry the preservative into the wood but vaporize after the pressure is released, leaving dry chemical deposited in the wood cell wall. Two such processes, the Dow and Cellon processes, use methylene chloride and butane, respectively, to dissolve penta. Because the solvents have a high vapor pressure, they rapidly volatilize from the wood, leaving the penta behind. The main advantage of these processes is the absence of surface oils that make painting difficult or mar the appearance. One disadvantage seems to be an increased susceptibility to the development of surface decay when the wood is used in ground contact.

POSTTREATMENT CLEANING

At the conclusion of the pressure period, some treaters heat wood in oil-type preservatives for several hours to force out excess preservative. Steaming also can be used to clean the wood surface after the pressure process. These heating or steaming periods reduce the amount of excessive preservative and decrease the potential for unsightly bleeding in service.

4.4 SPECIFYING TREATED TIMBER FOR BRIDGES

Although properly used preservative treatments will provide a long service life for wood products, the manner in which a commodity is specified can have a significant impact on its performance. Factors related to treatment preparation, processes, and results must all be carefully considered and specified, not only to ensure performance, but also to protect the buyer against inferior products. This section discusses treatment specifications, standards, and design considerations related to timber bridge applications. Methods of specifying treated timber, including typical specifications, also are addressed.

SPECIFICATIONS AND STANDARDS

Specifications and standards for the preservative treatment of wood are maintained by the American Wood Preservers' Association (AWPA), the American Association of State Highway and Transportation Officials (AASHTO), the American Institute of Timber Construction (AITC), and the Federal Government. The AWPA standards'the most widely used and most comprehensive standards and are the recommended source of specifications and treating process procedures for sawn lumber, glulam, piling, and poles used for timber bridges. The AASHTO (M133), AITC (AITC 109), and Federal standards directly reference or closely parallel the AWPA standards.

The AWPA standards are prepared by technical groups that consist of wood treaters, users, and general interest parties who assemble technical information to develop recommendations for the use of treated wood in specific environments. They contain requirements for the composition of preservatives and solvents, penetration and retention for various species and uses, and analytical procedures to ensure that treatment requirements are met. Also included are limits for pressures, temperatures, and exposure times during conditioning and treatment to avoid conditions that adversely affect strength or other wood properties. The standards are results oriented and are generally stated as minimums or acceptable levels over a designated range of values. This flexibility is intended to permit the purchaser and treater some latitude in meeting treatment requirements for specific applications without damaging the wood.

A book of AWPA standards is published annually and is available at nominal cost from AWPA (see Table 16-10 for address). The book is divided into five basic categories consisting of (1) preservative standards (P-standards), (2) commodity standards (C-standards), (3) analytical methods (A-standards), (4) miscellaneous standards (M-standards), and (5) conversion factors and correction tables (F-standards). The standards in these five groups are cross referenced and address a wide variety of timber products, many of which are not related to bridge applications. A list of those most applicable to timber bridges is given in Table 4-2. Although the standards may seem confusing at first glance, they contain a wealth of information and, with experience, are relatively simple to use. It is important that the designer obtain a current copy of these standards and become familiar with the contents prior to specifying treated timber.

DESIGN CONSIDERATIONS

Many of the design and performance considerations required for specifying treated timber for bridge applications were discussed in the preceding sections of this chapter. There are, however, several topics that continue to cause concern and deserve further emphasis before discussing treatment specifications. These topics include dimensional stability, surface appearance, and some special considerations for glulam.

Dimensional Stability

The primary purpose of wood preservatives is to protect timber members from decay and other deterioration. In addition to providing this protection, several of the oil-type preservatives, including creosote, creosote in petroleum oil, and penta or copper naphthenate in oil (Type A), provide added protection against the effects of weathering. Unlike waterborne preservatives or oil-type preservatives in volatile solvents, which afford little or no protection from moisture penetration, these heavier oil-type preservatives provide a water-resistant barrier on the wood surface.[10] Although they will not prevent splitting in members because of initial drying, they do reduce the susceptibility of the member to fluctuating moisture contents and associated dimensional changes and can reduce splitting and checking in service. This is an important consideration in timber bridges because checks provide avenues of entry for decay fungi and insects that would substantially reduce the service life of the structure.

The benefits of heavy oil preservatives are most pronounced in glulam members because of their large size. Glulam members are generally

Table 4-2. -Summary of AWPA Commodity Standards most applicable to bridges.

Preservative (P) Standards

Creosote P1
Creosote and Creosote Solutions P2
Creosote-Petroleum Oil Solution P3
Petroleum Oil for Blending with Creosote P4
Waterborne Preservatives P5
Creosote for Brush or Spray Treatment for Field Cuts P7
Oil-Borne Preservatives P8
Standard for Solvents for Organic Preservative Systems P9
Creosote-Pentachlorophenol Wood Preservative Solution P11
Creosote/Coal-Tar Solution for Use in Treatment of Marine (Coastal Waters) Piles and Timbers P12
Coal-Tar Creosote for Use in Treatment of Marine (Coastal Waters) Piles and Timbers P13

Commodity (C) Standards

All Timber Products, Pressure Treatment (General Requirements) C1
Lumber, Timbers, Bridge Ties, and Mine Ties, Pressure Treatment C2
Piles, Pressure Treatment C3
Poles, Pressure Treatment C4
Posts, Pressure Treatment C5
Wood for Highway Construction, Pressure Treatment C14
Material in Marine Construction, Pressure Treatment C18
Structural Glued-Laminated Members and Laminations Before Gluing, Pressure Treatment C28

Analysis Methods (A Standards)

Analysis of Creosote and Oil-Type Preservatives A1
Analyses of Waterborne Preservatives and Fire Retardant Formulations A2
Determining Penetration of Preservatives A3
Sampling Wood Preservatives A4
Analyses of Oil-Borne Preservatives A5
Determination of Water and Oil-Type Preservatives in Wood A6
Wet Ashing Procedure for Preparing Wood for Chemical Analysis A7
Qualitative Recovery of Creosote or Creosote/Coal-Tar Solution from Freshly Treated Piles, Poles,
   or Timber (Squeeze Method) A8
Analysis of Treated Wood and Treating Solutions by X-Ray Emission Spectroscopy A9
Analysis of CCA Treating Solutions and CCA Treated Wood by Colorimetry A10
Analysis of Treated Wood and Treating Solutions by Atomic Absorption Spectroscopy A11

Miscellaneous (M) Standards

Purchase of Treated Wood Products M1
Inspection of Treated Timber Products M2
Care of Pressure-Treated Wood Products M4
Glossary of Terms Used in Wood Preservation M5
Brands Used on Forest Products M6
Guideline for the Physical Inspection of Poles in Service M13
Miscellaneous Methods, Procedures and Information M17

From *AWPA Book of Standards.*[6] ©1986. Used by permission.

installed at a relatively low moisture content (less than 16 percent), and splitting and checking of the member because of initial moisture losses are minimal. However, without some protection to retard moisture absorption into the wood, members may split and check in service. Treatment with waterborne preservatives or penta in volatile solvents can lead to significant performance problems in glulam, as shown in Figure 4-12. With the exception of handrails or other components that may be subject to human or animal contact, or wood members that must be treated before they are glued, it is recommended that all bridge components be treated with creosote, creosote in petroleum oil, or penta or copper naphthenate in heavy oil (Type A) for best performance.

When waterborne treated members are used, the moisture content of the member after treating can also have an effect on dimensional stability. When timber is treated with waterborne preservatives, the wood becomes saturated with water, increasing the probability that seasoning checks and splits will develop as the member dries. It is recommended that all mem-

Figure 4-12.- Large glulam bridge members treated with waterborne preservatives (before or after gluing frequently check and split in service. With the exception of members that are subject to human or animal contact, all glulam used in bridge applications should be treated with oil-type preservatives.

bers treated with waterborne preservatives be dried after treatment. In most cases, drying to a moisture content of 19 percent is sufficient, but in very arid regions, lower moisture contents may be desirable. A number of recent studies have shown a significant posttreatment effect to be a direct result of the redrying after treatment. While stiffness has not been shown to be affected, some strength properties have been reduced. Recent modifications to the AWA standards for sawn lumber have restricted the posttreatment redrying temperature to no more than 190°F to minimize this potential problem.

Surface Appearance

In the past, users of treated wood were most concerned with performance, and there was less concern for such amenities as surface appearance. The recent environmental emphasis has changed this perspective, and the surface appearance and exudation or bleeding of oil-type preservatives have become important environmental issues. The most severe bleeding of treated wood members generally occurs along exterior beams or other components that are subjected to direct sunlight. The heating effect on these members can cause bleeding of preservatives that would otherwise not occur in shaded locations.

In most cases, the bleeding of oil-type preservatives in small quantities poses no harmful effects; however, bleeding should be minimized or eliminated whenever possible. Following are suggestions for improving the cleanliness of oil-type preservatives.

1. Specify the correct preservative retentions recommended in the appropriate AWPA standard for the type of material, use condition, and preservative. Retentions in excess of these levels increase the level of preservative in the wood, which may cause bleeding, and do not increase service life.

2. Use of empty-cell processes rather than the full-cell treating process for oil-type preservatives results in a lower level of preservative in the wood cell cavities and should be specified whenever possible. Empty-cell processes may not be feasible in situations when retentions greater than or equal to 20 lb/ft^3 for creosote are required.

3. When using creosote, use of clean creosote containing lower levels of xylene insolubles can reduce surface deposits.

4. Expansion baths (heating in preservative) at the conclusion of the treatment cycle and combinations of vacuum/steaming periods can reduce surface deposits and decrease bleeding once the wood is placed in service.

In addition to the above considerations, surface cleanliness also is related in some degree to the quality control and cleanliness of the treater. When the treating plant cylinder and pipes are kept free of sludges, surface residues and potential bleeding are reduced.

Special Considerations for Glulam

In most bridge applications, glulam is pressure treated after it is has been laminated (glued). In some cases, large members, such as arches, will not fit into treating cylinders after manufacture, and the individual laminations must be treated before gluing. Glulam can be manufactured from treated laminations when certain preservatives are used, specifically the water-borne preservatives or penta in light petroleum or volatile solvents (Type B, C, or D). When bridge members are treated before gluing, penta in light petroleum (Type C solvent) is recommended. Although penta in light petroleum is not as effective in protecting the member from moisture as the heavy oil preservatives, it does give limited surface protection and generally produces the best final results.

There recently has been some concern regarding glulam manufacture from treated laminations. In a December 1986 statement issued by the AITC, a policy was adopted by western laminators not to glue preservative-treated western species. Although this policy does not involve all laminators and does not affect Southern Pine species, the designer should verify industry capabilities before issuing designs or specifications that require preservative treatment before gluing.

SPECIFICATIONS FOR TREATED TIMBER

Treated timber must be properly and completely specified to obtain the required treatment for the intended application. For all types of treatments, specifications must include a preservative according to an AWPA P-standard and a treatment requirement (including preservative retention and penetration) in accordance with an AWPA C-standard. In addition, requirements for mechanical preparation and treating conditions may be desirable to ensure optimum preservative performance. These requirements vary for different component types and preservatives and generally include such items as pretreatment and posttreatment moisture contents, incising, prefabrication, treating procedures, and posttreatment steaming or cleaning.

The AWPA standards for treated timber in bridge applications are found in Standard C14, Wood for Highway Construction-Preservative Treatment by Pressure Process, and also in Standard C28, Standard for Preservative Treatment of Structural Glued Laminated Members and Laminations Before Gluing of Southern Pine, Pacific Coast Douglas Fir, Hemfir and Western Hemlock by Pressure Process. Both of these standards contain information related to treating requirements and preservative penetration and retention for various types of components, use conditions, and preservatives. Standard C14 gives specific requirements for sawn lumber, posts, poles, and piling but relies mainly on other AWPA standards for specific process requirements. Minimum preservative retentions from Standard C14 are shown in Table 4-3. Note that the retention for each preservative is specified for different components, such as sawn lumber, piles, and posts. The right column of the table specifies AWPA standard that gives additional treatment requirements for that type of component. For glulam, AWPA Standard C28 gives treating requirements for members treated before or after gluing. Retention requirements for glulam treated after gluing are shown in Table 4-4. Note that preservative retentions are based on the species of the laminations, not the type of component.

In most AWPA standards, minimum requirements for preservative retention are based on the type of material and the conditions where it will be used: aboveground, in ground contact, or in marine environments with exposure to borers. For wood used in bridges and other highway applications, aboveground conditions are generally not used and all components other than those subject to marine borers are treated to ground contact retentions. In Standard C14, one retention is specified regardless of whether the component is in ground contact or not (these retentions are approximately equal to ground contact requirements for sawn lumber specified in AWPA Standard C2). In Standard C28, retentions are specified for aboveground and ground contact; however, for bridge applications, the retentions specified for ground contact are normally used to provide retention levels comparable to those specified in Standard C14 for sawn lumber. Although much of a bridge will be out of ground or marine contact, it is important to recognize that some aboveground locations also are high-decay hazard environments. This is particularly true in the critical joint areas where moisture can collect and where decay is most likely to develop.

AWPA Standards C14 and C28 are designed to achieve 50 or more years of service life in most environments; however, additional requirements can be imposed when warranted by the needs of severe service. When additional retention or penetration requirements are considered, it is best to consult with specialists from a national treating organization, a university, or the USDA Forest Service, Forest Products Laboratory to ensure that such treatments are practical, safe, and worth the added costs. A listing of national treated timber organizations that provide assistance to users is given in Chapter 16.

Typical Treatment Specifications for Bridges

All information required to properly specify treated wood is found in the applicable AWPA standards. Additionally, the standards indicate which types of treatment are appropriate for various wood species and component types. The following sample specifications illustrate the information required to specify treated timber for several preservatives and commodity products. Additional requirements are included for treatment procedures,

Table 4-3.- Minimum preservative retentions for lumber, poles, and piling used for highway construction.

Material and Usage	Creosote[1]	Creosote-Coal Tar[1]	Creosote-Petroleum	Pentachlorophenol[2]	Pentachlorophenol, P9 Type E solvent	ACC	ACA	ACZA	CCA	PAS	AWPA standards
Lumber for Bridges, Structural Members, Decking, Cribbing, and Culverts											
Southern Pine, Coastal Douglas-fir, Western Hemlock, and Western Larch	12.0	12.0	12.0	0.60	NR[5]	NR	0.60	0.60	0.60	NR	C2
Structural Lumber in Salt Water											
Southern Pine	25.0	25.0	NR	NR	NR	NR	2.50	2.50	2.50	NR	C2
Coastal Douglas-fir, Hemlock	25.0	NR	NR	NR	NR	NR	2.50	2.50	2.50	NR	C2
Structural Lumber in Saltwater-Dual Treatment											
Southern Pine											
First treatment	NR	NR	NR	NR	NR	NR	1.50	1.50	1.50	NR	C2
Second treatment	20.0	20.0	NR	NR	NR	NR	NR	NR	NR	NR	C2
Coastal Douglas-fir, Western Hemlock											
First treatment	NR	NR	NR	NR	NR	NR	1.50	1.50	1.50	NR	C2
Second treatment	20.0	20.0	NR	NR	NR	NR	NR	NR	NR	NR	C2
Piles for Foundation, Land, or Fresh-Water Use											
Southern Pine, Ponderosa Pine, Jack Pine and Red Pine	12.0	12.0	12.0	0.60	NR	NR	0.80	0.80	0.80	NR	C3
Coastal Douglas-fir, Western Larch, Intermountain Douglas-fir, and Lodgepole Pine	17.0	17.0	17.0	0.85	NR	NR	1.00	1.00	1.00	NR	C3
Oak	6.0	6.0	6.0	0.30	NR	NR	NR	NR	NR	NR	C3
Posts, Fence, Guide, Sign, and Sight											
All Softwood Species											
Round, Half-Round, and Quarter Round[3]	8.0	8.0	8.0	0.40	NR	0.50	0.40	0.40	0.40	NR	C5
Sawn Four Sides	10.0	10.0	10.0	0.50	NR	0.62	0.50	0.50	0.50	NR	C2
Posts, Guardrail, Spacer Blocks[4]											
All Softwood Species											
Round	10.0	10.0	10.0	0.50	NR	NR	0.50	0.50	0.50	NR	C5
Sawn Four Sides	12.0	12.0	12.0	0.60	NR	NR	0.60	0.60	0.60	NR	C2
Poles, Lighting:											
Southern Pine, Ponderosa Pine	7.5	NR	NR	0.38	NR	NR	0.60	0.60	0.60	NR	C4
Red Pine	10.5	NR	NR	0.53	NR	NR	0.60	0.60	0.60	NR	C4
Coastal Douglas-fir	9.0	NR	NR	0.45	NR	NR	0.60	0.60	0.60	NR	C4
Jack Pine, Lodgepole Pine,	12.0	NR	NR	0.60	NR	NR	0.60	0.60	0.60	NR	C4
Western Red Cedar, Western Larch, Intermountain Douglas-fir	16.0	NR	NR	0.80	NR	NR	0.60	0.60	0.60	NR	C4
Handrails and Guardrails (not in contact with ground or water)											
All Softwood Species	8.0	NR	NR	0.40	0.40	0.25	0.25	0.25	0.25	0.40	C2

[1] When these preservatives are specified for materials to be used in salt water, the creosote-coal tar shall conform to AWPA Standard P12, and the creosote shall conform to AWPA Standard P13.

[2] Retention by lime ignition method. When copper pyridine method is used, multiply the results by 1.1 to convert to the lime ignition result.

[3] Where permitted in AWPA Standard C5.

[4] If spacer blocks are treated with other sawn material, the retention of the charge shall be determined by assay of borings taken from the other sawn material, unless each is sampled and assayed as an individual commodity.

[5] NR-Not recommended. Waterborne preservatives or pentachlorophenol in suitable solvents should be used where a dry surface is required or the material is to be painted.

From *AWPA Book of Standards.* ©1986. Used by permission.

Table 4-4. Minimum preservative retentions for glued-laminated timber treated after manufacture.

	Retention by assay (lb/ft^3), minimum			
	Southern Pine		Pacific coast Douglas-fir, hemfir, or western hemlock	
Treatment	Above-ground	Ground contact	Above-ground	Ground contact
Creosote	6.0	12.0	6.0	12.0
Creosote/Coal-Tar Solution	6.0	12.0	6.0	12.0
Creosote Petroleum	NR	NR	6.0	12.0
Pentachlorophenol	0.30	0.60	0.30	0.60

NR = Not recommended.
Refer to AWPA Standard C28 for table footnotes and requirements related to assay and penetration requirements.
From *AWPA* Book of Standards.[6]© 1986. Used by permission.

surface cleanliness, and moisture content for waterborne preservatives. These additional requirements are recommended but may be changed to meet specific design applications. For materials or use conditions other than those noted, sample specifications should be modified in accordance with AWPA Standards C14 and C28, and the applicable P-standards (preservative) listed in Table 4-2. For additional information on specifying treated timber, refer to AWPA Standard M1, Standard for the Purchase of Treated Wood Products.[6]

Creosote Treatment for Sawn Lumber
Sawn lumber shall be pressure treated using an empty-cell process with creosote conforming to AWPA Standard P1 to a minimum net retention of 12 lb/ft^3 in accordance with AWPA Standard C14. All members shall be fabricated before treatment and shall be free of excess preservative and solvent at the conclusion of the treating process.

Note: The same specification applies to glulam treated after gluing when AWPA Standard C14 is replaced by AWPA Standard C28.

Creosote Treatment for Douglas-Fir Foundation Piling in Land or Freshwater Use
Timber piling shall be incised and pressure-treated with creosote conforming to AWPA Standard P1 to a minimum net retention of 17 lb/ft^3 in the assay zone in accordance with AWPA Standard C14.

Note: Refer to AWPA Standard C14 for treating retentions for other species and piling used in salt water.

Creosote/Petroleum-Oil Treatment for Sawn Lumber
Sawn lumber shall be pressure treated using an empty-cell process with creosote/petroleum-oil solution conforming to AWPA Standard P3 to a minimum net retention of 12 lb/ft^3 in accordance with AWPA Standard

C14. All members shall be fabricated before treatment and shall be free of excess preservative and solvent at the conclusion of the treating process.

Note: The same specification applies to glulam treated after gluing when AWPA Standard C14 is replaced by AWPA Standard C28.

Penta in Petroleum-Oil (Type A) Treatment for Glulam Treated After Gluing

Glulam shall be pressure treated using an empty-cell process with pentachlorophenol conforming to AWPA Standard P8 in hydrocarbon solvent, Type A, conforming to AWPA Standard P9 to a minimum net retention of 0.60 lb/ft^3 in accordance with AWPA Standard C28. All members shall be fabricated before treatment and shall be free of excess preservative and solvent at the conclusion of the treating process.

Penta in Petroleum-Oil (Type C) Treatment for Laminations for Glulam Treated Before Gluing

Lumber laminations for glulam shall be pressure treated with pentachlorophenol conforming to AWPA Standard P8 in hydrocarbon solvent, Type C, conforming to AWPA Standard P9 to a minimum net retention of 0.60 lb/ft^3 in accordance with AWPA Standard C28.

CCA Treatment for Southern Pine Sawn Lumber Deck Planks

Sawn lumber planks shall be pressure treated with CCA conforming to AWPA Standard P5 to a minimum net retention of 0.60 lb/ft^3 in accordance with AWPA Standard C14. All members shall be fabricated before treatment and dried to a moisture content of 19 percent or less after treatment.

Note: CCA is used extensively for Southern Pine but is not recommended for Douglas-fir and other refractory species. These species are normally treated with ACA or ACZA.

ACZA Treatment for Douglas-Fir Sawn Lumber Guardrail Posts

Sawn lumber for guardrail posts shall be pressure treated with ACZA conforming to AWPA Standard P5 to a minimum net retention of 0.60 lb/ft^3 in the assay zone in accordance with AWPA Standard C14. All members shall be incised and fabricated before treatment and dried to a moisture content of 19 percent or less after treatment.

ACA Treatment for Western Hemlock Sawn Lumber Handrails

Sawn lumber for handrails shall be pressure treated with ACA conforming to AWPA Standard P5 to a minimum net retention of 0.25 lb/ft^3 in accordance with AWPA Standard C14. All members shall be incised and fabricated before treatment and dried to a moisture content of 19 percent or less after treatment.

While proper specifications help ensure proper treatment, wood is a variable material that does not always treat evenly. Inspection and quality control before, during, and after the treating process ensure that the material is suitable for the intended application. This inspection generally begins before treatment, when the untreated or white wood is inspected for grade, moisture content, stain or decay, and proper manufacture (cutting, boring, incising). Pieces with defects are rejected by the inspector based on end-use specifications. This point in the inspection is one of the most important because many defects are more easily seen in the white wood.

During the treatment procedure, the treater routinely removes samples of the treating solution for analysis to ensure adequate solution strength. In addition, the treating process is monitored by gauges to ensure compliance with the applicable AWPA standard. Following treatment, the material is again visually inspected to ensure that inadequate material did not slip through the white wood inspection. The inspector then removes a series of increment cores, at selected locations (depending on the commodity), from pieces in the charge. The depth of preservative penetration is measured either visually or by using chemical indicators to ensure that penetration requirements are met. Generally, a percentage of cores in each charge (usually 90 percent) must meet the requirements. If this does not occur, then all pieces in the charge are bored, and pieces not meeting the requirement must be retreated or rejected. The increment cores also are collected and returned to the laboratory where they are analyzed for preservative retention. Once again, failure to meet the retention requirement will lead to rejection of the charge.

Inspection of treated timber can be performed internally through a regular inspection staff or by contract through a third party. Many government bodies that purchase large quantities of treated wood maintain inspection staffs; however, the quantity of timber purchased by most users is usually not sufficient to justify a full-time staff. In these cases, the use of independent third-party inspection can provide reliable quality control at a reasonable cost. The treating industry has developed a quality control and certification program for treated products to assist users in obtaining properly treated material. The program is administered by the American Wood Preservers Bureau (AWPB), which acts as an independent third-party organization that licenses a number of inspection agencies to provide in-plant and field inspections of wood treaters and their products. Agency inspectors are highly qualified technicians who qualify individual treating plants for participation in the program. They train personnel for internal quality control programs and independently collect samples of pressure-treated wood; samples are sent to the agency or bureau laboratory for analysis of preservative retention and penetration. Treaters participating in the program who maintain their product quality are authorized to certify

A   Year of treatment
B   American Wood Preservers Bureau trademark or
    trademark of the AWPB certified agency
C   The preservative used for treatment
D   The applicable American Wood Preservers
    Bureau quality standard
E   Trademark of the AWPB certified agency
F   Proper exposure conditions
G   Treating company and plant location
H   Dry or KDAT if applicable

Figure 4-13.- Typical quality mark and nomenclature for wood treated in accordance with AWPB quality standards (courtesy of the American Wood Preservers Bureau). Used by permission.

their products with an AWPB quality mark (stamp or tag), which indicates that the product meets the specified standard (Figure 4-13). Additional information, including participating treaters and certified inspectors, may be obtained from AWPB at the address given in Table 16-10.

Although the AWPB is the largest nationwide organization for inspecting and certifying treated material, there are other qualified organizations and individuals that perform this service. For example, the Southern Pine Inspection Bureau administers an inspection and certification program for Southern Pine dimension lumber treated with waterborne preservatives. Regardless of the inspection organization or individual used, the user should always require that each piece of treated material be legibly ink stamped (waterborne preservatives only), branded, or tagged as evidence of inspection to certify compliance with treating standards. Examples of brands used for this purpose are given in AWPA Standard M6, Brands Used on Forest Products.[6]

4.6 SELECTED REFERENCES

1.   American Association of State Highway and Transportation Officials. 1982. AASHTO materials: part 1, specifications. Washington, DC: American Association of State Highway and Transportation Officials. 1094 p.

4-34

2. American Institute of Timber Construction. 1984. Standard for preservative treatment of structural glued laminated timber. AITC 109-84. Englewood, CO: American Institute of Timber Construction. 4 p.

3. American Institute of Timber Construction. 1985. Timber construction manual. 3d ed. New York: John Wiley and Sons, Inc. 836 p.

4. American Society of Civil Engineers. 1986. Evaluation and upgrading of wood structures: case studies. New York: American Society of Civil Engineers. 111 p.

5. American Wood Preservers Bureau. 1979. How you can be certain you're getting the pressure-treated wood you've specified. Arlington, VA: American Wood Preservers Bureau. 4 p.

6. American Wood-Preservers' Association. 1986. Book of standards. Stevensville, MD: American Wood-Preservers' Association. 240 p.

7. Blew, J.O., Jr. 1961. What can be expected from treated wood in highway construction. Rep. No. 2235. Madison, WI: U.S. Department of Agriculture, Forest Service, Forest Products Laboratory. 16 p.

8. Eslyn, W.E.; Clark, J.W. 1979. Wood bridges-decay inspection and control. Agric. Handb. 557. Washington, DC: U.S. Department of Agriculture, Forest Service. 32 p.

9. Gjovik, L.R.; Baechler, R.H. 1977. Selection, production, procurement and use of preservative-treated wood, supplementing Federal Specification TT-W-571. Gen. Tech. Rep. FPL-15. Madison, WI: U.S. Department of Agriculture, Forest Service, Forest Products Laboratory. 37 p.

10. McCutcheon, W.J.; Tuomi, R.L. 1973. Procedure for design of glued-laminated orthotropic bridge decks. Res. Pap. FPL 210. Madison, WI: U.S. Department of Agriculture, Forest Service, Forest Products Laboratory. 42 p.

11. Miller, D.J.; Graham, R.D. 1963. Treatability of Douglas-fir from western United States. Proceedings of the American Wood Preservers' Association. Vol. 59: 218-222.

12. Morrell, J.J.; Helsing, G.G.; Graham, R.D. 1984. Marine wood maintenance manual: a guide for proper use of Douglas fir in marine exposure. Res. Bull. 48. Corvallis, OR: Oregon State University, Forest Research Laboratory. 62 p.

13. Perrin, P.W. 1978. Review of incising and its effects on strength and preservative treatment of wood. Forest Products Journal 28(9): 27-33.

14. Southern Forest Products Association. 1983. Freedom of design with pressure-treated southern pine. New Orleans, LA: Southern Forest Products Association. 13 p.

15. Thomas, C.E. 1984. Consumer protection. Arlington, VA: American Wood Preservers' Association. 4 p.

16. U.S. Department of Agriculture, Forest Service, Forest Products Laboratory. 1981. Wood: Its structure and properties. Wangaard, F.F., ed. Clark C. Heritage Memorial Series on Wood. Compilation of educational modules; 2d Clark C. Heritage Memorial Workshop; 1980, August; Madison, WI. University Park, PA: The Pennsylvania State University. Vol. 1. 465 p.

17. U.S. Department of Agriculture. 1987. Wood handbook: Wood as an engineering material. Agric. Handb. No. 72. Madison, WI: U.S. Department of Agriculture, Forest Service, Forest Products Laboratory. 466 p.

18. Western Wood Products Association. 1984. Pressure treated lumber. WWPA spec point A-6. Portland, OR: Western Wood Products Association. 2 p.

CHAPTER 5

BASIC TIMBER DESIGN CONCEPTS FOR BRIDGES

5.1 INTRODUCTION

For thousands of years, timber bridges and other timber structures were built primarily by trial and error and rule of thumb. Designs were based on past experience, and little concern was given to efficient material usage or economy. As the complexity of structures increased, more attention was focused on the importance of accurate engineering methods. Research was undertaken to develop design criteria for wood with the same level of accuracy and reliability available for other engineering materials. As a result, developments in timber design have advanced substantially in this century. Although wood is orthotropic and differs in many respects from other materials, wood structures are designed using many of the same equations of mechanics developed for isotropic materials. Variations in material properties from growth characteristics, manufacturing, and use conditions are compensated for by material grading and stress adjustments applied in the design process. Timber design may seem confusing at first, but with experience it is no more difficult than design with other materials.

This chapter provides an overview of basic design concepts for sawn lumber and glulam used in bridge design. It includes specification requirements and methods for designing beams, tension members, columns, combined axial and bending members, and connections. Applications of these concepts to design situations are given in examples for each member and connection type. More detailed design related to specific bridge types is covered in Chapters 7, 8, and 9.

The discussions and examples in this chapter are based on a number of referenced specifications that were current at the time of publication. The reader is cautioned to verify these requirements against the most recent edition of the specifications before designing a bridge. In no case should the information presented in this chapter be considered a substitute for the most current design specifications.

5.2 DESIGN SPECIFICATIONS AND STANDARDS

The primary specifications for bridge design in the United States are the *Standard Specifications for Highway Bridges,* adopted and published by the American Association of State Highway and Transportation Officials (AASHTO). [1] These specifications are published intermittently and are revised annually through the issuance of interim specifications. They address all areas of bridge design, including geometry, loading, and design requirements for materials. AASHTO specifications are used extensively as the standard for bridge design and are the primary reference for the timber design requirements, procedures, and recommendations addressed in this manual.

The majority of the timber design requirements in AASHTO are based on the *National Design Specification for Wood Construction (NDS)*.[26] The NDS is the most widely recognized general specification for timber design and is published periodically by the National Forest Products Association. The specification includes design requirements and tabulated design values for sawn lumber, glulam, and timber piles. Although the NDS does not specifically address detailed bridge design, it does serve as the basis for the timber design concepts and requirements used for bridges. Notation of the NDS as the source of design requirements in this chapter reflects references in AASHTO that specify the NDS as the most current source of timber design information for bridges (AASHTO 13.1.1).

In addition to the NDS, AASHTO periodically references the specifications, standards, and technical publications of the American Institute of Timber Construction (AITC). AITC is the national technical trade association of the glulam industry and is responsible for numerous specifications and technical publications addressing fabrication, design, and construction of glulam. AITC also publishes *AITC 117-Design Standard Specifications for Structural Glued Laminated Timber of Softwood Species (AITC 117-Design)*, which is the source of tabulated values for glulam.[4]

Timber design requirements for bridges may differ from those commonly used for buildings and other structures. Although the requirements in AASHTO are based on the NDS and other referenced specifications and standards, modifications have been incorporated in AASHTO to address specific bridge requirements. The designer should become familiar with the content and requirements of current AASHTO, NDS, and AITC specifications. Copies of these specifications and other noted references are available from the parent organizations at the addresses listed in Table 16-10.

5.3 DESIGN METHODS AND VALUES

Timber bridges are designed according to the principles of engineering mechanics and strength of materials, assuming the same basic linear elastic theory applied to other materials. The method used for design is the allowable stress design method, which is similar to service load design for structural steel. In this method, stresses produced by applied loads must be less than or equal to the allowable stresses for the material. A design method called load and resistance factor design (LRFD) is used for timber design in other countries, but not in the United States. Progress is being made toward development of such a method in the United States; however, adoption is several years away.

As discussed in Chapter 3, wood strength and stiffness vary with species, growth characteristics, loading, and conditions of use. As a result, one set of allowable design values for all species and design situations would result in very uneconomical design in most cases. Conversely, tabulated values for all potential conditions would result in so many tables that they would be unusable. Rather than using either of these approaches, timber design is based on published tabulated values that are intended for one set of standard conditions. When these conditions differ from those of the design application, the tabulated values are adjusted by modification factors to arrive at the allowable values used for each design. This approach produces more realistic design values for a specific situation. In general terms, the basic timber design sequence is as follows:

1. Compute load effects and select an initial member size and species.

2. Compute the *applied stress* from applied loads.

3. Obtain the *tabulated stress* published for the specific material.

4. Determine appropriate modification factors and other adjustments required for actual use conditions.

5. Adjust the tabulated stress to arrive at the *allowable stress* used for design.

6. Compare applied stress to allowable stress. The design is satisfactory when applied stress is less than or equal to allowable stress.

SYMBOLS AND ABBREVIATIONS

Timber design uses standard symbols to denote the types of stresses for strength properties. These symbols consist of a stress symbol to designate the type of stress (applied, tabulated, or allowable), followed by a lower case subscript to denote the specific strength property (bending, shear, tension, and so forth). The symbols used for this purpose are shown in Table 5-1. For example, applied, tabulated, and allowable bending stresses are designated f_b, F_b, and F_b', respectively. The same type of designation without the strength property subscript applies to modulus of elasticity, where E denotes the tabulated value and E' denotes the allowable value. For glulam, an additional subscript of x or y may be included to designate

Table 5-1.- Stress symbols for timber components.

Stress symbol	Definition
f	Applied stress from loading
F	Tabulated stress from the applicable design specifications
F'	Allowable stress for design (tabulated stress adjusted by all applicable modification factors)
F''	Intermediate stress for calculating the tabulated stress for some beams or columns

Property subscript	Definition
b	Bending
v	Horizontal shear
t	Tension parallel to grain
c	Compression parallel to grain
$c\perp$	Compression perpendicular to grain
g	End grain in bearing

values about the x-x or y-y axis of the member (the x-x axis for glulam is always parallel to the wide face of the laminations). For example, F_{bx} is the tabulated bending stress about the x-x axis. In the absence of such a subscript, it is assumed that stresses act about the x-x axis.

TABULATED DESIGN VALUES

Tabulated design values for sawn lumber and glulam are based on testing and grading processes discussed in Chapter 3. These values represent the maximum permissible values for specific conditions of use and normally require adjustments for actual design conditions. In this sense, tabulated values should be viewed only as the basis or starting point for determining the allowable values to be used for design. An abbreviated summary of tabulated values for sawn lumber and glulam is published in AASHTO; however, these values do not include all species and grades and may not be current. For this reason, AASHTO requires that tabulated values comply with those specified in the most current edition of the NDS or AITC specifications (AASHTO 13.1.1 and 13.2.2). The source of tabulated values for sawn lumber is *Design Values for Wood Construction*, which is an integral part of the NDS, but is published as a separate volume. Tabulated values for glulam are given in *AITC 117-Design*. These NDS and AITC specifications represent the most comprehensive and current source of design information and include tabulated values for the following properties:

Bending (F_b)
Horizontal shear (F_v)
Tension parallel to grain (F_t)
Compression parallel to grain (F_c)
Compression perpendicular to grain ($F_{c\perp}$)

End grain in bearing (F_g)
Modulus of elasticity (E)

Tabulated Values for Sawn Lumber

Tabulated values for visually graded and machine stress rated (MSR) sawn lumber are published in the NDS based on the grading rules established by seven grading agencies. Separate tables are included for visually graded sawn lumber, MSR lumber, and end grain in bearing. The values are valid for sawn lumber used in dry applications under normal loading conditions (both of these conditions are discussed later for modification factors). In addition, each table contains an extensive set of footnotes for adjusting values to specific use conditions.

Visually Graded Sawn Lumber

Design values for visually graded sawn lumber are specified in Table 4A of the NDS. A portion of this table is shown in Table 5-2. The table gives tabulated values for F_b, F_t, F_v, F_c, $F_{c\perp}$, and E based on the species, size classification, and commercial grade of the lumber. When using the table, the following considerations will help interpret tabulated values:

1. Wood species may be specified as an individual species or a species combination. When species combinations are used, the individual species of the combination are listed in the Table 4A table of contents.

2. The grading rules agencies for each species are noted in the far right column of the tables. When grading rules for the same species differ among agencies, tabulated values are given separately for each grading agency.

3. Tabulated values for each species are based on the grade and size classification. Although commercial grade designations may be the same, tabulated values can vary among size classifications. For example, the tabulated values for grade No. 1 in the Beams and Stringers (B&S) size classification are not necessarily the same as those for No. 1 in the Posts and Timbers (P&T) size classification.

4. For all dimension lumber that is 2 to 4 inches thick, grading rules and commercial-grade nomenclature are standardized. When sawn lumber is thicker than 4 inches, grades are not standardized, and tabulated values for the same species, size, and grade of member may vary among grading agencies. In situations where conflicting tabulated values are given for different agencies, the designer must either specify the grading rules agency or use the lower tabulated values.

5. The availability of sawn lumber in the species, grade, and size classifications in Table 4A of the NDS may be geographically limited. The designer should verify availability before specifying a particular species, size, or grade.

Table 5-2. —Typical tabulated values for visually graded sawn lumber.

Species and Commercial Grade	Size Classification	Design values in pounds per square inch							Grading rules agency
		Extreme fiber in bending "F_b"		Tension parallel to grain "F_t"	Horizontal shear "F_v"	Compression perpendicular to grain "$F_{c\perp}$"	Compression parallel to grain "F_c"	Modulus of elasticity "E"	
		Single-member uses	Repetitive-member uses						
COTTONWOOD (Surfaced dry or surfaced green. Used at 19% max. m.c.)									
Stud	2" to 3" thick 2" to 4" wide	525	600	300	65	320	350	1,000,000	NHPMA (See footnotes 1–12)
Construction	2" to 4" thick 4" wide	675	775	400	65	320	650	1,000,000	
Standard		375	425	225	65	320	525	1,000,000	
Utility		175	200	100	65	320	350	1,000,000	
DOUGLAS FIR-LARCH (Surfaced dry or surfaced green. Used at 19% max. m.c.)									
Dense Select Structural		2450	2800	1400	95	730	1850	1,900,000	
Select Structural		2100	2400	1200	95	625	1600	1,800,000	
Dense No. 1		2050	2400	1200	95	730	1450	1,900,000	
No. 1	2" to 4" thick 2" to 4" wide	1750	2050	1050	95	625	1250	1,800,000	
Dense No. 2		1700	1950	1000	95	730	1150	1,700,000	
No. 2		1450	1650	850	95	625	1000	1,700,000	
No. 3		800	925	475	95	625	600	1,500,000	
Appearance		1750	2050	1050	95	625	1500	1,800,000	WCLIB
Stud		800	925	475	95	625	600	1,500,000	WWPA
Construction	2" to 4" thick 4" wide	1050	1200	625	95	625	1150	1,500,000	
Standard		600	675	350	95	625	925	1,500,000	
Utility		275	325	175	95	625	600	1,500,000	
Dense Select Structural		2100	2400	1400	95	730	1650	1,900,000	(See footnotes 1–12 and 20)
Select Structural		1800	2050	1200	95	625	1400	1,800,000	
Dense No. 1	2" to 4" thick 5" and wider	1800	2050	1200	95	730	1450	1,900,000	
No. 1		1500	1750	1000	95	625	1250	1,800,000	
Dense No. 2		1450	1700	775	95	730	1250	1,700,000	
No. 2		1250	1450	650	95	625	1050	1,700,000	
No. 3		725	850	375	95	625	675	1,500,000	
Appearance		1500	1750	1000	95	625	1500	1,800,000	
Stud		725	850	375	95	625	675	1,500,000	
Dense Select Structural		1900	—	1100	85	730	1300	1,700,000	
Select Structural	Beams and Stringers	1600	—	950	85	625	1100	1,600,000	
Dense No. 1		1550	—	775	85	730	1100	1,700,000	
No. 1		1300	—	675	85	625	925	1,600,000	
No. 2		875	—	425	85	625	600	1,300,000	WCLIB
Dense Select Structural		1750	—	1150	85	730	1350	1,700,000	
Select Structural	Posts and Timbers	1500	—	1000	85	625	1150	1,600,000	(See footnotes 1–12 and 20)
Dense No. 1		1400	—	950	85	730	1200	1,700,000	
No. 1		1200	—	825	85	625	1000	1,600,000	
No. 2		750	—	475	85	625	700	1,300,000	
Select Dex	Decking	1750	2000	—	—	625	—	1,800,000	
Commercial Dex		1450	1650	—	—	625	—	1,700,000	
Dense Select Structural		1850	—	1100	85	730	1300	1,700,000	
Select Structural	Beams and Stringers	1600	—	950	85	625	1100	1,600,000	
Dense No. 1		1550	—	775	85	730	1100	1,700,000	
No. 1		1350	—	675	85	625	925	1,600,000	
Dense No. 2		1000	—	500	85	730	700	1,400,000	
No. 2		875	—	425	85	625	600	1,300,000	WWPA
Dense Select Structural		1750	—	1150	85	730	1350	1,700,000	
Select Structural	Posts and Timbers	1500	—	1000	85	625	1150	1,600,000	
Dense No. 1		1400	—	950	85	730	1200	1,700,000	
No. 1		1200	—	825	85	625	1000	1,600,000	
Dense No. 2		800	—	550	85	730	550	1,400,000	
No. 2		700	—	475	85	625	475	1,300,000	(See footnotes 1–13 and 20)
Selected Decking	Decking	—	2000	—	—	—	—	1,800,000	
Commercial Decking		—	1650	—	—	—	—	1,700,000	
Selected Decking	Decking	—	2150	(Surfaced at 15% max. m.c. and used at 15% max. m.c.)			—	1,900,000	
Commercial Decking		—	1800				—	1,700,000	

Refer to the latest edition of the NDS for a complete and current listing of tabulated values and footnote explanations. From NDS;[24] © 1986. Used by permission.

MSR Lumber

For MSR lumber, tabulated values are derived by nondestructive stiffness testing of individual pieces that are 2 inches thick or less. Values are specified in Table 4B of the NDS for F_b, F_t, F_c, and E based on the grade designation and size classification of lumber (Table 5-3). Tabulated stresses for F_v and $F_{c\perp}$ are as specified in NDS Table 4A for No. 2 visually graded sawn lumber of the appropriate species.

End Grain in Bearing

The NDS contains a separate table of tabulated stress for end grain in bearing, F_g. These values are specified in Table 2B of the main NDS volume and pertain only to end-grain bearing parallel to grain on a rigid surface. The stresses are given for each species based on member size and use conditions and apply to both visually graded and MSR lumber.

Table 5-3. -Typical tabulated values for MSR sawn lumber.

| Grade designation[11] | Grading rules agency (see footnotes 1,2,3,4) | Size classification | Design values in pounds per square inch[10] | | Tension parallel to grain "F_t" | Compression parallel to grain "F_c" | Modulus of elasticity "E" |
| | | | Extreme fiber in bending "F_b"[8] | | | | |
			Single-member uses	Repetitive-member uses			
900f-1.0E	3,4		900	1050	350	725	1,000,000
1200f-1.2E	1,2,3,4,		1200	1400	600	950	1,200,000
1350f-1.3E	2,3,4		1350	1550	750	1075.[12]	1,300,000
1450f-1.3E	1,3,4		1450	1650	800	1150	1,300,000
1500f-1.3E	2		1500	1750	900	1200	1,300,000
1500f-1.4E	1,2,3,4		1500	1750	900	1200	1,400,000
1650f-1.4E	2		1650	1900	1020	1320	1,400,000
1650f-1.5E	1,2,3,4		1650	1900	1020	1320	1,500,000
1800f-1.6E	1,2,3,4	Machine rated lumber	1800	2050	1175	1450	1,600,000
1950f-1.5E	2	2" thick	1950	2250	1375	1550	1,500,000
1950f-1.7E	1,2,4	or less	1950	2250	1375	1550	1,700,000
2100f-1.8E	1,2,3,4	All	2100	2400	1575	1700	1,800,000
2250f-1.6E	2	Widths	2250	2600	1750	1800	1,600,000
2250f-1.9E	1,2,4		2250	2600	1750	1800	1,900,000
2400f-1.7E	2		2400	2750	1925	1925	1,700,000
2400f-2.0E	1,2,3,4		2400	2750	1925	1925	2,000,000
2550f-2.1E	1,2,4		2550	2950	2050	2050	2,100,000
2700f-2.2E	1,2,3,4		2700	3100	2150	2150	2,200,000
2850f-2.3E	2,4		2850	3300	2300	2300	2,300,000
3000f-2.4E	1,2		3000	3450	2400	2400	2,400,000
3150f-2.5E	2		3150	3600	2500	2500	2,500,000
3300f-2.6E	2		3300	3800	2650	2650	2,600,000
900f-1.0E	1,2,3		900	1050	350	725	1,000,000
900f-1.2E	1,2,3	See	900	1050	350	725	1,200,000
1200f-1.5E	1,2,3	footnote	1200	1400	600	950	1,500,000
1350f-1.8E	1,2	5	1350	1550	750	1075	1,800,000
1500f-1.8E	3		1500	1750	900	1200	1,800,000
1800f-2.1E	1,2,3		1800	2050	1175	1450	2,100,000

Refer to the latest edition of the NDS for a complete and current listing of tabulated values and footnote explanations. From the NDS;[24] © 1986. Used by permission.

Tabulated Values for Glued-Laminated Timber (Glulam)

Tabulated values for glulam are specified in *AITC 117-Design*. Separate tables are included for bending combinations, axial combinations, and end grain in bearing. Values are given for western species and Southern Pine made with either visually graded or E-rated lumber based on dry-use conditions (moisture content of 16 percent or less) and normal duration of load. Tabulated values for a specific combination symbol of glulam are standardized and are not subject to variations in grading rules or fabrication processes.

Bending Combinations

For bending combinations, tabulated values are given in Table 1 of *AITC 117-Design*. The combination symbols in this table are for members consisting of four or more laminations, stressed primarily in bending with loads applied perpendicular to the wide faces of the laminations *(x-x* axis). The table also includes tabulated values for axial loading and bending with loads applied parallel to the wide faces of the laminations *(y-y* axis); however, the axial combinations are usually better suited for these loading conditions. A limited number of combination symbols, taken from Table 1 from *AITC 117-Design*, are shown in Table 5-4. The first two columns of the table give the combination symbol and species of the member. The remainder of the table is divided into three parts based on the type and direction of applied stress. Columns 3 to 8 contain stresses for members loaded in bending about the *x-x* axis (the most common case). For this condition, stresses for F_b and $F_{c\perp}$ are specified separately for the tension and compression zones of the member. These stresses may be the same for both zones (balanced combination) or may differ significantly. Columns 9 to 13 are for members loaded in bending about the *y-y* axis where stresses in the tension and compression zones are equal. Columns 14 to 16 are for members loaded axially or with a combination of axial and bending loads. The intended use and limitations for groups of combinations are also noted in the table.

Axial Combinations

Tabulated values for axial combinations are specified in Table 2 of *AITC 117-Design*. The combinations in this table are intended primarily for members loaded axially or in bending with loads applied parallel to the wide faces of the laminations *(y-y* axis). The table also includes tabulated values for loading perpendicular to the wide faces of the laminations *(x-x* axis), but bending combinations are usually better suited for this condition. A limited number of combination symbols, taken from Table 2 from *AITC 117-Design*, are shown in Table 5-5. The table is organized in three sections based on the type and direction of applied stresses, as in Table 5-4. Tabulated values depend on the number of laminations and are given for members consisting of 2, 3, and 4 or more laminations. For all axial combinations, strength properties are balanced about the neutral axis, and tabulated stresses for F_b and $F_{c\perp}$ are equal in the tension and compression zones.

Table 5-4.- Typical tabulated values for glulam bending combinations.

		Bending About X-X Axis						Bending About Y-Y Axis					Axially Loaded		
		Loaded Perpendicular to Wide Faces of Laminations						Loaded Parallel to Wide Faces of Laminations							
		Extreme Fiber in Bending F_bx		Compression Perpendicular to Grain, F_cⱢx						Horizontal Shear F_vy				
Combination Symbol[d]	Species Outer Laminations/Core Laminations[e]	Tension Zone Stressed in Tension[i,v]	Compression Zone Stressed in Tension[g]	Tension Face	Compression Face	Horizontal Shear F_vx	Modulus of Elasticity, E_x ×10⁶psi	Extreme Fiber in Bending, F_by[r]	Compression Perpendicular to Grain, F_cⱢy	Horizontal Shear F_vy	(For members with multiple piece laminations which are not edge glued)	Modulus of Elasticity, E_y ×10⁶psi	Tension Parallel to Grain, F_t	Compression Parallel to Grain, F_c	Modulus of Elasticity, E ×10⁶psi
		psi	psi	psi	psi	psi		psi	psi	psi	psi		psi	psi	
1	2	3	4	5	6	7	8	9	10	11	12	13	14	15	16

Visually Graded Western Species															
The following four combinations are not balanced and are for either dry or wet use.															
16F-V1	DF/WW		560[h,i]	560[h,i]	140[s,w]	1.3[x]	950	255	130[s,w]	65[s,w]	1.1[x]	675	975	1.1[x]	
16F-V2	HF/HF		500[i]	375[i]	155	1.4	1250	375	135	70	1.3	875	1300	1.3	
16F-V3	DF/DF	1600	800	560[h,i]	560	165	1.5	1450	560	145	75	1.5	950	1550	1.5
16F-V8	DFS/DFS		650	500	165	1.2	1200	500	145	75	1.1	825	1350	1.1	
The following two combinations are intended for straight or slightly cambered members for dry use and industrial appearance.[k]															
16F-V4	DF/N3WW		650	560[h]	90[i,s,w]	1.5[x]	900	255	130[s,w]	65[s,w]	1.3[x]	650	600	1.3[x]	
16F-V5	HF/N3DF	1600	800	650	560[h]	90[m]	1.6	1000	470	135	70	1.5	750	875	1.5
The following two combinations are balanced and are intended for members continuous or cantilevered over supports and provide equal capacity in both positive and negative bending.															
16F-V4	DF/DF		560[h,i]	560[h]	165	1.5	1450	560	145	75	1.4	950	1550	1.5	
16F-V5	HF/HF	1600	1600	375[i]	375[i]	155	1.4	1200	375	135	70	1.3	850	1350	1.3
The following seven combinations are not balanced and are for either dry or wet use.															
20F-V1	DF/WW		650	560[h]	140[s,w]	1.4[k]	1000	255	130[s,w]	65[s,w]	1.2[x]	750	1000	1.2[x]	
20F-V2	HF/HF		500[i]	375[i]	155	1.5	1200	375	135	70	1.4	950	1350	1.4	
20F-V3	DF/DF		650	560[h]	165	1.6	1450	560	145	75	1.5	1000	1550	1.5	
20F-V4	DF/DF	2000	1000	590[h,i]	560[h]	165	1.6	1450	560	145	75	1.6	1000	1550	1.6
20F-V10	DF/HF		650	560	155	1.5	1300	375	135	70	1.4	950	1500	1.4	
20F-V11	DFS/DFS		650	500	165	1.3	1400	500	145	75	1.1	900	1400	1.1	
20F-V12	AC/AC		560	560	190	1.5	1200	470	165	80	1.4	900	1500	1.4	
The following two combinations are intended for straight or slightly cambered members for dry use and industrial appearance.[k]															
20F-V5	DF/N3WW		650	560[h]	90[i,s,w]	1.6[x]	1000	255	135[s,w]	70[s,w]	1.3[x]	750	725	1.3[x]	
20F-V6	DF/N3DF	2000	1000	650	560[h]	90[m]	1.6	1000	470	135	70	1.5	775	900	1.5
The following three combinations are balanced and are intended for members continuous or cantilevered over supports and provide equal capacity in both positive and negative bending.															
20F-V7	DF/DF		650	650	165	1.6	1450	560	145	75	1.6	1000	1600	1.6	
20F-V8	DF/DF	2000	2000	590[h,i]	590[h,i]	165	1.7	1450	560	145	75	1.6	1000	1600	1.6
20F-V9	HF/HF		500[i]	500[i]	155	1.5	1400	375	135	70	1.4	975	1400	1.4	
The following five combinations are not balanced and are for either dry or wet use.															
22F-V1	DF/WW		650	560[h]	140[s,w]	1.6[x]	1050	255	130[s,w]	65[s,w]	1.3[x]	850	1100	1.3[x]	
22F-V2	HF/HF		500[i]	500[i]	155	1.5	1250	375	135	70	1.4	950	1350	1.4	
22F-V3	DF/DF	2200	1100	650	560[i]	165	1.7	1450	560	145	75	1.6	1050	1500	1.6
22F-V4	DF/DF		650	560[h]	165	1.7	1450	560	145	75	1.6	1050	1550	1.6	
22F-V10	DF/DFS		650	560[h]	165	1.6	1600	500	145	75	1.3	1000	1400	1.3	
The following two combinations are intended for straight or slightly cambered members for dry use and industrial appearance.[x]															
22F-V5	DF/N3WW		650	560[h]	90[i,s,w]	1.6[x]	1100	255	135[s,w]	75[s]	1.4[k]	800	725	1.4[x]	
22F-V6	DF/N3DF	2200	1100	650	560[h]	90[m]	1.7	1250	470	135	75	1.6	900	925	1.6
The following three combinations are balanced and are intended for members continuous or cantilevered over supports and provide equal capacity in both positive and negative bending.															
22F-V7	DF/DF		650	650	165	1.8	1450	560	145	75	1.6	1100	1650	1.6	
22F-V8	DF/DF	2200	2200	590[h,i]	590[h,i]	165	1.7	1450	560	145	75	1.6	1050	1650	1.6
22F-V9	HF/HF		500[i]	500[i]	165	1.5	1250	375	135	70	1.4	975	1400	1.4	
The following six combinations are not balanced and are for either dry or wet use															
24F-V1	DF/WW		650	650	140[s,w]	1.7[x]	1250	255	135[s,w]	70[s]	1.4[x]	950	1300	1.4[x]	
24F-V2	HF/HF		500[i]	500[i]	155	1.5	1250	375	135	70	1.4	950	1300	1.4	
24F-V3	DF/DF	2400	1200	650	560[h]	165	1.8	1500	560	145	75	1.6	1100	1600	1.6
24F-V4	DF/DF		650	650	165	1.8	1500	560	145	75	1.6	1100	1650	1.6	
24F-V5	DF/HF		650	650	155	1.7	1350	375	140	70	1.5	1100	1450	1.5	
24F-V11	DF/DFS		650	560[h]	165	1.7	1600	500	145	75	1.4	1150	1700	1.4	

Refer to the latest edition of *AITC 117—Design* for a complete and current listing of tabulated values and footnote explanations. From *AITC 117—Design*[4] © 1987. Used by permission.

Table 5-5. Typical tabulated values for glulam axial combinations.

Combination Symbol	Species[d]	Grade[e]	Modulus of Elasticity E x10^6 psi	Compression Perpendicular to Grain F_{c⊥}[f,n] psi	Axially Loaded — Tension Parallel to Grain F_t, 2 or More Lams. psi	Axially Loaded — Compression Parallel to Grain F_c, 4 or More Lams. psi	2 or 3 Lams. psi	Bending about Y-Y Axis — Extreme Fiber in Bending F_by, 4 or More Lams. psi	3 Lams. psi	2 Lams. psi	Horizontal Shear F_vy, 4 or More Lams. (For members with multiple piece lams)[m] psi	Horizontal Shear F_vy, 4 or More Lams. psi	3 Lams. psi	2 Lams. psi	Bending About X-X Axis — Extreme Fiber in Bending F_bx, 2 Lams. to 15 in. Deep psi	4 or More Lams.[q] psi	Horizontal Shear F_vx, 2 or More Lams. psi
1	2	3	4	5	6	7	8	9	10	11	12	13	14	15	16	17	18
								Visually Graded Western Species									
1	DF	L3	1.5	560	900	1550	1200	1450	1250	1000	75	145	135	125	1250	1500	165
2		L2	1.7	560	1250	1900	1600	1800	1600	1300	75	145	135	125	1700	2000	165
3		L2D	1.8	650	1450	2300	1850	2100	1850	1550	75	145	135	125	2000	2300	165
4		L1CL	1.9	590	1400	2100	1900	2200	2000	1650	75	145	135	125	1900	2200	165
5		L1	2.0	650	1600	2400	2100	2400	2100	1800	75	145	135	125	2200	2400	165
6		N3C	1.4	470	350	875	550	550	550	550	60	120	115	105	450	—	140
7		N3M	1.5	560	900	1550	700	1450	1250	1000	75	145	135	125	1000	1600	165
8		N2	1.6	560	1000	1550	1150	1600	1550	1300	75	145	135	125	1350	1850	165
9		N2D	1.8	660	1150	1800	1350	1850	1800	1500	75	145	135	125	1600	2100	165
10		N1	1.8	560	1300	1950	1450	1950	1750	1500	75	145	135	125	1750	2100	165
11		N1D	2.0	650	1500	2300	1700	2300	2100	1750	75	145	135	125	2100	2400	165
12		SS	1.8	560	1400	1950	1650	2100	1950	1650	75	145	135	125	1900	2200	165
13		SSD	2.0	650	1600	2300	1950	2400	2300	1950	75	145	135	125	2200	2400	165
14	HF	L3	1.3	375	800	1100	975	1200	1050	850	70	135	130	115	1100	1300	155
15		L2	1.4	375	1050	1350	1300	1500	1350	1100	70	135	130	115	1450	1700	155
16		L1	1.6	375	1200	1500	1450	1750	1550	1300	70	135	130	115	1600	1900	155
17		L1D	1.7	500	1400	1750	1700	2000	1850	1550	70	135	130	115	1900	2200	155
18		N3	1.3	375	425	900	575	700	700	700	70	135	130	115	575	—	155
19		N2	1.4	375	850	1300	975	1350	1300	1100	70	135	130	115	1150	1350	155
20		N1	1.6	375	975	1450	1250	1550	1500	1250	70	135	130	115	1350	1550	155
21		SS	1.6	375	1100	1450	1350	1750	1650	1400	70	135	130	115	1500	1750	155
22	WW	L3	1.0	255	525	850	675	800	700	550	60	120	115	105	725	850	140
23		N3	1.0	255	275	625	450	450	450	450	60	120	115	105	400	—	140
24		N2	1.1	255	550	900	700	900	875	725	60	120	115	105	775	900	140
25		N1	1.2	255	650	1000	875	1050	1000	850	60	120	115	105	875	1050	140
26		SS	1.2	255	750	1000	1000	1150	1100	925	60	120	115	105	1000	1150	140
59	DFS	L3	1.1	500	800	1400	1050	1200	1050	850	75	145	135	125	1050	1250	165
60		L2	1.3	500	1050	1750	1400	1750	1550	1150	75	145	135	125	1450	1700	165
61		L1	1.5	660	1350	2200	1850	2000	1800	1500	75	145	135	125	1850	2200	165
69	AC	L3	1.3	470	700	1150	1150	1000	875	700	80	165	160	140	1000	1150	190
70		L2	1.4	470	1000	1450	1550	1250	1100	925	80	165	160	140	1350	1550	190
71		L1D	1.7	560	1250	1900	2050	1650	1500	1250	80	165	160	140	1700	2000	190
72		L1S	1.7	560	1250	1900	2050	1650	1500	1250	80	165	160	140	1700	2000	190

Refer to the latest edition of *AITC 117—Design* for a complete and current listing of tabulated values and footnote explanations. From *AITC 117—Design*.[v] ©1987. Used by permission.

End Grain in Bearing

Tabulated stress for end grain in bearing parallel to grain (F_g) is given in Annex A of *AITC 117-Design*. Annex A consists of Tables A-1 and A-2, which specify F_g for bending combinations and axial combinations, respectively. In both tables, F_g is specified by a combination symbol where member bearing is on the full cross section and where bearing is on a partial cross section.

ADJUSTMENTS TO TABULATED DESIGN VALUES

Tabulated values for sawn lumber and for glulam are based on the standard conditions noted in the applicable design tables. When actual use conditions vary from these standard conditions, tabulated values must be adjusted to compensate for (1) differences between the assumptions used to establish tabulated values and actual use conditions, (2) variations in wood behavior related to the type of stress or member orientation, and (3) differences between the physical or mechanical behavior of wood and that of an ideal material assumed in most equations of engineering mechanics.

Requirements for adjusting tabulated values are given in the text of the design specifications (AASHTO, NDS, and *AITC 117-Design*) and as footnotes to tabulated values. The type and magnitude of the adjustments, as well as the manner in which they are applied, vary with the type of material, strength property, and design application. Most adjustments are applied as modification factors that are multiplied by the tabulated values. These modification factors are designated by the letter C, followed by a subscript to denote the type of modification. They include the following:

C_M moisture content factor	C_L lateral stability of beams factor
C_D duration of load factor	C_P lateral stability of columns factor
C_t temperature factor	C_R fire-retardant treatment factor
C_f form factor	C_c curvature factor
C_F size factor	C_i interaction stress factor

Modification factors are applied to tabulated values only, not to applied stresses or loads. In most cases they are cumulative; however, in some cases the more restrictive value of two factors is used. A summary of the applicability of modification factors to various wood properties is given in Table 5-6. The factors C_c and C_t apply to curved and taper-cut glulam beams, respectively, and are not discussed in this chapter. Refer to the AITC *Timber Construction Manual* for additional information on these factors.[6]

Moisture Content Factor C_M)

The strength and stiffness of wood decrease as moisture content increases. To compensate for this effect, tabulated values are adjusted by C_M. This factor, which is also referred to as a wet-use factor or condition-of-use

Table 5-6.- Applicability of modification factors for strength properties and modulus of elasticity.

Design value	Modification factor[1]							
	Duration of load[2] C_D	Moisture content C_M	Temperature C_t	Fire retardant C_R	Size factor C_F	Stability of beams C_L	Form factor C_f	Stability of columns C_P
Bending, F_b	X	X	X	X	X	X	X	—
Tension parallel to grain, F_t	X	X	X	X	—	—	—	—
Compression parallel to grain, F_c	X	X	X	X	—	—	—	X
Compression perpendicular to grain, $F_{c\perp}$	—	X	X	X	—	—	—	—
Horizontal shear, F_v	X	X	X	X	—	—	—	—
End-grain bearing, F_g	X	X	X	X	—	—	—	—
Modulus of elasticity, E	—	X	X	X	—	—	—	—

[1] Factors are not always cumulative.
[2] The duration of load factor for impact does not apply to members pressure-impregnated with preservative salts to the heavy retentions required for marine exposure, or to sawn lumber treated with fire-retardant chemicals.
X = modification factor is applicable.
— = modification factor does not apply.

factor, is applicable to all tabulated values for strength and modulus of elasticity. It adjusts values for changes in strength and stiffness and compensates for variations in cross section caused by shrinkage.

Application of C_M differs for sawn lumber and glulam. For sawn lumber, tabulated values are based on the moisture content specified for each species in the NDS tables. With the exception of Southern Pine and Virginia Pine-Pond Pine, adjustment by C_M is applied when the moisture content of the member in service is expected to exceed 19 percent. For Southern Pine and Virginia Pine-Pond Pine, the C_M adjustment is not required because tabulated values are given in the design tables for three in-service moisture contents. These tabulated values already include the C_M adjustment, and no further adjustment for moisture is required. Values of C_M for all other lumber species are given in the footnotes to the design tables and depend on the member size and specific strength property (Table 5-7).

For glulam, all tabulated values in *AITC 117-Design* are based on a moisture content in service of 16 percent or less. When the moisture content in service is expected to be 16 percent or higher, tabulated values must be multiplied by the wet-use factors given in the design tables. Factor C_M for glulam depends on the strength property only and is independent of species, combination symbol, and member size. Values of C_M for glulam are given in Table 5-7.

In most applications, bridge members are exposed to the weather and should be adjusted by C_M for wet-use conditions. In cases where beams are protected by a waterproof deck, design for dry conditions may be appropriate, as discussed in Chapter 7.

Table 5-7. - Values of the moisture content factor C_M for sawn lumber and glulam.

Property	C_M values						
	F_b	F_t	F_c	$F_{c\perp}$	F_v	F_g	E
Sawn lumber; all species except Southern Pine and Virginia Pine-Pond Pine[a]							
All thicknesses surfaced dry or surfaced green and used at 19% maximum moisture content	1.00	1.00	1.00	1.00	1.00	1.00	1.00
Nominal 4 inches or less in thickness, surfaced green or dry and used at a moisture content greater than 19%	0.86	0.84	0.70	0.67	0.97	—[b]	0.97
Nominal 4 inches or less in thickness, surfaced green or dry and used at a moisture content of 15% or less[c]	1.08	1.08	1.17[d]	1.00	1.05	—[b]	1.05[d]
Nominal 5 inches and thicker used where moisture content exceeds 19%	1.00	1.00	0.91	0.67	1.00	—[b]	1.00
Glulam							
Used at moisture contents of 16% or less (dry conditions of use)	1.00	1.00	1.00	1.00	1.00	1.00	1.00
Used at moisture contents greater than 16% (wet conditions of use)	0.80	0.80	0.73	0.53	0.875	0.57	0.833

[a] Refer to the NDS[24] for adjusted tabulated values for Southern Pine and Virginia Pine-Pond Pine.
[b] Use tabulated values for wet-use conditions given in Table 2 of the NDS.[26]
[c] Refer to the NDS[24] for decking graded to WWPA rules that is surfaced at 15 percent maximum moisture content and used where the moisture content will exceed 15 percent for an extended period of time.
[d] For Redwood, use 1.15 for compression parallel to grain and 1.04 for modulus of elasticity.

Duration of load Factor (C_D)

Wood is capable of withstanding much greater loads for short durations than for long periods. This is particularly significant in bridge design where short-term increased loads from vehicle overloads, wind, earthquake, or railing impact must be considered. The tabulated values for sawn lumber and glulam are based on an assumed *normal* duration of load. In this case, a normal duration of load is based on the expectation that members will be stressed to the maximum stress level (either continuously or cumulatively) for a period of approximately 10 years, stressed to 90 percent of the maximum design level continuously for the remainder of the life of the structure, or both. This maximum stress is assumed to occur during the life of the member as a result of either continuous loading or a series of shorter duration loads that total 10 years. When the maximum design loads act for durations that are shorter or longer than these assumed durations, tabulated stresses are adjusted by C_D (Table 5-8). Factor C_D applies to tabulated strength properties but does not apply to compression perpendicular to grain $(F_{c\perp})$ or modulus of elasticity *(E)*. In most bridge

Table 5-8.   Modification factors for duration of load.

Load duration	Duration of load factor C_D
2 months (as for snow and ice)	1.15
7 days (as for snow and ice)	1.25
Wind or earthquake	1.33
5 minutes (rail loads only)	1.65[a]

[a] The duration of load factor for impact does not apply to members pressure-impregnated with preservative salts to the heavy retentions required for marine exposure, or sawn lumber treated with fire-retardant chemicals.

From *AASHTO* Section 13.2.5.1:[1] © 1983. Used by permission.

applications, the permanent load of the structure is small in relation to vehicle loads, and a decrease in tabulated stresses for permanent loading is not necessary

The stresses produced in bridge members are commonly the result of a combination of loads rather than a single load (Chapter 6). For a combination of loads of different durations, C_D for the entire group is the single value associated with the shortest load duration. When applying C_D the designer must recognize that for a given combination of loads, the most restrictive allowable stress may result from a partial combination involving loads of longer duration. The individual loads in a load combination must be evaluated in various combinations, with the value of C_D depending on the load of shortest duration for that combination. This is accomplished by progressively eliminating the load of shortest duration from the group and applying C_D for the load of next-shortest duration. In other words, the resulting size or capacity of a member required for a load combination must not be less than that required for a partial combination of the longer-duration loads. Application of C_D is discussed in more detail in Appendix B of the NDS and in Chapter 6. Duration of load is generally not applicable in bridge design, except for the design of railing systems.

Temperature Factor (C_t)

The strength and stiffness of wood increases as it cools and decreases as it warms. These changes in strength because of temperature occur immediately and depend on the magnitude of the temperature change and the moisture content of the wood. For temperatures up to approximately 150 °F, the immediate effects of strength loss are reversible, and the member will essentially recover its initial strength levels as the temperature is lowered. Prolonged exposure to temperatures higher than 150 °F may cause a permanent and irreversible loss in member strength.

Tabulated design values for sawn lumber and glulam assume that members will be used in normal temperature applications and may occasionally be heated to temperatures up to 150 °F. This applies to most bridge design situations. In cases where a member may be periodically exposed to elevated temperatures, humidity is generally low, and the increase in member strength that results from reduced moisture tends to offset the reduction in strength that results from temporary temperature increases. The design specifications do not require a mandatory adjustment to tabulated values for temperature effects, and as a general rule, none are warranted. In cases where members will be exposed to prolonged temperatures in excess of 150 °F, or will be used at very low temperatures for the entire design life, the modification factor, C_t, given in Table 5-9, may be applied at the discretion of the designer.

Table 5-9. - Temperature factor C_t given as a percentage increase or decrease in design values for each 1 °F decrease or increase in temperature.

Property	Moisture[1] content	Cooling[2] below 68 °F (Min. -100 °F)	Heating above 68 °F (Max. 150 °F)
Modulus of elasticity and tension parallel to grain	0%	+0.09%	−0.11%
	12%	+0.13%	−0.13%
	24%	+0.38%	−0.15%
Other properties and fastenings[2]	0%	+0.14%	−0.19%
	12%	+0.24%	−0.38%
	24%	+0.84%	−0.57%

[1] In-service (equilibrium) moisture content at design temperature.
[2] The effect of low temperatures on the ductility of metal fasteners should be considered.

From the NDS;[26] © 1986. Used by permission.

Fire-Retardant Treatment Factor (C_R)

Fire-retardant treatments are seldom used on bridge members and are unnecessary in most applications. For those situations where fire-retardant chemicals are considered necessary, tabulated values must be adjusted by the fire-retardant treatment factor C_R. The value for this factor depends on specific strength properties and is different for sawn lumber and glulam. C_R is given for sawn lumber in Table 2A of the NDS (Table 5-10). The basis for these values and treatment qualifications are outlined in Appendix Q of the NDS. C_R for glulam depends on the species and treatment combinations involved. The effects on strength properties must be determined for each treatment. However, indications are that 10 to 25 percent reductions in bending strength are applicable.[46] The treatment manufacturer should be contacted for more specific C_R values for glulam based on the specific material and design application.

Table 5-10.- Fire-retardant treatment factor for structural lumber.

Property	C_r
Extreme fiber in bending	0.85
Tension parallel to grain	0.80
Horizontal shear	0.90
Compression perpendicular to grain	0.90
Compression parallel to grain	0.90
Modulus of elasticity	0.90
Fastener design loads	0.90

From the NDS;"© 1986. Used by permission.

Size Factor (C_F)

Tabulated bending stresses are based on a square or rectangular member 12 inches deep in the direction of applied loads. For member depths greater than 12 inches, F_b must be adjusted by C_F, as computed by

$$C_F = \left(\frac{12}{d}\right)^{1/9}$$
(5-1)

where d is the member depth in inches.

For sawn lumber, C_F does not apply to MSR lumber or to visually graded lumber 2 to 4 inches thick used edgewise. For glulam, the C_F value computed by the above equation is based on a uniformly distributed load on a simply supported beam with a span to depth ratio $L/d = 21$. In most bridge applications, these assumptions result in reasonable accuracy as variations in loading and L/d result in relatively small deviations in the size factor. In cases where greater accuracy is warranted, C_F may be adjusted for other L/d ratios or loading conditions by the percentages in Table 5-11.

The effect of the size factor for both sawn lumber and glulam is to reduce the tabulated bending stress for members more than 12 inches deep. For members less than 12 inches deep, footnotes to design tables allow an increase in bending stress for sawn lumber members 2 to 4 inches thick used flatwise,[24] and glulam members loaded parallel to the wide faces of the laminations.[4] C_F is generally cumulative with other modification factors, but is normally not cumulative with the lateral stability of beams factor, C_L (see Sections 5.4 and 5.7).

Equation 5-1, used for computing size factor, is being reevaluated for glulam, and alternate forms of the equation are being considered by several industry-related technical committees. Thus, the designer should be aware of the potential for future revisions and refer to the latest editions of the NDS and AITC 117-Design for current requirements.

Table 5-11. —Adjustments to C_l for various span-to-depth ratios and loading conditions.

Span-to-depth ratio $(L/d)^a$	% change
7	+6.3
14	+2.3
21	0
28	−1.6
35	−2.8

Loading condition for simply supported beams	% change
Single concentrated load	+7.8
Uniform load	0
Third point load	−3.2

^{a} Use straight line interpolation for other L/d ratios.

From *AITC 117—Design;*[4] © 1987. Used by permission.

Lateral Stability of Beams Factor (C_L)

The lateral stability of beams factor, C_L, is applied to some bending members where the compressive stress in bending must be limited to prevent lateral buckling. Additional details on the use of C_L are discussed in Section 5.4.

Form Factor (C_f)

Tabulated bending stresses are based on members with a square or rectangular cross section loaded normal to one or more faces. For other member shapes, specifically round or diamond sections, stresses must be modified by the form factor, C_f. C_f does not apply to rectangular or square members and is not commonly used in bridge applications. Refer to the NDS for additional information on the use of C_f.

Lateral Stability of Columns Factor (C_P)

The lateral stability of columns factor, C_P, is applied to some compression members where the compressive stress must be limited to prevent lateral buckling. Additional details on the use of C_P are discussed in Section 5.6.

5.4 BEAM DESIGN

A beam is a structural component with loads applied transversely to the longitudinal axis. In bridge design, beams are the most frequently used structural components. The three most common bridge beams are girders, stringers, and floorbeams. Girders are large beams (normally glulam) that provide primary superstructure support, most often in beam-type superstructures. Stringers are longitudinal beams that support the bridge deck.

They are generally smaller than girders, but there is no clear size definition for either. Floorbeams are transverse beams that directly support the bridge deck or support longitudinal stringers that support the deck. In addition to girders, stringers, and floorbeams, other bridge components are designed as beams, including components of the deck and railing systems.

Beam design involves the analysis of member strength, stability, and stiffness for four basic criteria: (1) bending (including lateral stability), (2) deflection, (3) horizontal shear, and (4) bearing. Of these four criteria, bending, deflection, and shear can directly control member size, while bearing will influence the design of supports. Initial beam design is normally based on bending, then checked for deflection and shear. After an appropriate beam size is determined, bearing stresses are checked at supports to ensure sufficient bearing area.

Beam design requirements discussed in this section are limited to straight or slightly curved (cambered) solid rectangular beams of constant cross-sectional area. Refer to the NDS for design requirements for other beam configurations and shapes and for beams with notches or cutouts. The design of beams loaded in combined bending and axial tension or compression is discussed in Section 5.7.

DESIGN FOR BENDING

Beam design must consider the strength of the material in bending and the potential for lateral buckling from induced compressive stress. For positive and negative bending, compression stress occurs in the top and bottom portions of the beam, respectively. Single, simple spans are subjected to positive bending moments only, while multiple continuous spans and cantilevers will be subjected to both positive and negative moments. This distinction is particularly important for stability considerations, and also when the allowable stresses for positive and negative bending are different, as in some combination symbols of glulam beams.

Initial beam design is somewhat of a trial-and-error process. A beam size is first estimated, and applied stress is computed and checked against the allowable stress in bending. After a suitable beam is determined from strength requirements, it must be verified for lateral stability.

Applied Stress

Applied bending stress in timber beams is determined by the standard formulas of engineering mechanics assuming linear elastic behavior. Stress at extreme fiber in bending, f_b is computed by

$$f_b = \frac{M}{S} \tag{5-2}$$

where M = moment due to applied loads (in-lb), and

S = section modulus of the beam (in^3).

Section modulus values for standard sizes of sawn lumber and glulam are given in Chapter 16.

Lateral Stability and Beam Slenderness

Beams develop compressive stress from induced bending forces. If compression areas are not restrained from lateral movement and rotation, the member may buckle laterally at a bending stress considerably lower than that normally allowed for the material. The potential for lateral buckling depends on the magnitude of applied loads, beam dimensions, and the effectiveness and frequency of lateral restraint. Lateral stability is most critical in long slender beams with a high depth-to-width ratio. It is not critical in beams where the width of the beam exceeds its depth.

One of the primary factors affecting beam lateral stability is the distance between points of lateral support along the beam length. In bridge applications, lateral support is generally provided by cross frames, solid wood diaphragms, or framing connections that prevent beam rotation and lateral displacement (Figure 5-1). The distance between such points of lateral support is termed the unsupported length, or ℓ_u. When the compression edge is continuously supported along its length, ℓ_u is zero. For all other configurations, ℓ_u is simply the distance between cross frames, diaphragms, or bracing that prevent beam rotation and lateral displacement.

The basis for stability design in beams is the beam slenderness factor C_s, given by

$$C_s = \sqrt{\frac{\ell_e d}{b^2}} \leq 50 \tag{5-3}$$

where ℓ_e = effective beam length (in.),

d = beam depth (in.), and

b = beam width (in.).

The effective beam length ℓ_e in Equation 5-3 depends on the beam configuration and loading condition (Figure 5-2). For a single-span beam with a concentrated load at the center, ℓ_u is computed by

$$\ell_e = 1.37\ell_u + 3d \tag{5-4}$$

For a single-span beam with a uniformly distributed load, ℓ_e is computed by

$$\ell_e = 1.63\ell_u + 3d \tag{5-5}$$

For a single-span beam, or cantilever beam, with any load, ℓ_e is computed by

Figure 5-1. - Cross frames fabricated from steel angles are commonly used to provide lateral support for large glulam bridge beams.

$$\ell_e = 1.84\ \ell_u \qquad\qquad \text{when } \ell_u/d \geq 14.3 \qquad\qquad (5\text{-}6)$$

$$\ell_e = 1.63\ \ell_u + 3d \qquad\qquad \text{when } \ell_u/d < 14.3 \qquad\qquad (5\text{-}7)$$

Equations for computing ℓ_e for other beam configurations and loading conditions are given in the NDS. For single-span or cantilever beams, Equations 5-6 and 5-7 give slightly conservative results for any loading condition and are often used in bridge applications where several concentrated loads are positioned on the span.

Example 5-1 - Beam slenderness factor

A 10-3/4- by 48-inch glulam beam spans 60 feet and supports the three concentrated loads shown below. Lateral beam support is provided by transverse bracing located at the beam ends and at the third points. Compute the beam slenderness factor, C_s.

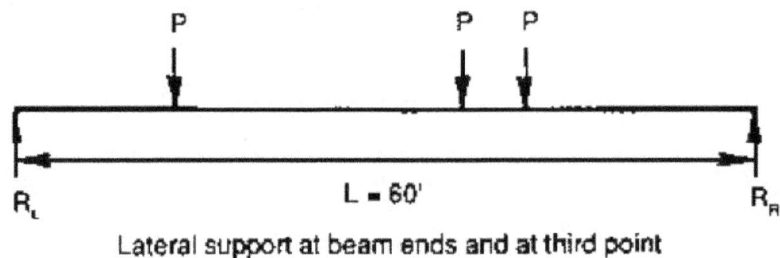

Lateral support at beam ends and at third point

5-20

Location of lateral beam support where transverse movement and rotation are prevented by cross frames or diaphragms

$\ell_e = 1.37\ell_u + 3d$

Single-span beam with concentrated load at center

$\ell_e = 1.63\ell_u + 3d$

Single-span beam with uniform load

$\ell_e = 1.84\ell_u$ if $\ell_u/d \geq 14.3$
$\ell_e = 1.63\ell_u + 3d$ if $\ell_u/d < 14.3$

Single-span or cantilever beam with any loading condition

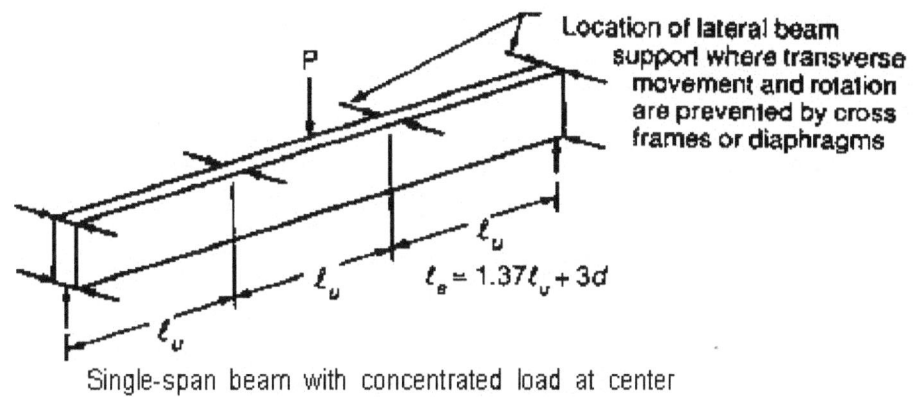

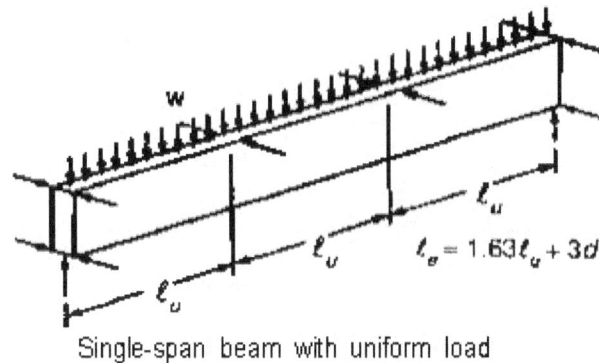

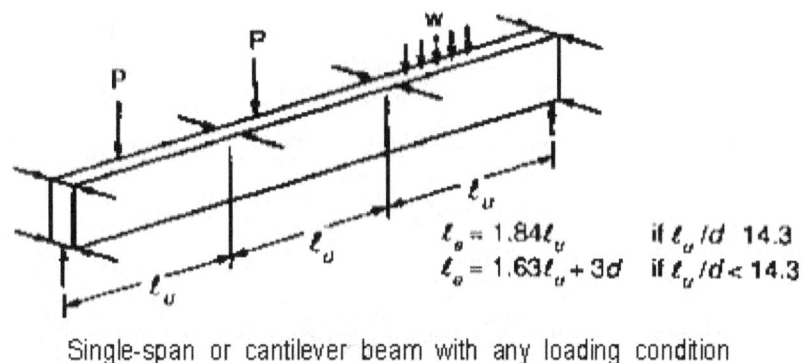

Figure 5-2.—Effective beam length, ℓ_e, for various loading conditions on single-span beams. Refer to the NDS™ for equations for other beam loads and configurations.

Solution

Lateral support is equally spaced along the beam, giving an unsupported length ℓ_u of 20 feet. Because the beam is loaded with three concentrated loads, the effective beam length ℓ_e will be computed by Equation 5-6 or 5-7, depending on the ratio of the unsupported length to the beam depth:

$$\frac{\ell_u}{d} = \frac{20(12 \text{ in/ft})}{48} = 5.0$$

5.0 < 14.3, so Equation 5-7 applies:

$$\ell_e = 1.63\ell_u + 3d = (1.63)(20)(12 \text{ in/ft}) + (3)(48) = 535.20$$

The slenderness factor is computed by Equation 5-3:

$$C_s = \sqrt{\frac{\ell_e d}{b^2}} = \sqrt{\frac{535.2(48)}{(10.75)^2}} = 14.9 \le 50$$

This example illustrates a typical case where transverse bracing is equally spaced and the value of C_s applies to all portions of the beam. In cases where the distance ℓ_u varies substantially along the beam length, C_s should be checked for each unsupported length. With few exceptions, however, C_s for the center portion of the beam, where bending stress is highest, will normally control.

Allowable Stress

The allowable bending stress in beams is controlled either by the size factor C_F, which limits bending stress in tension zone, or by lateral stability, which limits bending stress in the compression zone. *Adjustments for the size factor and lateral stability are not cumulative.* Therefore, the designer must compute allowable bending stress based on both criteria separately, and the lowest value obtained is used for design. In most bridge beams, allowable bending stress is controlled by C_F rather than stability. In addition, beam stability cannot be evaluated until an initial member size is selected. Therefore, it is most convenient and practical to assume that the size factor controls allowable bending stress and to initially design the beam based on the allowable stress given by

$$F_b' = F_b C_D C_M C_R C_t C_F \tag{5-8}$$

Values of C_F are normally included in tables of section properties for glulam bending combinations (see Tables 16-3 and 16-4). In addition, most glulam tables include C_F as a noted adjustment to the section modulus. This adjusted value, $S_x C_F$, is included for convenience and facilitates design by adjusting for C_F during initial member selection (see Example 5-3).

After a satisfactory beam size and grade are determined based on the allowable bending stress given by Equation 5-8, the beam must be checked for lateral stability. Criteria for allowable bending stress related to lateral stability are based on beam slenderness for the following three ranges:

$$0 < C_s \le 10 \quad \text{Short Beam}$$

$$10 < C_s \le C_k \quad \text{Intermediate Beam}$$

$$C_k < C_s \le 50 \quad \text{Long Beam}$$

where C_k is a slenderness factor defined later for intermediate beams.

5-22

Short Beams

In short beams with C_s of 10 or less, capacity of the member is controlled by the wood strength in bending rather than by lateral stability. In this case, the size factor is the controlling modification factor, and the allowable bending stress computed by Equation 5-8 is used for design.

Intermediate Beams

Intermediate beams have C_s greater than 10, but less than C_k determined by

$$C_k = 0.811\sqrt{E'/F_b''}$$

(5-9)

where C_k = the largest value of C_s at which the intermediate beam equation applies,

$E' = EC_M C_R C_t$ (lb/in²), and

$F_b'' = F_b C_D C_M C_R C_t$ (lb/in²).

In intermediate beams, failure can occur in bending or by torsional buckling from lateral instability. The controlling mode is indicated by the lateral stability of beams factor C_L given by

$$C_L = \left[1 - \frac{1}{3}\left(\frac{C_s}{C_k}\right)^4\right]$$

(5-10)

If C_L is less than C_F, bending stress is controlled by stability, and C_L is the controlling modification factor. The allowable bending stress is computed by

$$F_b' = F_b C_D C_M C_R C_t C_L$$

(5-11)

If C_L is greater than C_F, bending stress is controlled by strength, and the allowable stress computed by Equation 5-8 is used for design.

Equation 5-9 for lateral stability was developed from theoretical analyses and beam verification tests and is based on the modulus of elasticity of the member. For visually graded sawn lumber, tabulated E values are based on the average modulus of elasticity for the grade and species of material and represent a coefficient of variation of approximately 0.25. For glulam with six or more laminations, the coefficient of variation is 0.10 (less than half that for visually graded sawn lumber). To account for this reduced variability, the NDS allows the designer to use the following modified equation for C_k (Equation 5-12), which more accurately reflects the characteristics of glulam:

$$C_k = 0.956\sqrt{E'/F_b''}$$

(5-12)

5-23

This equation provides the same factor of safety at the 5-percent exclusion value for glulam that is provided for visually graded sawn lumber with a 0.25 coefficient of variation. Although use of Equation 5-12 is optional, it represents a more realistic approach to glulam beam design and is recommended for bridge applications. For additional information on low-variability equations for glulam beams, refer to Appendix O of the NDS [26] and the *AITC Timber Construction Manual.* [6]

Long Beams

Long beams have a slenderness ratio greater than C_s but less than or equal to 50. In long beams, bending stress is controlled by lateral stability rather than strength, and the allowable stress is computed using

$$F_b' = \frac{0.438E'}{(C_s)^2}$$ (5-13)

For glulam beams, the following low-variability equation may be used in lieu of Equation 5-13:

$$F_b' = \frac{0.609E'}{(C_s)^2}$$ (5-14)

Example 5-2- Beam design based on bending; sawn lumber beam

A sawn lumber beam spans 15 feet center-to-center of bearings and supports a uniform load of 350 lb/ft in addition to its own weight. The beam is laterally supported by blocking placed at the beam ends and at 5-foot intervals along the beam length. Determine the required beam size based on bending, assuming the following:

1. Normal load duration under wet-use conditions (lumber moisture content will exceed 19-percent in service); adjustments for temperature *(C$_t$)* and fire-retardant treatment *(C$_R$)* are not required.

2. The beam is surfaced (S4S) Douglas Fir-Larch.

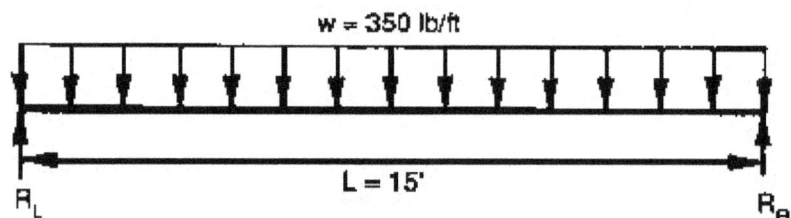

w = 350 lb/ft

L = 15'

R_L R_R

Solution

Beam design is somewhat of a trial-and-error process that starts with either an estimated beam size or a selected lumber species and grade. In this example, Douglas Fir-Larch, visually graded No. 1 in the Joist and Plank size classification is initially selected. The tabulated bending stress and modulus of elasticity for this species and grade are obtained from Table 4A of the NDS:

$$F_b = 1,500 \ lb/in^2$$

$$E = 1,800,000 \ lb/in^2$$

An initial section modulus based on applied moment and tabulated bending stress is computed as follows:

$$M = \frac{wL^2}{8} = \frac{(350)(15)^2}{8} \approx 9,844 \ ft\text{-}lb$$

Rearranging Equation 5-2,

$$S = \frac{M}{F_b} = \frac{(9,844)(12 \ in/ft)}{1,500} \approx 78.8 \ in^3$$

From lumber section properties in Table 16-2, a nominal beam size is selected with a section modulus slightly greater than the required 78.8 in³. The closest standard nominal size appears to be 4 inches by 14 inches with the following properties:

$b = 3.5$ in.

$d = 13.25$ in.

$S = 102.41$ in³

Beam weight = 16.1 lb/ft (based on a unit weight for wood of 50 lb/ft³)

The allowable bending stress is computed using the applicable modification factors given in Equation 5-8. The size factor, C_r is not applicable because it only applies to sawn lumber beams that are more than 4 inches thick. In this case, Equation 5-8 becomes

$$F_b' = F_b C_M$$

From Table 5-7, $C_M = 0.86$, and

$$F_b' = F_b C_M = 1,500(0.86) = 1,290 \ lb/in^2$$

Next, the applied bending stress is revised to reflect the beam weight of 16.1 lb/ft:

$$M = \frac{wL^2}{8} = \frac{(350 + 16.1)(15)^2}{8} = 10,297 \ ft\text{-}lb$$

By Equation 5-2,

$$f_b = \frac{M}{S} = \frac{(10,297 \ ft\text{-}lb)(12 \ in/ft)}{102.41 \ in^3} = 1,207 \ lb/in^2$$

$f_b = 1{,}207$ lb/in$^2 < F_b' = 1{,}290$ lb/in^2, so the initial beam is satisfactory in bending. The beam must next be checked for lateral stability.

For lateral support at 5-foot intervals,

$$\ell_u = 5 \text{ ft} = 60 \text{ in.}$$

By Equation 5-5 for a single-span beam with a uniformly distributed load,

$$\ell_e = 1.63\ell_u + 3d = (1.63)(60) + (3)(13.25) = 137.55$$

By Equation 5-3,

$$C_s = \sqrt{\frac{\ell_e d}{b^2}} = \sqrt{\frac{(137.55)(13.25)}{(3.5)^2}} = 12.20$$

The value $C_s = 12.20$ is greater than 10, so further stability calculations are required. From Table 5-7, C_M for modulus of elasticity is 0.97, and

$$E' = EC_M = 1{,}800{,}000(0.97) = 1{,}746{,}000$$

By Equation 5-9,

$$C_k = 0.811\sqrt{\frac{E'}{F_b''}} = 0.811\sqrt{\frac{1{,}746{,}000}{1{,}500\,(0.86)}} = 29.84$$

$10 < C_s = 12.20 < C_k = 29.84$, so the beam is classified in the intermediate slenderness range. By Equation 5-10,

$$C_L = 1 - \frac{1}{3}\left(\frac{C_s}{C_k}\right)^4 = 1 - \frac{1}{3}\left(\frac{12.20}{29.84}\right)^4 = 0.99$$

The allowable bending stress based on lateral stability is computed by Equation 5-11 using the modification factor C_L:

$$F_b' = F_b C_M C_L = 1{,}500(0.86)(0.99) = 1{,}277 \text{ lb/in}^2$$

$f_b = 1{,}207$ lb/in$^2 < F_b' = 1{,}277$ lb/in^2, so the beam size, species, and grade are satisfactory in bending.

Summary
Based on bending only, the beam will be a nominal 4-inch by 14-inch surfaced Douglas Fir-Larch beam, visually graded No. 1 in the Joists and Planks (J&P) size classification. The applied bending stress, f_b, is 1,207 lb/in^2. The allowable bending stress, F_b', is 1,277 lb/in^2 and is controlled by lateral stability.

Example 5-3 - Beam design based on bending; glulam beam

A glulam beam spans 50 feet center-to-center of bearings and supports a moving concentrated load of 20,000 pounds. Determine the required beam size based on bending for cases where: (A) the beam is laterally supported at the ends and at the third points, and (B) the beam is laterally supported at the ends only. The following assumptions apply:

1.   Normal load duration under wet-use conditions (glulam moisture content will exceed 16-percent in service); adjustments for temperature (C_t) and fire-retardant treatment (C_R) are not applicable.

2.   The glulam beam is manufactured from visually graded Southern Pine, combination symbol 24F-V2.

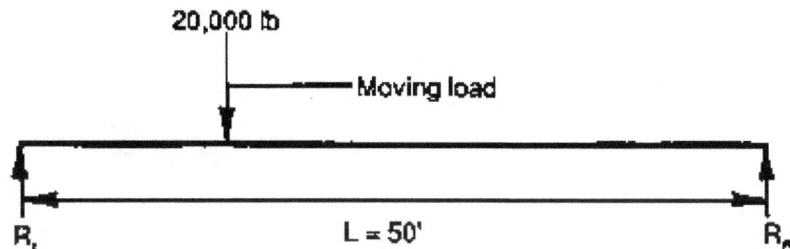

20,000 lb

Moving load

R_L $L = 50'$ R_R

Case A: Lateral support is provided at beam ends and
            at third points
Case B: Lateral support is provided at beam ends only

Solution

The first step in the design process is to determine the required beam size based on bending stress, adjusted by the size factor, C_r. The suitability of the initial beam size is then checked for each of the two conditions of lateral support.

Tabulated values for bending and modulus of elasticity are obtained from *AITC 117-Design*. Respective values for the moisture content modification factor are obtained from Table 5-7:

$$F_{bx} = 2,400 \text{ lb/in}^2 \qquad C_M = 0.80$$

$$E_x = 1,700,000 \text{ lb/in}^2 \qquad C_M = 0.833$$

The maximum applied moment is computed with the moving load positioned at the span centerline:

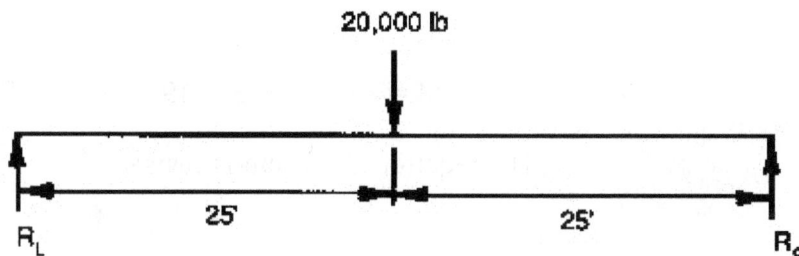

20,000 lb

R_L 25'   25' R_R

$$M = \frac{PL}{4} = \frac{(20,000)(50)}{4} = 250,000 \text{ ft-lb}$$

An initial beam size is determined using procedures similar to those used for sawn lumber beam design. For glulam, however, the size factor, C_F, is included as a noted adjustment to the section modulus $(S_x C_F)$ in Table 16-4. By Equation 5-8,

$$F_b' = F_{bx} C_M C_F$$

Assuming that the applied bending stress equals the allowable bending stress, Equation 5-2 is rearranged to compute the required value of $S_x C_F$ directly:

$$f_b = F_b' = F_{bx} C_M C_F = \frac{M}{S_x} \qquad \text{or} \qquad S_x C_F = \frac{M}{F_{bx} C_M}$$

Based on the moment from the concentrated load only, an initial value of $S_x C_F$ is computed:

$$S_x C_F = \frac{M}{F_{bx} C_M} = \frac{(250,000)(12 \text{ in/ft})}{(2,400)(0.80)} = 1,563 \text{ in}^3$$

From Table 16-4, an initial beam size is selected that provides an $S_x C_F$ value slightly greater than 1,563 in³. It is usually most convenient to find the closest $S_x C_F$ to that required, then increase the beam depth by one or two laminations to account for the beam dead load. In this case, a 6-3/4-inch by 41-1/4-inch beam is chosen with the following properties:

$$S_x C_F = 1,668.9 \text{ in}^3$$

Beam weight = 96.7 lb/ft (based on a unit weight of 50 lb/ft³)

Moment from the beam weight is computed and added to that from the concentrated load:

$$\text{Beam } M = \frac{wL^2}{8} = \frac{96.7(50)^2}{8} = 30,219 \text{ ft-lb}$$

M = 250,000 + 30,219 = 280,219 ft-lb

The required $S_x C_F$ value is revised:

$$S_x C_F = \frac{(280,219)(12 \text{ in/ft})}{(2,400)(0.80)} = 1,751 \text{ in}^3$$

From Table 16-4, a revised beam size of 6-3/4 inches by 42-5/8 inches is selected with the following properties:

$S_x = 2,044$ in^3

$C_F = 0.87$

Beam weight = 99.9 lb/ft (based on a unit weight of 50 lb/ft^3)

Moment from beam weight is revised and the applied bending stress is computed:

$$\text{Beam } M = \frac{wL^2}{8} = \frac{99.9(50)^2}{8} = 31,219 \text{ ft-lb}$$
$$M = 250,000 + 31,219 = 281,219 \text{ ft-lb}$$
$$f_b = \frac{M}{S_x} = \frac{281,219(12 \text{ in/ft})}{2,044} = 1,651 \text{ lb/in}^2$$

Allowable bending stress is computed by Equation 5-8:

$$F_b' = F_{bx} C_M C_F = 2,400(0.80)(0.87) = 1,670 \text{ lb/in}^2$$

$f_b = 1,651$ lb/in^2 < $F_b' = 1,670$ lb/in^2, so the beam is satisfactory in bending, assuming that the size factor controls. The beam is next checked for lateral stability.

Case A: Lateral support at beam ends and at third points

For lateral support at the beam ends and at the third points, the unsupported beam length is equal to one-third the span length:

$$\ell_u = \frac{50}{3} = 16.67 \text{ ft} = 200 \text{ in.}$$

Because the maximum moment is produced with the moving load at midspan, the effective beam length is computed using Equation 5-4:

$$\ell_e = 1.37\ell_u + 3d = 1.37(200) + 3(42.63) = 401.89 \text{ in.}$$

By Equation 5-3,

$$C_r = \sqrt{\frac{\ell_e d}{b^2}} = \sqrt{\frac{401.89\,(42.63)}{(6.75)^2}} = 19.39$$

The value of C_r is greater than 10, so lateral stability must be checked further. By equation 5-12 for low-variability material,

$$E' = E_x C_M = 1,700,000(0.83) = 1,416,100 \text{ lb/in}^2$$

$$F_b'' = F_{bx} C_M = 2,400(0.80) = 1,920 \text{ lb/in}^2$$

$$C_k = 0.956\,\frac{E'}{F_b''} = 0.956\,\frac{1,416,100}{1,920} = 25.96$$

$C_r = 19.39 < C_k = 25.96$, so the beam is in the intermediate beam slenderness range.

By Equation 5-10,

$$C_L = 1 - \frac{1}{3}\left(\frac{C_r}{C_k}\right)^4 = 1 - \frac{1}{3}\left(\frac{19.39}{25.96}\right)^4 = 0.90$$

$C_L = 0.90 > C_r = 0.87$, so the size factor reduction is more severe and controls the allowable bending stress. The selected beam size is therefore satisfactory in bending.

Case B: Lateral support at beam ends only

With lateral support at the beam ends only, the unsupported beam length equals the span length:

$$\ell_u = 50 \text{ ft} = 600 \text{ in.}$$

By Equation 5-4,

$$\ell_e = 1.37\,\ell_u + 3d = 1.37(600) + 3(42.63) = 949.89 \text{ in.}$$

By Equation 5-3,

$$C_r = \sqrt{\frac{\ell_e d}{b^2}} = \sqrt{\frac{949.89\,(42.63)}{(6.75)^2}} = 29.81$$

The previously computed value $C_k = 25.96$ is unchanged. In this case, however, $C_k = 25.96 < C_r = 29.81$, so the beam is in the long-beam slenderness range and lateral stability controls design. By low-variability Equation 5-14,

$$F_b' = \frac{0.609\,E'}{(C_s)^2} = \frac{0.609(1,416,100)}{(29.81)^2} = 970 \text{ lb/in}^2$$

$f_b = 1,651$ lb/in$^2 > F_b' = 970$ lb/in^2, so the beam must be redesigned. Using a modified form of Equation 5-2, with the previously computed moment (based on the previous beam size):

$$S_x = \frac{M}{F_b'} = \frac{281,219\,(12 \text{ in/ft})}{970} = 3,479 \text{ in}^3$$

From Table 16-4, a revised beam size of 8-1/2 inches by 50-7/8 inches is selected with the following properties:

$S_x = 3,666.7$ in^3

Beam weight = 150.2 lb/ft (based on a unit weight of 50 lb/ft^3)

Moment from beam weight is revised and bending stress is computed:

$$\text{Beam } M = \frac{wL^2}{8} = \frac{150.2(50)^2}{8} = 46,938 \text{ ft-lb}$$
$$M = 250,000 + 46,938 = 296,938 \text{ ft-lb}$$
$$f_b = \frac{M}{S_x} = \frac{296,938\,(12 \text{ in/ft})}{3,666.7} = 972 \text{ lb/in}^2$$

$F_b' = 970$ lb/in$^2 < f_b = 972$ lb/in^2, but the difference of 2 lb/in^2, or approximately 0.20 percent, is insignificant and the beam size is acceptable.

Summary

Based on bending only, the required size and bending stress for 24F-V2 Southern Pine beams are as follows:

Case A: With lateral support at beam ends and at third points

Beam size = 6-3/4 in. by 42-5/8 in.

$$f_b = 1,651 \text{ lb/in}^2$$

$$F_b' = 1,670 \text{ lb/in}^2$$

Case B: With lateral support at beam ends only

Beam size = 8-1/2 in. by 50-7/8 in.

$$f_b = 972 \text{ lb/in}^2$$

$$F_b' = 970 \text{ lb/in}^2$$

This example illustrates the effect of lateral support on beam size requirements. When support along the span is eliminated, the required beam size increases substantially. Additional requirements on the placement and design of lateral support for bridge beams are discussed in Chapter 7.

DESIGN FOR DEFLECTION

Deflection is the relative deformation that occurs in a beam as it is loaded. Deflection in timber beams results from bending and shear, but shear deformations are small in comparison to bending deformations and are normally not considered. Deflection does not seriously affect the strength of a beam, but it can affect the serviceability and appearance of bridge members and the performance of fasteners.

The length of time a load acts on a member influences its long-term deflection. When loads of relatively short duration are applied, deformation occurs immediately and remains at a relatively constant level for the duration of loading. When the load is removed, the member recovers elastically to the original unloaded position. For permanent loads (dead loads), initial elastic deformation is immediate, but members also develop an additional time-dependent, nonrecoverable deformation. This time-dependent deformation, known as creep, develops at a slow but persistent rate and is more pronounced for members seasoned in place or subject to variations in moisture content and temperature. Creep does not endanger the safety of the beam, but it can influence the performance, serviceability, and appearance of a structure when it is ignored in design. Thus, the two types of deflection considered in timber bridge design are: elastic deflection, and inelastic deflection, or creep.

Deflection Equations

Timber beam deflections are computed by the same engineering methods used for isotropic, elastic materials. Standard equations based on these methods are available in many engineering textbooks and manuals for numerous beam configurations and loading conditions.[6,27] Two of the most commonly used equations for simple beams are given below in Equations 5-15 and 5-16. Additional equations for more specific bridge applications and loads are discussed in Chapters 7, 8, and 9.

For a simply supported beam with one concentrated load at the center of the span:

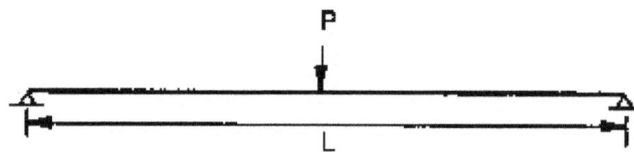

$$\Delta = \frac{PL^3}{48E'I} \qquad (5\text{-}15)$$

For a simply supported beam with a uniform load:

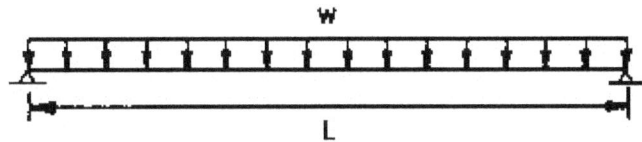

$$\Delta = \frac{5wL^4}{384E'I} \qquad (5\text{-}16)$$

where P = magnitude of a single concentrated load (lb),

w = magnitude of uniform load (lb/in),

L = beam span (in.),

$E' = EC_M C_t C_n$ (lb/in²), and

I = moment of inertia about the axis of bending (in⁴).

Note that the modification factor for duration of load, C_D, does not apply to E.

Deflection equations such as 5-15 and 5-16 can be used to accurately predict elastic beam deflections. For permanent load deflections, however, it is necessary to increase computed values to compensate for the long-term effects of creep. The magnitude of the increase depends on the type of material and the moisture content of the member at installation. A 50-percent increase in dead load deflection is normally sufficient for glulam and seasoned sawn lumber, while a 100-percent increase is more appropriate for unseasoned lumber (refer to Appendix F of the NDS for additional discussions on dead load deflection increases for creep).

Deflection Criteria

AASHTO specifications do not give deflection criteria for timber bridge members, and selection of an appropriate deflection limit is a matter of designer judgment. The acceptable deflection for a member will depend on specific use requirements and may vary among beam types within the same structure. Deflections in bridge members are important for serviceability, performance, and aesthetics and should not be ignored. From a structural viewpoint, large deflections cause fasteners to loosen and brittle materials, such as asphalt pavement, to crack and break. In addition, members that sag below a level plane present a poor appearance and can give the public a perception of structural inadequacy. Deflections from moving vehicle loads also produce vertical movement and vibrations that annoy motorists and alarm pedestrians.

Bridge deflection is normally expressed as a fraction, the denominator of which is obtained by dividing the beam span in inches by the computed deflection in inches. A deflection of *L/500*, for example, indicates a deflection equal to one five-hundredth of the beam span. The larger the denominator, the smaller the deflection. A brief literature search of bridge-related specifications and publications produced maximum recommended applied-load deflection values ranging from *L/200* to *L/1,200*. For general beam design discussed in this chapter, the recommended maximum deflections for timber beams are as follows:

1. For applied (short-term) loads, the maximum deflection should not exceed *L/360*.

2. For the combination of applied loads and dead load, the maximum deflection should not exceed *L/240*, where the portion of the total deflection from dead load is increased to account for creep.

Additional considerations and recommendations for deflection in timber bridge components are discussed in more detail in Chapters 7, 8, and 9.

Camber

Camber is circular or parabolic upward curvature built into a glulam beam, opposite to the direction of deflection. It is intended to offset dead load deflection and creep and is introduced during the manufacturing process. It is not feasible to camber sawn lumber beams. The amount of camber for bridge beams depends on the length and number of spans. For single spans shorter than approximately 50 feet, camber should be a minimum of 1.5 to 2.0 times the immediate (elastic) dead load deflection, plus one-half the applied load deflection.[6] For single beam spans equal to or longer than 50 feet and multiple-span beams of any span, camber should be a minimum of 1.5 to 2.0 times the immediate dead load deflection (multiple-span bridge beams are normally cambered for dead loads only to obtain acceptable riding qualities for vehicle traffic).

Camber is specified by the designer as a vertical centerline offset to the horizontal line between points of bearing (Figure 5-3). The glulam manufacturer will determine an appropriate radius of curvature based on offset distances and fabrication limitations. On multiple-span continuous beams, camber may vary along the beam and should be specified for each span segment. More specific information on cambering practices and limitations can be obtained from glulam manufacturers and the AITC.

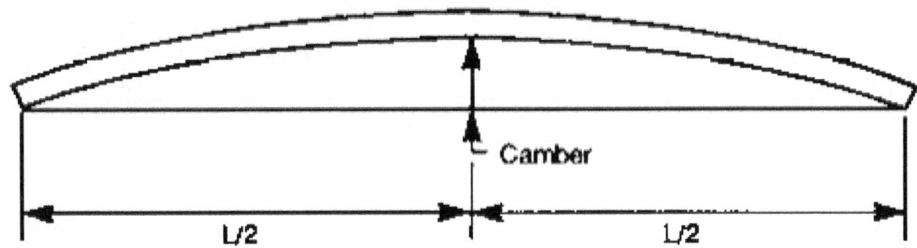

Figure 5-3.- Camber for glulam beams is specified as an upward vertical offset at the span centerline.

Example 5-4- Beam deflection and camber

For the glulam beam of Example 5-3, Case A, determine the deflection from the 20,000-pound moving load and the camber required to offset deflection from the beam weight. The beam spans 50 feet, measures 6-3/4 inches by 42-5/8 inches, and is manufactured from visually graded Southern Pine, combination symbol 24F-V2.

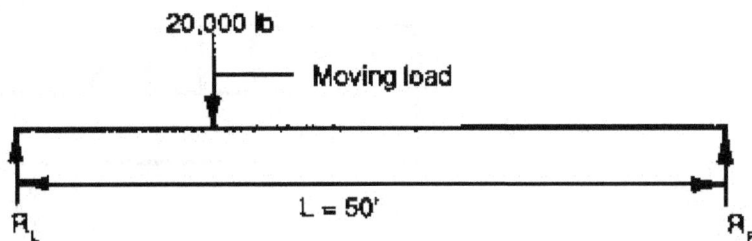

Solution:

The tabulated modulus of elasticity for a 24F-V2 Southern Pine beam is obtained from *AITC 117-Design:*

$$E_x = 1,700,000 \text{ lb/in}^2$$

The allowable modulus of elasticity is computed using the applicable C_M value from Table 5-7:

$$E' = E_x C_M = 1,700,000(0.833) = 1,416,100 \text{ lb/in}^2$$

From Table 16-4 for a 6-3/4-inch by 42-5/8-inch Southern Pine beam:

$$I_x = 43,562.8 \text{ in}^4$$

Beam weight = 99.9 lb/ft (based on a beam weight of 50 lb/ft^3)

Deflection for the 20,000-pound moving load is computed with the load at midspan by Equation 5-15:

5-35

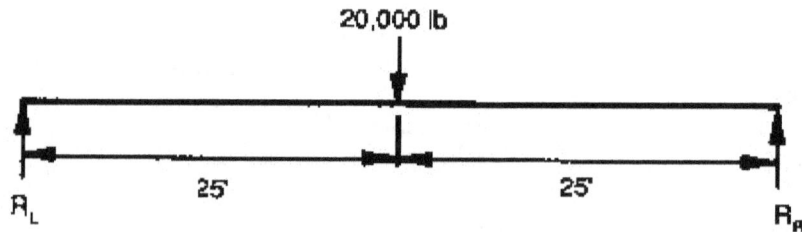

$$\Delta = \frac{PL^3}{48\ E'\ I_x} = \frac{20,000\left[50\left(12\ \text{in}/\text{ft}\right)\right]^3}{48\left(1,416,100\right)\left(43,562.8\right)} = 1.46\ \text{in}.$$

Expressing the deflection as a ratio of the bridge span,

$$\Delta = \frac{L}{\left[50\ \text{ft}\left(12\ \text{in}/\text{ft}\right)\right]/1.46\ \text{in}.} = \frac{L}{411}$$

$L/411 < L/360$, so deflection is acceptable.

For the beam weight of 99.9 lb/ft, deflection is computed by Equation 5-16:

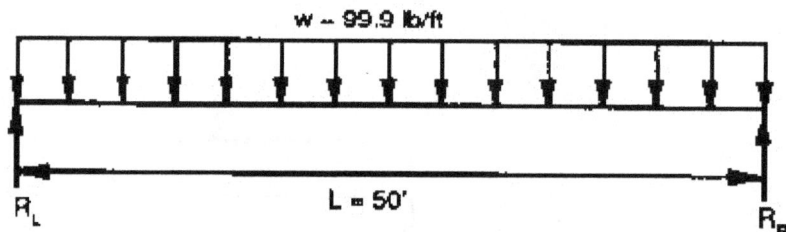

$$\Delta = \frac{5wL^4}{384\ E'\ I_x} = \frac{5(99.9)\left[50\left(12\ \text{in}/\text{ft}\right)\right]^4}{384(1,416,100)(43,562.8)(12\ \text{in}/\text{ft})} = 0.23\ \text{in}.$$

Camber of approximately 1/2-inch will be specified at centerline, which is approximately twice the beam dead load deflection.

DESIGN FOR SHEAR

Beams develop internal shear forces that act perpendicular and parallel to the longitudinal beam axis. In timber beams, horizontal shear rather than vertical shear will always control design. As discussed in Chapter 3, horizontal shear forces produce a tendency for the upper portion of the beam to slide in relation to the lower portion of the beam, with shear stresses acting parallel to the grain of the member. The maximum intensity of horizontal shear in rectangular beams occurs at the neutral axis and is proportional to the vertical shear force, V. In bridge applications, horizontal shear generally controls beam design only on relatively short, heavily loaded spans.

Shear requirements in AASHTO and the NDS apply at or near the supports for solid beams constructed of such materials as sawn lumber, glulam, or mechanically laminated lumber. Shear design for built-up components containing load-bearing connections at or near supports, such as between a web and chord, must be based on tests or other techniques.

Applied Stress

The applied stress in horizontal shear depends on the magnitude of the vertical shear and the area of the beam. Applied stress in square or rectangular timber beams is computed by Equation 5-17:

$$f_v = \frac{3V}{2bd} = \frac{1.5V}{A} \tag{5-17}$$

where f_v = unit stress in horizontal shear (lb/in^2),

V = vertical shear force (lb),

b = beam width at the neutral axis (in.),

d = beam depth (in.), and

A = beam cross-sectional area (in^2).

Equation 5-17 does not apply (1) at notches or joints, (2) in regions where the beam is supported by fasteners, or (3) when hanging loads are located at or near the supports. For these conditions, refer to AASHTO and the NDS.

The magnitude off, given by Equation 5-17 is based on the value of the vertical shear force, V. Unlike the situation in other construction materials, where the maximum vertical shear is computed at the face of the supports, in timber beams the maximum intensity of horizontal shear is produced by the maximum vertical shear force occurring at some distance from the support. This distance depends on the type of applied loading; different distances are used for moving loads and for stationary loads.

Current AASHTO requirements (AASHTO 13.3.1) specify that horizontal shear in beams from moving (vehicle) loads be computed from the maximum vertical shear *(V)* occurring at a distance from the support equal to three times the beam depth *(3d,* or the span quarter point *(L/4),* whichever is less (Figure 5-4). The moving loads are positioned on the beam to produce the maximum vertical shear at this location (Chapter 6). For stationary loads (such as dead load), vertical shear is computed at a distance from the support equal to the beam depth, *d,* and all loads occurring within the distance *d* from the supports are neglected. For sawn lumber, shear design requirements given in the NDS vary somewhat based on the beam configuration, loading condition, and wood species. Refer to the latest edition of the NDS for additional shear criteria for sawn lumber.

5-37

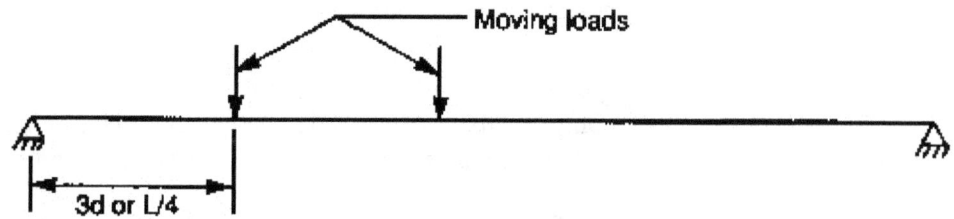

For moving loads, the loads are positioned to produce the maximum vertical shear at the lesser of 3d or L/4 from the support

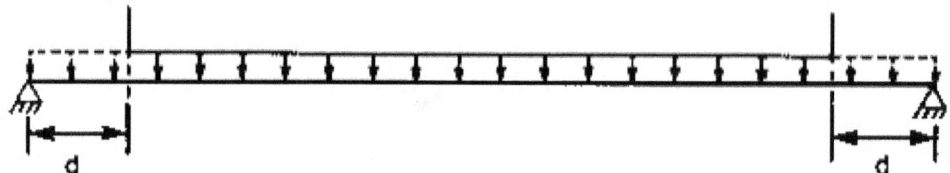

For stationary loads, such as dead load, the maximum vertical shear is computed at a distance d from the supports and all loads occurring within a distance d from the supports are neglected

Figure 5-4.- Locations for determining the maximum vertical shear (V) for timber beams.

Although the bases for shear design requirements are widely accepted, specific requirements for computing V are somewhat controversial and vary among design specifications. Research is currently under way to develop more accurate design criteria for shear, and the designer should remain familiar with the most current requirements and the potential for future revision.

Allowable Stress
The allowable stress in horizontal shear is computed by

$$F_v' = F_v C_D C_M C_R C_t \tag{5-18}$$

Individual sawn lumber members have a much higher potential for strength-reducing characteristics that reduce the ability of the member to resist horizontal shear. In glulam, most strength-reducing characteristics are excluded at fabrication and any that remain are dispersed throughout the individual laminations in the section. For sawn lumber, strength-reducing characteristics are not dispersed, and members are more susceptible to the development of checks and splits caused by variations in moisture content. As a result, tabulated values of F_v for sawn lumber are considerably lower than those for glulam because they are based on the worst-case assumption that members are split for their entire length. In situations where the length of split, or size of check or shake, can be estimated with reasonable certainty, the tabulated horizontal shear stress can be increased by the shear stress modification factors given in footnotes to the NDS Table 4A (Table 5-12). Application of this factor to specific design situations and materials is left to designer judgment, but the 2.0

increase is commonly used for mechanically laminated lumber and dimension lumber with loads applied perpendicular to the wide face. Additional information on application of the shear stress modification factor is discussed in Chapters 7 and 8.

Table 5-12.- Shear stress modification factor for sawn lumber.

Length of split on wide face of 2" lumber (nominal):	Multiply tabulated "F_v" value by:
No split	2.00
1/2 x wide face	1.67
3/4 x wide face	1.50
1 x wide face	1.33
1-1/2 x wide face or more	1.00

Length of split on wide face of 3" and thicker lumber (nominal):	Multiply tabulated "F_v" value by:
No split	2.00
1/2 x narrow laace	1.67
1 x narrow face	1.33
1-1/2 x narrow face or more	1.00

Size of shake[a] in 3" and thicker lumber (nominal):	Multiply tabulated "F_v" value by:
No shake	2.00
1/6 x narrow face	1.67
1/3 narrow face	1.33
1/2 x narrow face or more	1.00

[a] Shake is measured at the end between lines enclosing the shake and parallel to the wide face.

Specific horizontal shear values may be established by use of this table when the length of split, or size of check or shake is known and no increase in them is anticipated. For California Redwood, Southern Pine, Virginia Pine-Pond Pine, and Yellow Poplar, refer to the NDS for specific values of F_v for which these adjustments apply.

From the NDS;[24] © 1986. Used by permission.

Example 5-5- Horizontal shear in a sawn lumber beam

Determine the adequacy of the beam in Example 5-2 for horizontal shear. The beam measures 4 inches by 14 inches and is surfaced Douglas Fir-Larch, visually graded No. 1 in the J&P size classification. It spans 15 feet and supports a uniform load of 350 lb/ft.

Solution

Tabulated horizontal shear stress for No.1 Douglas Fir-Larch is obtained from Table 4A of the NDS (note that the tabulated shear stress for lumber 2 to 4 inches thick is the same for all grades):

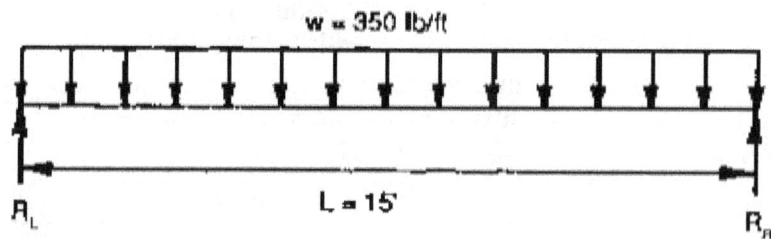

w = 350 lb/ft

L = 15'

R_L R_R

$F_v = 95 \ \text{lb/in}^2$

Allowable shear stress is computed by Equation 5-18 using the C_M value obtained from Table 5-7,

$$C_M = 0.97$$

$$F_v' = F_v C_M = 95(0.97) = 92 \ \text{lb/in}^2$$

The allowable stress in horizontal shear could be increased by the shear stress modification factor (Table 5-12) if the beam were free of shake, splits or checks, or if the length of such characteristics was known. For lumber bridge beams of this type, it is common for some beam checking to occur, however, its magnitude cannot be accurately predicted. Therefore, no adjustment by the shear stress modification factor will be used.

From Example 5-2, the beam weighs 16.1 lb/ft and has actual dimensions of 3.5 inches by 13.25 inches. The total load acting on the beam is equal to the 350 lb/ft applied load plus the beam weight of 16.1 lb/ft, for a total of 366.1 lb/ft. For a uniformly distributed load, the maximum vertical shear force, V, is computed at a distance from the support equal to the beam depth, d, and all loads acting within a distance d from the supports are neglected:

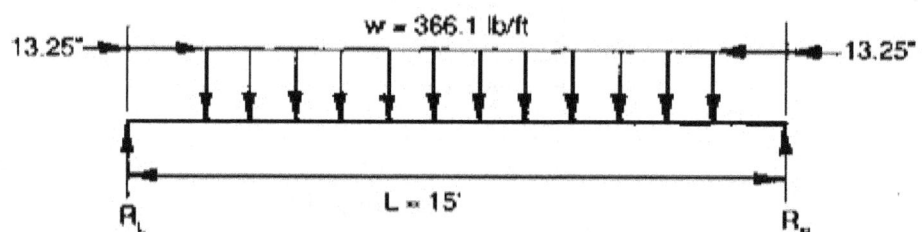

13.25"   w = 366.1 lb/ft   13.25"

L = 15'

R_L R_R

$$V = R_L = w\left(\frac{L}{2} - d\right) = 366.1 \left[\frac{15}{2} - \frac{13.25}{(12 \ \text{in/ft})}\right] = 2,342 \ \text{lb}$$

Horizontal shear stress is computed by Equation 5-17:

$$A = (3.5\text{ in.})(13.25\text{ in.}) = 46.38\text{ in}^2$$

$$f_v = \frac{1.5V}{A} = \frac{1.5(2,342)}{46.38} = 76\text{ lb/in}^2$$

$f_v = 76\text{ lb/in}^2 < F_v' = 92\text{ lb/in}^2$, so horizontal shear is acceptable

---

Example 5-6- Horizontal shear in a glulam beam.

Check the adequacy of the glulam beam in Example 5-3, Case A, for horizontal shear. The beam measures 6-3/4 inches by 42-5/8 inches and is manufactured from visually graded Southern Pine, combination symbol 24F-V2. It spans 50 feet and supports a moving concentrated load of 20,000 pounds.

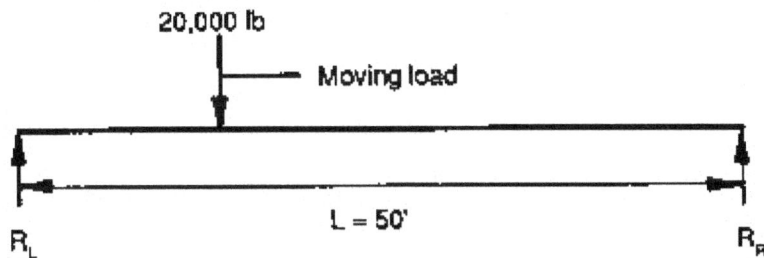

The tabulated stress for horizontal shear for a 24F-V2 beam is obtained from *AITC 117--Design*,

$$F_{vx} = 200\text{ lb/in}^2$$

Allowable shear stress is computed by Equation 5-18 using the applicable C_M value obtained from Table 5-7:

$$C_M = 0.875$$

$$F_v' = F_{vx}C_M = 200(0.875) = 175\text{ lb/in}^2$$

In this case the beam supports two loads; the uniform load from the beam weight and the moving concentrated load. Maximum vertical shear from the uniformly distributed beam weight is computed at a distance from the support equal to the beam depth, d, and all loads acting within a distance d from the supports are neglected. For the moving concentrated load, maximum vertical shear is computed at a distance from the support equal to three times the beam depth, $3d$, or the span quarter point, $L/4$, whichever is less.

For the uniformly distributed beam weight of 99.9 lb/ft and a beam depth of 42.63 inches,

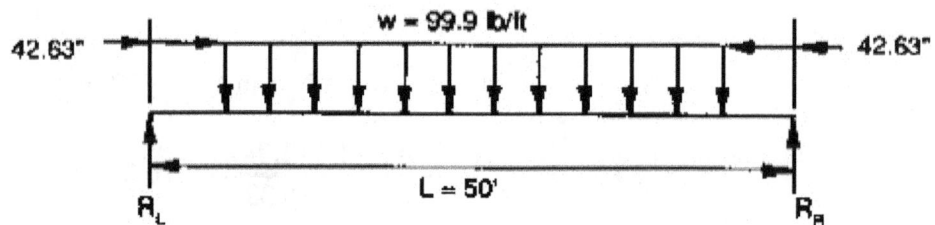

$$V = R_L = w\left(\frac{L}{2} - d\right) = 99.9\left[\frac{50}{2} - \frac{42.63}{(12\ \text{in/ft})}\right] = 2{,}143\ \text{lb}$$

For the moving concentrated load of 20,000 lb,

$$3d = 3\left(\frac{42.63}{12\ \text{in/ft}}\right) = 10.66\ \text{ft} \qquad \frac{L}{4} = \frac{50}{4} = 12.50\ \text{ft}$$

3d < L/4, so the maximum vertical shear from the 20,000-pound load is computed at a distance of 10.66 feet from the support:

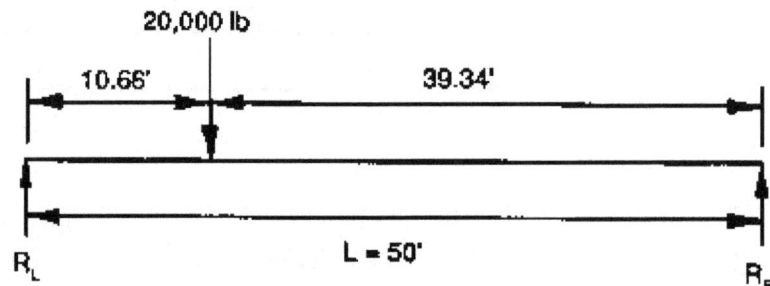

$$V = R_L = \frac{20{,}000\,(39.34)}{50} = 15{,}736\ \text{lb}$$

From Table 16-4, the cross-sectional area of a 6-3/4-inch by 42-5/8-inch Southern Pine glulam beam is 287.7 in². Applied stress is computed by Equation 5-17:

$$f_v = \frac{1.5V}{A} = \frac{1.5\,(2{,}143 + 15{,}736)}{287.7} = 93\ \text{lb/in}^2$$

$f_v = 93\ \text{lb/in}^2 < F_v' = 175\ \text{lb/in}^2$, so horizontal shear is acceptable.

DESIGN FOR BEARING

Reactions at beam supports produce bearing stress that acts perpendicular to or at an angle to the grain of the member. Bearing stress causes wood fibers to compress to a degree that depends on the magnitude of load and the area of bearing. The beam bearing area must be large enough to adequately transfer loads without causing the wood to compress or deform excessively.

Applied Stress

Applied bearing stress is computed by

$$f_{c\perp} = \frac{R}{A}$$
(5-19)

where $f_{c\perp}$ = unit stress in compression perpendicular to grain (lb/in²),

R = reaction or bearing force at the support (lb), and

A = net bearing area (in²).

When computing $f_{c\perp}$ at the end of a beam, no allowance is made for the fact that as the beam bends the pressure on the inner edge of the bearing is greater than that at the end of the beam.

Allowable Stress

The allowable stress for bearing perpendicular to grain is equal to the tabulated stress $F_{c\perp}$ adjusted by all applicable modification factors, except the duration of load factor, C_D, as computed by

$$F_{c\perp}' = F_{c\perp} C_M C_t C_R$$
(5-20)

When beam bearing is not perpendicular to grain (Figure 5-5), allowable stress must be computed for compression at an angle to the grain using the Hankinson Formula (Equation 5-21):

$$F_n' = \frac{F_s' F_{c\perp}'}{F_s' \sin^2(\theta) + F_{c\perp}' \cos^2(\theta)}$$
(5-21)

where F_n' = allowable stress in compression at an angle to the grain (lb/in²),

$F_s' = F_s C_M C_D C_t C_R$ (lb/in²),

$F_{c\perp}' = F_{c\perp} C_M C_t C_R$ (lb/in²), and

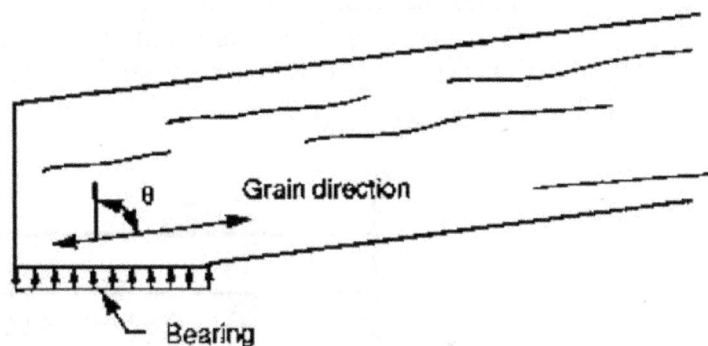

Figure 5-5. -- Beam bearing at an angle to the grain.

5-43

θ = angle between the direction of load and the direction of grain (degrees).

Values of $F_{c\perp}$ given in the NDS and *AITC 117-Design* apply to bearings of any length at beam ends and to all bearings 6 inches or more in length at other locations. Refer to the NDS for required adjustments in tabulated stress for bearings less than 6 inches long at locations between beam ends.

Example 5-7 - Beam bearing

For the glulam beam of Example 5-3, Case A, determine the required bearing length and the bearing stress in compression perpendicular to grain. The beam spans 50 feet center-to-center of bearings, is 6-3/4 inches wide and supports a moving concentrated load of 20,000 pounds. It is manufactured from visually graded Southern Pine, combination symbol 24F-V2.

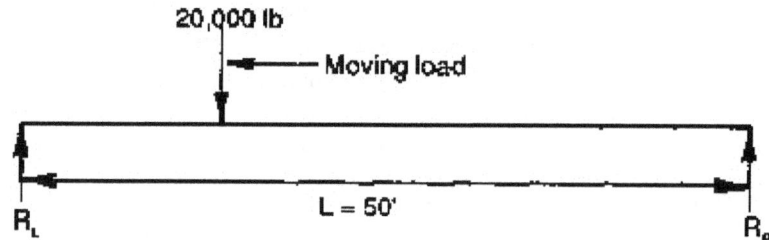

Solution

The tabulated stress in compression perpendicular to grain for a 24F-V2 Southern Pine beam is obtained from *AITC 117-Design*:

$$F_{c\perp x} = 650 \text{ lb/in}^2$$

The allowable compression perpendicular to grain is computed using Equation 5-20 and the applicable C_M value from Table 5-7:

$$F_{c\perp}' = F_{c\perp x}C_M = 650(0.53) = 345 \text{ lb/in}^2$$

The maximum reaction at the beam bearing is equal to the sum of the reactions from the moving concentrated load and the beam weight. The maximum reaction from the moving concentrated load occurs when the load is placed over one support:

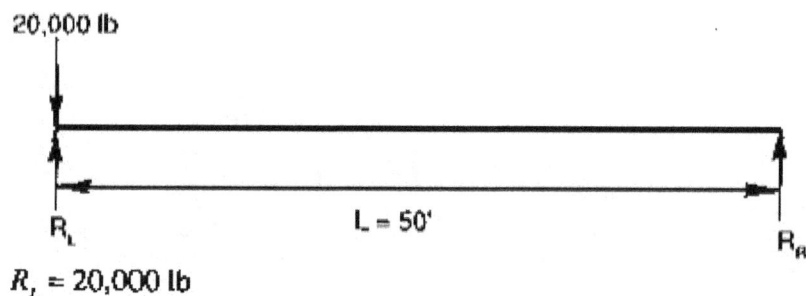

$R_L = 20,000 \text{ lb}$

The reaction from the beam weight is the same at both supports:

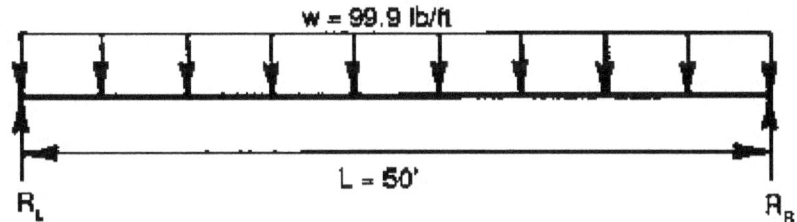

$$R_L = \frac{wL}{2} = \frac{99.9\,(50)}{2} = 2,498 \text{ lb}$$

Rearranging Equation 5-19, the minimum required bearing area is computed for the maximum reaction by substituting $F_{c\perp}'$ for $f_{c\perp}$:

$$A = \frac{R}{F_{c\perp}'} = \frac{(20,000 + 2,498)}{345} = 65.2 \text{ in}^2$$

For a beam width of 6-3/4 inches, the required bearing length is computed by dividing the bearing area by the bearing width:

$$\textbf{Bearing length} = \frac{A}{b} = \frac{65.30}{6.75} = 9.7 \text{ in.}$$

A bearing length of 10 inches is selected and applied stress is computed by Equation 5-19:

$$f_{c\perp} = \frac{P}{A} = \frac{(20,000 + 2,498)}{10(6.75)} = 333 \text{ lb/in}^2$$

$f_{c\perp} = 333 \text{ lb/in}^2 < F_{c\perp}' = 345 \text{ lb/in}^2$, so the bearing is satisfactory. For a center-to-center span of 50 feet, a beam length of 50 feet 10 inches will be required.

5.5 DESIGN OF TENSION MEMBERS

A tension member is a structural component loaded primarily in axial tension. In bridge design, tension members are used mostly as truss elements and occasionally as bracing (Figure 5-6). The direction of loading in tension members should always be parallel to the grain of the member. Timber is weak in tension perpendicular to the grain, and loading conditions that produce stress in this direction should be avoided. When loading conditions that induce tension perpendicular to the grain do exist, mechanical reinforcement must be designed to carry the load.

Discussions in this section apply to members loaded in axial tension only. Design criteria for members loaded in combined axial tension and bending are given in Section 5.7.

5-45

Figure 5-6.- Tension members in bridge applications are most common in trusses. This timber truss, located at Sioux Narrows, Ontario, Canada, spans 210 feet and is reputed to be the longest clear-span timber bridge in the world.

APPLIED STRESS

Applied stress in tension is computed by Equation 5-22:

$$f_t = \frac{P}{A} \tag{5-22}$$

where P = axial load applied to the member (lb), and

A = net cross-sectional area of the member (in²).

The net area, A, in Equation 5-22 is the gross area of the member minus the projected area of fastener holes or cuts that reduce the section. Requirements for determining net area for various fasteners are discussed in Section 5.8.

ALLOWABLE STRESS

Allowable stress in tension equals the tabulated stress for tension parallel to grain, F_t, adjusted by all applicable modification factors. This is computed by

$$F_t' = F_t C_D C_M C_F C_t \tag{5-23}$$

For sawn lumber, values of F_t for members 2 to 4 inches thick, and 5 inches and wider, apply to 5- and 6-inch widths only. When wider members are used, a reduction in tabulated stress ranging from 0.9 to 0.6 is required by footnotes to the NDS Table 4A. When glulam is used, the most economical tension members are generally selected from the axial combinations given in *AITC 117-Design.*

Example 5-8- Glulam tension member

A glulam truss member carries an axial tension load of 25,000 pounds. The ends of the member are attached to steel plates with a single row of 1-inch-diameter bolts aligned in the longitudinal direction. Design this truss member, assuming the following:

1. Normal load duration under wet-use conditions; adjustments for temperature *(C)* and fire-retardant treatment *(C_R)* are not applicable.

2. Bolt holes at member ends are 1/16 inch larger than the bolt diameter.

3. Glulam is manufactured from visually graded western species.

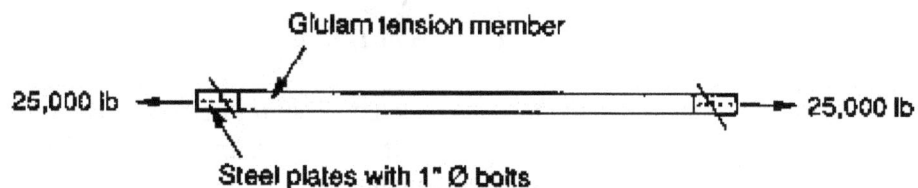

Glulam tension member

25,000 lb — Steel plates with 1" Ø bolts — 25,000 lb

Solution
The design of a tension member starts with either the selection of a glulam combination symbol or a standard member width. In this example, combination symbol No. 2 is selected and design will involve determining the required member size.

The tabulated stress for tension parallel to grain is obtained for combination symbol No. 2 from *AITC 117-Design:*

$$F_t = 1,250 \text{ lb/in}^2$$

The allowable stress for tension parallel to grain is computed by Equation 5-23 using the C_M value obtained from Table 5-7:

$$F_t' = F_t C_M = 1,250(0.80) = 1,000 \text{ lb/in}^2$$

Next, Equation 5-22 is rearranged to compute an initial member area based on the applied load and the allowable stress in tension parallel to grain:

$$A = \frac{P}{F_t'} = \frac{25,000}{1,000} = 25 \text{ in}^2$$

The required member depth is obtained for several standard glulam widths by dividing the required area by the standard width, then rounding the depth up to the next standard depth (based on a 1-1/2-inch lamination thickness for western species). For three standard glulam widths:

Member width	Minimum required depth	Depth rounded up to standard depth	Number of laminations
3-1/8 in.	$\dfrac{A}{b} = \dfrac{25}{3.125} = 8.00\,\text{in.}$	9 in.	6
5-1/8 in.	$\dfrac{A}{b} = \dfrac{25}{5.125} = 4.88\,\text{in.}$	6 in.	4
6-3/4 in.	$\dfrac{A}{b} = \dfrac{25}{6.75} = 3.70\,\text{in.}$	4.5 in.	3

Initial selection of a member width and depth is a matter of designer judgement and depends on size and economic considerations. In this case, the 5-1/8-inch width is selected and the gross member area is computed:

$$A_{GROSS} = b(d) = 5.125(6) = 30.75 \text{ in}^2$$

The net area used for design is equal to the gross area minus the projected area of bolt holes. Assuming that bolts pass through the narrow (5-1/8-inch) dimension,

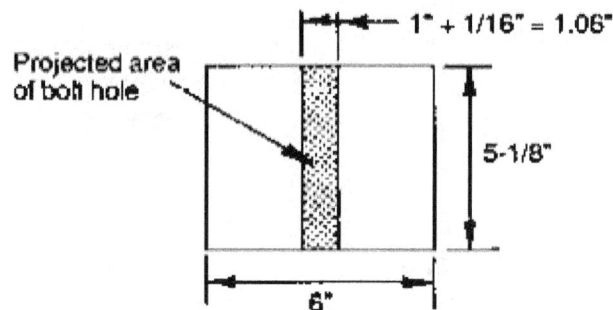

$$A_{BOLT} = (1.06 \text{ in.})(5.125 \text{ in.}) = 5.43 \text{ in}^2$$

$$A_{NET} = A_{GROSS} - A_{BOLT} = 30.75 - 5.43 = 25.32 \text{ in}^2$$

By Equation 5-22,

$$f_t = \frac{P}{A} = \frac{25,000}{25.32} = 987 \text{ in}^2$$

$f_t = 987$ in^2 < F_t' = 1,000 lb/in^2, so a 5-1/8-inch wide by 6-inch deep combination symbol No. 2 member is satisfactory.

A column is a structural component loaded primarily in axial compression parallel to its length. In bridge design, columns are used as supporting components of the substructure, truss elements, and bracing (Figure 5-7). The three general types of columns are simple solid columns, spaced columns, and built-up columns (Figure 5-8). Simple solid columns consist of a piece of sawn lumber or glulam. Spaced columns consist of two or more parallel pieces that are separated and fastened at the ends and at one or more interior points by blocking. Built-up columns consist of a number of solid members joined together with mechanical fasteners. The most common columns for timber bridges are simple solid columns constructed of sawn lumber, glulam (axial combinations), timber piles, or poles. Although spaced and built-up columns may be used for truss elements or other components, they are not common in modem bridge applications.

The column design requirements in this section are limited to simple solid columns of constant cross-sectional area. Loads are applied concentrically, and design is based on the stresses and instability from axial compression and end-grain bearing stress at column ends. Columns loaded in combined compression and bending are discussed in Section 5.7 of this chapter. For additional information on built-up, spaced, and tapered solid columns, refer to the *NDS* and the *AITC Timber Construction Manual*.

Figure 5-7.- Timber columns are common in bridge substructures such as these bents (photo courtesy Wheeler Consolidated, Inc.).

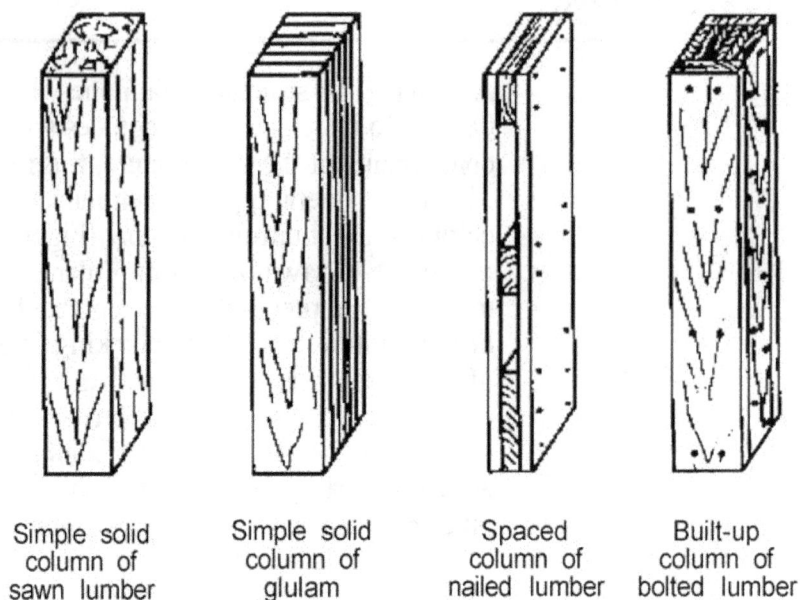

Figure 5-8. - General classes of timber columns.

| Simple solid column of sawn lumber | Simple solid column of glulam | Spaced column of nailed lumber | Built-up column of bolted lumber |

DESIGN FOR COMPRESSION

Compression in timber columns can induce failure by crushing the wood fibers or by lateral buckling (deformation). The first step in column design is to estimate an initial member size and compute applied stress (several iterations may be required to arrive at a suitable section). After an initial column size is selected, the column slenderness ratio is computed, which serves as the basis for design in compression. From the slenderness ratio, allowable stress is determined from equations given in the NDS and checked against the applied stress.

Applied Stress

Applied column stress in compression parallel to grain, f_c, is computed by

$$f_c = \frac{P}{A} \tag{5-24}$$

where P = the total compressive load supported by the column (lb), and

A = the cross-sectional area of the column (in^2).

The value of A used in Equation 5-24 depends on the location of fastener holes that reduce the column section. When the reduced section occurs at points of lateral support, failure occurs by wood crushing, and the gross column area is used without deductions for fastener holes. At locations away from points of lateral support, failure may occur by column buckling, and the net column area (gross column area minus fastener holes) is used. Refer to Section 5.8 for details on computing net area for different fastener types.

Column Slenderness Ratio

The slenderness ratio of a column provides a measure of the tendency of the column to fail by buckling from insufficient stiffness, rather than by crushing from insufficient strength. It is expressed as the ratio of the unsupported column length to its least radius of gyration and is computed for timber in the same manner as for other materials. For convenience in design, however, the slenderness ratio for square or rectangular simple solid columns is given in terms of the column cross-sectional dimension, rather than the radius of gyration, and is computed by

$$\text{Slenderness ratio} = \frac{\ell_e}{d} \tag{5-25}$$

where ℓ_e = effective column length (in.), and

d = cross-sectional dimension corresponding to ℓ_e (in.).

The effective column length in Equation 5-25 is the distance between two points along the column length at which the member is assumed to buckle in the shape of a sine wave. It is computed as the product of the unsupported column length and the effective buckling length factor given by

$$\ell_e = K_e \ell \tag{5-26}$$

where K_e = effective buckling length factor, and

ℓ = unbraced length between points of lateral support along the column length.

Values of K_e are given in Table 5-13 for various conditions of end fixity and lateral translation at column ends or intermediate points of lateral support. In most applications, timber columns with square-cut ends are fixed against translation but not rotation (approximately pinned connections), and the value of K_e is 1.0. Conditions may be encountered in design where restraint is more or less than this condition, and K_e must be adjusted accordingly based on designer judgment. Additional discussion on effective buckling length factors is given in Appendix N of the NDS.

The slenderness ratio provides an indication of the mode of failure and is the basis for determining the allowable design stress. If a column is loaded to failure by buckling, the buckling will always occur about the axis with the largest slenderness ratio. The task of the designer is to determine the controlling slenderness ratio for a given column configuration. For a rectangular column with the same unbraced length in both directions, the critical slenderness ratio can be determined by inspection (Figure 5-9 A). In this case, the column will obviously buckle about the weaker (y) axis, and that is the only slenderness ratio that must be computed (for buckling about the y axis the column deflects in the x direction). For column configurations where the unbraced length is not the same in both directions,

Table 5-13. - Effective buckling length factor, K_e.

Buckling modes						
Theoretical K_e value	0.5	0.7	1.0	1.0	2.0	2.0
Recommended design K_e when ideal conditions approximated	0.65	0.80	1.2	1.0	2.10	2.4
End condition code		Rotation fixed, translation fixed Rotation free, translation fixed Rotation fixed, translation free Rotation free, translation free				

the critical slenderness ratio cannot be determined by inspection and the designer must compute slenderness ratios for both directions (Figure 5-9 B). Depending on the spacing of lateral support, conditions may exist where the column design is controlled by buckling about the strong axis.

Allowable Stress

The allowable compressive stress for square or rectangular simple solid columns is computed from equations given in the NDS. These equations are based on the column slenderness for three ranges:

$$0 < \ell_e/d \quad 11 \qquad \text{Short Column}$$

$$11 < \ell_e/d \quad K \qquad \text{Intermediate Column}$$

$$K < \ell_e/d \quad 50 \qquad \text{Long Column}$$

where K is a slenderness factor defined later in this section for intermediate columns.

The NDS equations have been modified to incorporate the use of the column dimension (d) rather than the radius of gyration (r). They may be used for nonrectangular cross sections by substituting $3.46r$ for d ($\ell_e/3.46r$ is used in place of ℓ_e/d when determining the column-length class). For the special case of a round column, the NDS states that the load on a round column may be taken as the same as that for a square column of the same cross-sectional area. For round columns, the d used in determing the ℓ_e/d ratio is 0.866 times the diameter of the round column.

A. Column with equal unbraced lengths in both directions. The largest slenderness ratio $(\ell_e/d)_y$ can be determined by inspection.

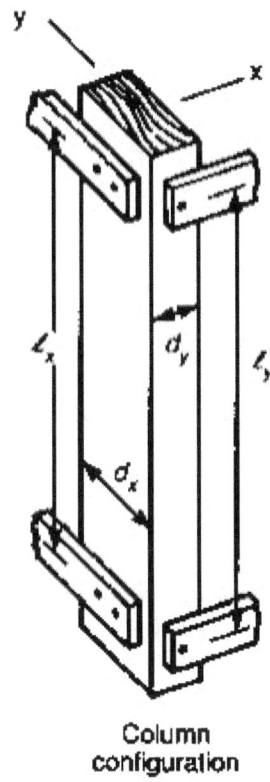

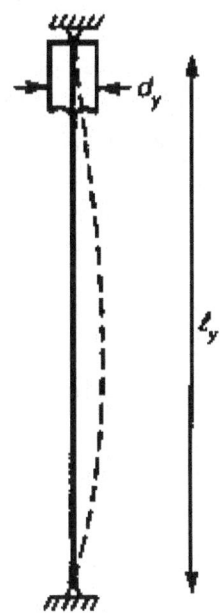

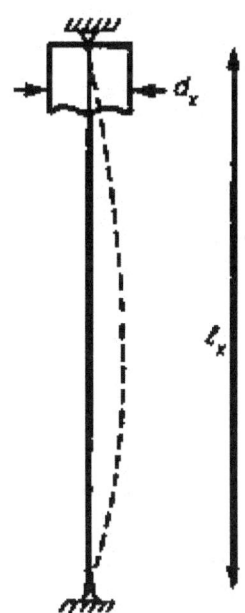

Column configuration

Buckling mode about the y-y axis $(\ell_e/d)_y$

Buckling mode about the x-x axis $(\ell_e/d)_x$

B. Column with different unbraced lengths for both axes. Both slenderness ratios $(\ell_e/d)_y$ and $(\ell_e/d)_x$ must be computed to determine the critical value.

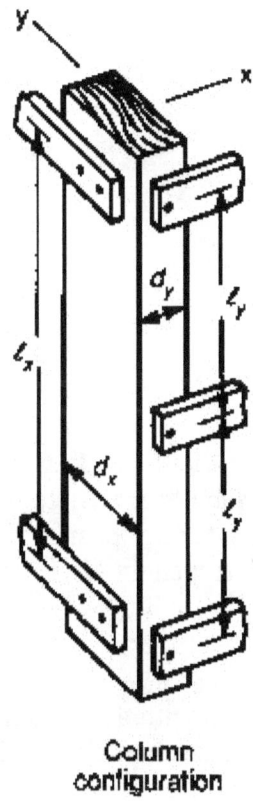

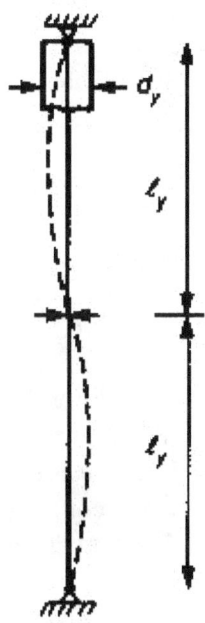

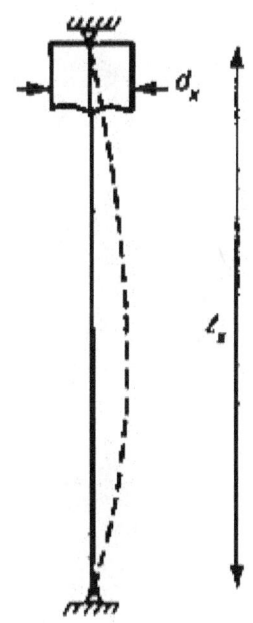

Column configuration

Buckling mode about the y-y axis $(\ell_e/d)_y$

Buckling mode about the x-x axis $(\ell_e/d)_x$

Figure 5-9.- Column slenderness ratios for columns with equal and unequal unbraced lengths.

5-53

Short Columns

Short columns are columns with a slenderness ratio of 11 or less. In short columns, the capacity of the member is controlled by the strength in compression parallel to grain, and failure always occurs by crushing of the wood fibers. Allowable stresses for short columns are equal to the tabulated stress in compression parallel to grain adjusted by applicable modification factors, as given by

$$F_c' = F_c C_D C_M C_R C_t \tag{5-27}$$

Intermediate Columns

Intermediate columns have a slenderness ratio greater than 11 but less than K as determined by

$$K = 0.671 \sqrt{\frac{E'}{F_c''}} \tag{5-28}$$

where K = minimum value of ℓ_e/d at which the column can be expected to perform as an Euler column,[6]

$$E' = E C_M C_R C_t \text{ (lb/in}^2\text{), and}$$

$$F_c'' = F_c C_D C_M C_R C_t \text{ (lb/in}^2\text{).}$$

In intermediate columns, failure can occur by crushing of the wood fibers or by lateral buckling, or both. The allowable stress for intermediate columns is the tabulated stress in compression parallel to grain adjusted by applicable modification factors, including the lateral stability of columns factor, C_p, and is computed by

$$F_c' = F_c C_p C_D C_M C_R C_t \tag{5-29}$$

where

$$C_p = 1 - \frac{1}{3}\left(\frac{\ell_e/d}{K}\right)^4 \tag{5-30}$$

In addition to Equation 5-29, the NDS gives optional column design adjustments for low variability materials (such as glulam) that are similar to those previously discussed for beams. For additional information on these equations, refer to Appendix G of the NDS and the AITC *Timber Construction Manual.*

Long Columns

Long columns are columns with a slenderness ratio greater than K and less than or equal to 50 (the maximum slenderness ratio allowed by the NDS for any column is 50). In long columns, the strength of the member is controlled by stiffness, and failure occurs by lateral buckling. The allowable design stress for long columns is given by

$$F_c' = \frac{0.30E'}{(\ell_e / d)^2}$$

(5-3 1)

DESIGN FOR BEARING

Column design must also consider bearing on the end grain of the member, given by

$$f_g = \frac{P}{A}$$

(5-32)

where

f_g = end-grain bearing stress from applied loads (lb/in²),

P = total applied load (lb), and

A = net area in bearing (in²).

The tabulated stress for end grain in bearing is specified in Table 2B of the NDS for sawn lumber and in Tables A-1 and A-2 of *AITC 117-Design* for glulam. The tabulated stress for sawn lumber is given for wet-service and dry-service conditions. For glulam, tabulated stress is for dry-service conditions and must be modified when the moisture content of the member is expected to exceed 16 percent in service (as in most bridge applications). Tabulated end-grain bearing stress is computed for sawn lumber and glulam as follows:

For sawn lumber,

$$F_g' = F_g C_D C_R C_t$$

(5-33)

For glulam,

$$F_g' = F_g C_M C_D C_R C_t$$

(5-34)

where

F_g' = allowable stress for end grain in bearing (lb/in²),

F_g = tabulated stress for end grain in bearing (lb/in²), and

C_M = moisture modification factor for glulam for end grain in bearing = 0.57.

When the bearing stress computed by Equation 5-32 exceeds 75 percent of the allowable stress computed by Equations 5-33 or 5-34, the NDS requires that the bearing be on a metal plate or strap, or on other durable, rigid, homogeneous material of adequate strength.

Example 5-9. - Column design; sawn lumber

A square, sawn lumber column is 6 feet high and supports a concentric load of 35,000 pounds. Lateral support for the column is provided by pinned connections at the column ends only. Design this column, assuming the following:

1. Normal load duration and wet-use conditions; adjustments for temperature (C_t) and fire-retardant treatment (C_R) are not required.

2. The column is S4S Douglas Fir-Larch, visually graded No. 1 to WCLIB rules in the Posts and Timbers (P&T) size classification.

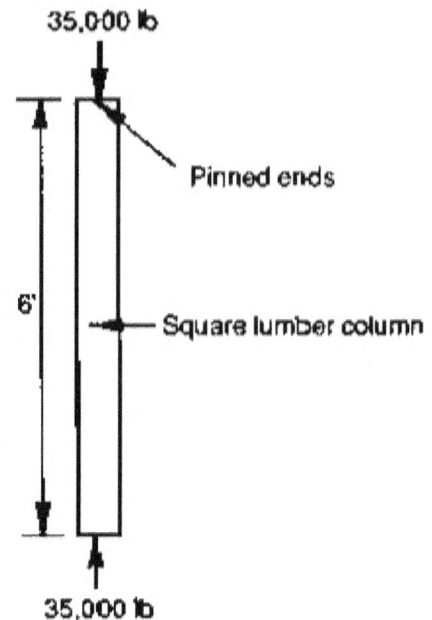

Solution
The first step in column design is to determine an initial column size. Since column dimensions are initially unknown, it is usually assumed that the column is in the short column slenderness range, and the allowable stress in compression parallel to grain is computed using Equation 5-27:

$$F_c' = F_c C_D C_M$$

From the NDS Table 4A for No. 1 Douglas Fir-Larch in the P&T size classification,

$$F_c = 1,000 \text{ lb/in}^2$$

From Table 5-7,

$$C_M = 0.91$$

Substituting values,

$$F_c' = F_c C_D C_M = 1,000(1.0)(0.91) = 910 \text{ lb/in}^2$$

An initial column area is obtained by dividing the applied load by F_c':

$$A = \frac{35,000 \text{ lb}}{910 \text{ lb/in}^2} = 38.5 \text{ lb/in}^2$$

From Table 16-2, the smallest square lumber size that meets the minimum area requirement is 8 inches by 8 inches, with the following properties:

$b = 7.5$ in.

$d = 7.5$ in.

$A = 56.25$ in^2

The column slenderness ratio must next be computed to determine the actual column slenderness range. The effective column length is computed by Equation 5-26 using an unbraced length of 6 feet and an effective buckling length factor, K_e of 1.0 for the pinned ends:

$$\ell_e = K_e \ell = 1.0 \ (6)(12 \ \text{in/ft}) = 72 \ \text{in.}$$

The column slenderness ratio is computed by Equation 5-25:

$$\text{Slenderness ratio} = \frac{\ell_e}{d} = \frac{72}{7.5} = 9.6$$

$\ell_e/d = 9.6 < 11.0$, so the column is in the short column slenderness range as initially assumed. Applied stress is computed by Equation 5-24:

$$f_c = \frac{P}{A} = \frac{35,000}{56.25} = 622 \ \text{lb/in}^2$$

$f_c = 622$ lb/in$^2 < F_c' = 910$ lb/in^2, so the column size is satisfactory.

Although normally not a controlling factor in column design, end grain in bearing stress should also be checked. From NDS Table 2B for wet-use Douglas Fir-Larch,

$$F_g = 1,340 \ \text{lb/in}^2$$

By Equation 5-33,

$$F_g' = F_g C_D = 1,340(\ 1.0) = 1,340 \ \text{lb/in}^2$$

$$0.75 F_g' = 0.75 \ (1,340) = 1,005 \ \text{lb/in}^2$$

Assuming a unit weight for wood of 50 lb/ft^3

$$\text{Column weight} = \frac{56.25 \ \text{in}^2}{144 \ \text{in}^2/\text{ft}^2} = (6 \ \text{ft})\left(50 \ \text{lb/ft}^2\right) = 117.2 \ \text{lb}$$

By Equation 5-32,

$$f_s = \frac{P}{A} = \frac{(117.2 + 35,000)}{56.25} = 624 \text{ lb/in}^2$$

$f_s = 624 \text{ lb/in}^2 < 0.75F_g' = 1,005 \text{ lb/in}^2$, so end-grain bearing is satisfactory, and bearing on a steel plate or other rigid, homogeneous material is not required.

Summary
The column will be nominal 8-inch by 8-inch surfaced Douglas Fir-Larch, visually graded No. 1 in the P&T size classification. The column is classified in the short column slenderness range and $f_c = 622 \text{ lb/in}^2 < F_c' = 910 \text{ lb/in}^2$. End-grain bearing stress is less than 75 percent of the allowable value, so special steel bearing plates are not required.

Example 5-10- Glulam column design

A glulam column is 17 feet long, 8-1/2 inches wide and 12-3/8 inches deep. Determine the column capacity for concentric loading when (A) the column is laterally supported at the ends only, and (B) the column is laterally supported at the ends and at midheight along the 12-3/8-inch dimension. The following assumptions apply:

1. Normal load duration under wet-use conditions; adjustments for temperature (C_t) and fire-retardant treatment (C_R) are not applicable.

2. Glulam is visually graded Southern Pine, combination symbol No. 47.

3. All support connections are pinned.

4. End-grain bearing is on a steel plate.

Solution
The procedure for determining the allowable load for each support condition will first involve computing the column slenderness range. From this, the allowable unit stress and load will be determined.

Tabulated values for compression parallel to grain and modulus of elasticity are obtained from *AITC 117--Design*. Respective values for the moisture content modification factor are obtained from Table 5-7:

$F_c = 1,900 \text{ lb/in}^2$ $C_M = 0.73$

$E = 1,400,000 \text{ lb/in}^2$ $C_M = 0.833$

Case A: Lateral support at ends only

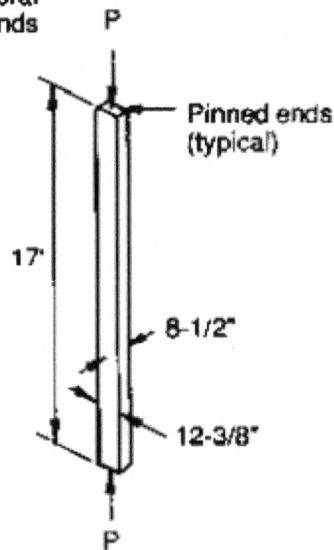

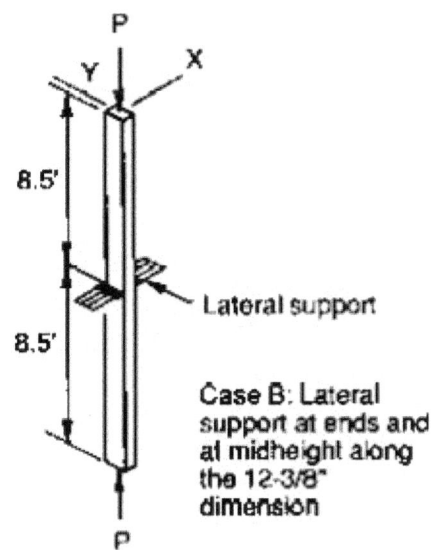

Pinned ends (typical)

17'

8-1/2"

12-3/8"

8.5'

8.5'

Lateral support

Case B: Lateral support at ends and at midheight along the 12-3/8" dimension

From Table 16-4, the area of an 8-1/2-inch by 12-3/8-inch glulam column is 105.2 in².

Case A: Lateral support at column ends only

With lateral support at the column ends only, the effective column length is computed using Equation 5-26. For an unbraced column length of 17 feet and a buckling length factor for pinned ends of 1.0,

$$\ell_e = K_e\ell = 1.0(17)(12 \text{ in/ft}) = 204 \text{ in.}$$

The column slenderness ratio is computed using Equation 5-25 with the least column dimension, $d = 8.5$ inches:

$$\text{Slenderness ratio} = \frac{\ell_e}{d} = \frac{204}{8.5} = 24$$

$\ell_e/d = 24 > 11$, so the column is in the intermediate or long slenderness range. The slenderness factor, K, is computed using Equation 5-28:

$$E' = EC_M = 1,400,000(0.833) = 1,166,200 \text{ lb/in}^2$$

$$F_c'' = F_cC_DC_M = 1,900(1.0)(0.73) = 1,387 \text{ lb/in}^2$$

$$K = 0.671\sqrt{\frac{E'}{F_c''}} = 0.671\sqrt{\frac{1,166,200}{1,387}} = 19.46$$

$\ell_e/d = 24 > K = 19.46$, so the column is in the long slenderness range. Allowable stress in compression parallel to grain is computed by Equation 5-31:

$$F_c' = \frac{0.30E'}{(\ell_e/d)^2} = \frac{0.30\,(1,166,200)}{(24)^2} = 607\ \text{lb/in}^2$$

The allowable load is the product of the column area and F_c':

$$P = A(F_c') = 105.2(607) = 63,856\ \text{lb}$$

Case B: Lateral support at column ends and at midheight along the 12-3/8-inch dimension

With lateral support at the column ends and at midheight along one axis, the slenderness ratio must be checked for both axes. About the x-x axis:

$$\ell = (17\ \text{ft})(12\ \text{in/ft}) = 204\ \text{in.}$$

$$K_e = 1.0$$

$$\ell_e = K_e\,(\ell) = 1.0\,(204) = 204\ \text{in.}$$

$$d = 12.38\ \text{in.}$$

$$\frac{\ell_e}{d} = \frac{204}{12.38} = 16.84$$

About the y-y axis:

$$\ell = \frac{17\ \text{ft}}{2}(12\ \text{in/ft}) = 102\ \text{in.}$$

$$K_e = 1.0$$

$$\ell_e = K_e\,(\ell) = 1.0\,(102) = 102\ \text{in.}$$

$$d = 8.5\ \text{in.}$$

$$\frac{\ell_e}{d} = \frac{102}{8.5} = 12.00$$

The largest slenderness ratio of 16.48 (about the x-x axis) will control design. By previous calculations $K = 19.46 > \ell_e/d = 16.48$, so the column is in the intermediate range.

The lateral stability of columns factor, C_p, is computed by Equation 5-30;

$$C_p = 1 - \frac{1}{3}\left(\frac{\ell_e/d}{K}\right)^4 = 1 - \frac{1}{3}\left(\frac{16.48}{19.46}\right)^4 = 0.83$$

Allowable stress in compression parallel to grain is computed by Equation 5-29:

$$F_c' = F_c C_p C_D C_M = 1,900(0.83)(1.0)(0.73) = 1,151\ \text{lb/in}^2$$

5-60

The allowable load is the product of the column area and F_c':

$$P = A(F_c') = 105.2(1,151) = 121,085 \text{ lb}$$

Check End-Grain Bearing

The tabulated stress for end grain in bearing is obtained from *AITC 117-Design:*

$$F_g = 2,300 \text{ lb/in}^2$$

The allowable stress is computed using Equation 5-34:

$$F_g' = F_g C_M C_D = 2,300(0.57)(1.0) = 1,311 \text{ lb/in}^2$$

$F_g' = 1,311 \text{ lb/in}^2$ is greater than previously computed values of F_c', so bearing stress will not control.

Summary

The allowable compression parallel to grain and maximum load for both column support cases are as follows:

Case A: Column laterally supported at ends only

$$F_c' = 607 \text{ lb/in}^2$$

Maximum allowable load = 63,856 lb

Case B: Column laterally supported at ends and at midheight along the 12-3/8-inch dimension

$$F_c' = 1,151 \text{ lb/in}^2$$

Maximum allowable load = 121,085 lb

This example illustrates the effect that lateral support can have on allowable column loading. When additional support is added at midheight, along the 12-3/8-inch dimension, the allowable load nearly doubles.

One or more loads acting on a column, beam, or other structural member may induce a combination of axial and bending stresses that occur simultaneously. In bridge design, combined loading most commonly occurs as axial compression and bending acting on supporting columns of the substructure (Figure 5-10). Even in columns designed for concentric loads, small eccentricities are created because of construction tolerances, slight member curvature, and material variations. Bending stress also occurs when columns are subjected to transverse loads from wind or earthquakes (see Chapter 6). Other conditions involving combined compression and bending or combined tension and bending are less common in bridge applications, but may occur in truss members or other components.

The design requirements discussed in this section are for combined axial tension or compression acting simultaneously with bending. It is assumed that bending occurs about one axis and that all loads are applied directly to the member. For cases involving axial loads with biaxial bending or loads acting through brackets attached to the member side, refer to references listed at the end of this chapter. [5,7,8,21,26,34]

GENERAL DESIGN CONSIDERATIONS

When members are subjected to simultaneous axial and bending loads, the resulting stress distribution is approximately the sum of the effects of the individual loads. In combined tension and bending, the effect is to reduce the compressive stress on one side of the member and increase the tensile stress on the other side. For combined compression and bending, tensile stress is reduced on one side and compressive stress is increased on the other. The case of combined compression and bending is critical because the higher compression increases the potential for lateral buckling of the member.

Combined stresses are evaluated using an interaction formula. In general terms, the interaction formula contains two expressions, one for the capacity in axial loading and one for the capacity in bending. In its basic form, the interaction formula is expressed by

$$\frac{f_a}{F_a'} + \frac{f_b}{F_b'} \leq 1.0 \tag{5-35}$$

where f_a = applied stress in tension f_t or compression f_c (lb/in²), and

F_a' = allowable stress in tension F_t' or compression F_c' (lb/in²).

Each of the expressions in Equation 5-35 can be thought of as representing the portion of the total member capacity taken by the respective axial or bending stress. The axial portion of the formula is the ratio of the applied axial stress to the allowable axial stress, assuming the member is loaded

Figure 5-10.- Members subjected to combined axial and bending forces are most common in bridge substructures. The vertical posts of this abutment support compressive loads from the superstructure and lateral loads from the earth pressure on the abutment wall.

with axial forces only. The bending portion is the ratio of the applied bending stress to the allowable bending stress, assuming the member is loaded with bending forces only. The sum of these expressions cannot exceed 1.0, or 100 percent of the member capacity.

When selecting a glulam member for combined axial and bending stresses, the designer should consider the relative magnitude of each type of stress. If tension or compression is the predominant stress, axial combinations are usually most economical. When bending is the predominant stress, bending combinations may be more appropriate.

COMBINED BENDING AND AXIAL TENSION

When members are loaded in combined axial tension and bending, the interaction equations that must be satisfied for design are given by

$$\frac{f_t}{F_t'} + \frac{f_b}{F_b'' C_F} \le 1.0 \tag{5-36}$$

$$\frac{f_b - f_t}{F_b'' C_L} \le 1.0 \tag{5-37}$$

where f_t = applied stress in axial tension computed by Equation 5-22 (lb/in^2),

$F_t' = F_t C_D C_M C_R C_t$ from Equation 5-23 (lb/in^2),

5-63

f_b = applied bending stress computed by Equation 5-2 (lb/in²), and

$$F_b'' = F_b C_D C_M C_R C_t \text{ from Equation 5-9 (lb/in²).}$$

In applying the interaction formulas, tension stress is computed for a tension member, as discussed in Section 5.5, and bending stress is computed for a beam, as discussed in Section 5.4. Considerations for tension are relatively straightforward; however, for bending, the member must be checked for strength in the tension zone and stability in the compression zone. In beam design, the size factor, C_F, applies to the tension side of the member where stresses from combined loading are greater than those from bending alone. As a result, C_F is always used as a modification factor in Equation 5-36. The lateral stability of beams factor, C_L, affects the compression side in bending where stresses from combined loading are reduced by the axial tension. When conditions of lateral support are such that the member is classified as an intermediate or long beam, and C_L rather than C_F controls beam design, the member must also meet the stability requirements given in Equation 5-37.

COMBINED BENDING AND AXIAL COMPRESSION

Members subjected to combined axial compression and bending are common in bridge design and are frequently referred to as beam columns. This type of loading is more critical than combined tension and bending because of the potential for lateral buckling and the additional bending stress created by the *P-delta* effect. The *P-delta* effect is produced when bending loads cause the axially loaded member to deflect along its longitudinal axis. When this occurs, an additional moment is generated by the axial load, P, acting over a lever arm equal to the deflected distance (Figure 5-11). The potential magnitude of the *P-delta* moment depends on the stiffness of the member and is not computed directly; however, the interaction equations for combined compression and bending include additional terms to compensate for this effect.

The exact analysis of a member with combined axial compression and bending can be a very time-consuming task and is most accurately determined by the secant formula. When timber members are considered, such an exacting analysis is generally not justified because of the material variability in modulus of elasticity and in strength properties and because of the degree of uncertainty in loading conditions. Rather than using a rigorous type of analysis, the NDS gives a simplified interaction formula for combined compression and bending that provides an accuracy well within an acceptable range for bridge applications. These equations are suitable for pin-end members of square or rectangular cross sections and are based on the following assumptions given in the NDS:

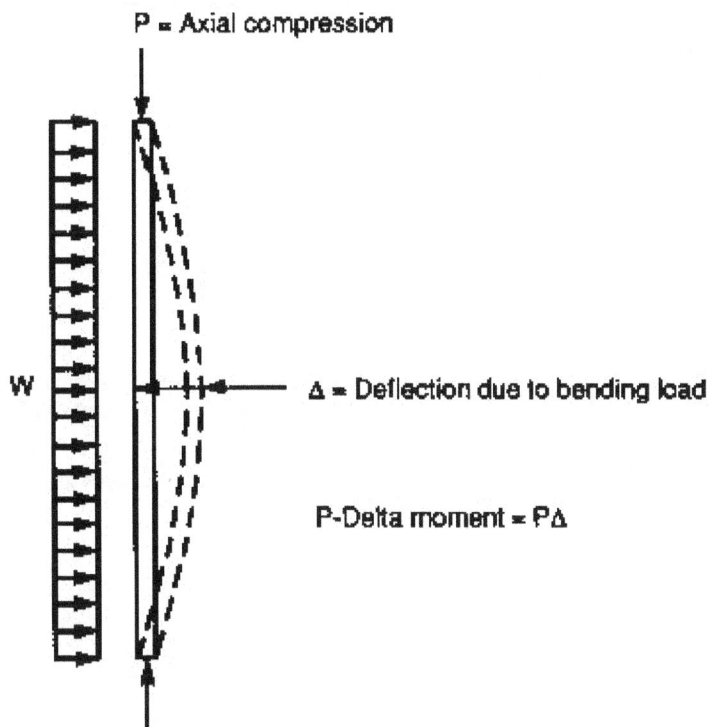

Figure 5-11.- P-delta effect on members loaded in combined axial compression and bending.

1. The stresses that cause a given deflection as a sinusoidal curve are the same as those for a beam with a uniform side load.

2. For a single concentrated side load, the stress under the load can be used, regardless of the position of the load with reference to the length of the column.

3. The stress to use with a system of side loads is the maximum stress from the system (some slight error on the side of overload will occur with large side loads near each end).

4. For columns with a slenderness ratio of 11 or less (short columns), the *P-delta* stress may be neglected.

The NDS interaction formula for combined compression and bending is given below by Equations 5-38 and 5-39. Appendix H of the NDS also gives eight modified forms of this equation for specified loading conditions that may be used at the option of the designer.

$$\frac{f_c}{F_c'} + \frac{f_b + f_c(6+1.5J)(e/d)}{F_b' - J(f_c)} \leq 1.0 \qquad (5\text{-}38)$$

and

$$= \frac{(\ell_e / d) - 11}{K - 11} \qquad 0 \leq J \leq 1.0 \qquad\qquad (5\text{-}39)$$

where f_c = applied stress in compression parallel to grain computed by Equation 5-24 (lb/in^2),

F_c' = allowable stress in compression parallel to grain by the applicable equations in Section 5.6 for the maximum slenderness ratio (ℓ_e / d), assuming the member is loaded in axial compression only (lb/in^2).

f_b = applied stress in bending from side loads or moments only, by Equation 5-2 (lb/in^2),

F_b' = allowable stress in bending computed by equations in Section 5.4, assuming the member is loaded in bending only (lb/in^2),

e = the eccentricity of an eccentrically applied axial load (in.),

d = the cross-sectional dimension of a rectangular or square column (in.),

J = a unitless convenience factor computed from the ℓ_e / d ratio in the plane of bending and limited to values between zero and 1.0, inclusively,

ℓ_e / d = for computing J, the column slenderness ratio of the member in the plane of bending, and

K = the smallest slenderness ratio ℓ_e / d at which the long column formula applies, from Equation 5-28.

The interaction Equations 5-38 and 5-39 are somewhat confusing at first glance, but become easier to use with experience. When applying the equations, five considerations will provide some clarification for various design applications. First, the compression terms f_c and F_c' are determined by the methods discussed in Section 5.6, in exactly the same manner as if the member was loaded in axial compression only.

Second, the term for bending stress f_b is applicable only when bending is from transverse loads or applied moments. When bending is from eccentric axial loads only, and no side loads or applied moments occur, f_b equals zero and Equation 5-38 becomes

$$\frac{f_c}{F_c'} + \frac{f_c(6 + 1.5J)(e/d)}{F_b' - J(f_c)} \leq 1.0 \qquad\qquad (5\text{-}40)$$

Third, the allowable bending stress F_b' is the tabulated bending stress adjusted by all applicable modification factors, assuming the member is loaded in bending only. In most applications, the more restrictive modification factor for size effect, C_F or lateral beam stability, C_L, applies; however, for combined compression and bending, both modification factors are applied cumulatively when the value of C_F is greater than 1.0. This will occur only when axial glulam combinations are less than 12 inches deep and are loaded in bending about the y-y axis (see *AITC 117-Design*). In all other cases, only the lowest value computed for C_F or C_L is applied as a modification factor to F_b.

Fourth, in the expression for eccentric loads e/d, d is the cross-sectional dimension of the member perpendicular to the axis about which bending is applied. When there are no eccentric axial loads, e/d equals zero and Equation 5-38 reduces to

$$\frac{f_c}{F_c'} + \frac{f_b}{F_b' - J(f_c)} \leq 1.0 \qquad (5\text{-}41)$$

Fifth, the J factor, whose value is limited between zero and 1.0, compensates for the effects of the P-delta moment. The column slenderness ratio used to determine J is always computed in the plane of bending. For column slenderness ratios of 11 or less (short columns), P-delta effects are ignored and the value J is zero. For ℓ_e/d values greater than K (long columns), the P-delta effects are greatest and J is at its maximum value of 1.0. When ℓ_e/d is greater than 11 but less than K, P-delta effects increase with the slenderness ratio and values of J vary linearly from zero to 1.0.

5.8 CONNECTIONS

A connection consists of two or more members joined with one or more mechanical fasteners. Connections are one of the most important considerations in timber bridge design because they provide continuity to the members as well as strength and stability to the system. The connections may consist entirely of wood members but frequently involve the connection of wood to steel or other materials. One advantage of wood as a structural material is the ease with which the members can be joined with a wide variety of fasteners. Progress in the past decade on fastener design and performance has led to reliable design criteria, allowing connections. to be designed with the same accuracy as other components of the structure.

This section discusses connection design for several types of fasteners commonly encountered in bridge construction. The types of connections and fasteners are discussed first, followed by basic design criteria and specific fastener requirements. The scope of coverage is limited to connections with two or three members, where fasteners are loaded perpendicular or parallel to their axis in the side grain of timber members. When fasteners are loaded at an angle to their axis, placed in wood end grain, or used in joints consisting of more than three members, refer to the NDS specifications for design criteria and requirements.

TYPES OF CONNECTIONS AND FASTENERS

There are two basic types of connections in timber bridges: lateral (shear) connections and withdrawal (tension) connections (Figure 5-12). In lateral connections, forces are transmitted by bearing stresses developed between the fastener and the members of the connection. A tight lateral connection also develops some strength by friction between members (at least when initially installed), but this effect is not considered in design. In withdrawal connections, the mechanism of load transfer depends on the type of fastener. For screw-type fasteners, load transfer is by a combination of friction and thread interaction between the fastener and the wood. For driven fasteners, such as nails, load transfer in withdrawal is entirely by friction developed between the fastener and the wood.

Selection of a fastener for a specific design application depends on the type of connection and the required strength capacity. Each connection must be designed to adequately transmit forces and provide good performance for the life of the structure without causing splitting, cracking, or deformation of the wood members. The five fastener types most commonly used for timber bridges are bolts, lag screws, timber connectors, nails or spikes, and drift bolts or pins (Figure 5-13). A brief description of each fastener is given below.

Bolts are the most common timber fastener for lateral connections where moderately high strength is required. They also are used in tension connections where loads are applied parallel to the bolt axis. Bolts used for bridge connections are standard machine bolts and should not be confused with machine screws, which have a much finer thread. Bolts are the only type of fastener that require nuts to maintain tightness of the connection.

Lag screws are pointed threaded fasteners with a square or hexagonal head that are placed in wood members by turning with a wrench. Although they provide a lower lateral strength than a comparable bolted connection, lag screws are advantageous when an excessive bolt length is required or when access to one side of a connection is restricted.

Timber connectors are steel rings or plates placed between members held by a bolt or lag screw. They are used in lateral connections only and provide the highest lateral strength of all fasteners because of the large bearing area provided by the connector.

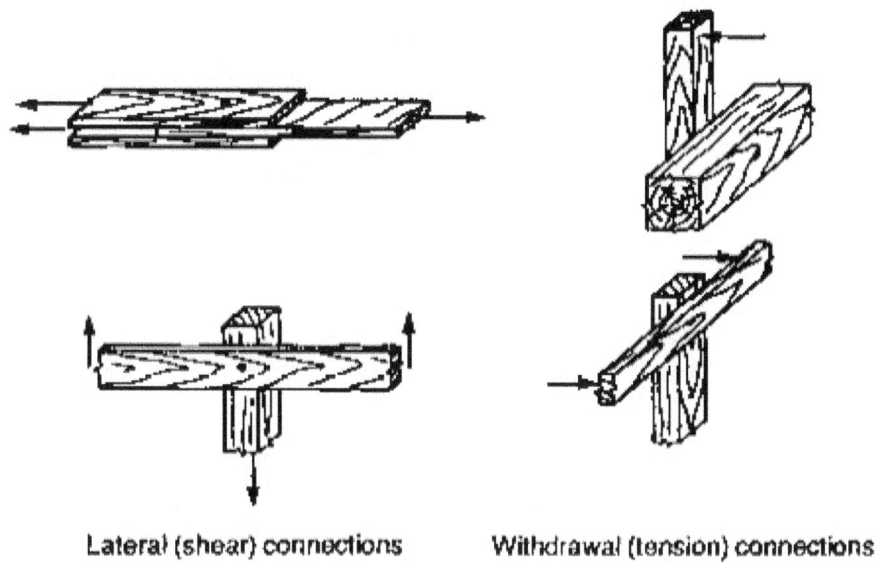

Lateral (shear) connections   Withdrawal (tension) connections

Figure 5-12.- Typical lateral and withdrawal connections for timber members.

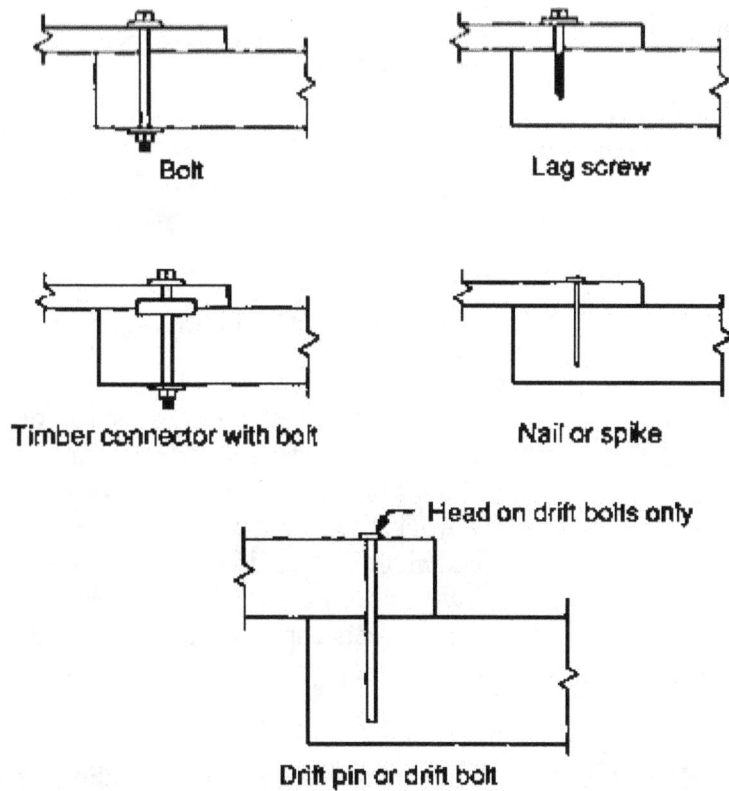

Bolt   Lag screw

Timber connector with bolt   Nail or spike

Head on drift bolts only

Drift pin or drift bolt

Figure 5-13. - Types of fasteners used for timber bridges.

Nails and spikes are driven fasteners used in bridges primarily for non-structural applications. They are more susceptible than other fasteners to loosening from vibrations and from dimensional changes in the wood caused by moisture content variations.

Drift bolts and drift pins are long unthreaded bolts or steel pins that are driven in prebored holes. Drift bolts have a head on one end, but drift pins have no head. In bridge applications, drift bolts and drift pins are used in lateral connections for large timber members. They are not suitable for withdrawal connections because of their low resistance to withdrawal loads.

When bolts or lag screws are used individually or with timber connectors, they must be provided with washers if the head or nut of the fastener is in wood contact. Washers distribute the load over a larger area to reduce stress and prevent wood crushing under the fastener head when the fastener is tightened. The three primary types of washers are cut washers, plate washers (round or square), and malleable iron washers (Figure 5-14). Cut washers are limited in application because they are thin and may bend from bearing forces. Malleable iron (MI) washers, intended only for timber connections, are most commonly used. Washers are not required when the head or nut of the fastener bears on a steel component; however, when steel components are used, they must be designed for adequate strength in accordance with AASHTO specifications for structural steel (AASHTO Section 10).

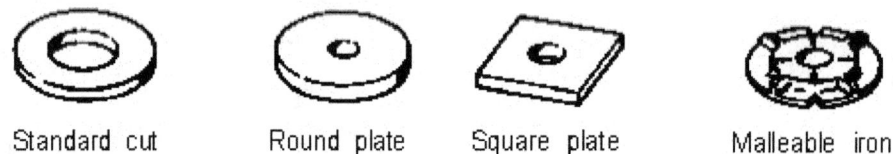

Standard cut   Round plate   Square plate   Malleable iron

Figure 5-14. - Common washer types for timber connections.

An important factor in connection performance and longevity is protection of the steel fasteners and hardware from corrosion. All steel components should be hot-dip galvanized in accordance with the applicable AASHTO specification M111 or M232. Such finishes as chrome and cadmium plating do not afford suitable protection for the exposure conditions encountered in bridges. When color is an important consideration, components can be painted after galvanizing or be coated with colored epoxy.

BASIC DESIGN CRITERIA

The strength of timber connections is usually controlled by the strength of the wood in bearing or withdrawal rather than by the strength of the fastener. As a result, connection design is affected by many of the same factors that influence the strength properties of wood. In addition to the type, number, and size of fasteners, connection strength depends on such factors as the wood species, direction and duration of load, and conditions of use.

Tabulated design values for different types of fasteners are given in the NDS. These values are based on one fastener, installed and used under specified conditions. Allowable design loads are determined by adjusting tabulated values with modification factors. When more than one fastener is used in a connection, the design value is the sum of the design values for the individual fasteners (for some fastener types, adjustments are required for multiple-fastener connections). It should be noted that the design criteria and tabulated values in the NDS are limited to connections involving the same type of fastener. Methods of analysis and test data for connections made with more than one type of fastener have not been developed.

The basic design procedures for connections are similar to those for structural components. For a given connection and fastener type, the designer must (1) compute fastener load requirements; (2) determine the tabulated value for one fastener based on the species group of the connected members; (3) apply modification factors to the tabulated value to reflect specific conditions; (4) adjust the modified value for lateral loading conditions other than parallel or perpendicular to grain, when applicable; (5) multiply the allowable design value for one fastener by the total number of fasteners in the connection; (6) compute the net section and verify the capacity of the members; and (7) detail the connection to ensure adequate fastener placement and performance.

Species Groups

The strength of timber connections is directly related to the species (density) of wood in which the fastener is installed. For lateral connections, wood species are divided into groups depending on the relative bearing capacity of the species for the specific fastener type. The three species groups consist of Groups 1 to 12 for bolts, Groups A to D for timber connectors, and Groups I to IV for lag screws, nails, spikes, drift bolts, and drift pins. There are no group designations for withdrawal connections, and design values are based on the specific gravity of the member. For both lateral and withdrawal connections, the species groups and specific gravities for sawn lumber (Table 5-14) and axial combinations of glulam (Table 5-15) apply to fasteners in the side grain at any location in the member. For bending combinations of glulam (Table 5-16), the species and grade of laminations vary for different locations in the member, and fastener groups and specific gravities are given separately for the tension face, side face, and compression face.

Modification Factors for Fasteners

Tabulated design values for fasteners are based on the strength of wood components assuming specific conditions of use. To adjust tabulated

Table 5-14.—Sawn lumber species groups for fastener design.

Species	Bolt load group	Timber connector load group[1]	Grouping for lag screws, nails, spikes, drift bolts and drift pins	
			Load group	Specific gravity[2]
Cedar, Northern White	12	D	IV	0.31
Cedars, Western[3]	9	D	IV	0.35
Coast Species	12	D	IV	0.39
Douglas Fir-Larch[3]	3	B	II	0.51
Douglas Fir-Larch (dense)	1	A	II	0.51[5]
Douglas Fir, South	6	C	III	0.48
Eastern Woods	12	D	IV	0.38
Fir, Balsam	11	D	IV	0.38
Hem-Fir[3]	8	C	III	0.42
Hemlock:				
Eastern-Tamarack[3]	8	C	III	0.45
Mountain	9	C	III	0.47
Western	8	C	III	0.48
Pine:				
Eastern White[3]	11	D	IV	0.38
Idaho White	11	D	IV	0.40
Lodgepole	10	C	III	0.44
Northern	9	C	III	0.46
Ponderosa[4]	11	C	III	0.49
Ponderosa-Sugar	11	C	III	0.42
Red[4]	11	C	III	0.42
Southern	3	B	II	0.55
Southern (dense)	1	A	II	0.55[5]
Western White	11	D	IV	0.40
Spruce:				
Eastern	10	C	III	0.43
Engelmann-Alpine Fir	12	D	IV	0.36
Sitka	10	C	III	0.43
Sitka, Coast	10	D	IV	0.39
Spruce-Pine-Fir	10	C	III	0.42
West Coast Woods (mixed species)	12	D	IV	0.35
White Woods (Western Woods)	12	D	IV	0.35

[1] When stress graded.

[2] Based on weight and volume when ovendry.

[3] Also applies when species name includes the designation "North."

[4] Applies when graded to NLGA rules.

[5] The specific gravity of dense lumber is slightly higher than for medium-grain lumber; however, the design values for this group are based on the average specific gravity of the species.

Load groups and specific gravities apply to all grades of that species unless otherwise noted.

This table contains a limited number of species. Refer to the NDS for a complete species listing.

From the NDS,[26] © 1986. Used by permission.

Table 5-15. - Glulam axial combination species groups for fastener design.

Combination symbol	Bolt group	Timber connector group	Lag screws, nails, spikes, drift bolts, and drift pins	
			Group	Specific gravity
Visually graded western species:				
1	3	B	II	0.51
2	3	B	II	0.51
3	1	A	II	0.51
4	3	B	II	0.51
5	1	A	II	0.51
Visually graded Southern Pine:				
46	3	B	II	0.55
47	3	B	II	0.55
48	1	A	II	0.55
49	3	B	II	0.55
50	1	A	II	0.55

Applicable to fasteners placed in any face of the member.

This table represents a partial listing of selected combination symbols. Refer to *AITC 117—Design*[4] and the *AITC Timber Construction Manual*[6] for a complete listing of all combination symbols. From *AITC 117—Design*.[4] © 1987. Used by permission.

values for actual design requirements, modification factors are applied to tabulated values in the same manner as those for strength properties. The modification factors for fasteners consist of the following:

C_M moisture content factor

C_D duration of load factor

C_t temperature factor

C_R fire-retardant treatment factor

C_e edge-distance factor

C_n end-distance factor

C_s spacing factor

C_g group action factor

C_s steel side-plate factor

C_{lo} lag-screw factor

A summary of fastener modification factors and their applicability to various fasteners are shown in Table 5-17.

Moisture Content Factor (C_M)

The moisture content of timber components affects joint strength in approximately the same manner as it affects other strength properties. For sawn lumber, moisture content must be considered at the time of fabrication (when the fastener is installed) and in service. For glulam, all laminations are dry when fabricated, and moisture effects are considered for in-use conditions only. Tabulated fastener values are based on fasteners that

Table 5-16.—Glulam bending combination species groups for fastener design.

Combination symbol	Tension face				Side face				Compression face			
	Bolt group	Timber conn. group	Lag screws, nails, spikes, drift bolts, and drift pins Group	Specific gravity	Bolt group	Timber conn. group	Lag screws, nails, spikes, drift bolts, and drift pins Group	Specific gravity	Bolt group	Timber conn. group	Lag screws, nails, spikes, drift bolts, and drift pins Group	Specific gravity
Visually graded western species												
16F-V3	3	B	II	0.51	3	B	II	0.51	3	B	II	0.51
16F-V6	3	B	II	0.51	3	B	II	0.51	3	B	II	0.51
20F-V3	1	A	II	0.51	3	B	II	0.51	3	B	II	0.51
20F-V7	1	A	II	0.51	3	B	II	0.51	1	A	II	0.51
24F-V4	1	A	II	0.51	3	B	II	0.51	1	A	II	0.51
24F-V8	1	A	II	0.51	3	B	II	0.51	1	A	II	0.51
Visually graded Southern Pine												
16F-V2	3	B	II	0.55	3	B	II	0.55	3	B	II	0.55
16F-V5	3	B	II	0.55	3	B	II	0.55	3	B	II	0.55
20F-V3	3	B	II	0.55	3	B	II	0.55	3	B	II	0.55
20F-V5	1	A	II	0.55	3	B	II	0.55	1	A	II	0.55
24F-V3	1	A	II	0.55	3	B	II	0.55	1	A	II	0.55
24F-V5	1	A	II	0.55	3	B	II	0.55	1	A	II	0.55

This table represents a partial listing of selected combination symbols. Refer to AITC 117—Design[4] and the AITC Timber Construction Manual[5] for a complete listing of all combination symbols.

Table 5-17.—Applicability of modification factors for fasteners.

Fastener type	Modification factor									
	Duration of load C_D	Moisture content C_M	Temperature C_t	Fire retardant C_R	Edge dist. C_e	End dist. C_n	Spacing C_s	Group action C_g	Steel side plate C_{st}	Lag screw C_{ss}
Timber connectors										
Split rings	X	X	X	X	X	X	X	X	—	X
Shear plates	X	X	X	X	X	X	X	X	X	X
Bolts	X	X	X	X	—	X	X	X	X	—
Lag screws	X	X	X	X	—	X	X	X	—	—
Nails and spikes	X	X	X	X	—	—	—	—	X	—
Drift bolts and drift pins	X	X	X	X	—	X	X	X	X	—

X = modification factor is applicable.

— = modification factor does not apply.

C_e, C_n, and C_s are not cumulative, and the most restrictive of the three values is used for design.

are installed and used in continuously dry conditions that do not exceed 19-percent moisture content for sawn lumber and 16-percent moisture content for glulam. For other conditions, tabulated values must be adjusted by C_M (Table 5-18). Note that C_M values for fasteners may vary from those used for other strength properties.

Duration of Load Factor (C_D)

Tabulated fastener values are for conditions where maximum loads are of normal duration. For other loading conditions, values are adjusted by the duration of load factor, C_D discussed in Section 5.3. The duration of load factor applies to wood members only and is not used for metal components. As a result, load increases due to application of C_D may be limited when the capacity of the connection is controlled by the strength of the steel connector rather than by the strength of the wood. This is discussed further in the following sections on fastener design.

Temperature Factor (C_t)

The strength of a wood connection is affected by temperature in the same manner as wood components. In unusual cases where connections will be subjected to prolonged temperatures in excess of 150 °F, fastener values should be adjusted by the temperature factor, C_t. Values and criteria for this factor are the same as those given in Section 5.3.

Fire-Retardant Treatment Factor (C_R)

Fire-retardant treatments are not common in bridge applications. However, when timber components are treated with fire-retardant chemicals, tabulated fastener values must be reduced by the fire-retardant treatment factor, C_R discussed in Section 5.3.

Table 5-18 Fastener load modification factors for moisture content, C_M.

| Type of fastener | Condition of wood[1] | | | | | |
| | Sawn lumber | | | Glulam | |
	At time of fabrication	In service	C_M	In service	C_M
Timber connectors[2]	Dry	Dry	1.0	Dry	1.0
	Partially seasoned[3]	Dry	Note 3	Wet	0.67
	Wet	Dry	0.8		
	Dry or wet	Partially seasoned or wet	0.67		
Bolts or lag screws	Dry	Dry	1.0	Dry	1.0
	Partially seasoned or wet[3]	Dry	See below	Wet	0.67
	Dry or wet	Exposed to weather	0.75		
	Dry or wet	Wet	0.67		
Drift bolts or pins - laterally loaded	Dry or wet	Dry	1.0	Dry	1.0
	Dry or wet	Partially seasoned, wet or subject to wetting and drying	0.70	Wet	0.7
Wire nails and spikes — Withdrawal loads	Dry	Dry	1.0	Dry	1.0
	Partially seasoned or wet	Will remain wet	1.0	Wet	0.25
	Partially seasoned or wet	Dry	0.25		
	Dry	Subject to wetting and drying	0.25		
— Lateral loads	Dry	Dry	1.0	Dry	1.0
	Partially seasoned or wet	Dry or wet	0.75	Wet	10.75
	Dry	Partially seasoned or wet	0.75		
Threaded, hardened steel nails	Dry or wet	Dry or wet	1.0	Dry	1.0
				Wet	1.0

[1] Condition of wood definitions applicable to fasteners are as follows:

"Dry" wood has a moisture content of 19% or less for sawn lumber and 16% or less for glulam.

"Wet" wood has a moisture content at or above the fiber saturation point (approximately 30%) for sawn lumber and above 16% for glulam.

"Partially seasoned" wood has a moisture content greater than 19%, but less than fiber saturation point.

"Exposed to weather" implies that the wood may vary in moisture content from dry to partially seasoned, but is not expected to reach the fiber saturation point at times when the joint is under full design load.

"Subject to wetting and drying" implies that the wood may vary in moisture content from dry to partially seasoned or wet, or vise versa, with consequent effects on the tightness of the joint.

[2] For timber connectors, moisture content limitations apply to a depth of 3/4 inch from the surface of the wood.

[3] When timber connectors, bolts, or laterally loaded lag screws are installed in wood that is partially seasoned at the time of fabrication, but that will be dry before full design load is applied, intermediate values may be used.

From NDS;[26] © 1986. Used by permission.

Moisture modification factors C_M for laterally loaded bolts and lag screws in sawn lumber seasoned in place[1]		
Arrangement of bolts or lag screws	Type of splice plate	C_M
One fastener only, or Two or more fasteners placed in a single line parallel to grain, or Fasteners placed in two or more lines parallel to grain with separate splice plates for each line.	Wood or metal	1.0
All other arrangements	Wood or metal	0.4

[1] Factors apply when wood is at or above the fiber saturation point (wet) at time of fabrication but dries to a moisture content of 19% or less (dry) before full design load is applied. For wood partially seasoned when fabricated, adjusted intermediate values may be used.

Edge-Distance Factor (C_e)

Edge distance is the distance from the center of a fastener to the edge of the member, measured perpendicular to grain (Figure 5-15). For loads applied perpendicular to the grain, the loaded edge is the edge toward which the load induced by the fastener acts. The unloaded edge is the opposite edge. Tabulated design values for bolts, lag screws, timber connectors, drift bolts, and drift pins are based on the full edge-distance requirements specified for the fastener. For timber connectors, it is permissible to reduce the edge distance provided the tabulated value for the connector is reduced by C_e (design tables for timber connectors include tabulated values reduced by C_e). The edge-distance factor is not cumulative with the end-distance factor (C_n) or the spacing factor (C_s). Of the three factors, the most restrictive value is used for design.

End-Distance Factor (C_n)

End distance is the distance from the center of a fastener to the end of the member (Figure 5-15). Tabulated values for bolts, lag screws, timber connectors, drift bolts, and drift pins are based on the full end-distance requirements for the fastener. Reduced end distances are permitted if the tabulated fastener value is reduced by C_n. End distance requirements and values of C_n for individual fasteners are discussed later in this section. The end-distance factor is not cumulative with edge-distance factor (C_e) or the spacing factor (C_s). Of the three factors, the most restrictive value is used for design.

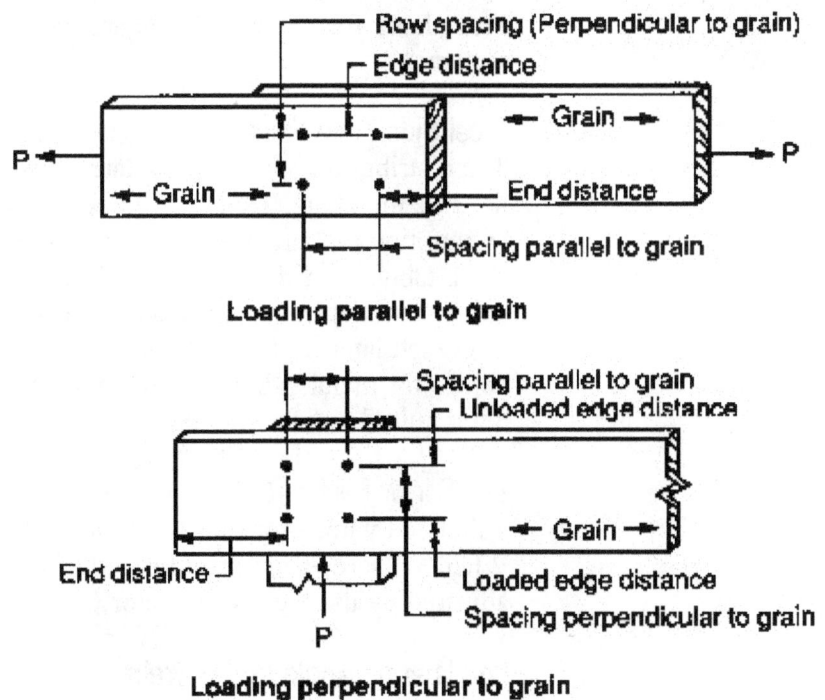

Figure 5-15.— Edge distance, end distance, and spacing for fasteners.

Spacing Factor (C_s)

Fastener spacing is the center-to-center distance between fasteners, measured parallel or perpendicular to grain (Figure 5-15). Tabulated design values for bolts, lag screws, timber connectors, drift bolts, and drift pins are based on minimum spacing requirements between fasteners. When spacings are less than the minimum, tabulated fastener values must be reduced by the spacing factor, C_s. Spacing requirements and values of C_s depend on the type of fastener and are discussed later in this section. The spacing factor is not cumulative with the edge-distance factor (C_e) or the end-distance factor. Of the three factors, the most restrictive value is used for design.

Group Action Factor (C_g)

A row of fasteners consists of two or more bolts, lag screws, timber connectors, drift bolts, or drift pins aligned in the direction of the applied load. When three or more of these fasteners are used in a row, the capacity of the connection is less than that computed by multiplying the value of an individual fastener by the total number of fasteners. To compensate for this effect, tabulated values for individual fasteners in the row are reduced by the group action factor, C_g. Values of C_g are given in Table 5-19 and are based on the gross areas of the members and the total number of fasteners in the row. It should be noted that the group action factor given in the NDS is applied to the group of fasteners acting as a whole. However, it also may be applied to individual fasteners, as presented here. Applying the factor to individual fasteners is more convenient for design and is more consistent with the application of other modification factors. Procedures for determining C_g are demonstrated in examples later in this section.

Steel Side-Plate Factor (C_{st})

The distribution of stress in a lateral connection depends on the material of the side members. Tabulated fastener values in the NDS are based on the assumption that all side members are wood. When steel side members are used, tabulated values for some fasteners may be increased by C_{st}. The value of C_{st} depends on the type of fastener and direction of loading and is discussed later in this section. For lag screws, a separate table of design values for metal side plates is given in the NDS, and adjustment by C_{st} is not required.

Lag-Screw Factor (C_{lb})

Tabulated values for timber connectors are based on a bolted connection. When lag screws are used instead of bolts, tabulated values must be adjusted by the lag-screw factor C_{lb}.

Loads at an Angle to the Grain

The strength of a laterally loaded wood connection for all fasteners other than nails and spikes depends on the direction of fastener bearing in relation to the grain of the members. Design values in the NDS are

Table 5-19. —Group action modification factor C_g for laterally loaded bolts, lag screws and timber connectors

Connections With Wood Side Plates

A_1/A_2 [2,3]	A_1 (in²) [4]	Number of fasteners in a row [1]										
		2	3	4	5	6	7	8	9	10	11	12
0.5	<12	1.00	0.92	0.84	0.76	0.68	0.61	0.55	0.49	0.43	0.38	0.34
	12 – <19	1.00	0.95	0.88	0.82	0.75	0.68	0.62	0.57	0.52	0.48	0.43
	19 – <28	1.00	0.97	0.93	0.88	0.82	0.77	0.71	0.67	0.63	0.59	0.55
	28 – <40	1.00	0.98	0.96	0.92	0.87	0.83	0.79	0.75	0.71	0.69	0.66
	40 – <64	1.00	1.00	0.97	0.94	0.90	0.86	0.83	0.79	0.76	0.74	0.72
	>64	1.00	1.00	0.98	0.95	0.91	0.88	0.85	0.82	0.80	0.78	0.76
1.0	<12	1.00	0.97	0.92	0.85	0.78	0.71	0.65	0.59	0.54	0.49	0.44
	12 – <19	1.00	0.98	0.94	0.89	0.84	0.78	0.72	0.66	0.61	0.56	0.51
	19 – <28	1.00	1.00	0.97	0.93	0.89	0.85	0.80	0.76	0.72	0.68	0.64
	28 – <40	1.00	1.00	0.99	0.96	0.92	0.89	0.86	0.83	0.80	0.78	0.75
	40 – <64	1.00	1.00	1.00	0.97	0.94	0.91	0.88	0.85	0.84	0.82	0.80
	>64	1.00	1.00	1.00	0.99	0.96	0.93	0.91	0.88	0.87	0.86	0.85

A_1 = gross cross-sectional area of the main member before boring or grooving.[5]
A_2 = sum of the cross-sectional areas of the side members before boring or grooving.[5]

Connections With Metal Side Plates

A_1/A_2	A_1 (in²)	Number of fasteners in a row [1]										
		2	3	4	5	6	7	8	9	10	11	12
2–12	5 – <8	1.00	0.78	0.64	0.54	0.46	0.40	0.35	0.30	0.25	0.20	0.15
	8 – <16	1.00	0.85	0.73	0.63	0.54	0.48	0.42	0.38	0.34	0.30	0.26
	16 – <24	1.00	0.91	0.83	0.74	0.66	0.59	0.53	0.48	0.43	0.38	0.33
	24 – <39	1.00	0.94	0.87	0.80	0.73	0.67	0.61	0.56	0.51	0.46	0.42
	39 – <64	1.00	0.96	0.92	0.87	0.81	0.75	0.70	0.66	0.62	0.58	0.55
	64 – <119	1.00	0.98	0.95	0.91	0.87	0.82	0.78	0.75	0.72	0.69	0.66
	119 – <199	1.00	0.99	0.97	0.95	0.92	0.89	0.86	0.84	0.81	0.79	0.78
12–18	17 – <24	1.00	0.94	0.88	0.81	0.74	0.67	0.61	0.55	0.49	0.43	0.37
	24 – <39	1.00	0.96	0.91	0.86	0.80	0.74	0.68	0.62	0.56	0.50	0.44
	39 – <64	1.00	0.98	0.94	0.90	0.85	0.80	0.75	0.70	0.67	0.62	0.58
	64 – <119	1.00	0.99	0.96	0.93	0.90	0.86	0.82	0.79	0.75	0.72	0.69
	119 – <199	1.00	1.00	0.98	0.96	0.94	0.92	0.89	0.86	0.83	0.80	0.78
	>199	1.00	1.00	1.00	0.98	0.97	0.95	0.93	0.91	0.90	0.88	0.87
18–24	40 – <64	1.00	1.00	0.96	0.93	0.89	0.84	0.79	0.74	0.69	0.64	0.59
	64 – <119	1.00	1.00	0.97	0.94	0.92	0.89	0.86	0.83	0.80	0.76	0.73
	119 – <199	1.00	1.00	0.99	0.98	0.96	0.94	0.92	0.90	0.88	0.86	0.85
	>199	1.00	1.00	1.00	1.00	0.98	0.96	0.95	0.93	0.92	0.92	0.91
24–30	40 – <64	1.00	0.98	0.94	0.90	0.85	0.80	0.74	0.69	0.65	0.61	0.58
	64 – <119	1.00	0.99	0.97	0.93	0.90	0.86	0.82	0.79	0.76	0.73	0.71
	119 – <199	1.00	1.00	0.98	0.96	0.94	0.92	0.89	0.87	0.85	0.83	0.81
	>199	1.00	1.00	0.99	0.98	0.97	0.95	0.93	0.92	0.90	0.89	0.89
30–35	40 – <64	1.00	0.96	0.92	0.86	0.80	0.74	0.68	0.64	0.60	0.57	0.55
	64 – <119	1.00	0.98	0.95	0.90	0.86	0.81	0.76	0.72	0.68	0.65	0.62
	119 – <199	1.00	0.99	0.97	0.95	0.92	0.88	0.85	0.82	0.80	0.78	0.77
	>199	1.00	1.00	0.98	0.97	0.95	0.93	0.90	0.89	0.87	0.86	0.85
35–42	40 – <64	1.00	0.95	0.89	0.82	0.75	0.69	0.63	0.58	0.53	0.49	0.46
	64 – <119	1.00	0.97	0.93	0.88	0.82	0.77	0.71	0.67	0.63	0.59	0.56
	119 – <199	1.00	0.98	0.96	0.93	0.89	0.85	0.81	0.78	0.76	0.73	0.71
	>199	1.00	0.99	0.98	0.96	0.93	0.90	0.87	0.84	0.82	0.80	0.78

A_1 = gross cross-sectional area of the main member before boring or grooving.[5]
A_2 = sum of the cross-sectional areas of metal side plates before drilling.

[1] When fasteners in adjacent rows are staggered, refer to the NDS[26] for requirements for determining the number of fasteners in a row.

[2] When A_1/A_2 > 1.0, use A_2/A_1.

[3] For values of A_1/A_2 between 0 and 1.0, interpolate or extrapolate from tabulated values.

[4] When A_1/A_2 > 1.0, use A_2 instead of A_1.

[5] When a wood member is loaded perpendicular to grain, its equivalent gross cross-sectional area for computing A_1 and A_2 is the product of the thickness of the member and the overall width of the fastener group.

From NDS.™ © 1986. Used by permission.

tabulated for loads acting parallel to grain *(P)* and perpendicular to grain *(Q)*. When the loads act at some intermediate angle (Figure 5-16), design values are computed using the Hankinson formula given by

$$N' = \frac{P'Q'}{P'\sin^2\theta + Q'\cos^2\theta} \qquad (5\text{-}42)$$

where

N' = allowable design value at an angle to the grain (lb),

P' = allowable value for the fastener parallel to grain (lb),

Q' = allowable value for the fastener perpendicular to grain (lb), and

θ = angle between the direction of load and the direction of grain (degrees).

For bolts, lag screws, drift bolts, and drift pins, the Hankinson formula is applied after tabulated values are adjusted by modification factors, and the value N' is the allowable fastener value used for design. For timber connectors, modification factors for distance and spacing are based on the angle of the load to the grain, and C, C_{ν} and C, are applied after N is computed by the Hankinson formula.

Member Capacity

The strength of a timber connection depends not only on the strength of the fasteners but also on the structural capacity of the connected members. As a part of the design process, the capacity of all members must be checked to ensure that factors related to the fasteners and connection

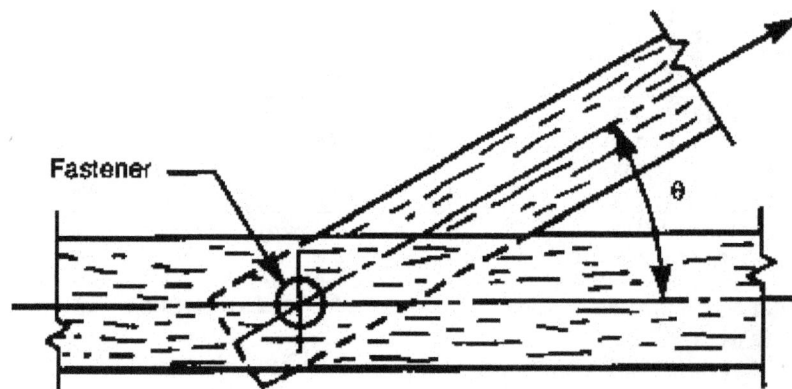

Figure 5-16. - Fastener loading applied at an angle to the grain.

configuration have not reduced the load-carrying capacity of the members. Connection-related factors that may affect member capacity include net area, eccentric loading, and shear capacity.

Net Area

The net area of a member is the cross-sectional area remaining after subtracting the area of material removed for fastener placement. The cross section where the net area is taken is called the critical section. Depending on the type of loading and size and placement of fasteners, the reduction in area for fasteners can significantly reduce member capacity. Requirements for determining net area vary among fasteners and are discussed in more detail later in this section.

Eccentricity

Eccentric loading is produced at connections when the resultant member forces are offset at the connection (Figure 5-17). Eccentricity in connections induces tension perpendicular to grain and can severely reduce the capacity of the members. The strength of eccentric connections is difficult to evaluate, and connections of this type must be avoided unless tests are employed in design to ensure that members can safely carry applied loads.

Shear Capacity

When fastener loads are applied transverse to beams or other components, the capacity of the member in horizontal shear may be reduced. Although not common in most bridge applications, this can occur when beams are supported entirely by fasteners, without bearing on another member, or when fastener loads are applied transverse to the beam (Figure 5-18). When conditions such as these are encountered, refer to the NDS for special provisions on computing horizontal shear in the member.

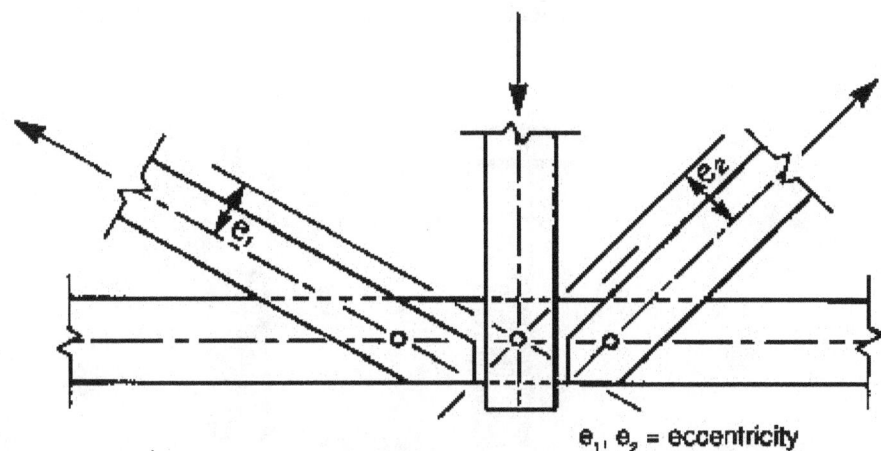

e_1, e_2 = eccentricity

Figure 5-17. - Example of an eccentrically loaded connection. Connections of this type can induce tension perpendicular to the grain and substantially reduce the capacity of connected members.

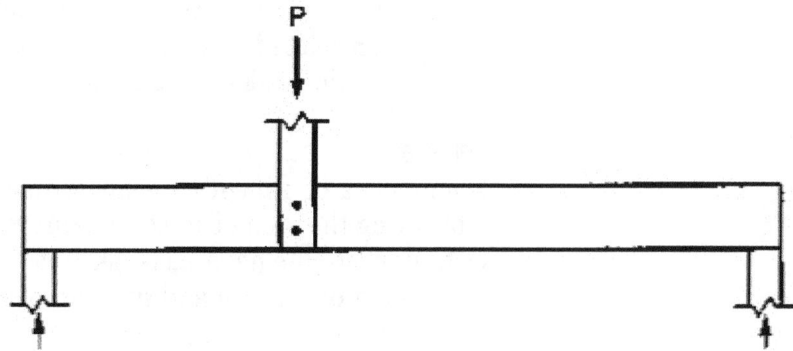

Figure 5-18. - Unsupported fastener loads applied transverse to the beam axis can reduce beam capacity in horizontal shear. Refer to the NDS ² for design requirements for this loading condition.

BOLTS

Bolts are the most common mechanical fasteners in timber bridge connections. They are used for lateral connections in double shear (three members) or single shear (two members) and in tension connections where the bolt is loaded parallel to its axis. The bolts most commonly used for timber connections conform to *ASTM Standard A307, Low-Carbon Steel Externally and Internally Threaded Standard Fasteners*. The allowable design stresses for these bolts are 20,000 lb/in² in tension and 10,000 lb/in² in shear. Bolts are generally available in diameters of 1/4 inch to 2 inches and lengths up to 24 inches or more in 1/2-inch increments. However, the designer should verify availability before specifying diameters over 1-1/4 inches or lengths over 16 inches. When long lengths are required, threaded rods conforming to ASTM A307 may be more practical.

Bolts are manufactured in a variety of types based on the configuration of the bolt head. The most common types are the hexagonal head, square head, dome head, and flat head (Figure 5-19). The standard hexagonal or square heads are used when the bolt head is in contact with wood or steel. More specialized bolts, such as the dome head and flat head, provide an

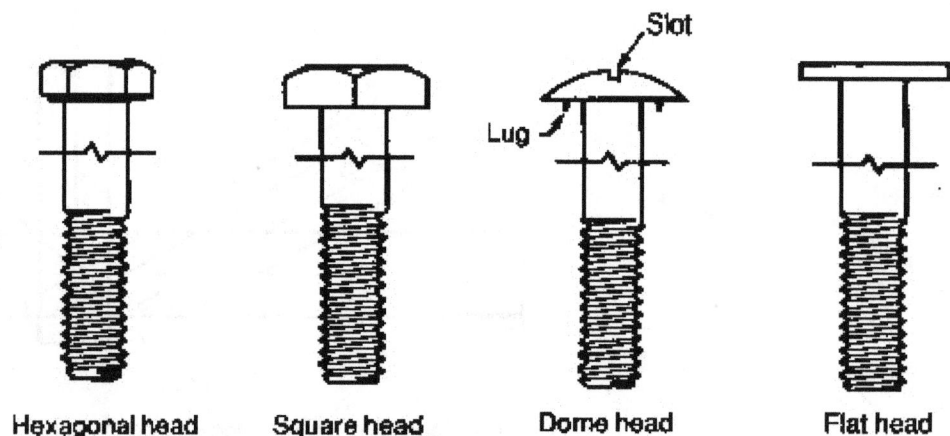

Hexagonal head **Square head** **Dome head** **Flat head**

Figure 5-19.- Bolt types used for timber connections.

increased head diameter and are used when the bolt head is in contact with wood. Bolts with dome heads also are referred to as economy bolts or mushroom bolts and may be slotted or provided with lugs to facilitate installation and tightening.

Net Area

The net area at a bolted connection is equal to the gross area of the timber member minus the projected area of the bolt holes at the section (bolt holes are typically 1/32 to 1/16 inch larger than the bolt diameter). For parallel-to-grain loading with staggered bolts, the nearest bolt in the adjacent row is considered to occur at the same critical section unless the parallel-to-grain spacing of bolts in each row is a minimum of eight times the bolt diameter (Figure 5-20). The required net area in tension and compression members is determined by dividing the total load transferred through the critical section by the applicable allowable stress $(F_t'$ or $F_c')$ for the species and grade of material used.

Design of Lateral Connections

The strength of laterally loaded, bolted connections is developed by bearing between the bolt and the wood (Figure 5-21). The capacity of the

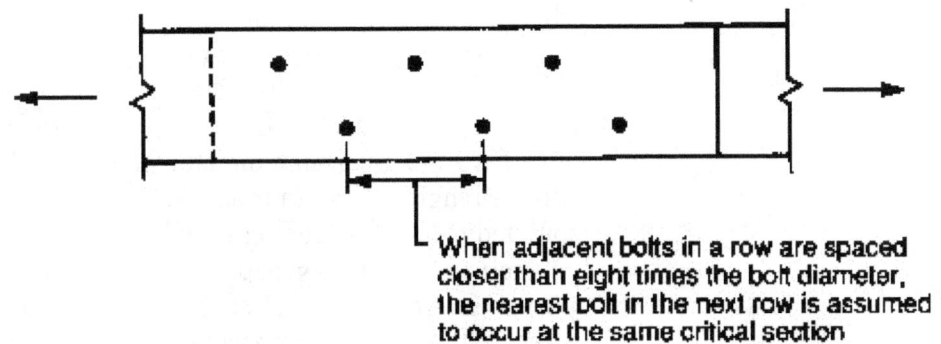

When adjacent bolts in a row are spaced closer than eight times the bolt diameter, the nearest bolt in the next row is assumed to occur at the same critical section

Figure 5-20. - Critical section for determining net area for staggered bolts loaded parallel to grain.

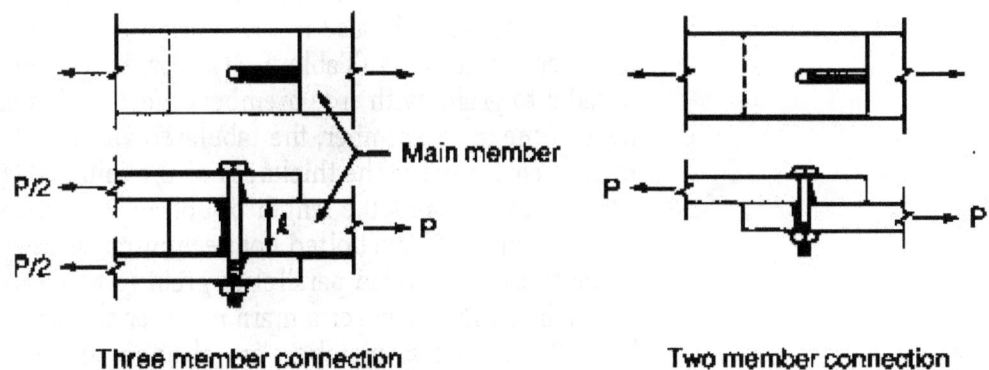

Three member connection

Two member connection

Figure 5-21. - Typical configuration and stress distribution for a laterally loaded bolted connection.

connection depends on the bearing strength of the wood and the slenderness ratio of the bolt. The slenderness ratio is defined as the length of the bolt in the main member (ℓ) divided by the bolt diameter (D). For bolted connections with low slenderness ratios, the bolt is relatively stiff, and the full bearing strength of the connection is developed. As the slenderness ratio increases, bolt stiffness is reduced, and bending may occur before full bearing strength is achieved, reducing the capacity of the connection.

The allowable value for one bolt is equal to the tabulated design value adjusted by all applicable modification factors and loading at an angle to the grain, when required. When more than one bolt is used, the allowable connection value is the sum of the design values of the individual bolts, adjusted by the group action factor, C_g. The applicable modification factors for loading parallel to grain (P) and perpendicular to grain (Q) are given for laterally loaded bolts by

$$P' = P C_D C_M C_t C_g C_n C_\Delta C_g C_{st} \tag{5-43}$$

$$Q' = Q C_D C_M C_t C_R C_n C_s C_g \tag{5-44}$$

Tabulated Design Values

Tabulated bolt design values are given in the NDS for one A307 bolt in a wood-to-wood, three-member connection where the side members are each a minimum of one-half the thickness of the main member (double shear). A portion of the NDS tables for several species groups is shown in Table 5-20. To determine the tabulated value for one bolt, enter the table with the length of bolt in main member and bolt diameter and read the tabulated values for loading parallel to grain and perpendicular to grain for the applicable species group. When joints have side pieces that are of a species different from that of the main member, the design value is the lesser of that obtained by assuming a comparable joint with all members the same species as the main member, or all members the same species as the side members.

Although tabulated values in the NDS are for a balanced three-member connection, the table also is used for other member thicknesses and two-member connections (Table 5-21). For three-member connections loaded parallel to grain, with side members that are less than one-half the thickness of the main member, the tabulated value is determined by assuming a main member twice the thickness of the thinnest side member. When steel side plates are used, the length of bolt is based on the thickness of the wood member. For a bolted connection consisting of two members of equal thickness loaded parallel to grain (single shear), the tabulated value is one-half that given for a main member the same thickness as the members. When the two members are of unequal thickness, the tabulated value is the lesser of one-half the tabulated value for the thicker member, or one-half the tabulated value for a piece twice the thickness of the thinner member. For a two-member connection consisting of one wood member

Table 5-20.—Tabulated design values for laterally loaded bolts.

				Species Group 1 DOUGLAS FIR-LARCH (Dense), SOUTHERN PINE (Dense)		Species Group 3 CALIFORNIA REDWOOD (Close grain), DOUGLAS FIR-LARCH, SOUTHERN PINE, SOUTHERN CYPRESS		Species Group 8 EASTERN HEMLOCK, TAMARACK, CALIFORNIA REDWOOD (Open grain), HEM-FIR, WESTERN HEMLOCK		Species Group 9 MOUNTAIN HEMLOCK, WESTERN CEDARS, NORTHERN PINE		Species Group 10 SPRUCE-PINE-FIR, SITKA SPRUCE, YELLOW POPLAR, LODGEPOLE PINE		Species Group 11 RED PINE, WESTERN WHITE PINE, PONDEROSA PINE-SUGAR PINE, EAST. WHITE PINE, EAST. SPRUCE, BALSAM FIR, IDAHO WHITE PINE	
Length of bolt in main member ℓ	Diameter of bolt D	ℓ/D	Projected area of bolt A=ℓ×D	Parallel to grain P	Perpendicular to grain Q	Parallel to grain P	Perpendicular to grain Q	Parallel to grain P	Perpendicular to grain Q	Parallel to grain P	Perpendicular to grain Q	Parallel to grain P	Perpendicular to grain Q	Parallel to grain P	Perpendicular to grain Q
2-1/2	1/2	5.00	1.250	1480	830	1260	720	1180	460	1100	500	1080	470	1010	340
	5/8	4.00	1.563	2140	950	1820	810	1620	520	1510	560	1410	530	1310	390
	3/4	3.33	1.875	2700	1050	2310	900	1990	580	1860	620	1700	590	1580	430
	7/8	2.66	2.188	3210	1160	2740	990	2330	630	2180	690	1980	650	1840	480
	1	2.50	2.500	3680	1270	3150	1080	2670	690	2490	750	2260	700	2110	520
3	1/2	6.00	1.500	1490	970	1270	860	1210	550	1130	590	1160	560	1080	410
	5/8	4.80	1.875	2290	1130	1960	970	1810	620	1690	670	1640	630	1520	470
	3/4	4.00	2.250	3080	1270	2630	1080	2340	690	2180	750	2030	700	1890	520
	7/8	3.43	2.625	3760	1390	3220	1190	2780	760	2600	820	2380	780	2210	570
	1	3.00	3.000	4390	1520	3750	1300	3200	830	2990	900	2710	850	2530	630
3-1/2	1/2	7.00	1.750	1490	1120	1270	980	1210	640	1130	690	1160	650	1080	480
	5/8	5.60	2.188	2320	1310	1980	1130	1890	720	1760	780	1790	740	1660	550
	3/4	4.67	2.625	3280	1470	2800	1260	2570	810	2400	870	2310	820	2150	610
	7/8	4.00	3.063	4190	1630	3580	1390	3180	890	2970	960	2760	900	2570	670
	1	3.50	3.500	5000	1770	4270	1520	3710	970	3460	1050	3170	990	2950	730
4	1/2	8.00	2.000	1490	1010	1270	1010	1210	700	1130	760	1160	720	1080	550
	5/8	6.40	2.500	2330	1410	1990	1290	1900	830	1770	900	1820	840	1690	620
	3/4	5.33	3.000	3340	1690	2850	1440	2690	920	2510	1000	2520	940	2350	690
	7/8	4.57	3.500	4440	1850	3790	1590	3470	1010	3240	1100	3090	1030	2880	770
	1	4.00	4.000	5470	2030	4670	1730	4150	1110	3880	1200	3600	1130	3360	840
4-1/2	5/8	7.20	2.813	2330	1440	1990	1400	1900	930	1770	1010	1820	950	1690	700
	3/4	6.00	3.375	3350	1830	2860	1620	2730	1040	2550	1120	2610	1060	2440	780
	7/8	5.14	3.938	4540	2110	3880	1790	3630	1140	3390	1240	3360	1160	3130	860
	1	4.50	4.500	5770	2280	4930	1950	4500	1250	4200	1350	3990	1270	3710	940
	1-1/4	3.60	5.625	7970	2670	6810	2280	5930	1460	5540	1580	5080	1490	4740	1100
5-1/2	5/8	8.80	3.438	2330	1390	1990	1410	1900	1010	1770	1090	1820	1030	1690	820
	3/4	7.33	4.125	3350	1930	2860	1880	2730	1260	2550	1360	2620	1280	2440	960
	7/8	6.29	4.813	4570	2400	3900	2180	3720	1400	3470	1510	3560	1420	3320	1050
	1	5.50	5.500	5930	2760	5070	2380	4820	1520	4500	1650	4550	1550	4240	1150
	1-1/4	4.40	6.875	8930	3260	7630	2790	6930	1780	6470	1930	6110	1820	5690	1350
7-1/2	5/8	12.00	4.688	2330	1260	1990	1260	1690	950	1770	1030	1820	960	1690	800
	3/4	10.00	5.625	3350	1820	2860	1820	2730	1320	2550	1430	2620	1340	2440	1110
	7/8	8.57	6.563	4560	2420	3900	2420	3720	1730	3470	1870	3560	1760	3320	1400
	1	7.50	7.500	5950	3090	5080	3030	4850	2060	4520	2230	4650	2100	4330	1570
	1-1/4	6.00	9.375	9310	4290	7950	3800	7580	2430	7070	2630	7260	2480	6770	1840
9-1/2	3/4	12.67	7.125	3350	1640	2860	1640	2730	1250	2550	1350	2620	1270	2440	1060
	7/8	10.86	8.313	4560	2270	3890	2270	3720	1660	3470	1790	3560	1690	3320	1410
	1	9.50	9.500	5950	2960	5080	2960	4850	2130	4530	2300	4650	2170	4330	1790
	1-1/4	7.60	11.875	9310	4510	7950	4450	7580	3030	7070	3280	7270	3090	6770	2330
	1-1/2	6.33	14.250	13420	6070	11470	5520	10930	3540	10200	3830	10470	3610	9760	2680
11-1/2	7/8	13.14	10.062	4560	2060	3900	2060	3700	1590	3460	1730	3570	1630	3330	1370
	1	11.50	11.500	5950	2770	5080	2770	4860	2040	4530	2210	4650	2060	4330	1730
	1-1/4	9.20	14.375	9310	4360	7960	4360	7590	3140	7080	3400	7270	3200	6780	2620
	1-1/2	7.67	17.250	13410	6210	11450	6140	10920	4210	10190	4550	10470	4280	9750	3240
13-1/2	1	13.50	13.500	5960	2530	5280	2530	4850	1970	4540	2140	4670	2020	4350	1680
	1-1/4	10.80	16.875	9300	4160	7950	4160	7590	3030	7060	3260	7260	3080	6770	2570
	1-1/2	9.00	20.250	13400	6040	11450	6040	10930	4340	10200	4700	10460	4420	9750	3600

Tabulated values, in pounds, are for one ASTM A307 bolt loaded in double shear in a three-member wood connection subjected to normal load duration and dry-use conditions. When high strength bolts are used (such as ASTM A325 bolts), values in this table can be used with slightly conservative results.

Use linear interpolation to determine design values for intermediate bolt lengths.

This table is limited to selected species and bolt lengths and is intended for illustrative purposes only. Refer to the current edition of the NDS for a more complete listing of tabulated values.

From the NDS;[26] ©1986. Used by permission.

Table 5-21. - Summary of requirements for determining tabulated bolt values for lateral connections.

A. Three-member joints

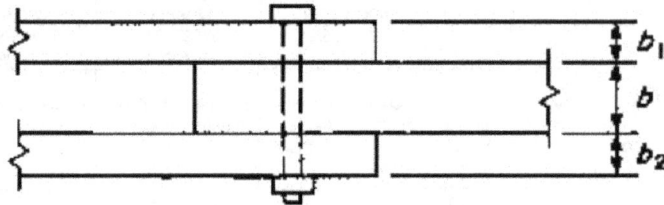

Wood side members

$b_1 = b_2 \geq b/2$ Use the tabulated value for main member thickness b.

$b_1 \leq b_2 < b/2$ Use the tabulated value for main member twice the thickness of b_1.

When side members are loaded at a different direction to the grain than the main member, the design value is the lesser of:

     a. the tabulated value for the main member, or;

     b. the tabulated value for a member twice the thickness of the side members and loaded in the same direction as the side members.

Steel side members

Use the tabulated value for the main (wood) member b for the direction of applied loading.

B. Two-member joints

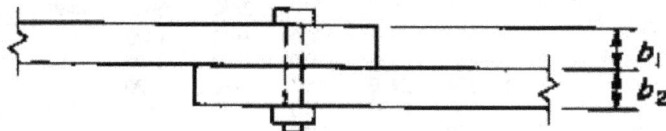

Wood side member

$b_1 = b_2$ Use one-half the tabulated value for a main member of thickness b_1.

$b_1 < b_2$ Use the lesser of one-half the tabulated value for main member of thickness = b_2 or $2b_1$.

When one member is loaded parallel-to-grain and the other is loaded at an angle to the grain, the design value is the lesser of:

     a. one-half the tabulated value for the thickness of the parallel to grain loaded member, or;

     b. the value obtained from application of the Hankinson formula (Equation 5-42) using one-half the tabulated parallel-to-grain and perpendicular-to-grain values for a member the thickness of the member loaded at an angle to the grain.

Steel side member

Use one-half the tabulated value for a member the thickness of the wood member for the direction of applied loading.

connected to a steel plate, the design load is one-half of the tabulated value of the thickness of the wood member.

Steel Side Plates

When steel rather than wood side plates are used for lateral connections, the tabulated design values for members loaded parallel to grain only may be increased by the steel side plate factor, C_{st} given below.

Sawn Lumber		Glulam	
Bolt diameter (in.)	C_{st}	Bolt diameter (in.)	C_{st}
≤1/2	1.75	All	1.25
3/4	1.63		
1	1.50		
1-1/4	1.38		
1-1/2	1.25		

Use linear interpolation to compute C_{st} for intermediate bolt diameters in sawn lumber. It should be noted that the values of C_{st} greater than 1.25 are currently being evaluated for sawn lumber and may be reduced in the future. In addition, AITC recommends that bolts used with steel side plates in glulam not exceed 1 inch in diameter.

Distance and Spacing Requirements

Tabulated bolt values are based on minimum distance and spacing requirements necessary to develop the full capacity of the connection. These requirements differ for parallel-to-grain loading and perpendicular-to-grain loading and are summarized in Table 5-22. When bolts are placed at the minimum dimension for full tabulated value, no reduction in capacity is required. For end distance and spacing parallel to grain only, the dimensions may be reduced provided the tabulated value is reduced by the modification factors C_{Δ} or $C_{\prime}$ For example, when a bolted tension connection is loaded parallel to gram, the minimum end distance to develop the full tabulated value is 7 times the bolt diameter. This distance may be reduced to an absolute minimum of 3.5 times the bolt diameter provided the tabulated value is reduced by 50 percent $(C_{\Delta} = 0.50)$. When reduced dimensions are used for any bolt in a group, the factors C_{Δ} or $C_{\prime}$ apply to all bolts in that group. Dimensions less than those given for reduced capacity are not permitted under any circumstances.

Distance and spacing requirements in the NDS are for loading parallel to grain and perpendicular to grain only. When loads act at an angle to the grain, bolt spacing and distance must be based on good engineering judgment. In this case, the gravity axis of the members should pass through the center of resistance of the bolt group to provide uniform stress in the main members and a uniform distribution of load to all bolts.

A. Loading parallel to grain

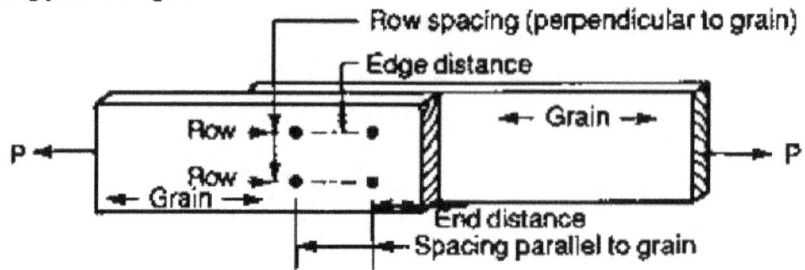

	Minimum dimension for full tabulated value	Minimum dimension for reduced value[1]
Edge distance		
$l/D \leq 6$	1.5D	N/A
$l/D > 6$	1.5D or 1/2 row spacing perpendicular to grain, whichever is greater	N/A
End distance		
Tension members	7D	3.5D ($C_n = 0.50$)
Compression member	4D	2D ($C_n = 0.50$)
Spacing		
Parallel to grain	4D	3D ($C_a = 0.75$)
Row spacing perpendicular to grain	1.5D	N/A

When steel members are used in connections, the spacing and distance
requirements are based on the requirements for the timber components,
not the steel components. As a practical consideration, the designer should
always check to ensure that spacing requirements are sufficient to place
washers without overlap.

Design of Tension Connections

In tension connections, the bolt is loaded in axial tension parallel to its
axis. This type of connection is common in bridge applications when rail
posts are bolted to curbs. The strength of a tension connection depends on
the bearing strength of the wood and the tensile strength of the bolt
(20,000 lb/in^2 for A307 bolts). The bearing stress under the washer must
not exceed the allowable stress for compression perpendicular to grain
($F_{c\perp}'$). To compute bearing stress, the bolt load is divided by the total
washer area minus the area of the bolt hole. Distance and spacing requirements for bolts loaded in tension only are not specified in the NDS and
should be based on designer judgment.

Table 5-22. - Summary of edge distance, end distance and spacing requirements for bolted connections. (Continued)

B. Loading perpendicular to grain

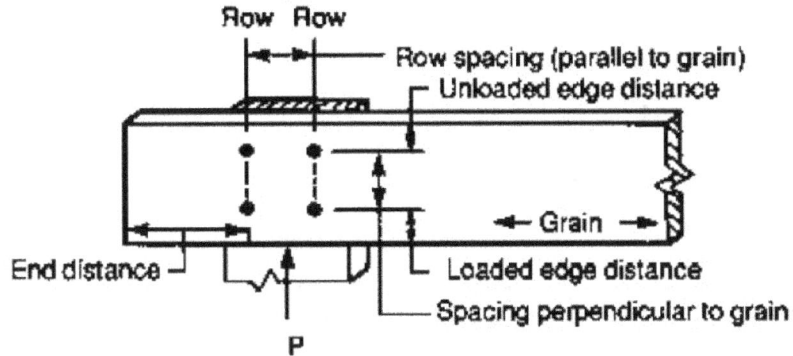

	Minimum dimension for full tabulated value	Minimum dimension for reduced value[1]
Edge distance		
Loaded edge	$4D$	N/A
Unloaded edge	$1.5D$	N/A
End distance	$4D$	$2D\,(C_n = 0.50)$
Spacing		
Row spacing parallel to grain[2,3]		
$\ell/D = 2$	$2.5D$	N/A
$\ell/D \geq 6$	$5D$	N/A
Perpendicular to grain	See note 4	See note 4

[1] For distances and spacings between the tabulated value and the reduced value use straight line interpolation to compute modification factor value.

[2] The spacing between rows of bolts shall not be more than 5 inches unless seperate splice plates are used for each row of bolts.

[3] For ℓ/D ratios between 2 and 6, spacing requirements are obtained by straight line interpolation.

[4] The spacing of bolts perpendicular to grain is limited by the spacing requirements of the attached member or members (whether of metal or of wood loaded parallel to grain).

All dimensions are measured from the center of the bolt hole.

Bolt Placement

The strength of a laterally loaded, bolted connection can be significantly affected by the diameter of the hole and the manner in which it is bored. When holes are too large, bearing is nonuniform, and the capacity of the connection is reduced. If holes are too small, the bolt cannot be inserted without driving, which may split the wood members. The NDS specifies that bolt holes be a minimum of 1/32 inch to a maximum of 1/16 inch larger than the bolt diameter. In some cases, it may be necessary to slightly enlarge the hole diameter slightly to compensate for galvanized coatings on large fasteners.

When bolts are installed in wood members, washers of the proper size or a steel plate or strap are required under all nuts and under square or hexagonal bolt heads. Nuts must be tightened so that member surfaces are brought into close contact without crushing the wood. Tabulated design values for bolts include an allowance for the loosening of nuts because of member shrinkage. However, when bolts are installed in unseasoned wood it is advisable to retighten connections at least every 6 months until the wood reaches equilibrium moisture content. Self-locking nuts are frequently used for decks and other components that may have a tendency to loosen because of vibrations from moving loads.

Example 5-11 - Lateral bolted connection parallel to grain

A tension splice in a timber truss joins two 2-inch by 6-inch side members to a 4-inch by 6-inch main member. Design the connection to develop the full capacity of the members, assuming the following:

1. Members will be exposed to weathering and carry loads of normal duration; adjustments for temperature (C_t) and fire-retardant treatment (C_p) are not required.

2. The connection is made with a single row of 1-inch-diameter bolts.

3. Lumber is dressed Southern Pine, visually graded No. 1 to SPIB rules.

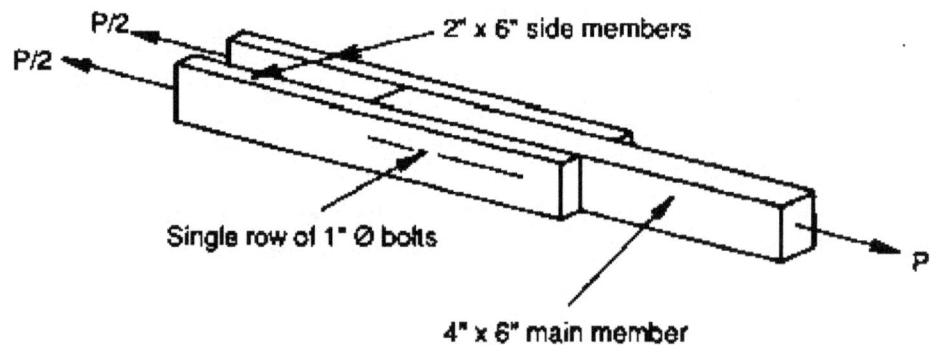

P/2   2" x 6" side members

Single row of 1" Ø bolts

4" x 6" main member

P

Solution

This connection involves a three-member configuration loaded in double shear. The design procedure will be to (1) compute the capacity of the lumber members, (2) determine the required number of bolts, and (3) detail the connection for minimum distance and spacing requirements.

Member Capacity

The tabulated stress for No. 1 Southern Pine in tension parallel to grain is obtained from NDS Table 4A. The NDS includes several tables for Southern Pine, and a value $F_t = 775$ lb/in^2 is selected from the table "surfaced green; used any condition" (footnotes to Table 4A specify use of this table when the moisture content in service is expected to exceed 19 percent). Further adjustment for moisture content is not required.

The allowable stress in tension parallel to grain is computed using Equation 5-23:

$$F_t' = F_t C_M = 775(1.0) = 775 \text{ lb/in}^2$$

The capacity of the connection depends on the net area of the lumber members. Gross section properties for nominal 2-inch by 6-inch and 4-inch by 6-inch lumber are obtained from Table 16-2.

For nominal 2-inch by 6-inch lumber	For nominal 4-inch by 6-inch lumber
$b = 1.5$ in.	$b = 3.5$ in.
$d = 5.5$ in.	$d = 5.5$ in.
$A = 8.25$ in^2	$A = 19.25$ in^2

Assuming that bolt holes are 1/16 inch larger than the bolt diameter, the net area of each member is equal to the gross area minus the projected area of the bolt holes:

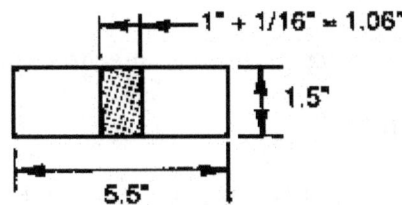

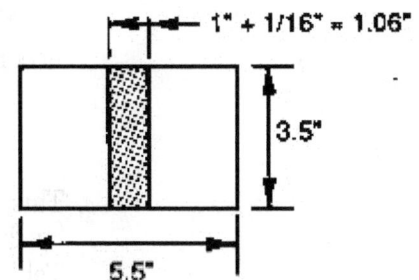

5-91

For two 2-inch by 6-inch members,

$$A_{NET} = 2 [8.25 \text{ in}^2 - (1.06 \text{ in.})(1.5 \text{ in.})] = 13.32 \text{ in}^2$$

For a single 4-inch by 6-inch member,

$$A_{NET} = 19.25 \text{ in}^2 - [(1.06 \text{ in.})(3.5 \text{ in.})] = 15.54 \text{ in}^2$$

Connection capacity will be limited by the smaller area of the two 2-inch by 6-inch members. The maximum connection load in tension, P_T, is equal to the net area times the allowable stress in tension parallel to grain:

$$P_T = A_{NET}(F_t') = 13.32(775) = 10,323 \text{ lb}$$

Number of Bolts

The next step is to determine the number of 1-inch-diameter bolts that are required to transfer the lateral load of 10,323 pounds. Because this is a three-member connection, tabulated bolt values can be read directly from the NDS bolt design tables (Table 5-20); however, the length of bolt in the main member, ℓ, must first be determined. In this case, the thickness of the side members (1.5 inches) is less than half the thickness of the main member (1.75 inches). From Table 5-21, the main member thickness used to determine the tabulated bolt value is equal to twice the thickness of the thinner side members:

$$\ell = 2(1.5 \text{ in.}) = 3 \text{ in.}$$

From Table 5-20 for a 1-inch-diameter bolt, Species Group 3, and a bolt length in main member of 3 inches,

$$P = 3,750 \text{ lb}$$

Assuming that adequate distance and spacing requirements can be met, the allowable load for one bolt loaded parallel to grain is given by Equation 5-43:

$$P' = PC_M C_g$$

From Table 5-18 for a bolted connection that is exposed to weathering:

$$C_M = 0.75$$

This connection will involve more than 2 bolts in a row and adjustment by the group action factor, C_g will be required. To determine C_g from Table 5-19, the number of bolts must be known. At this point, an estimate of the number of bolts is made by assuming adjustment by C_M only:

$$\text{Estimated number of bolts} = \frac{P_T}{P(C_M)} = \frac{10{,}323}{3{,}750\,(0.75)} = 3.7 \text{ bolts}$$

C_g will be determined for a row of 4 bolts. From Table 5-19 for connections with wood side plates:

$A_1 = 19.25$ in²

$A_2 = (2)(8.25 \text{ in}^2) = 16.50$ in²

$A_1/A_2 = 19.25/16.50 = 1.17 > 1.0$, so use A_2/A_1

$A_2/A_1 = 16.5/19.25 = 0.86$

The value A_2/A_1 is between the values 0.5 and 1.0 given in Table 5-19. Because $A_1/A_2 > 1.0$, A_2 is used instead of A_1. For $A_2 = 16.5$ and 4 bolts, linear interpolation between $C_g = 0.88$ for $A_2/A_1 = 0.50$ and $C_g = 0.94$ for $A_2/A_1 = 1.0$ gives a value $C_g = 0.92$. Using this factor, the allowable bolt load is computed by Equation 5-43:

$$P' = PC_M C_g = 3{,}750(0.75)(0.92) = 2{,}588 \text{ lb/bolt}$$

The required number of bolts is computed by dividing the maximum load by the allowable load per bolt:

$$\text{Required number of bolts} = \frac{P_T}{P'} = \frac{10{,}323}{2{,}588} = 3.99 \text{ or } 4 \text{ bolts}$$

Distance and Spacing Requirements
From Table 5-22, distance and spacing requirements for full connection capacity with loading parallel to grain are as follows:

Edge distance for $\ell/D \leq 6 = 1.5D = 1.5$ in.

End distance for tension members $= 7D = 7$ in.

Bolt spacing parallel to grain $= 4D = 4$ in.

All distance and spacing requirements for full load can be met; however, washer size should be checked to avoid potential overlapping. In most cases, malleable iron (MI) washers of the sizes given in Table 16-7 are used. For a 1-inch-diameter MI washer the outside washer diameter is 4 inches, which is the same distance required for bolt spacing parallel to grain. Spacing will be increased to 4-1/2 inches to allow for construction tolerances and washer placement.

Summary

The connection will be made with four 1-inch-diameter bolts to develop the member capacity of 10,323 pounds. Detailing is as follows:

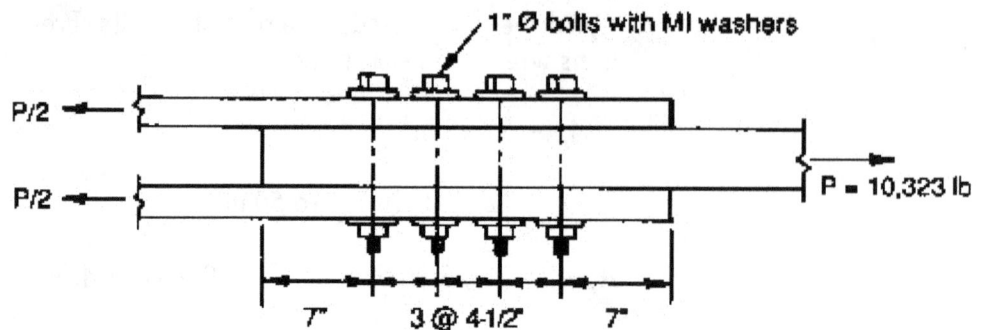

Example 5-12 - Lateral bolted connection perpendicular to grain

A 10-inch by 10-inch lumber curb is bolted along the edges of a 6-3/4-inch-thick transverse glulam deck. A transverse 5,000 pound load with a duration of load of 5 minutes is applied at the curb center. Determine the number of 7/8-inch-diameter bolts that are required to transfer the curb load to the deck, assuming the following:

1. Members will be exposed to weathering (wet-use conditions for glulam); adjustments for temperature (C_t) and fire-retardant treatment (C_R) are not required.

2. The glulam deck is combination symbol No. 2.

3. The curb is full-sawn Douglas Fir-Larch, visually graded No. 1 to WWPA rules.

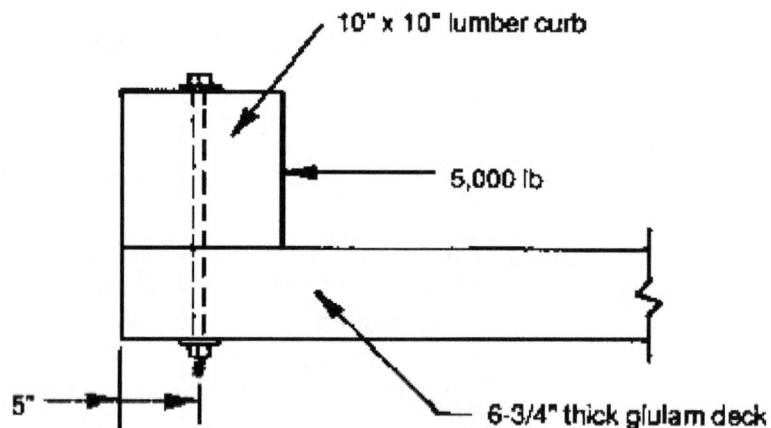

Solution

In this connection the curb is loaded perpendicular to grain while the deck is loaded parallel to grain. From Table 5-21, for a two-member connection with members loaded at different angles to the grain, the tabulated design value for one bolt is the lesser of the following:

1. one-half the tabulated parallel-to-grain value, P, for the thickness of the member loaded parallel to grain; or

2. one-half the tabulated perpendicular-to-grain value, Q, for the thickness of the member loaded perpendicular to grain (application of the Hankinson formula as stated in Table 5-21 is not necessary in this case because loading is perpendicular to grain rather than at some intermediate angle between 0 and 90 degrees).

Tabulated bolt values for loading perpendicular to grain are normally much lower than those for loading parallel to grain. Thus, the design sequence will be to (1) determine the number of bolts required for curb loading perpendicular to grain, (2) check the connection for deck loading parallel to grain, and (3) verify and detail distance and spacing requirements.

Curb Loading Perpendicular to Grain

The allowable design value for one bolt loaded perpendicular to grain is given by Equation 5-44. Assuming minimum distance and spacing requirements can be met, and substituting $Q/2$ for Q in this single-shear application, Equation 5-44 reduces to

$$Q' = \frac{Q}{2} C_D C_M$$

Using bolt design tables in the NDS (Table 5-20), the tabulated perpendicular to grain value, Q, is determined for one 7/8-inch-diameter bolt in Douglas Fir-Larch (Species Group 3), with a length of bolt in main member of 10 inches $(\ell = 10 \text{ inches})$. Table 5-20 does not include $\ell = 10$ inches, so interpolation is required.

For $\ell = 9$-1/2 in., $Q = 2{,}270$ lb

For $\ell = 11$-1/2 in., $Q = 2{,}060$ lb

By linear interpolation, for $\ell = 10$ in., $Q = 2{,}218$ lb

The duration of load factor for the 5-minute load duration is obtained from Table 5-8:

$$C_D = 1.65$$

The moisture-content factor for bolted sawn lumber exposed to weathering is obtained from Table 5-18:

$$C_M = 0.75$$

Substituting values into Equation 5-44, the allowable perpendicular-to-grain load for one 7/8-inch-diameter bolt is computed:

$$Q' = \frac{Q}{2} C_D C_M = \frac{(2,218)}{2}(1.65)(0.75) = 1,372 \text{ lb}$$

The required number of bolts is computed by dividing the applied load by the allowable load per bolt:

$$\text{Required number of bolts} = \frac{5,000 \text{ lb}}{1,372 \text{ lb/bolt}} = 3.64 = 4 \text{ bolts}$$

Deck Loading Parallel to Grain

The allowable design value for one bolt loaded parallel to grain is given by Equation 5-43. Assuming that minimum distance and spacing requirements can be met, and substituting P/2 for P in this single-shear application,

$$P' = \frac{P}{2} C_D C_M$$

From Table 5-15, glulam combination symbol No. 2 is in Species Group 3 for bolt design. As with curb loading, tabulated values in Table 5-20 do not include a bolt length in main member that matches the required length of ℓ = 6-3/4 inches. However, values for ℓ = 5-1/2 inches and ℓ = 7-1/2 inches are both 3,900 pounds, so the same value also applies to ℓ = 6-3/4 inches:

$$P = 3,900 \text{ lb}$$

From Table 5-18 for glulam used under wet-use conditions,

$$C_M = 0.67$$

Substituting into Equation 5-43,

$$P' = \frac{P}{2} C_D C_M = \frac{3,900}{2}(1.65)(0.67) = 2,156 \text{ lb}$$

P' = 2,156 lb > Q' = 1,372 lb, so curb loading perpendicular to grain will control design.

5-96

Distance and Spacing Requirements

The two most critical distances in this connection are the curb loaded edge distance and the deck end distance:

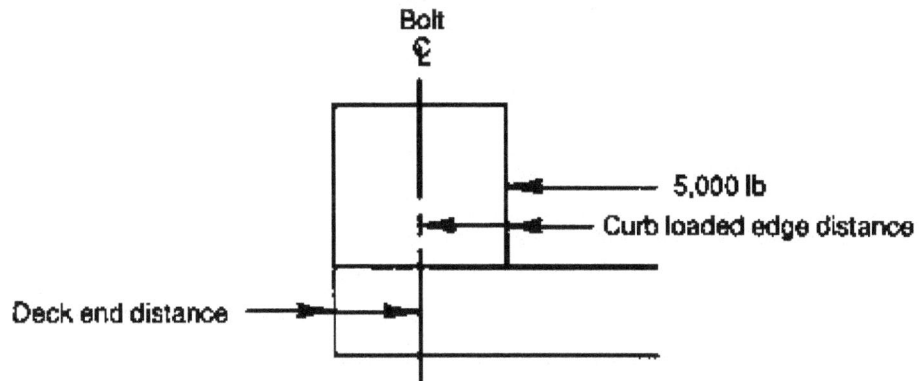

From Table 5-22, the minimum loaded edge distance for loading perpendicular to grain is four times the bolt diameter,

$$4D = 4(0.875 \text{ in.}) = 3.5 \text{ in.}$$

The actual loaded edge distance of 5 inches exceeds the minimum 3.5 inches, and is sufficient.

For loading parallel to grain, the minimum end distance for full capacity on the glulam deck is seven times the bolt diameter,

$$7D = 7(0.875) = 6.13 \text{ in.}$$

This value is greater than the 5 inches provided. The end distance can be reduced to a minimum value of $3.5D = 3.06$ inches, provided the allowable load is reduced by 50 percent $(C_n = 0.50)$. By linear interpolation for the actual end distance of 5 inches, $C_n = 0.82$, and the allowable load is revised as follows:

$$P' = \frac{P}{2} C_D C_M C_n = \frac{3,900}{2} (1.65)(0.67)(0.82) = 1,768 \text{ lb}$$

The revised value is still greater than $Q' = 1,372$ pounds, so a reduced end distance of 5 inches will not affect connection capacity.

From Table 5-22, the spacing of bolts parallel to grain on the curb is controlled by $5D = 4.38$ inches (based on $\ell/d = 10/.875 = 11.4$). From Table 16-7, the outside diameter of a 7/8-inch MI washer is 3.5 inches. Bolts will be spaced 4-1/2 inches apart to meet spacing requirements and allow for construction tolerance.

Summary

The connection will be made using four 7/8-inch-diameter bolts for a total capacity of 4(1,372) = 5,488 pounds. The bolts will be spaced 4-1/2 inches on-center and will be provided with malleable iron washers on each end.

LAG SCREWS

Lag screws are used in bridge applications for two-member connections loaded laterally in single shear (two members) or in withdrawal. The strength of a lag screw is less than that of a comparable bolt, but lag screws offer the advantage of being placed from one side of the connection. They are used primarily for convenience or when through bolts are undesirable or impractical. This occurs in connections where access for nut placement is restricted or when an excessively long bolt is required to fully penetrate the connection. Lag screws also may be used instead of spikes in nonstructural applications (such as timber wearing surface attachment) because they are less susceptible to loosening from vibrations and from dimensional changes in the wood.

Lag screws are manufactured of the same material as bolts, conforming to *ASTM Standard A307, Low-Carbon Steel Externally and Internally Threaded Standard Fasteners*. They have a square or hexagonal bolt head and require a washer when the screw head is in wood contact. A diagram of a typical lag screw is shown in Figure 5-22. The specified diameter of the screw corresponds to the diameter of the unthreaded shank portion. Nominal length is the distance from the base of the head to the tip of the threads. Lag screws are commonly available in stock diameters of 3/16 inch to 1-1/4 inch and nominal lengths up to 16 inches, in 1/2-inch increments. The length of the threaded portion varies with the length of the screw. Dimensions of common lag screws are given in Table 16-5.

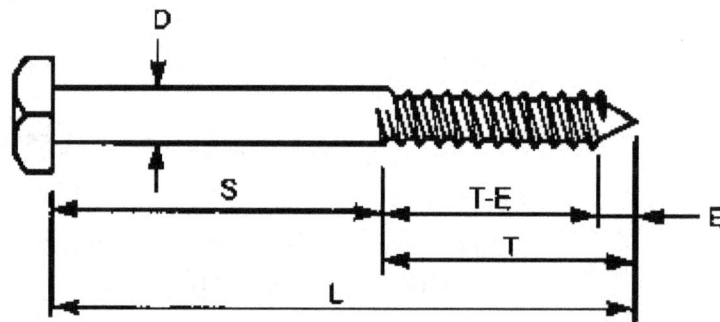

D = Nominal diameter or shank diameter
L = Nominal length
S = Length of shank
T = Length of thread
E = Length of tapered tip

Figure 5-22. - Lag screw configuration and nomenclature.

Net Area

Net area is computed for lag screws in the same manner as bolts with the same diameter as the shank diameter of the lag screw.

Design of Lateral Connections

The strength of a laterally loaded, lag screw connection is developed by bearing between the screw and the members, and the interaction of the threads in the main member (Figure 5-23). In bridge applications, these connections should be limited to applications where the screw is inserted into the side grain of the member, perpendicular to the wood fiber direction. Refer to the NDS for design criteria when end-grain connections cannot be avoided.

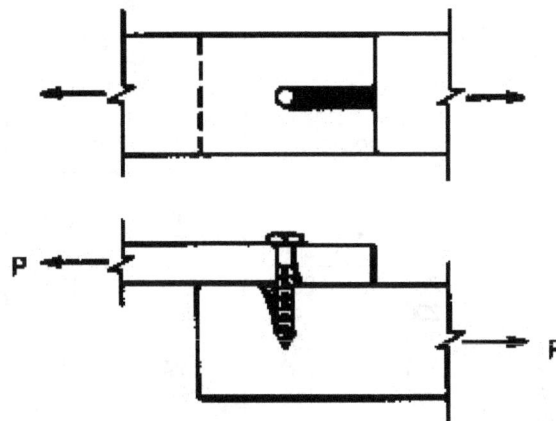

Figure 5-23. - Typical configuration and stress distribution for a laterally loaded lag screw connection.

The allowable value for one laterally loaded lag screw is equal to the tabulated value, adjusted by all applicable modification factors. When more than one lag screw is used, the allowable value for the connection is the sum of the allowable values for the individual fasteners, including adjustment by the group action factor, C_g. Equations 5-45 and 5-46 follow:

$$P' = PC_D C_M C_t C_g C_g C_g$$ (5-45)

$$Q' = QC_D C_M C_t C_g C_g C_g$$ (5-46)

If loads act at an angle to the grain, the allowable design value is computed using the Hankinson formula.

Tabulated Design Values

Tabulated values are specified in the NDS for one A307 lag screw loaded in single shear in a two-member joint. Unlike other fasteners, separate tables are included for connections with wood side pieces and connections with metal side pieces. Portions of the NDS tables for a limited number of lag screw lengths are given in Tables 5-23 and 5-24. For connections with wood side members, tabulated values are based on the thickness of the side members and the nominal length and diameter of the lag screw (Table 5-23). When side members are 1-1/2 or 2-1/2 inches thick, values are read directly from the table. The NDS does not include tabulated values for other side member thicknesses, but additional tabulated values for other side member thicknesses are given in the AITC *Timber Construction Manual.* [24]

Table 5-23.—Tabulated design values for laterally loaded lag screws with wood side plates.

Thickness of side member (inches)	Length of lag screw (inches)	Diameter of lag screw shank (inches)	Species Group							
			GROUP I		GROUP II		GROUP III		GROUP IV	
			Total lateral load per lag screw in single shear (pounds)		Total lateral load per lag screw in single shear (pounds)		Total lateral load per lag screw in single shear (pounds)		Total lateral load per lag screw in single shear (pounds)	
			Parallel to grain	Perpendicular to grain	Parallel to grain	Perpendicular to grain	Parallel to grain	Perpendicular to grain	Parallel to grain	Perpendicular to grain
1-1/2	4	1/4	200	200	170	170	130	130	100	100
		5/16	290	240	220	180	150	130	120	110
		3/8	330	250	250	190	180	140	140	110
		7/16	370	260	280	190	200	140	160	110
		1/2	390	250	290	190	210	140	170	110
		5/8	470	260	360	210	260	160	200	120
	6	1/4	270	260	230	220	210	200	180	180
		5/16	380	320	330	280	290	250	260	220
		3/8	490	370	420	320	370	280	300	230
		7/16	600	420	520	360	410	280	330	230
		1/2	700	460	600	390	430	280	340	220
		5/8	850	510	710	430	510	310	410	250
2-1/2	6	3/8	450	340	380	290	270	210	220	170
		7/16	590	410	440	310	320	220	250	180
		1/2	620	410	470	310	340	220	270	180
		5/8	730	440	550	330	390	240	320	190
		3/4	830	460	630	350	450	250	360	200
		7/8	950	490	720	370	510	270	410	210
		1	1060	530	800	400	570	290	460	230
	8	3/8	560	420	480	370	430	330	380	290
		7/16	730	510	630	440	560	390	450	320
		1/2	890	580	770	500	600	390	480	310
		5/8	1230	740	970	580	700	420	560	340
		3/4	1440	790	1090	600	780	430	630	340
		7/8	1610	840	1220	630	870	450	700	360
		1	1810	910	1370	690	980	490	790	390

Tabulated values are for normal load durations under dry service conditions.

This table contains a limited number of lag screw lengths and is intended for illustrative purposes only. Refer to the current edition of the NDS for a more complete listing of design values.

From the NDS; [26] © 1986. Used by permission.

Table 5-24. - Tabulated design values for laterally loaded lag screws with metal side plates up to 1/2-inch thick.

Length of lag screw (inches)	Diameter of lag screw shank (inches)	GROUP I Total lateral load per lag screw in single shear (pounds)		GROUP II Total lateral load per lag screw in single shear (pounds)		GROUP III Total lateral load per lag screw in single shear (pounds)		GROUP IV Total lateral load per lag screw in single shear (pounds)	
		Parallel to grain	Perpendicular to grain	Parallel to grain	Perpendicular to grain	Parallel to grain	Perpendicular to grain	Parallel to grain	Perpendicular to grain
4	1/4*	270	210	240	180	210	160	190	150
	5/16	410	280	350	240	290	200	230	160
	3/8	570	350	480	290	340	210	280	170
	7/16	730	410	550	310	390	220	310	180
	1/2	810	420	610	320	440	230	350	180
	5/8	980	470	740	360	530	250	430	200
6	5/16*	450	300	390	260	340	230	300	210
	3/8	630	390	550	330	490	300	430	260
	7/16	850	480	730	410	660	370	540	300
	1/2	1100	570	950	490	760	400	610	320
	5/8	1640	790	1290	620	920	440	740	350
	3/4	1990	870	1500	660	1070	470	860	380
8	7/16*	880	490	760	420	680	380	600	330
	1/2	1140	590	980	510	880	460	780	400
	5/8	1750	840	1510	720	1320	630	1060	510
	3/4	2470	1090	2130	940	1560	690	1250	550
	7/8	3260	1360	2480	1030	1770	470	1420	590
10	5/8*	1790	860	1550	740	1380	660	1220	590
	3/4	2550	1120	2200	970	1970	870	1630	720
	7/8	3430	1420	2960	1230	2340	970	1880	780
	1	4410	1770	3680	1470	2640	1050	2110	850
12	7/8	3490	1450	3020	1260	2700	1120	2320	960
	1	4520	1810	3900	1560	3260	1310	2620	1050
	1-1/8	5670	2270	4890	1960	3630	1450	2910	1170

Tabulated values are for normal load durations under dry service conditions.

The asterisk (*) indicates that greater lengths for the lag screw diameter do not provide higher loads.

This table contains a limited number of lag screw lengths and is intended for illustrative purposes only. Refer to the current edition of the NDS for a more complete listing of design values.

From the NDS;[26] © 1986. Used by permission.

For lag screw connections with metal side plates, tabulated values in the NDS are based on the nominal length and diameter of the lag screw (Table 5-24). Values for side plates up to 1/2 inch thick are read directly from the table. When side plates are thicker than 1/2 inch, tabulated values must be reduced in proportion to the reduced lag screw penetration, by linear interpolation of table values. Values in Table 5-24 have been adjusted by C_d and further adjustment by this factor is not required or permitted.

Distance and Spacing Requirements

End distance, edge distance, and spacing requirements for lag screws are the same as those for bolts of a diameter equal to the shank diameter of the lag screw (Table 5-22). Bolt modification factors for reduced distance and spacing also apply to lag screws.

Design of Withdrawal Connections

In withdrawal connections, lag screws develop their strength by the interaction of the threads with the wood. The capacity of the connection depends on the specific gravity of the wood and the length of penetration of the lag screw. As shown in Equation 5-47, the allowable value for one lag screw in axial withdrawal is equal to the tabulated value in withdrawal, P_w adjusted by all applicable modification factors:

$$P_w' = P_w C_D C_M C_t C_k \tag{5-47}$$

When more than one lag screw is used, the value for one screw is multiplied by the total number of screws in the connection.

In determining allowable withdrawal values, the washer bearing stress on wood members must be less than the allowable stress in compression perpendicular to grain $F_{c\perp}'$, as discussed for bolts. In addition, the allowable tensile strength of the lag screw at the net (root) section must not be exceeded. The strength of A307 lag screws in axial tension is developed when the penetration depth of the threaded portion is approximately 7 diameters for Group I species, 8 diameters for Group II species, 10 diameters for Group III species, and 11 diameters for Group IV species. When the penetration of the screw exceeds these values, connection strength is generally controlled by the tensile strength of the fastener.

Tabulated Design Values

Tabulated withdrawal values for lag screws are given in the NDS for one A307 lag screw loaded in withdrawal from side grain. A portion of the NDS table for a limited number of specific gravities is shown in Table 5-25. To determine the tabulated value for one lag screw, enter the table with the specific gravity of the member and read the value in pounds per inch of penetration given for the screw diameter. The tabulated value for one screw is computed by multiplying this value times the distance of

Table 5-25. - Tabulated design values for lag screws loaded in withdrawal.

Specific gravity G	Lag Screw Shank Diameter (in.)											
	1/4 0.250	5/16 0.3125	3/8 0.375	7/16 0.4375	1/2 0.500	9/16 0.5625	5/8 0.625	3/4 0.750	7/8 0.875	1 1.000	1-1/8 1.125	1-1/4 1.250
0.55	260	307	352	395	437	477	516	592	664	734	802	868
0.54	253	299	342	384	425	464	502	576	646	714	780	844
0.51	232	274	314	353	390	426	461	528	593	656	716	775
0.49	218	258	296	332	367	401	434	498	559	617	674	730
0.48	212	250	287	322	356	389	421	482	542	599	654	708
0.47	205	242	278	312	345	377	408	467	525	580	634	686
0.46	199	235	269	302	334	365	395	453	508	562	613	664
0.45	192	227	260	292	323	353	382	438	492	543	594	642
0.44	186	220	252	283	312	341	369	423	475	525	574	621
0.43	179	212	243	273	302	330	357	409	459	508	554	600
0.42	173	205	235	264	291	318	344	395	443	490	535	579

Tabulated values are for load in withdrawal in pounds per inch of penetration of the threaded portion of the screw into the side grain of the member holding the point; normal load duration under dry service conditions.

This table is limited to selected values for specific gravity and is intended for illustrative purposes only. Refer to the current edition of the NDS for a more complete listing of design values.

From the NDS; [26] © 1986. Used by permission.

thread penetration into the member. When determining thread penetration, the screw tip length is not included as a portion of the threads. Refer to Table 16-5 for lag screw thread and tip lengths.

Lag Screw Placement

Lag screws are installed in prebored lead holes of sufficient diameter and length to develop thread strength and prevent the wood from splitting as the screw is installed. This requires that holes be drilled in two diameters, one for the shank and one for the threads (Figure 5-24). The lead hole for the shank is 1/16 inch larger than the shank diameter and is bored to the depth of penetration of the shank. The lead hole diameter for the threaded portion, which is bored at least the length of the threads, is based on the species of the member receiving the point. The NDS requires that for the threaded portion, the lead hole be 65 to 85 percent of the shank diameter in Group I species, 60 to 75 percent in Group II species, and 40 to 70 percent in Group III and IV species (the larger percentile figure in each range applies to screws of greater diameters). Recommended prebore diameters for lag screws are given in Table 5-26. The effect of prebore diameter on the lag screw thread penetration is illustrated in Figure 5-25.

Lag screws must be provided with a washer of the proper size unless the head of the screw bears on steel. When installing lag screws, the threaded portion is inserted in the lead hole by turning with a wrench, not by driv-

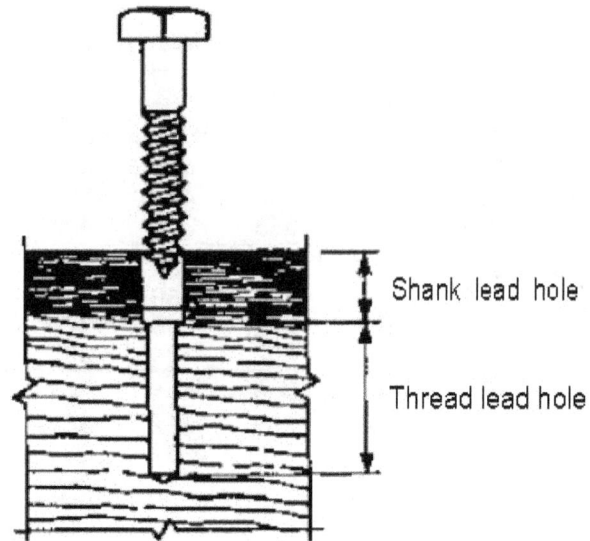

Figure 5-24.- Lead holes for lag screws are prebored in two diameters; one diameter for the shank and a smaller diameter for the threads.

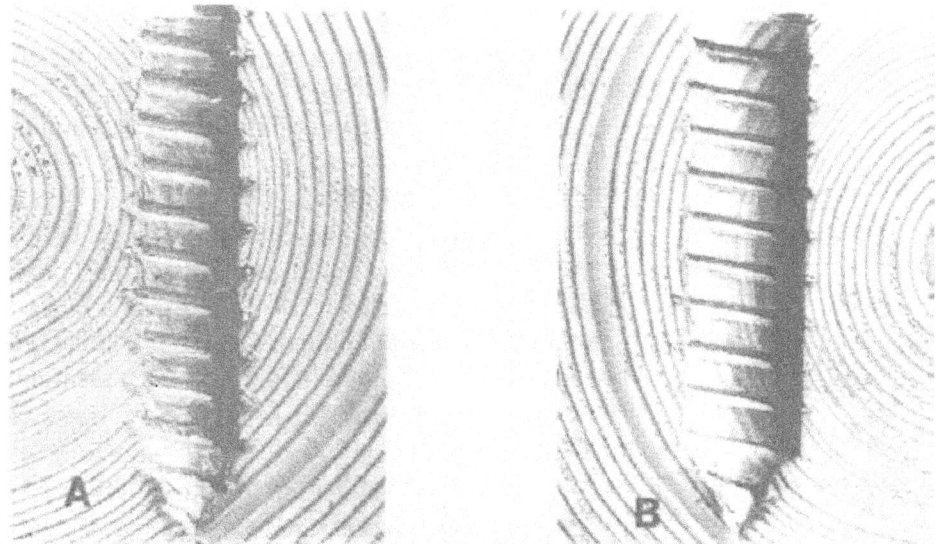

Figure 5-25. - (A) Clean-cut, deep penetration of thread made by a lag screw turned into a lead hole of proper size. (B) Shallow penetration of thread made by a lag screw turned into an oversized lead hole.

ing with a hammer. If screws are difficult to insert, soap or other lubricants can be placed on the screw to facilitate placement. In timber treated with an oil-type preservative, the preservative facilitates placement, and additional lubricants are normally not required.

5-104

Table 5-26. - Recommended lead hole diameters for lag screws.

Nominal diameter of lag screw (in.)	Shank (unthreaded) portion (in.)	Diameter of lead hole (in.)		
		Threaded portion		
		Group I species	Group II species	Groups III and IV species[1]
1/4	5/16	3/16	5/32	3/32
5/16	3/8	13/64	3/16	9/64
3/8	7/16	1/4	15/64	11/64
7/16	1/2	19/64	9/32	13/64
1/2	9/16	11/32	5/16	15/64
9/16	5/8	13/32	23/64	9/32
5/8	11/16	29/64	13/32	5/16
3/4	13/16	9/16	1/2	13/32
7/8	15/16	43/64	39/64	33/64
1	1-1/16	51/64	23/32	5/8
1-1/8	1-3/16	59/64	53/64	3/4
1-1/4	1-5/16	1-1/16	15/16	7/8

[1] When loaded primarily in withdrawal, lag screws of 3/8-inch diameter or less may be inserted into group III and IV species without a lead hole provided that spacings, end distances and edge distances are sufficient to prevent unusual splitting.

From the *AITC Timber Construction Manual.*[b] American Institute of Timber Construction, © 1985, John Wiley & Sons. Reprinted by permission.

Example 5-13 - Lateral lag screw connection with steel side plates

A beam bearing shoe consists of a pair of 1/2-inch-thick steel angles that are 12 inches long. Each angle is connected to a full-sawn 12-inch by 12-inch pile cap with two lag screws placed at the angle third points. Determine the required diameter and length of lag screws to resist a longitudinal beam load of 2,500 pounds per angle, assuming the following:

1. There is a 2-month duration of load (C_D= 1.15).

2. Members will be exposed to weathering; adjustments for temperature (C_t) and fire-retardant treatment (C_R) are not required.

3. The pile cap is Douglas Fir-Larch, visually graded No. 1 to WWPA rules.

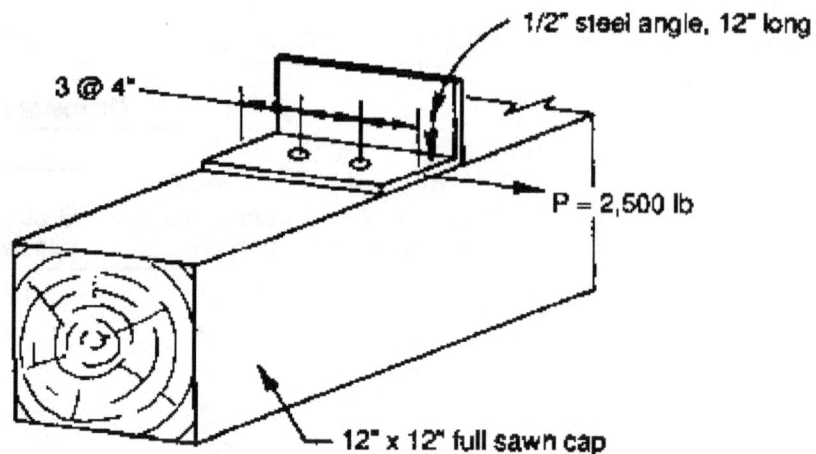

3 @ 4"

1/2" steel angle, 12" long

P = 2,500 lb

12" x 12" full sawn cap

Solution

The allowable load perpendicular to grain for one lag screw, Q', is given by Equation 5-46. Assuming that distance and spacing requirements can be met, Equation 5-46 for this case reduces to

$$Q' = QC_D C_M$$

To facilitate selection of a lag screw from NDS tables, the above equation is rearranged so that the required tabulated lag screw value, Q is computed directly:

$$Q = \frac{Q'}{C_D C_M}$$

The applied load of 2,500 pounds is resisted by two lag screws. Therefore, the minimum allowable load for one lag screw is one-half the applied load:

$$Q' = \frac{2,500}{2} = 1,250 \text{ lb}$$

From Table 5-18 for lag screws installed in sawn lumber exposed to weathering:

$$C_M = 0.75$$

Substituting values and solving for the required tabulated value perpendicular to grain:

$$Q = \frac{1,250}{1.15 (0.75)} = 1,449 \text{ lb}$$

Before entering design tables, limitations on lag screw length and diameter must be checked. For the 12-inch pile cap depth, lag screw length will be limited to 12-inches. Limitations on lag screw diameter are checked against distance and spacing requirements given in Table 5-22. For loading perpendicular to grain, the minimum loaded edge distance is four times the lag screw diameter. For the 4-inch distance provided by the angle configuration, requirements for loaded edge distance cannot be met if the lag screw diameter exceeds 1-inch. Thus, design requirements for lag screw selection are as follows:

Tabulated value for loading perpendicular to grain = $Q \geq 1{,}449$ lb

Lag screw diameter = $D \leq 1$ in.

Lag screw length = $L \leq 12$ in.

From Table 5-14, No. 1 Douglas Fir-Larch is in Species Group II for lag screw design. Entering Table 5-24 for laterally loaded lag screws with metal side plates, two possible lag screw sizes meet design requirements; a 10-inch-long by 1-inch-diameter lag screw with $Q = 1{,}470$ pounds, or a 12-inch-long by 1-inch-diameter lag screw with $Q = 1{,}560$ pounds. In this case the 10-inch-long lag screw is selected, but either screw is feasible depending on availability and relative economics.

Example 5-14 - Lag screw loaded in withdrawal

A 4-inch by 4-inch lumber railpost is attached to the side of a 10-1/2-inch-wide glulam beam with a 7/8-inch diameter by 10-inch-long lag screw with malleable iron washer. Determine the capacity of the connection in withdrawal, assuming the following:

1. Members will be exposed to weathering (wet-use conditions) and carry loads of normal duration; adjustments for temperature (C_t) and fire-retardant treatment (C_{rt}) are not required.

2. The railpost is rough-sawn Southern Pine, visually graded No. 1 to SPIB rules.

3. The glulam beam is combination symbol 24F-V5 Southern Pine.

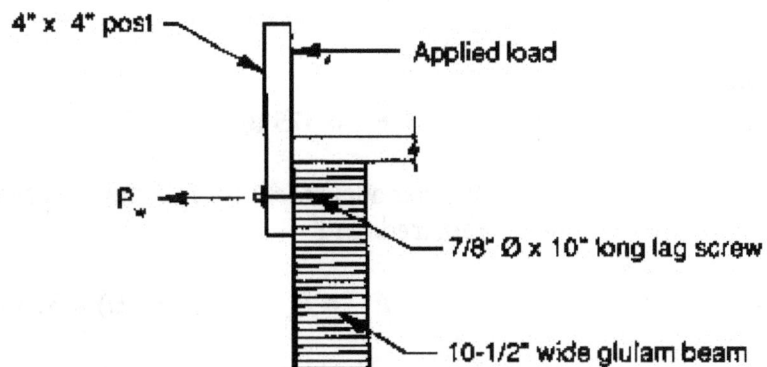

Solution
Capacity of this connection will be controlled either by the strength of the lag screw in withdrawal, or by bearing stress under the washer.

Lag Screw in Withdrawal

Using applicable modification factors from Equation 5-47, the allowable lag screw load in withdrawal is given as follows:

$$P_W' = P_W C_M$$

The strength in withdrawal depends on the length of penetration of the threaded portion minus tip length $(T - E)$ into the member receiving the point. From Table 16-5, dimensions for a 7/8-inch-diameter by 10-inch-long lag screw are as follows:

S = Length of shank = 4.75 in.

T = Length of thread = 5.25 in.

$T - E$ = Length of thread minus length of tip = 4.75 in.

From Table 5-16, the specific gravity of the side face of a 24F-V5 glulam beam is 0.55. Entering Table 5-25 with a specific gravity of 0.55, and a lag screw shank diameter of 7/8 inch,

$$P_w \text{ per inch of penetration} = P_w/in. = 664 \text{ lb}$$

P_w is obtained by multiplying $P_w/in.$ by the thread penetration, $T - E$,

$$P_w = P_w/in. \ (T - E) = 664 \ (4.75) = 3,154 \text{ lb}$$

From Table 5-18, $C_M = 0.67$ for lag screws in glulam under wet-use conditions, and

$$P'_W = P_W C_M = 3,154 \ (0.67) = 2,113 \text{ lb}$$

Check Washer Bearing Stress

From NDS Table 4A for No. 1 Southern Pine, surfaced green, used any condition:

$$F_{c\perp} = 375 \text{ lb/in}^2$$

Further adjustment for moisture content in excess of 19 percent is not required, and

$$F_{c\perp}' = F_{c\perp} C_M = 375(1.0) = 375 \text{ lb/in}^2$$

From malleable iron washer sizes in Table 16-7, the bearing area of a 7/8-inch-diameter washer is computed by subtracting the hole area from the total washer area:

$$A = A_{TOTAL} - A_{HOLE} = \pi(1.75)^2 - \pi(0.5)^2 = 9.62 - 0.79 = 8.83 \text{ in}^2$$

The allowable bearing load is equal to the bearing area times the allowable stress in compression perpendicular to grain:

$$\text{Bearing capacity} = AF_{c\perp}' = 8.83(375) = 3,311 \text{ lb}$$

Summary
The allowable capacity of the connection is limited by the withdrawal strength of the lag screw to 3,113 pounds.

TIMBER CONNECTORS

Timber connectors are round steel rings or plates embedded between members in precut grooves. When used with bolts or lag screws, they develop the highest strength in lateral loading of all fastener types. The two types of timber connectors most common in bridge applications are split rings and shear plates (Figure 5-26). Split rings are round steel rings with slightly tapered edges that wedge the connector in the precut grooves. They are manufactured in diameters of 2-1/2 inches and 4 inches from hot-rolled carbon steel meeting Society of Automotive Engineers Specification SAE-1010. As the name implies, the side of the ring is split to allow the connector to expand as it is placed in the groove. Shear plates are 2-5/8-inch or 4-inch-diameter round steel plates with a flange on one side. The 2-5/8-inch plates are pressed from hot-rolled steel meeting SAE-1010. The 4-inch plates are cast malleable iron manufactured to Grade 32510 of ASTM Standard A47. Typical dimensions for split rings and shear plates are given in Table 16-6.

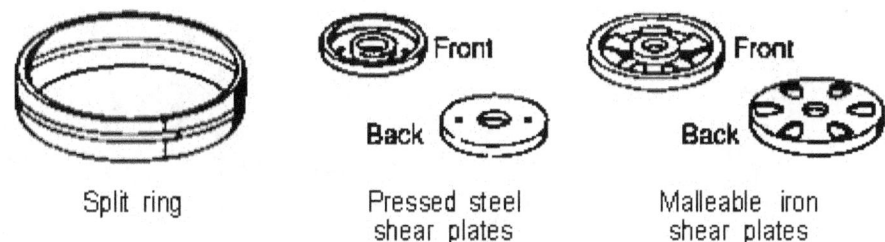

Split ring   Pressed steel shear plates   Malleable iron shear plates

Figure 5-26. - Types of timber connectors.

Timber connectors are used in lateral connections with a bolt or lag screw placed concentrically through the center of the connector (Figure 5-27). Split rings are limited to wood-to-wood connections where one ring is placed at each wood interface. Shear plates are best adapted for wood-to-metal connections but may be used back to back for wood connections; however, one split ring in wood connections is more economical than two shear plates. For both types of connectors, the bolt or lag screw is an integral part of the connector unit and serves to clamp the members together so that the connector functions effectively. For shear plates, the bolt also must transfer the shear across the member interface.

5-109

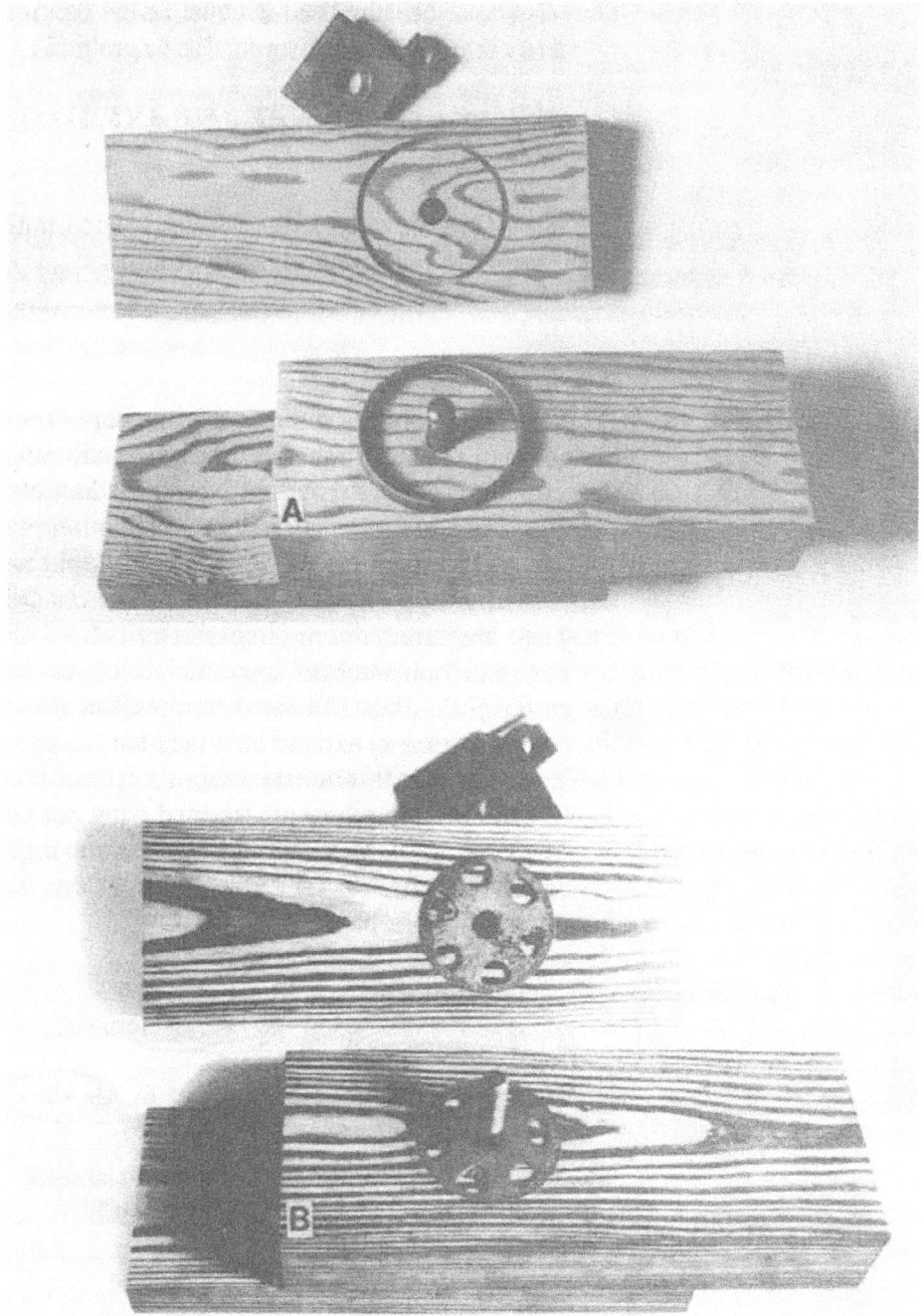

Figure 5-27.- Typical timber connector joints: (A) split ring connector between wood members. (B) Shear plates used back-to-back between wood members.

Figure 5-27. - Typical timber connector joints (continued): (C) Shear plate used between wood and steel members.

Net Area

The net area at a timber connection is the gross area of the member minus the projected area of the bolt holes and the projected area of the connector groove within the member (Figure 5-28). When connectors are staggered, adjacent connectors with a parallel to grain spacing equal to or less than one connector diameter are considered to occur at the same critical section. The required net area in tension and compression members is determined by dividing the total load transferred at the connection by the applicable allowable design stress, F_t' or F_c'.

Design of Lateral Connections

As with other types of lateral connections, the strength of timber connectors is developed by bearing between the connector and the wood (Figure 5-29). Design values for connectors are considerably higher than bolts or

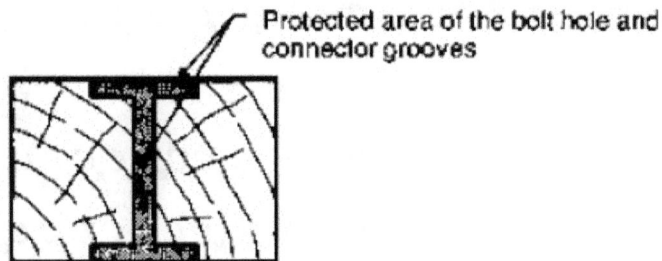

Protected area of the bolt hole and connector grooves

Figure 5-28. - The net area at a timber connector is equal to the gross area of the member minus the projected area of the bolt hole and connector grooves.

5-111

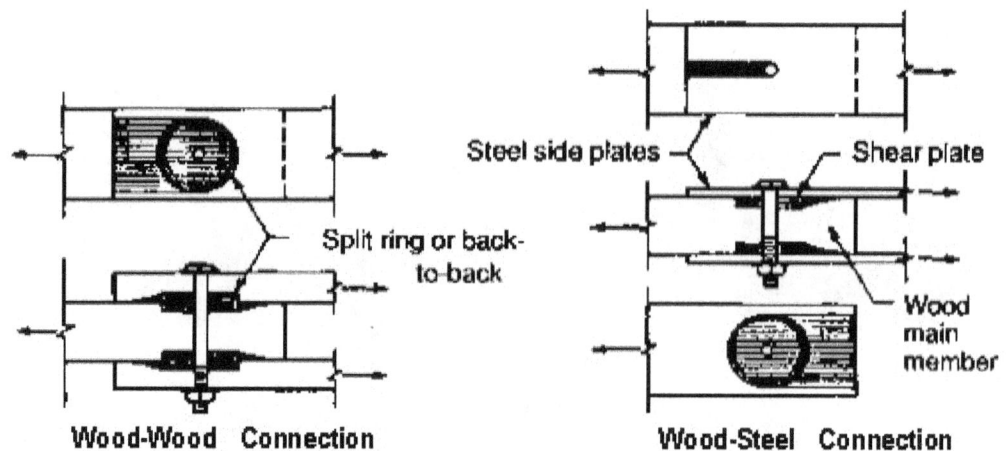

Figure 5-29. - Typical configuration and stress distribution for laterally loaded timber connectors.

lag screws because they bend less and provide more bearing area. With connectors, the inner surface of the ring bears against the inner core of wood, while the outer surface bears against the outer wall of the groove. Split rings are especially efficient because the tongue and groove split allows expansion, resulting in better load distribution in bearing. In most applications, connector capacity is controlled by the strength of the wood; however, for some shear plates, capacity may be controlled by the strength of the connector. In such cases, maximum design values are limited by the NDS.

The allowable value on one timber connector is equal to the tabulated value adjusted by all applicable modification factors. When several connectors are used, the design value is the sum of the individual connector values adjusted by the group action factor, C_g. Applicable modification factors for timber connectors are given by Equations 5-48 and 5-49 for split rings and by Equations 5-50 and 5-51 for shear plates:

For split rings,

$$P' = PC_D C_M C_t C_R C_{\mathit{d}} C_n C_s C_g C_{tb} \tag{5-48}$$

$$Q' = QC_D C_M C_t C_R C_{\mathit{d}} C_n C_s C_g C_{tb} \tag{5-49}$$

For shear plates,

$$P' = PC_D C_M C_t C_R C_{\mathit{d}} C_n C_s C_g C_{st} C_{tb} \leq P'_{MAX} \tag{5-50}$$

$$Q' = QC_D C_M C_t C_R C_{\mathit{d}} C_n C_s C_g C_{tb} \leq Q'_{MAX} \tag{5.51}$$

where P'_{MAX} and Q'_{MAX} are the maximum allowable values for shear plates, limited by the strength of the connector.

When timber connectors are loaded at an angle to the grain, modification factors for end distance (C_n), edge distance (C_j), and spacing (C_j) are based on the loading angle. As a result, these factors are applied after application of the Hankinson formula (Equation 5-42).

Tabulated Design Values

The tabulated NDS design values for split rings and shear plates are shown in Tables 5-27 and 5-28. The values are based on normal duration of load and dry-use conditions for one connector unit with an A307 bolt. For the purpose of determining tabulated values, one connector unit is defined as (1) one split ring in a wood-wood connection, (2) two shear plates back to back in a wood-wood connection, or (3) one shear plate in a wood-steel connection. In each case, the tabulated value is the load that occurs in single shear at the location of the connector, regardless of the total number of members in the connection.

To determine the tabulated value for either type of connector, enter the appropriate table with the connector diameter and read the tabulated value, by species group, based on the number of faces of the piece with connectors on the same bolt, and the net thickness of the thinnest member in contact with the connector. For loading perpendicular to grain, tabulated values are additionally based on the loaded edge distance of the connector. Values for intermediate member thicknesses and loaded edge distances are determined by linear interpolation.

When determining tabulated connector values, the following considerations apply:

1. Timber connectors cannot be used in members less than the minimum net thickness given in Tables 5-27 and 5-28.

2. The bolt diameter specified for each connector is the minimum diameter A307 bolt required to meet tabulated values. Increasing the bolt diameter is permissible but does not increase the tabulated values.

3. Maximum loads on shear plates (P'_{MAX} and Q'_{MAX}) shall not exceed the following:

    (a) 2,900 pounds for a 2-5/8-inch shear plate,

    (b) 4,400 pounds for a 4-inch shear plate with a 3/4-inch bolt, or

    (c) 6,000 pounds for a 4-inch shear plate with a 7/8-inch bolt.

    When tabulated values exceed P'_{MAX} or Q'_{MAX}, they are marked with an asterisk in Table 5-28.

Table 5-27.—Tabulated split ring design values.

Split-ring diam. (inches)	Bolt diam. (inches)	Number of faces of piece with connectors on same bolt	Net thickness of piece (inches)	Minimum edge distance (inches)	Loaded parallel to grain (0°) Design value per connector unit and bolt (pounds) Group A woods	Group B woods	Group C woods	Group D woods	Edge distance (inches) Unloaded edge, min.	Loaded edge	Loaded perpendicular to grain (90°) Design value per connector unit and bolt (pounds) Group A woods	Group B woods	Group C woods	Group D woods
2-1/2	1/2	1	1 min.	1-3/4	2630	2270	1900	1640	1-3/4	1-3/4 min.	1580	1350	1130	970
										2-3/4 or more	1900	1620	1350	1160
			1-1/2 or more	1-3/4	3160	2730	2290	1960	1-3/4	1-3/4 min.	1900	1620	1350	1160
										2-3/4 or more	2280	1940	1620	1390
		2	1-1/2 min.	1-3/4	2430	2100	1760	1510	1-3/4	1-3/4 min.	1460	1250	1040	890
										2-3/4 or more	1750	1500	1250	1070
			2 or more	1-3/4	3160	2730	2290	1960	1-3/4	1-3/4 min.	1900	1620	1350	1160
										2-3/4 or more	2280	1940	1620	1390
4	3/4	1	1 min.	2-3/4	4090	3510	2920	2520	2-3/4	2-3/4 min.	2370	2030	1700	1470
										3-3/4 or more	2840	2440	2040	1760
			1-1/2	2-3/4	6020	5160	4290	3710	2-3/4	2-3/4 min.	3490	2990	2490	2150
										3-3/4 or more	4180	3590	2990	2580
			1-5/8 or more	2-3/4	6140	5260	4380	3790	2-3/4	2-3/4 min.	3560	3050	2540	2190
										3-3/4 or more	4270	3660	3050	2630
		2	1-1/2 min.	2-3/4	4110	3520	2940	2540	2-3/4	2-3/4 min.	2480	2040	1700	1470
										3-3/4 or more	2980	2450	2040	1760
			2	2-3/4	4950	4250	3540	3050	2-3/4	2-3/4 min.	2870	2470	2050	1770
										3-3/4 or more	3440	2960	2460	2120
			2-1/2	2-3/4	5830	5000	4160	3600	2-3/4	2-3/4 min.	3380	2900	2410	2080
										3-3/4 or more	4050	3480	2890	2500
			3 or more	2-3/4	6140	5260	4380	3790	2-3/4	2-3/4 min.	3560	3050	2540	2190
										3-3/4 or more	4270	3660	3050	2630

Design values in pounds apply to one split ring and bolt in single shear when installed in seasoned wood that will remain dry in service and be subject to normal loading conditions.
From the NDS.[28] © 1966. Used by permission.

5-114

Table 5-28.—Tabulated shear plate design values.

Shear-plate diam. (inches)	Bolt diam. (inches)	Number of faces of piece with connectors on same bolt	Net thickness of piece (inches)	Minimum edge distance (inches)	Loaded parallel to grain (0°) Design value per connector unit and bolt (pounds) Group A woods	Group B woods	Group C woods	Group D woods	Edge distance (inches) Unloaded edge, min.	Loaded edge	Loaded perpendicular to grain (90°) Design value per connector unit and bolt (pounds) Group A woods	Group B woods	Group C woods	Group D woods
2-5/8	3/4	1	1-1/2 min.	1-3/4	3110*	2670	2220	2010	1-3/4	1-3/4 min. / 2-3/4 or more	1810 / 2170	1550 / 1860	1290 / 1550	1110 / 1330
		2	1-1/2 min.	1-3/4	2420	2080	1730	1500	1-3/4	1-3/4 min. / 2-3/4 or more	1410 / 1690	1210 / 1450	1010 / 1210	870 / 1040
			2	1-3/4	3190*	2730	2270	1960	1-3/4	1-3/4 min. / 2-3/4 or more	1850 / 2220	1590 / 1910	1320 / 1580	1140 / 1370
			2-1/2 or more	1-3/4	3330*	2860	2380	2060	1-3/4	1-3/4 min. / 2-3/4 or more	1940 / 2320	1660 / 1990	1380 / 1650	1200 / 1440
4	3/4 or 7/8	1	1-1/2 min.	2-3/4	4370	3750	3130	2700	2-3/4	2-3/4 min. / 3-3/4 or more	2540 / 3040	2180 / 2620	1810 / 2170	1550 / 1860
			1-3/4 or more	2-3/4	5090*	4360	3640	3140	2-3/4	2-3/4 min. / 3-3/4 or more	2950 / 3540	2530 / 3040	2110 / 2530	1810 / 2200
		2	1-3/4 min.	2-3/4	3390	2910	2420	2090	2-3/4	2-3/4 min. / 3-3/4 or more	1970 / 2360	1680 / 2020	1400 / 1680	1250 / 1410
			2	2-3/4	3790	3240	2700	2330	2-3/4	2-3/4 min. / 3-3/4 or more	2200 / 2640	1880 / 2260	1570 / 1880	1360 / 1630
			2-1/2	2-3/4	4310	3690	3080	2660	2-3/4	2-3/4 min. / 3-3/4 or more	2500 / 3000	2140 / 2550	1780 / 2140	1540 / 1850
			3	2-3/4	4630	4140	3450	2980	2-3/4	2-3/4 min. / 3-3/4 or more	2800 / 3360	2400 / 2880	2000 / 2400	1720 / 2060
			3-1/2 or more	2-3/4	5030*	4320	3600	3110	2-3/4	2-3/4 min. / 3-3/4 or more	2920 / 3500	2500 / 3000	2090 / 2510	1800 / 2160

Design values in pounds apply to one shear plate unit and bolt in single shear when installed in seasoned wood members that will remain dry in service and be subject to normal loading conditions.
Allowable values for shear plates (after adjustment by applicable modification factors) shall not exceed P'_{max} or Q'_{max} as given below:

a. 2-5/8 inch shear plate .... 2,900 lb.
b. 4-inch shear plate with 3/4 inch bolt .... 4,400 lb.
c. 4-inch shear plate with 7/8 inch bolt .... 6,000 lb.

Loads followed by an asterisk exceed P'_{max} or Q'_{max} but are necessary for proper determination of values for other angles of load to the grain.
From the NDS.[28] © 1986. Used by permission.

4. If concentric grooves for two sizes of split rings are cut in the member, rings must be installed in both grooves; however, the tabulated design value is that for the larger ring only.

Lag Screws

Tabulated values for timber connectors are based on a bolted connection. Lag screws may be used, provided the shank diameter of the lag is the same as specified for a bolt and provided the lag screw threads are cut rather than rolled (cut threads hold better). When lag screws are used instead of bolts, tabulated values must be adjusted by the lag screw factor C_{lb} given in Table 5-29.

Steel Side Plates

When steel rather than wood side plates are used, tabulated values may be increased for 4-inch shear plates loaded parallel to grain only (no increase is allowed for 2-5/8-inch shear plates or split rings). Values of C_{st} are given below. However, the adjusted load on any shear plate is limited to the maximum values P'_{MAX} and Q'_{MAX}.

Species Group	C_{st} for 4-inch shear plates loaded parallel to grain
A	1.18
B	1.11
C	1.05
D	1.00

Table 5-29. - Lag screw modification factor, C_{lb}

Connector size and type	Side plate	Penetration[1]	Penetration of lag screw into member receiving point (number of shank diameters) Fastener species group				C_{lb}
			I	II	III	IV	
2-1/2-inch split ring 4-inch split ring 4-inch shear plate	Wood or Metal	Standard	7	8	10	11	1.00
		Minimum	3	3-1/2	4	4-1/2	0.75
2-5/8-inch shear plate	Wood	Standard	4	5	7	8	1.00
		Minimum	3	3-1/2	4	4-1/2	0.75
2-5/8-inch shear plate	Metal	Standard and Minimum	3	3-1/2	4	4-1/2	1.00

[1] Use straight line interpolation for intermediate values.

From the NDS;[26] © 1986. Used by permission.

Distance and Spacing Requirements

Distance and spacing requirements for timber connectors loaded parallel to grain and perpendicular to grain are summarized in Table 5-30. These requirements are given as the minimum dimension for the full tabulated value and as the minimum dimension for reduced value, as previously discussed for bolts. It is recommended that the minimum dimensions for full tabulated value be used whenever possible. When space is not available, the minimum dimensions for a reduced value may be used provided tabulated values are reduced by the applicable modification factors for edge distance, C_d, end distance, C_N, or spacing, C_s. The edge-distance factor for the loaded edge is already factored into tabulated values, and further application of C_d is not required when tabulated minimum loaded edge-distance values are used. The modification factors C_d, C_N, and C_s are not cumulative, and the lowest value of the three is used. However, when end distance or spacing is reduced for any connector in a group, the lowest applicable factor applies to all connectors in the group. Modification factor values for intermediate dimensions are determined by straight-line interpolation.

When timber connectors are loaded at an angle to the grain of the member, refer to the NDS and the *AITC Timber Construction Manual* for distance and spacing requirements.

Connector Placement

All holes, grooves, and daps for timber connectors must be precision machined with special cutters for proper connector performance and assembly (Figure 5-30). Fabrication is best suited to a shop environment but can be done in the field when shop fabrication is not possible (Figure 5-31). The holes for bolts and lag screws are prebored in the manner previously discussed for the individual fasteners. Grooves and daps for split rings and shear plates must be appropriate for the type and size of connector. Connectors from different manufacturers may differ slightly in shape or cross section, and cutter heads must be specifically designed to accurately conform to the dimensions and shape of the par-

Figure 5-30. - Tools used for grooving wood for timber connectors.

Table 5-30. - Summary of edge distance, end distance, and spacing requirements for timber connectors.

Loading parallel to grain

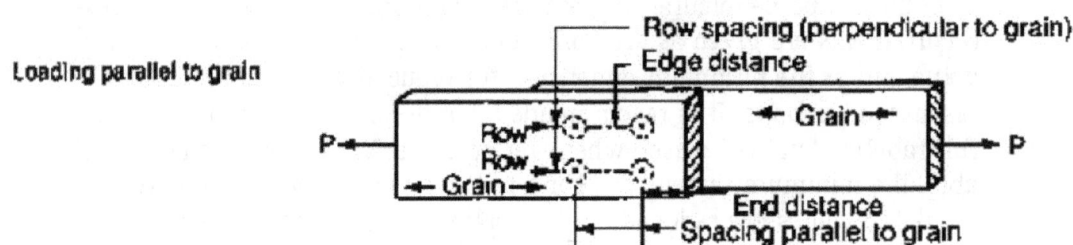

	2-1/2-inch split rings or 2-5/8-inch shear plates		4-inch split rings or shear plates	
	Minimum dimension for full tabulated value (inches)	Minimum dimension for reduced value[1] (inches)	Minimum dimension for full tabulated value (inches)	Minimum dimension for reduced value[1] (inches)
Edge distance	1-3/4	N/A[2]	2-3/4	N/A
End distance				
Tension members	5-1/2	3/4 $(C_n = 0.625)$	7	3-1/2 $(C_n = 0.625)$
Compression members	4	2-1/2 $(C_n = 0.625)$	5-1/2	3-1/4 $(C_n = 0.625)$
Spacing				
Parallel to grain	6-3/4	3-1/2 $(C_a = 0.5)$	9	5 $(C_a = 0.5)$
Row spacing perpendicular to grain	3-1/2	N/A	5	N/A

Loading perpendicular to grain

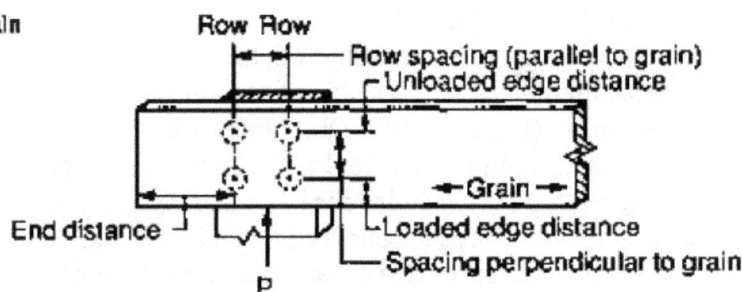

	2-1/2-inch split rings or 2-5/8-inch shear plates		4-inch split rings or shear plates	
	Minimum dimension for full tabulated value (inches)	Minimum dimension for reduced value[1] (inches)	Minimum dimension for full tabulated value (inches)	Minimum dimension for reduced value[1] (inches)
Edge distance				
Unloaded edge	1-3/4	N/A	2-3/4	N/A
Loaded edge[2]	2-3/4	1-3/4	3-3/4	2-3/4
End distance				
Tension members	5-1/2	2-3/4 $(C_n = 0.625)$	7	3-1/2 $(C_n = 0.625)$
Compression members	5-1/2	2-3/4 $(C_n = 0.625)$	7	3-1/2 $(C_n = 0.625)$
Spacing				
Row spacing parallel to grain	3-1/2	N/A	5	N/A
Perpendicular to grain	4-1/2	3-1/2 $(C_a = 0.5)$	5	5 $(C_a = 0.5)$

[1] For dimensions between the tabulated value and the reduced value use straight line interpolation to compute the modification factor value.
[2] See Table 5-27 and Table 5-28 for reduced design values for minimum loaded edge distance.
All dimensions are measured from the center of the connector.

Figure 5-31. - Field-grooving for a split ring connector for a curb-deck attachment. Field fabrication such as this requires field treating with wood preservative, as discussed in Chapter 12 (photo courtesy of Wheeler Consolidated, Inc.).

ticular connector used. The heavy 4-inch-diameter shear plates may sometimes be cast out-of-round and should be checked for dimensions and roundness before assembly. An out-of-round plate should not be forced into a round groove.

Bolts and lag screws installed with timber connectors must be provided with plate or malleable iron washers between the outside wood member and the head or nut of the fastener. Cut washers are not suitable for use with connectors and are not permitted. The minimum washer size for each type of connector is given in Table 16-6. When an outside member is a steel plate or shape, the washer may be omitted except when desirable to prevent bearing on the fastener threads.

Design values for timber connectors are based on the assumption that the faces of the members will be brought into tight contact when the connectors are installed. When timber connectors are installed in wood with a high moisture content, they should be checked periodically to ensure that shrinkage of the wood has not caused members to separate. It may be necessary to retighten connections as the wood dries.

Example 5-15 - Lateral split ring connection

A dressed 12-inch by 12-inch lumber curb is bolted along the edges of a 6-3/4-inch-thick transverse glulam deck. The curb serves at an attachment point for vehicular railing where a transverse reaction of 15,600 pounds is transfered at the center of the curb height. Determine the number of 4-inch-diameter split ring connectors that are required to transfer the curb load to the deck, assuming the following:

1. Members will be exposed to weathering (wet-use conditions) with a duration of load factor, C_D, of 1.65; adjustments for temperature (C_t) and fire-retardant treatment (C_R) are not required.

2. The glulam deck is combination symbol No. 2.

3. The curb is surfaced Douglas Fir-Larch, graded No. 1 to WWPA rules.

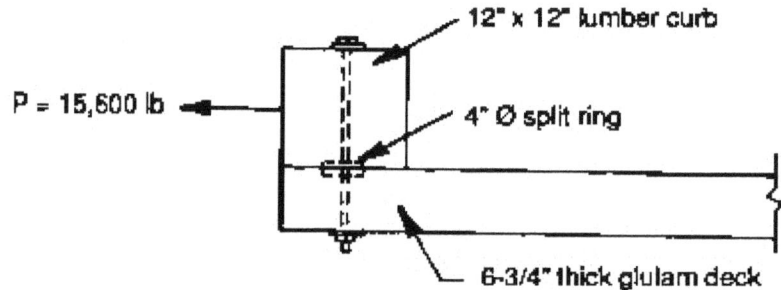

Solution

In this connection, the curb is loaded perpendicular to grain while the deck is loaded parallel to grain. Unlike bolted connections, tabulated values for timber connectors are based on a two-member (single shear) joint and values for loading parallel to grain and perpendicular to grain are read directly from tables in the NDS (Table 5-27). The procedure used here will be to design the connection based on perpendicular-to-grain loading (which normally controls), then check for parallel-to-grain loading.

Curb Loading Perpendicular to Grain

The allowable design value for one split ring loaded perpendicular to grain is given by Equation 5-49. Including possible modification factors for connector distance and spacing, the equation in this case becomes

$$Q' = QC_D C_M C_d C_s$$

From Table 5-14, Douglas Fir-Larch is in Load Group B for timber connector design. The tabulated value for one split ring loaded perpendicular to grain is obtained from Table 5-27. Entering that table for a 4-inch-diameter split ring, 3/4-inch bolt, one member face with a connector on the same bolt, member thickness greater than 1-5/8-inches, and a loaded edge distance greater than 3-3/4-inches:

$$Q = 3,660 \text{ lb}$$

From Table 5-18 for timber connectors used in partially seasoned or wet-condition sawn lumber:

$$C_M = 0.67$$

Minimum values of connector distance and spacing for full loading are obtained from Table 5-30:

Unloaded edge distance   2.75 in.

Loaded edge distance   3.75 in.

Spacing parallel to grain   5 in.

From Table 16-2, the width of a dressed 12-inch by 12-inch curb is 11.5 inches. Centering the connector on the curb provides a loaded and unloaded edge distance of 5.75 inches:

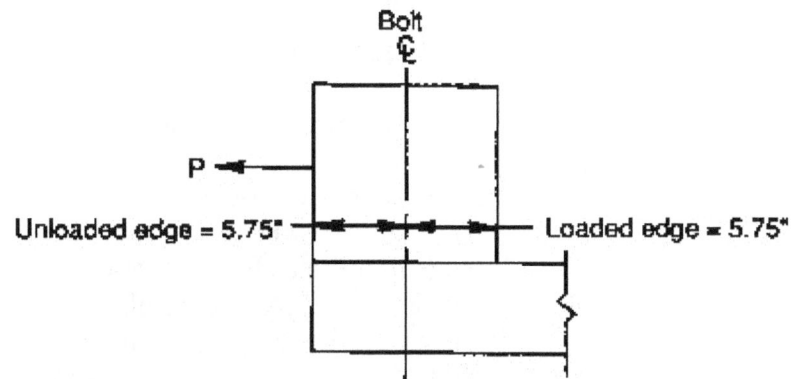

Using a minimum connector spacing of 5 inches, all distance and spacing requirements for full load are met and values of C_e and C_s each become 1.0.

The allowable load for one split ring is computed by substituting values into the equation for Q':

$$Q' = QC_D C_M C_e C_s = 3,660(1.65)(0.67)(1.0)(1.0) = 4,046 \text{ lb}$$

The required number of split rings is obtained by dividing the applied load by Q':

$$\text{Number of split rings} = \frac{15,600 \text{ lb}}{4,046 \text{ lb}} = 3.86 = 4$$

Deck Loading Parallel to Grain

Using the applicable modification factors for this case, the allowable load for one split ring loaded parallel to grain is given by Equation 5-48:

$$P' = PC_D C_M C_e C_\Delta C_t$$

From Table 5-15, glulam combination symbol No. 2 is in Load Group B for timber connector design. The tabulated value for one split ring loaded parallel to grain is obtained from Table 5-27 using the same table values previously used for loading perpendicular to grain:

$$P = 5,260 \text{ lb}$$

From Table 5-18 for timber connectors used in glulam under wet-use conditions:

$$C_M = 0.67$$

Minimum values of connector distance and spacing for full loading are obtained from Table 5-30:

Edge distance	2.75 in.
End distance (tension members)	7 in.
Spacing perpendicular to grain	5 in.

All distance and spacing requirements can be met with the exception of end distance, which is 5.75 inches rather than the 7 inches required for full load (end distance for parallel-to-grain loading is the same as the unloaded edge distance for perpendicular-to-grain loading). From Table 5-30, end distance can be reduced to a minimum of 3.5 inches provided the tabulated load is reduced by $C_e = 0.625$. Using linear interpolation for the 5.75-inch distance, $C_e = 0.87$.

Substituting values into the equation for P',

$$P' = PC_D C_M C_e C_\Delta C_t = 5,260(1.65)(0.67)(1.0)(0.87)(1.0) = 5,059 \text{ lb}$$

$P' = 5,059 \text{ lb} > Q' = 4,046$ pounds so connector capacity is controlled by loading perpendicular to grain.

Summary

The connection will be made using four 4-inch-diameter split rings with 3/4-inch-diameter bolts. The capacity of the connection is limited by curb loading perpendicular to grain to 16,184 pounds. The bolts will be spaced 5 inches on-center and will be provided with malleable iron washers at each end:

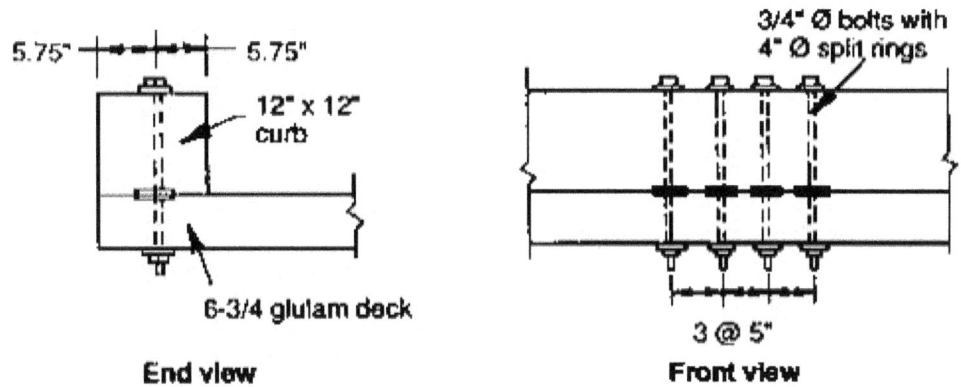

End view   **Front view**

Example 5-16 - Lateral shear-plate connection

A glulam tension member measures 3-inches wide by 5.5 inches deep. The end of the member is held between steel plates by two 3/4-inch-diameter bolts with four 4-inch-diameter shear plates. Determine the capacity of the connection, assuming the following:

1. Members will be exposed to weathering (wet-use conditions) and a normal duration of load; adjustments for temperature (C_t) and fire-retardant treatment (C_R) are not required.

2. The glulam is combination symbol No. 47.

3. The capacity of the steel plates is satisfactory.

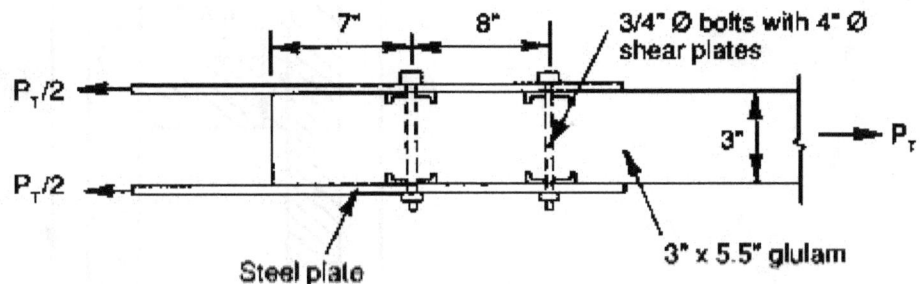

Solution

The capacity of this connection will be controlled either by the strength of the glulam member or the strength of the connectors. Glulam capacity will be computed first, followed by connector capacity.

Capacity of Glulam Member

The capacity of the glulam member in tension is equal to the allowable tensile stress times the net member area. The allowable stress in tension parallel to grain is computed using the applicable modification given by Equation 5-23:

$$F_t' = F_t C_D C_M$$

From *AITC 117-Design* for combination symbol No. 47,

$$F_t = 1,200 \text{ lb/in}^2$$

From Table 5-7,

$$C_M = 0.80$$

Substituting,

$$F_t' = F_t C_D C_M = 1,200(1.0)(0.80) = 960 \text{ lb/in}^2$$

The net area of the member is equal to the gross area minus the projected area of the shear plates and bolt hole. Dimensions of the shear plates and bolt hole are obtained from timber connector properties given in Table 16-6:

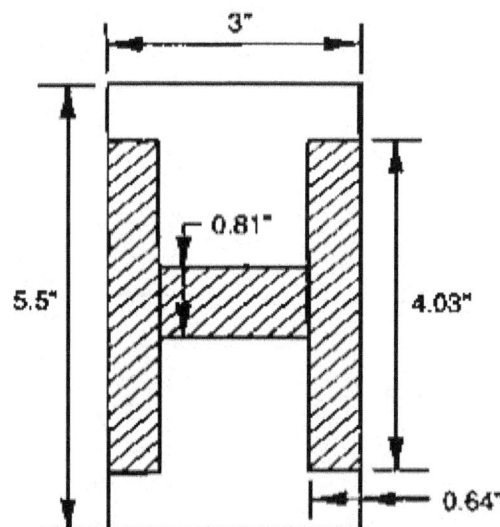

$$\text{Gross area} = (3 \text{ in.})(5.5 \text{ in.}) = 16.5 \text{ in}^2$$

$$\text{Shear plate area} = 2\,[(4.03 \text{ in.})(0.64 \text{ in.})] = 5.16 \text{ in}^2$$

$$\text{Bolt hole area} = [\,(3 \text{ in.}) - 2(0.64 \text{ in.})]\,(0.81 \text{ in}) = 1.39 \text{ in}^2$$

$$A_{NET} = 16.5 \text{ in}^2 - 5.16 \text{ in}^2 - 1.39 \text{ in}^2 = 9.95 \text{ in}^2$$

The member capacity equals the allowable stress times the net area:

$$\textbf{Member capacity} = F_t{'}(A_{NET}) = 960\,(9.95) = 9{,}552\ \textbf{lb}$$

Capacity of Shear Plates

The allowable design value for one shear plate loaded parallel to grain is given by Equation 5-50. Including possible modification factors for connector distance and spacing, the equation in this case becomes

$$P' = PC_D C_M C_e C_n C_s C_g C_{st}$$

From Table 5-15, glulam combination symbol No. 47 is in Load Group B for timber connector design. The tabulated value for one shear plate loaded parallel to grain is obtained from Table 5-28. Entering that table for a 4-inch-diameter shear plate, 3/4-inch-diameter bolt, two member faces with a connector on the same bolt, and a member thickness of 3 inches:

$$P = 4{,}140 \text{ lb}$$

From Table 5-18 for timber connectors used under wet-use conditions in glulam,

$$C_M = 0.67$$

Values of connector distance and spacing for full loading are obtained from Table 5-30:

Edge distance	2.75 in.
End distance (tension members)	7 in.
Spacing parallel to grain	9 in.

All distance and spacing requirements for full load are met except spacing parallel to grain, which is 8 inches instead of 9 inches. Spacing can be reduced to a minimum of 5 inches provided tabulated values are reduced by $C_s = 0.50$. By interpolation for an 8-inch spacing,

$$C_s = 0.88$$

For two connectors in a row, adjustment for group action is not required, and

$$C_g = 1.0$$

For steel side plates used with Group B species,

$$C_{st} = 1.11$$

Substituting values into the equation for P',

$$P' = PC_D C_M C_t C_n C_s C_g C_{st}$$

$$= 4,140(1.0)(0.67)(1.0)(1.0)(0.88)(1.0)(1.11) = 2,709 \text{ lb}$$

For four shear plates,

$$4(P') = 4(2709) = 10,836 \text{ lb}$$

Summary
The capacity of the connection is 9,552 pounds and is controlled by the capacity of the glulam member in tension parallel to grain.

NAILS AND SPIKES

Nails and spikes are the most common wood fastener for building construction. For bridge applications, however, their use is mostly limited to laminating lumber decks and attaching plank-wearing surfaces. Design is usually based on nailing schedules or specification requirements rather than on structural analysis, but an engineered design may be required in some situations. The primary disadvantage with nails and spikes is their susceptibility to loosening from vibrations or changes in moisture content. Withdrawal connections are not recommended, and discussions in this section are limited to lateral loading conditions only. Refer to the NDS for criteria on withdrawal connections.

Nails and spikes are available in a wide variety of lengths and diameters in four different types: box nails, common wire nails, common wire spikes, and threaded hardened-steel nails and spikes. Size is specified by penny-weight, or by diameter and length for larger spikes (Table 5-31). Spikes are longer and have a larger diameter than nails. Most nails and spikes are manufactured from low- or medium-carbon steel. Threaded hardened-steel nails and spikes are made of high-carbon steel wire that is heat treated and tempered to provide higher strength.

Table 5-31. - Typical sizes of nails and spikes.

Pennyweight	Length (in.)	Wire diameter (in.)			
		Box nails	Common wire nails	Threaded hardened-steel nails	Common wire spikes
6d	2	0.099	0.113	0.120	—
8d	2-1/2	0.113	0.131	0.120	—
10d	3	0.128	0.148	0.135	0.192
12d	3-1/4	0.128	0.148	0.135	0.192
16d	3-1/2	0.135	0.162	0.148	0.207
20d	4	0.148	0.192	0.177	0.225
30d	4-1/2	0.148	0.207	0.177	0.244
40d	5	0.162	0.225	0.177	0.263
50d	5-1/2	—	0.244	0.177	0.283
60d	6	—	0.263	0.177	0.283
70d	7	—	—	0.207	—
80d	8	—	—	0.207	—
90d	9	—	—	0.207	—
5/16	7	—	—	—	0.312
3/8	8-1/2	—	—	—	0.375

From the NDS;[26] ©1986. Used by permission.

Nail and spike classifications are based on the type of shank, whether smooth or deformed (Figure 5-32). Deformed shanks are generally spiral (helical) or ringed, but patterns may vary. Deformed shanks are used in most bridge applications because they provide greater withdrawal resistance and are less susceptible to loosening from vibrations or changes in wood moisture content.

Net Area

The net area at nailed or spiked connections is normally taken as the gross area of the member. When large-diameter spikes are used, the net area may be computed by subtracting the projected area of the fasteners, based on designer judgment.

Design of Lateral Connections

In laterally loaded nail and spike connections, the capacity of the connection is controlled by deformation (slip) rather than strength. As a result, design values are independent of the direction of loading with respect to the direction of grain. The allowable value for one nail or spike is the tabulated design value from the NDS adjusted by all applicable modification factors, as given by

$$P_N' = P_N C_D C_M C_t C_R C_{st}$$

(5-52)

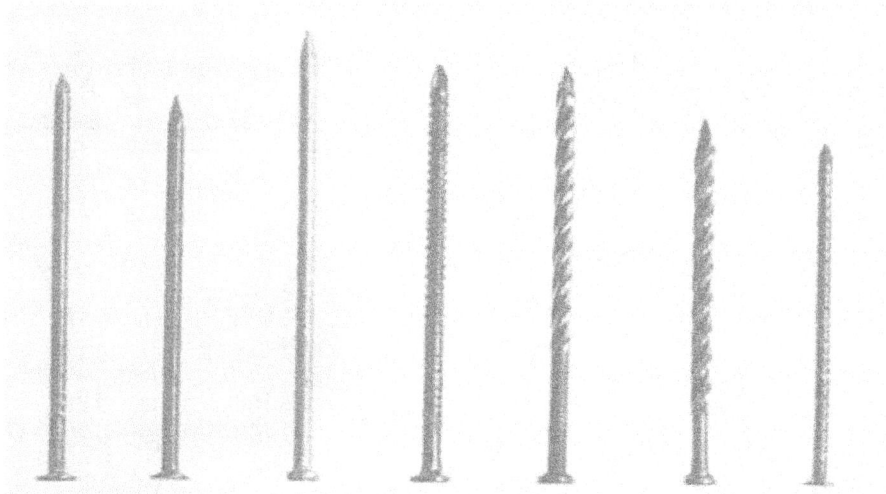

Figure 5-32. - Types of nails: (left to right), bright, smooth wire; cement coated; zinc-coated; annularly threaded; helically threaded; helically threaded and barbed; and barbed.

where P_n' is the allowable lateral load applied at any angle to the grain of the members.

When more than one nail or spike is used, the allowable value for the connection is the sum of the individual design values. Adjustment by the group action factor, C_g, is not required for nails and spikes.

Tabulated Design Loads

Tabulated lateral values for nails and spikes loaded at any angle to grain are given in Table 5-32. The values are for side-grain connections in seasoned wood and are based on the depth of penetration of the nail or spike into the member. In two-member connections, the penetration is measured in the member holding the point. For three-member connections, the penetration is measured in the center member. To determine the tabulated value, enter the table with the type and size of fastener and read horizontally across from the applicable species group (for connections with members of different species, use the higher numbered species group). For full tabulated value, penetration must be a minimum of 10 diameters in Group I species, 11 diameters in Group II species, 13 diameters in Group III species, and 14 diameters in Group IV species. The minimum penetration for any connection cannot be less than one-third of these values. For intermediate penetrations, values are determined by linear interpolation between zero and the tabulated value. However, values cannot be increased for penetrations greater than those required for full tabulated value.

Table 5-32. - Tabulated lateral load design values for nails and spikes.

Common Wire Nails

Pennyweight	6d	8d	10d	12d	16d	20d	30d	40d	50d	60d
Length	2	2-1/2	3	3-1/4	3-1/2	4	4-1/2	5	5-1/2	6
Diameter	0.113	0.131	0.148	0.148	0.162	0.192	0.207	0.225	0.244	0.263
10 Diameters	1.13	1.31	1.48	1.48	1.62	1.92	2.07	2.25	2.44	2.63
11 Diameters	1.24	1.44	1.63	1.63	1.78	2.11	2.28	2.48	2.68	2.89
13 Diameters	1.47	1.70	1.92	1.92	2.11	2.50	2.69	2.93	3.17	3.42
14 Diameters	1.58	1.83	2.07	2.07	2.27	2.69	2.90	3.15	3.42	3.68
Species Group I	77	97	115	116	133	172	192	218	246	275
Species Species Group II	63	78	94	94	108	139	155	175	199	223
Species Species Group III	51	64	77	77	88	114	127	144	163	182
Species Species Group IV	41	51	61	61	70	91	102	115	130	146

Threaded Hardened-Steel Nails and Spikes

Pennyweight	6d	8d	10d	12d	16d	20d	30d	40d	50d	60d	70d	80d	90d
Length	2	2-1/2	3	3-1/4	3-1/2	4	4-1/2	5	5-1/2	6	7	8	9
Diameter	0.120	0.120	0.135	0.135	0.148	0.177	0.177	0.177	0.177	0.177	0.207	0.207	0.207
10 Diameters	1.20	1.20	1.35	1.35	1.48	1.77	1.77	1.77	1.77	1.77	2.07	2.07	2.07
11 Diameters	1.32	1.32	1.49	1.49	1.63	1.95	1.95	1.95	1.95	1.95	2.28	2.28	2.28
13 Diameters	1.56	1.56	1.76	1.76	1.92	2.30	2.30	2.30	2.30	2.30	2.69	2.69	2.69
14 Diameters	1.68	1.68	1.89	1.89	2.07	2.48	2.48	2.48	2.48	2.48	2.90	2.90	2.90
Species Group I	77	97	116	116	133	172	172	172	172	172	218	218	218
Species Group II	63	78	94	94	108	139	139	139	139	139	176	176	176
Species Group III	51	64	77	77	88	114	114	114	114	114	144	144	144
Species Group IV	41	51	61	61	70	91	91	91	91	91	115	115	115

Common Wire Spikes

Pennyweight	10d	12d	16d	20d	30d	40d	50d	60d	5/16"	3/8"
Length	3	3-1/4	3-1/2	4	4-1/2	5	5-1/2	6	7	8-1/2
Diameter	0.192	0.192	0.207	0.225	0.244	0.263	0.283	0.283	0.312	0.375
10 Diameters	1.92	1.92	2.07	2.25	2.44	2.63	2.83	2.83	3.12	3.75
11 Diameters	2.11	2.11	2.28	2.48	2.68	2.89	3.11	3.11	3.43	4.13
13 Diameters	2.50	2.50	2.69	2.93	3.17	3.42	3.68	3.68	4.06	4.88
14 Diameters	2.69	2.69	2.90	3.15	3.42	3.68	3.96	3.96	4.37	5.25
Species Group I	172	172	192	218	246	275	307	307	356	468
Species Group II	139	139	155	176	199	223	248	248	288	379
Species Group III	114	114	127	144	163	182	203	203	235	310
Species Group IV	91	91	102	115	130	146	163	163	188	248

Diameters and lengths are in inches; loads are in pounds.
Design values are for lateral loads in single shear (two members) for nails and spikes penetrating not less than 10 diameters in Group I species, 11 diameters in Group II species, 13 diameters in Group III species, and 14 diameters in Group IV species, into the member holding the point. For other diameters and lengths refer to the *Wood Handbook.*[35]
From the NDS;[26] ©1986. Used by permission.

Steel Side Plates

When steel rather than wood side plates are used for lateral connections, the tabulated design values for nails and spikes may be increased by the steel side plate factor $(C_{st} = 1.25)$.

Distance and Spacing Requirements

End distance, edge distance, and spacing of nails and spikes should be sufficient to avoid unusual splitting of the wood. Although no criteria or dimensions are given in AASHTO or the NDS, the following criteria are given in the *Wood Handbook* [35] based on the diameter d of the nail or spike:

End distance (tension members)	*15d*
End distance (compression members)	*12d*
Edge distance	*10d*

Nail and Spike Placement

Nails and spikes are generally hand-driven but may be placed with power drivers for smaller diameters and lengths. They should be driven through the thinner member, into a thicker member, and be flush or countersunk to the member surface. Holes for large-diameter fasteners should be prebored to prevent the wood from splitting during placement. In such cases, the diameter of the lead hole must not exceed 0.90 times the fastener diameter for Group I species and 0.75 times the fastener diameter for Group II, III, and IV species. For deformed shanks, the diameter of the nail or spike may vary among types and manufacturers and should be verified before preboring lead holes.

Example 5-17 - Lateral nailed connection

A nominal 2-inch by 6-inch handrail is attached to a 6-inch by 6-inch post with common wire nails. The connection between the rail and post must be capable of resisting a downward force of 300 pounds. Determine the size and number of common wire nails that are required for the connection, assuming the following:

1. Members will be exposed to weathering (wet-use conditions) with a normal duration of load; adjustments for temperature (C_t) and fire-retardant treatment (C_R) are not required.

2. Lumber is surfaced Southern Pine.

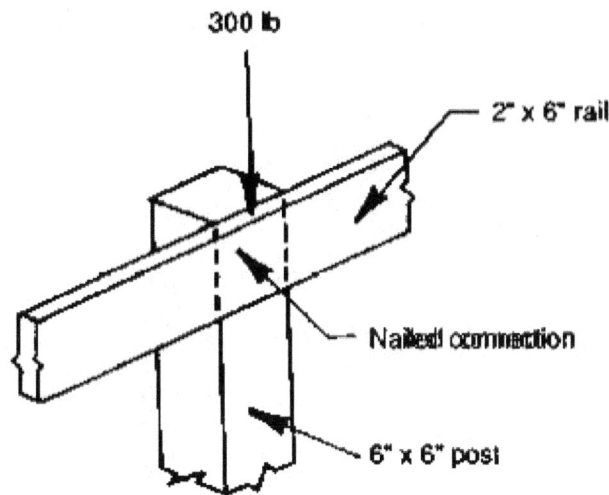

300 lb

2" x 6" rail

Nailed connection

6" x 6" post

Solution

In this connection, the rail is loaded perpendicular to grain while the post is loaded parallel to grain. For nailed connections, however, allowable loads are independent of load orientation to grain. The allowable load for one nail loaded in either direction is computed using Equation 5-52 with the applicable modification factors:

$$P_N' = P_N C_D C_M$$

The moisture modification factor for nailed connections is obtained from Table 5-18:

$$C_M = 0.75$$

Tabulated values for nails and spikes are given in the NDS (Table 5-32). From Table 5-14, Southern Pine is in Species Group II for nailed and spiked connections. To develop the full tabulated load in this species group, the nail must penetrate a minimum of 11 diameters $(11D)$ into the member holding the point (reduced penetration requires reduced load). In this case, the nail length minus $11D$ must not be less than the rail thickness of 1-1/2 inches. Using information from Table 5-32, nail sizes are evaluated to determine the minimum nail pennyweight for full penetration:

Pennyweight	Length (in.)	11D (in.)	Length − 11D (in.)
8d	2.5	1.44	1.06
10d	3	1.63	1.37
12d	3.25	1.63	1.62

A 12d nail is the minimum nail size that provides the required penetration for full load.

Substituting values for the allowable load on one nail,

$$P_N' = P_N C_D C_M = P_N(1.0)(0.75) = P_N(0.75)$$

Using tabulated values from Table 5-32 for nails 12d and larger, a table is compiled of allowable nail loads and the number of nails required:

Pennyweight	P_N (lb)	P_N' (lb)	# nails required
12d	94	70.5	4.3 = 5
16d	108	81.0	3.7 = 4
20d	139	104.3	2.9 = 3
30d	155	116.3	2.6 = 3

In any nailed connection it is desirable to use the minimum diameter and number of nails to minimize the potential for splitting. In this case, the 20d nails will be used because only three nails are required and the increase in diameter from 16d to 20d is small.

Summary
The connection will be made with three 20d nails for a connection capacity of 3(104.3 lb) = 313 lb.

DRIFT BOLTS AND DRIFT PINS

Drift pins and drift bolts are long, unthreaded steel rods that are driven in prebored holes for lateral connections in large timber members. Drift bolts have a head, for use with steel side plates and for convenience in driving, while drift pins have no head (Figure 5-33). In bridge applications, drift bolts and drift pins are used for connecting pile caps to timber piles or posts, or for attaching sawn lumber beams to their supporting cap or sill (Figure 5-34). Manufactured fasteners generally conform to ASTM A307, but pins of concrete reinforcing steel also are used. Because they have poor resistance in withdrawal, drift bolts and drift pins are not recommended for bridge connections subjected to significant withdrawal forces.

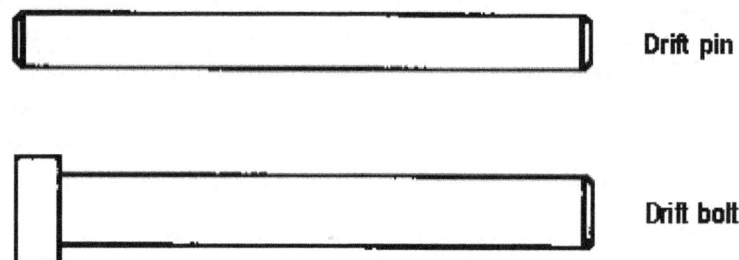

Drift pin

Drift bolt

Figure 5-33. - Typical drift pin and drift bolt.

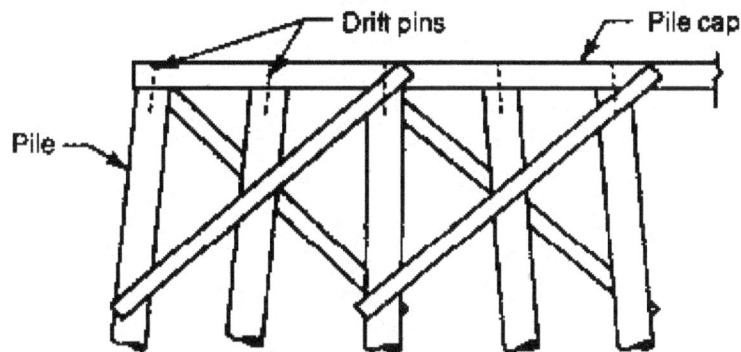

Figure 5-34. - Drift pins or drift bolts are normally used to connect large timber members such as a pile cap to piling.

There is little design information available on drift bolts or drift pins, and requirements for net area, end distance, edge distance, and spacing are taken to be the same as those for a bolt of the same diameter. The NDS specifies that lateral design values in wood side grain not exceed 75 percent of the design value for a comparable bolt of the same diameter and length in the main member. Fastener penetration is left to the judgment of the designer. Drift bolts and drift pins are driven in prebored holes that are 1/8 inch to 1/16 inch smaller in diameter than the fastener.

5.9 SELECTED REFERENCES

1. American Association of State Highway and Transportation Officials. 1983. Standard specifications for highway bridges. 13th ed. Washington, DC: American Association of State Highway and Transportation Officials. 394 p.

2. American Institute of Timber Construction. 1986. Bolts in glued laminated timber. AITC Tech. Note No. 8. Englewood, CO: American Institute of Timber Construction. 1 p.

3. American Institute of Timber Construction. 1977. Connections in glued timber. AITC Tech. Note No. 3. Englewood, CO: American Institute of Timber Construction. 17 p.

4. American Institute of Timber Construction. 1987. Design standard specifications for structural glued laminated timber of softwood species. AITC 117-87-Design. Englewood, CO: American Institute of Timber Construction. 28 p.

5. American Institute of Timber Construction. 1985. Spec-data sheet on structural glued laminated timber. Cat. No. 06170. Englewood, CO: American Institute of Timber Construction. 4 p.

6. American Institute of Timber Construction. 1985. Timber construction manual. 3d ed. New York: John Wiley and Sons, Inc. 836 p.

7. American Society of Civil Engineers. 1975. Wood structures, a design guide and commentary. New York: American Society of Civil Engineers. 416 p.

8. Breyer, D.E. 1980. Design of wood structures. New York: McGraw Hill. 542 p.

9. Brungraber, R.L. 1983. Timber design for the civil professional engineering exam. San Carlos, CA: Professional Publications. 202 p.

10. Canadian Wood Council. 1982. Canadian wood construction. Bolts, lag screws, and timber connectors. CWC datafile WJ-4. Ottawa, Can.: Canadian Wood Council. 28 p.

11. Canadian Wood Council. 1982. Canadian wood construction. Design-lumber and glued laminated lumber. CWC datafile WP-1. Ottawa, Can.: Canadian Wood Council. 32 p.

12. Canadian Wood Council. 1986. Canadian wood construction. Glued laminated timber design. CWC datafile WD-3. Ottawa, Can.: Canadian Wood Council. 36 p.

13. Canadian Wood Council. 1986. Canadian wood construction. Introduction to wood design. CWC datafile WD-1. Ottawa, Can.: Canadian Wood Council. 16 p.

14. Canadian Wood Council. 1986. Canadian wood construction. Lumber design. CWC datafile WD-2. Ottawa, Can.: Canadian Wood Council. 28 p.

15. Canadian Wood Council. 1982. Canadian wood construction. Lumber standards and design data. CWC datafile WP-3. Ottawa, Can.: Canadian Wood Council. 20 p.

16. Canadian Wood Council. Canadian wood construction. Nails, spikes and staples. CWC datafile WJ-2. Ottawa, Can.: Canadian Wood Council. 12 p.

17. Canadian Wood Council. 1982. Canadian wood construction. Structural glued laminated timber specification and design data. CWC datafile WP-4, Ottawa, Can.: Canadian Wood Council. 24 p.

18. Canadian Wood Council. 1986. Effect of datafile SP-1. Ottawa, Can.: Canadian Wood Council. 16 p.

19. Gerhards, C.C. 1977. Effect of duration and rate of loading on strength of wood and wood-based materials. Res. Pap. FPL-283. Madison, WI: U.S. Department of Agriculture, Forest Service, Forest Products Laboratory. 24 p.

20. Goodman, J.R. 1970. Orthotropic elastic properties of wood. Journal of the Structural Division, American Society of Civil Engineers 96(ST11): 2301-2319.

21. Gurfinkel, G. 1981. Wood engineering. 2d ed. Dubuque, IA: Kendall/Hunt Publishing Co. 552 p.

22. Hockaday, E. 1980. Preliminary study of duration of load for glulam wood bridges used in rural areas. St. Paul, MN: Weyerhaeuser Co. 4 p.

23. Milbradt, K.P. 1968. Timber structures. In: Gaylord, E.H., Jr.; Gaylord, C.N., eds. Structural engineering handbook. New York: McGraw Hill. Chapter 16.

24. National Forest Products Association. 1986. Design values for wood construction. A supplement to the national design specification for wood construction. Washington, DC: National Forest Products Association. 34 p.

25. National Forest Products Association. 1979. Heavy timber construction details. Wood Constr. Data No. 5. Washington, DC: National Forest Products Association. 28 p.

26. National Forest Products Association. 1986. National design specification for wood construction. Washington, DC: National Forest Products Association. 87 p.

27. National Forest Products Association. 1978. Wood structural design data. Washington, DC: National Forest Products Association. 240 p.

28. Smith, A.K. 1945. Timber connectors in highway structures. British Columbia, Can. Lumberman 29: 40-41, 104.

29. Somayaji, F.S. 1983. Design of timber beams: a graphical approach. Journal of Structural Engineering 109: 271-278.

30. Southern Forest Products Association. 1983. Freedom of design with pressure-treated Southern Pine. New Orleans, LA: Southern Forest Products Association. 13 p.

31. Southern Forest Products Association. 1985. Southern Pine use guide. New Orleans, LA: Southern Forest Products Association. 13 p.

32. Timber Engineering Co. 1978. Structural wood fasteners. TECO publ. no. 101. Washington, DC: Timber Engineering Co. 15 p

33. U.S. Department of Agriculture, Forest Service, Forest Products Laboratory. 1982. Wood as a structural material. Dietz, A.G.H.; Schaffer, E.L.; Gromala, D.S., eds. Clark C. Heritage Memorial Series on Wood. Compilation of educational modules; 2d Clark C. Heritage Memorial Workshop; 1980, August; Madison, WI. University Park, PA: The Pennsylvania State University. Vol. 2.282 p.

34. U.S. Department of Agriculture, Forest Service, Forest Products Laboratory. 1986. Wood: Engineering design concepts. Freas, A.D.; Moody, R.C.; Soltis, L., eds. Clark C. Heritage Memorial Series on Wood. Compilation of educational modules; 4th Clark C. Heritage Memorial Workshop; 1982, August; Madison, WI. University Park, PA: The Pennsylvania State University. Vol. 4. 600 p.

35. U.S. Department of Agriculture. 1987. Wood handbook: Wood as an engineering material. Agric. Handb. No. 72. Madison, WI: U.S. Department of Agriculture, Forest Service, Forest Products Laboratory. 466 p.

36. Western Wood Products Association. 1971. Dimensional stability. WWPA tech. guide TG-3. Portland, OR: Western Wood Products Association. 32 p.

37. Western Wood Products Association. 1984. Lumber specification information. WWPA spec point A-2. Portland, OR: Western Wood Products Association. 2 p.

38. Western Wood Products Association. 1983. Western woods use book. 3d ed. Portland, OR: Western Wood Products Association. 350 p.

CHAPTER 6

LOADS AND FORCES ON TIMBER BRIDGES

6.1 INTRODUCTION

A bridge must be designed to safely resist all loads and forces that may reasonably occur during its life. These loads include not only the weight of the structure and passing vehicles, but also loads from natural causes, such as wind and earthquakes. The loads may act individually but more commonly occur as a combination of two or more loads applied simultaneously. Design requirements for bridge loads and loading combinations are given in *AASHTO Standard Specifications for Highway Bridges* (AASHTO).[3] AASHTO loads are based on many years of experience and are the minimum loads required for design; however, the designer must determine which loads are likely to occur and the magnitudes and combinations of loads that produce maximum stress.

This chapter discusses AASHTO load fundamentals as they relate to timber bridges. Methods and requirements for determining the magnitude and application of individual loads are presented first, followed by discussions on loading combinations and group loads. Additional information on load application and distribution related to specific bridge types is given in succeeding chapters on design.

6.2 DEAD LOAD

Dead load is the permanent weight of all structural and nonstructural components of a bridge, including the roadway, sidewalks, railing, utility lines, and other attached equipment. It also includes the weight of components that will be added in the future, such as wearing surface overlays. Dead loads are of constant magnitude and are based on material unit weights given by AASHTO (Table 6-1). Note that the minimum design dead load for timber is 50 lb/ft^3 for treated or untreated material.

Dead loads are commonly assumed to be uniformly distributed along the length of a structural element (beam, deck panel, and so forth). The load sustained by any member includes its own weight and the weight of the components it supports. In the initial stages of bridge design, dead load is unknown and must be estimated by the designer. Reasonable estimates may be obtained by referring to similar types of structures or by using empirical formulas. As design progresses, members are proportioned and dead loads are revised. When these revised loads differ significantly from estimated values, the analysis must be repeated. Several revision cycles

Table 6-1. - Material dead load unit weights.

Material	Dead load (lb/ft³)
Timber (treated or untreated)	50
Steel of cast steel	490
Cast iron	450
Aluminum alloys	175
Concrete (plain or reinforced)	150
Pavement, other than wood block	150
Macadam or rolled gravel	140
Compacted sand, earth, gravel, or ballast	120
Loose sand, earth, and gravel	00
Cinder filling	60
Stone masonry	70

From AASHTO' 3.3.6; © 1983. Used by permission.

may be required before arriving at a final design. It is often best to compute the final dead load of one portion of the structure before designing its supporting members.

6.3 VEHICLE LIVE LOAD

Vehicle live load is the weight of the vehicles that cross the bridge. Each of these vehicles consists of a series of moving concentrated loads that vary in magnitude and spacing. As the loads move, they generate changing moments, shears, and reactions in the structural members. The extent of these forces depends on the number, weight, spacing, and position of the loads on the span. The designer must position vehicle live loads to produce the maximum effect for each stress. Once the locations for maximum stress are found, other positions result in lower stress and are no longer considered.

TERMINOLOGY

Vehicle live loads are generally depicted in diagrams that resemble trucks or other specialized vehicles. The terms used to describe these loads are defined below and shown in Figure 6-1.

Gross vehicle weight (GVW) is the maximum total weight of a vehicle.

Axle load is the total weight transferred through one axle.

Axle spacing is the center-to-center distance between vehicle axles. Axle spacing may be fixed or variable.

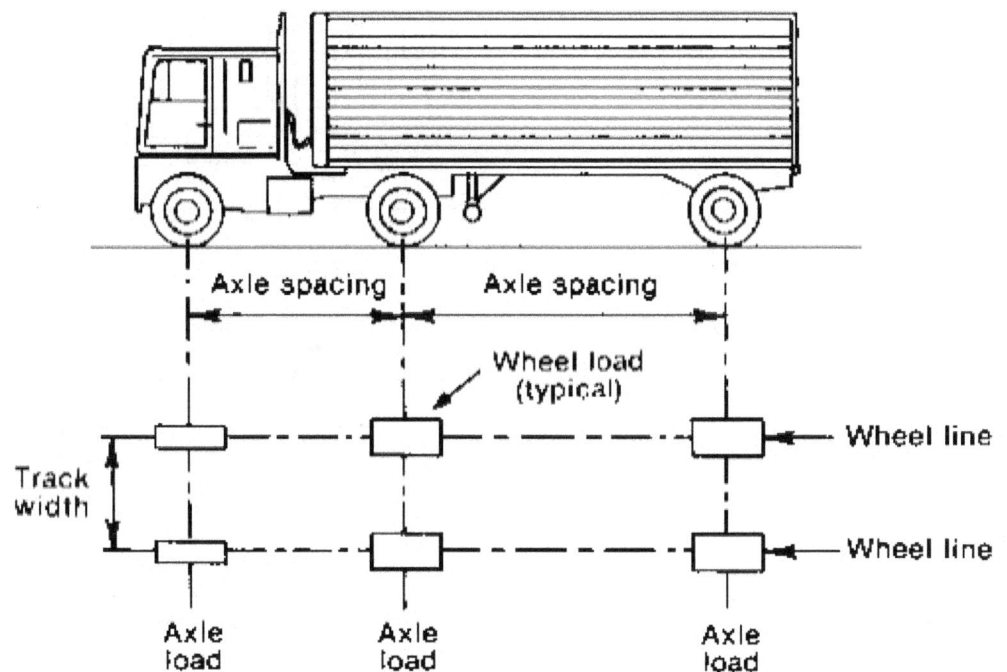

Figure 6-1. - Typical diagrams and terms for describing vehicle live loads used for bridge design.

Wheel load is one-half the axle load. Wheel loads for dual wheels are given as the combined weight of both wheels.

Wheel line is the series of wheel loads measured along the vehicle length. The total weight of one wheel line is equal to one-half the GVW.

Track width is the center-to-center distance between wheel lines.

STANDARD VEHICLE LOADS

AASHTO specifications provide two systems of standard vehicle loads, H loads and HS loads. Each system consists of individual truck loads and lane loads. Lane loads are intended to be equivalent in weight to a series of vehicles (discussed in the following paragraphs). The type of loading used for design, whether truck load or lane load, is that producing the highest stress. It should be noted that bridges are designed for the stresses and deflection produced by a standard highway loading, not necessarily the individual vehicles. The design loads are hypothetical and are intended to resemble a type of loading rather than a specific vehicle. Actual stresses produced by vehicles crossing the structure should not exceed those produced by the hypothetical design vehicles.

Truck Loads

There are currently two classes of truck loads for each standard loading system (Figure 6-2). The H system consists of loading H 15-44 and loading H 20-44. These loads represent a two-axle truck and are designated by the letter H followed by a number indicating the GVW in tons.

20-44	8,000 lbs	32,000 lbs
15-44	6,000 lbs	24,000 lbs

14'-0"

0.2W   W = Total weight of   0.8W
                truck and load

0.1 W   0.4 W

                                                      6'-0"

0.1 W   0.4 W

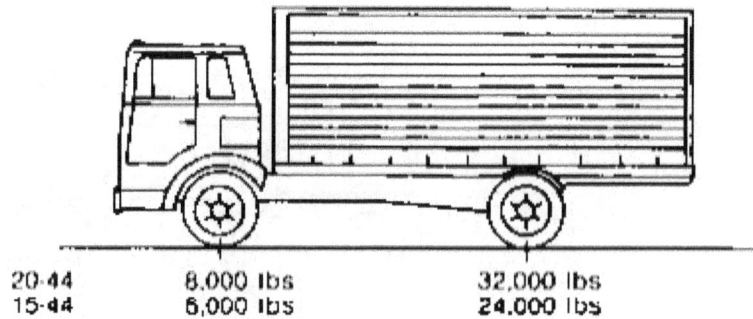

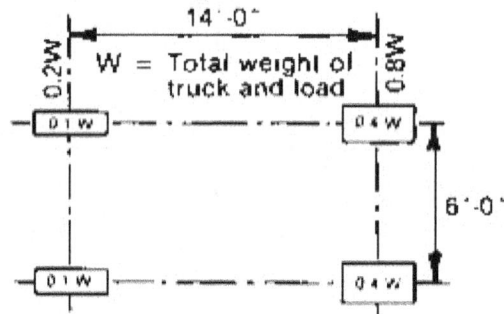

H 15-44 and H 20-44 trucks

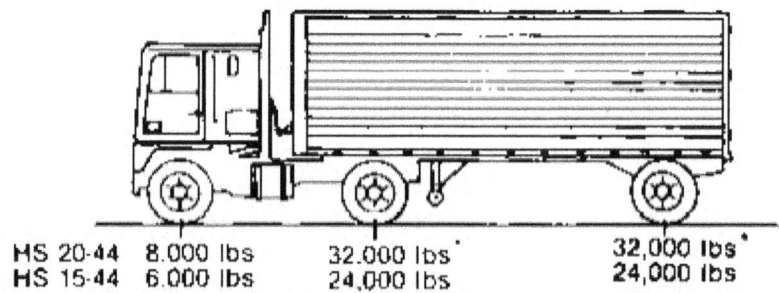

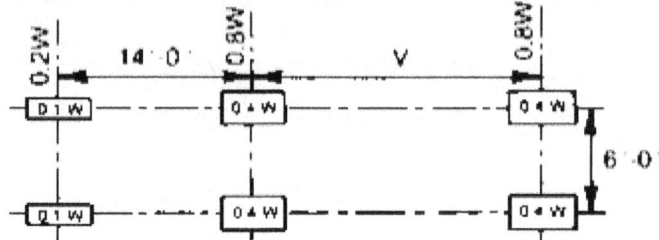

HS 20-44	8,000 lbs	32,000 lbs*	32,000 lbs*
HS 15-44	6,000 lbs	24,000 lbs	24,000 lbs

0.2W   14'-0"   0.8W   V   0.8W

0.1 W   0.4 W   0.4 W

                                                      6'-0"

0.1 W   0.4 W   0.4 W

W = Combined weight on the first two axles which
    is the same as for the corresponding H (M) truck.
V = Variable spacing—14 feet to 30 feet inclusive.
    Spacing to be used is that which produces
    maximum stresses.

HS 15-44 and HS 20-44 trucks

*In the design of timber floors and orthotropic steel decks (excluding
transverse beams) for H 20 and HS 20 loading one axle load of
24,000 pounds or two axle loads of 16,000 pounds each spaced 4 feet
apart may be used, whichever produces the greater stress, instead
of the 32,000 pound axle shown

Figure 6-2. - Standard AASHTO truck loads (from AASHTO Figures 3.7.6A and 3.7.7A;
© 1983. Used by permission).

The load designations also include a "-44" suffix to indicate the year that the load was adopted by AASHTO (1944). The weight of an H truck is assumed to be distributed two-tenths to the front axle and eight-tenths to the rear axle. Axle spacing is fixed at 14 feet and track width at 6 feet.

Truck loads for the HS system consist of loadings HS 15-44 and HS 20-44. These loads represent a two-axle tractor truck with a one-axle semitrailer and are designated by the letters HS, followed by a number indicating the gross weight in tons of the tractor truck. The configuration and weight of the HS tractor truck is identical to the corresponding H load. The additional semitrailer axle is equal in weight to the rear tractor truck axle and is spaced at a variable distance of 14 to 30 feet. The axle spacing used for design is that producing the maximum stress.

When H 20-44 and HS 20-44 loads are used for timber deck (floor) design, a modified form of standard loading is permitted by AASHTO. Instead of the 32,000-pound axle load specified for the standard trucks, one-axle loads of 24,000 pounds or two-axle loads of 16,000 pounds each, spaced 4 feet apart, may be used (AASHTO Figures 3.7.6A and 3.7.7A). Of the two options, the loading that produces the maximum stress is used design. These modified loads apply to the design of most timber decks, but do not apply to transverse beams, such as floorbeams (Chapter 8).

Lane Loads

Lane loads were adopted by AASHTO in 1944 to provide a simpler method of calculating moments and shears. These loads are intended to represent a line of medium-weight traffic with a heavy truck positioned somewhere in the line. Lane loads consist of a uniform load per linear foot of lane combined with a single moving concentrated load, positioned to produce the maximum stress (for continuous spans, two concentrated loads -- one placed in each of two adjoining spans -- are used to determine maximum negative moment). Both the uniform load and the concentrated loads are assumed to be transversely distributed over a 10-foot width.

AASHTO specifications currently include two classes of lane loads: one for H 20-44 and HS 20-44 loadings and one for H 15-44 and HS 15-44 loadings (Figure 6-3). The uniform load per linear foot of lane is equal to 0.016 times the GWV for H trucks or 0.016 times the weight of the tractor truck for HS trucks. The magnitude of the concentrated loads for shear and moment are 0.65 and 0.45 times those loads, respectively.

Modification to Standard Loads

There may be instances when the standard vehicle loads do not accurately represent the design loading required for a bridge. In such cases, AASHTO permits deviation from the standard loads provided they are obtained by proportionately changing the weights for both the standard truck and corresponding lane loads (AASHTO 3.7.2). The weights of the standard loads are increased or decreased, but the configuration and other requirements remain unchanged.

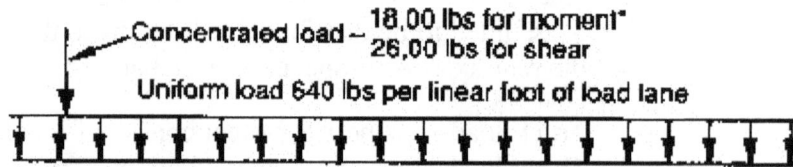

Concentrated load — 18,00 lbs for moment*
26,00 lbs for shear

Uniform load 640 lbs per linear foot of load lane

H 20-44 and HS 20-44 loading

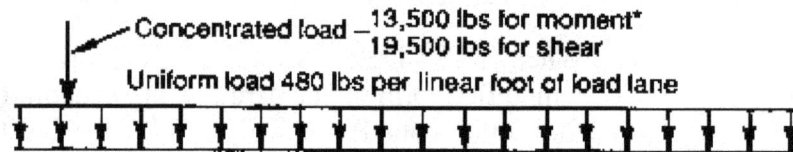

Concentrated load — 13,500 lbs for moment*
19,500 lbs for shear

Uniform load 480 lbs per linear foot of load lane

H 15-44 and HS 15-44 loading

*For computing maximum negative moment on continuous spans, two concentrated loads are used; one in each of two adjoining spans

Figure 6-3. - Standard AASHTO lane loads (from AASHTO' Figure 3.7.6B; © 1983. Used by permission).

Example 6-1 - Modified loading for standard AASHTO loads

Determine the AASHTO truck and lane loads for H 10-44 and HS 25-44 loadings.

Solution
H 10-44 Loading

The GVW of an H 10-44 truck load is 10 tons, or 20,000 pounds. From Figure 6-2, the GVW is distributed 20 percent to the front axle and 80 percent to the rear axle:

$$\text{Front axle load} = 0.20(\text{GVW}) = 0.20(20,000) = 4,000 \text{ lb}$$

$$\text{Rear axle load} = 0.80(\text{GVW}) = \mathbf{0.80(20,000) = 16,000 \ lb}$$

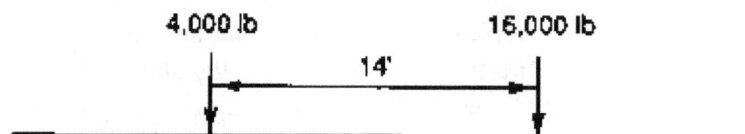

4,000 lb   16,000 lb

14'

For lane loading, the uniform load is 0.016 times the GVW:

$$\text{Uniform lane load} = 0.016(\text{GVW}) = 0.016(20,000) = 320 \text{ lb/ft}$$

Concentrated loads for moment and shear are 0.45 and 0.65 times the GVW, respectively:

$$\text{Concentrated load for moment} = 0.45(\text{GVW}) = 0.45(20,000) = 9,000 \text{ lb}$$

$$\text{Concentrated load for shear} = 0.65(\text{GVW}) = 0.65(20,000) = 13,000 \text{ lb}$$

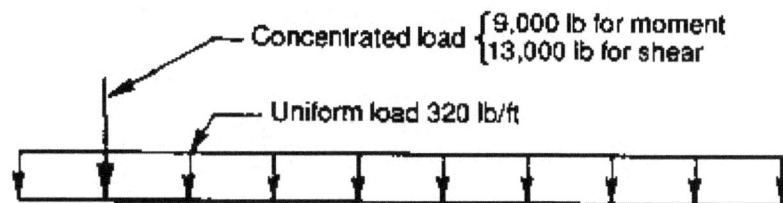

HS 25-44 Loading

For an HS 25-44 truck load, the weight of the tractor truck is 25 tons, or 50,000 pounds. From Figure 6-2, the weight is distributed 20 percent to the front axle and 80 percent each to the rear tractor truck axle and semi-trailer axle:

$$\text{Front axle load} = 0.20(50,000) = 10,000 \text{ lb}$$

$$\text{Rear tractor and semitrailer axle loads} = 0.80(50,000) = 40,000 \text{ lb}$$

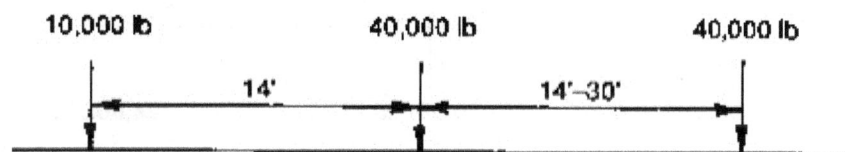

For lane loading, the uniform load is 0.016 times the weight of the tractor truck:

$$\text{Uniform lane load} = 0.016(50,000) = 800 \text{ lb/ft}$$

Concentrated loads for moment and shear are 0.45 and 0.65 times the weight of the tractor truck:

Concentrated load for moment = 0.45(50,000) = 22,500 lb

Concentrated load for shear = 0.65(50,000) = 32,500 lb

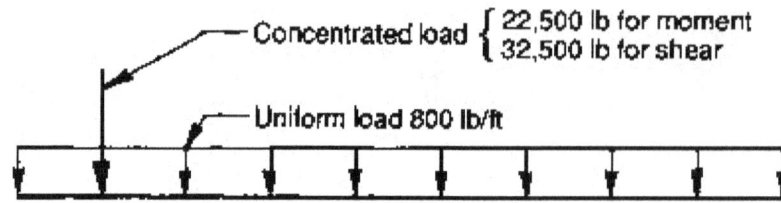

Alternate Military Loading

In addition to the standard loading systems, AASHTO also specifies an alternate military loading (AASHTO 3.7.4) that is used in some design applications discussed later in this section. This hypothetical loading consists of two 24,000-pound axles spaced 4 feet apart (Figure 6-4). There is no lane load for the alternate military loading.

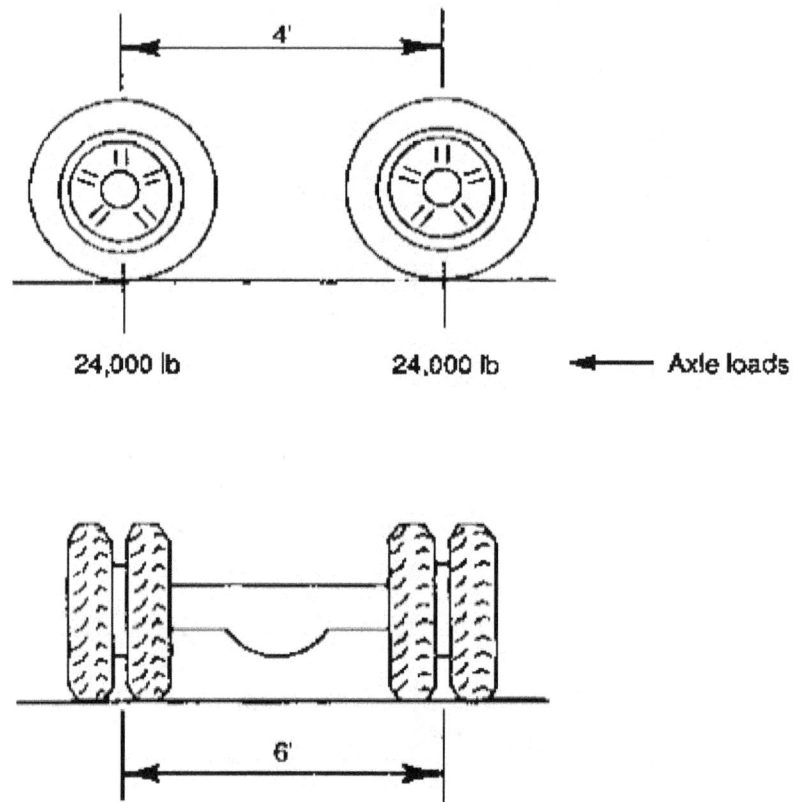

Figure 6-4. - AASHTO alternate military loading.

Overloads

An overload or permit load is a design vehicle that represents the maximum load a structure can safely support. It is generally a specialized vehicle that is not part of the normal traffic mix but must occasionally cross the structure. Although there are no standardized AASHTO overloads, many States and agencies have adopted standard vehicle overloads to meet the use requirements of their jurisdictions. Three of the overloads commonly used by the Forest Service are shown in Figure 6-5. In most cases, overloads are controlled or restricted from crossing bridges without a special permit.

U80 truck – GVW = 80 tons
(Axle loads are shown)

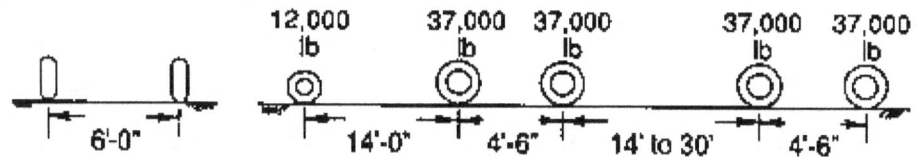

U102 truck– GVW = 102.5 tons
(Axle loads are shown)

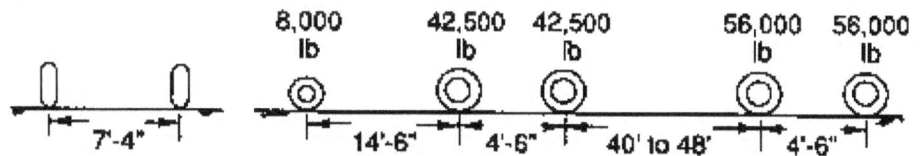

L90 tracked loader– GVW = 90 tons

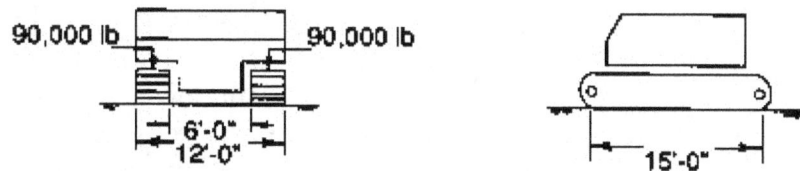

Figure 6-5. - Overload vehicles used by the USDA Forest Service.

APPLICATION OF VEHICLE LIVE LOAD

Vehicle live loads are applied to bridges to produce the maximum stress in structural components. The designer must determine the type of design loading and overload (when required), compute the absolute maximum vehicle forces (moment, shear, reactions, and so forth), and distribute those forces to the individual structural components. The first two topics are discussed in the remainder of this section. Load distribution to specific components depends on the configuration and type of structure; it is addressed in subsequent chapters on design.

Design Loading

Vehicle live loads used for design vary for different locations and are established by the agency having jurisdiction for traffic regulation and control. Bridges that support highway traffic are designed for heavy truck loads (HS 20-44 or HS 25-44). On secondary and local roads, a lesser loading may be appropriate. To provide a minimum level of safety, AASHTO specifications give the following minimum requirements for bridge loading:

1. Bridges that support interstate highways or other highways that carry or may carry heavy truck traffic are designed for HS 20-44 loading or the alternate military loading, whichever produces the maximum stress (AASHTO 3.7.4).

2. Bridges designed for less than H 20-44 loading also must be designed to support an infrequent heavy overload equal to twice the weight of the design vehicle. This increased load is applied in one lane, without concurrent loading in any other lane. The overload applies to the design of all affected components of the structure, except the deck (AASHTO 3.5.1). When an increased loading of this type is used, it is applied in AASHTO Load Group IA, and a 50-percent increase in design stress permitted by AASHTO (see discussions on load groups in Section 6.19).

Traffic Lanes

Vehicle live loads are applied in design traffic lanes that are 12 feet wide, measured normal to the bridge centerline (AASHTO 3.6). The number of traffic lanes depends on the width of the bridge roadway measured between curbs, or between rails when curbs are not used (AASHTO 2.1.2). Fractional parts of design lanes are not permitted; however, for roadway widths from 20 to 24 feet, AASHTO requires two design lanes, each equal to one-half the roadway width (this requirement generally does not apply for single-lane, low-volume bridges that require additional width for curve widening). For all other widths, the number of traffic lanes is equal to the number of full 12-foot lanes that will fit the roadway width.

Each traffic lane is loaded with one standard truck or one lane load, regardless of the bridge length or number of spans. The standard loads occupy a 10-foot width within the lane and are considered as a unit (Figure 6-6). Fractional parts of either type of load are not allowed. Traffic lanes and the vehicle loads within the lanes are positioned laterally on the bridge to produce the maximum stress in the member being designed, but traffic lanes cannot overlap. In the outside lanes, the load position in relation to the nearest face of the rail or curb depends on the type of component being designed. For deck design, the center of the wheel line is placed 1 foot from the railing or curb. For the design of supporting beams and other components, the center of the wheel line is placed 2 feet from the rail or curb. Vehicle positioning in traffic lanes is discussed in more detail in subsequent chapters on bridge design.

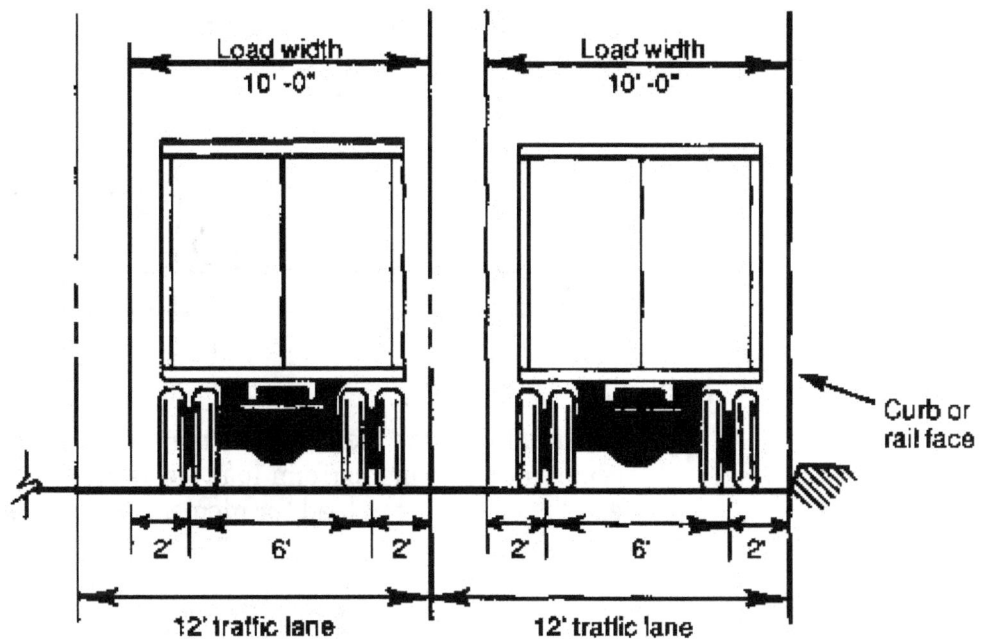

For deck design, the center of the wheel line is assumed to be positioned 1 foot from the nearest face of the curb or rail

Figure 6-6. – AASHTO traffic lanes. The 10-foot truck width is positioned laterally within the 12-foot traffic lane to produce the maximum stress in the component being designed.

Maximum Forces on Simple Spans

Maximum forces from vehicle live loads on simple spans depend on the position of the loads on the span. For lane loads, these positions are well defined and apply to all span lengths. For truck loads, general load positions are defined; however, the specific combination of wheel loads that produces the maximum forces may vary for different span lengths. When the span is less than or equal to the vehicle length (in some cases slightly greater than the vehicle length), the group of wheel loads that produces the maximum force must be determined by the designer. Some trial and error may be required when short spans are loaded with long vehicles with many axles. For truck loads with variable axle spacing, for example, the HS 15-44 and HS 20-44 loads, the minimum axle spacing always produces the maximum forces on simple spans.

General procedures for determining maximum vehicle live load forces on simple spans are discussed below and shown in Examples 6-2 and 6-3. Tables for computing maximum moment, vertical shear, and end reactions for standard truck and lane loads and selected overloads are given in Chapter 16. For additional information, refer to references listed at the end of this chapter.[18,24]

Maximum Moment

In most cases, the maximum moment on a simple span from a series of moving wheel loads occurs under the wheel load nearest the resultant *(R)*

6-11

of all loads when the resultant is the same distance on one side of the span centerline as the wheel load nearest the resultant is on the other side.

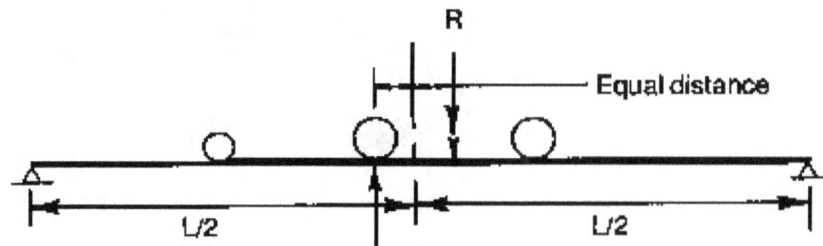

For lane loads, the maximum moment on a simple span occurs at the span centerline when the uniform load *(w)* is continuous over the span length and the concentrated load for moment *(P_M)* is positioned at the span centerline.

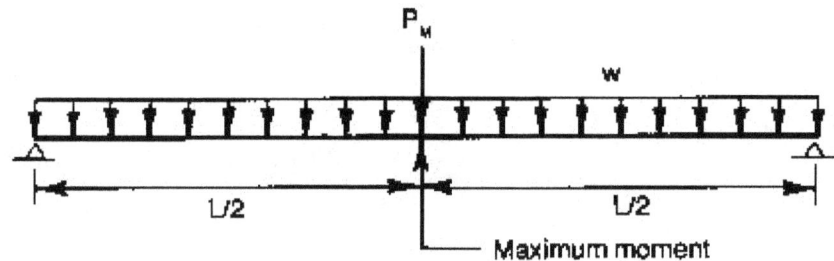

Maximum simple span moments for AASHTO vehicle loads are shown graphically in Figure 6-7. Truck loads control for simple spans less than 56.7 feet for H loads and 144.8 feet for HS loads (the alternate military loading controls over the HS 20-44 load on spans less than 41.3 feet). On longer spans, lane loads control.

Maximum Vertical Shear and End Reactions
The maximum vertical shear and end reactions for wheel loads on a simple span occur under the wheel over the support when the heaviest wheel (generally the rear wheel) is positioned at the support, with the remaining wheel loads on the span.

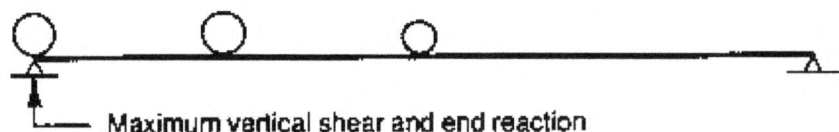

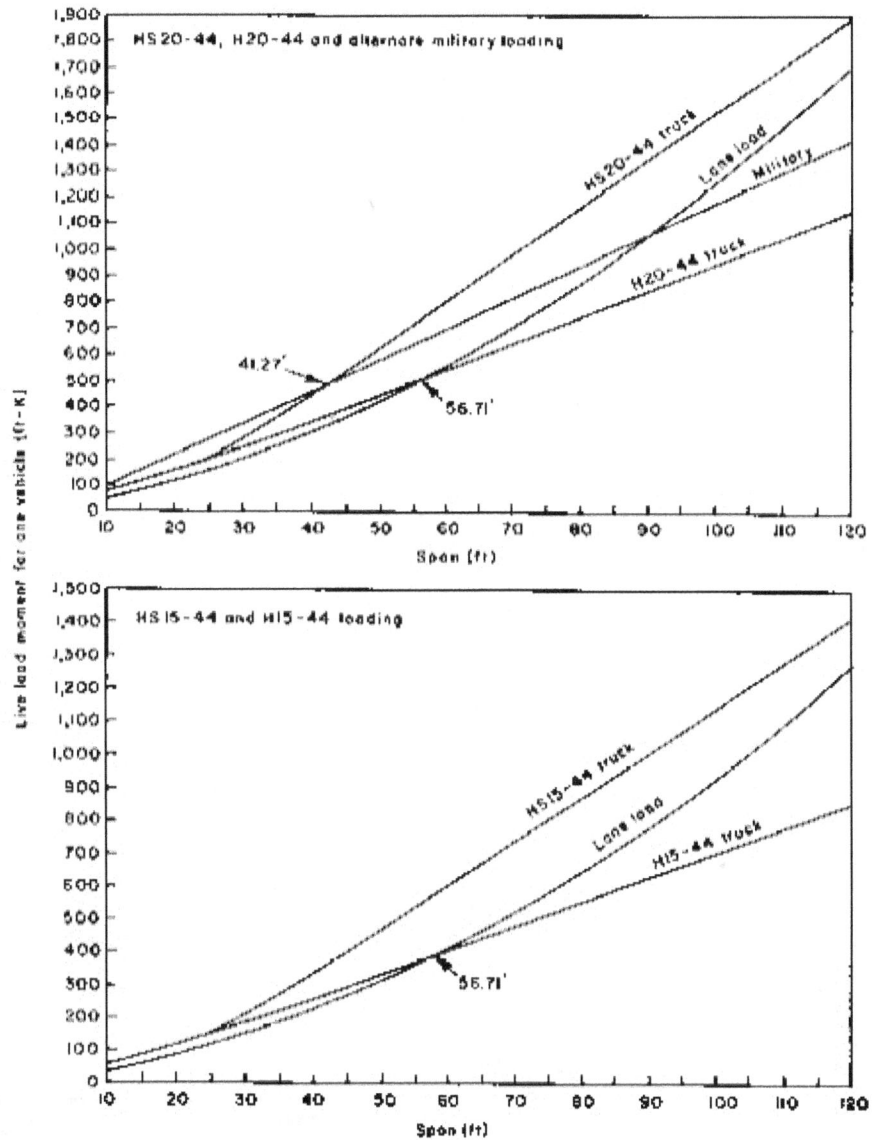

Figure 6-7. - Maximum moment on a simple span from one traffic lane of standard AASHTO vehicle loading.

The absolute maximum vertical shear and end reaction for lane loads occur when the uniform load is continuous and the concentrated load for shear (P_v) is positioned over the support.

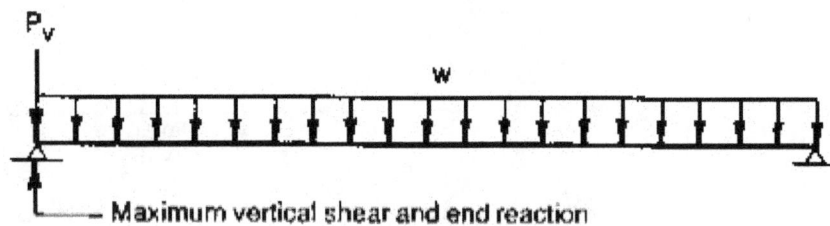

— Maximum vertical shear and end reaction

6-13

Maximum end reactions computed by these procedures are based on the bridge span measured center to center of bearings and are commonly tabulated in bridge design specifications and handbooks. Although they are technically correct for point bearing at span ends only, they do provide a very close approximation of the actual reaction for short bearing lengths. For very long bearing lengths, reactions should be computed based on the out-to-out span length with loads placed at the span end.

Maximum vertical shear and end reactions produced by AASHTO loads are shown graphically in Figure 6-8. Truck loads control maximum vertical shear and end reactions for simple spans less than 33.2 feet for H loads and 127.3 feet for HS loads (alternate military loading controls over HS 20-44 loading on spans less than 22 feet). On longer spans, lane loads control.

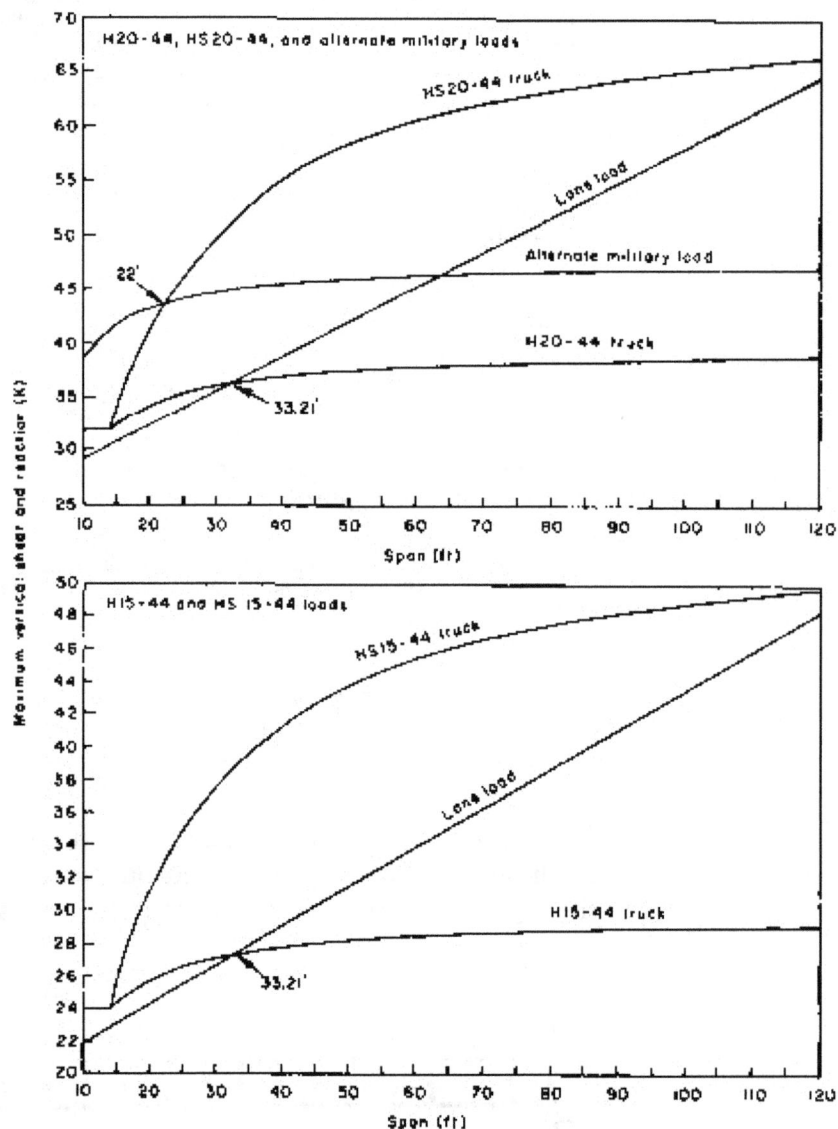

Figure 6-8. —Maximum vertical shear and end reactions on a simple span from one traffic lane of standard AASHTO vehicle loading.

6-14

Maximum Intermediate Vertical Shear

The maximum vertical shear at an intermediate point on a simple span is computed by positioning the loads to produce the maximum reaction at the support nearest the point. For truck loads, this generally occurs when the heaviest (rear) wheel load is placed over the point and no wheel loads occur on the shortest span segment between the point and the support.

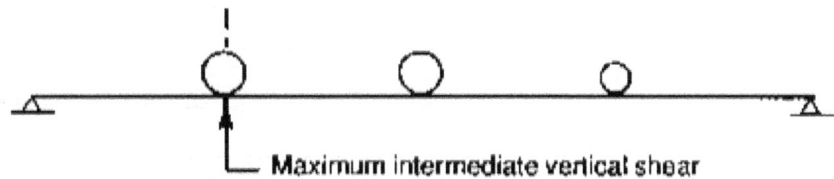

The maximum intermediate vertical shear for lane loads is produced by using a discontinuous uniform load with the concentrated load for shear (P_v) positioned at the point where shear is computed.

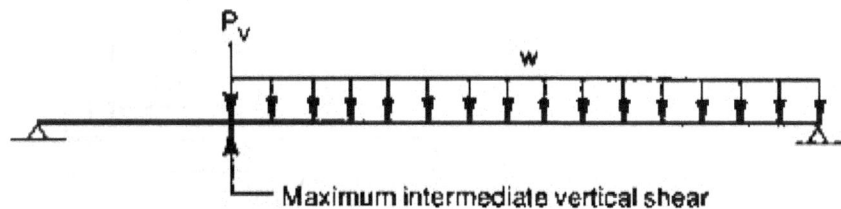

Example 6-2 - Maximum vehicle forces on a simple span; H 15-44 loading

For one lane of H 15-44 loading on a 62-foot simple span, determine the (1) maximum moment, (2) maximum reactions, and (3) maximum vertical shear at a distance 10 feet from the supports.

Solution

From Figure 6-2, the H 15-44 truck load consists of one 6,000-pound axle and one 24,000-pound axle with an axle spacing of 14 feet:

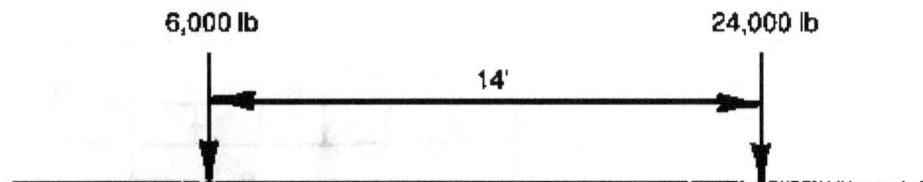

From Figure 6-3, H 15-44 lane loading consists of a uniform load of 480 lb/ft and a concentrated load of 13,500 pounds for moment and 19,500 pounds for shear.

Maximum Moment

Maximum moment from truck loading will be computed first. The distance (x) of the load resultant from the 24,000-pound axle is determined

6-15

by summing moments about the 24,000-pound axle and dividing by the gross vehicle weight:

$$x = \frac{(14 \text{ ft})(6,000 \text{ lb})}{(6,000 \text{ lb} + 24,000 \text{ lb})} = 2.8 \text{ ft}$$

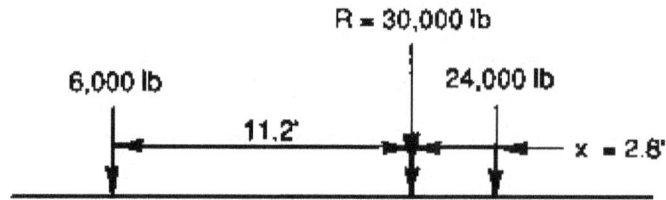

Maximum moment occurs under the 24,000-pound axle when the span centerline bisects the distance between the load resultant and the axle load:

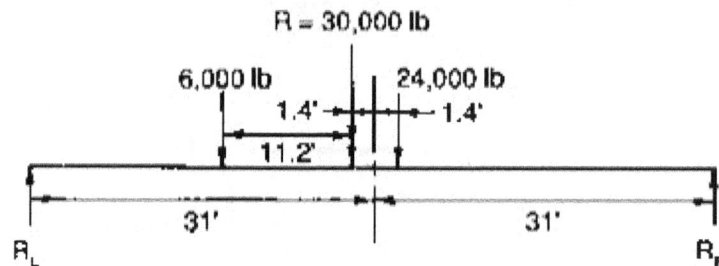

$$R_R = \frac{(30,000 \text{ lb})(31 \text{ ft} - 1.4 \text{ ft})}{62 \text{ ft}} = 14,322 \text{ lb}$$

$$M_{MAX} = (14,322 \text{ lb})(31 \text{ ft} - 1.4 \text{ ft}) = 423,931 \text{ ft-lb}$$

For lane loading, the concentrated load for moment is positioned at the span centerline:

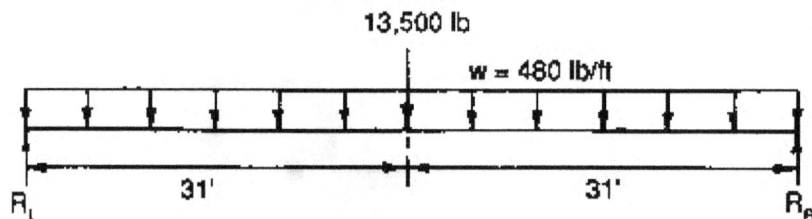

$$M_{MAX} = \frac{wL^2}{8} + \frac{PL}{4} = \frac{480(62)^2}{8} + \frac{13,500(62)}{4} = 439,890 \text{ ft-lb}$$

439,890 ft-lb > 423,931 ft-lb, so lane loading produces maximum moment.

Maximum Reactions

For truck loading, the maximum reaction is obtained by positioning the 24,000-pound axle over the support:

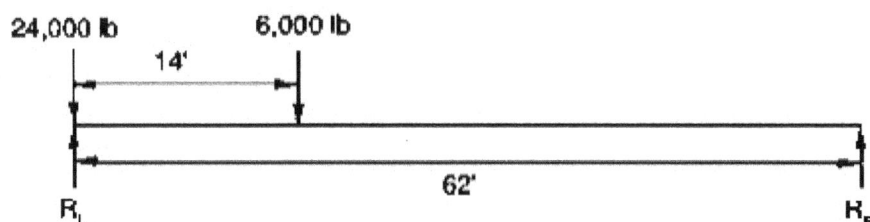

$$R_{MAX} = R_L = \frac{(6,000 \text{ lb})(62 \text{ ft} - 14 \text{ ft}) + (24,000 \text{ lb})(62 \text{ ft})}{62 \text{ ft}} = 28,645 \text{ lb}$$

For lane loading, the maximum reaction is obtained by placing the concentrated load for shear over the support:

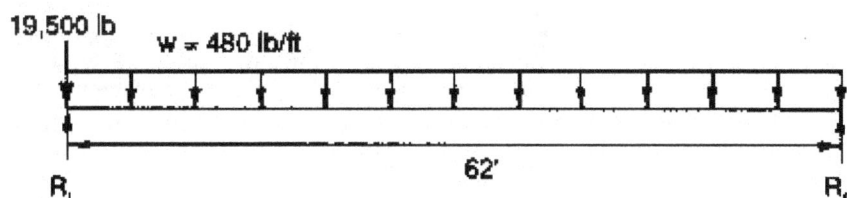

$$R_{MAX} = R_L = \frac{(480 \text{ lb/ft})(62 \text{ ft})}{2 \text{ ft}} + 19,500 \text{ lb} = 34,380 \text{ lb}$$

34,380 lb > 28,645 lb, so lane loading also produces the maximum reaction.

Maximum Vertical Shear 10 feet from the Support

For truck loading, the maximum vertical shear 10 feet from the support is obtained by positioning the 24,000-pound axle 10 feet from the support:

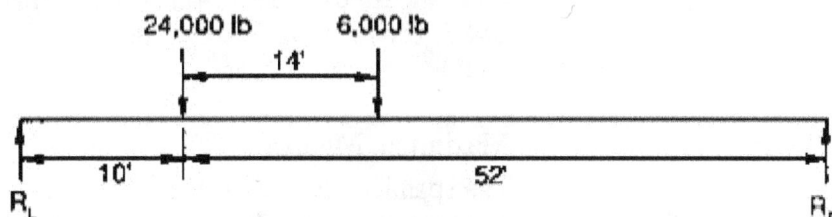

$$V_{MAX} = R_L = \frac{(6,000 \text{ lb})(52 \text{ ft} - 14 \text{ ft}) + (24,000 \text{ lb})(52 \text{ ft})}{62 \text{ ft}} = 23,806 \text{ lb}$$

For lane loading, maximum vertical shear is obtained using a partial uniform load with the concentrated load for shear positioned 10 feet from the support:

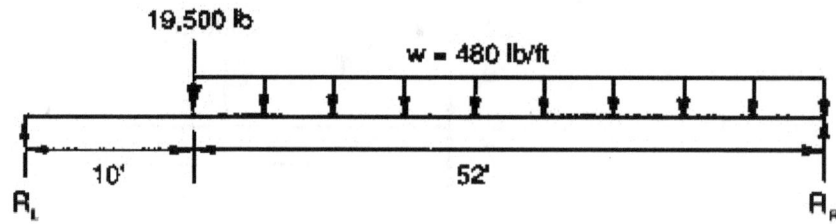

$$V_{MAX} = R_L = \frac{(480\ lb/ft)(52\ ft)(52\ ft/2) + (19,500\ lb)(52\ ft)}{62} = 26,822\ lb$$

26,822 lb > 23,806 lb and lane loading again controls maximum loading.

Example 6-3 - Maximum vehicle forces on a simple span; HS 20-44 loading

Determine the absolute maximum moment and reactions for one lane of HS 20-44 loading on a 23-foot simple span.

Solution

From Figure 6-2, the HS 20-44 truck load consists of one 8,000-pound axle and two 32,000-pound axles with a variable axle spacing of 14 to 30 feet. For this simple span application, the minimum axle spacing of 14 feet produces maximum forces:

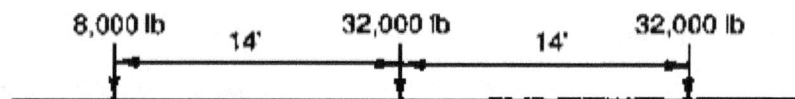

From Figure 6-3, HS 20-44 lane loading consists of a uniform load of 640 lb/ft and a concentrated load of 18,000 pounds for moment and 26,000 pounds for shear.

Maximum Moment

The span length of 23 feet is less than the vehicle length, so the maximum moment from truck loading will be produced by a partial vehicle configuration. For the two 32,000-pound axles,

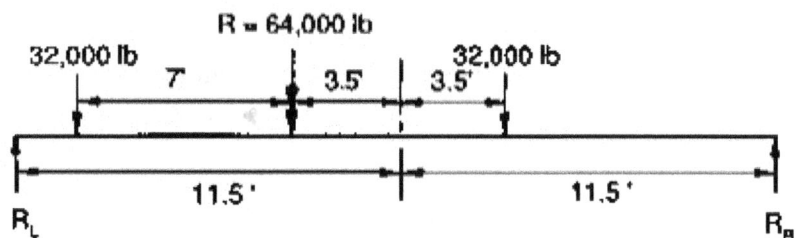

$$R_R = \frac{(64{,}000\ \text{lb})(11.5\ \text{ft} - 3.5\ \text{ft})}{23\ \text{ft}} = 22{,}261\ \text{lb}$$

$$M_{MAX} = (22{,}261\ \text{lb})(11.5\ \text{ft} - 3.5\ \text{ft}) = 178{,}088\ \text{ft-lb}$$

For a single 32,000-pound axle at the span centerline,

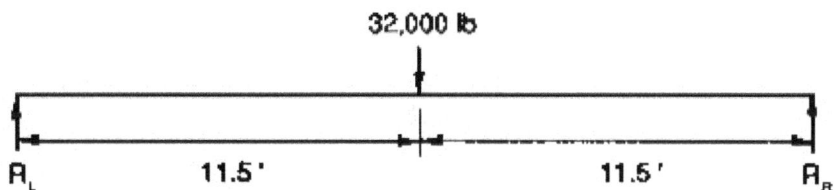

$$R_R = \frac{32{,}000\ \text{lb}}{2} = 16{,}000\ \text{lb}$$

$$M_{MAX} = (11.5\ \text{ft})(16{,}000\ \text{lb}) = 184{,}000\ \text{ft-lb}$$

In this case, maximum moment is controlled by a single axle at the span centerline, rather than by both axles positioned for maximum moment. This usually occurs when one axle is located close to a support. For HS truck loads, the single axle configuration will control maximum moment for spans up to approximately 23.9 feet.

For lane loading, maximum moment is produced when the concentrated load for moment is positioned at the span centerline:

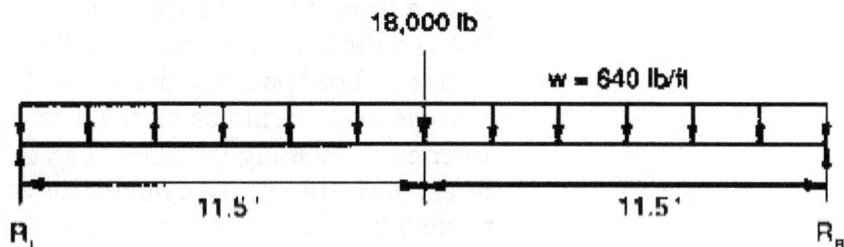

$$M_{MAX} = \frac{wL^2}{8} + \frac{PL}{4} = \frac{640(23)^2}{8} + \frac{18,000(23)}{4} = 145,820 \text{ ft-lb}$$

The maximum moment of 184,000 ft-lb is produced by truck loading with a single 32,000-pound axle positioned at the span centerline.

Maximum Reaction

For truck loading, the maximum reaction is obtained by positioning the rear 32,000-pound axle over the support (the front axle is off the span):

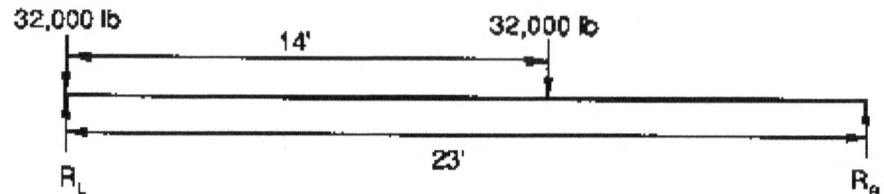

$$R_{MAX} = R_L = \frac{(32,000 \text{ lb})(23 \text{ ft} - 14 \text{ ft}) + (32,000 \text{ lb})(23 \text{ ft})}{23 \text{ ft}} = 44,522 \text{ lb}$$

For lane loading, the concentrated load for shear is placed over the support:

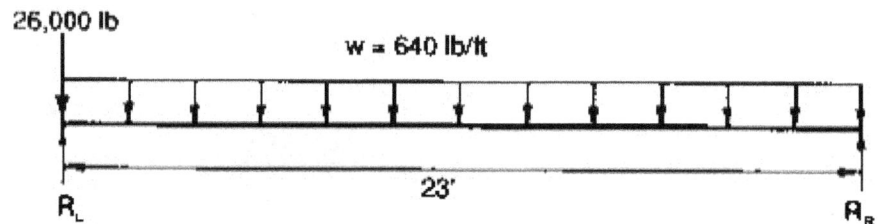

$$R_{MAX} = R_L = \frac{(640 \text{ lb/ft})(23 \text{ ft})}{2} + 26,000 \text{ lb} = 33,360 \text{ lb}$$

44,522 lb > 33,360 lb, so truck loading also produces the maximum reaction.

Maximum Forces on Continuous Spans

Maximum vehicle live load forces on continuous spans depend on the number, length, and stiffness of individual spans. In contrast to the case of simple spans, for continuous spans the designer must consider both positive and negative moments, as well as shear and reactions at several locations. Load positions are not well defined, and it is not always obvious how the loads should be placed. Historically, load positions have been determined by using influence diagrams or through trial and error. In recent years, inexpensive microcomputer programs have become the primary tool for determining maximum force envelopes. A detailed discussion of influence diagrams and other methods is beyond the scope of this chapter. For additional information, refer to references at the end of this chapter or other structural analysis publications.

Reduction in Load Intensity

The probability of the maximum vehicle live load occurring simultaneously in all traffic lanes of a multiple-lane structure decreases as the number of lanes increases. This is recognized in AASHTO specifications, and a reduction in vehicle live load is allowed in some cases (AASHTO 3.12.1). When the maximum stresses are produced in any member by loading a number of traffic lanes simultaneously, the percentages of the live loads given in Table 6-2 are used for design.

Table 6-2. - Reduction in load intensity for simultaneous lane loading.

Number of traffic lanes loaded simultaneously	Percent of vehicle live load used for design
One or two lanes	100
Three lanes	90
Four or more lanes	75

From AASHTO³ 3.12.1; © 1983. Used by permission.

6.4 DYNAMIC EFFECT (IMPACT)

A moving vehicle produces stresses in bridge members that are greater than those produced by the same loads applied statically. This increase in stress is from dynamic effects resulting from (1) the force of the vehicle striking imperfections in the roadway, (2) the effects of sudden loading, and (3) the vibrations of the vehicle or bridge-vehicle system. In bridge design, the word *impact* is used to denote the incremental stress increase from moving vehicle loads. In most contexts, impact denotes one body striking another. However, in bridge design, it refers to the total dynamic effect of moving loads.

AASHTO specifications require that an allowance for impact be included in the design of some structures. This allowance is expressed as an impact factor and is computed as a percentage increase in vehicle live load stress. Because of timber's ability to absorb shock and loads of short duration, AASHTO does not require an impact factor for timber bridges (AASHTO 3.8.1). However, when main components are made of steel or concrete, the impact factor may apply to the design of that member. Refer to AASHTO specifications for requirements related to application of the impact factor for materials other than timber.

Longitudinal forces develop in bridges when crossing vehicles accelerate or brake. These forces are caused by the change in vehicle momentum and are transmitted by the tires to the bridge deck. The magnitude of the longitudinal force depends on the vehicle weight, the rate of acceleration or deceleration, and the coefficient of friction between the tires and the deck surface. The most severe loading is produced by a braking truck and is computed, using physics, by

$$F_L = \mu \left(\frac{W}{g} \frac{dV}{dT} \right) \qquad (6\text{-}1)$$

where F_L = the longitudinal force transferred to the bridge (lb),

W = the weight of the vehicle (lb),

g = the acceleration due to gravity (32.2 ft/sec^2),

dV = the change in vehicle velocity (ft/sec),

dT = the time required for velocity change (sec), and

μ = the friction factor of the tires on the bridge deck.

The magnitude of the longitudinal force given by Equation 6.1 can vary substantially, depending on the physical condition of the vehicle and deck surface. The friction factor, μ, is a function of vehicle velocity and varies from 0.01 to 0.90, depending on the air pressure and type of tires, amount of tire tread, and roadway conditions. Additionally, and perhaps of more significance, is the rate of vehicle deceleration, dV/dT. In stops from high speeds, vehicle deceleration depends more on the condition of the braking system than on the friction between the tires and the roadway.

In view of the variables affecting the actual longitudinal force F_L, AASHTO specifies an approximate longitudinal force LF based on vehicle loads (AASHTO 3.9.1). A longitudinal force equal to 5 percent of the live load is applied in all lanes carrying traffic in the same direction. When a bridge is likely to become one directional in the future, all lanes are loaded. The live load used to compute longitudinal force is the uniform lane load plus the concentrated load for moment. Values of the longitudinal force for one traffic lane are shown in Figure 6-9.

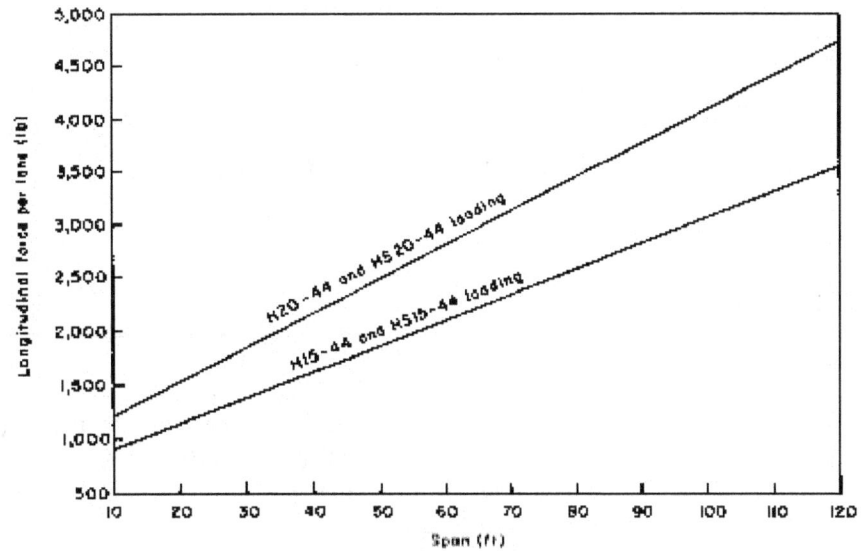

Figure 6-9. - Longitudinal force for one traffic lane of standard AASHTO vehicle loading.

The longitudinal force is applied in the center of the traffic lane at an elevation 6 feet above the bridge deck (Figure 6-10). The force acts horizontally in the direction of traffic and is positioned longitudinally on the span to produce maximum stress. When the maximum stress in any member is produced by loading a number of traffic lanes simultaneously, the longitudinal forces may be reduced for multiple-lane loading as permitted for vehicle live load (Table 6-2).

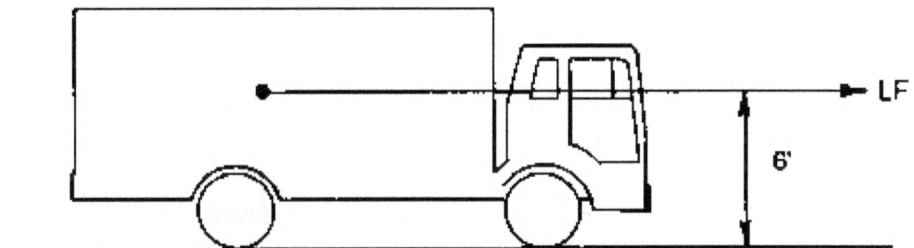

Figure 6-10. - Application of the vehicle longitudinal force.

Longitudinal forces are distributed to the structural elements of a bridge through the deck. For superstructure design, the forces generate shear at the deck interface and produce moments and axial forces in longitudinal beams. Application of the force 6 feet above the deck also produces a longitudinal overturning effect resulting in vertical reactions at bearings. In most cases, longitudinal forces have little effect on timber superstructures, but they may have a substantial effect on the substructure. When substructures consist of bents or piers, the forces produce shear and moment in supporting members. These forces are most critical at the base of high substructures when longitudinal movement of the superstructure can occur at expansion bearings or joints. Bearings on timber bridges are generally fixed, and members are restrained against longitudinal sidesway. In this case, forces on bents or piers are reduced by load transfer through the superstructure to the abutments.

6.6 CENTRIFUGAL FORCE

When a vehicle moves in a curvilinear path, it produces a centrifugal force that acts perpendicular to the tangent of the path (Figure 6-11). In bridge design, this force must be considered when the bridge is horizontally curved, when a horizontally curved deck is supported by straight beams, or when a straight bridge is used on a curved roadway. Situations of this type are not common for timber bridges, but may occur in some applications (Figure 6-12).

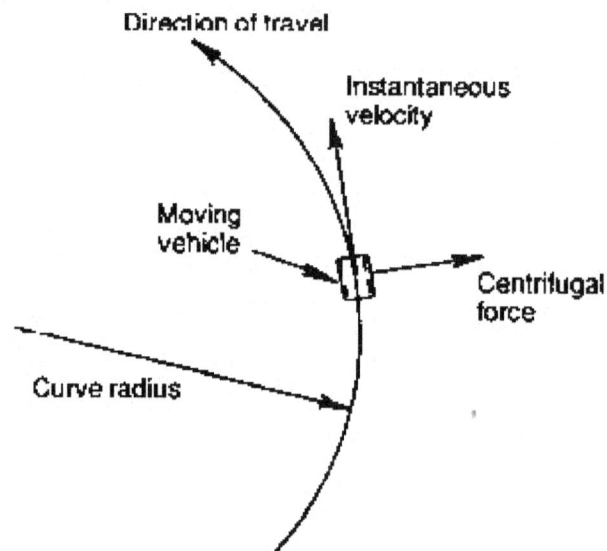

Figure 6-11. - Centrifugal force produced by a vehicle moving on a curved path.

Centrifugal force depends on vehicle weight and velocity as well as the curve radius. Magnitude of the force is given in AASHTO as a percentage of vehicle live load applied in each traffic lane (AASHTO 3.10.1), as given by

$$C = 0.00117S^2D = \frac{6.68S^2}{R} \tag{6-2}$$

where C = the centrifugal force in percent of live load,

          S = the design speed (mph),

Figure 6-12. - A timber bridge with sharply curved approach roadways. Trucks crossing the bridge can produce centrifugal forces that affect the bridge superstructure and substructure.

D = the degree of curve, and

R = the radius of the curve (ft).

The live load used to compute centrifugal force is the vehicle truck load (lane loads are not used). Traffic lanes in both directions are loaded with one truck in each lane, placed in a position to produce the maximum force. The force is applied 6 feet above the centerline of the roadway surface and acts horizontally, away from the curve (Figure 6-13). When roadway superelevation is provided, the centrifugal force is resolved into horizontal and vertical components.

Centrifugal forces are most significant for bridges that have high design speeds and small radii curves, or are supported by substructures with tall columns. For substructure design, centrifugal forces can produce large moments and shears in supporting members, particularly tall piers or columns. Additionally, they generate a transverse overturning effect on the superstructure that results in vertical forces at the reactions. For superstructure design, centrifugal forces produce transverse shear at the deck interface. For longitudinally rigid decks that are adequately attached to supporting beams, these forces are resisted in the plane of the deck and transferred to bearings by transverse bracing. When timber decks are considered, many configurations are not longitudinally rigid, and transverse loads can generate torsion in beams between points of transverse bracing.

6-25

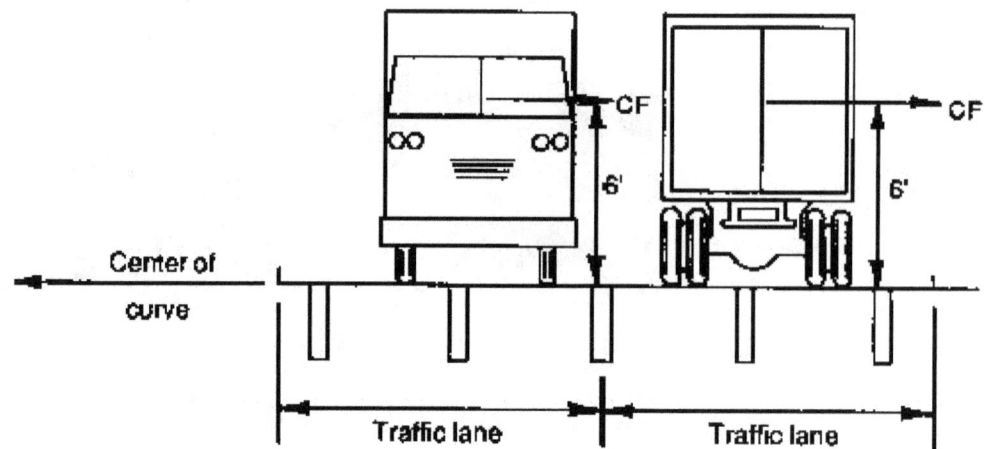

Figure 6-13. - Application of the vehicle centrifugal force.

6.7 WIND LOAD

Wind loads are caused by the pressure of wind acting on the bridge members. They are dynamic loads that depend on such factors as the size and shape of the structure, the velocity and angle of the wind, and the shielding effects of the terrain. For design purposes, AASHTO specifications give wind loads as uniformly distributed static loads. This simplified loading is intended for rigid structures that are not dynamically sensitive to wind, that is, structural design is not controlled by wind loads. With very few exceptions, timber bridges are included in this category. For structures that are highly sensitive to dynamic effects (bridges with long flexible members or suspension bridges), a more detailed analysis is required. Wind-tunnel tests may be appropriate when significant uncertainties about structural behavior exist.

Wind loads are applied to bridges as horizontal loads acting on the superstructure and substructure and as vertical loads acting upward on the deck underside. The magnitude of the loads depends on the component of the structure and the base wind velocity used for design. Wind loads given in AASHTO are based on an assumed base wind velocity of 100 miles per hour (mph) (AASHTO 3.15). In some cases, a lower or higher velocity is permitted when precise local records or permanent terrain features indicate that the 100-mph velocity should be modified. When the base wind velocity is modified, the specified loads are changed in the ratio of the square of the design wind velocity to the square of the 100-mph wind velocity.

SUPERSTRUCTURE LOADS Superstructures are designed for wind loads that are applied directly to the superstructure (W) and/or those that act on the moving vehicle live load (WL). The magnitude of these loads varies for different loading combinations (AASHTO 3.15.1). In general, the full wind load acts directly on the structure when vehicle live loads are not present. When live loads are present, the wind load on the structure is reduced 70 percent, and an additional wind load acting on the moving vehicle live load is applied simultaneously (see Section 6.19).

Loads Applied Directly to the Superstructure

Wind loads acting directly on the bridge superstructure (W) are applied as uniformly distributed loads over the exposed area of the structure (Figure 6-14). The exposed area is the sum of areas of all members, including the deck, curbs, and railing, as viewed in elevation at 90 degrees to the longitudinal bridge axis. The magnitude of the uniform load for beam (girder) superstructures is 50 lb/ft^2 of exposed area, but not less than 300 lb/lin ft (AASHTO 3.15.1.1.1). For trusses and arches, the wind load is 75 lb/ft^2 of exposed area, but only for trusses not less than 300 lb/lin ft in the plane of the windward chord and 150 lb/lin ft in the plane of the leeward chord. The wind loads for all superstructure types are applied horizontally, at right angles to the longitudinal bridge axis.

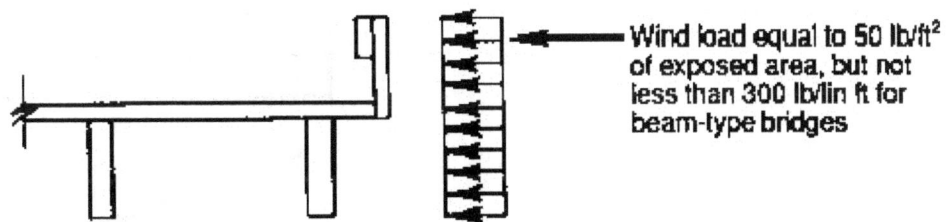

Figure 6-14. - Wind load applied to the bridge superstructure.

Loads Applied to the Vehicle Live Load

Wind loads acting on the moving vehicle live load (WL) are applied along the span length as a horizontal line load of 100 lb/lin ft. The loads are applied horizontally at right angles to the longitudinal bridge axis, 6 feet above the roadway surface (Figure 6-15).

Wind loads on the superstructure are laterally distributed to structural members and the bearings by the deck and transverse bracing. The loads produce transverse forces that develop shear at the deck interface and bearings, axial forces in the bracing, and small moments in beams or other supporting members. Wind loads generally have little or no effect on main superstructure components, but are considered in the design of transverse bracing and bearings.

SUBSTRUCTURE LOADS

Substructures are designed for wind loads transmitted to the substructure by the superstructure, and those applied directly to the exposed area of the substructure (AASHTO 3.15.2). Both loads act in a horizontal plane, but are applied at various skew angles to the structure. The skew angle is measured from the perpendicular to the longitudinal bridge axis (Figure 6-16). The angle used for design is that which produces the greatest stress in the substructure.

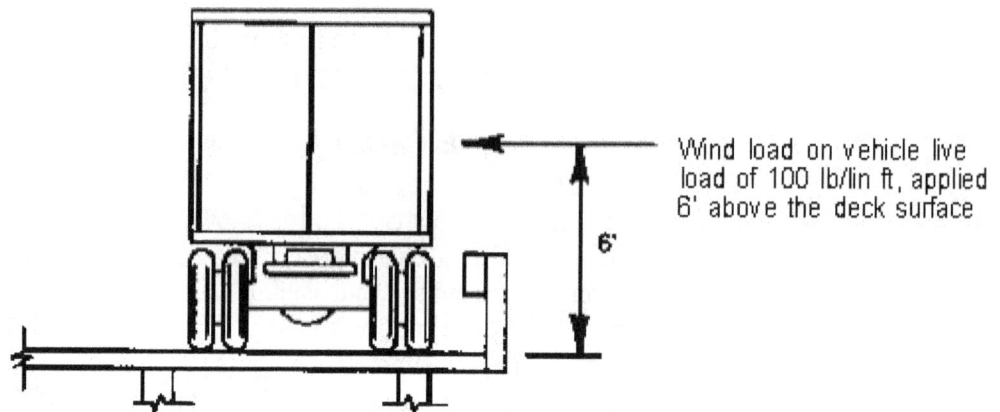

Figure 6-15. - Application of wind load acting on the vehicle live load.

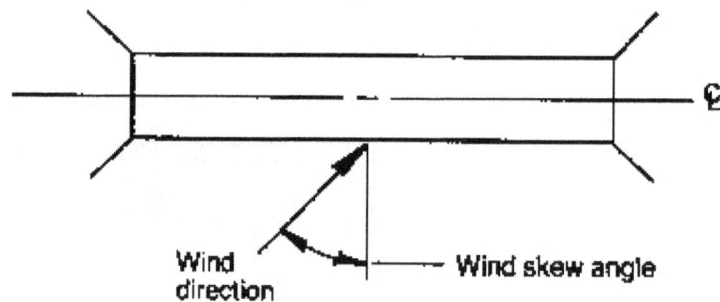

Figure 6-16. - Wind skew angle for substructure design.

Loads Transmitted to the Substructure by the Superstructure

Wind loads transmitted to the substructure by the superstructure include the loads acting directly on the superstructure (W) and those acting on the moving vehicle live load (WL). Both loads are applied simultaneously in the lateral and longitudinal directions (Table 6-3). Wind loads acting directly on the superstructure are applied at the center of gravity of the exposed superstructure area. Loads acting on the moving live load are applied 6 feet above the deck.

For beam and deck bridges with a maximum span length of 125 feet or less, which includes most timber bridges, AASHTO contains special provisions for superstructure wind loads transmitted to the substructure (AASHTO 3.15.2.1.3). Instead of the more precise loading given above, these structures may be designed for the following loads without further consideration for skew angles:

Wind load on structure (W): 50 lb/ft^2, transverse, and 12 lb/ft^2, longitudinal, both applied simultaneously

Wind load on live load (WL): 100 lb/lin ft, transverse, and 40 lb/lin ft, longitudinal, both applied simultaneously

Table 6-3. - Wind loads transmitted to the substructure by the superstructure.

| Skew angle of wind (deg) | Wind load on the superstructure (lb/ft²) | | | | Wind load on the moving vehicle live load (lb/ft) | |
| | Trusses | | Beams (girders) | | | |
	Lateral load	Longitudinal load	Lateral load	Longitudinal load	Lateral load	Longitudinal load
0	75	0	50	0	100	0
15	70	12	44	6	88	12
30	65	28	41	12	82	24
45	47	41	33	16	66	32
60	24	50	17	19	34	38

From AASHTO[3] 15.2.1.3; © 1983. Used by permission.

Loads Applied Directly to the Substructure

Wind loads applied directly to the substructure are 40 lb/ft² of exposed substructure area (AASHTO 3.15.2.2). The force for skewed wind directions is resolved into components perpendicular to the end and front elevations of the substructure. The component acting perpendicular to the end elevation acts on the exposed area seen in the end elevation. The component acting perpendicular to the front elevation acts on the exposed area seen in the front elevation and is applied simultaneously with the wind loads from the superstructure.

Wind loads acting on the substructure generate lateral and longitudinal forces that produce the same effects previously discussed for centrifugal and longitudinal forces. They are most significant for continuous or multiple-span structures supported by high piers or bents.

OVERTURNING FORCE

AASHTO specifications (AASHTO 3.15.3) require that the wind forces tending to overturn a bridge be computed in some loading combinations (Load Groups II, III, V, and VI discussed in Section 6.19). When overturning is considered, the wind loads applied to the superstructure and substructure are assumed to act perpendicular to the longitudinal bridge centerline. In addition, a vertical wind load is applied upward at the windward quarter point of the transverse superstructure width (Figure 6-17). This vertical wind load (W) is equal to 20 lb/ft² of deck and sidewalk area as seen in the plan view. When applied in load combinations where vehicle live loads are present (Load Groups III and IV), the vertical force is reduced to 6 lb/ft² of deck and sidewalk area.

6-29

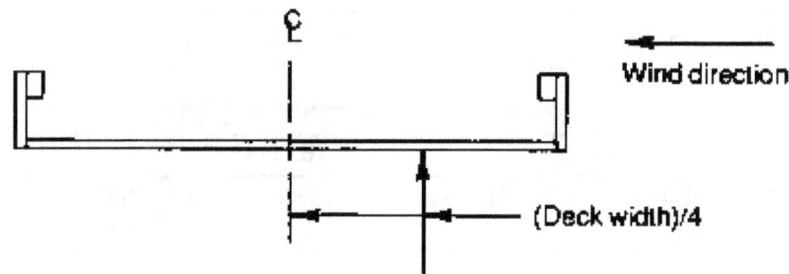

Upward wind force equal to 20 lb/ft²of deck
and sidewalk area, applied at the windward
1/4 point of the deck width

Figure 6-17. - Wind load overturning force.

6.8 EARTHQUAKE FORCES

When earthquakes occur, bridges can be subject to large lateral displacements from the ground movement at the base of the structure. In many areas of the United States, the risk of earthquakes is low, while in others, it is high. Large earthquakes, such as those that occurred in San Francisco in 1906 and Alaska in 1964, induce strong structure motions that can last up to 1 minute or more. Smaller earthquakes also can produce significant motion, although the duration of movement is shorter. Bridge failures in earthquakes generally occur by shaking that causes the superstructure to fall off the bearings, displacement or yielding of tall supporting columns, or settlement of the substructure caused by a strength loss in the soil from ground vibrations (Figure 6-18). Earthquake or seismic analysis is concerned primarily with ensuring that the bearings and substructure are capable of resisting the lateral forces generated by movement of the superstructure. The objective of seismic analysis is not to design the structure to resist all potential loads with no damage, but to minimize damage to a level below that associated with failure.

Bridge earthquake loads depend on a number of factors, including the earthquake magnitude, the seismic response of soil at the site, and the dynamic response characteristics (stiffness and weight distribution) of the structure. An exact analysis is complex and requires specific seismic data for the site. For timber bridges, the most appropriate method of analysis is generally the equivalent static force method given in AASHTO 3.21.1. Using this simplified procedure, which is intended for structures with supporting members of approximately equal stiffness, the earthquake force (EQ) applied as an equivalent static force at the structure's center of mass, is computed as

$$EQ = (C)(F)(W) \qquad (6\text{-}3)$$

Figure 6-18. - Earthquake damage to a timber trestle highway bridge that occurred during the Alaska earthquake of 1964 (photo courtesy of the Alaska Department of Transportation and Public Facilities).

where   EQ = equivalent static horizontal force applied at the center of gravity of the structure (lb),

C = combined response coefficient,

F = framing factor (1.0 for structures where single columns or piers resist the horizontal forces, 0.80 for structures where continuous frames resist horizontal forces applied along the frame), and

W = total dead load weight of the structure (lb).

The combined response coefficient C in Equation 6-3 can be computed directly from equations given in AASHTO when seismic data are available for the site. In many cases, such data are not available, and C is determined from graphs based on the natural period of vibration *(T)* of the structure, the expected rock acceleration *(A)*, and the depth of alluvium to rocklike material at the site. Graphs for determining C for depths of alluvium to rocklike material of 0 to 10 feet and 11 to 80 feet are shown in Figure 6-19 (see AASHTO 3.21.2 for greater depths). To use the graphs, the designer must determine the applicable values of T and A:

6-31

$$T = 0.32\sqrt{\frac{W}{P}}$$

(6-4)

where T = period of vibration of the structure (sec), and

P = total uniform force required to cause a 1-inch maximum horizontal deflection of the structure (lb).

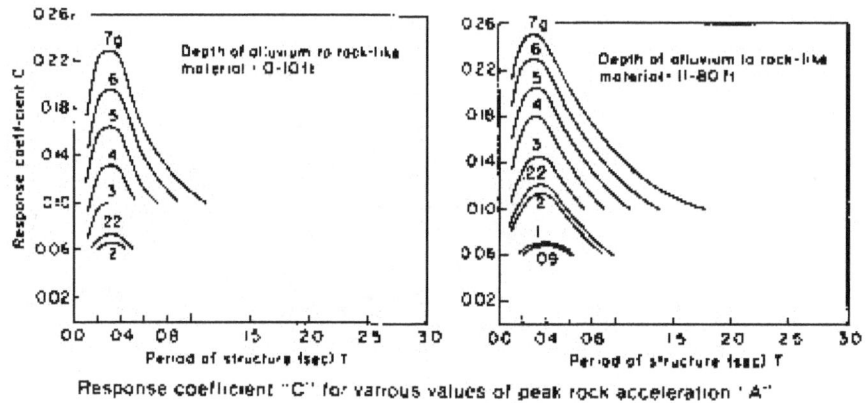

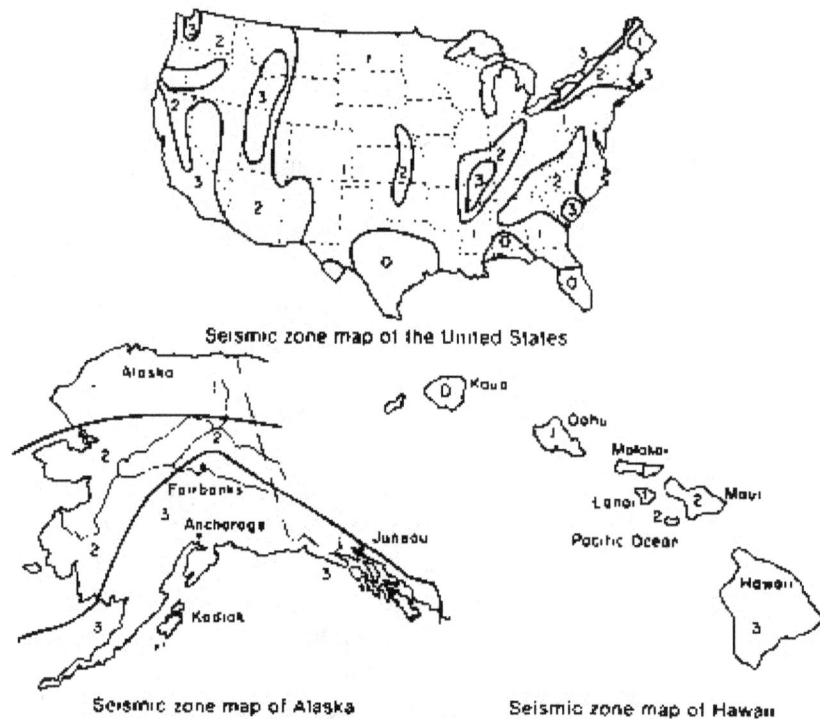

Figure 6-19. - Combined response coefficients and seismic zones used for computing earthquake loads by the equivalent static force method (from AASHTO'3.21.1; © 1983. Used by permission).

When maximum expected rock acceleration maps are not available for the specific site, the following values for A should be used based on the site zone from the seismic risk maps given in Figure 6-19:

Zone 1	$A = 0.09g$
Zone 2	$A = 0.22g$
Zone 3	$A = 0.50g$

In addition to the equivalent static method given in the AASHTO bridge specifications, AASHTO has also published a much more comprehensive *Guide Specifications for Seismic Design of Highway Bridges.* This guide, which may be used in lieu of the equivalent static force method, gives several methods of analysis based on a number of factors related to the location and type of structure. For single-span bridges, no seismic analysis is required; however, the connections between the bridge span and the abutments must be designed to longitudinally and transversely resist the dead load reaction at the abutment multiplied by the acceleration coefficient, A, at the site. In addition, expansion ends (which are generally not required on timber bridges) must meet minimum bearing length requirements given in the specifications. The AASHTO guide specifications present a good approach to seismic analysis and include commentary and design examples. Their use is currently optional but highly recommended.

6.9 SNOW LOAD

Snow loads should be considered when a bridge is located in an area of potentially heavy snowfall. This can occur at high elevations in mountainous areas with large seasonal accumulations. Snow loads are normally negligible in areas of the United States that are below 2,000 feet elevation and east of longitude 105°W, or below 1,000 feet elevation and west of longitude 105°W. In other areas of the country, snow loads as large as 700 lb/ft² may be encountered in mountainous locations.

AASHTO specifications do not require consideration of snow loads except under special conditions (AASHTO 3.3.2). The effects of snow are assumed to be offset by an accompanying decrease in vehicle live load. This assumption is valid for most structures, but is not realistic in areas where snowfall is significant. When prolonged winter closure of a road makes snow removal impossible, the magnitude of snow loads may exceed those from vehicle live loads (Figure 6-20). Loads also may be notable when plowed snow is stockpiled or otherwise allowed to accumulate. The applicability and magnitude of snow loads are left to designer judgment.

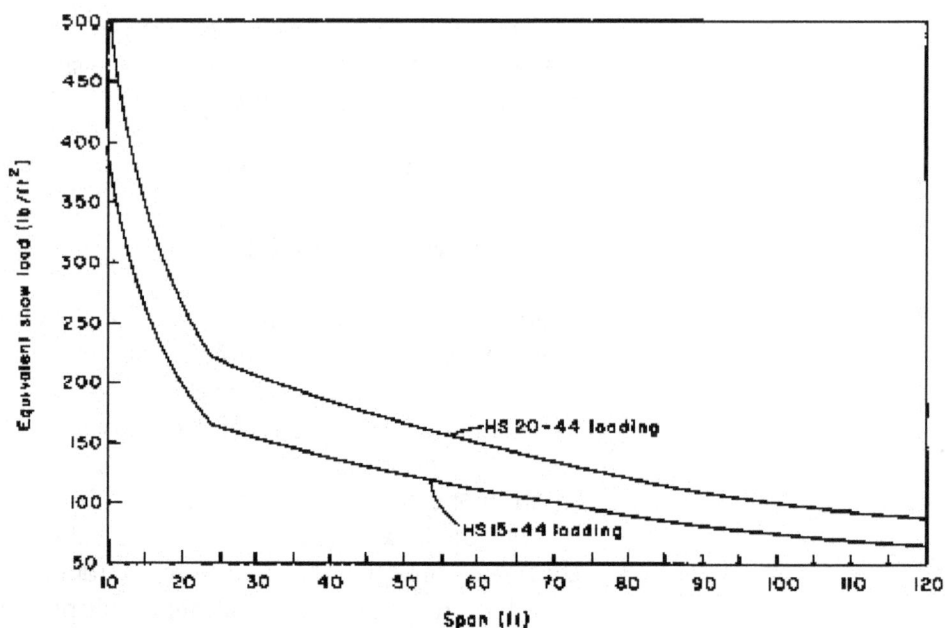

Figure 6-20. - Equivalent snow load required to produce the same moment as one truck load.

Snow loads vary from year to year and depend on the depth and density of snow pack. The depth used for design should be based on a mean recurrence interval or the maximum recorded depth. Density is based on the degree of compaction. The lightest accumulation is produced by fresh snow falling at cold temperatures. Density increases when the snow pack is subjected to freeze-thaw cycles or rain. Probable densities for several snow pack conditions are as follows:[9]

Condition of snow pack	Probable density (lb/ft³)
Freshly fallen	6
Accumulated	19
Compacted	31
Rain on snow	31

Estimated snow load can be determined from historical records or other reliable data. General information on ground snow loads is available from the National Weather Service, from State and local agencies, and in ANSI A58.1.[9] Snow loads in mountainous areas are subject to extreme variations, and determining the extent of these loads should be based on local experience or records, rather than generalized information.

The effect of snow loads on a bridge structure is influenced by the pattern of snow accumulation. Windblown snow drifts may produce unbalanced loads considerably greater than those from uniformly distributed loads. Drifting is influenced by the terrain, structure shape, and other features that cause changes in the general wind flow. Bridge components, such as railing, can serve to contain drifting snow and cause large accumulations to develop.

6-34

6.10 THERMAL FORCE

Thermal forces develop in bridge members that are restrained from movement and are subjected to temperature change. The magnitude of the thermal force depends on the member length, the degree of temperature change, and the coefficient of thermal expansion for the material. Like other solid materials, timber expands when heated and contracts when cooled; however, the thermal expansion for timber is only one-tenth to one-third that for other common construction materials (Chapter 3). As a result, thermal forces can be induced at connections or other locations where timber is used in conjunction with other materials that are more sensitive to temperature. In most bridge applications, thermal forces in timber members are insignificant and are commonly ignored. When members are very long, are subjected to extreme temperature changes, or are used in conjunction with other materials, consideration of thermal forces and/or provisions for expansion and contraction are left to the judgment of the designer.

6.11 UPLIFT

Uplift is an upward vertical reaction produced at the supports of continuous-span superstructures. It develops under certain combinations of bridge configuration and loading that generate forces acting to lift the superstructure from the substructure. Uplift forces may develop in continuous-span timber bridges where short spans are adjacent to longer spans (Figure 6-21).

Uplift forces are transmitted from the superstructure to the substructure by anchor bolts or tension ties at the bearings. The strength of the connections and the mass or anchorage of the substructure must be sufficient to resist these forces. AASHTO specifications require that the calculated uplift at any support be resisted by members designed for the largest force obtained under the following two conditions (AASHTO 3.17.1):

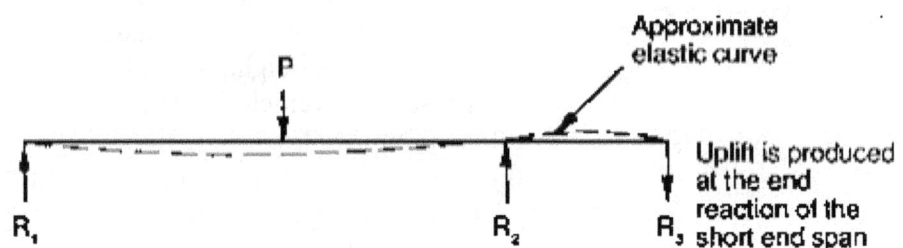

Figure 6-21. - Uplift force on a continuous span superstructure.

1.   100 percent of the calculated uplift caused by any loading or loading combination in which the vehicle live load (including impact, when applicable) is increased by 100 percent

2.   150 percent of the computed uplift at working load level from any applicable loading combination

The allowable stress in anchor bolts in tension or other elements of the structure stressed under these conditions may be increased by 150 percent.

6.12 EARTH PRESSURE

Earth pressure is the lateral pressure generated by fill material acting on a retaining structure. In bridge design, it is most applicable in the design of substructures, primarily abutments and retaining walls (Figure 6-22). Earth pressures also may be transmitted to the superstructure when back-walls or endwalls are directly supported by superstructure ends; however, in most design applications, earth pressure is significant in substructure design only.

The magnitude of earth pressure depends on the physical properties of the soil, the interaction at the soil-structure interface, and the deformations in the soil-structure system. For routine bridge design, active earth pressures are generally computed using Rankine's formula, a somewhat simplified procedure employing an equivalent fluid pressure. The fill material is assumed to act as a fluid of known weight, and the forces acting on the structure are computed from the triangular distribution of fluid pressure (Figure 6-23 A). AASHTO specifications require that a minimum equivalent fluid weight of 30 lb/ft³ be used for retaining structures (AASHTO 3.20.1). In practice, an equivalent fluid weight of 35 or 36 lb/ft³ is more commonly used (sandy backfill with a unit weight of approximately 120 lb/ft³). These fluid weights assume that fill material is free draining and that no significant hydrostatic forces exist. When hydro-static forces may be generated, the equivalent fluid weight must be increased.

The earth pressure acting on a retaining structure is increased when ve-hicle live loads occur in the vicinity of the structure. When vehicle traffic can come within a horizontal distance from the top of a retaining structure equal to one-half its height, a live load surcharge of 2 feet of fill is added to compensate for vehicle loads (AASHTO 3.20.3). The resulting load distribution on the structure is trapezoidal (Figure 6-23 B). This additional load is not required when a reinforced concrete approach slab supported at one end by the bridge is provided.

Earth pressures can vary significantly, based on soil conditions at the site and the type and complexity of the structure. In some cases, a more

Figure 6-22. - Bulging in an abutment retaining wall caused by earth pressure.

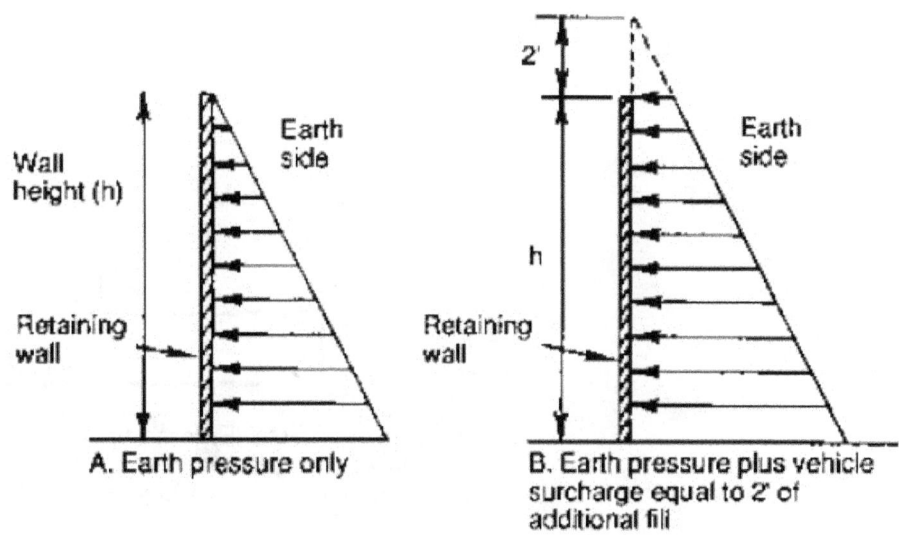

Figure 6-23. - Distribution of earth pressure on retaining structures.

sophisticated analysis than that required in AASHTO is warranted. References listed at the end of this chapter provide more detailed information on the application of soil mechanics to the design of abutments and retaining walls.[1228]

6.13 BUOYANCY

Buoyancy is the resultant of the upward surface forces acting on a submerged body (Figure 6-24). It is considered in bridge design when a portion of the structure is submerged or is located below the water table. Buoyancy is equal in magnitude to the weight of fluid displaced, or 62.4 lb/ft³ for water. Its effect is to reduce the weight of the substructure, which may result in smaller footing or pier sizes and a more economical design; however, buoyancy also reduces the ability of the substructure to resist uplift from vertical or lateral (overturning) loads. When combined with significant longitudinal or transverse moments, the effects of buoyancy could result in a larger footing. In either case, buoyancy is most significant in the design of massive footings or piers where dead load is a considerable percentage of the total load. In most timber bridge applications, the ratio of dead load to live load is small, and the effects of buoyancy are generally of little or no significance.

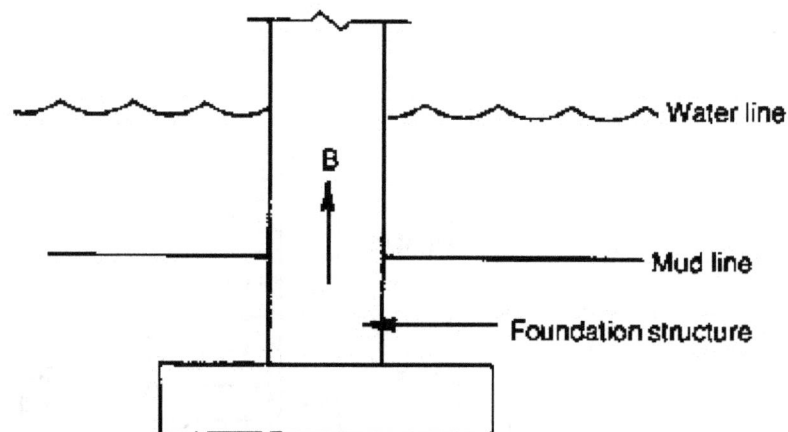

The buoyancy force (B) is an upward vertical force equal in magnitude to the volume of the structure below the water line times the unit weight of water (62.4 lb/ft³)

Figure 6-24. - Buoyancy forces on a submerged substructure.

Stream currents produce forces acting on piers, bents, and other portions of the structure located in moving water. These forces produce pressure against the submerged structure and are computed as a function of stream velocity (AASHTO 3.18.1) as

$$P = KV^2 \tag{6-5}$$

where P = stream flow pressure (lb/ft^2),

V = water velocity (ft/sec), and

K = a constant for the shape of the pier (1-3/8 for square ends, 1/2 for angle ends where the angle is 30° or less, 2/3 for round ends).

The stream flow pressure computed by Equation 6-5 is applied to the area of the substructure over the estimated stream depth (Figure 6-25). Although stream velocity varies with depth, a constant velocity for the full depth provides sufficiently accurate results. The pressures act to slide or overturn the structure and are most significant on large piers or bents located in deep, fast-moving streams or rivers.

Forces associated with streams depend on a number of factors that must be thoroughly investigated for each site. In general, hydraulic parameters for flow velocity and depth are based on the 50- or 100-year occurrence interval. For many streams, flow records and other data have been established to provide this information. When such data are not available, estimated flow should be based on local experience or the best judgment of the designer.

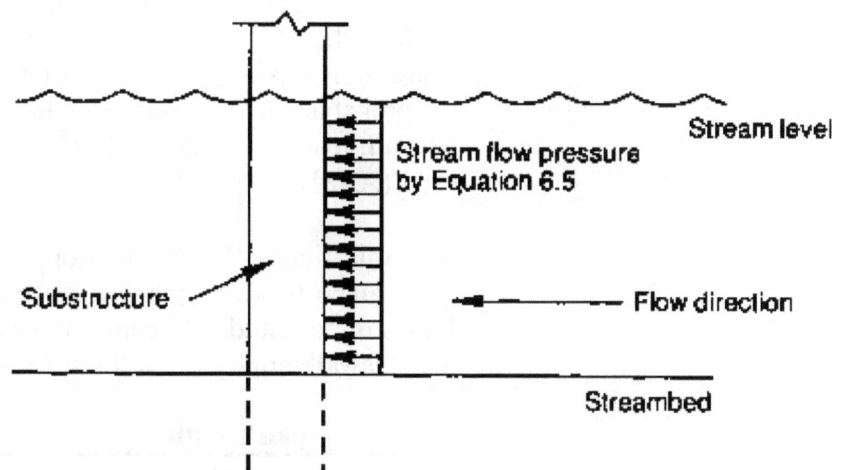

Figure 6-25. - Application of stream flow pressure on a submerged substructure.

6.15 ICE FORCE

In areas of cold climate, substructures located in streams or other bodies of water may be subjected to ice forces. These forces result from (1) the dynamic force of floating ice sheets and floes striking the structure; (2) the static ice pressure from thermal movement of continuous ice sheets on large bodies of water; (3) the static pressure produced by ice jams forming against the structure; and (4) the static vertical forces caused by fluctuating water levels when piers are frozen into ice sheets.

Ice forces are difficult to predict and depend on a number of factors including the thickness, strength, and movement of ice, as well as the configuration of the structure. AASHTO specifications give guidelines for computing dynamic ice forces on piers (AASHTO 3.18.2); however, definitive recommendations for static forces are not practical because of variations in local conditions. When ice formation is possible, potential forces should be determined by specialists using field investigations, published records, past experience, and other appropriate means. Consideration should be given to the probability of extreme rather than average conditions. Additional information on ice forces is given in AASHTO and references listed at the end of this chapter.[9,15]

6.16 SIDEWALK LIVE LOAD

Sidewalks are provided on vehicle bridges to allow concurrent use of the structure by pedestrians, bicycles, and other nonhighway traffic. Sidewalks are subjected to moving live loads that vary in magnitude and position, just as do vehicle live loads. For design purposes, AASHTO gives sidewalk live loads as uniformly distributed static loads that are applied vertically to the sidewalk area (Figure 6-26). The magnitude of the load depends on the component of the structure and the length of sidewalk it supports. When a member supports a long section of sidewalk, the probability of maximum loading along the entire length is reduced. As a result, loads vary and are based on the type of member and sidewalk span (AASHTO 3.14.1).

Sidewalk floors, floorbeams (longitudinal or transverse), and their immediate supports are designed for a live load of 85 lb/ft^2 of sidewalk area. Loads on longitudinal beams, arches, and other main members supporting the sidewalk are based on the sidewalk span:

Span length	Sidewalk load
Up to 25 ft	85 lb/ft^2
25 ft to 100 ft	60 lb/ft^2

Sidewalk loads are applied as uniformly distributed loads acting over the exposed sidewalk area.

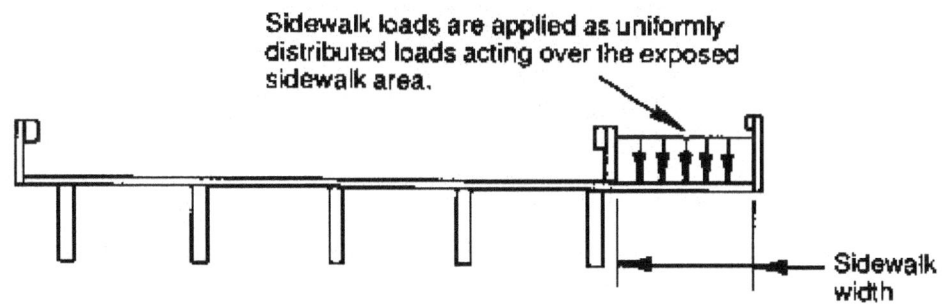

Sidewalk width

Figure 6-26. - Application of side walk loads.

When the span length exceeds 100 feet, the design live load is determined as

$$P = \left(30 + \frac{3,000}{L}\right)\left(\frac{55 - W}{50}\right) \le 60 \text{ lb/ft}^2 \qquad (6\text{-}6)$$

where

P = load per square foot of sidewalk area (lb/ft^2),

L = loaded length of sidewalk (ft), and

W = sidewalk width (ft).

It should be noted that sidewalk loads given in AASHTO are intended for conditions where loading is primarily pedestrian and bicycle traffic. If sidewalks will be used by maintenance vehicles, horses, or other heavier loads, the designer should increase the design loading accordingly.

Sidewalk loads are distributed to structural components in a manner similar to dead load. The load supported by any member is computed from the tributary area of sidewalk it supports. If bridges have cantilevered sidewalks on both sides, one or both sides should be fully loaded, whichever produces the maximum stress. In cases where the maximum design load in an outside longitudinal beam results from a combination of dead load, sidewalk live load, and vehicle live load, AASHTO allows a 25-percent increase in allowable design stresses, provided the beam is of no less carrying capacity than would be required if there were no sidewalks (AASHTO Table 3.22.1A).

6.17 CURB LOADS

Curbs are provided on bridges to guide the movement of vehicle wheels and protect elements of the structure from wheel impact. When traffic railing is provided, curbs may be included as a part of the rail system and are frequently used to connect rail posts to the deck. On low-volume roads with relatively slow design speeds, barrier curbs are sometimes used instead of traffic railing to delineate the roadway edge and inhibit slow-moving vehicles from leaving the structure (Figure 6-27). In both cases, curb loading is from vehicle impact applied either directly to the curb or through the rail system.

AASHTO specifications give curb loading requirements based on the interaction of the curb and traffic railing (AASHTO 3.14.2). When curbs are used without railing, or are not an integral part of a traffic railing system, the minimum design load consists of a transverse line load of 500 lb/lin ft of curb applied at the top of the curb, or at an elevation 10 inches above the floor if the curb is higher than 10 inches (Figure 6-28). When curbs are connected with traffic railing to form an integral system, the design loads applied to the curb are those produced by the railing loads (see Chapter 10).

Figure 6-27—Barrier curbs on a timber bridge. Such curbs are sometimes used instead of railing on single-lane, low-volume bridges.

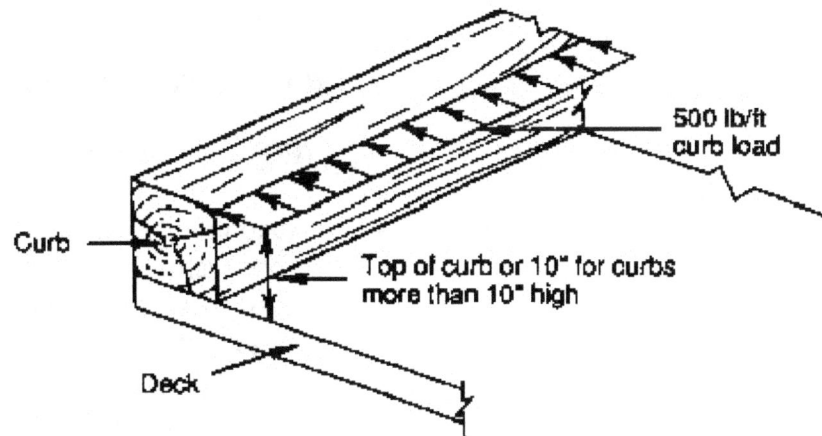

Figure 6-28 - Application of curb loads when the curb is not integral with the vehicular railing system.

6.18 OTHER LOADS

In addition to the minimum AASHTO load requirements discussed in this chapter, timber bridges may be subjected to other loads during construction and in service. Consideration should be given to loads resulting from transportation, handling, and erection, especially when long, slender beams or columns are considered. Because these loads are difficult to quantify, they are left to the judgment of the designer and must be based on specific information for each project.

6.19 LOAD COMBINATIONS

Timber bridges may be subjected to any of the loads and forces previously discussed. In practice, these loads seldom act individually, but normally occur as a combination of loads acting simultaneously. The designer must determine which loads are applicable to the design of a structure and the combination of loads that produce the maximum stress in each bridge component.

Load combinations for bridge design are based on load groups given in AASHTO (AASHTO 3.22) for service-load design (load-factor design is currently not applicable for timber). These load groups consist of a number of individual loads that are assumed to act simultaneously on a particular bridge component. Each load group is computed using the following equation and the load group numbers and factors given in Table 6-4:

$$GROUP(N) = \gamma[\beta_D(D) + \beta_L(L+I) + \beta_C(CF) + \beta_E(E) + \beta_B(B) \tag{6-7}$$

$$+ \beta_S(SF) + \beta_W(W) + \beta_{WL}(WL) + \beta_L(LF) + \beta_R(R+S+T)$$

$$+ \beta_{EQ}(EQ) + \beta_{ICE}(ICE)]$$

where N = load group number,
γ = load factor from Table 6-4,
β = load coefficient from Table 6-4,
D = dead load,
L = vehicle live load,
I = vehicle live load impact (not applicable to timber),
E = earth pressure,
B = buoyancy,
W = wind load on the structure,
WL = wind load acting on the vehicle live load,
LF = longitudinal force from vehicle live load,
CF = centrifugal force,
R = rib shortening,
S = shrinkage,
T = temperature,
EQ = earthquake force,
SF = stream flow pressure, and
ICE = ice pressure.

Table 6-4. - AASHTO load group coefficients for service load design of timber bridges.

	1	2	3	3A	4	5	6	7	8	9	10	11	12	13	14
Group						β factors									
(N)	γ	D	$(L+I)_n$	$(L+I)_p$	CF	E	B	SF	W	WL	LF	R+S+T	EQ	ICE	%[a]
I	1.0	1	1	0	1	β_E	1	1	0	0	0	0	0	0	100
IA	1.0	1	2	0	0	0	0	0	0	0	0	0	0	0	150
IB	1.0	1	0	1	1	β_E	1	1	0	0	0	0	0	0	—[b]
II	1.0	1	0	0	0	1	1	1	1	0	0	0	0	0	125
III	1.0	1	1	0	1	β_E	1	1	0.3	1	1	0	0	0	125
IV	1.0	1	1	0	1	β_C	1	1	0	0	0	1	0	0	125
V	1.0	1	0	0	0	1	1	1	1	0	0	1	0	0	140
VI	1.0	1	1	0	1	β_E	1	1	0.3	1	1	1	0	0	140
VII	1.0	1	0	0	0	1	1	1	0	0	0	0	1	0	133
VIII	1.0	1	1	0	1	1	1	1	0	0	0	0	0	1	140
IX	1.0	1	0	0	0	1	1	1	1	0	0	0	0	1	150
X	1.0	1	1	0	0	β_E	0	0	0	0	0	0	0	0	100 Culvert

[a] Percentage of allowable unit stress used for design. No increase is permitted for members or connections carrying wind loads only.

[b] Percentage = $\dfrac{\text{Maximum unit stress (operating rating)}}{\text{Allowable unit stress}} \times 100 = 133\%$ for timber bridges.

$(L+I)_n$ = Live load plus impact for AASHTO H or HS loading.

$(L+I)_p$ = Live load plus impact consistent with the overload criteria of the operating agency.

β_E = 1.0 and 0.5 for lateral loads on rigid frames (check both to see which governs).

β_E = 1.0 for vertical and lateral loads on all other structures.

From AASHTO[3] Table 3.22.1A. © 1983. Used by permission

For service load design, the load factor for all load groups is 1.0 and the requirements of Equation 6-7 can by read directly from values specified in Table 6-4. The relative magnitude of each load within a group is determined by the ß factor in columns 2 through 13. When the ß factor for an individual load is zero, that load is not considered in the load group. For example, Load Group III consists of the dead load, vehicle live load, centrifugal force, earth pressure (factored by the applicable beta factor), buoyancy, stream force, 30 percent of the wind load on the structure, wind load on the vehicle live load, and the longitudinal force. Although each of these loads is assumed to act simultaneously in the load group, the applicability of any load for a specific structure is left to the judgment of the designer. If an individual load is not applicable, the ß factor for that load is zero, regardless of the ß factor given in Table 6-4.

The concept of load groups is based on the assumption that a number of loads willoccur simultaneously on the structure. To compensate for the small probability that all loads will act together at their maximum intensities, an increase in allowable design stresses is permitted for most groups. These increases are based on the premise that the possibility of all loads acting at the same time is small enough to justify a reduction in the factor of safety. Percentages of allowable stresses for each load group are given in column 14 of Table 6-4, with the following two exceptions:

1. When a member loaded in any load group is subjected to wind load only, no increase in allowable stress is permitted.

2. For overloads considered in Load Group IB, the design stresses are a percentage of allowable stresses computed as the ratio of the maximum unit stress allowed at the operating rating level given in the *AASHTO Manual for Maintenance Inspection of Bridges*[2] and the allowable unit stress. For timber components, this ratio is 133 percent.

For timber bridges, increases in allowable stresses for load groups are cumulative, with modifiers for duration of load. Because duration of load adjustments reflect the material properties of timber, they should not be confused with increases based on load probability. The total increase in allowable unit stress for timber components is that given for the load group plus the applicable factor for duration of load discussed in Chapter 5.

Each component of a bridge superstructure and substructure must be proportioned to safely withstand all load group combinations that are applicable to the structure. Different load groups will control the design of different parts of the structure. Load Groups I, II, and III are most applicable for bridge superstructures and substructures; Load Groups IV, V, and VI are for arches and frames; and Load Groups VII, VIII, and IX are for substructures. Load Group X is for culvert design only and is not used for bridges. To determine the controlling load groups, the designer must determine which individual loads are applicable and compute magnitudes and effects of these loads; however, it is not necessary to investigate all group loads for all bridges. In most cases, it is evident by inspection that only a few loadings are likely to control the design of any single type of structure or component. In general, the following three load groups are most applicable for timber bridges:

Superstructures: Load Group I; Load Group IB when overloads are considered

Abutments: Load Groups I and III; Load Group IB when overloads are considered; Load Group VII when earthquake loads are applicable

Piers: Load Groups I, II, and III; Load Group IB when overloads are considered; Load Group VIII when ice loads are applicable

6.20 SELECTED REFERENCES

1. American Association of State Highway and Transportation Officials. 1976. AASHTO manual for bridge maintenance. Washington, DC: American Association of State Highway and Transportation Officials. 251 p.
2. American Association of State Highway and Transportation Officials. 1983. Manual for maintenance inspection of bridges. Washington, DC: American Association of State Highway and Transportation Officials. 50 p.
3. American Association of State Highway and Transportation Officials. 1983. Standard specifications for highway bridges. 13th ed. Washington, DC: American Association of State Highway and Transportation Officials. 394 p.
4. American Association of State Highway and Transportation Officials. 1984. A policy on geometric design of highways and streets. Washington, DC: American Association of State Highway and Transportation Officials. 1,087 p.
5. American Association of State Highway and Transportation Officials. 1984. Guide specifications for seismic design of highway bridges.- Washington, DC: American Association of State Highway and Transportation Officials. 107 p.
6. American Institute of Timber Construction. 1985. Timber construction manual. 3d ed. New York: John Wiley and Sons, Inc. 836 p.
7. American National Standards Institute. 1982. Minimum design loads for buildings and other structures. ANSI A58.1. New York: American National Standards Institute. 100 p.

8. American Society of Civil Engineers. 1982. Evaluation, maintenance, and upgrading of wood structures. Freas, A., ed. New York: American Society of Civil Engineers. 428 p.

9. American Society of Civil Engineers. 1980. Loads and forces on bridges. Preprint 80-173, 1980 American Society of Civil Engineers national convention; 1980 April 14-18; Portland, OR. New York: American Society of Civil Engineers. 73 p.

10. American Society of Civil Engineers. 1961. Wind forces on structures. Final report of the task committee on wind forces. Pap. 3269. American Society of Civil Engineers Transactions 126(2): 1124-1198.

11. Bell, L.C.; Yoo, C.H. 1984. Seminar on fundamentals of timber bridge construction. Course notes; 1984 May 22-25; Auburn University, AL. Auburn University. [150 p.].

12. Bruner, R.F.; Coyle, H.M.; Bartoskewitz, R.E. 1983. Cantilever retaining wall design. Res. Rep. 236-2F. College Station, TX: Texas Transportation Institute. 179 p.

13. Grubb, M.A. 1984. Horizontally curved I-girder bridge analysis: V-load method. In: Bridges and foundations. Trans. Res. Rec. 982. Washington, DC: National Academy of Sciences, National Research Council, Transportation Research Board: 26-36.

14. Gurfinkel, G. 1981. Wood engineering. 2d ed. Dubuque, IA: Kendall/Hunt Publishing Co. 552 p.

15. Haynes, F.D. 1983. Ice forces on bridge piers. Hanover, NH: U.S. Army CRREL. 16 p.

16. Heins, C.P.; Lawrie, R.A. 1984. Design of modem concrete highway bridges. New York: John Wiley and Sons. 635 p.

17. Kozak, J.J.; Leppmann, J.F. 1976. Bridge engineering. In: Merrit, F.S., ed. Standard handbook for civil engineers. New York: McGraw-Hill Co. Chapter 17.

18. McCormac, J.C. 1975. Structural analysis. 3d ed. New York: Intext Educational Publishers. 603 p.

19. Milbradt, K.P. 1968. Timber structures. In: Gaylord, E.H., Jr.; Gaylord, C.N., eds. Structural engineering handbook. New York: McGraw Hill. Chapter 16.

20. Ministry of Transportation and Communications. 1983. Ontario highway bridge design code. Downsview, ON, Can.: Ministry of Transportation and Communications. 357 p.

21. Ministry of Transportation and Communications. 1983. Ontario highway bridge design code commentary. Downsview, ON, Can.: Ministry of Transportation and Communications. 279 p.

22. Ministry of Transportation and Communications. 1986. Ontario highway bridge design code updates. Downsview, ON, Can.: Ministry of Transportation and Communications. 28 p.

23. State of California, Department of Transportation. 1986. Bridge design specifications manual. Sacramento, CA: State of California, Department of Transportation. 379 p.

24. State of California, Department of Transportation. 1983. Bridge design practice-load factor. Sacramento, CA: State of California, Department of Transportation. 619 p.

25. State of California, Department of Transportation. 1985. Revisions of the standard specifications for highway bridges relating to seismic design. Sacramento, CA: State of California, Department of Transportation. 85 p.

26. State of Wisconsin, Department of Transportation. 1979. Bridge manual. Madison, WI: State of Wisconsin, Department of Transportation. [350 p.]

27. U.S. Department of Agriculture, Forest Service, Northern Region. 1985. Bridge design manual. Missoula, MT: U.S. Department of Agriculture, Forest Service, Northern Region. 299 p.

28. U.S. Department of Agriculture, Forest Service, Pacific Northwest Region. 1979. Retaining wall design guide. Portland, OR: U.S. Department of Agriculture, Forest Service, Pacific Northwest Region. [300 p.]

29. United States Steel Corp. 1965. Highway structures design handbook. Pittsburgh, PA: United States Steel Corp. Vol. 2. 597 p.

CHAPTER 7

DESIGN OF BEAM SUPERSTRUCTURES

7.1 INTRODUCTION

Beam superstructures consist of a series of longitudinal timber beams supporting a transverse timber deck. They are constructed of glulam or sawn lumber components and have historically been the most common and most economical type of timber bridge (Figure 7-1). For the past 20 years, beam bridges have been constructed almost exclusively from glulam because of the greater size and better performance characteristics it provides compared with sawn lumber systems. Sawn lumber bridges are still used to a limited degree on local public roads and private road systems with low traffic volumes.

This chapter addresses design considerations and requirements for beam superstructures and is divided into two parts. Part I deals with glulam systems and includes the design of glulam beams and transverse glulam deck panels. Part II covers sawn lumber systems and includes the design of lumber beams and transverse nail-laminated and plank decks. In both parts, deck design is limited to transverse and configurations only. Applications involving longitudinal decks on beam superstructures are discussed in Chapter 8. Railing systems and wearing surfaces for beam bridges are covered in Chapters 10 and 11, respectively.

7.2 DESIGN CRITERIA AND DEFINITIONS

The material presented in this chapter is based on the 1983 edition of the *AASHTO Standard Specifications for Highway Bridges* (AASHTO), including interim specifications through 1987.[1] When specific design requirements or criteria are not addressed by that specification, recommendations are based on referenced standards and specifications or commonly accepted design practice. Because AASHTO specifications are periodically revised to reflect new developments in bridge design, the designer should refer to the latest edition for the most current requirements. This chapter is not intended to serve as a substitute for current specifications.

General design criteria used in this chapter are summarized below. Additional criteria related to specific component design are given in the applicable sections.

Figure 7-1. - Beam superstructures constructed of (A) glulam timber and (B) sawn lumber.

DESIGN PROCEDURES AND EXAMPLES

Sequential design procedures and examples are included in this chapter to familiarize the designer with the requirements for beam bridges. Design procedures are intended to outline basic requirements and present applicable design equations and aids. The order of the procedures is based on the most common sequence used in design and may vary for different applications. Examples are based on more specific site requirements, and criteria are noted for each example.

LOADS

Loads are based on the AASHTO load requirements discussed in Chapter 6. Beam and deck design procedures are limited to AASHTO Group I loads where design is routinely controlled by a combination of structure dead load and vehicle live load. Vehicle live loads are standard AASHTO loads consisting of H 15-44, H 20-44, HS 15-44, and HS 20-44 vehicles. Overloads are considered in the design examples in AASHTO Group IB, where allowable stresses are increased by 33 percent, as discussed in Chapter 6.

For deck design, AASHTO special provisions for HS 20-44 and H 20-44 loads apply, and a 12,000-pound wheel load is used unless otherwise noted (AASHTO Figures 3.7.6A and 3.7.7A). In most cases, deck design aids include the dead load of a 3-inch asphalt wearing surface. These aids can be used with reasonable accuracy for other common wearing surfaces since wearing-surface dead load normally has little effect on beam or deck design.

MATERIALS

Tabulated values for sawn lumber are taken from the 1986 edition of the NDS. [37,38] Species used are Douglas Fir-Larch and Southern Pine, but the principles of design apply to wood of any species group. For glulam, tabulated values are taken from the 1987 edition of *AITC 117--Design.* [5] Material specifications are given by combination symbol; however, glulam can also be specified by required design values in a format similar to that given in *AITC 117--Design.* Visually graded combination symbols are recommended, with provisions for E-rated substitution at the option of the manufacturer. *All timber components are assumed to be pressure-treated with an oil-type preservative prior to fabrication, as discussed in Chapter 4.*

LIVE LOAD DEFLECTION

AASHTO specifications do not include design criteria or guidelines for beam or deck live load deflection. The recommendations in this chapter are based on field experience and common design practice as noted for the specific component. Although it is highly recommended that these deflection guidelines be followed, deflection criteria should be based on specific design circumstances and are left to designer judgment.

CONDITIONS OF USE
Tabulated values for timber components must be adjusted for specific use conditions by all applicable modification factors discussed in Chapter 5. The following criteria have been used in this chapter.

Duration of Load. Beam and deck design for combined dead load and vehicle live load are based on a normal duration of load (that is, design stresses at the maximum allowable level do not exceed a cumulative total of 10 years). Therefore, equations for allowable design values do not include the duration of load factor, C_D.

Moisture Content. With the exception of glulam beams covered by a watertight deck, all stresses in bridge components are adjusted for wet-use conditions. Based on recommendations of the AITC,[7] covered glulam beams are designed for dry-condition stresses with the exception of compression perpendicular to grain at supports, where wet-condition stress is recommended. This is based on the assumption that a watertight deck sufficiently protects glulam beams and that superficial surface wetting does not cause significant increases in beam moisture content except at supports.

Temperature Effects and Fire-Retardant Treatment. Conditions requiring adjustments for temperature or fire-retardant treatment are rare in bridge applications. Design equations in this chapter do not include modification factors for temperature effect, C_t, or fire-retardant treatment, C_R.

PART I:
GLUED-LAMINATED TIMBER (GLULAM) SYSTEMS

7.3 GENERAL

Glued-laminated beam bridges consist of a series of transverse glulam deck panels supported on straight or slightly curved beams (Figure 7-2). They are the most practical for clear spans of 20 to 100 feet and are widely used on single-lane and multiple-lane roads and highways. Glulam has proved to be an excellent material for beam bridges because members are available in a range of sizes and grades and are easily adaptable to a modular or systems concept of design and construction. Although glulam can be custom fabricated in many shapes and sizes, the most economical structure uses standardized components in a repetitious arrangement, an approach that is particularly adaptable to bridges (Figure 7-3).

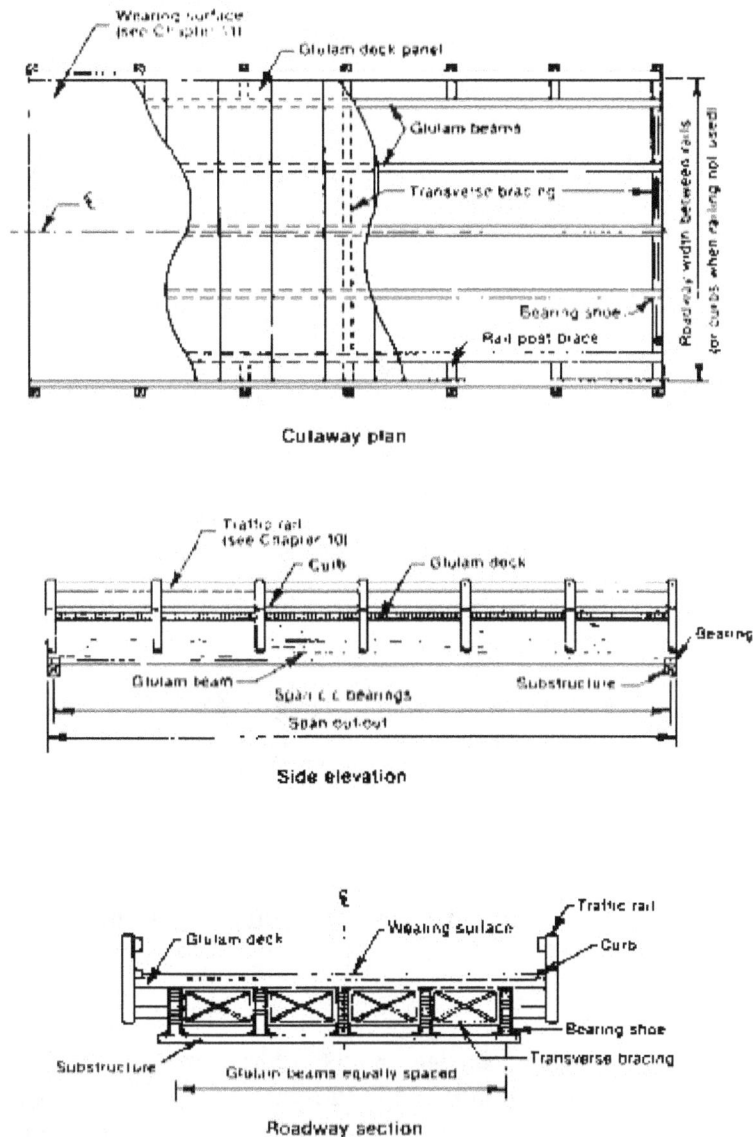

Figure 7-2. - Typical glulam beam bridge configuration.

The following three sections address design considerations, procedures, and details for glulam beam bridges. Beams and beam components are discussed first, followed by transverse glulam deck panels.

7.4 DESIGN OF BEAMS AND BEAM COMPONENTS

Beams are the principal load-carrying components of the bridge super-structure. They must be proportioned to resist applied loads and meet serviceability requirements for deflection. The total beam system consists of three primary components: beams, transverse bracing, and bearings.

Figure 7-3.- Glulam beam bridge, 290 feet long in Tioga County, New York. This bridge was completely prefabricated in standardized components that were bolted together at the project site (photo courtesy of Weyerhaeuser Co.).

Each of these components is designed individually to perform specific functions. Together they interact to form the structural framework of the bridge.

BEAM DESIGN

Glulam bridge beams are horizontally laminated members designed from the bending combinations given in Table 1 of *AITC 117--Design*. These combinations provide the most efficient beam section where primary loading is applied perpendicular to the wide face of the laminations. The quality and strength of outer laminations are varied for different combination symbols to provide a wide range of tabulated design values in both positive and negative bending.

Glulam beams offer substantial advantages over conventional sawn lumber beams because they are manufactured in larger sizes, provide improved dimensional stability, and can be cambered to offset dead load deflection: Beams are available in standard widths ranging from 3 to 14-1/4 inches (Table 7-1) and in depth multiples of 1-1/2 inches for western species and 1-3/8 inches for Southern Pine. Beam length is usually limited by treating and transportation considerations to a practical maximum of 110 to 120 feet, but longer members may be feasible in some areas. Tables of standard glulam section properties are given in Chapter 16.

Table 7-1. - Standard glulam beam widths.

Nominal width (in.)	Net finished width (in.)	
	Western species	Southern Pine
4	3-1/8	3
6	5-1/8	5
8	6-3/4	6-3/4
10	8-3/4	8-1/2
12	10-3/4	10-1/2
14	12-1/4	—
16	14-1/4	—

Live Load Distribution

Methods for determining the maximum moment, shear, and reactions for truck and lane loads were discussed in Chapter 6. For beam superstructures, the designer must also determine the portion of the total load that is laterally distributed to each beam. The ability of a bridge to laterally distribute loads to individual beams depends on the transverse stiffness of the structure as a unit and is influenced by the type and configuration of the deck and the number, spacing, and size of beams. Load distribution may also be influenced by the type and spacing of beam bracing or diaphragms, but the effect of these components is not considered for determining load distribution.

In view of the complexity of the theoretical analysis involved in determining lateral wheel-load distribution, AASHTO specifications give empirical methods for longitudinal beam design. The fractional portion of the total vehicle load distributed to each beam is computed as a distribution factor (DF) expressed in wheel lines (WL) per beam. The magnitude of the design forces is determined by multiplying the distribution factor for each beam by the maximum force produced by one wheel line of the design vehicle (moment, shear, reaction, and so forth). The procedures for determining distribution factors for longitudinal beams depend on the type of force and are specified separately for moment, shear, and reactions.

Distribution for Moment

When computing bending moments in longitudinal beams (AASHTO 3.23.2), wheel loads are assumed to act as point loads. Lateral distribution is determined by empirical methods based on the position of the beam relative to the transverse roadway section. Different criteria are given for outside beams and for interior beams; however, AASHTO requires that the load distributed to an outside beam not be less than that distributed to an interior beam.

The distribution factor for moment in outside beams is determined by computing the reaction of the wheel lines at the beam, assuming the deck acts as a simple span between beams (Figure 7-4). Wheel lines in the outside traffic lane are positioned laterally to produce the maximum reaction at the beam, but wheel lines are not placed closer than 2 feet from the face of the traffic railing or curb (Chapter 6). The distribution factor for moment for interior beams is computed from empirical formulas based on deck thickness, beam spacing, and the number of traffic lanes (Table 7-2). For glulam decks 6 inches or more in nominal thickness, these equations are valid up to the maximum beam spacing specified in the table. When the average beam spacing exceeds the maximum, the distribution factor is the reaction of the wheel lines at the beam, assuming the flooring between beams acts as a simple span (Figure 7-5). In this case, wheel lines are laterally positioned in traffic lanes to produce the maximum beam reaction (wheel lines in adjacent traffic lanes are separated by 4 feet).

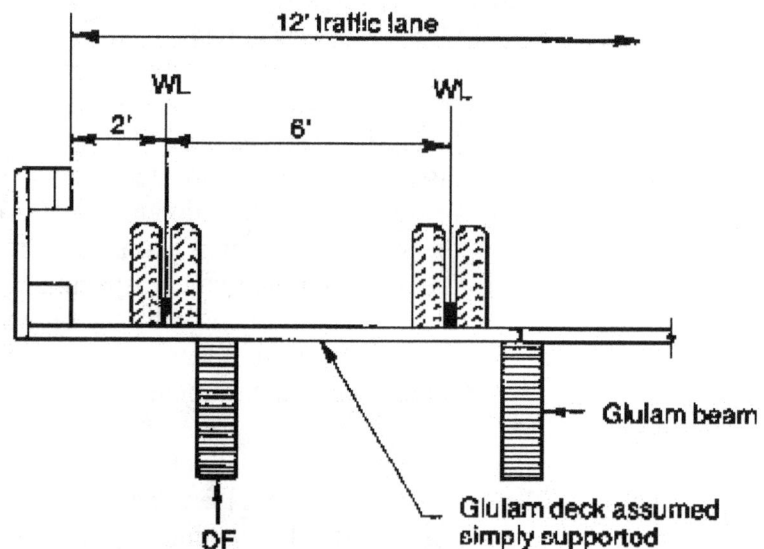

Figure 7-4. - Wheel load distribution factor to outside beams, assuming the deck acts as a simple span between supporting beams.

Table 7-2 - Interior beam live load distribution factors for glulam beams with transverse glulam decks.

| Nominal deck thickness (in.) | DF for moment (wheel lines/beam) | |
	Bridges designed for one traffic lane	Bridges designed for two or more traffic lanes
4	$S/4.5$	$S/4.0$
≥ 6	$S/6.0$	$S/5.0$
	If S exceeds 6 ft, use footnote a.	If S exceeds 7.5 ft, use footnote a.

[a] In this case, the distribution factor for each beam is the reaction of the wheel lines, assuming the deck between beams to act as a simple beam.

S = average beam spacing (ft).

From AASHTO[1] Table 3.23.1; ©1983. Used by permission.

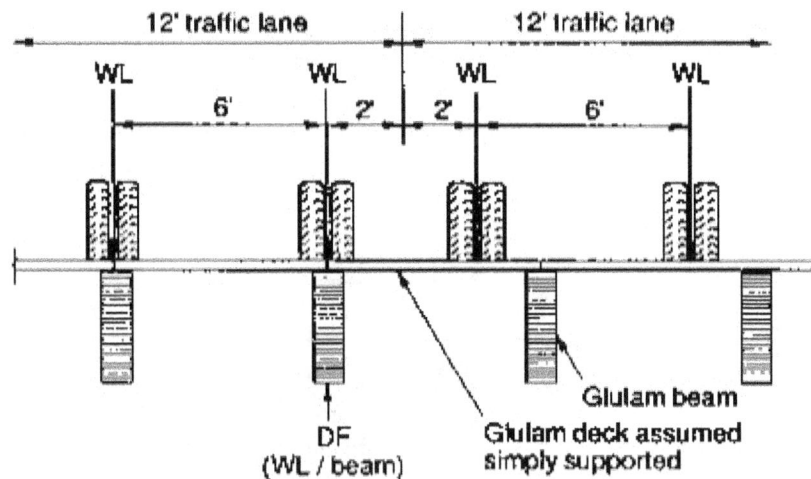

Figure 7-5. - Wheel load distribution factor to interior beams, assuming the deck acts as a simple span between supporting beams.

Distribution for Shear

Live-load horizontal shear in glulam beams (AASHTO 13.3.1) is computed from the maximum vertical shear occurring at a distance from the support equal to three times the beam depth (3*d*) or the span quarter point *(L*/4), whichever is less (Figure 7-6). Lateral shear distribution at this point is computed as one-half the sum of 60 percent of the shear from the undistributed wheel lines and the shear from the wheel lines distributed laterally as specified for moment. For undistributed wheel lines, one wheel line is assumed to be carried by one beam. These requirements are expressed as

$$V_{LL} = 0.5[(0.6V_{LU}) + V_{LD}] \qquad (7\text{-}1)$$

where
V_{LL} = distributed live-load vertical shear used to compute horizontal shear (lb),

V_{LU} = maximum vertical shear from an undistributed wheel line (lb), and

V_{LD} = maximum vertical shear from the vehicle wheel lines distributed laterally as specified for moment (lb).

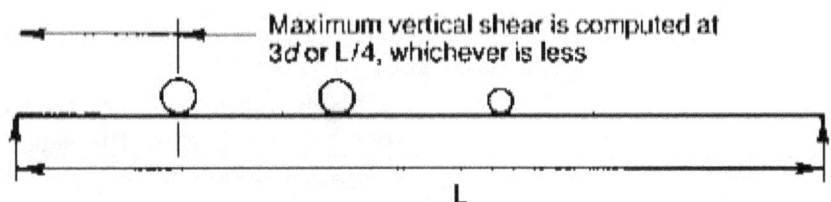

Figure 7-6. - Live load horizontal shear in timber beams is based on the maximum vertical shear occurring at a distance from the support equal to three times the beam depth (3d), or the span quarter point (L/4), whichever is less.

7-9

Distribution for Reactions

Live load distribution for end reactions (AASHTO 3.23.1) is computed assuming no longitudinal distribution of wheel loads. The DF for outside and interior beams is determined by computing the reaction of the wheel lines at the beam, assuming the deck acts as a simple span between beams (Figures 7-4 and 7-5).

Example 7-1 - Live load distribution on a multiple-lane beam bridge

---

A two-lane beam bridge with a 28-foot roadway width spans 52 feet. The superstructure consists of a 6-3/4-inch glulam deck supported by 5 glulam beams, symmetrically spaced at 6-feet on center. Determine the distributed live load moment, shear and reactions for an HS 20-44 design vehicle. Assume an initial beam depth of 43-1/2 inches for shear distribution.

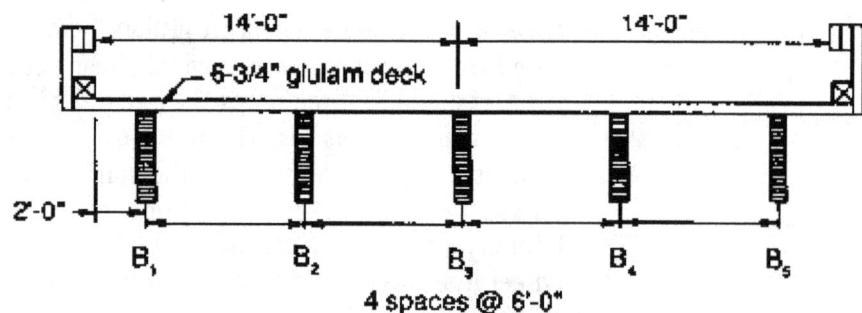

Solution

The designer must determine the distribution factors for interior and outside beams and the magnitude of the maximum forces produced by one wheel line of the design vehicle. The product of applicable DF and wheel line force provides the design value for each beam.

Distribution for Moment

The moment distribution factor for interior beams is determined from Table 7-2 based on the deck thickness, number of traffic lanes, and beam spacing. For a 6-3/4-inch glulam deck, two-lane bridge, and 6-foot beam spacing,

$$DF = \frac{S}{5.0} = \frac{6.0}{5.0} = 1.20 \text{ WL/beam}$$

For outside beams, the DF is computed by assuming the deck acts as a simple span between beams. The center of the wheel load is placed 2 feet from the face of the railing, and the outside beam reaction is computed, in wheel lines, by statics:

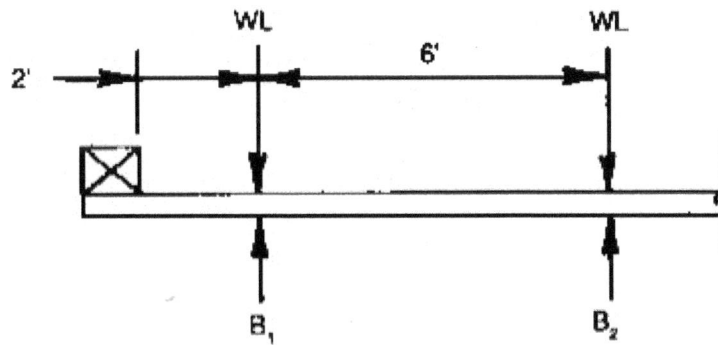

The maximum reaction results in a DF of 1.0 WL/beam; however, the DF to outside beams cannot be less than that to interior beams. Therefore, the DF to both outside and interior beams is 1.2 WL/beam.

From Table 16-8, or computations discussed in Chapter 6, the maximum moment for one wheel line of an HS 20-44 truck on a 52-foot span is 331.77 ft-k. The distributed live load moment for interior and outside beams is

$$M_{LL} = (1.20 \text{ WL/beam})(331.77 \text{ ft-k}) = 398.12 \text{ ft-k}$$

Distribution for Shear

Live load shear distribution is computed by Equation 7-1 using the same distribution factors used for moment. The first step is to compute the maximum vertical shear occurring at the lesser of $3d$ or $L/4$ from the support:

$$3d = \frac{(3)(43.5)}{12 \text{ in/ft}} = 10.88 \text{ ft} \qquad \frac{L}{4} = \frac{52}{4} = 13.0 \text{ ft}$$

$3d$ = 10.88 ft controls.

The maximum vertical shear for an undistributed wheel line (V_{LU}) is computed by placing the heaviest axle 10.88 feet from the support as discussed in Chapter 6:

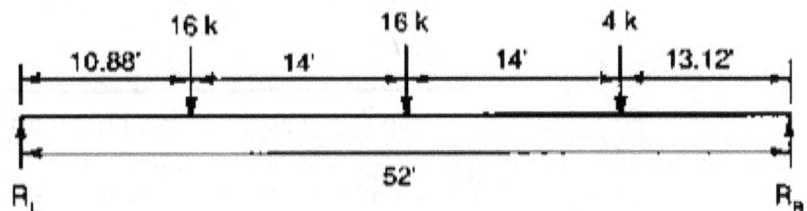

$$V_{LU} = R_L = \frac{(13.12)(4\text{k}) + (27.12)(16\text{k}) + (41.12)(16\text{k})}{52} = 22.01 \text{ k}$$

Because the moment DF is the same for interior and outside beams, the distributed shear for interior and outside beams will also be the same. By Equation 7-1,

$$V_{LD} = DF(V_{LU}) = 1.20(22.01) = 26.41 \text{ k}$$

$$V_{LL} = 0.50[(0.60V_{LU}) + V_{LD}] = 0.50[(0.60)(22.01) + (26.41)] = 19.81 \text{ k}$$

Distribution for Reactions

The reaction distribution factors to interior and outside beams are computed by assuming the deck acts as a simple span between beams. In this case, the vehicle track width of 6 feet equals the beam spacing, and the maximum DF for interior beams is 1.0:

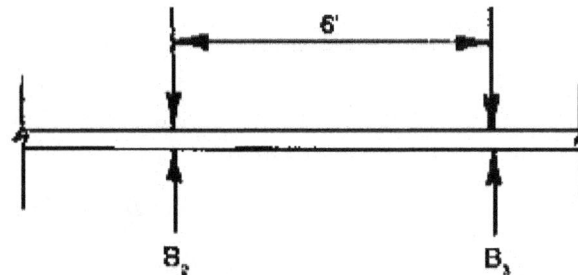

For outside beams, the DF also equals 1.0 as initially computed for moment.

From Table 16-8, the maximum reaction for one wheel line of an HS 20-44 truck on a 52-foot span is 29.54 k. The distributed reaction for interior and outside beams is

$$R_{LL} = (1.0 \text{ WL/beam})(29.54 \text{ k}) = 29.54\text{k}$$

Summary

	Interior beams	Outside beams
Moment	398.12 ft-k	398.12 ft-k
Shear	19.81 k	19.81 k
Reaction	29.54 k	29.54 k

Example 7-2 - Live load distribution on a single-lane beam bridge

A single-lane beam bridge with a 14-foot roadway width spans 32 feet. The superstructure consists of a 5-1/8-inch glulam deck supported by 3 glulam beams, symmetrically spaced at 5 feet on center. Determine the distributed live load moment, shear, and reactions for an HS 15-44 design vehicle. Assume an initial beam depth of 30 inches for shear distribution.

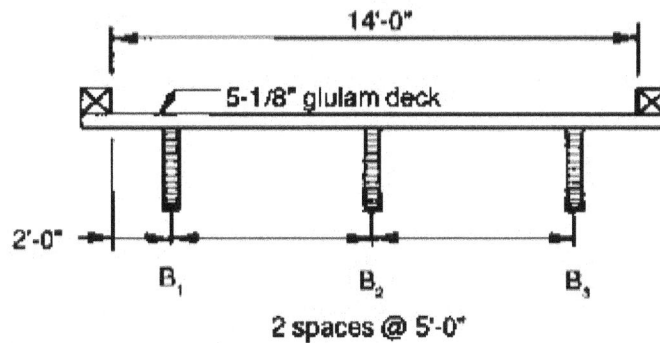

14'-0"

5-1/8" glulam deck

2'-0"

B₁   B₂   B₃

2 spaces @ 5'-0"

Solution

Distribution for Moment

Moment distribution to the interior beam is determined from Table 7-2:

$$DF = \frac{S}{6.0} = \frac{5.0}{6.0} = 0.83 \text{ WL/beam}$$

For outside beams, the distribution factor is computed by assuming the deck acts as a simple span between beams:

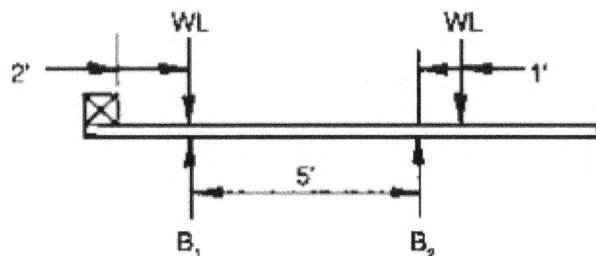

WL   WL

2'   1'

5'

B₁   B₂

By examination, the DF to the outside beam is 1.0 WL/beam.

From Table 16-8, the maximum moment for one wheel line of an HS 15-44 truck on a 32-foot span is 117.19 ft-k. The distributed live load moments for interior and outside beams are

Interior beam $M_{LL} = (0.83 \text{ WL/beam})(117.19 \text{ ft-k}) = 97.27 \text{ ft-k}$

Outside beam $M_{LL} = (1.0 \text{ WL/beam})(117.19 \text{ ft-k}) = 117.19 \text{ ft-k}$

Distribution for Shear

Shear distribution is computed by Equation 7-1 based on the maximum vertical shear at the lesser of $3d$ or $L/4$ from the support:

$$3d = \frac{(3)(30)}{12 \text{ in/ft}} = 7.5 \text{ ft} \qquad \frac{L}{4} = \frac{32}{4} = 8.0 \text{ ft}$$

$3d = 7.5$ ft controls.

The maximum vertical shear 7.5 feet from the support is computed for one wheel line:

7-13

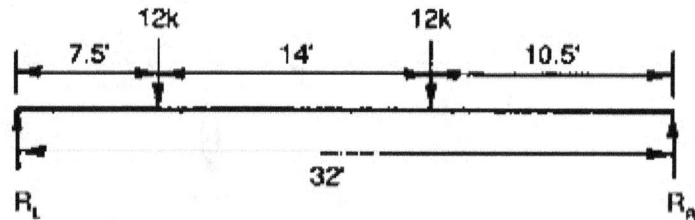

$$V_{LU} = R_L = \frac{(17.5\ ft)(24\ k)}{32\ ft} = 13.13\ k$$

By Equation 7-1,

Interior beam $V_{LL} = 0.50\ [(0.60V_{LU}) + V_{LO}]$

$$= 0.50\ [(0.60)(13.13) + (0.83)(13.13)] = 9.39\ k$$

Outside beam $V_{LL} = 0.50\ [(0.60V_{LU}) + V_{LO}]$

$$= 0.50\ [(0.60)(13.13) + (1.0)(13.13)] = 10.50\ k$$

Distribution for Reactions

Distribution factors for reactions are computed by assuming the deck acts as a simple span between beams. For interior beams, the wheel line is placed 2 feet from the curb face and moments for span B_2-B_3 are summed about B_3:

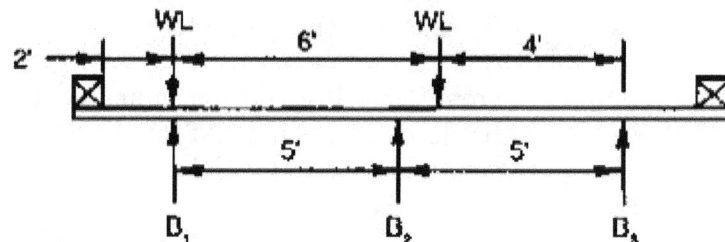

$$DF = \frac{(4\ ft)(WL)}{5\ ft} = 0.80\ WL$$

For outside beams, the distribution factor is the same as that obtained for moment, DF = 1.0 WL/beam.

From Table 16-8, the maximum reaction for one wheel line of an HS 15-44 truck on a 32-foot span is 19.13 k. The distributed reactions for interior and outside beams are

Interior beam R_{LL} (0.80 WL/beam)(19.13 k) = 15.30 k

Outside beam R_{LL} = (1.0 WL/beam)(19.13 k) = 19.13 k

7-14

Summary

	Interior beam	Outside beams
Moment	97.27 ft-k	117.19 ft-k
Shear	9.39 k	10.50 k
Reaction	15.30 k	19.13 k

Beam Configuration

One of the most influential factors on the overall economy and performance of a glulam bridge is the beam configuration. For a given roadway width, the number and spacing of beams can affect size and strength requirements for beam and deck elements and significantly influence the cost for material, fabrication, and construction. The number of combinations of beam size and spacing is potentially infinite, and the designer must select the most economical combination that provides the required structural capacity and meets serviceability requirements for deflection. In most situations, beam configuration is based on an economic evaluation influenced by three factors: (1) site restrictions, (2) deck thickness and performance, and (3) live load distribution to the beams.

Site Restrictions

Efficient beam design favors a relatively narrow, deep section. In some cases, the optimum beam depth may not be practical because of vertical clearance restrictions at the site. In these situations beam depth is limited, and the number of beams must be increased to achieve the same capacity provided by fewer, deeper beams. The most common configuration for such low-profile beam bridges uses a series of closely spaced beam groups (Figure 7-7). In most cases, however, the longitudinal deck designs discussed in Chapters 8 and 9 will provide a more economical design. Additional information on low-profile beam configurations is given in references listed at the end of this chapter.[7,62]

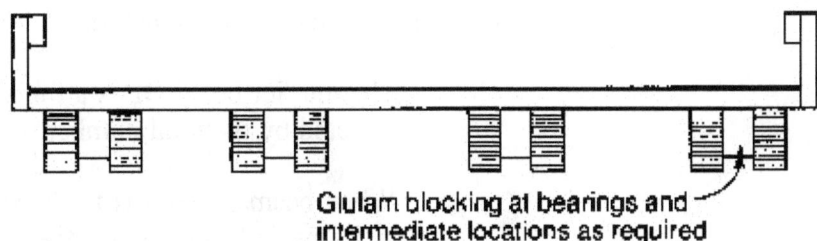

Glulam blocking at bearings and intermediate locations as required

Figure 7-7. - Typical low-profile glulam beam configuration.

Deck Thickness and Performance

Deck thickness and performance vary with the spacing of supporting beams. As beam spacing increases, the stress and deflection of the deck increase, resulting in greater deck thickness, strength, or stiffness requirements. The thickness of glulam deck panels is based on standard member sizes that increase in depth in 1-1/2- to 2-inch increments. As a result, the load-carrying capacity and stiffness of a panel is adequate for a range of beam spacings. For example, a 6-3/4-inch deck panel is used when the computed deck thickness is between 5-1/8 and 6-3/4 inches. The largest effect of beam spacing on the deck occurs when the panel thickness must be increased to the next thicker panel; for example, from 6-3/4 to 8-3/4 inches. On the other hand, considerable savings may be realized when the next smaller deck thickness can be used.

In general, the most practical and most economical beam spacing for transverse glulam decks supporting highway loads is between 4.5 and 6.5 feet. The maximum recommended deck overhang, measured from the centerline of the exterior beam to the face of the curb or railing, is approximately 2.5 feet. These values are based on deck stress and deflection considerations that may vary slightly for different panel combination symbols and configurations.

Live Load Distribution

In beam design, the magnitude of the vehicle live load supported by each beam is directly related to the distribution factor computed for that beam. The higher the distribution factor, the greater the load the beam must support. Thus, the value of the DF gives a good indication of relative beam size and grade requirements for different configurations.

The relationship between the distribution factor for moment and beam spacing is illustrated for a 24-foot-wide roadway and three equally spaced beam configurations in Figure 7-8. The concepts shown for this configuration are also applicable to other roadway widths and beam configurations. The graph shows the moment distribution factor, DF, for interior and outside beams as a function of center-to-center beam spacing, S. Solid curves for outside beams represent the feasible range in spacing where the deck overhang is between 1 and 2.5 feet. The dashed portion of the curves identifies beam spacings where the overhang is greater than 2.5 feet. The following points should be noted:

1. The interior beam DF is a function of beam spacing and is not affected by the total number of beams.

2. When beam spacing is to the right of the intersection of interior and outside beam curves, the interior beam DF controls for all beams and outside beams must be designed for the higher interior beam DF.

3. When beam spacing is to the left of the curve intersection, the DF for outside beams is greater than for interior beams. In this case, the load supported by each beam is based on the respective DF for

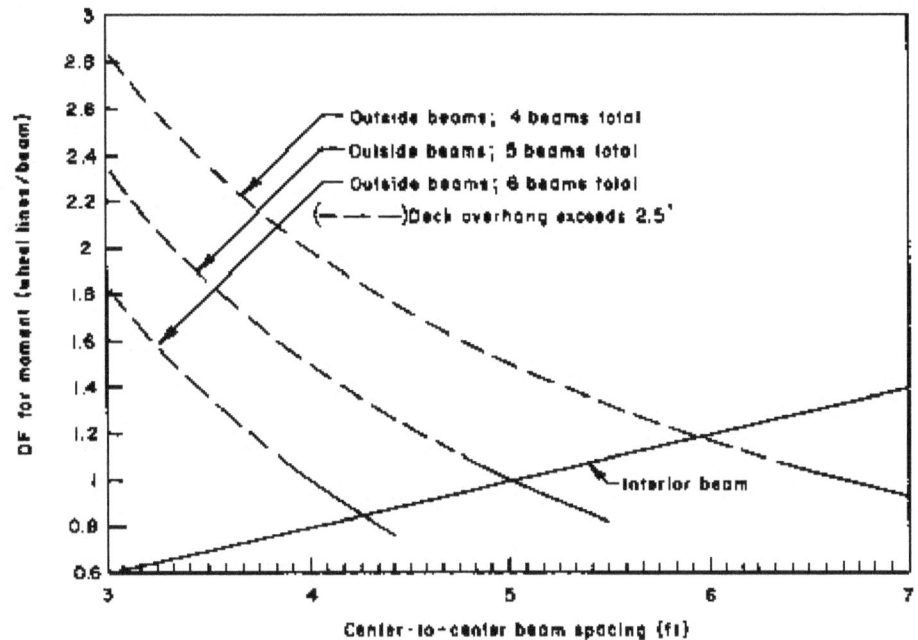

Figure 7-8. - Effects of beam configuration on the vehicle live load distribution factor (DF); roadway width of 24 feet; transverse glulam deck, 6 inches or more in nominal thickness.

that beam; that is, exterior beams support a greater portion of the load than interior beams.

4. The DF for each beam decreases as the total number of beams increases (beam spacing decreases). At the intersection of interior and outside beam curves for the five- and six-beam configurations, the distribution factors are 1.00 WL/beam and 0.85 WL/beam, respectively. For the four-beam configuration, beam spacing is limited by deck overhang restrictions, and the minimum DF of 1.27 WL/beam is controlled by interior beams.

As a general rule, beam spacing to the right of the curve intersection is the least economical because all beams must be designed for the higher DF required for interior beams. Spacing should be kept at or to the left of the intersection to achieve maximum economy. For wide bridges with many interior beams it may be beneficial to use a spacing left of the curve intersection that provides a lower DF for interior beams; however, all beams are normally designed to be the same depth, and a reduced DF for interior beams is not economical unless it allows the use of the next-lower standard beam width.

Exclusive of site restrictions, beam configuration should be based on economic and performance considerations for the deck and beam components. These considerations will vary depending on material prices,

7-17

availability at the time of construction, and transportation and construction costs. The recommended beam configurations used in this chapter are given in Table 7-3.

Table 7-3.—Recommended beam spacing for glulam beams with transverse glulam decks.

Roadway width (ft)[a]	Number of beams	Beam spacing (ft)	Deck overhang (ft)[b]	Moment DF Interior beams[c]	Moment DF outside beams[d]
Single-lane bridges					
14	3	5.5	1.5	0.92	0.92
16	3	6.0	2.0	1.00	1.00
Double-lane bridges					
24	5	5.0	2.0	1.00	1.00
26	5	5.5	2.0	1.10	1.10
28	5	6.0	2.0	1.20	1.20
34	6	6.0	2.0	1.20	1.20

[a] Measured face to face of railings, or of curbs when railing is not used.

[b] Measured from centerline of outside beam to face of railing or curbs.

[c] For glulam decks 6 inches or more in nominal thickness (S/6 for single-lane; S/5 for two or more lanes).

[d] Computed assuming the deck acts as a simple span between beams, but not less than the interior beam DF.

Beam Design Procedures

Beam design is an interactive process that follows the same basic procedures discussed in Chapter 5. A combination symbol is selected and the beam is designed for bending, deflection, shear, and bearing requirements. Design is routinely controlled by a combination of dead load and vehicle live load given in AASHTO Load Groups I or IB (Chapter 6). Transverse or longitudinal loads may be significant in some cases and should also be checked.

Basic design procedures for glulam bridge beams are summarized in the following steps. The sequence assumes a typical case, where bending or deflection controls design. On short, heavily loaded spans, shear may control design, and the sequence should be modified. For clarity, design procedures are limited to one beam of a simple-span structure loaded with dead load and a standard AASHTO vehicle live load. Application of these procedures is illustrated in Examples 7-3 and 7-4, following the procedures.

1. Define basic configuration and design criteria.

Define the longitudinal and transverse bridge configuration, including the following:

    a.   Span length *L* measured center-to-center of bearings

    b.   Roadway width measured face-to-face of railings or curbs (AASHTO 2.1.2)

    c.   Number of traffic lanes (Chapter 6)

    d.   Number and spacing of beams

    e.   Deck and railing/curb configuration

Identify design vehicles (including overloads), other applicable loads, and AASHTO load combinations discussed in Chapter 6. Also note design requirements for live load deflection and any restrictions on beam depth or other design criteria.

2. Select beam combination symbol.

An initial beam combination symbol is selected from the visually graded bending combinations given in Table 1 of *AITC 117-Design*. Combination symbols that are commonly used for bridges are given in Table 7-4. Select a species and combination symbol and note tabulated values in bending (F_{bx}), compression perpendicular to grain $(F_{c\perp x})$, horizontal shear (F_{vx}), and modulus of elasticity (E_x).

Table 7-4. - Glulam bending combination symbols commonly used for bridge beams.

Beam configuration	Western species combination symbols	Southern Pine combination symbols
Single span	24F-V3	24F-V2
	24F-V4	24F-V3
		24F-V6
Continuous spans	24F-V8	24F-V5

3. Determine deck dead load and dead load moment.

Compute the deck dead load supported by each beam, including the weight of the deck, wearing surface, railing, and other attached components (lb/ft). Refer to Chapter 6 for procedures and material weights used for dead load calculations. When deck thickness is unknown, use an estimated thickness of 6-3/4 inches. Estimates of rail dead loads can be made from typical designs shown in Chapter 10. Minor differences between estimated and actual deck and rail dead loads normally have an insignificant effect on beam design, but should be verified and revised during the design process.

For the usual case of a uniformly distributed deck dead load, dead load moment is computed as

7-19

$$M_{DL} = \frac{w_{DL}L^2}{8} \tag{7-2}$$

where M_{DL} = dead load moment (in-lb),

w_{DL} = uniform deck dead load (lb/in), and

L = beam span (in).

When the deck dead load is not uniformly distributed, dead load moment should be computed by statics for the specific loading condition.

4. Determine live load moment.

Live load moments are computed for interior and outside beams by multiplying the maximum moment for one wheel line of the design vehicle by the applicable moment distribution factors. Tables of maximum vehicle live load moments for standard AASHTO loads and selected overloads on simple spans are given in Table 16-8.

5. Determine beam size based on bending.

Allowable bending stress in beams is controlled by the largest reduction in tabulated stress resulting from application of the size factor, C_F, or the lateral stability of beams factor, C_L (Chapter 5). The allowable bending stress in bridge beams is normally controlled by C_F, rather than C_L. Thus, initial beam size is estimated based on the deck dead load moment and vehicle live load moment, assuming the size factor controls allowable bending stress (beam dead load moment is unknown at this point). This is computed as

$$S_x C_F = \frac{M}{F_b'} \tag{7-3}$$

where $S_x C_F$ = required beam section modulus adjusted by the size factor, C_F (in^3),

M = applied dead load and live load bending moment (in-lb),

$F_b' = F_{bx} C_M$ (lb/in^2), and

C_M = moisture content factor for bending = 0.80.

An initial beam size can be selected from the $S_x C_F$ values given in glulam section property tables in Chapter 16, but it is usually more convenient to use Figure 7-9. By entering the graph with the required $S_x C_F$ value, the required beam depth for standard beam widths can be readily obtained. Beam design generally favors a relatively narrow, deep section with a depth-to-width ratio between 4:1 and 6:1.

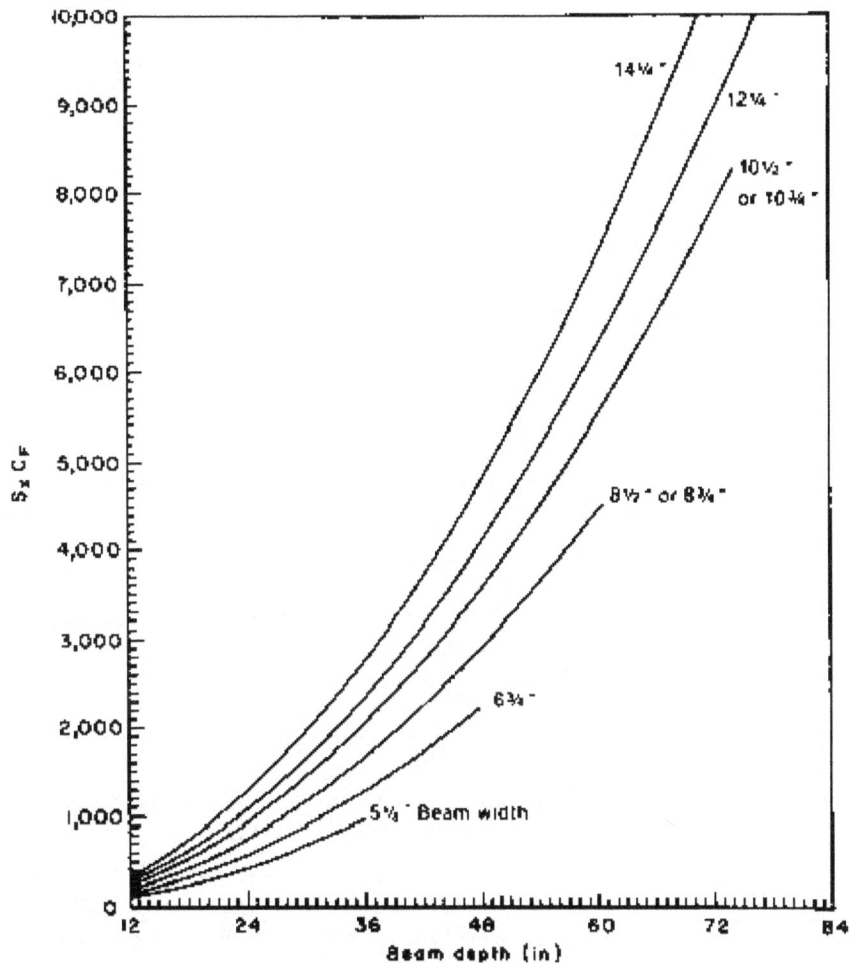

Figure 7-9. - Approximate adjusted section modulus (S_xC_F) versus beam depth for standard glulam beam widths.

After an initial beam size is selected, beam dead load moment is computed for the estimated beam size and added to the deck dead load and live load moments. A revised beam size is selected using the same procedures for initial beam selection. This interactive process is continued until a satisfactory beam size is finalized. Applied stress is then computed for the member using

$$f_b = \frac{M}{S_x} \tag{7-4}$$

This stress must not be greater than the allowable stress from

$$F_b' = F_{bx}C_MC_F \tag{7-5}$$

Allowable bending stress may be increased by a factor of 1.33 for overloads in AASHTO Load Group IB.

7-21

Beam size based on bending stress must next be checked for lateral stability. Criteria for lateral stability are based on the frequency of lateral support provided by transverse bracing between beams. Transverse bracing should be provided at each bearing for all spans and at intermediate intervals for spans greater than 20 feet. Maximum intermediate spacing is 25 feet, but bracing is generally spaced at equal intervals over the beam span (lateral bracing configurations are discussed later in this section).

Determine the spacing of transverse bracing and compute allowable bending stress based on stability from the low-variability equations given in Chapter 5. If stability controls over the size factor, it is generally most economical to reduce the unsupported beam length by adding additional bracing. When this is not practical, the beam must be redesigned for the lower stress required for stability.

6. Check live load deflection.

Vehicle deflections are computed from standard methods of engineering analysis. Deflection coefficients for standard AASHTO loads on simple spans are given in Table 16-8.

The distribution of deflection to bridge beams depends on the transverse deck stiffness. On single-lane bridges with glulam decks, it is generally assumed that the deflection produced by one vehicle (two wheel lines) is resisted equally by all beams. On multiple-lane structures, deflection can be distributed using the distribution factor for beam moment, or by assuming that all beams equally resist the deflection produced by the simultaneous loading of one vehicle in each traffic lane. For glulam decks, deflection in multiple-lane bridges is usually distributed using the DF for beam moment.

Compute beam live load deflection and compare it with maximum deflection criteria for the structure. When actual deflection exceeds acceptable levels, the beam moment of inertia, I, must be increased. Deflections are important in timber bridges and must be limited for proper performance and serviceability. Excessive deflections loosen connections and cause asphalt wearing surfaces to crack or disintegrate. Criteria for maximum deflection are based on designer judgment, but should not exceed $L/360$. When the structure supports a pedestrian walkway or will be paved with asphalt, a further reduction in deflection is desirable.

7. Check horizontal shear.

Dead load horizontal shear is based on the maximum vertical shear occurring a distance from the support equal to the beam depth, d. Compute the dead load vertical shear for interior and outside beams, neglecting loads acting within a distance d from the supports:

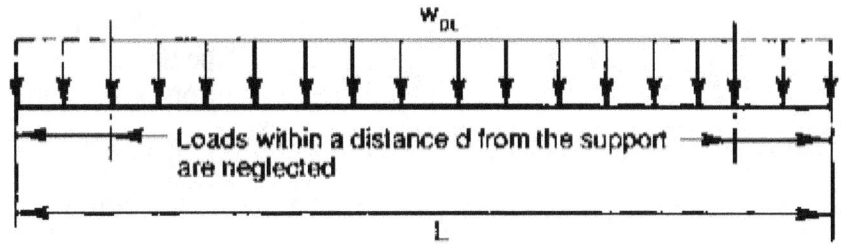

Loads within a distance d from the support are neglected

$$V_{DL} = w_{DL}\left(\frac{L}{2} - d\right) \tag{7-6}$$

where V_{DL} = vertical dead load shear at a distance d from the support (lb) and

    w_{DL} = uniform dead load supported by the beam (lb/in).

Live load vertical shear is computed at the lesser distance of $3d$ or $L/4$ by Equation 7-1. Applied stress in horizontal shear must not be greater than the allowable stress, as given by

$$f_v = \frac{1.5V}{A} \le F_v' = F_{vx} C_M \tag{7-7}$$

where $V = V_{DL} + V_{LL}$ (lb),

    A = beam cross-sectional area (in^2), and

    C_M = moisture content factor for shear = 0.875.

Allowable shear stress may be increased by a factor of 1.33 for overloads in AASHTO Load Group IB.

When $f_v \le F_v'$, the beam is adequately proportioned for horizontal shear. If $f > F_v'$, the beam is insufficient in shear and the cross-sectional area must be increased.

8. Check lateral and longitudinal loads.

The applicability and magnitude of lateral and longitudinal loads, such as wind load, longitudinal force, and centrifugal force will vary among different structures. Loads should be computed and applied to affected members in accordance with the AASHTO load groups discussed in Chapter 6. Stresses from AASHTO loading combinations may be increased by stress adjustments for duration of load and those allowed by the specific load group, when applicable.

9. Determine bearing length and stress.

Bearing area at beam reactions must be sufficient to limit stress to an allowable level. Compute the dead load reaction, R_{DL} at each beam (dead load of beam, deck, wearing surface, railing, and so forth). Compute the live load reaction, R_{LL}, at each beam by multiplying the maximum reaction for one wheel line by the applicable distribution factor for reactions. Maximum reactions for one wheel line of standard AASHTO loads are given in Table 16-8.

For a given beam width, the minimum bearing length must not be less than that computed by

$$\text{Required bearing length} = \frac{(R_{DL} + R_{LL})}{bF_{c\perp}'} \tag{7-8}$$

where R_{DL} = dead load reaction (lb),

R_{LL} = distributed live load reaction (lb),

b = beam width (in), and

$F_{c\perp}' = F_{c\perp x} C_M \ (\text{lb/in}^2).$

Minimum required bearing lengths for the usual $F_{c\perp x}$ = 650 lb/in^2 are given in Figure 7-10.

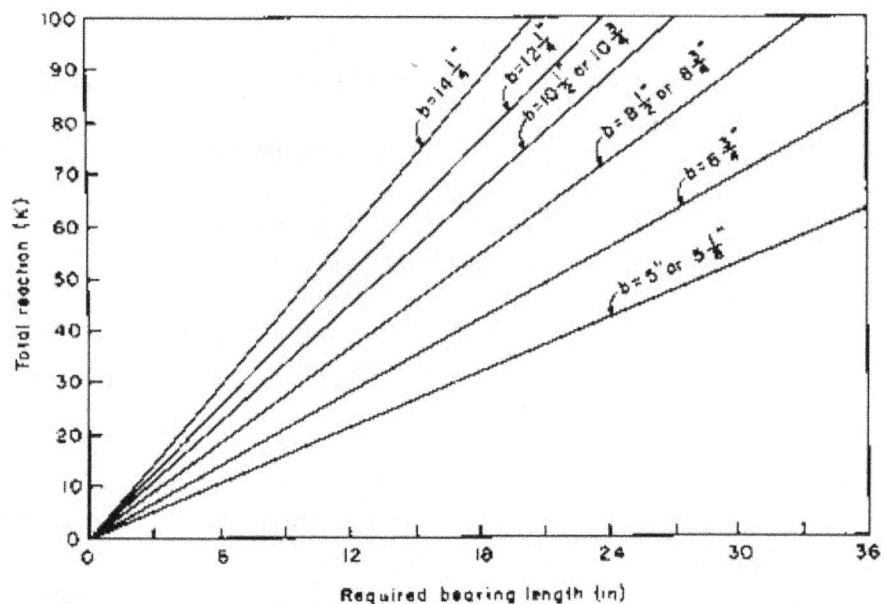

Figure 7-10.—Approximate adjusted minimum bearing length for glulam beams based on an allowable compression perpendicular to grain, $F_{c\perp}'$, of 344.5 lb/in^2 ($F_{c\perp}' = F_{c\perp x} C_M$ = 650 lb/in^2 (0.53)).

Values of $F_{c\perp x}$ in *AITC 117-Design* are based on a deformation limit of 0.04 inch and are not subject to increases for duration of load. An increase in allowable stress for overloads may result in additional nonrecoverable deformation at the bearings and is left to designer judgment.

Compute the applied stress at bearings using

$$f_{c\perp} = \frac{R_{DL} + R_{LL}}{A}$$

(7-9)

where A is the bearing area in square inches. This stress must not be greater than $F_{c\perp}'$ computed for Equation 7-8. When bearing is on an inclined surface, refer to Chapter 5 for methods for computing bearing stress.

10. Determine camber.

Camber is based on the span length and configuration of beams. For beams with spans greater than 50 feet, camber is generally 1.5 to 2.0 times the computed dead load deflection (Chapter 5). For spans less than 50 feet, camber is 1.5 to 2.0 times the dead load deflection plus one-half the vehicle live load deflection. Regardless of span, camber on multiple-span beams is normally based on dead load deflections only in order to obtain acceptable riding qualities.

Camber for single-span beams is specified as a vertical offset at the beam centerline. On multiple-span continuous beams, camber may vary along the beam and should be specified at the center of each span segment.

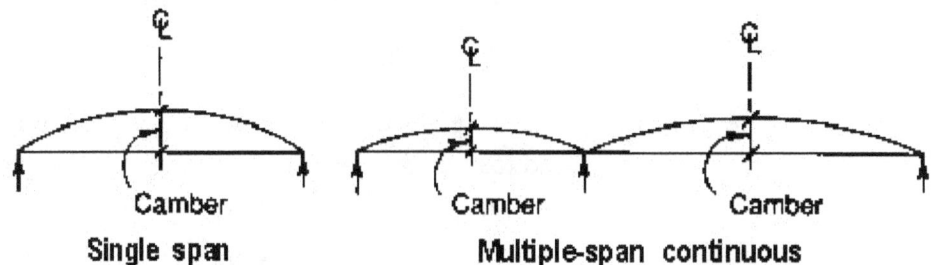

Single span   Multiple-span continuous

Example 7-3 - Glulam beam design; two-lane highway loading

A deteriorated bridge on a state highway is to be removed and replaced with glulam beam bridge. The new superstructure will be placed on the existing substructure where the span measured center-to-center of bearings is 94 feet. It will carry two traffic lanes and have a roadway width of 24 feet. Design the supporting beams for the structure, assuming the following:

1. A watertight glulam deck constructed of 5-1/8-inch-thick panels with a 3-inch asphalt wearing surface (including allowance for future overlay)

2. AASHTO Load Croup I loading with HS 20-44 vehicles

3. Vehicular railing with an approximate dead load of 45 lb/ft

4. Beams manufactured from visually graded western species

Solution

From the given information, a configuration of five beams spaced 5 feet on center is obtained from Table 7-3. Total deck width is increased 6 inches on each edge to account for rail width and attachment (Chapter 10).

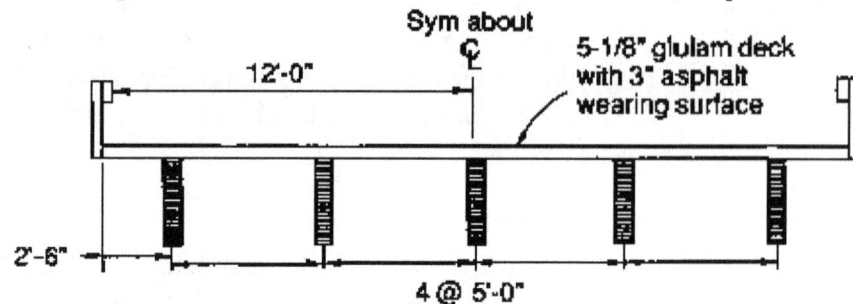

Select a Beam Combination Symbol

A beam combination symbol 24F-V4 manufactured from visually graded western species is selected from *AITC 117--Design*. Tabulated values are as follows:

$$F_{bx} = 2,400 \text{ lb/in}^2$$

$$F_{c\perp x} = 650 \text{ lb/in}^2$$

$$F_{vx} = 165 \text{ lb/in}^2$$

$$E_x = 1,800,000 \text{ lb/in}^2$$

Determine Deck Dead Load and Dead Load Moment

Dead load of the deck and wearing surface is computed in lb/ft² based on unit weights of 50 lb/ft³ for timber and 150 lb/ft³ for asphalt pavement:

$$DL = \frac{(5.125 \text{ in.})(50 \text{ lb/ft}^3)}{12 \text{ in./ft}} + \frac{(3 \text{ in.})(150 \text{ lb/ft}^3)}{12 \text{ in./ft}} = 58.9 \text{ lb/ft}^2$$

The dead load applied to each beam is equal to the tributary deck width supported by the beam. In this case, interior beams support 5 feet of deck width. Exterior beams also support 5 feet of deck plus 45 lb/ft of rail dead load.

For interior beams,

$$\text{Deck } w_{DL} = (5.0 \text{ ft})(58.9 \text{ lb/ft}^2) = 294.5 \text{ lb/ft}$$

$$\text{Deck } M_{DL} = \frac{w_{DL}L^2}{8} = \frac{(294.5 \text{ lb/ft})(94 \text{ ft})^2}{8} = 325,275 \text{ ft-lb}$$

For outside beams,

$$\text{Deck } w_{DL} = (294.5 \text{ lb/ft}) + 45 \text{ lb/ft} = 339.5 \text{ lb/ft}$$

$$\text{Deck } M_{DL} = \frac{(339.5 \text{ lb/ft})(94 \text{ ft})^2}{8} = 374{,}978 \text{ ft-lb}$$

Determine Live Load Moment

From Table 7-3, the moment DF = 1.0 WL/beam for interior and outside beams. From Table 16-8, the maximum moment for one wheel line of an HS 20-44 truck on a 94-foot span is 708.09 ft-k.

$$M_{LL} = M(\text{DF}) = (708.09 \text{ ft-k})(1.0)(1{,}000 \text{ lb/k}) = 708{,}090 \text{ ft-lb}$$

Determine Beam Size Based on Bending

An initial beam section modulus is computed based on the deck dead load and live load moments (beam dead load is unknown). Because the deck is watertight and beams are protected from direct exposure, dry condition allowable stress is used for bending $(C_M = 1.0)$.

$$F_b' = F_{bx}C_M C_F = (2{,}400 \text{ lb/in}^2)(1.0)(C_F)$$

For interior beams,

$$M = \text{Deck } M_{DL} + M_{LL} = 325{,}275 + 708{,}090 = 1{,}033{,}365 \text{ ft-lb}$$

By Equation 7-3,

$$S_x C_F = \frac{M}{F_b'} = \frac{(1{,}033{,}365 \text{ ft-lb})(12 \text{ in/ft})}{2{,}400 \text{ lb/in}^2} = 5{,}167 \text{ in}^3$$

For outside beams,

$$M = \text{Deck } M_{DL} + M_{LL} = 374{,}978 + 708{,}090 = 1{,}083{,}068 \text{ ft-lb}$$

$$S_x C_F = \frac{M}{F_b'} = \frac{(1{,}083{,}068 \text{ ft-lb})(12 \text{ in/ft})}{2{,}400 \text{ lb/in}^2} = 5{,}415 \text{ in}^3$$

Section modulus requirements differ slightly for interior and outside beams because of the greater load carried by the outside beams. In this case, equal beam depth is desired for even bearing, and beam design will be based on the more severe requirements for outside beams.

Entering Figure 7-9 with an outside beam value $S_x C_F = 5{,}415 \text{ in}^3$, an initial beam size of 12-1/4 by 57 inches is selected. From glulam section properties in Table 16-3,

$$S_x C_F = 5{,}579 \text{ in}^3$$

$$\text{Beam } w_{DL} = 242.4 \text{ lb/ft}$$

Beam dead load moment is computed and S_xC_r revised:

$$\text{Beam } M_{DL} = \frac{w_{DL}L^2}{8} = \frac{242.4\,(94\text{ ft})^2}{8} = 267{,}731\text{ ft-lb}$$

$$\text{Revised } S_xC_r = \frac{M}{F_b'} = \frac{(1{,}083{,}068 + 267{,}731)\,(12\text{ in/ft})}{2{,}400\text{ lb/in}^2} = 6{,}754\text{ in}^3$$

From Table 16-3, a revised beam size of 12-1/4 by 64-1/2 inches is selected with the following section properties:

$$A = 790.1\text{ in}^2$$

$$S_xC_r = 7{,}046.3\text{ in}^3$$

$$S_x = 8{,}493.8\text{ in}^3$$

$$C_r = 0.83$$

$$I_x = 273{,}927\text{ in}^4$$

Beam $w_{DL} = 274.3$ lb/ft

Applied moment is revised and bending stress is computed:

$$\text{Beam } M_{DL} = \frac{w_{DL}L^2}{8} = \frac{274.3\,(94\text{ ft})^2}{8} = 302{,}964\text{ ft-lb}$$

$$M = 302{,}964 + 1{,}083{,}068 = 1{,}386{,}032\text{ ft-lb}$$

$$f_b = \frac{M}{S_x} = \frac{1{,}386{,}032\,(12\text{ in/ft})}{8{,}493.8\text{ in}^3} = 1{,}958\text{ lb/in}^2$$

$$F_b' = F_{bx}C_M C_F = 2{,}400(1.0)(0.83) = 1{,}992\text{ lb/in}^2$$

$f_b = 1{,}958$ lb/in^2 < $F_b' = 1{,}992$ lb/in^2, so a 12-1/4 by 64-1/2-inch beam is satisfactory in bending.

Check bending stress in interior beams:

$$M = \text{Beam } M_{DL} + (\text{Deck } M_{DL} + M_{LL}) = 302{,}964 + 1{,}033{,}365 = 1{,}336{,}329\text{ ft-lb}$$

$$f_b = \frac{M}{S_x} = \frac{1{,}336{,}329\,(12\text{ in/ft})}{8{,}493.8} = 1{,}888\text{ lb/in}^2 < F_b' = 1{,}992\text{ lb/in}^2$$

When there is a difference of 200 lb/in^2 or more between beams with the lowest bending stress and the allowable bending stress, a lower glulam combination symbol should be considered. In this case, the difference between interior beam f_b and F_b' is only 104 lb/in^2, so the 12-1/4 by 64-1/2-inch member will be used for all beams.

7-28

The beam must next be checked for lateral stability. Assuming a maximum 25-foot spacing between points of lateral support, transverse bracing will be provided at the beam ends and at the quarter points:

$$\ell_u = \frac{L}{4} = \frac{94}{4} = 23.5 \text{ ft} \quad \text{and} \quad \frac{\ell_u}{d} = \frac{23.5\,(12 \text{ in/ft})}{64.5} = 4.37 < 14.3$$

By Equation 5-7,

$$\ell_e = 1.63\ell_u + 3d = 1.63(23.5)(12 \text{ in/ft}) + 3(64.5) = 653.16 \text{ in}$$

By Equation 5-3,

$$C_s = \sqrt{\frac{\ell_e d}{b^2}} = \sqrt{\frac{653.16(64.5)}{(12.25)^2}} = 16.76 < 50$$

$C_s >$ 10, so further stability calculations are required. As with bending stress, dry conditions of use are assumed for E, and

$$E' = E_x C_M = 1{,}800{,}000(\ 1.0) = 1{,}800{,}000 \text{ lb/in}^2$$

By low-variability Equation 5-12,

$$F_b'' = F_{bx}C_M = 2{,}400(1.0) = 2{,}400 \text{ lb/in}^2$$

$$C_k = 0.956\sqrt{\frac{E'}{F_b''}} = 0.956\sqrt{\frac{1{,}800{,}000}{2{,}400}} = 26.18$$

$C_s = 16.76 < C_k = 26.18$, so the beam is in the intermediate slenderness range. By Equation 5-10,

$$C_L = 1 - \frac{1}{3}\left(\frac{C_s}{C_k}\right)^4 = 1 - \frac{1}{3}\left(\frac{16.76}{26.18}\right)^4 = 0.94$$

$C_L = 0.94 > C_F = 0.83$, so strength rather than stability controls allowable bending stress.

Check Live Load Deflection

Live load deflection is checked by assuming that deflection is distributed in the same manner as bending: one beam resists the deflection produced by one wheel line. From Table 16-8, the deflection coefficient for one wheel line of an HS 20-44 truck on a 94-foot simple span is 1.02×10^{12} lb-in^3.

$$\Delta_{LL} = \frac{1.02 \times 10^{12}}{E' I_x} = \frac{1.02 \times 10^{12}}{(1{,}800{,}000)\,(273{,}927)} = 2.07 \text{ in.} = L/545$$

$L/545 < L/360$, so live load deflection is acceptable.

Check Horizontal Shear

From bending calculations, the total dead load for outside beams is 339.5 lb/ft for the deck and railing and 274.3 lb/ft for the beam, for a total of 613.8 lb/ft. Neglecting loads within a distance $d = 64.5$ inches from the supports, dead load vertical shear is computed by Equation 7-6:

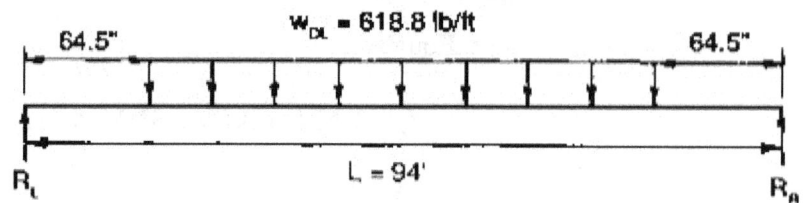

$$V_{DL} = w_{DL}\left(\frac{L}{2} - d\right) = 613.8\left(\frac{94}{2} - \frac{64.5}{12 \text{ in/ft}}\right) = 25,549 \text{ lb}$$

Live load vertical shear is computed from the maximum vertical shear occurring at the lesser of $3d$ or $L/4$ from the support:

$$3d = \frac{3(64.5)}{12 \text{ in/ft}} = 16.13 \text{ ft} \qquad \frac{L}{4} = \frac{(94)}{4} = 23.5 \text{ ft}$$

$3d = 16.13$ feet controls, and maximum vertical shear is determined at that location for one wheel line of an HS 20-44 truck:

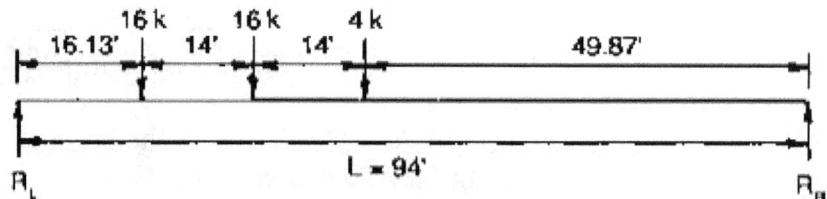

$$V_{LU} = R_L = \frac{(49.87 \text{ ft})(4 \text{ k}) + (63.87 \text{ ft})(16 \text{ k}) + (77.87 \text{ ft})(16 \text{ k})}{94 \text{ ft}}$$

$$V_{LU} = 26.25 \text{ k} = 26,250 \text{ lb}$$

For a moment DF to outside beams of 1.0,

$$V_{LD} = V(\text{Moment DF}) = 26,250(1.0) = 26,250 \text{ lb}$$

$$V_{LL} = 0.50\left[(0.6 \, V_{LU}) + V_{LD}\right]$$

$$= 0.50\left[(0.6)(26,250) + 26,250\right] = 21,000 \text{ lb}$$

Total vertical shear $= V_{DL} + V_{LL} = 25,549 + 21,000 = 46,549 \text{ lb}$

Stress in horizontal shear is computed by Equation 7-7:

$$f_v = \frac{1.5V}{A} = \frac{1.5(46,549)}{790.1} = 88 \text{ lb/in}^2$$

7-30

$$F_v' = F_{vx}(C_M) = (165)(1.0) = 165 \text{ lb/in}^2$$

$F_v' = 165 \text{ lb/in}^2 > f_v = 88 \text{ lb/in}^2$, so the beam is satisfactory in horizontal shear.

Determine Bearing Length and Stress

Although the watertight deck is assumed to protect the beams from exposure, bearings are subject to wetting from runoff and debris accumulations that trap water. Therefore, bearings will be designed using wet-condition stress in compression perpendicular to grain.

From Table 5-7, $C_M = 0.53$, and

$$F_{c\perp}' = F_{c\perp x}(C_M) = 650(0.53) = 345 \text{ lb/in}^2$$

For a unit dead load $w_{DL} = 613.8$ lb/ft to outside beams,

$$R_{DL} = \frac{w_{DL}L}{2} = \frac{(613.8)(94)}{2} = 28,849 \text{ lb}$$

For a 2-foot deck overhang and beam spacing of 5 feet, the reaction DF is 1.0 WL/beam for interior and outside beams. From Table 16-8, the maximum reaction for one wheel line of an HS 20-44 truck on a 94-foot span is 32.43 k = 32,430 lb:

$$R_{LL} = R(DF) = 32,430(1.0) = 32,430 \text{ lb}$$

By Equation 7-8 (or by Figure 7-10),

$$\textbf{Required bearing length} = \frac{(R_{DL} + R_{LL})}{b(F_{c\perp}')} = \frac{28,849 + 32,430}{12.25(345)} = 14.5 \text{ in.}$$

A bearing length of 18 inches is selected. For an out-to-out beam length of 95-1/2 feet, reactions are revised and applied stress is computed by Equation 7-9:

$$R_{DL} = \frac{613.8(95.5)}{2} = 29,309 \text{ lb}$$

$$f_{c\perp} = \frac{R_{DL} + R_{LL}}{A} = \frac{29,309 + 32,430}{12.25(18)}$$

$$= 280 \text{ lb/in}^2 < F_{c\perp}' = 345 \text{ lb/in}^2$$

Determine Camber

Dead load deflection is computed by Equation 5-16:

$$\Delta_{DL} = \frac{5wL^4}{384E'I_x} = \frac{5(613.8)[(94)(12 \text{ in./ft})]^4}{384(1,800,000)(273,927)(12 \text{ in./ft})} = 2.19 \text{ in.}$$

Using camber slightly greater than twice the dead load deflection, a minimum midspan offset of 5 inches will be specified.

Summary

The superstructure will consist of five 12-1/4-inch-wide by 64-1/2-inch-deep glulam beams, 95-1/2 feet long, with a distance center to center of bearings of 94 feet. Transverse bracing will be provided for lateral support at the bearings and at the beam quarter points. The glulam will be specified as visually graded western species conforming to combination symbol 24F-V4, or may be specified by required stresses as outlined in *AITC 117--Design*.

Stresses and deflection are as follows:

	Interior beams	Outside beams
f_b	1,888 lb/in²	1,958 lb/in²
F_b'	1,992 lb/in²	1,992 lb/in²
Δ_{LL}	2.07 in. = $L/545$	2.07 in. = $L/545$
f_v	< Outside beam	88 lb/in²
F_v'	165 lb/in²	165 lb/in²
$f_{c\perp}$	< Outside beam	280 lb/in²
$F_{c\perp}'$	345 lb/in²	345 lb/in²

Example 7-4 - Glulam beam design; single-lane with overload

A new bridge on a local rural road will span 48 feet center-to-center of bearings. It will carry one traffic lane and have a roadway width of 14 feet. Design the supporting glulam beams for the structure, assuming

1. a nonwatertight deck constructed of 6-3/4-inch glulam panels with a 4-inch rough-sawn lumber wearing surface;

2. AASHTO Load Group I loading with an H 20-44 vehicle and AASHTO Group IB loading with a U80 overload (Figure 6-5);

3. a 12- by 12-inch rough-sawn brush curb along each deck edge; and

4. beams manufactured from visually graded Southern Pine.

Solution

A configuration of three beams spaced 5-1/2 feet on center is obtained from Table 7-3. Deck width is increased 1 foot on each edge to account for the brush curb:

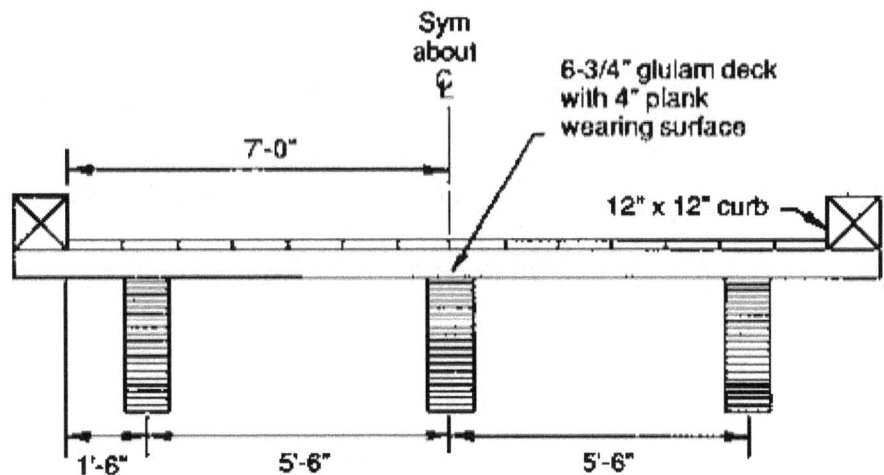

Select a Beam Combination Symbol

A beam combination symbol 24F-V2 is selected from *AITC 117--Design*. Tabulated values are as follows:

$$F_{bx} = 2,400 \text{ lb/in}^2 \qquad C_M = 0.80$$

$$F_{c \perp x} = 650 \text{ lb/in}^2 \qquad C_M = 0.53$$

$$F_{vx} = 200 \text{ lb/in}^2 \qquad C_M = 0.875$$

$$E_x = 1,700,000 \text{ lb/in}^2 \quad C_M = 0.833$$

Compute Deck Dead Load and Dead Load Moment

Deck and wearing surface dead loads are computed as follows:

$$\text{Deck DL} = \frac{(6.75 \text{ in.})(50 \text{ lb/ft}^3)}{12 \text{ in./ft}} = 28.1 \text{ lb/ft}^2$$

$$\text{Wearing surface DL} = \frac{(4 \text{ in.})(50 \text{ lb/ft}^3)}{12 \text{ in./ft}} = 16.7 \text{ lb/ft}^2$$

The interior beam supports a 5.5-foot width of deck and wearing surface, while exterior beams support 5.25 feet of deck, 4.25 feet of wearing surface, and 50 lb/ft of curb.

For interior beams,

$$w_{DL} = (5.5)(28.1 + 16.7) = 246.4 \text{ lb/ft}$$

$$M_{DL} = \frac{w_{DL} L^2}{8} = \frac{(246.4)(48)^2}{8} = 70,963 \text{ ft-lb}$$

For outside beams,

$$w_{DL} = (5.25)(28.1) + (4.25 \text{ ft})(16.7) + 50 \text{ lb/ft} = 268.5 \text{ lb/ft}$$

$$M_{DL} = \frac{(268.5)(48)^2}{8} = 77,328 \text{ ft-lb}$$

Determine Live Load Moment

From Table 7-3, the moment DF = 0.92 WL/beam for interior and outside beams. Maximum live load moments per wheel line are obtained from Table 16-8 and are multiplied by the moment DF:

$$H\ 20\text{-}44\ \ M_{LL} = 0.92\ (212,820 \text{ ft-lb}) = 195,794 \text{ ft-lb}$$

$$U80\ \ M_{LL} = 0.92\ (572,590 \text{ ft-lb}) = 526,783 \text{ ft-lb}$$

Determine Beam Size Based on Bending

For a U80 overload, the tabulated bending stress can be increased 33 percent in AASHTO Load Group IB. Comparing the U80 moment to the lesser H 20-44 moment,

$$\frac{526,783}{1.33} = 396,077 > 195,794$$

so the U80 will control bending.

$$F_b' = F_{bx}(1.33)C_M C_F = 2,400(1.33)(0.80)(C_F) = 2,554(C_F) \text{ lb/in}^2$$

For outside beams,

$$M = \text{Deck } M_{DL} + M_{LL} = 77,328 + 526,783 = 604,111 \text{ ft-lb}$$

$$S_x C_F = \frac{M}{F_b'} = \frac{604,111 \text{ ft-lb}\ (12 \text{ in/ft})}{2,554 \text{ lb/in}^2} = 2,838 \text{ in}^3$$

Entering Figure 7-9 with a value $S_x C_F = 2,838$ in^3, an approximate beam size of 8-1/2 by 51 inches is selected. From Table 16-4,

$$S_x C_F = 2,965.5 \text{ in}^3$$

Beam $w_{DL} = 146.1$ lb/ft

$$\text{Beam } M_{DL} = \frac{w_{DL} L^2}{8} = \frac{146.1\,(48)^2}{8} = 42,077 \text{ ft-lb}$$

Revising section modulus requirements,

$$S_x C_F = \frac{M}{F_b'} = \frac{(604,111 + 42,077)\,(12 \text{ in/ft})}{2,554 \text{ lb/in}^2} = 3,036 \text{ in}^3$$

From Table 16-4, a revised beam size of 8-1/2 by 50-7/8 inches is chosen with the following section properties:

$$A = 432.4 \text{ in}^2$$

$$S_x C_r = 3{,}123.0 \text{ in}^3$$

$$S_x = 3{,}666.7 \text{ in}^3$$

$$C_r = 0.85$$

$$I_x = 93{,}271.9 \text{ in}^4$$

Beam $w_{DL} = 150.2$ lb/ft

$$\text{Beam } M_{DL} = \frac{w_{DL}L^2}{8} = \frac{150.2(48)^2}{8} = 43{,}258 \text{ ft-lb}$$

$$M = 43{,}258 + 604{,}111 = 647{,}369 \text{ ft-lb}$$

$$f_b = \frac{M}{S_x} = \frac{647{,}369(12 \text{ in/ft})}{3{,}666.7 \text{ in}^3} = 2{,}119 \text{ lb/in}^2$$

$$F_b' = F_{bx}(1.33)C_M C_F = 2{,}554(0.85) = 2{,}171 \text{ lb/in}^2$$

$f_b < F_b'$, therefore an 8-1/2 by 50-7/8-inch outside beam is sufficient in bending.

Check U80 bending stress in interior beams:

$$M = M_{DL} + M_{LL} = (43{,}258 + 70{,}963) + 526{,}783 = 641{,}004 \text{ ft-lb}$$

$$f_b = \frac{M}{S_x} = \frac{641{,}004(12 \text{ in/ft})}{3{,}666.7 \text{ in}^3} = 2{,}098 \text{ lb/in}^2$$

The difference between interior beam f_b and F_b' is only 73 lb/in^2, so an 8-1/2- by 50-7/8-inch 24F-V2 will be used for all beams.

Check outside beam bending stresses for the H 20-44 load:

$$F_b' = F_{bx}C_M C_F = 2{,}400(0.80)(0.85) = 1{,}632 \text{ lb/in}^2$$

$$M = M_{DL} + M_{LL} = (43{,}258 + 77{,}328) + 195{,}794 = 316{,}380 \text{ ft-lb}$$

$$f_b = \frac{M}{S_x} = \frac{316{,}380(12 \text{ in/ft})}{3{,}666.7 \text{ in}^3} = 1{,}035 \text{ lb/in}^2 < F_b' = 1{,}632 \text{ lb/in}^2$$

Check lateral stability assuming lateral support at beam ends and centerspan:

$$\ell_u = \frac{L}{2} = \frac{48}{2} = 24 \text{ ft} \qquad \frac{\ell_u}{d} = \frac{24 \, (12 \text{ in/ft})}{50.88} = 5.66 < 14.3$$

$$\ell_e = 1.63\ell_u + 3d = 1.63(24)(12 \text{ in/ft}) + 3(50.88) = 622.08$$

$$C_s = \sqrt{\frac{\ell_e d}{b^2}} = \sqrt{\frac{(622.08)(50.88)}{(8.5)^2}} = 20.93 < 50$$

$C_s > 10$, so further stability calculations are required.

$$E' = E_x C_M = 1,700,000(0.833) = 1,416,100 \text{ lb/in}^2$$

By low-variability Equation 5-12,

$$F_b'' = F_{bx} C_M = 2,400(0.80) = 1,920 \text{ lb/in}^2$$

$$C_k = 0.956\sqrt{\frac{E'}{F_b''}} = 0.956\sqrt{\frac{1,416,100}{1,920}} = 25.96$$

$C_s = 20.93 < C_k = 25.96$, so the beam is in the intermediate slenderness range. By Equation 5-10,

$$C_L = 1 - \frac{1}{3}\left(\frac{C_s}{C_k}\right)^4 = 1 - \frac{1}{3}\left(\frac{20.93}{25.96}\right)^4 = 0.86$$

$C_L = 0.86 > C_r = 0.85$, so strength rather than stability controls the allowable bending stress and an 8-1/2- by 50-7/8-inch beam is satisfactory.

Check Live Load Deflection

Live load deflection for this single-lane configuration will be checked by assuming deflection is equally resisted by all beams. Criteria for the H 20-44 vehicle will be a maximum deflection of $L/360$. For the U80 overload, no criteria will apply, but deflection will be computed for reference.

For the H 20-44 vehicle, the deflection coefficient from Table 16-8 for one wheel line on a 48-foot simple span is 7.40 x 10^{10} lb-in^3. Deflection is computed by assuming that all beams equally resist the deflection produced by one truck (two wheel lines):

$$\Delta_{LL} = \frac{2\,(7.40 \times 10^{10})}{E_x (C_M)(3) I_x} = \frac{1.48 \times 10^{11}}{1,700,000(0.833)(3)(93,272.9)} = 0.37 \text{ in.}$$

0.37 in. = $L/1,557 < L/360$ allowed.

For the U80 vehicle, the deflection coefficient from Table 16-8 for one wheel line is 2.35 x 10^{11} lb-in^3, and

$$\Delta_{LL} = \frac{2(2.35 \times 10^{11})}{E_x (C_M)(3)I_x} = \frac{4.70 \times 10^{11}}{1,700,000 (0.833)(3)(93,272.9)} = 1.19 \text{ in.}$$

which is approximately equal to $L/484$.

Live load deflection is acceptable.

Check Horizontal Shear

From bending calculations the total outside-beam dead load is 268.5 lb/ft for the deck and curb and 150.2 lb/ft for the beam, for a total load of 418.7 lb/ft. Neglecting loads within a distance of $d = 50\text{-}7/8$ inches from the supports, dead load vertical shear is computed by Equation 7-6:

$$V_{DL} = w_{DL}\left(\frac{L}{2} - d\right) = 418.7\left(\frac{48}{2} - \frac{50.88}{12 \text{ in/ft}}\right) = 8,274 \text{ lb}$$

Live load vertical shear is computed from the maximum vertical shear occurring at the lesser of $3d$ or $L/4$ from the support:

$$3d = \frac{3(50.88)}{12 \text{ in/ft}} = 12.72 \text{ ft} \qquad \frac{L}{4} = \frac{(48)}{4} = 12 \text{ ft}$$

$L/4 = 12$ feet controls, and maximum vertical shear is computed at that location for one wheel line of a U80 truck:

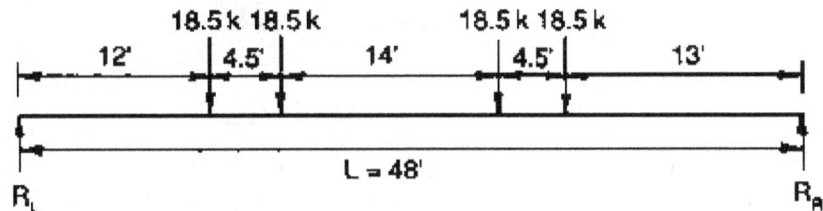

$$V_{LU} = R_L \frac{(13\text{ft} + 4.5\text{ft} + 14/2\text{ft})(74\text{k})}{48\text{ft}} = 37.77\text{k} = 37,770\text{lb}$$

$$V_{LD} = V(DF) = 37,770(0.92) = 34,748 \text{ lb}$$

$$V_{LL} = 0.50 [(0.6 V_{LU}) + V_{LD}]$$

$$= 0.50 [(0.6)(37,770) + 34,748] = 28,705 \text{ lb}$$

$$V = V_{DL} + V_{LL} = 8,274 + 28,705 = 36,979 \text{ lb}$$

Stress in horizontal shear is computed by Equation 7-7:

$$f_v = \frac{1.5V}{A} = \frac{1.5(36,979)}{432.4} = 128 \text{ lb/in}^2$$

$$F_v' = F_{vx}(1.33)(C_M) = (200)(1.33)(0.875) = 233 \text{ lb/in}^2$$

$f_v = 128 \text{ lb/in}^2 < F_v' = 233 \text{ lb/in}^2$, so horizontal shear is acceptable.

For reference, check shear for H 20-44 loading.

For truck loading,

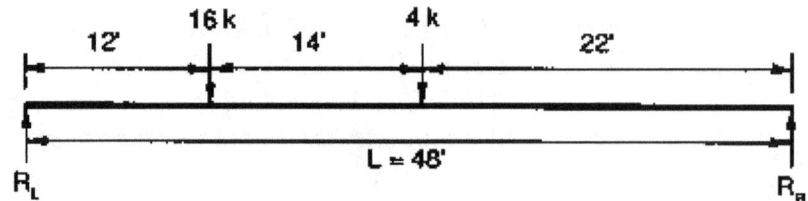

$$V_{LU} = R_L = \frac{(22 \text{ ft})(4 \text{ k}) + (36 \text{ ft})(16 \text{ k})}{48 \text{ ft}} = 13.83 \text{ k} = 13,833 \text{ lb}$$

For one-half lane loading (one wheel line),

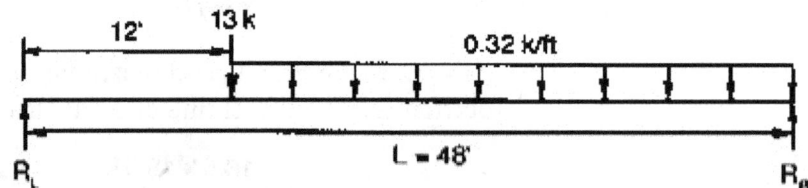

$$V_{LU} = R_L \frac{(36 \text{ ft})(13 \text{ k}) + (36 \text{ ft})(0.32 \text{ k/ft})(18 \text{ ft})}{48 \text{ ft}}$$

$$= 14.07 \text{ k} = 14,070 \text{ lb}$$

H 20-44 shear stress is computed for the controlling lane load:

$$V_{LD} = V(\text{DF}) = 14,070(0.92) = 12,944 \text{ lb}$$

$$V_{LL} = 0.50 [(0.6 \, V_{LU}) + V_{LD}]$$

$$= 0.50 [(0.6)(14,070) + 12,944] = 10,693 \text{ lb}$$

$$V = V_{DL} + V_{LL} = 8,274 + 10,693 = 18,967 \text{ lb}$$

$$f_v = \frac{1.5V}{A} = \frac{1.5(18,967)}{432.4} = 66 \text{ lb/in}^2$$

$$F_v' = F_{vx}(C_M) = (200)(0.875) = 175 \text{ lb/in}^2 > 66 \text{ lb/in}^2$$

The beam is satisfactory in horizontal shear.

7-38

Determine Bearing Length and Stresses

Bearing design will be based on the heavier U80 loading without the 33-percent stress increase for overloads.

$$F_{c\perp}' = F_{c\perp s}(C_M) = 650(0.53) = 345 \text{ lb/in}^2$$

For a unit dead load $w = 418.7$ lb/ft,

$$R_{DL} = \frac{wL}{2} = \frac{(418.7)(48)}{2} = 10,049 \text{ lb}$$

Reaction distribution factors are computed by placing the wheel line two feet from the curb face. For this single-lane bridge, one vehicle position is used for interior and outside beam distribution factors:

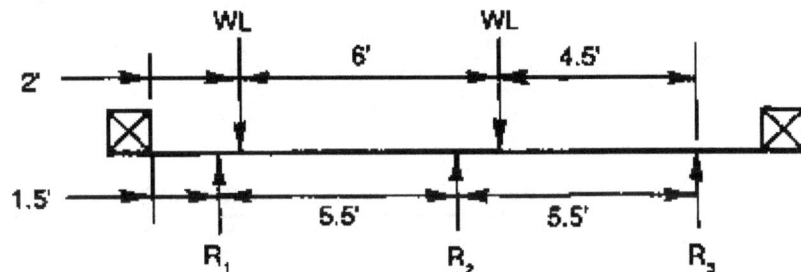

Assuming that the deck acts as a simple span between supports,

$$\text{Interior beam reaction DF} = \frac{(4.5 \text{ ft})(WL)}{5.5 \text{ ft}} = 0.82 \text{ WL/beam}$$

$$\text{Outside beam reaction DF} = \frac{(5 \text{ ft})(WL)}{5.5 \text{ ft}} = 0.91 \text{ WL/beam}$$

From Table 16-8, the maximum reaction for one wheel line of a U80 vehicle on a 48-foot span is 57,650 pounds. For the controlling outside beams,

$$R_{LL} = 57,650(0.91) = 52,461 \text{ lb}$$

$$\text{Required bearing length} = \frac{R_{DL} + R_{LL}}{b(F_{c\perp}')} = \frac{10,049 + 52,461}{8.5(345)} = 21.3 \text{ in.}$$

A bearing length of 24 inches will be used for an out-to-out beam length of 50 feet:

$$R_{DL} = \frac{wL}{2} = \frac{(418.7)(50)}{2} = 10,468 \text{ lb}$$

$$f_{c\perp} = \frac{R_{DL} + R_{LL}}{A} = \frac{10,468 + 52,461}{8.5(24)} = 308 \text{ lb/in}^2 = 345 \text{ lb/in}^2$$

Determine Camber

Dead load deflection is computed by Equation 5-16:

$$\Delta_{DL} = \frac{5wL^4}{384E'I_x} = \frac{5(418.7)[(48)(12 \text{ in./ft})]^4}{384(1,416,100)(93,272)(12 \text{ in./ft})} = 0.38 \text{ in.}$$

Camber of 1 inch will be specified at centerline, which is approximately 2-1/2 times the dead load deflection.

Summary

The superstructure will consist of three 8-1/2 by 50-7/8-inch glulam beams, 50 feet long, with a distance center to center of bearings of 48 feet. Transverse bracing will be provided for lateral support at the bearings and at midspan. The glulam will be specified as visually graded Southern Pine conforming to combination symbol 24F-V2, or may be specified by required stresses as outlined in *AITC 117--Design*. Stresses and deflection for controlling outside beams are as follows:

	H 20-44 loading	U80 loading
f_b	1,035 lb/in^2	2,119 lb/in^2
F_b'	1,632 lb/in^2	2,171 lb/in^2
Δ_{LL}	0.37 in. = $L/1,557$	1.19 in. = $L/484$
f_v	66 lb/in^2	128 lb/in^2
F_v'	175 lb/in^2	233 lb/in^2
$f_{c\perp}$	< U80	308 lb/in^2
$F_{c\perp}'$	345 lb/in^2	345 lb/in^2

DESIGN OF TRANSVERSE BRACING

Beams must be transversely braced to provide lateral strength and rigidity to the members. In bridge applications, beam bracing is provided to maintain the relative spacing of beams during construction and in service, laterally support the beam compression zone, and distribute lateral loads such as wind and centrifugal loads from the superstructure to the bearings. It is recommended by AASHTO that transverse bracing be provided at the bearings for all span lengths and at intermediate locations for spans longer than 20 feet. The spacing of intermediate bracing is based on requirements for lateral beam support, but should not exceed 25 feet. Although some lateral beam support and load distribution are also provided by the deck, these effects vary with the type of deck attachment and are normally neglected in design.

Bracing for glulam beams generally consists of cross frames or diaphragms placed normal to the longitudinal beam axes and stepped for skewed crossings (Figure 7-11). Cross frames are constructed of welded steel angles, a minimum of 5/16 inch thick, that are galvanized after fabrication (Figure 7-12). They are economical, lightweight, and are completely prefabricated for easy field erection. Cross frame design is based on the design requirements for structural steel given in AASHTO specifications. The size of the steel angles must be sufficient to resist applied loads and provide sufficient width for attachment bolts and hardware. Diaphragms are solid glulam blocks placed vertically between the beams (Figure 7-13). In most cases, the beams are held against the diaphragms by steel tie rods that pass through the beams on alternate sides of the diaphragm. Diaphragms are more effective in laterally distributing wheel loads to beams, but diaphragms are heavier and more difficult to erect than cross frames.

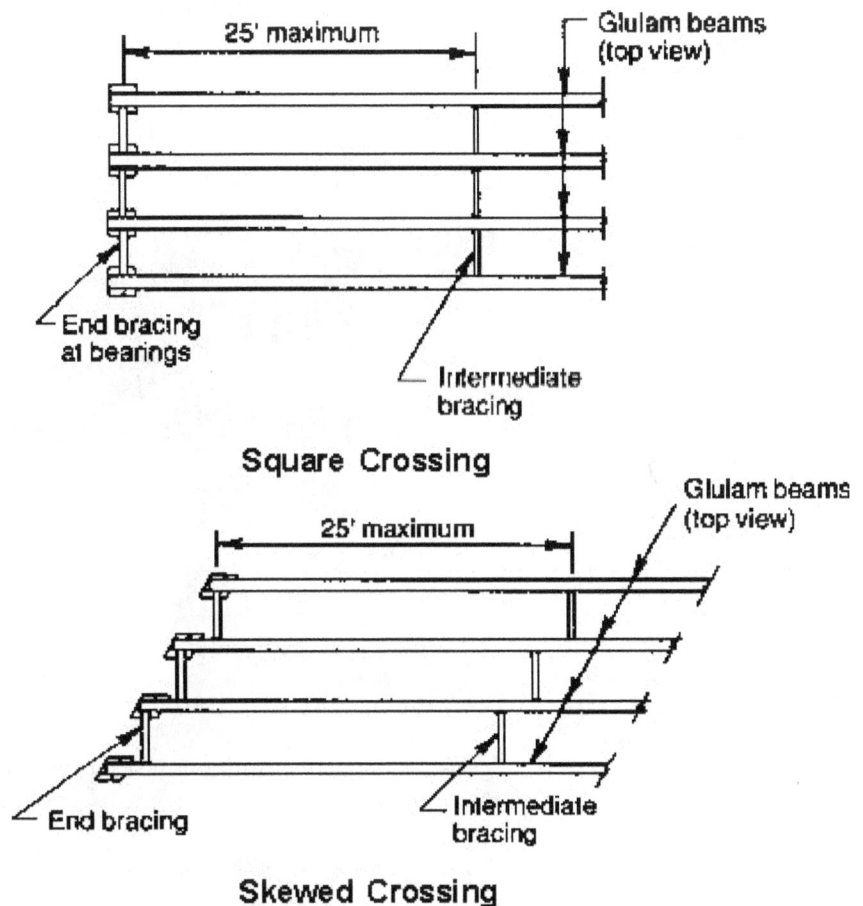

Square Crossing

Skewed Crossing

Figure 7-11. - Transverse bracing configurations for glulam beams.

Figure 7-12. - Transverse beam bracing constructed of welded-steel cross frames (photo courtesy of Tim Chittenden, USDA Forest Service).

Figure 7-13. - Transverse beam bracing constructed of solid glulam diaphragms.

Cross frames and diaphragms are designed to be as deep as practical to provide support for lateral loads over the entire beam depth. They are typically designed for the most severe loading at the bearings and the same configuration is used at intermediate points, although loading at these locations may be somewhat less. The top of the bracing should be 2 to 5 inches below the deck to ensure air circulation and clearance from deck attachment hardware. The lower beam connection should be inside the outer tension zone of the beam, which is generally considered to be the lower 10 percent of the beam depth (in areas of negative bending, this applies to the beam top). Bracing at bearings should extend to the top of the bearing shoe but not conflict with bearing anchor-bolt placement. Bolted connections between the bracing and the beam should also permit minor vertical movement of the beam from variations in moisture content. Two or more bolts rigidly connecting bracing to a beam at widely spaced points can restrain vertical beam shrinkage and may cause splitting, if shrinkage occurs.

DESIGN OF BEARINGS

Bearings support the bridge beams and transmit vertical, longitudinal, and transverse loads from the superstructure to the substructure. The two general types of bearings used are fixed bearings and expansion bearings. Fixed bearings are designed to prevent beam movement in the longitudinal direction. Expansion bearings allow longitudinal movement and are used when the superstructure will expand or contract because of thermal changes or deflection. Both types of bearings prevent transverse movement but allow small beam rotations at the support. For most timber bridges, longitudinal movement is insignificant, and fixed bearings are used. Nevertheless, expansion bearings may be required for exceptionally long spans or when thermal movement of other material such as steel or concrete must be considered.

A typical bearing for timber beams consists of four components: bearing shoe, bearing pad, beam attachment bolts, and anchor bolts (Figure 7-14). Design of these components is based on the direction and magnitude of loads transmitted by the superstructure. The bearing must be capable of distributing vertical loads from dead load and vehicle live load (including uplift when applicable), and lateral loads from sources such as wind, seismic forces, centrifugal forces, and vehicle braking.

Bearing Shoe

The bearing shoe is a bracket constructed of a welded steel plate or angles that connects the beams to the substructure (Figure 7-15). The plate configuration includes a base plate and is most commonly used for spans of approximately 50 feet or more. The angle configuration may be used for longer spans but is generally most suited for spans shorter than approximately 50 feet. A base plate for the angle configuration is optional, but is commonly used when bearing is on a timber cap or sill.

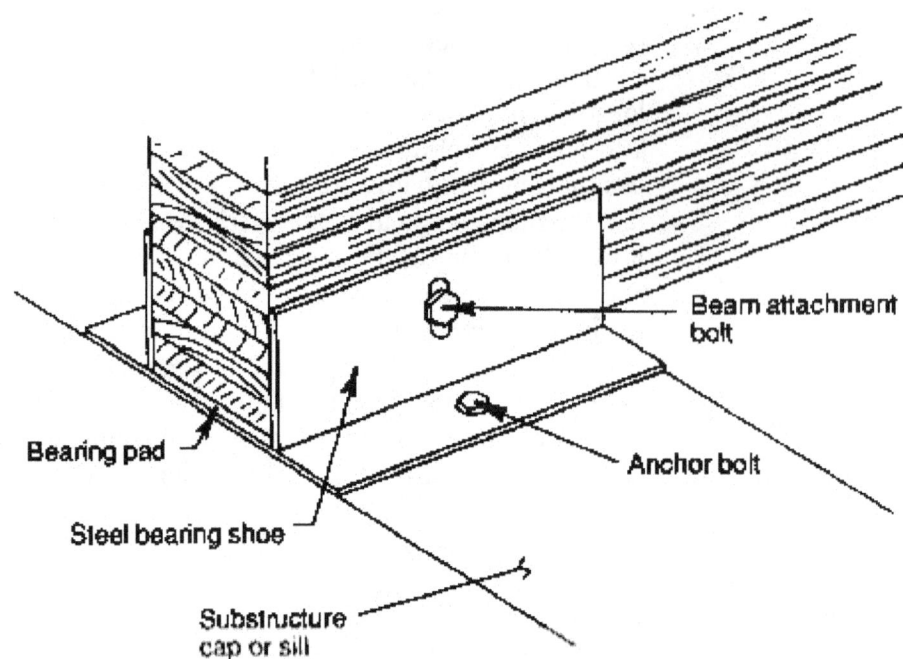

Figure 7-14. - Typical fixed-bearing configuration for glulam beams.

The size of the bearing shoe depends on the beam size and required length of bearing. Minimum length is the required beam bearing length. The width between side plates is the beam width plus 1/4 inch. The height of the side plates must be sufficient to resist transverse loads and locate the beam attachment bolt a minimum or four times, but preferably five times, the bolt diameter above the base of the beam. When the bearings are subject to uplift, the minimum height of the attachment bolt is seven times the bolt diameter.

Bearing Pad

A bearing pad is a thin pad of elastomeric rubber (usually neoprene) placed between the beam and the support. For timber bridges, the purpose of the pad is to allow slight movement and rotation of the beam through deformation of the pad, provide a smooth bearing surface and compensate for irregularities in the bearing surfaces, and elevate the beam above the sill or cap where water may collect.

Bearing pad size depends on the bearing area of the beam. Pads are equal in length to the beam bearing length and are 1/4 inch narrower than the beam width. Pad thickness depends on the type of bearing, whether fixed or expansion. For fixed bearings, pads are typically 1/2 inch thick for spans shorter than 50 feet, and 3/4 to 1 inch thick for spans longer than 50 feet. For expansion bearings, pad thickness is based on the anticipated

7-44

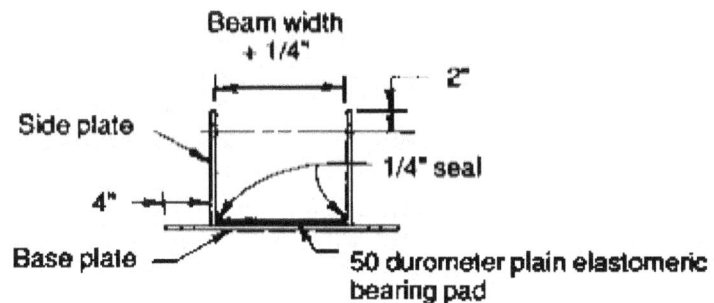

Beam width
+ 1/4"

2"

Side plate

1/4" seal

4"

Base plate

50 durometer plain elastomeric
bearing pad

Steel plate bearing shoe

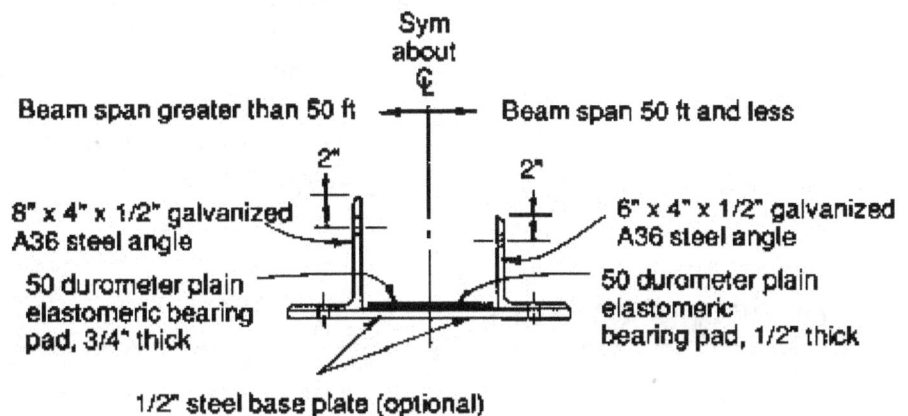

Sym
about
℄

Beam span greater than 50 ft

Beam span 50 ft and less

2"

2"

8" x 4" x 1/2" galvanized
A36 steel angle

6" x 4" x 1/2" galvanized
A36 steel angle

50 durometer plain
elastomeric bearing
pad, 3/4" thick

50 durometer plain
elastomeric
bearing pad, 1/2" thick

1/2" steel base plate (optional)

Steel angle bearing shoe

Figure 7-15. - Typical bearing shoe details for glulam beams.

movement of the superstructure, and must be based on design criteria
given in AASHTO for elastomeric bearings (AASHTO Section 14).
In both cases, a pad with nominal 50 or 60 durometer hardness is
recommended.

Beam Attachment Bolts

Beam attachment bolts connect the beams to the bearing shoe and transmit
longitudinal and uplift forces from the superstructure. Minimum recommended bolt diameters are 3/4 inch for spans up to approximately 50 feet
and 1 inch for spans longer than 50 feet. For most designs, one bolt at the
center of the bearing length is adequate; however, the number and diameter of bolts should be based on the magnitude and direction of applied
loads.

Beam attachment bolts are placed in round holes bored through the beam before preservative treatment. Holes in the bearing shoe are slotted or round depending on the type of bearing and direction of vertical forces. For fixed bearings without uplift, holes are generally slotted vertically to allow for construction tolerances and permit the beam to rotate slightly at the support. When fixed bearings are subjected to uplift, holes are round. For expansion bearings, holes are slotted horizontally to allow longitudinal beam movement.

Anchor Bolts

Anchor bolts transmit vertical and lateral loads from the bearing shoe to the substructure. On steel and concrete substructures, anchor bolts are normally machine bolts or studs. On timber substructures, lag screws may be used. Anchor bolts are typically placed through round holes in the bearing shoe, but slotted holes may be used at the option of the designer to allow for construction tolerances.

The number and diameter of anchor bolts depends on load magnitude and bolt capacity. As a minimum, two bolts are provided at each bearing, one on each side of the beam. Recommended minimum diameters are 3/4 inch for spans 50 feet or shorter and 1 inch for spans longer than 50 feet. Additional bolts or increased bolt diameters may be required depending on the magnitude of transmitted loads.

7.5 DESIGN OF GLULAM DECKS

Glulam decks are constructed of panels manufactured of vertically laminated lumber. The panels are placed transverse to the supporting beams, and loads act parallel to the wide face of the laminations. The two basic types of glulam decks are the noninterconnected deck and the doweled deck (Figure 7-16). Noninterconnected decks have no mechanical connection between adjacent panels. Doweled decks are interconnected with steel dowels to distribute loads between adjacent panels. Both deck types are stronger and stiffer than conventional nail-laminated lumber or plank decks, resulting in longer deck spans, increased spacing of supporting beams, and reduced live load deflection. Additionally, glulam panels can be placed to provide a watertight deck, protecting the structure from the deteriorating effects of rain and snow.

Glulam decks are manufactured from visually graded western species or Southern Pine sawn lumber using the same lumber grade throughout. Any of several axial combination symbols in Table 2 of *AITC 117--Design* may be used. The three most frequently used combination symbols for each species are listed in Table 7-5. Combination symbols with a tabulated bending stress of 1,800 lb/in^2 or less are the most economical and most commonly used.

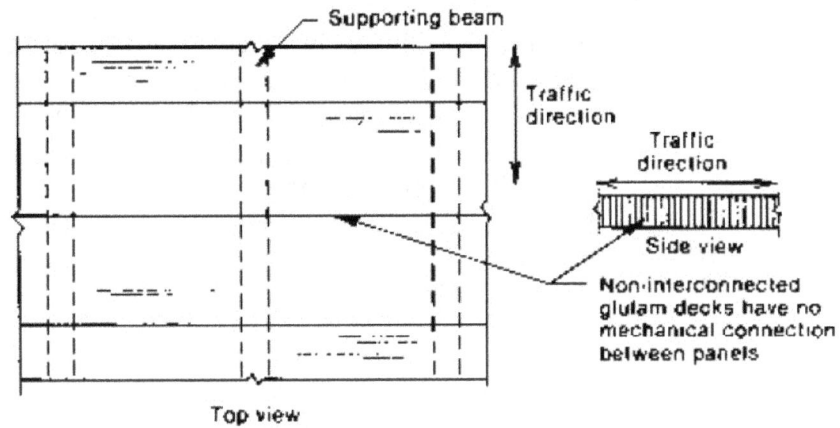

Non-interconnected glulam deck

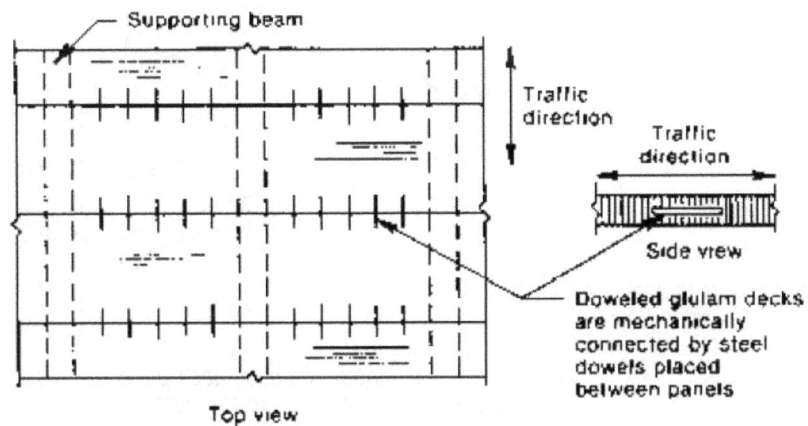

Doweled glulam deck

Figure 7-16. - Configurations for noninterconnected and doweled glulam decks.

Table 7-5 - Glulam axial combination symbols commonly used for bridge decks.

Western species			Southern Pine		
Combination symbol	Grade	F_s	Combination symbol	Grade	F_s
1	L3	1,450	46	N3M	1,450
2	L2	1,800	47	N2M	1,750
3	L2D	2,100	48	N2D	2,000

Glulam decks are generally 5-1/8 inches (5 inches for Southern Pine) or 6-3/4 inches thick. Increased thicknesses up to 14-1/4 inches are available, but are seldom required (design aids in this section are limited to decks 8-3/4 inches thick or less). Panel width is a multiple of 1-1/2 inches, the net width of the individual lumber laminations. The practical width of panels ranges from approximately 30 to 55 inches; however, the designer should check local manufacturing and treating limitations before specifying widths over 48 inches. Panels can be manufactured in any specified length to be continuous across the structure. It is common practice to vary adjacent panel lengths to provide a drainage opening under curbs (Figure 7-17).

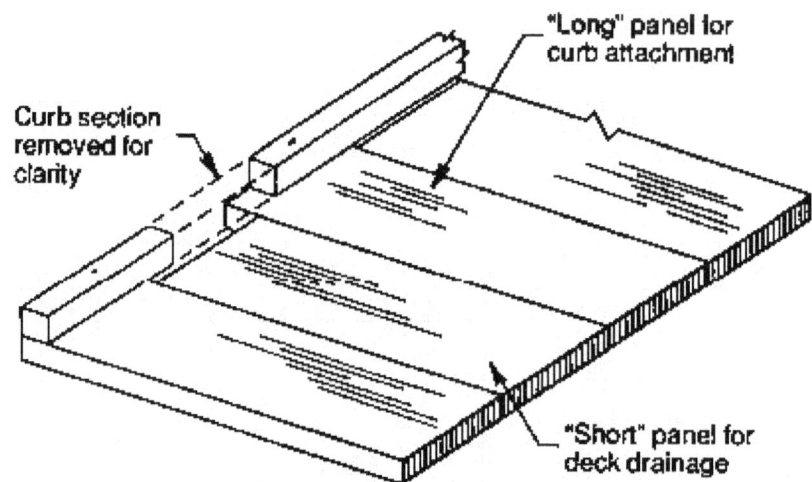

Figure 7-17. - The length of glulam deck panels may be varied between adjacent panels to provide a drainage opening under the curb.

The performance and economy of glulam deck panels can be significantly affected by the configuration and materials specified in design. The most economical design is one that uses a modular-type system with two or three standardized panels in a repetitious arrangement. Panel width and configuration are usually based on criteria for curb or railing systems (Chapter 10). When the bridge length is not evenly divisible by the selected panel width, odd-width panels are placed on the approach ends of the deck.

NONINTERCONNECTED GLULAM DECKS

Noninterconnected glulam decks are the most widely used type of glulam deck in modern timber bridge construction (Figure 7-18). They are economical, require little fabrication, and are easy to install with unskilled labor and without special equipment. Because the panels are not connected to one another, each panel acts individually to resist the stresses and deflection from applied loads.

Figure 7-18. - (A) Noninterconnected glulam deck being placed (photo courtesy of LamFab Wood Structures, Inc.). (B) Completed glulam deck is prepared for paving (photo courtesy of Ron Vierra, USDA Forest Service).

Design Procedures

Noninterconnected glulam decks are designed using an interactive procedure, similar to that previously discussed for beams. The deck is assumed to act as a simple span between beams and is designed for the stresses acting in the direction of the deck span, and deflection. Stresses occurring in the direction perpendicular to the span are not critical and are not considered in design.

The basic design procedures for noninterconnected glulam decks are given in the following steps. The sequence assumes that panels are initially designed for bending, then checked for deflection and shear. Although deflection rather than bending stress usually controls in most applications, the acceptable level of deflection is established by the designer and may vary for different applications.

1. Define the deck span, design loads, and panel size.

The effective deck span, s, is the clear distance between supporting beams plus one-half the width of one beam, but not greater than the clear span plus the panel thickness (AASHTO 3.25.1.2). Panel width and length are based on considerations previously discussed.

The deck design load is the maximum wheel load of the design vehicle. For H 20-44 and HS 20-44 loads, AASHTO special provisions for timber decks apply, and a 12,000-pound wheel load is used instead of the standard 16,000-pound wheel load. As a result, the maximum wheel load for all standard AASHTO vehicles (H 15-44, HS 15-44, H 20-44 and HS 20-44) is 12,000 pounds.

2. Estimate deck thickness.

Deck thickness, t, must be estimated for initial calculations. It is generally most practical to start with a 6-3/4-inch deck (an initial estimate of deck thickness based on bending or deflection can also be made from Tables 7-8 and 7-9 presented later in this section).

3. Determine. wheel distribution widths and effective deck section properties.

In the direction of the deck span, the wheel load is assumed to be uniformly distributed over a width, b_t (AASHTO 3.25.1.1), as computed by

$$b_t = \sqrt{0.025P} \tag{7-10}$$

where b_t = wheel load distribution width in the direction of the deck span (in) and

P = maximum wheel load (lb).

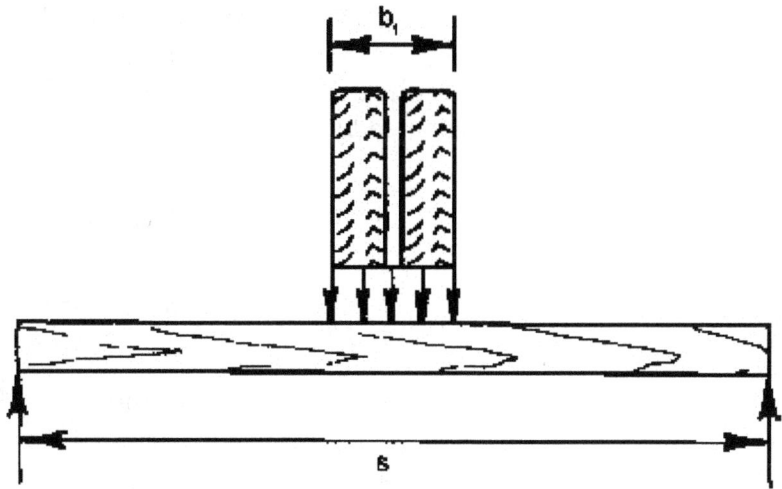

For a 12,000-pound wheel load, $b_t = 17.32$ inches.

In the direction perpendicular to the deck span, the wheel load is distributed over an effective width, b_d, equal to the deck thickness, t, plus 15 inches, but not greater than the deck panel width (AASHTO 3.25.1.1):

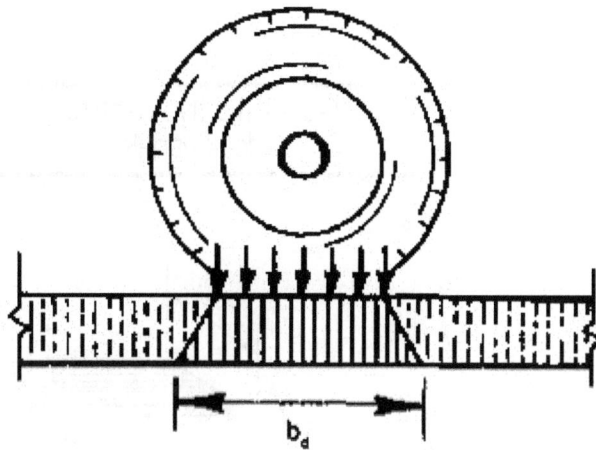

$$b_d = t + 15 \leq \text{actual panel width} \qquad (7\text{-}11)$$

where $b_d =$ wheel load distribution width perpendicular to the deck span (in.) and

$t =$ deck thickness (in.)

The effective deck section, defined by a deck width, b_d, and thickness, t, is designed as a beam to resist the loads and deflection produced by one wheel line of the design vehicle. Effective deck section properties are computed by

$$A = \text{effective deck area (in}^2) = b_d t \tag{7-12}$$

$$S_y = \text{effective deck section modulus (in}^3) = \frac{b_d t^2}{6} \tag{7-13}$$

$$I_y = \text{effective deck moment of inertia (in}^4) = \frac{b_d t^3}{12} \tag{7-14}$$

Effective deck section properties for common deck thicknesses are given in Table 7-6.

Table 7-6. - Effective deck section properties for noninterconnected glulam deck panels.

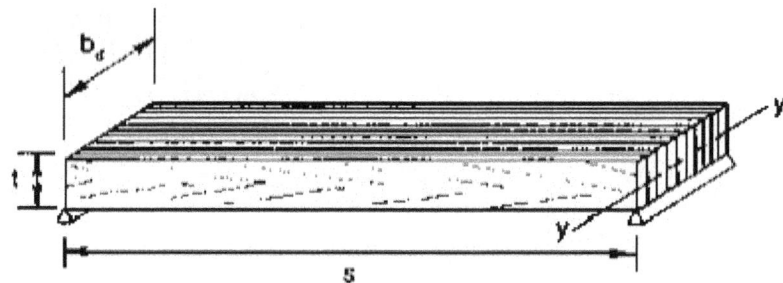

t	b_d	A	S_Y	I_Y
(in.)	(in.)	(in²)	(in³)	(in⁴)
5	20	100.00	83.33	208.33
5-1/8	20.13	103.17	88.01	225.75
6-3/4	21.75	146.81	165.16	557.43
8-1/2	23.50	199.75	282.98	1,202.66
8-3/4	23.75	207.81	303.06	1,325.89

4. Determine dead load moment.

Uniform dead load moment for the effective deck section can be computed:

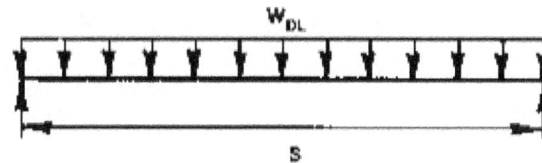

$$M_{DL} = \frac{w_{DL} \, s^2}{8} \tag{7-15}$$

where M_{DL} = deck dead load moment (in-lb),

w_{n} = dead load of the deck and wearing surface over the wheel load distribution width, b_d(lb/in), and

s = effective deck span (in.).

When a portion of the dead load is not uniformly distributed (as when the deck supports utility lines or other components), dead load moment from these nonuniform loads is computed by assuming the deck acts as a simple span, and the moment from the additional loading is added to M_{n} computed by Equation 7-15.

5. Determine live load moment.

Compute the maximum vehicle live load moment by assuming that the deck acts as a simple span between beams. Wheel loads are positioned laterally on the span to produce the maximum moment using the same procedures discussed in Chapter 6 for a moving series of loads.

For one traffic lane, the maximum moment for a standard 12,000-pound wheel load and 6-foot-track width depends on the effective deck span, s. When the effective deck span is greater than 17.32 inches, but less than or equal to 122 inches $(17.32 < s \leq 122)$, maximum moment is produced when a single wheel load is positioned at the span centerline, and is computed as follows:

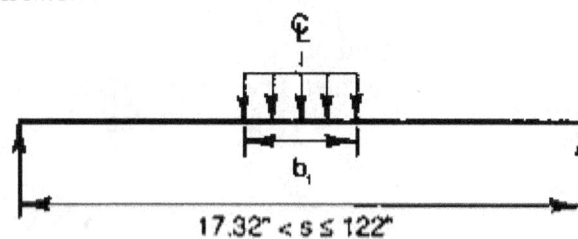

$$M_{LL} = 3,000s - 25,983 \tag{7-16}$$

where M_{LL} is the maximum live moment (in-lb).

When the effective deck span is greater than 122 inches ($s > 122$), the maximum moment is produced when both wheel loads are on the span. Maximum moment occurs under the wheel load closest to the span center-line when the span centerline bisects the centroid of the wheel loads and the adjacent wheel load, and is computed as follows:

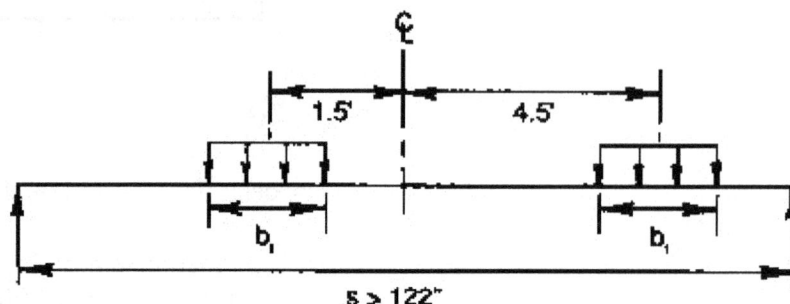

$$M_{LL} = 6,000s + \frac{7,776,000}{s} - 457,983 \tag{7-17}$$

6. Compute bending stress and select a deck combination symbol.

When deck panels are continuous over two spans or less, bending stress is based on simple span moments and is computed by

$$f_b = \frac{M}{S_y} \tag{7-18}$$

where $M = M_{LL} + M_{DL}$ computed for a simple span (in-lb).

When the deck is continuous over more than two spans, the maximum bending moment is 80 percent of that computed for a simple span to account for span continuity (AASHTO 3.25.4), and is computed by

$$f_b = \frac{0.8M}{S_y} \tag{7-19}$$

Select a panel combination symbol from Table 2 of *AITC 117-Design* that provides the required bending stress. The most common combination symbols are No. 2 for western species ($F_{by} = 1,800$ lb/in²) and No. 47 for Southern Pine ($F_{by} = 1,750$ lb/in²). The applied bending stress, f_b, must not exceed F_b' for the selected combination symbol, computed by

$$F_b' = F_{by}C_F C_{du} \tag{7-20}$$

where F_{by} = tabulated bending stress from Table 2 of *AITC 117-Design* (lb/in²) and

C_F = size factor for panels less than 12 inches thick:

t (in.)	C_s
5 or 5-1/8	1.10
6-3/4	1.07
8 or 8-3/4	1.04

F_b' computed by Equation 7-20 is given in Table 7-7 for common values of F_{by}. Allowable bending stress may be increased by a factor of 1.33 for overloads in AASHTO Load Group IB.

Table 7-7. - Values of F_b' for glulam deck panels.

Deck t (in.)	Allowable bending stress, F_b' (lb/in^2)[a]				
	F_{by}=1,450	F_{by}=1,750	F_{by}=1,800	F_{by}=2,000	F_{by}=2,100
5-1/8 [b]	1,276.0	1,540.0	1,584.0	1,760.0	1,848.0
6-3/4	1,241.2	1,498.0	1,540.8	1,712.0	1,797.6
8-3/4 [c]	1,206.4	1,456.0	1,497.6	1,664.0	1,747.2

[a] $F_b' = F_{by}C_M C_F$

[b] Also applies to t = 5 inches for Southern Pine.

[c] Also applies to t = 8 1/2 inches for Southern Pine.

If $f_b \leq F_b'$, the initial deck thickness and combination symbol are satisfactory in bending. When F_b is significantly lower than F_b', a thinner deck or lower grade combination symbol may be more economical; however, no changes in the panel thickness or combination symbol should be made until the live load deflection is determined.

If $f_b > F_b'$, the deck is insufficient in bending and the deck thickness or grade must be increased, or the effective deck span reduced. If deck thickness or span is changed, the design sequence must be repeated. In some cases, it may be more economical to increase deck thickness to the next higher standard size, rather than use a higher-grade combination symbol. The designer should check local availability and prices for different panel thicknesses and combination symbols before specifying panels with F_b greater than 1,800 lb/in^2 for visually graded western species or 1,750 lb/in^2 for visually graded Southern Pine.

Approximate maximum spans based on bending for noninterconnected glulam decks continuous over more than two spans are given in Table 7-8.

Table 7-8. - Approximate maximum effective span for noninterconnected transverse glulam deck panels based on bending; deck continuous over more than two spans; loading from a 12,000-pound wheel load plus the deck dead load, including a 3-inch asphalt wearing surface; b_t = 15 inches + deck thickness.

F_{by} (lb/in^2)	Approximate maximum deck span (in.)		
	t = 5 in. or t = 5-1/8 in.	t = 6-3/4 in.	t = 8-1/2 in. or t = 8-3/4 in.
1,450	52	91	>120
1,750	61	107	>120
1,800	65	109	>120
2,000	68	120	>120
2,100	74	>120	>120

$F_b' = F_{by}C_M C_F$ as given in Table 7-7.

7. Check live load deflection.

Live load deck deflection is computed by standard methods of engineering analysis, assuming the deck to be a simple span between beams. For standard AASHTO trucks, with 12,000-pound wheel loads and a 6-foot track width, equations for maximum deflection on a simple span are as follows:

For effective spans greater than 17.32 inches, but less than or equal to 110 inches ($17.32 < s < 110$), maximum live load deflection occurs with one wheel load positioned at the span centerline and is computed as follows:

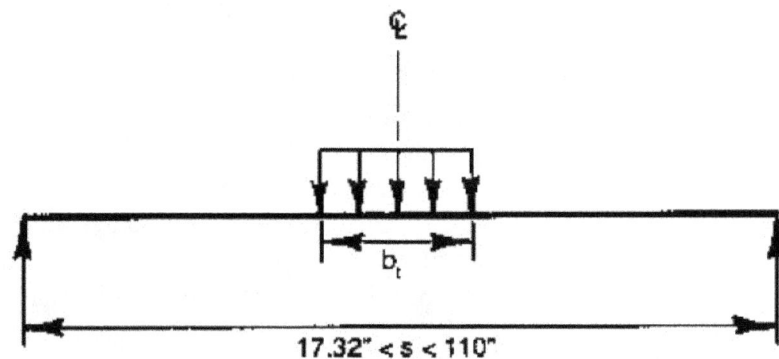

$$\Delta_{LL} = \frac{1.80}{E'I_y}\left(138.8s^3 - 20,780s + 90,000\right) \tag{7-21}$$

where $E' = EC_M$ (lb/in^2).

When the effective deck span is greater than or equal to 110 inches ($s \geq 110$), maximum live load deflection is obtained when both wheel loads are centered on the span and is computed as follows:

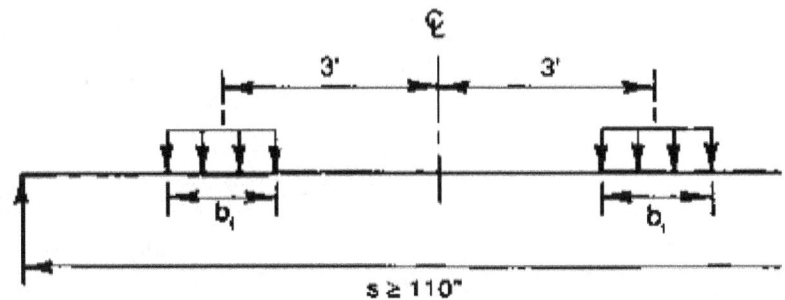

$$\Delta_{LL} = \frac{1}{E'I_y}\left(500s^3 + 90.5s^2 - 3,967,074s + 98,663,396\right) \qquad (7\text{-}22)$$

When the deck is continuous over more than two spans, the maximum deflection is 80 percent of that computed for a simple span to account for deck continuity. In this case, values obtained from Equations 7-21 or 7-22 may be multiplied by 0.80. Deflection coefficients for standard 12,000-pound wheel load(s) on decks continuous over more than two spans are given in Figure 7-19.

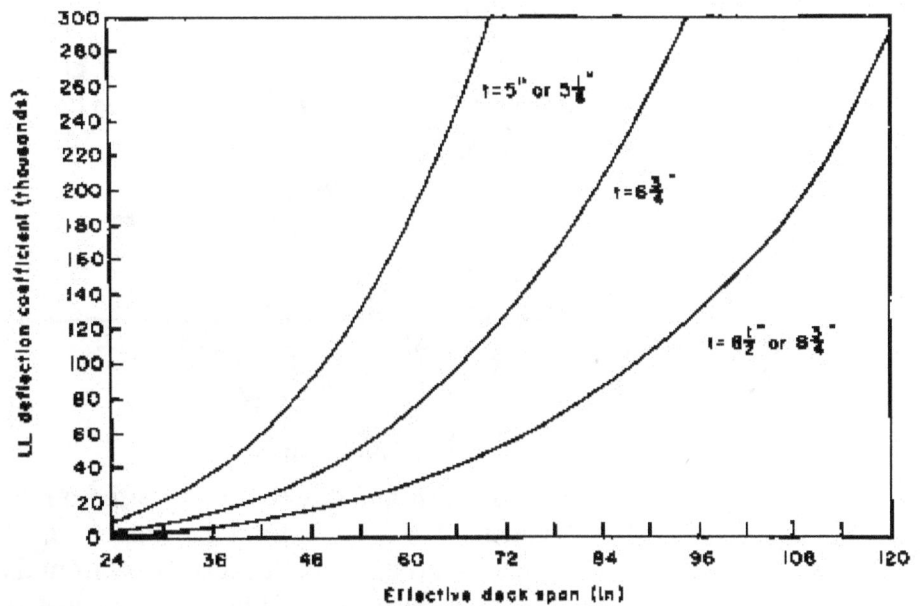

Figure 7-19. - Vehicle live load deflection coefficients for 12,000-pound wheel load(s) on a transverse, noninterconnected glulam deck that is continuous over more than two spans. Divide the deflection coefficient by E' to obtain the deck deflection in inches.

Requirements for live load deflection in glulam decks are not included in AASHTO specifications, and the acceptable deflection limit is left to designer judgment. Deck deflection is important because it directly influences the performance and serviceability of the deck, wearing surface, and mechanical connections. When deflections are large, vertical movement of the panel causes vibrations in the structure and rotation of the deck panel about the beam. This can cause bolts or other connections to loosen and asphalt wearing surfaces to crack. Deck movement can also be alarming to users, especially pedestrians.

The maximum recommended live load deflection for noninterconnected glulam panels is 0.10 inch. This limit was derived from research and field observations related to panel attachment and asphalt wearing surface performance. [62] Deflection will control over bending in most design applications, but panel spans remain within the acceptable range of recommended beam spacings previously discussed. Based on this criterion, maximum effective deck spans for live load deflection are shown in Table 7-9. A further reduction in deflection for deck panels supporting pedestrian walkways or an asphalt wearing surface is desirable.

Table 7-9. - Approximate maximum effective span for noninterconnected transverse glulam deck panels based on a maximum vehicle live load deflection of 0.10 inch; deck continuous over more than two supports; loading from a 12,000-pound wheel load; b_d = 15 inches + deck thickness.

		Approximate maximum deck span (in.)		
E (lb/in²)	E' (lb/in²)	t = 5 in. or t = 5-1/8 in.	t = 6-3/4 in.	t = 8-1/2 in. or t = 8-3/4 in.
1,300,000	1,082,900	50	68	91
1,400,000	1,166,200	51	70	94
1,500,000	1,249,500	53	72	95
1,700,000	1,416,100	56	75	99
1,800,000	1,499,400	57	76	101

$E' = EC_M = 0.833E$.

8. Check horizontal shear.

Horizontal shear for dead load is based on the maximum vertical shear occurring at a distance from the support equal to the deck thickness, t. Loads occurring within the distance t from the supports are neglected. Horizontal shear for dead load is computed as follows:

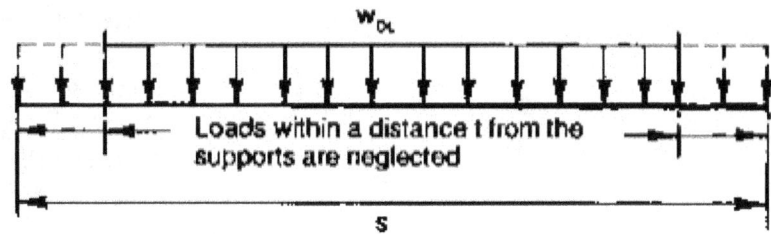

$$V_{DL} = w_{DL}\left(\frac{s}{2} - t\right) \tag{7-23}$$

where V_{DL} = dead load vertical shear (lb).

Live load vertical shear is computed by placing the edge of the wheel load distribution width, b_t, a distance t from the support.

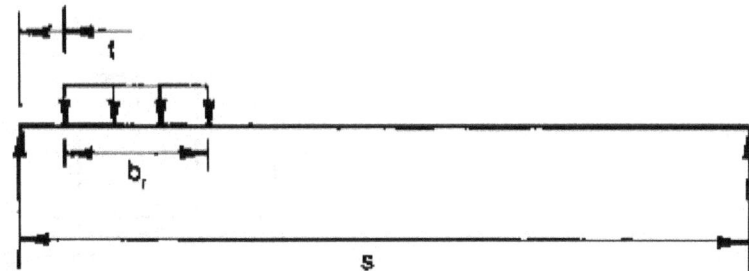

Applied stress in horizontal shear is based on a different effective panel width than that used for bending and deflection. Current AASHTO specifications (interims through 1987) allow the stress to be distributed over the full panel width (AASHTO 13.3.1). AITC has recently recommended a more conservative distribution width of 15 inches plus twice the deck thickness, but not greater than the panel width. In either case, shear stress is normally not a controlling factor in glulam panel design. The distribution width used in this chapter follows the AITC recommendations. Either convention may be used based on designer judgment.

Horizontal shear stress is computed using

$$f_v = \frac{1.5V}{A_v} \tag{7-24}$$

where $\qquad V = V_{DL} + V_{LL}$ (lb), and

$$A_v = t(15 + 2t) \leq t(\text{panel width})(\text{in}^2)$$

Allowable shear stress is computed using

$$F_v' = F_v C_M \tag{7-25}$$

where $\qquad F_v'$ = allowable horizontal shear stress (lb/in²),

$\qquad\qquad F_v$ = tabulated shear stress from Table 2, *AITC 117--Design* (lb/in²), and

7-59

C_M = wet-use factor for shear = 0.875.

Values of F_v within each species group are the same for the various combination symbols commonly used for glulam decks. For western species, F_v = 145 lb/in², while for Southern Pine, F_v = 175 lb/in². When $f_v > F_v'$ the only options are to increase the deck thickness or reduce the effective deck span.

9. Check overhang.

The deck overhang at exterior supports is checked using an effective span measured to the centerline of the outside beam, minus one-fourth the beam width. For vehicle live load stresses and deflection, the wheel load is positioned with the load centroid 1 foot from the face of the railing or curb.

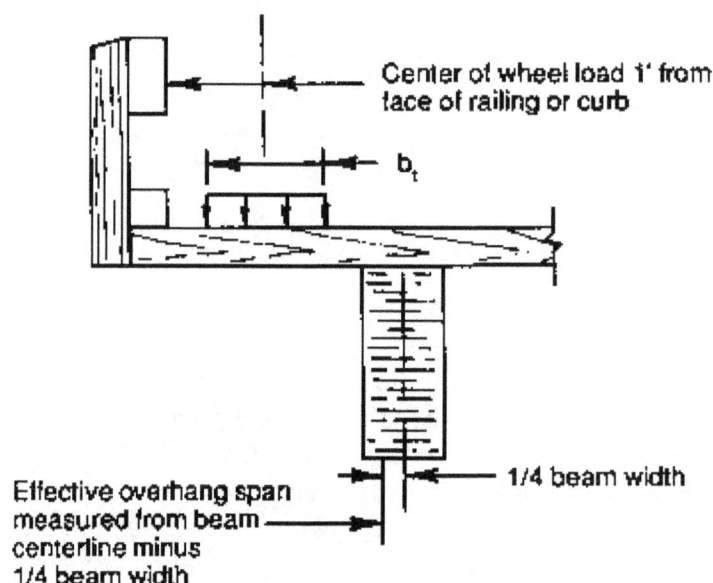

Deck stress in bending and horizontal shear must be within allowable values previously determined.

Example 7-5 - Noninterconnected glulam deck with highway loading

Design a noninterconnected glulam deck for the beam superstructure of Example 7-3. The superstructure has a 24-foot roadway that carries two lanes of AASHTO HS 20-44 loading. Support is provided by five 12-1/4-inch-wide glulam beams that are spaced 5 feet on center and are 95-1/2 feet long. The following assumptions apply:

1. glulam deck panels are manufactured from visually graded Southern Pine;

2. rail system dead load is 300 pounds at each post with a maximum post spacing of 7 feet; and

3. deck live load deflection is limited to approximately 0.10 inch.

7-60

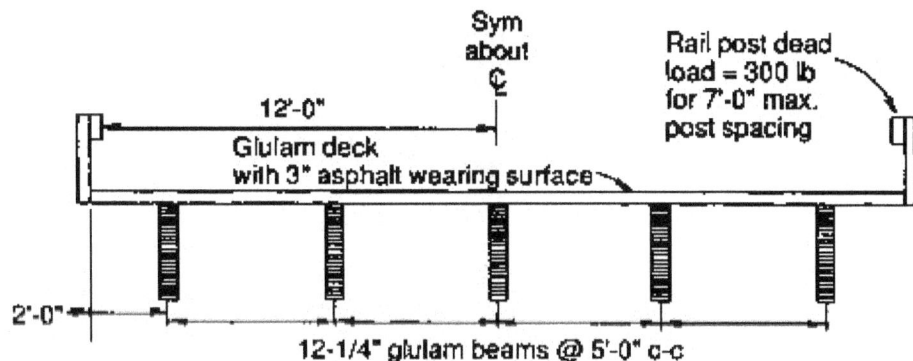

Solution

Determine the Deck Span, Design Loads, and Panel Size

The deck span is the clear distance between supporting beams plus one-half the width of one beam, but not greater than the clear span plus the panel thickness:

Clear distance between beams = 60 in. - 12.25 in. = 47.75 in.

$$s = 47.75 \text{ in. } + \frac{12.25}{2} = 53.88 \text{ in.}$$

If a 5-inch deck is used, s will be limited by the clear span plus deck thickness to 47.75 inches + 5 inches = 52.75 inches. For other deck thicknesses, $s = 53.88$ inches will control.

For HS 20-44 loading, AASHTO special provisions apply and the deck will be designed for a 12,000-pound wheel load. Panel width for an out-to-out bridge length of 95-1/2-feet will be based on an alternating repetition of panels to allow standardized panel configurations. In this case, 46-3/4-inch-wide panels will be used with two 41-1/4-inch-wide panels at each end (one of the end panels will be trimmed 3/4 inch before pressure treatment). Rail posts will be placed at the center of end panels and at the center of every second panel:

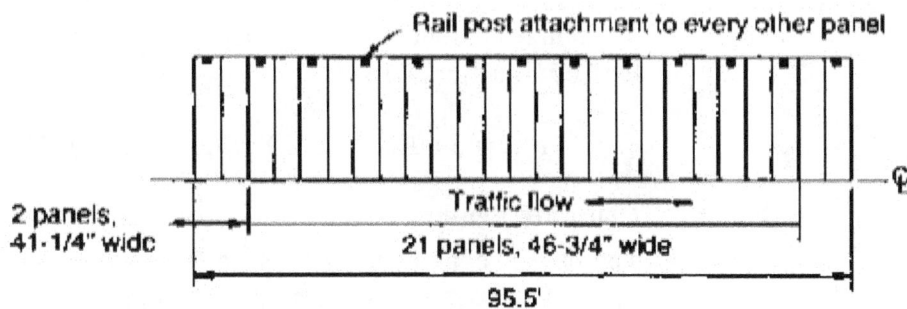

Estimate Deck Thickness

From approximate maximum deck spans given in Tables 7-8 and 7-9, an initial deck thickness of 5 inches is selected. The effective span used for design will therefore be controlled by the clear span plus deck thickness to 52.75 inches.

Determine Wheel Distribution Widths and Effective Deck Section Properties

In the direction of the deck span,

$$b_t = \sqrt{0.025P} = \sqrt{0.025(12,000)} = 17.32 \text{ in.}$$

Normal to the deck span,

$$b_d = t + 15 = 5 + 15 = 20 \text{ in.}$$

Effective deck section properties from Table 7-6 are

$$A = 100 \text{ in}^2$$

$$S_y = 83.33 \text{ in}^3$$

$$I_y = 208.33 \text{ in}^4$$

Determine Deck Dead Load

For a 5-inch deck and 3-inch asphalt wearing surface, dead load unit weight and moment over the effective distribution width of 20 inches are computed as follows:

$$w_{DL} = (20 \text{ in.})\frac{\left[(5 \text{ in.})(50 \text{ lb/ft}^3) + (3 \text{ in.})(150 \text{ lb/ft}^3)\right]}{1728 \text{ in}^3/\text{ft}^3} = 8.1 \text{ lb/in.}$$

$$M_{DL} = \frac{w_{DL}s^2}{8} = \frac{8.1(52.75)^2}{8} = 2,817 \text{ in-lb}$$

Determine Live Load Moment

For an effective deck span less than 122 inches, maximum live load moment is computed for a (6-foot track width and 12,000-pound wheel load by Equation 7-16:

$$M_{LL} = 3,000s - 25,983 = 3,000(52.75) - 25,983 = 132,267 \text{ in-lb}$$

Compute Bending Stress and Select a Deck Combination Symbol

The deck is continuous over more than two spans, so bending stress is based on 80 percent of the simple span moment:

$$M = M_{DL} + M_{LL} = 2,817 + 132,267 = 135,084 \text{ in-lb}$$

$$f_b = \frac{0.80M}{S_y} = \frac{0.80(135,084)}{83.33} = 1,297 \text{ lb/in}^2$$

From Table 7-7, F_{by} = 1,750 lb/in² is the closest value for a Southern Pine combination that will meet bending requirements. An initial combination symbol No. 47 is selected, and the following values are obtained from *AITC 117-Design*:

$$F_{by} = 1,750 \text{ lb/in}^2 \qquad C_M = 0.80$$

$$F_{vy} = 175 \text{ lb/in}^2 \qquad C_M = 0.875$$

$$E_y = 1,400,000 \text{ lb/in}^2 \qquad C_M = 0.833$$

By Equation 7-20 (or Table 7-7),

$$F_b' = F_{by}C_F C_M = 1,750(1.1)(0.80) = 1,540 \text{ lb/in}^2$$

f_b = 1,297 lb/in² < F_b' = 1,540 lb/in², so a 5-inch combination symbol No. 47 panel is satisfactory in bending.

Check Live Load Deflection

Maximum deflection is computed for a 12,000-pound wheel load and 6-foot track width by Equation 7-21:

$$E' = E_y C_M = 1,400,000(0.833) = 1,166,200 \text{ lb/in}^2$$

$$\Delta_{LL} = \frac{1.80}{E'I_y}\left(138.8s^3 - 20,780s + 90,000\right)$$

$$\Delta_{LL} = \frac{1.80\left[(138.8)(52.75)^3 - (20,780)(52.75) + 90,000\right]}{1,166,200(208.33)} = 0.14 \text{ in.}$$

The deck is continuous over more than two spans, so 80 percent of the simple span deflection is used to account for span continuity:

$$\Delta_{LL} = 0.80(0.14) = 0.11 \text{ in.}$$

The computed deflection of 0.11 inch is slightly greater than 0.10 inch, but the difference of 0.01 inch is considered insignificant and deflection is acceptable.

Check Horizontal Shear

Dead load vertical shear is computed at a distance t from the support. By Equation 7-23 for w_{DL} = 8.1 lb/in,

$$V_{DL} = w_{DL}\left(\frac{s}{2} - t\right) = 8.1\left(\frac{52.75}{2} - 5\right) = 173.1 \text{ lb}$$

Live load vertical shear is computed by placing the edge of the wheel load distribution width *(b)* a distance t from the support. The resultant of the 12,000-pound wheel load acts through the center of the distribution width and V_{LL} is computed by statics:

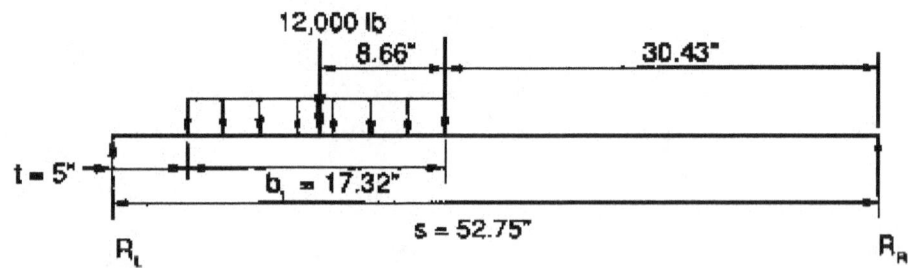

$$V_{DL} = R_L \frac{12{,}000 \text{ lb}(8.66 \text{ in.} + 30.43 \text{ in.})}{52.75 \text{ in.}} = 8{,}893 \text{ lb}$$

By Equation 7-24,

$$V = V_{DL} + V_{LL} = 173.1 + 8{,}893 = 9{,}066 \text{ lb}$$

$$A_v = t(15 + 2t) = 5(15 + 10) = 125 \text{ in}^2$$

$$f_v = \frac{1.5V}{A_v} = \frac{1.5(9{,}066)}{125} = 109 \text{ lb/in}^2$$

By Equation 7-25,

$$F_v' = F_{vy}C_M = 175(0.875) = 153 \text{ lb/in}^2$$

$f_v = 109 \text{ lb/in}^2 < F_v' = 153$, so the panel is satisfactory in horizontal shear.

Check Overhang

Bending and shear stresses are checked in the deck overhang by positioning the wheel load centroid 1 foot from the rail face. Moments are computed using an effective span measured from the load to the beam centerline, minus one-fourth the beam width:

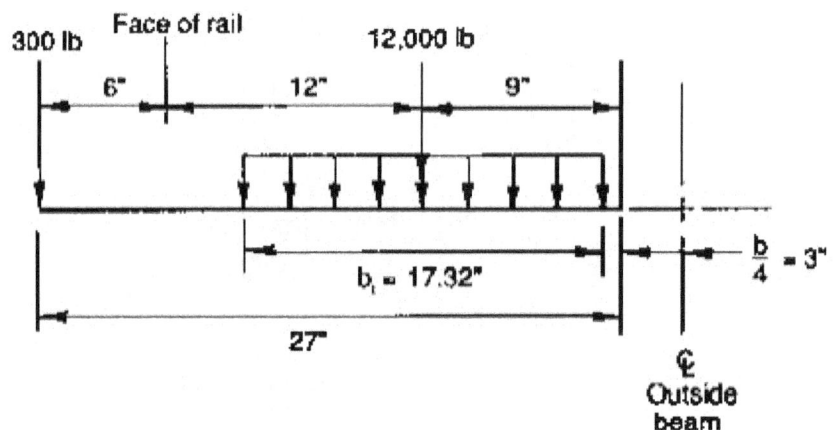

$$M_{LL} = (9 \text{ in.})(12{,}000 \text{ lb}) = 108{,}000 \text{ in-lb}$$

$$\text{Rail } M_{DL} = (27 \text{ in.})(300 \text{ lb}) = 8{,}100 \text{ in-lb}$$

7-64

$$\text{Deck } M_{DL} = (8.1 \text{ lb/in })(27 \text{ in.})\left(\frac{27 \text{ in.}}{2}\right) = 2,952 \text{ in-lb}$$

$$M = M_{LL} + M_{DL}$$

$$= 108,000 + 8,100 + 2,952 = 119,052 \text{ in-lb}$$

$$f_b = \frac{M}{S_y} = \frac{119,052}{83.3} = 1,429 \text{ lb/in}^2 \; < 1,540 \text{ lb/in}^2$$

The overhang is satisfactory in bending.

Horizontal shear in the overhang is based on the maximum vertical shear occurring a distance from the beam centerline equal to one-fourth the beam width plus the deck thickness. Loads acting within this distance from the beam centerline are neglected. The distributed wheel load is equal to the wheel load divided by the distribution width:

$$\text{Distributed wheel load} = \frac{12,000 \text{ lb}}{17.32 \text{ in.}} = 692.8 \text{ lb/in}$$

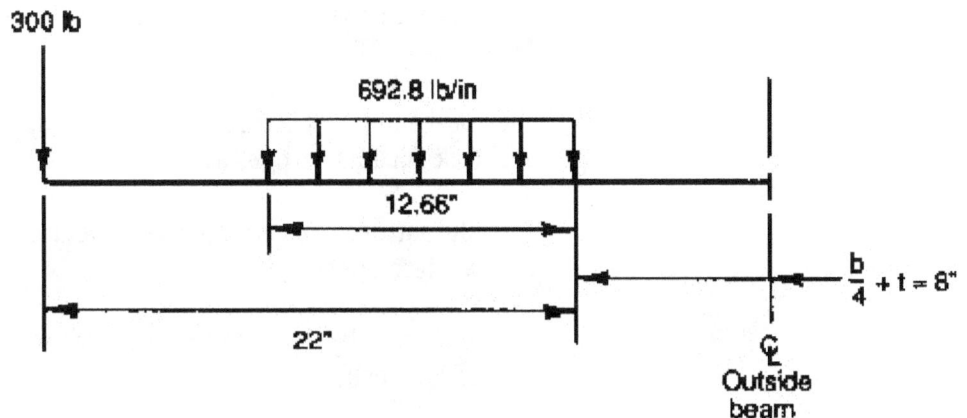

$$V_{LL} = (12.66 \text{ in.})(692.8 \text{ lb/in}) = 8,771 \text{ lb}$$

$$\text{Rail } V_{DL} = 300 \text{ lb}$$

$$\text{Deck } V_{DL} = (8.1 \text{ lb/in})(22 \text{ in.}) = 178 \text{ lb}$$

$$V = V_{LL} + V_{DL} = 8,771 + 300 + 178 = 9,249 \text{ lb}$$

$$f_v = \frac{1.5V}{A_v} = \frac{1.5(9,249)}{125} = 111 \text{ lb/in}^2 \; < 153 \text{ lb/in}^2$$

The overhang is satisfactory.

Summary

The deck will consist of 5-inch-thick noninterconnected glulam panels, 25 feet long. A total of 25 panels are required: 21 panels that are 46-3/4 inches wide and 4 panels that are 41-1/4 inches wide (one end panel will be trimmed 3/4 inch before treatment). Deck panels will be manufactured from visually graded Southern Pine, combination symbol No. 47. Stresses and deflection are as follows:

	Center spans	Overhang
f_b	1,297 lb/in^2	1,429 lb/in^2
F_b'	1,540 lb/in^2	1,540 lb/in^2
Δ_{LL}	0.11 in.	—
f_v	109 lb/in^2	111 lb/in^2
F_v'	153 lb/in^2	153 lb/in^2

Example 7-6 - Noninterconnected glulam deck; single-lane with overload

---

A glulam beam superstructure has a 14-foot-wide roadway that carries one lane of AASHTO H 20-44 loading with a U80 overload. Support is provided by three 8-1/2-inch-wide glulam beams that are spaced 5 feet 6 inches on center and are 52 feet long. Design a noninterconnected glulam deck for this bridge, assuming

1. glulam deck panels are manufactured from visually graded western species;

2. a 12-inch by 12-inch lumber curb is continuous along each edge of the deck;

3. the deck is covered with a rough-sawn, 4-inch-thick plank wearing surface; and

4. deck live load deflection is limited to approximately 0.10 inch for H 20-44 loads. No deflection criteria apply to the U80 overload.

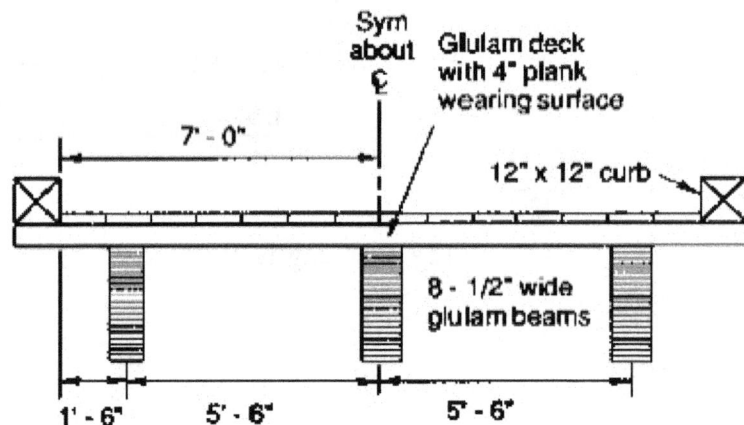

Solution

Determine the Deck Span, Design Loads, and Panel Size

Clear distance between beams = 66 in - 8.50 in = 57.50 in

$$s = 57.50 \text{ in} + \frac{8.50}{2} = 61.75 \text{ in.}$$

For H 20-44 loading, AASHTO special provisions apply and a 12,000-pound wheel load is used for design. From Figure 6-5, the U80 wheel load weight is 18,500 pounds.

Panel width for an out-to-out bridge length of 52 feet will be 48 inches. End panels and alternating interior panels will be 16 feet long for curb attachment. Other panels will be 14 feet long:

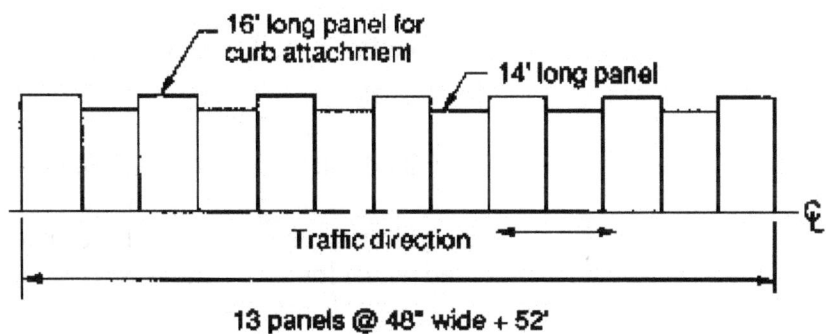

Estimate Deck Thickness

An initial deck thickness of 6-3/4 inches will be used. Although it is anticipated that U80 loading will control design, stresses will be computed for both vehicles for future reference.

Determine Wheel Distribution Widths and Effective Deck Section Properties

In the direction of the deck span,

$$\text{H 20-44 } b_t = \sqrt{0.025P} = \sqrt{0.025(12,000)} = 17.32 \text{ in.}$$

$$\text{U80 } b_t = \sqrt{0.025(18,500)} = 21.51 \text{ in.}$$

Normal to the deck span,

$$b_d = t + 15 = 6.75 + 15 = 21.75 \text{ in.}$$

Effective deck section properties from Table 7-6 are

$$A = 146.81 \text{ in}^2$$

$$S_y = 165.16 \text{ in}^3$$

$$I_y = 557.43 \text{ in}^4$$

Determine Deck Dead Load

For a 6-3/4-inch deck and a 4-inch plank wearing surface, dead load unit weight and moment over the effective distribution width of 21.75 inches are computed as follows:

$$w_{DL} = 21.75 \text{ in.} \frac{\left[(6.75 + 4 \text{ in.})(50 \text{ lb/in}^3)\right]}{1,728 \text{ in}^3/\text{ft}^3} = 6.8 \text{ lb/in.}$$

$$M_{DL} = \frac{w_{DL} s^2}{8} = \frac{6.8 (61.75)^2}{8} = 3,241 \text{ in-lb}$$

Determine Live Load Moment

By Equation 7-16,

$$\text{H 20-44 } M_{LL} = 3,000s - 25,983 = 3,000(61.75) - 25,983 = 159,267 \text{ in-lb}$$

U80 moment is computed at the span centerline by centering the distributed wheel load.

$$\text{U80 distributed wheel load} = \frac{18,500 \text{ lb}}{21.51 \text{ in.}} = 860.1 \text{ lb/in.}$$

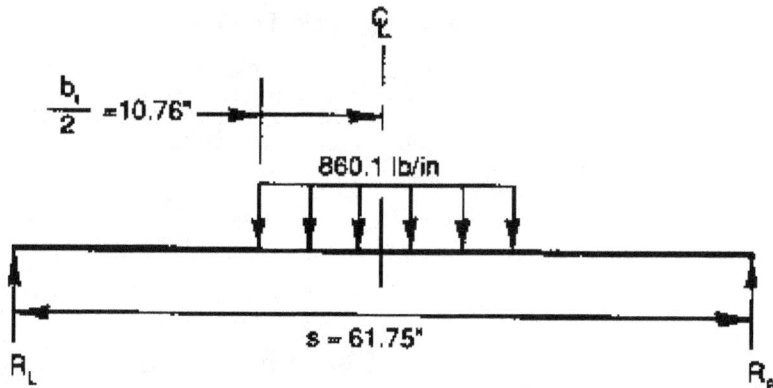

$$R_L = \frac{18,500 \text{ lb}}{2} = 9,250 \text{ lb}$$

$$\text{U80 } M_{LL} = \left[(9,250 \text{ lb})\frac{(61.75 \text{ in.})}{2}\right] - \left[(9,250 \text{ lb})\frac{(10.76 \text{ in.})}{2}\right]$$
$$= 235,829 \text{ in-lb}$$

Dividing U80 moment by the allowable overload increase of 1.33 and comparing the value to H 20-44 moment indicates the controlling vehicle:

$$\frac{235,829 \text{ in-lb}}{1.33} = 177,315 > 159,267$$

so the U80 will control bending.

Compute Bending Stress and Select a Deck Combination Symbol

The deck is continuous over two spans, so the reduction in bending stress for continuity is not applicable.

$$\text{U80 } M = M_{DL} + M_{LL} = 3,241 + 235,829 = 239,070 \text{ in-lb}$$

$$f_b = \frac{M}{S_y} = \frac{239,070}{165.16} = 1,448 \text{ lb/in}^2$$

From Table 7-7 for an approximate $F_b' = 1,448/1.33 = 1,089$ lb/in^2, an F_{by} of 1,450 lb/in^2 is the closest value for a western species combination symbol. An initial combination symbol No. 1 is selected, and the following values are obtained from *AITC 117-Design*:

$F_{by} = 1,450$ lb/in^2 $\qquad$ $C_M = 0.80$

$F_{vy} = 145$ lb/in^2 $\qquad$ $C_M = 0.875$

$E_y = 1,500,000$ lb/in^2 $\qquad$ $C_M = 0.833$

By Equation 7-20,

$$\text{U80 } F_b' = F_{by}(1.33)C_F C_M = 1,450(1.33)(1.07)(0.80) = 1,651 \text{ lb/in}^2$$

$f_b = 1,448$ lb/in$^2 < F_b' = 1,651$ lb/in^2, so a 6-3/4-inch combination symbol No. 1 panel is satisfactory in bending.

Check H 20-44 loading:

$$M = M_{DL} + M_{LL} = 3,241 + 159,267 = 162,508 \text{ in-lb}$$

$$f_b = \frac{M}{S_y} = \frac{162,508}{165.16} = 984 \text{ lb/in}^2$$

$$F_b' = F_{by}C_F C_M = 1,450(1.07)(0.80) = 1,241 \text{ lb/in}^2$$

$f_b = 984$ lb/in$^2 < F_b' = 1,241$ lb/in^2, so the deck is satisfactory for H 20-44 loading.

Check Live Load Deflection

Maximum H 20-44 deflection for a panel continuous over two spans is computed by Equation 7-21:

$$E' = E_y C_M = 1,500,000(0.833) = 1,249,500 \text{ lb/in}^2$$

$$\Delta_{LL} = \frac{1.80}{E' I_y}\left(138.8s^3 - 20,780s + 90,000\right)$$

$$\Delta_{LL} = \frac{1.80\left[(138.8)(61.75)^3 - 20,780(61.75) + 90,000\right]}{(1,249,500)(557.43)} = 0.08 \text{ in.}$$

0.08 inch is less than 0.10 inch, so deck deflection is acceptable.

Check Horizontal Shear

Dead load vertical shear is computed by Equation 7-23 for $w_{DL} = 6.8$ lb/in.:

$$V_{DL} = w_{DL}\left(\frac{s}{2} - t\right) = 6.8\left(\frac{61.8}{2} - 6.75\right) = 164.1 \text{ lb}$$

Live load vertical shear is computed by placing the edge of the wheel load distribution width (b_t) a distance t from the support.

For U80 loading,

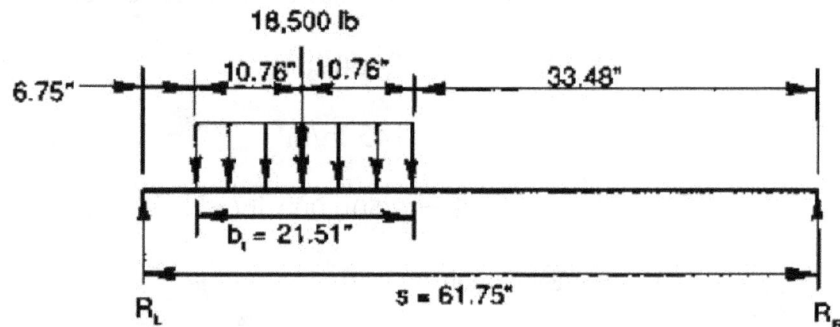

$$V_{LL} = R_L = \frac{(18,500 \text{ lb})(10.76 \text{ in.} + 33.48 \text{ in.})}{61.75 \text{ in.}} = 13,254 \text{ lb}$$

$$V = V_{DL} + V_{LL} = 164.1 + 13,254 = 13,418 \text{ lb}$$

$$A_v = t(15 + 2t) = 6.75[15 + (2)(6.75)] = 192.38 \text{ in}^2$$

$$f_v = \frac{1.5V}{A_v} = \frac{1.5(13,418)}{192.38} = 105 \text{ lb/in}^2$$

$$F_v' = F_{vy}(1.33)C_M = 145(1.33)(0.875) = 169 \text{ lb/in}^2 > 105 \text{ lb/in}^2$$

For H 20-44 loading,

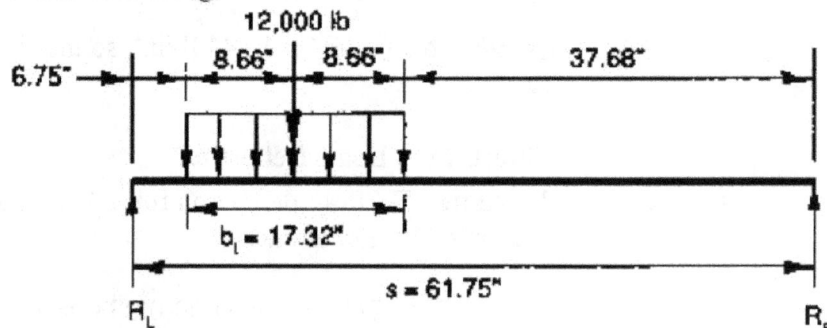

$$V_{LL} = R_L = \frac{12,000 \text{ lb}(8.66 \text{ in.} + 37.68 \text{ in.})}{61.75 \text{ in.}} = 9,005 \text{ lb}$$

$$V = V_{DL} + V_{LL} = 164.1 + 9,005 = 9,169 \text{ lb}$$

$$f_v = \frac{1.5V}{A_v} = \frac{1.5(9,169)}{192.38} = 71 \text{ lb/in}^2$$

$$F_v' = F_{vy}C_M = 145(0.875) = 127 \text{ lb/in}^2 > 71 \text{ lb/in}^2$$

The panel is satisfactory in horizontal shear for both vehicles.

Check Overhang

Bending and shear stresses are checked in the deck overhang by positioning the wheel load centerline 1 foot from the curb face, which is 6 inches from the outside beam centerline. Moments are computed using an effective span measured from the load to the beam centerline, minus one-fourth the beam width. Horizontal shear is based on the maximum vertical shear occurring a distance from the beam centerline equal to one-fourth the beam width plus the deck thickness. Loads acting within this distance from the beam centerline are neglected for shear.

Dead load of the 12- by 12-inch curb, and the distributed dead load of the deck and wearing surface, is computed for the U80 distribution width $b_d = 21.75$ in:

$$\text{Curb DL} = \frac{21.75 \text{ in.}}{12 \text{ in./ft}}(50 \text{ lb/ft}) = 90.6 \text{ lb}$$

$$\text{Deck } w_{DL} = (21.75 \text{ in.})\left[\frac{(6.75 \text{ in.})(50 \text{ lb/ft}^3)}{1,728 \text{ in}^3/\text{ft}^3}\right] = 4.3 \text{ lb/in.}$$

$$\text{Wearing surface } w_{DL} = (21.75 \text{ in.})\left[\frac{(4 \text{ in.})(50 \text{ lb/ft}^3)}{1,728 \text{ in}^3/\text{ft}^3}\right] = 2.5 \text{ lb/in.}$$

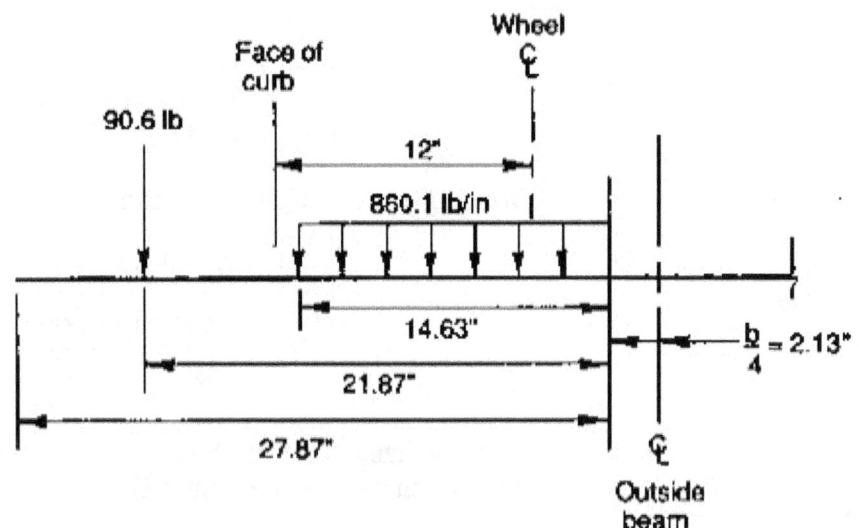

Summing moments at a point $b/4 = 2.13$ inches from the outside beam centerline,

$$M_{LL} = (14.63 \text{ in.})(860.1 \text{ lb/in.})\left(\frac{14.63 \text{ in.}}{2}\right) = 92,047 \text{ in-lb}$$

$$\text{Curb } M_{DL} = (21.87 \text{ in.})(90.6 \text{ lb}) = 1,981 \text{ in-lb}$$

$$\text{Deck } M_{DL} = (4.3 \text{ lb/in.})(27.87 \text{ in.})\left(\frac{27.87 \text{ in.}}{2}\right) = 1,670 \text{ in-lb}$$

$$\text{Wearing surface } M_{DL} = (2.5 \text{ lb/in.})(15.87 \text{ in.})\left(\frac{15.87 \text{ in.}}{2}\right) = 314.8 \text{ in-lb}$$

$$M = M_{LL} + M_{DL} = 92,047 + 1,981 + 1,670 + 314.8$$
$$= 96,013 \text{ in-lb}$$

$$f_b = \frac{M}{S_y} = \frac{96,013}{165.16} = 581 \text{ lb/in}^2 < \text{U80} F_b' = 1,651 \text{ lb/in}^2$$

Horizontal shear is computed at a distance of $b/4 + t = 8.88$ inches from the outside beam centerline:

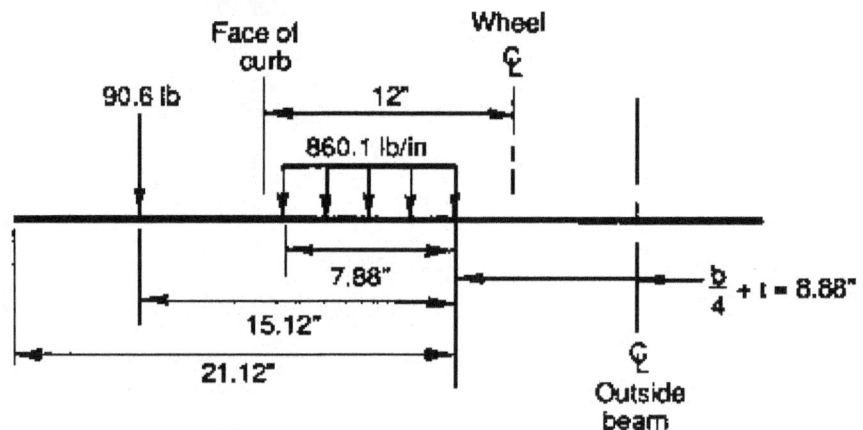

$$V_{LL} = (7.88 \text{ in.})(860.1 \text{ lb/in.}) = 6,778 \text{ lb}$$

$$\text{Curb } V_{DL} = 90.6 \text{ lb}$$

$$\text{Deck } V_{DL} = (4.3 \text{ lb/in.})(21.12 \text{ in.}) = 90.8 \text{ lb}$$

$$\text{Wearing surface } V_{DL} = (2.5 \text{ lb/in.})(9.12 \text{ in.}) = 22.8 \text{ lb}$$

$$V = V_{LL} + V_{DL} = 6,778 + 90.6 + 90.8 + 22.8 = 6,982 \text{ lb}$$

$$f_v = \frac{1.5V}{A_v} = \frac{1.5(6,982)}{192.38} = 54 \text{ lb/in}^2 < \text{U80} F_v' = 169 \text{ lb/in.}$$

The overhang is satisfactory for U80 loading with low stress levels. Further checks for the lighter H 20-44 loading are not required.

Summary

The deck will consist of 6-3/4-inch noninterconnected glulam panels that are 48 inches wide. A total of 13 panels are required: 7 panels 16 feet long and 6 panels 14 feet long. Panels will be manufactured from visually graded western species combination symbol No. 1. Stresses and deflections are as follows:

	H 20-44 loading	U80 loading
Center spans		
f_b	984 lb/in²	1,448 lb/in²
F_b'	1,241 lb/in²	1,651 lb/in²
Δ_{LL}	0.08 in.	—
f_v	71 lb/in²	105 lb/in²
F_v'	127 lb/in²	169 lb/in²
Overhang		
f_b	< U80	581 lb/in²
F_b'	< U80	1,651 lb/in²
f_v	< U80	54 lb/in²
F_v'	< U80	169 lb/in²

DOWELED GLULAM DECKS   Doweled glulam decks consist of a series of glulam deck panels interconnected at the panel joints with steel dowels (Figure 7-20). The dowels transfer loads between panels and reduce relative displacements and rotations between adjacent panels. As a result, doweled decks generally have lower live load deflections and may result in longer deck spans or thinner panels than noninterconnected decks. These advantages can be significant in some cases but may not be sufficient to offset the increased costs required for dowel installation.

The suitability of a doweled deck for a specific application depends on the design requirements of the structure and the economics of fabrication and construction. Doweled panels are more expensive than noninterconnected decks because they require precise fabrication for proper installation and performance. As a general rule, they are most practical when an asphalt wearing surface is used and the deflection at the panel joints must be limited to prevent cracking. However, it may be more cost effective to use a noninterconnected deck and limit deflections by using a thicker deck or decreased deck span. When paving is not planned, noninterconnected panels will generally provide the most economical deck.

Design Procedures

Doweled deck design is basically a two-part process involving separate criteria for the glulam panels and interconnecting dowels. First, the glulam panels are designed for the primary moment, shear, and deflection acting between beams in the x direction, parallel to the length of the laminations

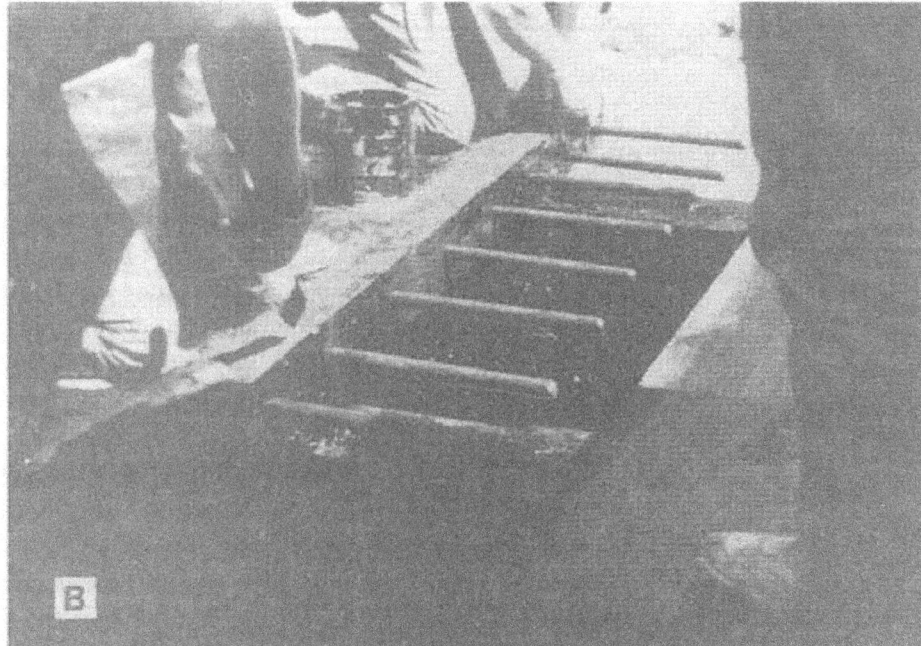

Figure 7-20 - Construction of a doweled glulam deck. The panels are (A) lifted into position and (B) interconnected with steel dowels (photos courtesy of Steve Bunnell, USDA Forest Service).

(Figure 7-21). These strength computations are based on the maximum unit stress acting in the panels. Second, the size and spacing of the dowels are determined from the average secondary moment and shear acting parallel to the supporting beams in the y direction, perpendicular to the length of the laminations. These computations assume that the dowels provide deck continuity for the length of the bridge.

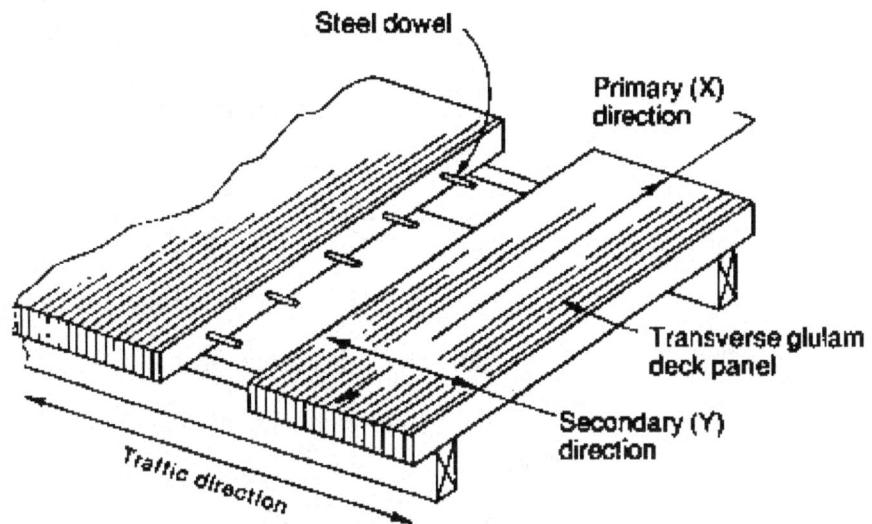

Figure 7-21. - *Primary and secondary directions for doweled glulam deck panels.*

Basic design procedures for doweled glulam decks are given below in a sequential order used for most design applications. The procedures were adopted by AASHTO in 1975 based on research conducted at the USDA Forest Service, Forest Products Laboratory.[29,30] They are based on experimental and analytical analyses of the deck as an orthotropic plate, acting as a simple span between two supports. The procedures were developed for single wheel loads of 12,000 pounds and 16,000 pounds and are valid for effective spans of 122 inches or less for standard track widths of 6 feet.

1. Define the deck span, design loads, and panel size.

The effective deck span, s, is the clear distance between supporting beams plus one-half the width of one beam, but not greater than the clear span plus the panel thickness (AASHTO 3.25.1.2). The maximum effective span for doweled decks designed by these procedures is 122 inches. Panel configuration should be based on the same considerations previously discussed for noninterconnected glulam decks.

The design load for doweled decks is the maximum wheel load of the design vehicle. Special AASHTO provisions for HS 20-44 and H 20-44 loads on timber decks *do not* apply to doweled decks designed in accordance with these procedures. Wheel loads for standard AASHTO trucks are 16,000 pounds for HS 20-44 and H 20-44, and 12,000 pounds for HS 15-44 and H 15-44.

2. Estimate deck thickness.

Deck thickness, t, must be estimated for initial calculations. Use a minimum thickness of 5-1/8 inches (5 inches for Southern Pine) for HS 15-44 and H 15-44 loads (12,000-pound wheel load) and 6-3/4 inches for HS 20-44 and H 20-44 loads (16,000-pound wheel load).

3. Compute the primary dead load moment and vertical shear.

Dead load moment and shear are based on the unit dead load, DL, of the deck and wearing surface, including allowance for future wearing surface overlays. Primary dead load moment is computed at the effective span centerline by

$$M_{DL} = \frac{DL\,s^2}{1152} \tag{7-26}$$

where M_{DL} = primary dead load moment (in-lb/in), and

$\quad\quad$ DL = dead load of the deck and wearing surface (lb/ft²).

Primary dead load vertical shear is computed at a distance t from the support by

$$R_{DL} = \frac{DL}{144}\left(\frac{s}{2} - t\right) \tag{7-27}$$

where R_{DL} = primary dead load vertical shear (in-lb/in).

4. Determine primary live load moment and vertical shear.

Primary live load moment and vertical shear are computed directly, assuming the deck to act as a simple span between supporting beams (AASHTO 3.25.1.3):

$$M_x = P\left[(0.51 \log_{10} s) - K\right] \tag{7-28}$$

$$R_x = 0.034P \tag{7-29}$$

where $\quad$ M_x = primary live load bending moment (in-lb/in),

$\quad\quad$ P = design wheel load (lb),

$\quad\quad$ K = design constant based on the wheel load contact area, and

$\quad\quad$ R_x = primary live load vertical shear (lb/in).

Design values for P, K, and R_x for standard highway loads are given in Table 7-10.

Table 7-10. - Design values for primary live load moment and shear for doweled glulam deck panels.

Vehicle Type	P (lb)	K*	R_x (lb/in)
HS 20-44 and H 20-44	16,000	0.51	544
HS 15-44 and H 15-44	12,000	0.47	408

* For wheel loads greater than 16,000 pounds, K = 0.51 may be used with slightly conservative results.

5. Select a panel combination symbol and compute allowable stresses.

Select an axial combination symbol from Table 2 of *AITC 117--Design* based on the same selection criteria given for noninterconnected panels. Compute allowable stresses for bending and horizontal shear by adjusting tabulated values by all applicable modification factors:

$$F_b' = F_{by} C_F C_M \tag{7-30}$$

$$F_v' = F_v C_M \tag{7-31}$$

F_b' and F_v' may be increased by a factor of 1.33 for overloads in AASHTO Load Group IB.

6. Compute required deck thickness.

Deck thickness is based on the most restrictive requirements for primary moment or horizontal shear, but the nominal deck thickness cannot be less than 6 inches (actual thickness of 5-1/8 inches for western species or 5 inches for Southern Pine) (AASHTO 3.25.1.1). The minimum required deck thickness is obtained from

$$t = \sqrt{\frac{6(M_x + M_{DLx})}{F_b'}} \tag{7-32}$$

$$t = \frac{3(R_x + R_{DLx})}{2F_v'} \tag{7-33}$$

whichever is the largest (AASHTO 3.25.1.3).

When the deck is continuous over more than two spans, M_{DLx} and M_x used in Equation 7-32 are 80 percent of the simple-span values computed by Equations 7-26 and 7-28 to account for the effects of span continuity.

The required deck thickness may be computed for several combination symbols to obtain the most economical panel. When the required deck thickness varies significantly from the estimated thickness, dead load moment, M_{DLx}, and vertical shear, R_{DLx}, must be revised.

7. Check live load deflection.

Maximum live load deflection in the primary direction is computed by

$$\Delta_{LL} = \frac{0.51 Ps(s-10)}{E't^3} \qquad (7\text{-}34)$$

where $E' = E_y C_M$.

When the deck is continuous over more than two spans, the live load deflection is 80 percent of the deflection computed by Equation 7-34 to account for span continuity.

The recommended deflection limits for doweled glulam decks are the same as those previously discussed for noninterconnected glulam decks. Maximum effective deck spans based on an allowable deck deflection of 0.10 inch are given in Table 7-11 for decks continuous over more than two spans.

Table 7-11. - Approximate maximum effective span for doweled transverse glulam deck panels based on a maximum vehicle live load deflection of 0.10 inch; deck continuous across more than two spans.

		Approximate maximum deck span (in.)		
E (lb/in²)	E'(lb/in²)	$t = 5$ in. or $t = 5\text{-}1/8$ in.	$t = 6\text{-}3/4$ in.	$t = 8\text{-}1/2$ in. or $t = 8\text{-}3/4$ in.
		12,000-lb wheel load		
1,300,000	1,082,900	58	88	>110
1,400,000	1,166,200	60	91	>110
1,500,000	1,249,500	64	94	>110
1,700,000	1,416,100	68	100	>110
1,800,000	1,499,400	70	103	>110
		16,000-lb wheel load		
1,300,000	1,082,900	51	77	>110
1,400,000	1,166,200	53	80	>110
1,500,000	1,249,500	57	82	>110
1,700,000	1,416,100	60	87	>110
1,800,000	1,499,400	61	90	>110

$E' = EC_M = 0.833 E.$

8. Compute secondary moment and shear.

Requirements for the number and size of dowels are based on the secondary live load moment and shear (AASHTO 3.25.1.4). Equations for computing these values depend on the effective deck span, s.

When the effective deck span is less than or equal to 50 inches ($s \leq 50$),

$$M_y = \frac{Ps}{1,600}(s - 10) \tag{7-35}$$

$$R_y = \frac{6Ps}{1,000} \tag{7-36}$$

where M_y = secondary live load moment (in-lb), and

   R_y = secondary live load shear (lb).

When the effective deck span is more than 50 inches ($s > 50$),

$$M_y = \frac{Ps}{20}\frac{(s - 30)}{(s - 10)} \tag{7-37}$$

$$R_y = \frac{P}{2s}(s - 20) \tag{7-38}$$

9. Determine required size and spacing of steel dowels.

The number of dowels required for each deck span is based on the dowel diameter and properties given in Table 7-12. Select a dowel diameter and compute the required number of dowels using

$$n = \frac{1,000}{\sigma_{PL}}\left(\frac{R_y}{R_D} + \frac{M_y}{M_D}\right) \tag{7-39}$$

where n = number of steel dowels required for each deck span,

   σ_{PL} = proportional limit stress for timber, perpendicular to grain (1,000 lb/in^2 for Douglas Fir-Larch and Southern Pine),

   R_D = dowel shear capacity from Table 7-12 (lb), and

   M_D = dowel moment capacity from Table 7-12 (in-lb).

The required number of dowels from Equation 7-39 is given for standard AASHTO highway loads in Figure 7-22. Dowel placement is shown in Figure 7-23.

Table 7-12. - Properties and required lengths of steel dowels for doweled glulam deck panels.

Dowel diameter (in.)	Shear capacity R_D (lb)	Moment capacity M_D (in-lb)	Steel stress coefficients		Required dowel length (in.)
			C_R (1/in²)	C_m (1/in³)	
1/2	600	850	36.9	81.5	8.5
5/8	800	1,340	22.3	41.7	10.0
3/4	1,020	1,960	14.8	24.1	11.5
7/8	1,260	2,720	10.5	15.2	13.0
1	1,520	3,630	7.75	10.2	14.5
1-1/8	1,790	4,680	5.94	7.15	15.5
1-1/4	2,100	5,950	4.69	5.22	17.0
1-3/8	2,420	7,360	3.78	3.92	18.0
1-1/2	2,770	8,990	3.11	3.02	19.5

10. Check dowel stress.

Applied stress in the steel dowels must not exceed the allowable stress computed by

$$\sigma_A = 0.8 F_y \tag{7-40}$$

$$\sigma = \frac{1}{n}\left(C_R R_y + C_m M_y\right) \tag{7-41}$$

where σ_A = allowable steel stress in bending (AASHTO Table 10.32.1A) (lb/in²),

σ = dowel stress from applied loads (lb/in²),

F_y = minimum specified yield point of the steel dowels (lb/in²), and

C_R, C_m = steel stress coefficients from Table 7-12.

When $\sigma > \sigma_A$, stress in the steel dowels exceeds allowable values and the dowel diameter must be increased.

11. Check deck overhang.

There are no analysis criteria given in AASHTO for checking dowel deck stresses in the overhang at outside beams. Although slightly conservative, it is recommended that overhangs be checked using the same criteria previously discussed for noninterconnected decks, using an effective panel distribution width of 15 inches plus twice the deck thickness (15 + 2t).

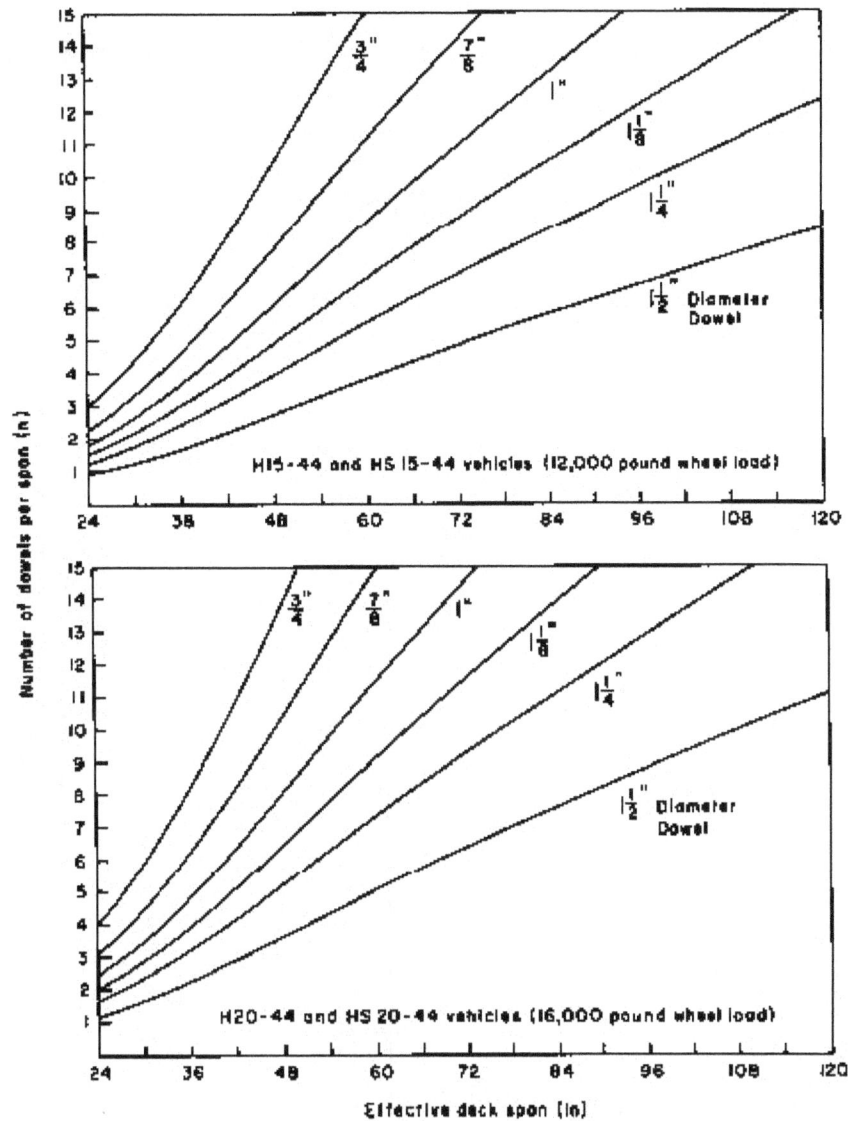

Figure 7-22. - Number of dowels required for each effective span of a doweled glulam deck.

Example 7-7 - Doweled glulam deck with highway loading

A glulam beam bridge spans 71 feet 6 inches out to out and carries two traffic lanes of HS 20-44 loading on a 28-foot-wide roadway. Support is provided by five 12- 1/4-inch-wide glulam beams spaced 6 feet on center, Design a doweled glulam deck for the beam superstructure, assuming

1. glulam deck panels are visually graded western species;

2. rail system dead load is 150 pounds at each post with a maximum post spacing of 6 feet;

3. the deck will be surfaced with 3 inches of asphalt (includes future overlay); and

4. deck live load deflection is limited to approximately 0.10 inch.

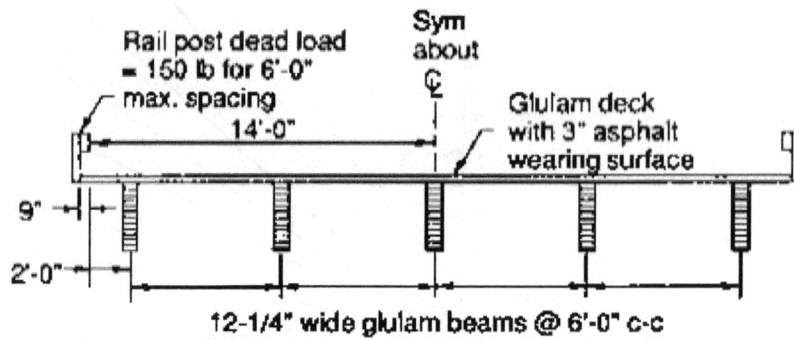

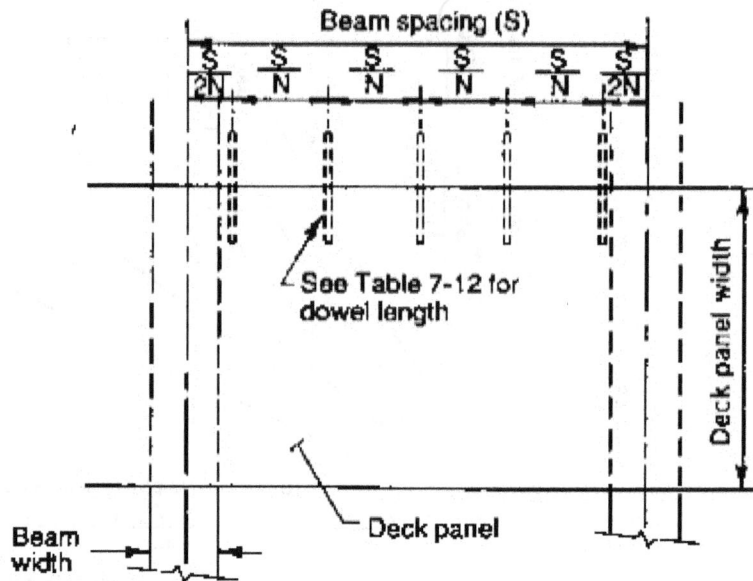

Rail post dead load
= 150 lb for 6'-0"
max. spacing

Sym about ₵

14'-0"

Glulam deck with 3" asphalt wearing surface

9"

2'-0"

12-1/4" wide glulam beams @ 6'-0" c-c

Beam spacing (S)

$\frac{S}{2N}$ $\frac{S}{N}$ $\frac{S}{N}$ $\frac{S}{N}$ $\frac{S}{N}$ $\frac{S}{2N}$

See Table 7-12 for dowel length

Deck panel width

Beam width

Deck panel

Panel top view

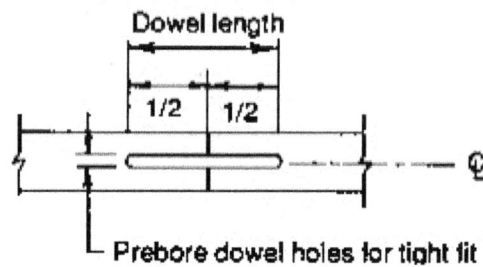

Dowel length

1/2   1/2

₵

Prebore dowel holes for tight fit

Panel section through dowel

Figure 7-23. - Dowel placement requirements for glulam deck panels.

Solution

Determine Deck Span, Design Loads, and Panel Size

Clear distance between beams = 72 in.- 12.25 in.= 59.75 in.

$$s = 59.75 \text{ in.} + \frac{12.25}{2} = 65.88 \text{ in.}$$

If a 5-1/8-inch deck is used, s will be limited by the clear span plus deck thickness to $59.75 + 5$-$1/8 = 64.88$ inches. For other deck thicknesses, $s = 65.88$ inches will control.

For HS 20-44 loading on doweled decks, AASHTO special wheel load provisions do not apply, and the deck will be designed for a 16,000-pound wheel load. Panel length will be increased 1-1/2 feet over the roadway width for curb/rail attachment. Panel width for an out-to-out bridge length of 71 feet 6 inches will be 66 inches with a railpost attachment centered on each panel (local availability of deck panels in this width may be limited by manufacturing or treating limitations and should be verified).

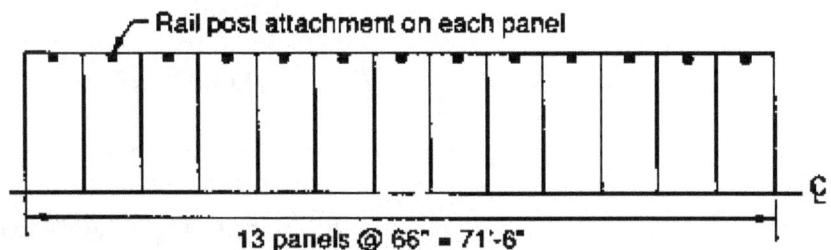

Rail post attachment on each panel

13 panels @ 66" = 71'-6"

Estimate Deck Thickness

For HS 20-44 loading, an initial panel thickness of 6-3/4 inches will be used. For this deck thickness, $s = 65.88$ inches.

Compute Primary Dead Load Moment and Vertical Shear

For a 6-3/4-inch deck and 3-inch asphalt wearing surface:

$$DL = \frac{6.75 \text{ in.}}{12 \text{ in.}/\text{ft}}\left(50 \text{ lb}/\text{ft}^3\right) + \frac{3 \text{ in.}}{12 \text{ in.}/\text{ft}}\left(150 \text{ lb}/\text{ft}^3\right) = 65.6 \text{ lb}/\text{ft}^2$$

By Equation (7-26),

$$M_{DLx} = \frac{DLs^2}{1,152} = \frac{65.6(65.88)^2}{1,152} = 247.2 \text{ in-lb}/\text{in.}$$

By Equation 7-27,

$$R_{DLx} = \frac{DL}{144}\left(\frac{s}{2} - t\right) = \frac{65.6}{144}\left(\frac{65.88}{2} - 6.75\right) = 11.9 \text{ lb}/\text{in.}$$

Determine Primary Live Load and Vertical Shear

From Table 7-10 for HS 20-44 loading, $P = 16,000$ lb and $K = 0.51$.

By Equation 7-28,

$$M_x = P\left[(0.51 \log_{10} s) - K\right] = 16,000\left[(0.51 \log_{10} 65.88) - 0.51\right]$$

$$= 6,681 \text{ in-lb/in}$$

By Equation 7-29 (or Table 7-10),

$$R_x = 0.034P = 0.034(16,000 \text{ lb}) = 544 \text{ lb/in}$$

Select a Panel Combination Symbol and Compute Allowable Stresses

From *AITC 117--Design*, combination symbol No. 1 is selected with the following tabulated values:

$$F_{by} = 1,450 \text{ lb/in}^2 \qquad C_M = 0.80$$

$$F_{vy} = 145 \text{ lb/in}^2 \qquad C_M = 0.875$$

$$E_y = 1,500,000 \text{ lb/in}^2 \qquad C_M = 0.833$$

Allowable stresses are computed:

$$F_b' = F_{by}C_F C_M = 1,450(1.07)(0.80) = 1,241 \text{ lb/in}^2$$

$$F_v' = F_{vy}C_M = 145(0.875) = 127 \text{ lb/in}^2$$

$$E' = E_y C_M = 1,500,000(0.833) = 1,249,500 \text{ lb/in}^2$$

In this case, the deck is continuous over more than two spans and 80 percent of the simple span moments are used to account for span continuity. Minimum required deck thickness based on bending is computed by Equation 7-32:

$$t = \sqrt{\frac{6\left(M_x + M_{DLx}\right)}{F_b'}} = \sqrt{\frac{6\,(0.80)\,(6,681 + 247.2)}{1,241}} = 5.2 \text{ in.}$$

Minimum required deck thickness based on shear is computed by Equation 7-33:

$$t = \frac{3\left(R_x + R_{DLx}\right)}{2F_v'} = \frac{3\,(544 + 11.94)}{2(127)} = 6.6 \text{ in.}$$

A 6-3/4-inch deck exceeds the minimum 6.6-inch thickness required for shear and is satisfactory.

Check Live Load Deflection

Because the deck is continuous over more than two spans, live load deflection is 80 percent of that computed by Equation 7-34:

$$\Delta_{LL} = (0.80)\frac{0.51 Ps(s-10)}{E't^3}$$

$$= (0.80)\frac{0.51(16,000)(65.88)(65.88-10)}{1,249,500(6.75)^3} = 0.06 \text{ in.}$$

The actual deflection of 0.06 inch is less than the maximum allowable of 0.10 inch, so deck deflection is acceptable.

Compute Secondary Moment and Shear

$s = 65.88$ in. > 50, so secondary moment and shear are computed by Equations 7-37 and 7-38, respectively:

$$M_y = \frac{Ps}{20}\frac{(s-30)}{(s-10)} = \frac{16,000(65.88)}{20}\frac{(65.88-30)}{(65.88-10)} = 33,841 \text{ in-lb}$$

$$R_y = \frac{P}{2s}(s-20) = \frac{16,000}{2(65.88)}(65.88-20) = 5,571 \text{ lb}$$

Determine the Required Size and Spacing of Steel Dowels

An estimated number of dowels for various dowel diameters is obtained from Figure 7-11. For an effective deck span of 65.88 inches, the required number of dowels for each deck span varies from approximately 13 for 1-inch-diameter dowels to 6 for 1-1/2-inch-diameter dowels. The 1-1/2-inch-diameter dowels are selected, and the required number of dowels is confirmed by Equation 7-39 based on the dowel shear and moment capacity given in Table 7-12:

$$n = \frac{1,000}{\sigma_{PL}}\left(\frac{R_y}{R_D} + \frac{M_y}{M_D}\right) = \frac{1,000}{1,000}\left(\frac{5,571}{2,770} + \frac{33,841}{8,990}\right) = 5.8 \text{ dowels}$$

Six 1-1/2-inch-diameter dowels per deck span is satisfactory.

From Table 7-12, a minimum dowel length of 19.5 inches is required. The dowel layout obtained from Figure 7-23 is as follows:

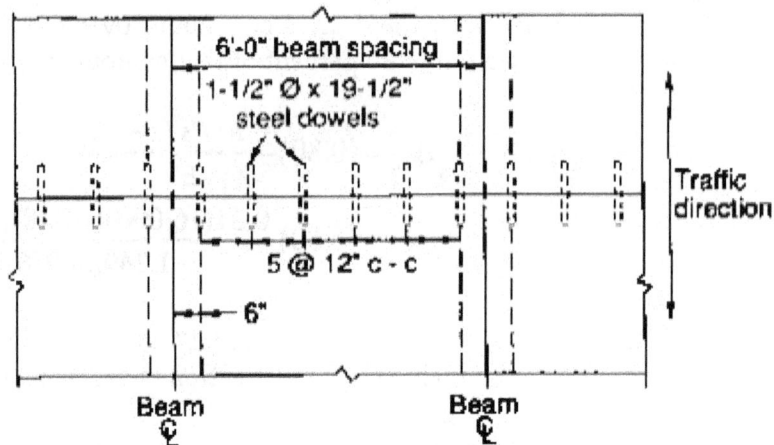

Check Dowel Stress

Assuming A36 steel dowels $(F_y = 36,000 \text{ lb/in}^2)$, allowable dowel stress is computed by Equation 7-40:

$$\sigma_A = 0.80 F_y = 0.80(36,000) = 29,000 \text{ lb/in}^2$$

Applied dowel stress is computed by Equation 7-41 based on previously computed values of R_y and M_y and coefficients given in Table 7-12:

$$\sigma = \frac{1}{n}\left(C_R R_y + C_m M_y\right) = \frac{1}{6}\left[3.11\,(5,571) + 3.02\,(33,841)\right]$$
$$= 19,921 \text{ lb/in}^2$$

$29,000 \text{ lb/in}^2 > 19,921 \text{ lb/in}^2$, so dowel stress is acceptable.

Check Overhang

Stresses in the deck overhang are checked in the same manner as for noninterconnected glulam decks, but an increased wheel load distribution for bending of 15 inches plus twice the deck thickness $(15 + 2t)$ is used for doweled decks. In this case, the deck is thicker and the distribution width greater than the deck overhang previously checked in Example 7-5. Refer to that example for procedures.

Summary

The deck will consist of 13 combination symbol No. 1 glulam panels that are 6-3/4 inches thick, 66 inches wide and 29-1/2 feet long. Panels will be interconnected with 1-1/2-inch-diameter A36 steel dowels, 19-1/2 inches long. The dowels will be spaced 12 inches on center along the deck panel edges.

Example 7-8 - Doweled glulam deck with highway loading

An old steel truss is structurally deficient and will be rehabilitated for HS 15-44 loads. As part of the rehabilitation, the existing concrete deck will be removed and replaced with transverse doweled glulam panels. The bridge is 74 feet 3 inches long (out to out) and carries two traffic lanes on a roadway width of approximately 23 feet. Deck support is provided by six steel beams with 7-inch flange widths, spaced 4-1/2 feet on center. Design a doweled glulam deck for this structure, assuming

1. glulam deck panels are manufactured from visually graded Southern Pine;

2. the deck will be surfaced with 3 inches of asphalt (includes future overlay); and

3. deck live load deflection is limited to approximately 0.10 inch.

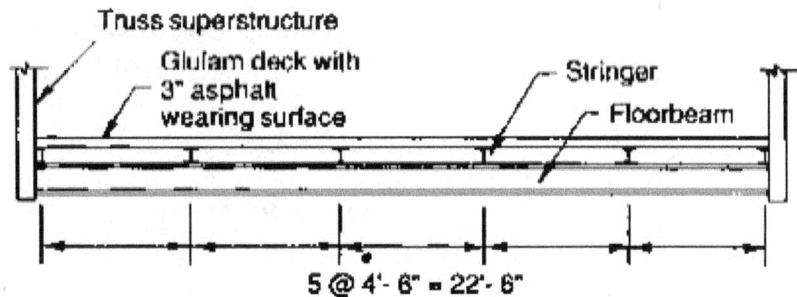

Solution

Determine Deck Span, Design Loads, and Panel Size

Clear distance between beams = 54 in. - 7 in. = 47 in.

$$s = 47 \text{ in.} + \frac{7}{2} = 50.50 \text{ in.}$$

For HS 15-44 loads, the deck will be designed for a 12,000-pound wheel load. Panel width for an out-to-out bridge length of 74 feet 3 inches will be 49-1/2 inches. Panel length will equal the roadway width of 23 feet.

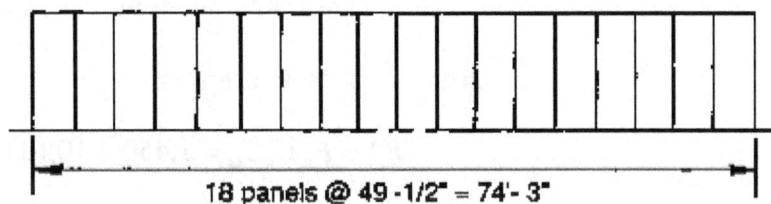

18 panels @ 49-1/2" = 74'-3"

Estimate Deck Thickness

For HS 15-44 loading, an initial panel thickness of 5 inches is selected.

Compute Primary Dead Load Moment and Vertical Shear

For a 5-inch deck and 3-inch asphalt wearing surface,

$$DL = \frac{5 \text{ in.}}{12 \text{ in.} / \text{ft}} \left(50 \text{ lb}/\text{ft}^3\right) + \frac{3 \text{ in.}}{12 \text{ in.} / \text{ft}} \left(150 \text{ lb}/\text{ft}^3\right) = 58.3 \text{ lb}/\text{ft}^2$$

By Equation 7-26,

$$M_{DLx} = \frac{DLs^2}{1152} = \frac{58.3(50.50)^2}{1152} = 129.1 \text{ in-lb/in.}$$

By Equation 7-27,

$$R_{DLx} = \frac{DL}{144}\left(\frac{s}{2} - t\right) = \frac{58.3}{144}\left(\frac{50.50}{2} - 5\right) = 8.2 \text{ lb/in.}$$

Determine Primary Live Load and Vertical Shear

From Table 7-10 for HS 15-44 loading, $P = 12{,}000$ pounds and $K = 0.47$.

By Equation 7-28,

$$M_x = P\left[(0.51 \log_{10} s) - K\right] = 12{,}000\left[(0.51 \log_{10} 50.50) - 0.47\right]$$

$$= 4{,}784 \text{ in-lb/in.}$$

By Equation 7-29 (or Table 7-10),

$$R_x = 0.034\, P = 0.034(12{,}000 \text{ lb}) = 408 \text{ lb/in.}$$

Select a Panel Combination Symbol and Compute Allowable Stresses

From *AITC 117--Design*, combination symbol No. 46 is selected with the following tabulated values:

$$F_{by} = 1{,}450 \text{ lb/in}^2 \qquad\qquad C_M = 0.80$$

$$F_{vy} = 175 \text{ lb/in}^2 \qquad\qquad C_M = 0.875$$

$$E_y = 1{,}300{,}000 \text{ lb/in}^2 \qquad\qquad C_M = 0.833$$

Allowable stresses are computed:

$$F_b' = F_{by} C_F C_M = 1{,}450(1.10)(0.80) = 1{,}276 \text{ lb/in}^2$$

$$F_v' = F_{vy} C_M = 175(0.875) = 153 \text{ lb/in}^2$$

7-88

$$E' = E_y C_M = 1,300,000(0.833) = 1,082,900 \text{ lb/in}^2$$

Using 80 percent of the simple span moments, minimum required deck thickness based on bending is computed by Equation 7-32:

$$t = \sqrt{\frac{6\left(M_x + M_{DLx}\right)}{F_b'}} = \sqrt{\frac{6(0.80)(4,784 + 129.1)}{1,276}} = 4.3 \text{ in.}$$

Minimum required deck thickness based on shear is computed by Equation 7-33:

$$t = \frac{3\left(R_x + R_{DLx}\right)}{2F_v'} = \frac{3(408 + 8.2)}{2(153)} = 4.1 \text{ in.}$$

A 5-inch deck meets minimum deck thickness requirements for moment and shear.

Check Live Load Deflection

Live load deflection is 80 percent of that computed by Equation 7-34 to account for span continuity:

$$\Delta_{LL} = (0.80)\frac{0.51 Ps(s - 10)}{E't^3}$$

$$= (0.80)\frac{0.51(12,000)(50.50)(50.50 - 10)}{1,082,900\,(5)^3} = 0.07 \text{ in.}$$

Deck deflection is less than the maximum allowable of 0.10 inch.

Compute Secondary Moment and Shear

$s = 50.50$ inches > 50, so secondary moment and shear are computed by Equations 7-37 and 7-38, respectively:

$$M_y = \frac{Ps}{20}\frac{(s - 30)}{(s - 10)} = \frac{12,000\,(50.50)}{20}\frac{(50.50 - 30)}{(50.50 - 10)} = 15,337 \text{ in-lb}$$

$$R_y = \frac{P}{2s}(s - 20) = \frac{12,000}{2(50.50)}(50.50 - 20) = 3,624 \text{ lb}$$

Determine the Required Size and Spacing of Steel Dowels

From Figure 7-22, a 1-inch dowel diameter is selected. The required number of dowels is computed by Equation 7-39:

$$n = \frac{1,000}{\sigma_{PL}}\left(\frac{R_y}{R_D} + \frac{M_y}{M_D}\right) = \frac{1,000}{1,000}\left(\frac{3,624}{1,520} + \frac{15,337}{3,630}\right) = 6.6 \text{ dowels}$$

Seven dowels 14.5 inches long will be used for each deck span. Spacing from Figure 7-23 is slightly adjusted to the closest 1/4 inch:

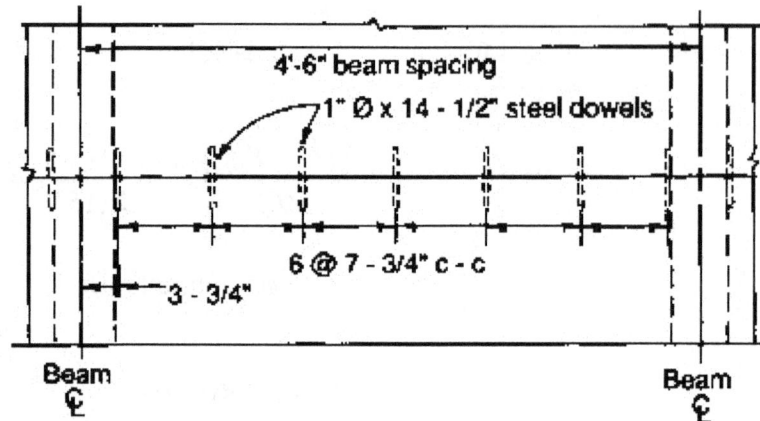

Check Dowel Stress

For A36 steel dowels, allowable dowel stress is computed by Equation 7-40:

$$\sigma_A = 0.80F_y = 0.80(36,000) = 29,000 \text{ lb/in}^2$$

Applied dowel stresses are computed by Equation 7-41:

$$\sigma = \frac{1}{n}\left(C_R R_y + C_m M_y\right) = \frac{1}{7}\left[7.75\,(3,364) + 10.20\,(15,337)\right]$$
$$= 26,360 \text{ lb/in}^2$$

29,000 lb/in^2 > 26,360 lb/in^2, so dowel stress is acceptable.

Summary

The deck will consist of 18 glulam deck panels that are 5 inches thick, 49-1/2 inches wide, and 23 feet long. Panels will be manufactured from visually graded Southern Pine, combination symbol No. 46. Panels will be interconnected with 1-inch-diameter by 14-1/2-inch-long A36 steel dowels, placed between panels at 7-3/4 inches on center.

Glulam decks are attached to supporting beams with mechanical fasteners such as bolts and lag screws. The attachments must securely hold the panels and transmit longitudinal and transverse forces from the deck to the beams. They should also be easy to install and maintain and be adjustable for construction tolerances in deck alignment. The most desirable connection requires no field fabrication where holes or cuts made after preservative treatment increase susceptibility to decay.

The performance of deck attachments is affected primarily by live load deflection in the panels. Deflections cause attachments to loosen from vibrations and from panel rotation about the support. The larger the deflection, the more significant the effects. Acceptable panel deflection is difficult to quantify and should be based on the best judgment of the designer. Recommended maximum deck deflections given in preceding discussions should provide acceptable attachment performance.

Some of the common attachment configurations for glulam panels on timber or steel beams are discussed below. The attachments are sufficient to resist vertical loads, longitudinal forces from vehicle braking, and transverse forces from wind on the vehicle. A decreased spacing may be required when centrifugal forces are applied. Although the attachments also provide a varying degree of lateral beam support, such support is currently not recognized in design.

Attachment to Glulam Beams

Glulam decks are placed directly on glulam beams without material at the deck-beam interface. Material such as roofing felt placed between the deck and beam is not recommended because the material can decompose with age and hold moisture, enhancing conditions for decay. Deck panels are attached to beams with bolted brackets that connect to the beam side, or with lag screws that are placed through the deck and into the beam top. The bracket configuration uses a cast aluminum alloy bracket (Weyco bracket) that bolts through the deck and connects to the beam in a routed slot (Figure 7-24). It includes small teeth that firmly grip the deck and beam but do not penetrate through the preservative treatment. This bracket, which is available from a number of glulam suppliers and manufacturers, is the preferred attachment for glulam beams because it provides a tight connection, does not alter the preservative effectiveness, and is easily tightened in service.

When panels are attached with lag screws, the screws are placed through the panel and into beam tops (Figure 7-25). It is impractical to drill beam lead holes before pressure treatment; therefore, holes must be field bored and treated before placing the screws. Lag screw attachments are not recommended because the field boring increases the susceptibility to beam and deck decay, and they are not accessible for tightening if the deck is paved.

Figure 7-24. - Aluminum deck bracket for attaching glulam decks to glulam beams.

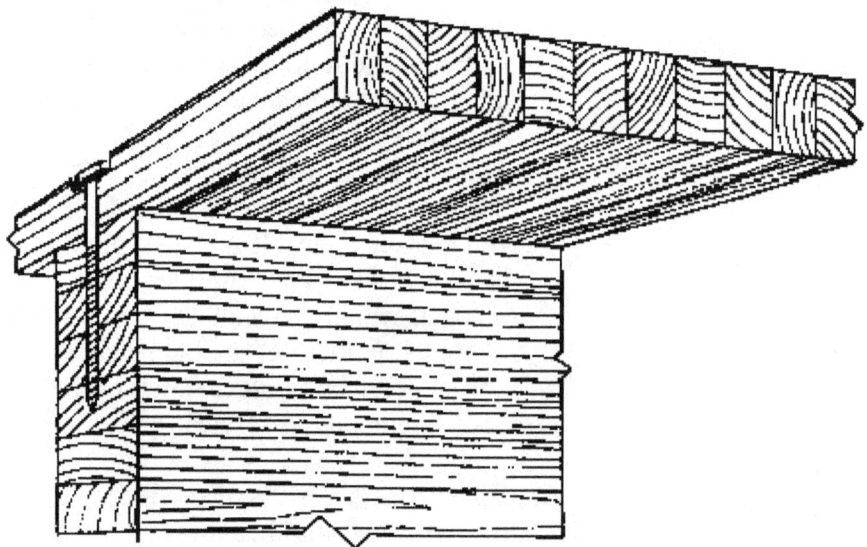

Figure 7-25. - Lag screw connection for attaching glulam decks to glulam beams.

Attachment to Steel Beams

Glulam decks are used on steel beams in new construction and rehabilitation of existing structures. Panels are placed directly on the beams with no special treatment to the top beam flange; however, when panels are placed on unpainted weathering steel beams (AASHTO M 222), a corrosion coating on the top flange should be considered to reduce the potential for steel corrosion at the panel-flange interface. The most suitable attachment for steel beams is a bracket connection that bolts through the panel and over the top beam flange. Through-bolting of the panel directly to the flange is not recommended because it allows little or no tolerance for placement or minor panel movements from variations in moisture content or thermal expansion of the steel.

The most common attachments for glulam panels on steel beams are the C-clip and angle bracket. A C-clip is a galvanized, forged-steel bracket that bolts through the panel and over the top beam flange (Figure 7-26). The clip is provided with small teeth on the deck side to prevent rotation of the bracket without penetrating the preservative envelope. C-clips are commercially available from several glulam suppliers and manufacturers and are suitable for use on beam flanges of approximately 3/4 inch or less. For thicker flanges, the angle bracket is used. Angle brackets are galvanized steel brackets fabricated from standard A36 steel angles (Figure 7-27). They are similar in connection and performance to C-clips, but can be fabricated locally. Angle clips are cut from standard 1/4- or 5/16-inch angle stock and leg dimensions can be varied for any flange thickness.

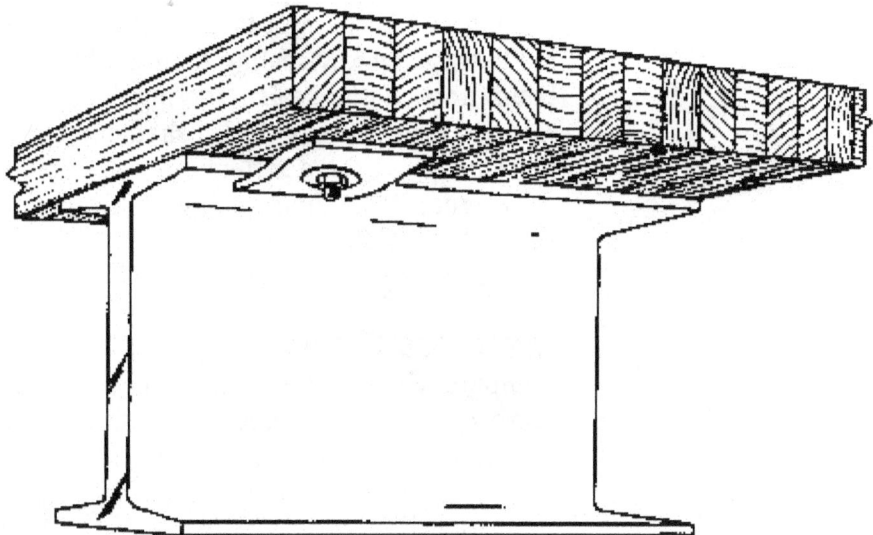

Figure 7-26. - C-clip for attaching glulam decks to steel beams.

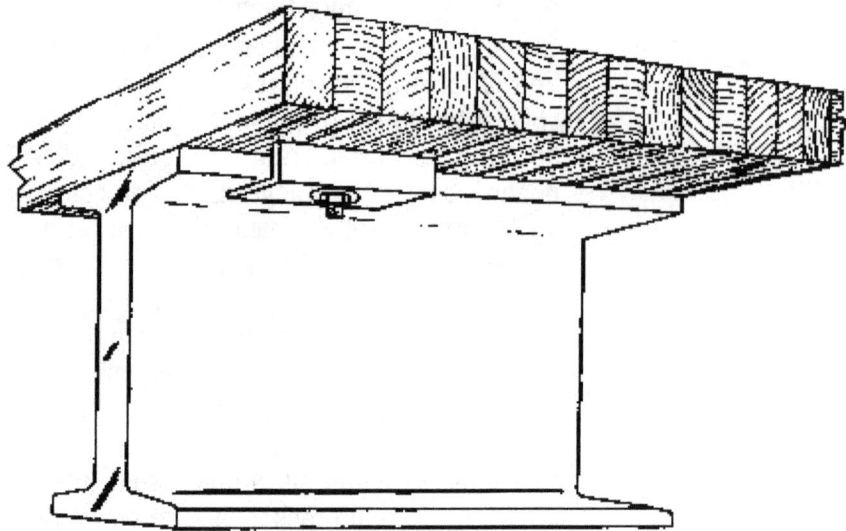

Figure 7-27. - Steel angle bracket for attaching glulam decks to steel beams.

ADDITIONAL DETAILS AND CONSIDERATIONS FOR GLULAM DECKS

Design details for fabrication and placement of bridge components can influence performance and should be suited to specific project needs. Several common details used with glulam deck panels are discussed below. The applicability of these details will vary for different projects and is left to designer judgment.

Transverse Joint Configuration

A bridge deck should provide a watertight roof over beams and other components of the superstructure. Glulam panels are especially suited for this purpose because of their relatively large size. Glulam decks can be made watertight by sealing the joint between adjacent panels with a bituminous mastic sealer (roofing cement is commonly used). It is recommended that the sealer be brushed or spread on panel edges just before placement, but some sealers can be poured into the joint after panels are set (Figure 7-28). Joint sealing is inexpensive and can contribute significantly to long structure life. It is strongly recommended for all panel configurations.

Dimensional Stability

Although glulam exhibits a much higher dimensional stability than sawn lumber, it can be affected by substantial changes in moisture content. The magnitude and effects of moisture changes are greatly reduced when panels are treated with oil-type preservatives and protected with a watertight asphalt wearing surface.[7,29] Cases involving problems with dimensional stability are not common; however, the designer should be aware of the potential for swelling or shrinkage as well as the steps to reduce or eliminate their effects.

Figure 7-28. - Bituminous sealer is spread on the edges of glulam deck panels to water-proof the panel joints.

The biggest adjustment in moisture content normally occurs during the first 2 years after construction when the panels reach equilibrium moisture content with the environment. After equilibrium is reached, subsequent changes in moisture content from seasonal variations occur gradually and have a relatively minor effect on the member. Glulam is manufactured at a moisture content of 16 percent or less, which may be reduced slightly when treated with oil-type preservatives. The panel moisture content is also affected by storage conditions between manufacture and installation. When installed in arid regions, some checking of panel ends may occur as panels dry and subsequently shrink in service. In such locations, shrinkage can be reduced if a lower panel moisture content is specified when the material is ordered. As discussed in Chapter 3, maximum moisture contents as low as 10 percent may be specified for glulam based on designer judgment. Although lower moisture contents will slightly increase costs, the potential for panel shrinkage can be greatly reduced.

In contrast to shrinkage, swelling may occur when dry panels (moisture content less than 16 percent) are installed in wet or humid areas without the protection of a watertight wearing surface. There has been at least one case where significant swelling occurred in panels protected with an asphalt wearing surface, although this condition is very rare. Swelling can cause breaks in the wearing surface, substructure backwalls, curbs, and railing depending on the magnitude of the moisture changes and the bridge span. Little can be done to increase panel moisture content for installation. In cases where the bridge is over 50 feet long, and the deck moisture content is expected to exceed 18 percent (as when unpaved decks are used in warm, humid climates), a transverse joint or gap of approximately

7-95

1/2 inch between every third or forth panel will allow the necessary room for potential expansion. If the deck is not paved and if beams are designed for wet-condition stresses, the gap can be left open, based on designer judgment. A preferable solution is to seal the gap with metal flashing or commercial joint material that will allow some panel movement.

Nosing Angles

Steel nosing angles are placed on the edge of end panels to minimize damage from vehicle impact and abrasion. They are used when approach roads are unpaved or when the potential for vehicle impact exists. The angles are generally galvanized and are attached to the deck with lag screws.

Figure 7-29. - Steel nosing angle placed across an unpaved deck to reduce damage from vehicle impact and abrasion.

PART II: SAWN LUMBER SYSTEMS

7.6 GENERAL

Sawn lumber beam bridges consist of a series of closely spaced lumber beams supporting a transverse nail-laminated or plank deck (Figure 7-30). For AASHTO highway loads, they are most practical for clear spans up to approximately 25 feet, when sawn lumber in the required sizes is available. Longer crossings are made with a series of single spans, usually in a trestle arrangement. Lumber beam bridges are among the oldest and simplest of all bridge types and were widely used in the United States through the 1950's. Their use has declined significantly over the past 20 years because of the popularity of glulam and its increased member size and improved performance. It has also become increasingly difficult to obtain sawn lumber beams in the sizes and grades typically required for bridges.

Figure 7-30. - Typical sawn lumber beam bridge with a transverse nail-laminated deck.

The following sections address design considerations, procedures, and details for sawn lumber beam bridges with transverse nail-laminated or plank decks. Although design with sawn lumber differs from glulam because of smaller member sizes and the wider variety of species and grades, many of the concepts are the same. When possible, reference will be made to previous material discussed for glulam.

As with other beam superstructures, sawn lumber beam systems consist of beams, transverse bracing, and bearings. Design considerations and procedures are addressed in that order.

BEAM DESIGN

Sawn lumber beams are designed from the species and grades of visually graded lumber given in Table 4A of the NDS.[37] Although any species can be used provided it is treatable with preservatives, most bridges are constructed from Douglas Fir-Larch or Southern Pine because of the high strength and availability of these species.

Douglas Fir-Larch beams are generally available in widths up to 16 inches, depths up to 24 inches, and lengths up to 40 feet. There may be a substantial price premium for larger sizes, however, and 6- to 8-inch widths up to 16 inches deep are normally most economical. Beams are most efficiently designed from the Beams and Stringers (B&S) size classification where tabulated bending stress, F_b, is based on loads applied to the narrow face of the member (Beams and Stringers are sawn lumber of rectangular cross section, 5 or more inches thick with the width more than 2 inches greater than the thickness). Grades for bridge beams in this classification are normally No. 1 or Select Structural. Beams can also be specified from the Posts and Timbers (P&T) size classification but these sizes generally do not provide the most efficient section in bending (Posts and Timbers are sawn lumber of square or approximately square cross-section, 5 by 5 inches and larger, with the width not more than 2 inches greater than thickness). When P&T sizes are graded to B&S requirements, design values for the applicable B&S grades may be used.

For Southern Pine, beams are generally available in widths up to 10 inches, depths up to 12 inches, and lengths up to 24 feet. Grades for bridge beams are normally Dense Structural 72 or Dense Structural 65 in the 2-1/2 inches and thicker size classification. Southern Pine does not follow many of the conventions and standards used for other species, and the designer should carefully check design tables for footnotes. Beams are generally specified from the table noted "surfaced green; used any conditions." Values in this table have been adjusted for wet-use conditions and further adjustment by C_M is not required.

Bridge beams can be specified as surfaced (S4S), rough-sawn, or full-sawn (Chapter 3). Rough- or full-sawn lumber should be edge planed (S2E) to ensure an even depth for all members. When design is based on rough- or full-sawn sizes, the applicable moisture content and size used for design must be clearly indicated on the specifications and drawings.

Live Load Distribution

Vehicle live load distribution criteria for moment, shear, and reactions in sawn lumber beams follow the same basic criteria previously discussed for glulam. However, because the distribution factors for moment are based on the relative deck stiffness, different interior beam DF equations are required for the various decks used on lumber beams. Empirical equations from AASHTO for computing interior beam distribution factors for plank and nail-laminated lumber decks are given in Table 7-13. Examples of live load distribution for sawn lumber beams are included in examples later in this section.

Table 7-13. - Interior beam live load distribution factors for plank and nail-laminated timber decks.

| Deck type[a] | DF for moment (wheel lines/beam) | |
	Bridges designed for one traffic lane	Bridges designed for two or more traffic lanes
Plank	$S/4.0$	$S/3.75$
Nail-laminated; 4 in. thick or multiple layer floors over 5 in. thick[b]	$S/4.5$	$S/4.0$
Nail-laminated; 6 in. or more thick	$S/5.0$	$S/4.25$
	If S exceeds 5 ft, use footnote c.	If S exceeds 6.5 ft, use footnote c.

[a] Deck thickness is based on nominal thickness.

[b] Multiple layer floors consist of two or more layers of planks, each layer being laid at an angle to the other.

[c] In this case, the distribution factor for each beam is the reaction of the wheel lines, assuming the deck between beams to act as a simple beam.

S = average center-to-center beam spacing (feet).

From AASHTO[1] Table 3.23.1; © 1983. Used by permission.

Beam Configuration

The number and spacing of beams can affect the overall economy and performance of the sawn lumber bridges in many of the same ways previously discussed for glulam. The effects are normally less pronounced, however, because beam spacing is often controlled primarily by strength requirements and material availability. Because of the large number of species, grades, and sizes of lumber beams, specific recommendations on bridge beam configuration are impractical. In general terms, the designer should first check material availability, then try several configurations to determine the most economical combination that meets strength and stiffness requirements.

Site restrictions are normally not a problem with sawn lumber beams because beams are not available in large depths. Deck considerations can influence beam spacing, although to a lesser degree than for glulam. Nominal 4-inch-thick plank decks are feasible for spacings up to approximately 20 inches, while nail-laminated decks are practical for spans up to approximately 38 inches for nominal 4-inch decks and 72 inches for nominal 6-inch decks. The most significant deck effect on beam spacing is at the break between a 4-inch and a 6-inch nail-laminated deck where cost savings for the thinner deck may be greater than the increased cost for closer beam spacing.

Perhaps the most important consideration in lumber beam configuration is the live load distribution to outside beams. The most suitable design is one where moment distribution factors are approximately equal for all beams, interior and outside. This allows the use of one beam size and grade across the width of the structure. The outside beam distribution factor is controlled by limiting the deck overhang so that the reaction at the beam in wheel lines does not exceed the interior beam DF given in Table 7-13.

Beam Design Procedures

Design procedures for sawn lumber beams follow the same basic procedures used for glulam timber. Minor differences in procedures and criteria are illustrated in the following examples.

Example 7-9 - Lumber beam design; two-lane HS 15-44 loading

A lumber beam bridge is required to span 17 feet center to center of bearings and support two lanes of HS 15-44 loading over a roadway width of 24 feet. The deck is nominal 4-inch-thick nail-laminated lumber with a full sawn 3-inch timber wearing surface. Design the beam system for this structure, assuming

1. beam spacing is limited by deck requirements to a maximum of 26 inches;

2. a curb and vehicular railing are provided with an approximate dead load of 60 lb/ft;

3. all lumber except the wearing surface is dressed (S4S);

4. beams are visually graded Douglas Fir-Larch;

5. beam live load deflection must not exceed *L/360*; and

6. AASHTO requirements for Load Group IA do not apply.

7-100

Solution

From the given information, an initial configuration of 13 beams spaced 24 inches on center is selected. The face of the rail is aligned with the outside beam centerline with an additional 10-inch deck extension for the curb and rail attachment:

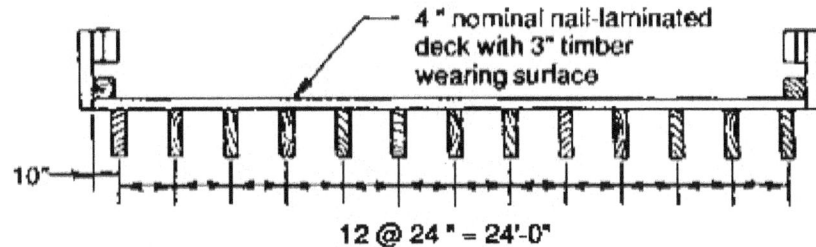

Select Lumber Species and Grade

From NDS Table 4A, an initial beam species and grade are selected as Douglas Fir-Larch, visually graded No. 1 in the Beams and Stringers (B&S) size classification (WWPA rules). Tabulated values are as follows:

$$F_b = 1,350 \text{ lb/in}^2$$

$$F_v = 85 \text{ lb/in}^2$$

$$F_{c\perp} = 625 \text{ lb/in}^2$$

$$E = 1,600,000 \text{ lb/in}^2$$

Compute Deck Dead Load and Dead Load Moment

Dead load of the deck (3-1/2 inches actual thickness) and wearing surface is computed as

$$DL = \frac{(3.5 \text{ in.} + 3 \text{ in.})(50 \text{ lb/ft}^3)}{12 \text{ in./ft}} = 27.1 \text{ lb/ft}^2$$

For interior beams, each beam supports a tributary deck width of 2 feet:

$$\text{Deck } w_{DL} = 2 \text{ ft } (27.1 \text{ lb/ft}^2) = 54.2 \text{ lb/ft}$$

$$\text{Deck } M_{DL} = \frac{w_{DL} L^2}{8} = \frac{54.2 (17)^2}{8} = 1,958 \text{ ft-lb}$$

For outside beams, each beam supports 1 foot of combined deck and wearing surface, 10 inches (0.83 feet) of deck only and 60 lb/ft of curb and railing:

$$\text{Deck } w_{DL} = (1 \text{ ft})(27.1 \text{ lb/ft}^2) + 0.83 \text{ ft}\left[\frac{(3.5 \text{ in.})(50 \text{ lb/ft}^3)}{12 \text{ in./ft}}\right] = 60 \text{ lb/ft}$$

$$= 99.2 \text{ lb/ft}$$

$$\text{Deck } M_{DL} = \frac{w_{DL}L^2}{8} = \frac{99.2\,(17)^2}{8} = 3{,}584 \text{ ft-lb}$$

Compute Live Load Moment

The equation for the interior beam moment DF is obtained from Table 7-13:

$$\text{Interior beam DF} = \frac{S}{4} = \frac{2}{4} = 0.50 \text{ WL/beam}$$

The outside beam moment DF is computed by positioning the wheel line 2 feet from the rail face, assuming the deck acts as a simple span between beams. In this case, the rail face is aligned with the outside beam center-line and the wheel line is directly over the first interior beam:

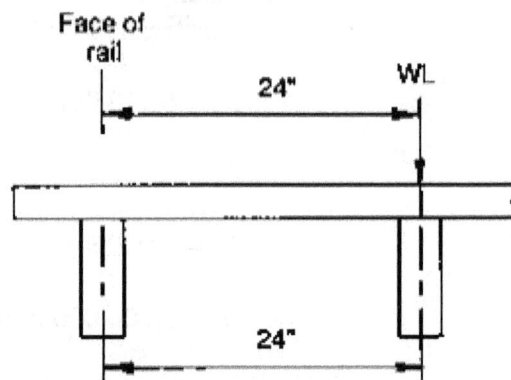

The moment DF to outside beams is technically zero; however, AASHTO requires that the DF to outside beams not be less than that to interior beams. The moment DF is therefore 0.50 WL/beam.

From Table 16-8, the maximum moment for one wheel line of an HS 15-44 truck on a 17-foot span is 51 ft-k. The design live load moment is computed by multiplying the maximum moment for one wheel line by the moment DF:

$$M_{LL} = M(\text{DF}) = 51(0.50)(1{,}000 \text{ lb/k}) = 25{,}500 \text{ ft-lb}$$

Determine Beam Size Based on Bending

The allowable stress in bending is equal to tabulated stress adjusted by all applicable modification factors. In this case

$$F_b' = F_b C_M C_F$$

At this point the beam size, dead load, C_M and C_F are unknown. Assuming a beam dead load of 50 lb/ft, an initial interior beam size is computed based on the tabulated bending stress:

$$\text{Estimated Beam } M_{DL} = \frac{w_{DL}\, L^2}{8} = \frac{50\,(17)^2}{8} = 1,806 \text{ ft-lb}$$

Using the inside beam $M_{\Omega} = 1,958$ ft-lb,

$$M = (\text{Beam } M_{DL} + \text{Deck } M_{DL}) + M_{LL} = (1,806 + 1,958) + 25,500$$

$$= 29,264 \text{ ft-lb}$$

$$S = \frac{M}{F_b} = \frac{29,264\,(12 \text{ in./ft})}{1,350} = 260.13 \text{ in}^3$$

From Table 16-2, an initial interior beam size of 6 by 18 inches is selected with the following properties:

$b = 5\text{-}1/2$ in.	$S = 280.73$ in^3
$d = 17\text{-}1/2$ in.	$I = 2,456.38$ in^4
$A = 96.25$ in^2	$w_{DL} = 33.4$ lb/ft

Modification factors and the allowable bending stress are computed as follows:

From Table 5-7, $C_M = 1.0$ for lumber 5 inches or thicker.

$$C_F = \left(\frac{12}{d}\right)^{1/9} = \left(\frac{12}{17.5}\right)^{1/9} = 0.96$$

$$F_b' = F_b C_M C_F = 1350(1.0)(0.96) = 1,296 \text{ lb/in}^2$$

Bending stress is computed based on the actual beam dead load:

$$\text{Beam } M_{DL} = \frac{w_{DL}\, L^2}{8} = \frac{33.4\,(17)^2}{8} = 1,207 \text{ ft-lb}$$

$$M = (\text{Beam } M_{DL} + \text{Deck } M_{DL}) + M_{LL}$$

$$= (1,207 + 1,958) + 25,500 = 28,665 \text{ ft-lb}$$

$$f_b = \frac{M}{S} = \frac{28,665\,(12 \text{ in/ft})}{280.73} = 1,225 \text{ in}^2$$

$f_b = 1,225$ lb/in^2 < $F_b' = 1,296$ lb/in^2, so 6- by 18-inch beams are satisfactory in bending for interior beams. Checking outside beams:

$$M = M_{DL} + M_{LL} = (1{,}207 + 3{,}584) + 25{,}500 = 30{,}291 \text{ ft-lb}$$

$$f_b = \frac{M}{S} = \frac{30{,}291(12 \text{ in/ft})}{280.73} = 1{,}295 \text{ lb/in}^2$$

$f_b = 1{,}295 \text{ lb/in}^2 < Fb' = 1{,}296 \text{ lb/in}^2$, so outside beams are satisfactory in bending.

The beams must next be checked for lateral stability. Transverse bracing (blocking) will be provided at the beam ends and the span centerline:

$$\ell_u = \frac{L}{4} = \frac{17}{2} = 8.5 \text{ ft} \quad \text{and} \quad \frac{\ell_u}{d} = \frac{8.5(12 \text{ in/ft})}{17.5} = 5.83$$

By Equation 5-7,

$$\ell_e = 1.63\ell_u + 3d = 1.63(8.5)(12 \text{ in/ft}) + 3(17.5) = 218.76$$

By Equation 5-3,

$$C_s = \sqrt{\frac{\ell_e d}{b^2}} = \sqrt{\frac{218.76(17.5)}{(5.5)^2}} = 11.25 < 50$$

$C_s > 10$, so further stability calculations are required:

$$E' = EC_M = 1{,}600{,}000(1.0) = 1{,}600{,}000 \text{ lb/in}^2$$

By Equation 5-9,

$$F_b'' = F_b C_M = 1{,}350(1.0) = 1{,}350 \text{ lb/in}^2$$

$$C_k = 0.811\sqrt{\frac{E'}{F_b''}} = 0.811\sqrt{\frac{1{,}600{,}000}{1{,}350}} = 27.92$$

$C_s = 11.25 < C_k = 27.92$, so the beam is in the intermediate slenderness range. By Equation 5-10,

$$C_L = 1 - \frac{1}{3}\left(\frac{C_s}{C_k}\right)^4 = 1 - \frac{1}{3}\left(\frac{11.25}{27.92}\right)^4 = 0.99$$

$C_L = 0.99 > C_r = 0.96$; therefore, strength rather than stability controls allowable bending stress.

Check Live Load Deflection

Live load deflection is checked by assuming deflection is distributed the same as bending; one beam resists the deflection produced by 0.50 wheel lines. From Table 16-8, the deflection coefficient for one wheel line of an HS 15-44 truck on a 17-foot simple span is 2.12×10^9 lb-in^3.

$$\Delta_{LL} = \frac{0.50\left(2.12 \times 10^9\right)}{E'I} = \frac{0.50\left(2.12 \times 10^9\right)}{(1,600,000)(2,456.38)} = 0.27 \text{ in.} = L/756$$

$L/756 < L/360$, so deflection is acceptable.

Check Horizontal Shear

From bending calculations, outside beam dead load is 99.2 lb/ft for the deck and railing and 33.4 lb/ft for the beam, for a total of 132.6 lb/ft. Neglecting loads within a distance of $d = 17.5$ inches from the supports, dead load vertical shear is computed by Equation 7-6:

$$V_{DL} = w_{DL}\left(\frac{L}{2} - d\right) = 132.6\left(\frac{17}{2} - \frac{17.5}{12 \text{ in/ft}}\right) = 934 \text{ lb}$$

Live load vertical shear is computed at the lesser of $3d$ or $L/4$ from the support:

$$3d = \frac{3(17.5)}{12 \text{ in/ft}} = 4.38 \text{ ft} \qquad \frac{L}{4} = \frac{(17)}{4} = 4.25 \text{ ft}$$

$L/4 = 4.25$ feet controls, and the maximum vertical shear is determined at that location for one wheel line of an HS 15-44 truck:

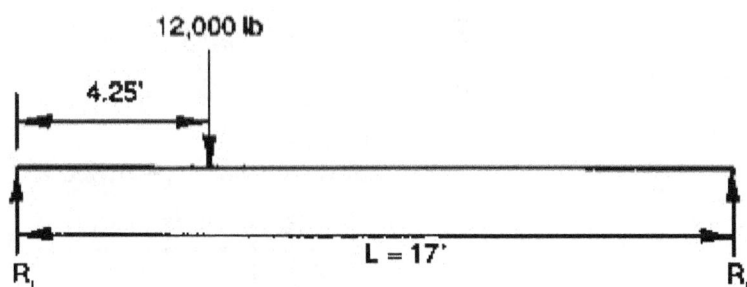

$$V_{LU} = R_L = \frac{12,000 \text{ lb } (17 \text{ ft} - 4.25 \text{ ft})}{17 \text{ ft}} = 9,000 \text{ lb}$$

For a moment DF to outside beams of 0.50,

$$V_{LD} = 9,000(0.50) = 4,500 \text{ lb}$$

By Equation 7-1,

$$V_{LL} = 0.50 \left[(0.6 V_{LW}) + V_{LD} \right]$$

$$= 0.50 \left[(0.6)(9,000) + 4,500 \right] = 4,950 \text{ lb}$$

$$V = V_{DL} + V_{LL} = 934 + 4,950 = 5,884 \text{ lb}$$

$$f_v = \frac{1.5 \, V}{A} = \frac{1.5 \, (5,884)}{96.25} = 92 \text{ lb/in}^2$$

$$F_v' = F_v(C_M) \text{ (shear stress modification factor)}$$

Without the shear stress modification factor,

$$F_v' = F_v(C_M) = 85(1.0) = 85 \text{ lb/in}^2$$

Without an increase in allowable stress by the shear stress modification factor (Table 7-17), the beam is overstressed by approximately 7 lb/in². It is reasonable to assume that some splitting of the beam may occur as it seasons; however, a full-length split assumed by no stress increase is unlikely. A slight increase in allowable stress of approximately 10 percent is considered appropriate in this case. This is a matter of designer judgment that must be specifically addressed in each case.

$$F_v' = 85(1.0)(1.10) = 94 \text{ lb/in}^2$$

$f_v = 92$ lb/in² $< F_v' = 94$ lb/in², so the beam is acceptable in horizontal shear.

Determine Bearing Length and Stress

From Table 5-7, $C_M = 0.67$, and

$$F_{c\perp}' = F_{c\perp}(C_M) = 625(0.67) = 419 \text{ lb/in}^2$$

For a unit dead load $w_{DL} = 132.6$ lb/ft to outside beams,

$$R_{DL} = \frac{w_{DL} \, L}{8} = \frac{(132.6)\,(17)}{8} = 1,127 \text{ lb}$$

The live load reaction DF is determined as the reaction at the beam, assuming the deck acts as a simple span between supports. For a 24-inch beam spacing, the maximum reaction is 1.0 WL/beam. From Table 16-8, the maximum reaction for one wheel line of an HS 15-44 truck on a 17-foot span is 14.12 k = 14,120 lb:

$$R_{LL} = R(DF) = 14,120(1.0) = 14,120 \text{ lb}$$

By Equation 7-8,

$$\text{Required bearing length} = \frac{R_{DL} + R_{LL}}{b\left(F_{c\perp}'\right)} = \frac{1{,}127 + 14{,}120}{5.5\,(419)} = 6.6 \text{ in.}$$

A bearing length of 7 inches will be used, for an out-to-out beam length of 17 feet 7 inches. Applied stress is computed by Equation 7-9:

$$f_{c\perp} = \frac{R_{DL} + R_{LL}}{A} = \frac{1{,}127 + 14{,}120}{5.5\,(7)} = 396 \text{ lb/in}^2$$

Summary

The superstructure will consist of thirteen 6- by 18-inch dressed lumber beams spaced 24 inches on center. The beams will be 17 feet 7 inches long and span a distance of 17 feet measured center to center of bearings. Transverse blocking will be provided for lateral support at the bearings and at the span centerline. Lumber will be specified as Douglas Fir-Larch in the B & S size classification, visually graded No. 1 or better to WWPA rules. Stresses and deflection are as follows:

	Interior beams	Outside beams
f_b	1,225 lb/in^2	1,295 lb/in^2
F_b'	1,296 lb/in^2	1,296 lb/in^2
Δ_{LL}	0.27 in. = $L/756$	0.27 in. = $L/756$
f_v	< Outside beam	92 lb/in^2
F_v'	94 lb/in^2	94 lb/in^2
$f_{c\perp}$	< Outside beam	396 lb/in^2
$F_{c\perp}'$	419 lb/in^2	419 lb/in^2

Example 7-10 - Lumber beam design; single-lane H 10-44 loading

---

A farmer wants to construct a bridge over a small creek to access additional acreage. Based on a study of the site, an 11-foot span, measured center-to-center of bearings, will be adequate. The bridge must be capable of supporting farming equipment that closely resembles an AASHTO H 10-44 truck. The required roadway width is approximately 10-1/2 feet with 6- by 6-inch curbs installed along each edge. Design the beam system for this structure, assuming

1. the beams and curbs are full-sawn Douglas Fir-Larch;

2. the transverse timber deck is constructed of surfaced 4-inch planks, with no wearing surface;

3. beam spacing is limited by deck span capabilities to approximately 14 inches; and

4. live load deflection and AASHTO Load Group IA loading need not be considered.

Solution

For an AASHTO H 10-44 truck the GVW is 10 tons distributed 20 percent to the front axle and 80 percent to the rear axle (Example 6-1). The vehicle configuration for one wheel line is as follows:

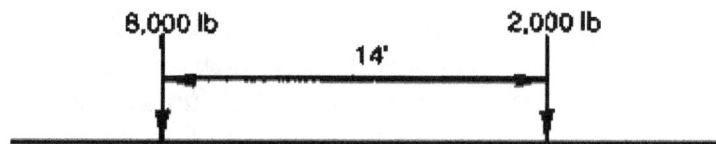

Because this bridge spans a short crossing, it is anticipated that shear will control beam design. The design procedure will be to size the beams based on horizontal shear, then check for bending. An initial configuration of 11 beams spaced 12 inches on center is selected:

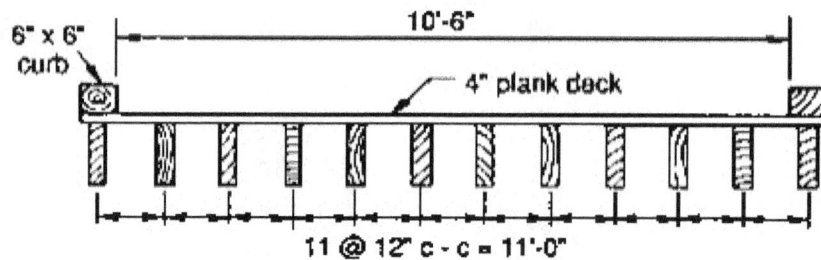

Compute Deck Dead Load

Interior beams support 1 foot of deck width. Outside beams support a more severe loading from a 9-inch deck width plus the 6- by 6-inch curb:

$$\text{Deck } w_{DL} = \left(\frac{(3.5 \text{ in.})(9 \text{ in.}) + (6 \text{ in.})(6 \text{ in.})}{144 \text{ in}^2 / \text{ft}^2} \right)(50 \text{ lb/ft}^2) = 23.4 \text{ lb/ft}$$

Compute Live Load Distribution Factors

Live load distribution for shear is based on the distribution factors used for moment. Assuming the deck acts as a simple span between beams, placing the wheel line 2 feet from the face of the curb results in no live load distribution to outside beams. Therefore, the moment DF for interior and outside beams will be controlled by interior beams. From Table 7-13 for a single-lane plank deck,

$$\text{Moment DF} = \frac{S}{4} = \frac{1}{4} = 0.025 \ \text{WL/beam}$$

Determine Beam Size Based on Horizontal Shear

From NDS Table 4A for visually graded Douglas Fir-Larch, there are two tabulated shear values given for different size classifications. For all grades in the J&P size classification (lumber 2 to 4 inches thick), $F_v = 95$ lb/in^2. For all grades in the B&S size classification, $F_v = 85$ lb/in^2. The smaller 4-inch material is selected as a first choice.

Starting with a 4- by 12-inch full-sawn beam, section properties required for shear are computed:

$$b = 4 \text{ in.}$$

$$d = 12 \text{ in.}$$

$$A = 4 \text{ in. } (12 \text{ in.}) = 48 \text{ in}^2$$

$$w_{DL} = \frac{48 \text{ in}^2}{144 \text{ in}^2/\text{ft}^2}\left(50 \text{lb/ft}^3\right) = 16.7 \text{ lb/ft}$$

Dead load vertical shear is computed for combined deck and beam dead load by Equation 7-6:

$$V_{DL} = w_{DL} = \left(\frac{L}{2} - d\right) = (23.4 + 16.7)\left(\frac{11}{2} - \frac{12}{12 \text{ in./ft}}\right) = 180.5 \text{ lb}$$

Live load vertical shear is computed from the maximum vertical shear occurring at the lesser of $3d$ or $L/4$ from the support:

$$3d = 3\left(\frac{12}{12 \text{ in./ft}}\right) = 3 \text{ ft} \qquad \frac{L}{4} = \frac{11}{4} = 2.75 \text{ ft}$$

$L/4 = 2.75$ feet controls, and the maximum vertical shear is determined at that point for one wheel line of an H 10-44 truck:

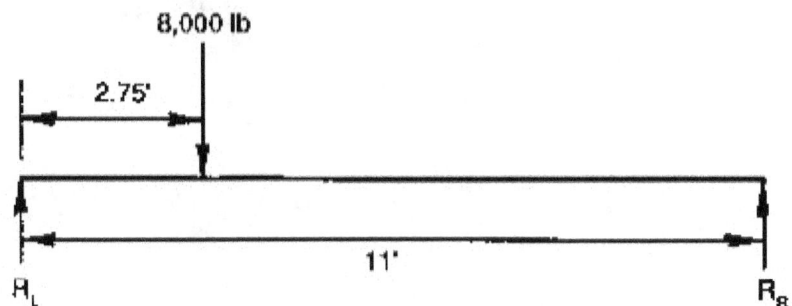

7-109

$$V_{LU} = R_L = \frac{(8{,}000 \text{ lb})(11 \text{ ft} - 2.75 \text{ ft})}{11 \text{ ft}} = 6{,}000 \text{ lb}$$

For a moment DF to outside beams of 0.25,

$$V_{LD} = 6{,}000(0.25) = 1{,}500 \text{ lb}$$

$$V_{LL} = 0.50\left[(0.6V_{LU}) + V_{LD}\right] = 0.50\left[(0.6)(6{,}000) + 1{,}500\right] = 2{,}550 \text{ lb}$$

$$V = V_{DL} + V_{LL} = 180.5 + 2{,}550 = 2{,}731 \text{ lb}$$

By Equation 5-18,

$$F_v' = F_v(C_M) \text{ (shear stress modification factor)}$$

From Table 5-7, $C_M = 0.97$ for wet-condition use. Because it is likely that some beam splitting may occur as the material seasons, the shear stress modification factor (Table 7-17) will be limited to 1.0 based on designer judgment.

$$F_v' = (95 \text{ lb/in}^2)\,(0.97)(1.0) = 92 \text{ lb/in}^2$$

Rearranging Equation 5-17, the required beam area is computed:

$$A = \frac{1.5V}{F_v} = 1.5\,\frac{(2{,}731)}{92.15} = 44.45 \text{ in}^2$$

44.45 in² < 48 in², so a 4- by 12-inch beam is satisfactory with the following applied stress:

$$f_v = \frac{1.5V}{A} = \frac{1.5(2{,}731)}{48 \text{ in}^2} = 85 \text{ lb/in}^2$$

Check Bending and Select Beam Grade

For a 4- by 12-inch full-sawn beam,

$$S = \frac{bd^2}{6} = \frac{4\,(12)^2}{6} = 96 \text{ in}^3$$

$$I = \frac{bd^3}{12} = \frac{4\,(12)^3}{12} = 576 \text{ in}^4$$

$$\text{Beam } M_{DL} = \frac{w_{DL}L^2}{8} = \frac{(16.7 \text{ ft})(11 \text{ ft})^2}{8} = 252.6 \text{ ft-lb}$$

$$\text{Deck } M_{DL} = \frac{(23.4 \text{ ft}) (11 \text{ ft})^2}{8} = 353.9 \text{ ft-lb}$$

$$\text{Total } M_{DL} = 252.6 + 353.9 = 606.5 \text{ ft-lb}$$

For an H 10-44 truck on an 11-foot span, maximum live load moment occurs when the 8,000-pound wheel load is positioned at the span centerline:

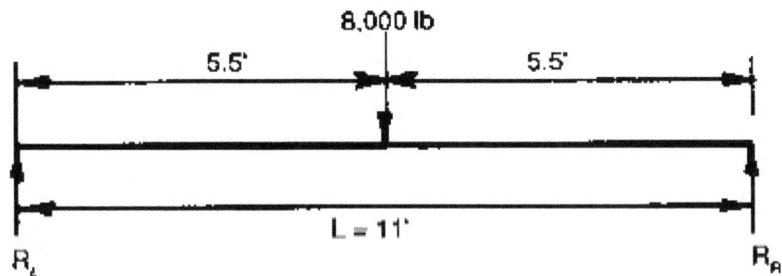

$$M_{LL}/\text{WL} = R_L \frac{L}{2} = 4,000 \frac{11}{2} = 22,000 \text{ ft-lb}$$

Applying the moment DF = 0.25, applied bending stress is computed:

$$M_{LL} = 0.25(22,000 \text{ ft-lb}) = 5,500 \text{ ft-lb}$$

$$M = M_{DL} + M_{LL} = 606.5 + 5,500 = 6,107 \text{ ft-lb}$$

$$f_b = \frac{M}{S} = \frac{6,107 \, (12 \text{ in/ft})}{96} = 763 \text{ lb/in}^2$$

From NDS Table 4A, No, 2 Douglas Fir-Larch is selected with the following tabulated values:

$F_b = 1,250 \text{ lb/in}^2$	$C_M = 0.86$
$F_v = 95 \text{ lb/in}^2$	$C_M = 0.97$
$F_{c\perp} = 625 \text{ lb/in}^2$	$C_M = 0.67$
$E = 1,700,000 \text{ lb/in}^2$	$C_M = 0.97$

$$F_b' = F_b C_M C_r = 1250(0.86)(1.0) = 1,075 \text{ lb/in}^2$$

$f_b = 763 \text{ lb/in}^2 < F_b' = 1,075 \text{ lb/in}^2$, so the beam is satisfactory in bending. The beam is next checked for lateral stability. Because of the very short span, transverse bracing (blocking) will be provided at the beam ends only:

$$\ell_x = L = 11 \text{ ft} \quad \text{and} \quad \frac{\ell_u}{d} = \frac{11(12 \text{in/ft})}{12} = 11$$

By Equation 5-7,

$$\ell_e = 1.63\ell_u + 3d = 1.63(11)(12 \text{ in.}/\text{ft}) + 3(12) = 251.16$$

By Equation (5-3),

$$C_s = \sqrt{\frac{\ell_e d}{b^2}} = \sqrt{\frac{251.16(12)}{(4)^2}} = 13.72 < 50$$

$$E' = EC_M = 1,700,000(0.97) = 1649,000 \text{ lb/in}^2$$

By Equation 5-9,

$$F_b'' = F_b C_M = 1,250 \, (0.86) = 1,075 \text{ lb/in}^2$$

$$C_k 0.811 \sqrt{\frac{E'}{F_b''}} = 0.811 \sqrt{\frac{1,649,000}{1,075}} = 31.76$$

$C_s = 13.72 < C_k = 31.76$; therefore the beam is in the intermediate slenderness range. By Equation 5-10,

$$C_L = 1 - \frac{1}{3}\left(\frac{C_s}{C_k}\right)^4 = 1 - \frac{1}{3}\left(\frac{13.72}{31.76}\right)^4 = 0.99$$

$C_L = 0.99 < C_r = 1.0$, so stability controls over strength and allowable bending stress must be adjusted by C_L:

$$F_b' = F_b C_M C_L = 1250(0.86)(0.99) = 1,064 \text{ lb/in}^2$$

$F_b' = 1,064 \text{ lb/in}^2 > f_b = 763 \text{ lb/in}^2$, so the 4- by 12-inch No. 2 beams are satisfactory.

Determine Bearing Length and Stresses

Allowable stress in compression perpendicular to grain is computed by Equation 5-20:

$$F_{c\perp}' = F_{c\perp}(C_M) = 625(0.67) = 419 \text{ lb/in}^2$$

For a unit dead load of 23.4 lb/ft for the deck and 16.7 lb/ft for the beams,

$$R_{DL} = \frac{w_{DL} L}{2} = \frac{(23.4 + 16.7)(11)}{2} = 220.6 \text{ lb}$$

Assuming the deck acts as a simple span over the 12-inch beam spacing, the reaction DF is 1.0 WL/beam. The reaction for one wheel line of an H 10-44 truck is computed and multiplied by the reaction DF:

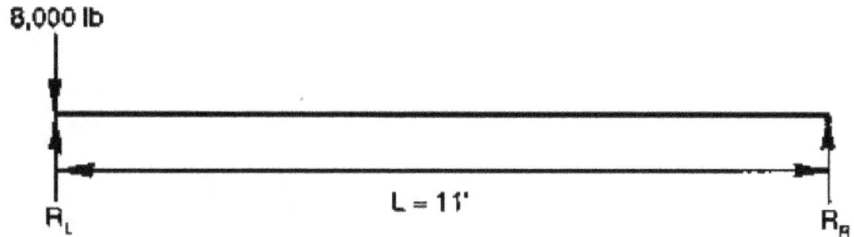

$$R_L = 8,000 \text{ lb}$$

$$R_{LL} = R(DF) = 8,000(1.0) = 8,000 \text{ lb}$$

By Equation 7-8,

$$\text{Required bearing length} = \frac{R_{DL} + R_{LL}}{b\left(F_{c\perp}'\right)} = \frac{220.6 + 8,000}{4(419)} = 4.9 \text{ in.}$$

A bearing length of 6 inches will be used, for an out-to-out beam length of 11 feet 6 inches Applied stress is computed by Equation 7-9:

$$f_{c\perp} = \frac{R_{DL} + R_{LL}}{A} = \frac{220.6 + 8,000}{4(6)} = 343 \text{ lb/in}^2$$

Summary

The superstructure will consist of twelve 4- by 12-inch full-sawn lumber beams, 11 feet 6 inches long, spaced 12 inches on center. Transverse blocking will be provided for lateral support at the bearings. Stresses based on No. 2 Douglas Fir-Larch in the J&P size classification are as follows:

	Interior beams	Outside beams
f_b	< Outside beams	763 lb/in^2
F_b'	1,075 lb/in^2	1,075 lb/in^2
f_v	< Outside beams	85 lb/in^2
F_v'	92 lb/in^2	92 lb/in^2
$f_{c\perp}$	< Outside beams	343 lb/in^2
$F_{c\perp}'$	419 lb/in^2	419 lb/in^2

Design of Transverse Bracing

Transverse bracing for sawn lumber beams is normally provided by lumber blocks placed between the beams (Figure 7-31). Blocks should be positioned as close as practical to the beam top and preferably extend the entire beam depth. They are generally 4 inches thick for beams up to 12 inches wide, and 6 inches thick for wider beams. As a minimum, blocks should be placed at both bearings, and at centerspan for span lengths over 20 feet.

Figure 7-31. - Lumber blocks placed as transverse bracing for sawn lumber beams.

An examination of existing lumber beam bridges will show that the number of different block attachments has been limited only by designer imagination. Two of the most common attachments used in recent years are steel brackets attached to the beam sides and rods placed through the beams. The simplest brackets are prefabricated steel joist or beam hangers commonly used in building construction (Figure 7-32). These hangers, which are nailed or spiked to the beams and blocks, are available in a variety of standard sizes for members up to 6 inches wide and 16 inches deep. They are relatively inexpensive, simple to install, and provide adequate performance. For the rod configuration, a 3/4-inch-diameter steel rod is placed continuously through all beams across the structure width (Figure 7-33). Lumber blocks are then toenailed to adjacent beams and connected to the rod with 3/16-inch driven staples. This system provides the added advantage of tying all beams together, but it requires additional fabrication and materials and is normally more difficult to erect than other systems.

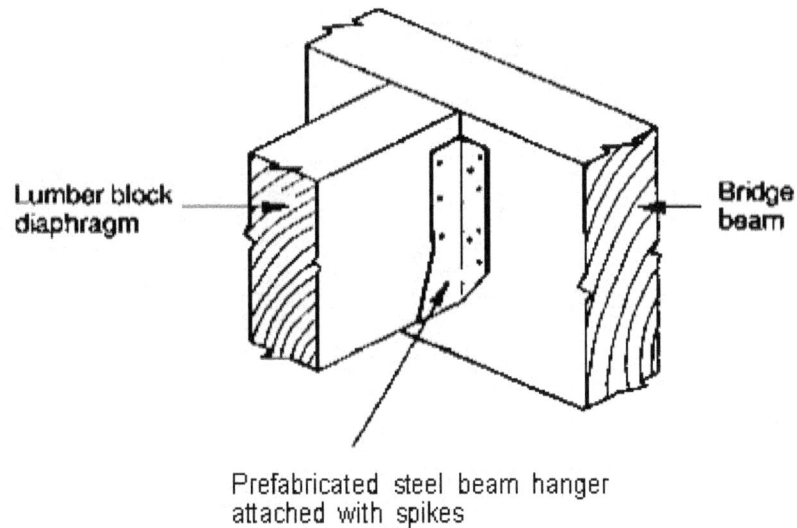

Figure 7-32. - Lumber block diaphragm configuration using steel beam hangers.

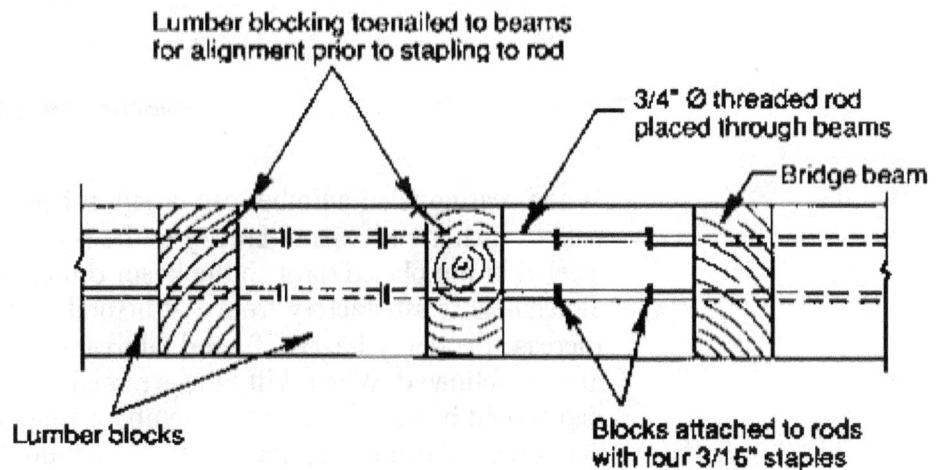

Figure 7-33. - Lumber block diaphragm configuration using steel rods and driven staples.

Design of Bearings

Bearings for sawn lumber beams must provide sufficient area for compression and must be able to transfer longitudinal and transverse loads from the superstructure to the substructure. The design considerations for glulam beams also apply to lumber beams, although some details are often modified because of the smaller beam size. The most suitable bearing is generally the steel bearing shoe arrangement. For sawn lumber applications, the shoe is constructed of standard steel angles with one beam attachment bolt and two anchor bolts, one for each angle (Figure 7-34).

Because of the smaller beam sizes, the base plate and bearing pad used for glulam are normally not required for sawn lumber beams, but may be provided at the option of the designer.

Figure 7-34. - Steel angle bearing attachment for sawn lumber beams.

When bearing is on a timber cap or sill, it has been common practice in the past to anchor each beam directly to the support with a 1/2- to 3/4-inch steel drift pin placed through the beam center. Although this type of attachment is satisfactory from a structural standpoint, it can significantly increase the decay hazard if good fabrication and construction practices are not followed. When drift pins are used, lead holes in the beams and cap should be bored before the members are pressure-treated with preservatives. When this is not practical, field-bored holes must be thoroughly treated with preservatives before placing the pin (Chapter 12).

7.8 NAIL-LAMINATED DECKS

Transverse nail-laminated decks consist of a series of dimension lumber laminations placed on edge and nailed together on their wide faces (Figure 7-35). The deck is constructed by progressively nailing laminations to the preceding section to form a continuous surface over the bridge length. Nail-laminated decks are similar in arrangement to glulam, but load transfer between laminations is done mechanically by nails rather than by glue. The laminations are generally nominal 2 by 4 or 2 by 6 sawn

Figure 7-35.—(A) Edge view of a transverse nail-laminated lumber deck. (B) Top view comparison of a nail-laminated lumber deck (right) and glulam deck (left).

lumber for spans up to approximately 6 feet under standard AASHTO highway loads. Nail-laminated decks have been widely used on timber and steel superstructures for more than 40 years. Their popularity has declined significantly since the introduction of glulam panels.

The performance of nail-laminated decks depends on the effectiveness of the nails in transferring loads between adjacent laminations. Loose nails lead to reduced load distribution and increased deck deflection. This typically causes laminations to separate and asphalt paving to deteriorate. Although the static strength of a loose deck may remain high, deck serviceability under dynamic vehicle loads is greatly reduced. Looseness is normally caused by two factors, high deck deflections and dimensional changes from moisture variations. Deflections can be controlled in design, but have frequently been neglected in the past. Moisture effects have a somewhat lesser effect that deflection and depend on local environmental conditions and the degree of exposure to weathering. Dimensional stability of nail-laminated decks is improved when seasoned, edge-grain lumber is used and the deck is protected by a watertight wearing surface (Chapter 11).

Nail-laminated decks are economical and are easily constructed with locally available materials. When properly designed, they provide acceptable performance on low- to moderate-volume bridges that are not subjected to heavy highway loads. They do not provide a service life comparable to properly designed glulam panels because the nails penetrate the preservative layer of the wood, making it more susceptible to decay. In areas where de-icing chemicals are used, the chemicals may also corrode the nails over time.

DESIGN PROCEDURES

Nail-laminated decks are designed using the same basic procedures previously discussed for noninterconnected glulam panels. An initial species and grade of lumber lamination is selected, and deck thickness is determined based on bending. Live load deflection and horizontal shear are then checked.

The design procedures given below are for continuous nail-laminated decks constructed of 2-inch nominal sawn lumber, 4 to 6 inches deep. A continuous nail-laminated deck is one in which all laminations are nailed to the previous laminations (see AASHTO 3.25.1.1 for design criteria for nail-laminated decks constructed as noninterconnected panels). The criteria apply to all deck spans and loading conditions, but design aids are limited to standard AASHTO vehicle loads on effective deck spans of 72 inches or less. Examples 7-11 and 7-12, which follow the procedures, illustrate their application to deck design.

1. Define deck span, configuration, and design loads.

The effective deck span s is the clear distance between supporting beams plus one-half the width of one beam. The deck width is equal to the roadway width plus additional width required for curb and rail systems (Chapter 10). Whenever possible, lumber laminations should be continuous (one piece) for the entire deck width. On multiple-lane decks where sawn lumber is not available in the required lengths, butt joints should be placed at the center of the support, with joints for adjacent laminations staggered on different supports (Figure 7-36).

The design live load on nail-laminated decks is the maximum wheel load of the design vehicle. For standard AASHTO H 20-44 and HS 20-44 loads, special provisions for timber decks apply and a 12,000-pound wheel load is used for all four standard AASHTO truck loads.

2. Estimate deck thickness.

Deck thickness must be estimated for initial calculations. The following values provide a reasonable estimate of the maximum deck span for standard AASHTO vehicle loads.

Initial deck thickness (in.)	Maximum effective span (in.)
3-1/2	30
4	38
5-1/2	67
6	72

Deck thicknesses of 3-1/2 and 4 inches are based on the depths of dimension and full-sawn 2 by 4 lumber, respectively. Thicknesses of 5-1/2 and 6 inches are based on the same relative depths for 2 by 6 lumber.

Initial deck thickness may also be estimated for a known species and grade of lumber based on bending, deflection, or shear by Tables 7-15, 7-16, and 7-18 presented later in this section.

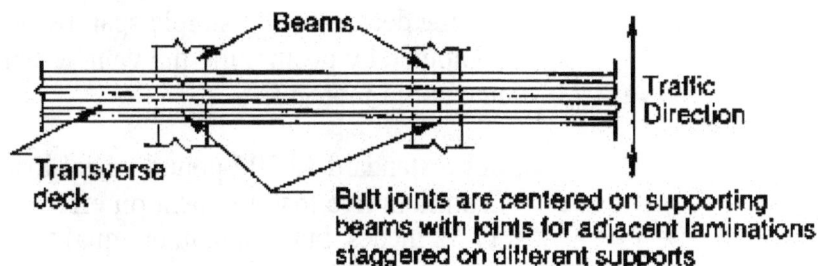

Figure 7-36. - Joint placement for transverse nail-laminated lumber decks.

3. Determine wheel distribution widths and effective deck section properties.

In the direction of the deck span, the wheel load, P, is assumed to be a uniformly distributed load acting over a width, b_t (AASHTO 3.25.1):

$$b_t = \sqrt{0.025P} \tag{7-42}$$

For a 12,000-pound wheel load, $b_t = 17.32$ inches.

In the direction normal to the deck span, the wheel load distribution width, b_d, is equal to 15 inches plus the deck thickness, t (AASHTO 3.25.1.1), as computed by

$$b_d = 15 + t \tag{7-43}$$

The deck is designed as a beam of width b_d and depth t. Effective section properties are computed by the same equations used for noninterconnected glulam decks, and are given in Table 7-14 for nominal 2 by 4 and 2 by 6 sawn lumber decks.

Table 7-14. - Effective deck section properties for continuous transverse nail-laminated decks.

t (in.)	b_d (in.)	A (in^2)	S (in^3)	I (in^4)
3-1/2	18.5	64.75	37.77	66.10
4	19.0	76.00	50.67	101.33
5-1/2	20.5	112.75	103.35	284.22
6	21.0	126.00	126.00	378.00

4. Compute dead load, dead load moment, and live load moment.

Deck dead load, dead load moment, and live load moment are computed in the same manner as for noninterconnected glulam decks. The uniform dead load moment for the effective deck section is determined by assuming the deck acts as a simple span between supports. Live load moment is computed by positioning the vehicle wheel load on the span to produce the maximum moment.

For a standard 12,000-pound wheel load and 6-foot-track width, the maximum live load moment on effective deck spans greater than 17.32 inches, but less than or equal to 122 inches ($17.32 < s < 122$), is given by

$$M_{LL} = 3,000s - 25,983 \tag{7-44}$$

where M_{LL} is the maximum live load moment (in-lb).

5. Compute bending stress and select a lamination species and grade.

For decks continuous over two spans or less, bending stress is based on the simple span moment, computed by

$$f_b = \frac{M}{S} \tag{7-45}$$

where $M = M_{DL} + M_{LL}$ computed for a simple span (in-lb) and

S = section modulus of the effective deck section (in³).

For decks continuous over more than two spans, bending stress is based on 80 percent of simple span moment to account for deck continuity and is computed by

$$f_b = \frac{0.8M}{S} \tag{7-46}$$

After f_b is computed, a species and grade of sawn lumber is selected based on the size classification for the estimated deck thickness. Allowable bending stress is computed by adjusting the tabulated stress by all applicable modification factors (for nail-laminated decks, the tabulated bending stress listed in the NDS Table 4A for repetitive member use may be used):

$$F_b' = F_b C_M \tag{7-47}$$

The allowable stress computed by Equation 7-47 may be increased by a factor of 1.33 for overloads in AASHTO Load Group IB.

If $f_b \leq F_b'$, the lamination size, species, and grade are satisfactory in bending. If f_b is substantially lower than F_b', it may be more economical to select a lower-grade material or reduce the deck thickness.

If $f_b > F_b'$, the lamination is insufficient in bending and the grade of sawn lumber or the deck thickness must be increased. If the thickness is increased, revise calculations starting at step 2.

Table 7-15 gives approximate maximum spans based on bending for nail-laminated decks continuous over more than two spans.

Table 7-15. - Approximate maximum effective span for continuous transverse nail-laminated decks based on bending; deck continuous across over more than two spans; loading from a 12,000-pound wheel load plus the deck dead load; b_d= 15 inches + deck thickness.

F_b (lb/in²)	F_b' (lb/in²)	Maximum deck span (in.)			
		t = 3-1/2 in.	t = 4 in.	t = 5-1/2 in.	t = 6 in.
1,100	946	23	29	49	58
1,150	989	24	29	51	60
1,200	1,032	25	30	53	62
1,250	1,075	26	31	54	64
1,300	1,118	26	32	56	66
1,350	1,161	27	33	58	69
1,400	1,204	28	34	60	71
1,450	1,247	28	35	62	73
1,500	1,290	29	36	64	75
1,550	1,333	30	37	65	77
1,600	1,376	30	38	67	80
1,650	1,419	31	38	69	82
1,700	1,462	32	39	71	84
1,750	1,505	32	40	73	86
1,800	1,548	33	41	74	88
1,850	1,591	34	42	76	91
1,900	1,634	34	43	78	93
1,950	1,677	35	44	80	95
2,000	1,720	36	45	82	97

$F_b' = F_b C_u = F_b(0.86)$.

6. Check live load deflection.

Live load deck deflection is computed by the standard methods of engineering analysis, assuming the deck behaves elastically as a simple beam between supports. The maximum deflection for a standard 12,000-pound wheel load on deck spans greater than 17.32 inches, but less than 110 inches, is given by

$$\Delta_{LL} = \frac{1.80}{E'I}\left(138.8s^3 - 20,780s + 90,000\right) \qquad (7\text{-}48)$$

where I is the effective moment of inertia of the effective deck section of width b_d and depth t.

When the deck is continuous over more than two spans, the deflection computed by Equation 7-48 may be multiplied by 0.80 to account for span continuity. Deflection coefficients for decks that are continuous over more than two spans are given in Figure 7-37.

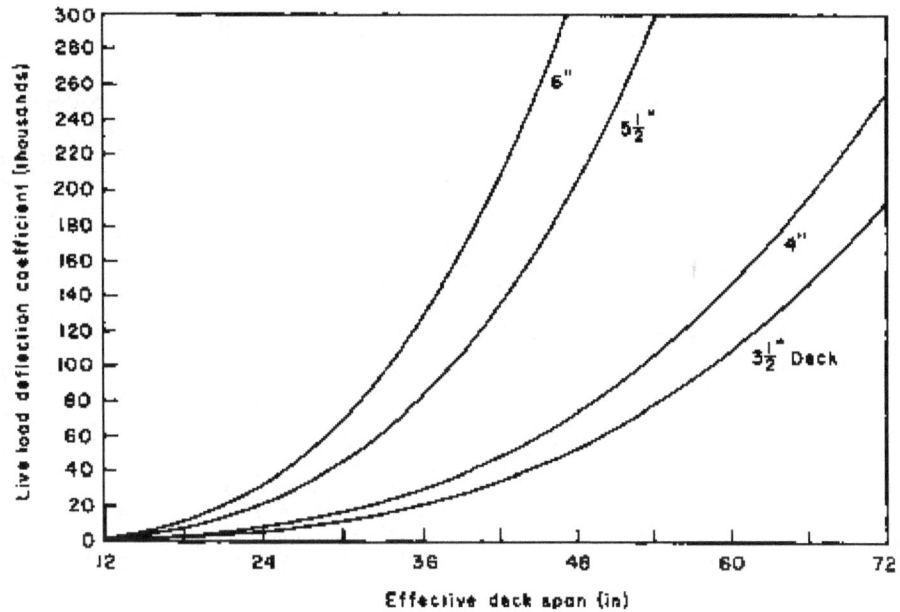

Figure 7-37. - Vehicle live load deflection coefficients for 12,000-pound wheel load(s) on a continuous, transverse nail-laminated lumber deck that is continuous over more than two spans. Divide the deflection coefficient by E' to obtain the deck deflection in inches.

Deflection is an important consideration in nail-laminated deck design and must be limited to ensure deck and wearing surface performance. The maximum acceptable deflection should be based on the type and volume of traffic and the type of wearing surface. The maximum recommended deflection is $s/500$, where s is the effective deck span. Based on this limit, maximum effective deck spans for a 12,000-pound wheel load are given in Table 7-16. When the computed live load deflection exceeds acceptable limits, the lumber grade must be increased to provide a higher E value, or the deck thickness must be increased.

7. Check horizontal shear.

Horizontal shear is based on the maximum vertical shear occurring at a distance from the support equal to the deck thickness, t. Dead load vertical shear, V_{DL} is determined by

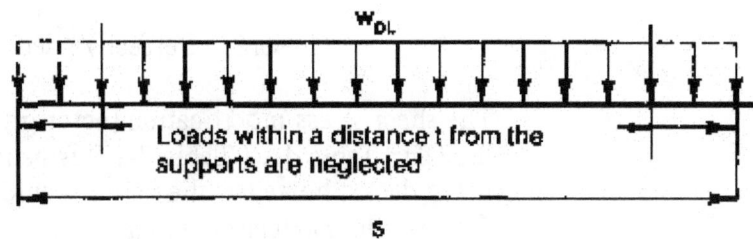

7-123

Table 7-16. - Approximate maximum effective span for continuous transverse nail-laminated decks based on a maximum vehicle live load deflection of $s/500$; deck continuous over more than two spans; loading from a 12,000-pound wheel load; b_d = 15 inches + deck thickness.

E (lb/in²)	E' (lb/in²)	Maximum deck span (in.)			
		t = 3-1/2 in.	t = 4 in.	t = 5-1/2 in.	t = 6 in.
1,500,000	1,455,000	33	40	62	>72
1,600,000	1,552,000	34	41	67	>72
1,700,000	1,649,000	35	42	69	>72
1,800,000	1,746,000	36	44	71	>72

$E' = EC_M = 0.97\, E$

$$V_{DL} = w_{DL}\left(\frac{s}{2} - t\right) \tag{7-49}$$

Live load vertical shear is determined by placing the edge of the wheel load distribution width, b, a distance, t, from the support.

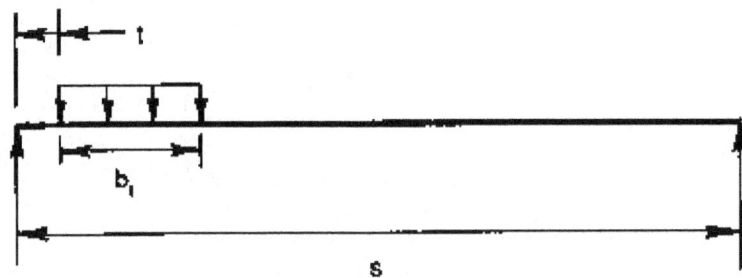

Applied stress in horizontal shear must be less than or equal to the allowable stress for the laminations, as computed by

$$f_v = \frac{1.5V}{A} \leq F_v' = F_v C_M \text{ (shear stress modification factor)} \tag{7-50}$$

where $\quad V = V_{DL} + V_{LL}$ (lb) and

$\quad\quad\quad\quad A$ = area of effective deck section (in²).

The shear stress modification factor given for sawn lumber in footnotes to the NDS Table 4A (Table 7-17) is generally taken as 2.0 for nail-laminated decks; however, the value should be based on designer judgment for the specific application and material.

7-124

Table 7-17. - Shear stress modification factor for sawn lumber.

Length of split on wide face of 2" lumber (nominal):	Multiply tabulated "F_v" value by:
No split	2.00
1/2 x wide face	1.67
3/4 x wide face	1.50
1 x wide face	1.33
1-1/2 x wide face or more	1.00

Length of split on wide face of 3" and thicker lumber (nominal):	Multiply tabulated "F_v" value by:
No split	2.00
1/2 x narrow faace	1.67
1 x narrow face	1.33
1-1/2 x narrow face or more	1.00

Size of shake* in 3" and thicker lumber (nominal):	Multiply tabulated "F_v" value by:
No shake	2.00
1/6 x narrow face	1.67
1/3 narrow face	1.33
1/2 x narrow face or more	1.00

* Shake is measured at the end between lines enclosing the shake and parallel to the wide face.

Specific horizontal shear values may be established by use of this table when the length of split, or size of check or shake, is known and no increase in them is anticipated. For California Redwood, Southern Pine, Virginia Pine-Pond Pine, and Yellow Poplar, refer to the NDS for specific values of F_v for which these adjustments apply.

From the NDS;[36] © 1986. Used by permission.

If $f_v > F_v'$, the deck does not have sufficient strength in horizontal shear and either F_v must be increased by selecting another grade or species of lamination or f_v must be reduced by increasing the deck thickness. For most species, tabulated values for horizontal shear do not increase substantially as grade increases, and increasing deck thickness is the only option. Maximum effective spans for continuous nail-laminated decks based on shear criteria are given in Table 7-18.

8. Check overhang.

The deck overhang at outside beams is checked for strength using an effective deck span measured to the centerline of the support, minus one-fourth of the beam width. For vehicle live load stresses, the wheel load is positioned with the load centroid 1 foot from the face of the railing or curb, as previously discussed for noninterconnected glulam decks. Deck stresses in bending and shear must be within allowable values previously computed.

9. Determine nail size and placement pattern.

Laminations are nailed with galvanized common wire nails or threaded hardened-steel nails of sufficient length to penetrate 2.5 laminations. For 1-1/2-inch laminations, *20d* (4-inch) nails are used. For full-sawn 2-inch laminations, *40d* (5-inch) nails are sufficient. Nails are placed on approximately 9-inch centers near the top and bottom edges of the lamination.[22,60] The placement pattern is staggered over three successive laminations as shown in Figure 7-38.

Table 7-18. - Approximate maximum effective span for continuous transverse nail-laminated decks based on horizontal shear; loading from a 12,000-pound wheel load plus the deck dead load; b_d = 15 inches + deck thickness.

F_v (lb/in²)	F_v' (lb/in²)	Maximum deck span (in.)			
		t = 3-1/2 in.	t = 4 in.	t = 5-1/2 in.	t = 6 in.
100	194	40	68	>72	>72
95	185	36	57	>72	>72
90	175	32	48	>72	>72
85	165	30	41	>72	>72
80	155	27	36	>72	>72
75	146	25	33	>72	>72

$F_v' = F_v C_M$ (shear stress modification factor) = F_v(0.97)(2.0). The 2.0 shear stress modification factor assumes no splitting of the deck laminates across the wheel load distribution width, b_d.

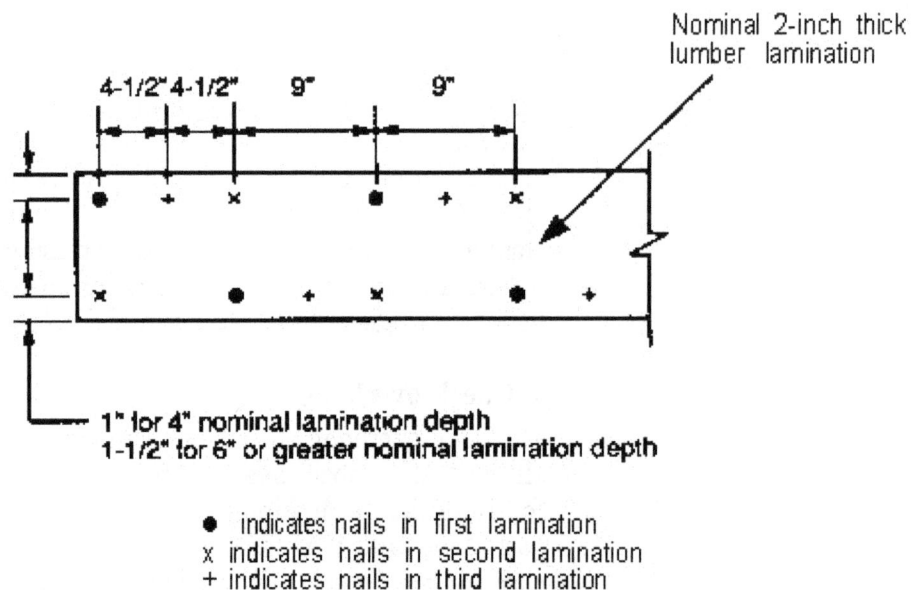

Figure 7-38. - Nail placement pattern for transverse nail-laminated lumber decks.

7-126

Design a transverse continuous nail-laminated lumber deck for the beam superstructure of Example 7-9. The superstructure has a two-lane, 24-foot roadway that carries AASHTO HS 15-44 loading. Support is provided by surfaced 6- by 18-inch lumber beams, spaced 24 inches on center. The out-to-out bridge span is 17 feet 7 inches. The following assumptions apply:

1. Deck laminations are visually graded Southern Pine.

2. The deck is provided with a full-width lumber wearing surface of full-sawn planks, 3 inches thick.

3. Deck live load deflection must be limited to $s/500$.

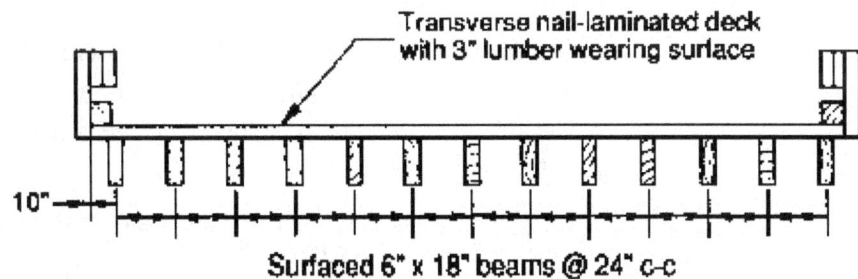

Solution

Define the Deck Span, Configuration, and Design Loads

The effective deck span is the clear distance between supporting beams plus one-half the width of one beam, but not greater than the clear span plus the deck thickness:

Clear distance between beams = 24 in. - 5.5 in. = 18.50 in.

$$s = 18.5 \text{ in.} + \frac{5.5 \text{ in.}}{2} = 21.25 \text{ in.}$$

The deck will be thicker than 2.75 inches, so $s = 21.25$ inches will control design.

For HS 15-44 loading the design load is one 12,000-pound wheel. Laminations will be continuous across the deck width in lengths of 25 feet 8 inches (25.67 feet).

Estimate Deck Thickness

An initial deck thickness of 4 inches (3.5 inches actual) is selected.

Determine Wheel Distribution Widths and Effective Deck Section Properties

In the direction of the deck span,

$$b_t = \sqrt{0.025P} = \sqrt{0.025\,(12,000)} = 17.32 \text{ in.}$$

Normal to the deck span,

$$b_d = 15 + t = 15 + 3.5 = 18.5 \text{ in.}$$

Effective deck section properties from Table 7-14 are

$$A = 64.75 \text{ in}^2$$

$$S = 37.77 \text{ in}^3$$

$$I = 66.10 \text{ in}^4$$

Compute Dead Load, Dead Load Moment, and Live Load Moment

For a 3.5-inch deck and 3-inch timber wearing surface, the dead load unit weight and moment over the effective distribution width of 18.5 inches are computed:

$$w_{DL} = (18.5 \text{ in.})\frac{(13.5 \text{ in.} + 3 \text{ in.})(50 \text{ lb/ft}^3)}{1,728 \text{ in}^3/\text{ft}^3} = 3.5 \text{ lb/in.}$$

$$M_{DL} = \frac{w_{DL}s^2}{8} = \frac{3.5(21.25)^2}{8} = 197.6 \text{ in-lb}$$

Live load moment is computed by Equation 7-44:

$$M_{LL} = 3,000s - 25,983 = 3,000\,(21.25) - 25,983 = 37,767 \text{ in-lb}$$

Compute Bending Stress and Select a Lamination Species and Grade

The deck is continuous over more than two spans, so bending stress is based on 80 percent of the simple span moment:

$$M = M_{DL} + M_{LL} = 196.6 + 37,767 = 37,964 \text{ in-lb}$$

$$f_b = \frac{0.80M}{S} = \frac{0.80(37,964)}{37.7} = 804 \text{ lb/in}^2$$

From NDS Table 4A, No.2 Southern Pine in the size classification 2 to 4 inches thick, 2 to 4 inches wide is selected from the table "surfaced dry used at 19% m.c." For wet-use conditions (>19 percent), NDS Table 4A footnotes require that tabulated values be taken from the Southern Pine table "surfaced green used any condition." These values are adjusted for moisture content and further application of C_M is not required:

$F_b = 1,300$ lb/in^2 (repetitive member use)

$F_v = 85$ lb/in^2

$E = 1,400,000$ lb/in^2

$F_b' = F_b C_M = 1,300(1.0) = 1,300$ lb/in^2

$f_b = 804$ lb/in$^2 < F_b' = 1,300$ lb/in^2, so a 4-inch nominal deck is satisfactory in bending. Although the allowable stress is considerably higher than the applied stress, No. 2 is the lowest grade of structural lumber that meets stress requirements.

Check Live Load Deflection

The deck is continuous over more that two spans, so deflection is 80 percent of the simple span deflection computed by Equation 7-48 (or by Figure 7-37):

$$E' = EC_M = 1,400,000 \ (1.0) = 1,400,000 \text{ lb/in}^2$$

$$\Delta_{LL} = 0.80 \left[\frac{1.80}{EI} \left(138.8s^3 - 20,780s + 90,000 \right) \right]$$

$$\Delta_{LL} = 0.80 \left[\frac{1.80 \left[(138.8)(21.25)^3 - 20,780 (21.25) + (90,000) \right]}{1,400,000 (66.10)} \right]$$

$$= 0.02 \text{ in.}$$

0.02 inch $= s/1,063 < s/500$, so live load deflection is acceptable.

Check Horizontal Shear

Dead load vertical shear is computed at a distance t from the support by Equation 7-49:

$$V_{DL} = w_{DL} \left(\frac{s}{2} - t \right) = 3.5 \left(\frac{21.25}{2} - 3.5 \right) = 24.9 \text{ lb}$$

Live load vertical shear is computed by placing the edge of the wheel load distribution width (b_t) a distance t from the support. The resultant of the 12,000-pound wheel load acts through the center of the distribution width and V_{LL} is computed by statics:

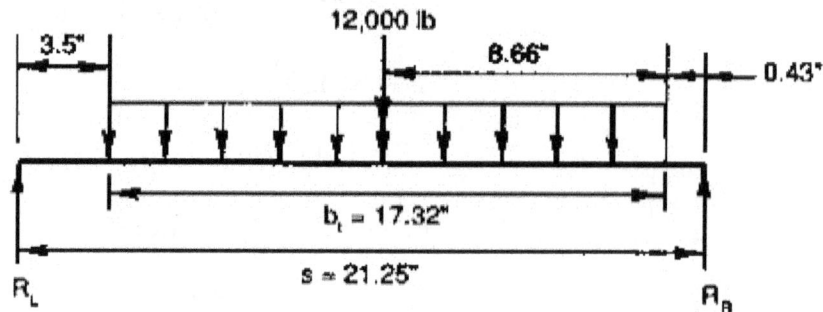

$$V_{LL} = R_L = \frac{(12,000 \text{ lb})(8.66 \text{ in.} + 0.43 \text{ in.})}{21.25 \text{ in.}} = 5,133 \text{ lb}$$

$$V = V_{DL} + V_{LL} = 24.9 + 5,133 = 5,158 \text{ lb}$$

$$f_v = \frac{1.5V}{A} = \frac{1.5(5,158)}{64.75} = 119 \text{ lb/in}^2$$

By Equation 7-50,

$$F_v' = F_v C_{ts} \text{(shear stress modification factor)}$$

For nail-laminated lumber treated with oil-type preservatives, a shear stress modification factor of 2.0 is applicable (Table 7-17):

$$F_v' = 85(1.0)(2.0) = 170 \text{ lb/in}^2$$

$f_v = 119 \text{ lb/in}^2 < F_v' = 170 \text{ lb/in}^2$, so the deck is satisfactory in horizontal shear.

Summary

The deck will consist of 141 surfaced 2- by 4-inch lumber laminations that are 25 feet 8 inches long. The laminations will be nailed together and to the beams using the nailing pattern shown in Figures 7-38 and 7-39. The lumber will be No. 2 or better Southern Pine (surfaced dry), visually graded to SPIB rules. Stresses and deflection are as follows:

$$f_b = 804 \text{ lb/in}^2$$

$$F_b' = 1,300 \text{ lb/in}^2$$

$$\Delta_{LL} = 0.02 \text{ in.} = L/1,063$$

$$f_v = 119 \text{ lb/in}^2$$

$$F_v' = 170 \text{ lb/in}^2$$

An existing bridge spans 38 feet out-to-out and is supported by three steel wide flange beams, spaced 5 feet on center. The roadway width of 12 feet carries one lane of AASHTO HS 20-44 loading. The existing concrete deck is to be removed and replaced with a continuous transverse nail-laminated lumber deck with a 4-inch-thick plank wearing surface. Design the deck for this structure, assuming the following:

1. All lumber is surfaced (S4S) visually graded Douglas Fir-Larch.

2. The beam top flange width is 12 inches.

3. Deck live load deflection is limited to *s/500*.

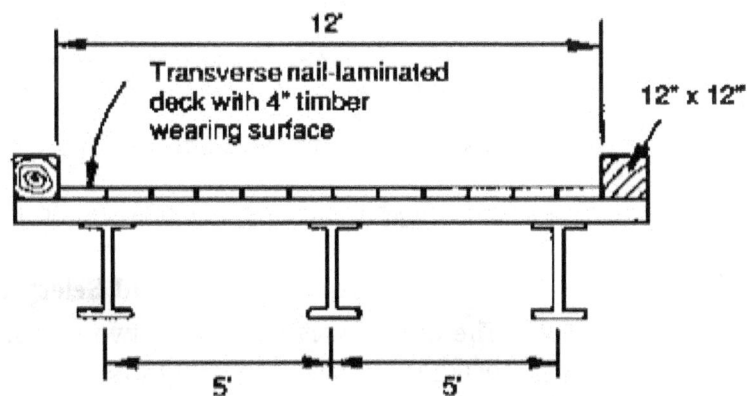

Solution

Define the Deck Span, Configuration, and Design Loads

Clear distance between beams = 60 in - 12 in = 48 in

$$s = 48 \text{ in.} + \frac{12}{2} \text{ in.} = 54 \text{ in.}$$

For HS 20-44 loading, AASHTO special wheel load provisions apply and the deck will be designed for a 12,000-pound wheel load. Laminations will be continuous across the deck width in lengths of 14 feet.

Estimate Deck Thickness

An initial deck thickness of 6 inches (5.5 inches actual) is selected. Deck span will be controlled by the clear distance plus deck thickness:

$$s = 48 \text{ in.} + 5.5 \text{ in.} = 53.5 \text{ in.}$$

Determine Wheel Distribution Widths and Effective Deck Section Properties

$$b_t = \sqrt{0.025P} = \sqrt{0.025(12,000)} = 17.32 \text{ in.}$$

7-131

$$b_d = 15 + t = 15 + 5.5 = 20.5 \text{ in.}$$

From Table 7-14,

$$A = 112.75 \text{ in}^2$$

$$S = 103.35 \text{ in}^3$$

$$I = 284.22 \text{ in}^4$$

Compute Dead Load, Dead Load Moment, and Live Load Moment

For a 5.5-inch deck and 3.5-inch timber wearing surface over the effective distribution width of 20.5 inches,

$$w_{DL} = (20.5 \text{ in.}) \left[\frac{(5.5 \text{ in.} + 3.5 \text{ in.})(50 \text{ lb/ft}^3)}{1,728 \text{ in}^3/\text{ft}^3} \right] = 5.3 \text{ lb/in.}$$

$$M_{DL} = \frac{w_{DL} s^2}{8} = \frac{5.3 (53.50)^2}{8} = 1,896 \text{ in-lb}$$

Live load moment is computed by Equation 7-44:

$$M_{LL} = 3,000s - 25,983 = 3,000(53.50) - 25,983 = 134,517 \text{ in-lb}$$

Compute Bending Stress and Select a Lamination Species and Grade

The deck is continuous over two spans, so the 80-percent reduction in bending for span continuity does not apply.

$$M = M_{DL} + M_{LL} = 1,896 + 134,517 = 136,413 \text{ in-lb}$$

$$f_b = \frac{M}{S} = \frac{136,413}{103.35} = 1,320 \text{ lb/in}^2$$

From NDS Table 4A, visually graded No. 1 Douglas Fir-Larch in the J&P size classification is selected. Tabulated values are as follows:

$F_b = 1,750 \text{ lb/in}^2$ (repetitive uses) $C_M = 0.86$

$F_v = 95 \text{ lb/in}^2$ $C_M = 0.97$

$E = 1,800,000 \text{ lb/in}^2$ $C_M = 0.97$

$$F_b' = F_b C_M = 1,750(0.86) = 1,505 \text{ lb/in}^2$$

$f_b = 1,320 \text{ lb/in}^2 < F_b' = 1,505 \text{ lb/in}^2$, so a 6-inch nominal deck is satisfactory in bending.

Check Live Load Deflection

Maximum deflection is computed by Equation 7-48 (or Figure 7-37):

$$E' = EC_M = 1,800,000(0.97) = 1,746,000 \text{ lb/in}^2$$

$$\Delta_{LL} = \frac{1.80}{E'\,I}\left(138.8s^3 - 20,780s + 90,000\right)$$

$$\Delta_{LL} = \frac{1.80\left[(138.8)(53.5)^3 - 20,780(53.5) + (90,000)\right]}{1,649,000(284.22)} = 0.07 \text{ in.}$$

0.07 in.= s/764 < s/500, so live load deflection is acceptable.

Check Horizontal Shear

For $w_{DL} = 5.3$ lb/in.,

$$V_{DL} = w_{DL}\left(\frac{s}{2} - t\right) = 5.3\left(\frac{53.5}{2} - 5.5\right) = 112.6 \text{ lb}$$

For a 12,000-pound wheel load,

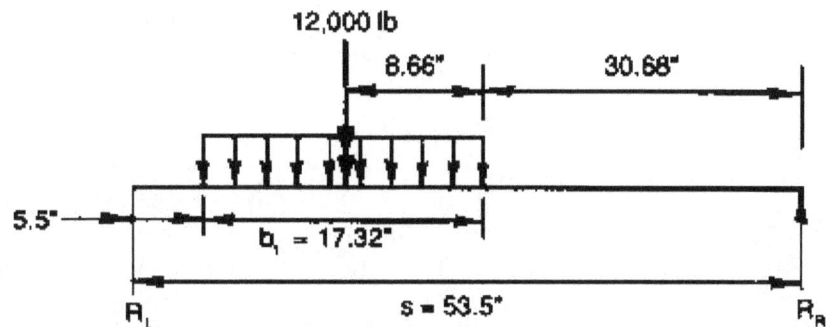

$$V_{LL} = R_L = \frac{(12,000 \text{ lb})(8.66 \text{ in.} + 30.68 \text{ in.})}{53.5 \text{ in.}} = 8,824 \text{ lb}$$

$$V = V_{DL} + V_{LL} = 112.6 + 8,824 = 8,937 \text{ lb}$$

$$f_v = \frac{1.5V}{A} = \frac{1.5(8,937)}{112.75} = 119 \text{ lb/in}^2$$

$$F_v' = F_v C_M (\text{shear stress modification factor})$$

Using a shear stress modification factor of 2.0 (Table 7-17),

$$F_v' = 95(0.97)(2.0) = 184.30 \text{ lb/in}^2$$

$f_v = 119$ lb/in^2 < $F_v' = 184$ lb/in^2, so the deck is satisfactory in horizontal shear.

Summary

The deck will consist of 304 surfaced 2-inch by 6-inch lumber laminations, 14 feet long. The laminations will be nailed as shown in Figures 7-38 and 7-39. The lumber will be No. 1 or better Douglas Fir-Larch, visually graded to WCLIB rules. Stresses and deflection are as follows:

$$f_b = 1{,}320 \text{ lb/in}^2$$

$$F_b' = 1{,}505 \text{ lb/in}^2$$

$$\Delta_{LL} = 0.07 \text{ in.} = L / 764$$

$$f_v = 119 \text{ lb/in}^2$$

$$F_v' = 184 \text{ lb/in}^2$$

DECK ATTACHMENT

Nail-laminated decks can be placed on timber or steel beams using several attachment configurations. For timber beams, the most common attachment is to nail the laminations to beam tops as the deck is constructed. Every other lamination is toenailed to every other beam with nails the same size as those used for laminating. When this method is used, the NDS recommends that toenails be driven at an angle of approximately 30 degrees with the piece and started approximately one-third the length of the nail from the edge of the piece (Figure 7-39). Although nailing provides satisfactory performance from a structural standpoint, the nails penetrate the beam top and increase susceptibility to decay. A more suitable connection is achieved using bolted bracket attachments like those used for glulam panels. On steel beams, nail-laminated decks can be attached with bolted C-clip or angle-clip attachments previously discussed. Another method of attachment involves a thin steel plate (or sheet) connector that fits over the top beam flange and is nailed to the lamination (Figure 7-40).

7.9 PLANK DECKS

Transverse plank decks consist of a series of sawn lumber planks placed flatwise across supporting beams (Figure 7-41). The planks are normally 10 or 12 inches wide and 4 inches thick, although a minimum plank thickness of 3 inches is allowed by AASHTO (AASHTO 13.9.4.1). Plank decks are used primarily on low-volume or special-use roads. They are not suitable for asphalt pavement because of large live load deflections and movements from moisture changes in the planks. In addition, plank decks are normally not practical in applications where traffic railing is required to meet full AASHTO standards (Chapter 10).

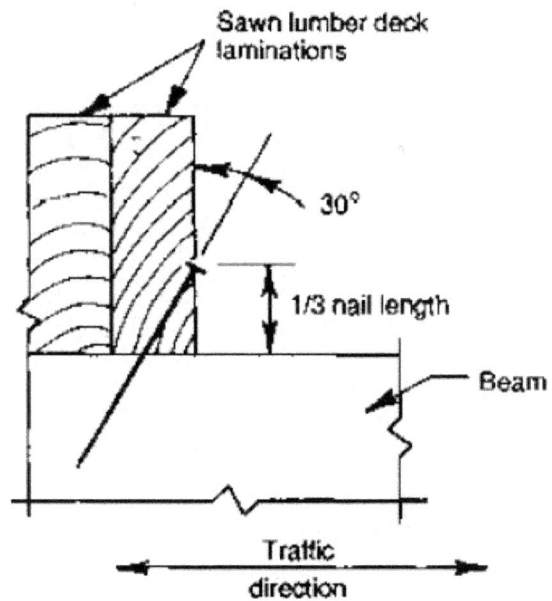

Figure 7-39. - Recommended toenail placement for attaching transverse lumber laminations to timber beams.

Figure 7-40. - Steel plate deck attachment for nail-laminated lumber decks on steel beams. The thin steel plate is placed over the top beam flange and is nailed to the lumber laminations during deck construction.

Figure 7-41. - Transverse plank deck on a single-lane, low-volume road (photo courtesy of Wheeler Consolidated, Inc.).

The performance of plank decks can be improved when edge-grain rather than flat-grain lumber is used (Chapter 3). In edge-grain material, dimensional changes from moisture result in fairly uniform changes in plank width and depth. For flat-grain material, dimensional changes depend on the orientation of growth rings, and swelling or shrinking can cause planks to cup. If edge-grain lumber is not available, flat-grain lumber should be placed with the bark side up so any cupping that occurs will be downward, rather than upward where water can be trapped. When green (unseasoned) planks are used, they should be placed with a tight joint between planks. When seasoned planks are used, a small gap of 1/4 to 1/2 inch should be left between planks to allow for potential swelling as the moisture content of the planks increases.

Planks are attached to supporting beams with galvanized spikes that are 1/4 to 3/8 inch in diameter and approximately twice as long as the deck is thick. Two spikes are placed in each plank at each beam. Resistance to withdrawal is improved if spikes are driven at a slight angle rather than vertically into the beam.

DESIGN PROCEDURES

Design procedures for transverse plank decks are fundamentally the same as those previously given for nail-laminated decks. Instead of a wheel load distribution width, however, wheel loads on plank decks are assumed to be distributed over the plank width (AASHTO 3.25.1.1). Because of the relatively short-span capabilities of plank decks, design is often controlled by horizontal shear rather than bending.

Design procedures for plank decks are illustrated in the following example. Approximate maximum spans for plank decks based on bending and shear are given in Tables 7-19 and 7-20.

Table 7-19. - Approximate maximum effective span for transverse plank decks based on bending; deck continuous over more than two spans; loading from a 12,000-pound wheel load plus the deck dead load; wheel-load distribution width equals the plank width.

| | | Maximum effective span (In.) | | | |
| | | Dimension lumber[a] | | Full-sawn lumber[b] | |
F_b	F_b'	4 by 10	4 by 12	4 by 10	4 by 12
1,900	1,814	22	26	28	32
1,850	1,766	22	25	28	32
1,800	1,718	22	24	27	31
1,750	1,623	21	24	27	30
1,700	1,623	21	24	26	30
1,650	1,527	21	23	26	29
1,600	1,527	20	23	25	28
1,550	1,480	20	22	25	28
1,500	1,432	19	22	24	27
1,450	1,384	19	21	24	27
1,400	1,336	19	21	23	26
1,350	1,289	18	20	22	25
1,300	1,240	18	20	22	25
1,250	1,193	18	20	21	24
1,200	1,146	17	19	21	23
1,150	1,098	17	19	20	23
1,100	1,050	16	18	20	22
1,050	1,002	16	18	19	22
1,000	955	16	17	19	21

[a] Plank sizes for 4 by 10 and 4 by 12 dressed lumber are 3-1/2 inches by 9-1/4 inches and 3-1/2 inches by 11-1/4 inches, respectively.

[b] Plank sizes for 4 by 10 and 4 by 12 full-sawn lumber are 4 inches by 10 inches and 4 inches by 12 inches, respectively.

$F_b' = F_b C_M$(modification factor for flatwise use) $= F_b(0.86)(1.11)$.

Table 7-20. - Approximate maximum effective span for transverse plank decks based on horizontal shear; loading from a 12,000-pound wheel load plus the deck dead load; wheel load distribution width equal to plank width.

		Maximum Effective Span (in.)			
		Dimension lumber[a]		Full-sawn lumber[b]	
F_v	F_v'	4 by 10	4 by 12	4 by 10	4 by 12
100	194	18	21	22	26
95	184	17	20	21	24
90	175	17	19	20	23
85	165	16	18	19	22
80	155	15	18	19	21
75	146	15	17	18	20
70	136	14	16	17	19
65	126	14	15	16	18
60	116	13	15	15	17

[a] Plank sizes for 4 by 10 and 4 by 12 dressed lumber are 3-1/2 inches by 9-1/4 inches and 3-1/2 inches by 11-1/4 inches, respectively.

[b] Plank sizes for 4 by 10 and 4 by 12 full-sawn lumber are 4 inches by 10 inches and 4 inches by 12 inches, respectively.

$F_v' = F_v C_M$ (Shear stress modification factor) $= F_v (0.97)(2.0)$.

Example 7-13 - Transverse plank deck design; single-lane HS 15-44 loading

A longitudinal lumber beam superstructure carries AASHTO HS 15-44 loading and consists of a series of nominal &inch-wide lumber beams spaced 24 inches center-to-center. Design a transverse plank deck for this bridge assuming the following:

1. The deck is provided with a full-width lumber wearing surface constructed of nominal 2-inch planks.

2. All lumber, including the wearing surface, is dressed (S4S) Douglas Fir-Larch.

3. Deck live load deflection must be limited to *s/500*.

7-138

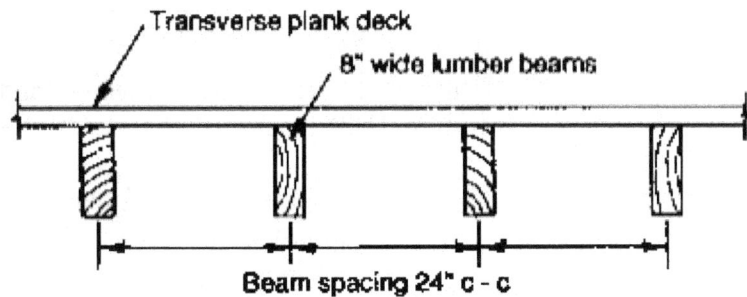

Transverse plank deck

8" wide lumber beams

Beam spacing 24" c - c

Solution

Define the Deck Span, Configuration, and Design Loads

The deck span is the clear distance between supporting beams plus one-half the width of one beam, but not greater than the clear span plus the deck thickness. From Table 16-2, the actual width of a dressed 8-inch-wide beam is 7.50 inches:

Clear distance between beams = 24 in. - 7.5 in. = 16.5 in.

$$s = 16.5 \text{ in.} + \frac{7.5}{2} \text{ in.} = 20.25 \text{ in.}$$

If a nominal 4-inch-thick plank is used (3.5 inches actual thickness), the deck span will be limited by the clear span plus the deck thickness:

$$s = 16.5 \text{ in.} + 3.5 \text{ in.} = 20 \text{ in.}$$

For HS 15-44 loading, the deck will be designed for a 12,000-pound wheel load.

Estimate Plank Size and Determine Section Properties

Plank decks are generally constructed of 4- by 10-inch or 4- by 12-inch lumber. In this case, a dressed 4- by 12-inch plank is selected. Section properties are obtained from Table 16.2:

$b = 11.25$ in.

$d = 3.50$ in.

$A = 39.38$ in^2

$S = 22.97$ in^3

$I = 40.20$ in^4

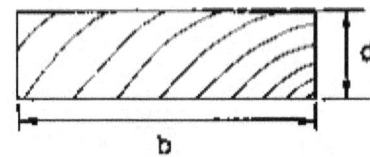

Determine Wheel Distribution Widths

In the direction of the deck span, the wheel load is distributed over the tire width given by Equation 7-42:

7-139

$$b_t = \sqrt{0.025P} = \sqrt{0.025\,(12,000)} = 17.32 \text{ in.}$$

Normal to the deck span, the wheel load is distributed over the plank width of 11.25 inches.

Compute Dead Load, Dead Load Moment, and Live Load Moment

For a 3.5-inch deck and 1.5-inch timber wearing surface, the dead load is computed for the plank width:

$$w_{DL} = (11.25 \text{ in.}) \frac{\left[(3.5 \text{ in.} + 1.5 \text{ in.})(50 \text{ lb}/\text{ft}^3)\right]}{1,728 \text{ in}^3/\text{ft}^3} = 1.6 \text{ lb}/\text{in.}$$

$$M_{DL} = \frac{w_{DL}\,s^2}{8} = \frac{1.6(20)^2}{8} = 80 \text{ in-lb}$$

Live load moment is computed by Equation 7-44:

$$M_{LL} = 3,000s - 25,983 = 3,000(20) - 25,983 = 34,017 \text{ in-lb}$$

Compute Bending Stress and Select Plank Species and Grade

The deck is continuous over more than two spans, so bending stress is based on 80 percent of the simple span moment:

$$M = M_{DL} + M_{LL} = 80 + 34,017 = 34,097 \text{ in-lb}$$

$$f_b = \frac{0.80M}{S} = \frac{0.80\,(34,097)}{22.97} = 1,188 \text{ lb}/\text{in}^2$$

From Table 4A of the NDS, No. 2 Douglas Fir-Larch in the J&P size classification is chosen with the following tabulated values:

$$F_b = 1,250 \text{ lb/in}^2 \qquad\qquad C_M = 0.86$$

$$F_v = 95 \text{ lb/in}^2 \qquad\qquad C_M = 0.97$$

$$E = 1,700,000 \text{ lb/in}^2 \qquad\qquad C_M = 0.97$$

Footnotes to the NDS tabulated values also specify that bending stress may be increased by a factor of 1.11 for flatwise use:

$$F_b' = F_b C_M(1.11) = 1,250(0.86)(1.11) = 1,193 \text{ lb/in}^2$$

$f_b = 1,188$ lb/in$^2 < F_b' = 1,193$ lb/in^2, so the plank size and grade are satisfactory in bending.

Check Horizontal Shear

Dead load vertical shear is computed at a distance t from the support. By Equation 7-49 for $w_{DL} = 1.6$ lb/in,

$$V_{DL} = w_{DL}\left(\frac{s}{2} - t\right) = 1.6\left(\frac{20}{2} - 3.5\right) = 10.4 \text{ lb}$$

Live load vertical shear is computed by placing the edge of the wheel load distribution width (b_t) a distance t from the support. In this case, the remaining span is less than b, and the wheel load is converted to a uniform load:

$$w_{LL} = \frac{P}{b_t} = \frac{12,000}{17.32} = 692.8 \text{ lb/in.}$$

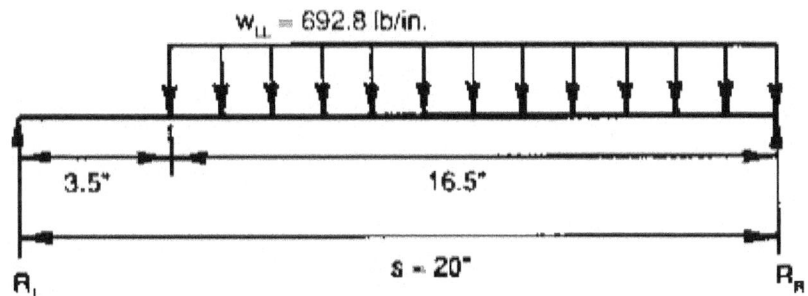

$$V_{LL} = R_L = \frac{(692 \text{ lb/in.})(16.5 \text{ in.})(16.5 \text{ in.}/2)}{20 \text{ in.}} = 4,715 \text{ lb}$$

$$V = V_{DL} + V_{LL} = 10.4 + 4,715 = 4,725 \text{ lb}$$

$$f_v = \frac{1.5 V}{A} = \frac{1.5 (4,725)}{39.38} = 180 \text{ lb/in}^2$$

By Equation 7-50,

$$F_v' = F_v C_M \text{ (shear stress modification factor)}$$

For planks treated with oil-type preservatives, a 2.0 shear stress modification factor is used (Table 7-17):

$$F_v' = 95(0.97)(2.0) = 184 \text{ lb/in}^2$$

$f_v = 180$ lb/in^2 $< F_v' = 184$ lb/in^2, so the deck is satisfactory in horizontal shear.

Check Live Load Deflection

Maximum deflection for a 12,000-pound wheel load and 6-foot track width on a simple span is computed by Equation 7-48. Because the deck is continuous over more than two spans, 80 percent of the simple span deflection is used to account for span continuity:

$$E' = EC_M = 1{,}700{,}000(0.97) = 1{,}649{,}000 \text{ lb/in}^2$$

$$\Delta_{LL} = (0.80)\left[\frac{1.80}{E'I}\left(138.8s^3 - 20{,}780s + 90{,}000\right)\right]$$

$$\Delta_{LL} = (0.80)\left[\frac{1.80\left[(138.8)(20)^3 - 20{,}780(20) + (90{,}000)\right]}{1{,}649{,}000(40.20)}\right] = 0.02 \text{ in.}$$

A deflection of 0.02 inch $= s/1{,}000 < s/500$, so live load deflection is acceptable.

Summary

The deck will consist of surfaced 4-inch by 12-inch Douglas Fir-Larch planks, visually graded No. 2 or better in the J&P size classification. Stresses and deflection are as follows:

$$f_b = 1{,}188 \text{ lb/in}^2$$

$$F_b' = 1{,}193 \text{ lb/in}^2$$

$$\Delta_{LL} = 0.02 \text{ in.} = s/1{,}000$$

$$f_v = 180 \text{ lb/in}^2$$

$$F_v' = 184 \text{ lb/in}^2$$

7.10 SELECTED REFERENCES

1. American Association of State Highway and Transportation Officials. 1983. Standard specifications for highway bridges. 13th ed. Washington, DC: American Association of State Highway and Transportation Officials. 394 p.

2. American Association of State Highway and Transportation Officials. 1983. Manual for maintenance inspection of bridges. Washington, DC: American Association of State Highway and Transportation Officials. 50 p.

3. American Association of State Highway and Transportation Officials. 1982. AASHTO materials: pt. 1, specifications. Washington, DC: American Association of State Highway and Transportation Officials. 1094 p.

4. American Institute of Timber Construction in conjunction with the Virginia Highway Research Council. [1973]. Typical timber bridge design and details. Englewood, CO: American Institute of Timber Construction. 10 p.

5. American Institute of Timber Construction. 1987. Design standard specifications for structural glued laminated timber of softwood species. AITC 117-87-Design. Englewood, CO: American Institute of Timber Construction. 28 p.

6. American Institute of Timber Construction. 1974. Glulam bridge systems plans and details. Englewood, CO: American Institute of Timber Construction. 16 p.

7. American Institute of Timber Construction. 1988. Glulam bridge systems. Vancouver, WA: American Institute of Timber Construction. 33 p.

8. American Institute of Timber Construction. 1973. Modem timber highway bridges, a state of the art report. Englewood, CO: American Institute of Timber Construction. 79 p.

9. American Institute of Timber Construction. 1985. Timber construction manual. 3d ed. New York: John Wiley and Sons, Inc. 836 p.

10. American Society of Civil Engineers. 1975. Wood structures, a design guide and commentary. New York: American Society of Civil Engineers. 416 p.

11. American Wood-Preservers' Association. 1987. Book of standards. Stevensville, MD: American Wood-Preservers' Association. 240 p.

12. Anderson, L.O.; Heebink, T.B.; Oviatt, A.E. 1971. Construction guide for exposed wood decks. Portland, OR: U.S. Department of Agriculture, Forest Service, Pacific Northwest Forest and Range Experiment Station. 78 p.

13. Barnhart, J.E. 1986. Ohio's experiences with treated timber for bridge construction. In: Trans. Res. Rec. 1053. Washington, DC: National Academy of Sciences, National Research Council, Transportation Research Board: 56-58.

14. Bell, L.C.; Yoo, C.H. 1984. Seminar on fundamentals of timber bridge construction. Course notes; 1984 May 22-25; Auburn University, AL. Auburn University. [150 p.].

15. Better Roads. 1976. Glulam helping to solve America's bridge problem. Better Roads 46(5): 36-37.

16. Bohannan, B. 1972. FPL timber bridge deck research. Journal of the Structural Division, American Society of Civil Engineers 98(ST3): 729-740.

17. Boomsliter, G.P.; Cather, C.H.; Worrell, D.T. 1951. Distribution of wheel loads on a timber bridge floor. Res. Bull. 24. Morgantown, WV: West Virginia University, Engineering Experiment Station. 31 p.

18. Canadian Institute of Timber Construction. 1970. Modem timber bridges, some standards and details. 3d ed. Ottawa, Can.: Canadian Institute of Timber Construction. 48 p.

19. Commonwealth of Pennsylvania, Department of Transportation. 1984. Standard plans for low cost bridges. Series BLC-540, timber spans. Pub. No. 130. [Pittsburgh, PA]: Commonwealth of Pennsylvania, Department of Transportation. 28 p.

20. Erickson, E.C.O.; Romstad, K.M. 1965. Distribution of wheel loads on timber bridges. Res. Pap. FPL 44. Madison, WI: U.S. Department of Agriculture, Forest Service, Forest Products Laboratory. 62 p.

21. Freas, A.D. 1952. Laminated timber permits flexibility of design. Civil Engineering 22(9): 173-175.

22. Gurfinkel, G. 1981. Wood engineering. 2d ed. Dubuque, IA: Kendall/Hunt Publishing Co. 552 p.

23. Gutkowski, R.M.; McCutcheon, W.J. 1984. Comparative performance of experimental timber bridges. Madison, WI: U.S. Department of Agriculture, Forest Service, Forest Products Laboratory. 27 p.

24. Gutkowski, R.M.; Williamson, T.G. 1983. Timber bridges: state-of-the-art Journal of Structural Engineering 109(9): 2175-2191.

25. Hale, C.Y. 1975. Field test of a 40-ft span, two-lane Weyerhaeuser panelized wood bridge. Rep. No. RDR 045-1092. Tacoma, WA: Weyerhaeuser Co. 22 p.

26. Hockaday, E. 1980. Preliminary study of duration of load for glulam wood bridges used in rural areas. St. Paul, MN: Weyerhaeuser Co. 4p.

27. Leviasky, S. 1982. Timber bridges. In: Arches and short span bridges. New York: Chapman and Hall Ltd. 48-83.

28. Manesh, A.A.S. 1977. Design formulas for slab-stringer bridges. Morgantown, WV: West Virginia University, Department of Civil Engineering. 53 p.

29. McCutcheon, W.J.; Tuomi, R.L. 1973. Procedure for design of glued-laminated orthotropic bridge decks. Res. Pap. FPL 210. Madison, WI: U.S. Department of Agriculture, Forest Service, Forest Products Laboratory. 42 p.

30. McCutcheon, W.J.; Tuomi, R.L. 1974. Simplified design procedure for glued-laminated bridge decks. Res. Pap. FPL 233. Madison, WI: U.S. Department of Agriculture, Forest Service, Forest Products Laboratory. 8 p.

31. Mielke, K.F. 1977. Experimental project for glued-laminated timber deck panels on highway bridges. Juneau, AK: State of Alaska, Department of Highways. [50 p.].

32. Ministry of Transportation and Communications. 1986. Ontario highway bridge design code updates. Downsview, ON, Can.: Ministry of Transportation and Communications. 28 p.

33. Ministry of Transportation and Communications. 1983. Ontario highway bridge design code. Downsview, ON, Can.: Ministry of Transportation and Communications. 357 p.

34. Ministry of Transportation and Communications. 1983. Ontario highway bridge design code commentary. Downsview, ON, Can,: Ministry of Transportation and Communications. 279 p.

35. Muchmore, F.W. 1986. Designing timber bridges for long life. In: Trans. Res. Rec. 1053. Washington, DC: National Academy of Sciences, National Research Council, Transportation Research Board: 12-17.

36. Nagy, M.M.; Trebett, J.T.; Wellburn, G.V. 1980. Log bridge construction handbook. Vancouver, Can.: Forest Engineering Research Institute of Canada. 421 p.

37. National Forest Products Association. 1986. Design values for wood construction. A supplement to the national design specification for wood construction. Washington, DC: National Forest Products Association. 34 p.

38. National Forest Products Association. 1986. National design specification for wood construction. Washington, DC: National Forest Products Association. 87 p.

39. Nowak, A.S.; Taylor, R.J. 1986. Ultimate strength of timber deck bridges. In: Trans. Res. Rec. 1053. Washington, DC: National Academy of Sciences, National Research Council, Transportation Research Board: 26-30.

40. Ou, F.L. 1986. An overview of timber bridges. In: Trans. Res. Rec. 1053. Washington, DC: National Academy of Sciences, National Research Council, Transportation Research Board: 1-12.

41. Parry, J.D. 1986. A prefabricated modular timber bridge. In: Trans. Res. Rec. 1053. Washington, DC: National Academy of Sciences, National Research Council, Transportation Research Board: 49-55.

42. Sanders, W.W. 1984. Distribution of wheel loads on highway bridges. National Cooperative Highway Research Program Synthesis of Highway Practice. No. 3. Washington, DC: National Academy of Sciences, National Research Council, Transportation Research Board. 22 p.

43. Sanders, W.W., Jr. 1980. Load distribution in glulam timber highway bridges. Report ISU-ERI-AMES-80124. Ames, IA: Iowa State University, Engineering Research Institute. 21 p.

44. Sanders, W.W., Jr.; Elleby, H.A. 1970. Distribution of wheel loads on highway bridges. Cooperative Highway Research Program. Rep. 83. Washington, DC: National Academy of Sciences, National Research Council, Highway Research Board. 56 p.

45. Scales, W.H. 1959. Standard treated timber bridges. In: Standardization of highway bridges. Bull. No. 244. Washington, DC: American Road Builders' Association: 22-26.

46. Scarisbrick, R.G. 1976. Laminated timber logging bridges in British Columbia. Journal of the Structural Division, American Society of Civil Engineers 102(ST1): [10 p.].

47. Selbo, M.L. 1966. Laminated bridge decking (progress report). Madison, WI: U.S. Department of Agriculture, Forest Service, Forest Products Laboratory. 32 p.

48. Selbo, M.O.; Knauss, A.C.; Worth, H.E. 20 years of service prove durability of pressure-treated glulam bridge members. Wood Preserving News 44(3): 5-8.

49. Southern Forest Products Association. 1985. Southern Pine use guide. New Orleans, LA: Southern Forest Products Association. 13 p.

50. Sprinkel, M.M. 1978. Glulam timber deck bridges. VHTRC 79-R26. Charlottesville, VA: Virginia Highway and Transportation Research Council. 33 p.

51. Sprinkel, M.M. 1985. Prefabricated bridge elements and systems. National Cooperative Highway Research Program, Synthesis of Highway Practice 119. Washington, DC: National Academy of Sciences, National Research Council, Transportation Research Board. 75 p.

52. Stacey, W.A. 1935. The design of laminated timber bridge floors. Wood Preserving News 13(4): 44-46, 55-56.

53. Stanton, J.F.; Roeder, C.W. 1982. Elastomeric bearing design. Nat. Coop. Highw. Res. Program Rep. 248. Washington, DC: National Academy of Sciences, National Research Council, Transportation Research Board. 82 p.

54. Stone, M.F. [1975]. New concepts for short span panelized bridge design of glulam timber. Tacoma, WA: Weyerhaeuser Co. 8 p.

55. Sunset Foundry Company, Inc. [1970]. Deck brackets for treated timber bridges. AIA File 17F. Kent, WA: Sunset Foundry. 4 p.

56. Tuomi, R.L. 1972. Advancements in timber bridges through research and engineering. In: Proceedings, 13th annual Colorado State University bridge engineering conference; Ft. Collins, CO. Colorado State University: 34-61.

57. Tuomi, R.L. 1976. Erection procedure for glued-laminated timber bridge decks with dowel connectors. Res. Pap. FPL 263. Madison, WI: U.S. Department of Agriculture, Forest Service, Forest Products Laboratory. 15 p.

58. Tuomi, R.L.; McCutcheon, W.J. 1973. Design procedure for glued laminated bridge decks. Forest Products Journal 23(6): 36-42.

59. U.S. Department of Agriculture, Forest Service, Northern Region. 1985. Bridge design manual. Missoula, MT: U.S. Department of Agriculture, Forest Service, Northern Region. 299 p.

60. U.S. Department of Transportation, Federal Highway Administration. 1979. Standard plans for highway bridges. Timber bridges. Washington, DC: U.S. Department of Transportation, Federal Highway Administration. Vol. 3. 19 p.

61. Weyerhaeuser Company. 1976. Weyerhaeuser bridge deck bracket. SL-495. Tacoma, WA: Weyerhaeuser Co. 4 p.

62. Weyerhaeuser Company. 1980. Weyerhaeuser glulam wood bridge systems. Tacoma, WA: Weyerhaeuser Co.; 1980. 114 p.

63. Weyerhaeuser Company. 1975. Weyerhaeuser panelized bridge system for secondary roadways, highways & footbridges. SL-1318. Tacoma, WA: Weyerhaeuser Co. 4 p.

64. Weyerhaeuser Company. 1974. Weyerhaeuser panelized bridge system. SL-1318. Tacoma, WA: Weyerhaeuser Co. 4 p.

CHAPTER 8

DESIGN OF LONGITUDINAL DECK SUPERSTRUCTURES

8.1 INTRODUCTION

Longitudinal deck superstructures consist of a glulam or nail-laminated lumber deck placed over two or more substructure supports (Figure 8-1). The lumber laminations are placed parallel to traffic, and loads are applied parallel to the wide face of the laminations. The deck provides all structural support for the roadway, without the aid of beams or other components. In most configurations, however, transverse stiffener beams are connected to the deck underside to distribute loads laterally across the bridge width. Longitudinal deck bridges provide a low profile that makes them especially suitable for short-span applications where clearance below the structure is limited. The same basic configuration can also be used over transverse floorbeams for the construction or rehabilitation of other superstructure types.

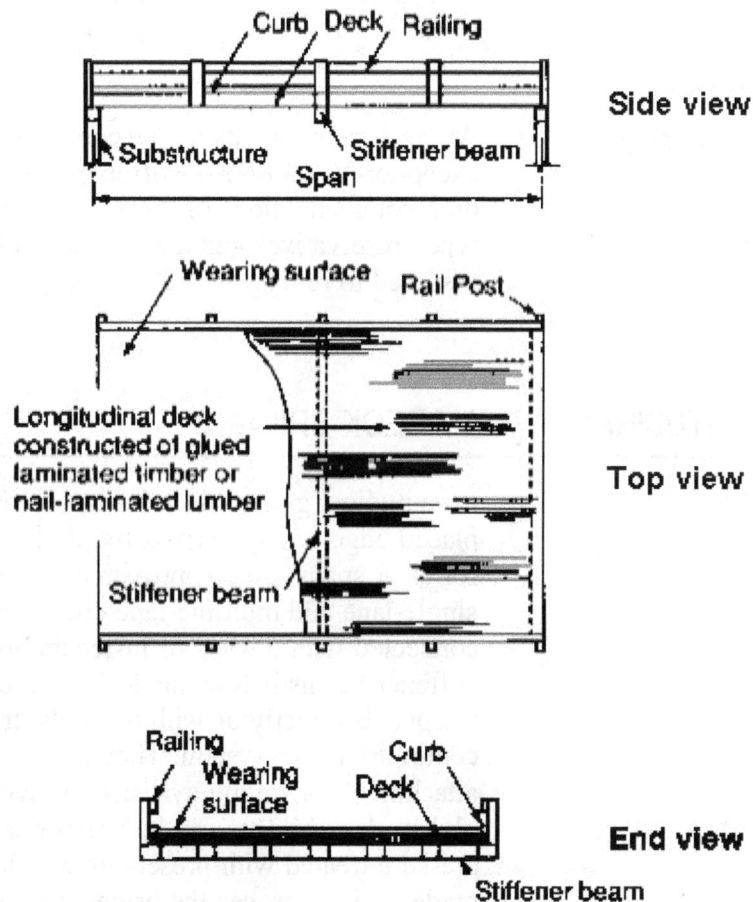

Figure 8-1. - Typical configuration for a single-lane longitudinal deck bridge.

This chapter discusses the design requirements and considerations for longitudinal deck bridges constructed of glulam and nail-laminated sawn lumber. Railing and wearing surfaces for longitudinal decks are addressed in Chapters 10 and 11, respectively.

8.2 DESIGN CRITERIA AND DEFINITIONS

The design requirements addressed in this chapter are based on the 1983 edition of the AASHTO Standard Specifications for Highway Bridges, including interim specifications through 1987.[1] The criteria related to design procedures and examples, loads, materials, live load deflection, and conditions of use are the same as those given for beam superstructures in Chapter 7, with the following exceptions.

LOADS

Longitudinal decks are designed for the maximum forces and deflection produced by the design vehicle, assuming that wheel loads act as point loads in the direction of the deck span (AASHTO 3.25.2.3). AASHTO special provisions for reduced wheel loads for H 20-44 and HS 20-44 trucks *do not* apply to longitudinal decks.

CONDITIONS OF USE

All deck components are designed using wet-condition stresses with the exception of transverse stiffener beams for watertight glulam decks. Based on recommendations of AITC, stiffener beams that are treated with oil-type preservatives and are located under a watertight glulam deck are assumed to remain within the range of dry-use conditions.[4]

8.3 LONGITUDINAL GLULAM DECK BRIDGES

Longitudinal glulam deck bridges consist of a series of glulam panels placed edge to edge across the deck width (Figure 8-2). They are practical for clear spans up to approximately 35 feet and are equally adaptable to single-lane and multiple-lane crossings. The panels are usually not inter-connected with dowels or fasteners but are provided with transverse stiffener beams below the deck. These stiffener beams, which are bolted to the panels directly or with brackets, transfer loads between panels and give continuity to the system. They are also frequently used as a point of attachment for railing systems. As with glulam beam bridges, longitudinal glulam deck bridges can be prefabricated in a modular system that is pressure treated with preservatives after all required cuts and holes are made. This improves the bridge economy and longevity and reduces field erection time.

Figure 8-2. - Longitudinal glulam deck bridges. (A) Panel placement during construction of a multiple-span bridge. (B) Typical single-span bridge configuration (photo courtesy of Dave Nordenson, USDA Forest Service).

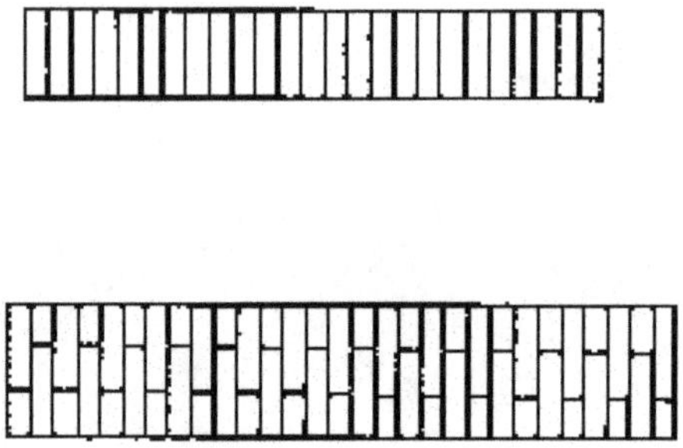

Single-piece laminations are used
for deck thicknesses up to
10-1/2 inches for Southern Pine and
10-3/4 inches for western species.

Multiple-piece laminations are
required for deck thicknesses of
12-1/4 inches and 14-1/4 inches.

Panel end views

Figure 8-3. - Laminating patterns for longitudinal glulam deck panels.

Longitudinal glulam decks are manufactured from visually graded axial combinations specified in Table 2 of *AITC 117--Design.*[3] Combination symbols with a tabulated bending stress, F_b, of 1,800 lb/in^2 or less are most economical and are most commonly used. Panels are 42 to 54 inches wide in increments equal to the net lamination thickness (1-1/2 inches for western species and 1-3/8 inches for Southern Pine). They can be manufactured in any length subject to local pressure treating and transportation restrictions. Deck thicknesses of 5-1/8, 6-3/4, 8-3/4, and 10-3/4 inches for western species and 5, 6-3/4, 8-1/2, and 10-1/2 inches for Southern Pine are manufactured from full-width laminations (Figure 8-3). Thicknesses of 12-1/4 and 14-1/4 inches are also available but require multiple-piece laminations, which normally must be edge glued to meet design requirements in horizontal shear. Unglued edge joints may also be used, but the tabulated horizontal shear values for panels with unglued joints is approximately 50 percent of that for comparable panels with glued joints.

The design criteria for longitudinal deck bridges were developed from research conducted at Iowa State University (ISU).[14,27,28] The primary emphasis of the ISU studies dealt with the lateral live load distribution characteristics for deck panel design. Empirical methods for stiffener-beam design were also developed based on limitations placed on design parameters within the load distribution studies. Additional experimental data obtained by ISU subsequent to development of the load distribution criteria should eventually provide a basis for more explicit stiffener-beam design criteria, rather than the empirical methods currently used.

DESIGN PROCEDURES

Deck panels for longitudinal glulam superstructures are designed as individual glulam beams of rectangular cross section. The portion of the vehicle wheel line distributed to each panel is computed as a Wheel Load Fraction (WLF) that is similar in application to the distribution factors used for beam design. The bending, deflection, shear, and reactions dis-

8-4

tributed to each panel are assumed to be resisted by the entire panel cross section.

Sequential design procedures for longitudinal glulam deck bridges are given in the following steps. These procedures are based on ISU research and are valid for panels that are 3-1/2 to 4-1/2 feet wide and are provided with transverse stiffener beams. The basic sequence is to (1) estimate a panel thickness and width, (2) determine loads and load distribution criteria, (3) select an initial panel combination symbol based on bending, and (4) check the suitability of the panel in deflection and shear. The process is iterative in nature if panel dimensions are changed at any point during the design process. After a suitable panel size and grade are determined, stiffener beams and bearings are designed.

1. Define deck geometric requirements and design loads.

    a.   Define geometric requirements for bridge span and width. The effective deck span, L, is the distance measured center-to-center of the bearings. Deck width is the roadway width plus any additional width required for curb and railing systems.

    b.   Identify design vehicles (including overloads) and other applicable loads and AASHTO load combinations discussed in Chapter 6. Note design requirements for live load deflection and other site-specific requirements for geometry or loading.

2. Estimate panel thickness and width and compute section properties.

Deck thickness and width must be estimated for initial calculations. Approximate maximum deck spans that may be used for estimating an initial deck thickness are shown in Table 8-1.

Panel width depends on the out-to-out structure width. Panels are 42 to 54 inches wide in multiples of 1-1/2 inches for western species or 1-3/8 inches for Southern Pine. The panels are normally designed to be of equal width, obtained by dividing the bridge width by a selected number of panels.

Based on the estimated panel dimensions, properties are computed for the panel cross section as follows:

Table 8-1. - Approximate maximum spans for longitudinal glulam deck bridges for purposes of estimating deck thickness.

Deck thickness (in.)	HS 15-44 loads		HS 20-44 loads	
	Simple spans (ft)	Continuous multiple spans (ft)	Simple spans (ft)	Continuous multiple spans (ft)
5 or 5-1/8	8	9	6	7
6-3/4	12	14	10	12
8-1/2 or 8-3/4	18	20	15	18
10-1/2 or 10-3/4	23	25	21	23
12-1/4	26	30	24	27
14-1/4	30	33	27	31

Spans listed in this table are generally limited by a live load deflection of $L/360$. Longer spans may be possible with an increased deflection, subject to designer judgment.

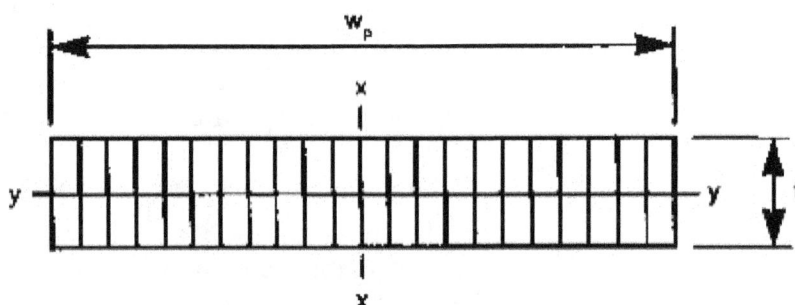

$$A = \text{panel area (in}^2) = w_p t \tag{8-1}$$

$$S_y = \text{section modulus of the panel (in}^3) = \frac{w_p t^2}{6} \tag{8-2}$$

$$I_y = \text{moment of inertia of the panel (in}^4) = \frac{w_p t^3}{12} \tag{8-3}$$

where w_p = panel width (in.), and

t = panel thickness (in.).

3. Compute panel dead load.

Compute the uniform dead load, w_{DL}, of the deck and wearing surface in lb/ft (or lb/in) of panel length using the unit material weights given in Chapter 6. Typical deck dead loads for various panel widths are given in Table 8-2. When railings and curbs are supported by transverse stiffener

Table 8-2. - Typical dead loads for longitudinal glulam deck panels.

Panel width (in.)	Dead load of deck only (lb/ft)[a]					
	t = 5-1/8	t = 6-3/4	t = 8-3/4	t = 10-3/4	t = 12-1/4	t = 14-1/4
42.0	74.7	98.4	127.6	156.8	178.6	207.8
43.5	77.4	102.0	132.2	162.4	185.0	215.2
45.0	80.1	105.5	136.7	168.0	191.4	222.7
46.5	82.7	109.0	141.3	173.6	197.8	230.1
48.0	85.4	112.5	145.8	179.2	204.2	237.5
49.5	88.1	116.0	150.4	184.8	210.5	244.9
51.0	90.8	119.5	154.9	190.4	216.9	252.3
52.5	93.4	123.0	159.5	196.0	223.3	259.8
54.0	96.1	126.6	164.1	201.6	229.7	267.2

Panel width (in.)	Dead load of deck plus a 3-inch asphalt wearing surface (lb/ft)[a]					
	t = 5-1/8	t = 6-3/4	t = 8-3/4	t = 10-3/4	t = 12-1/4	t = 14-1/4
42.0	206.0	229.7	258.9	288.0	309.9	339.1
43.5	213.3	237.9	268.1	298.3	321.0	351.2
45.0	220.7	246.1	277.3	308.6	332.0	363.3
46.5	228.1	254.3	286.6	318.9	343.1	375.4
48.0	235.4	262.5	295.8	329.2	354.2	387.5
49.5	242.8	270.7	305.1	339.5	365.2	399.6
51.0	250.1	278.9	314.3	349.7	376.3	411.7
52.5	257.5	287.1	323.6	360.0	387.4	423.8
54.0	264.8	295.3	332.8	370.3	398.4	435.9

[a] For 5-, 8-1/2-, and 10-1/2-inch deck thicknesses, respective values listed for 5-1/8-, 6-3/4-, and 10-3/4-inch deck thicknesses can be used with slightly conservative results.

beams, their dead load is normally assumed to be equally distributed to all panels. When railings and curbs are attached to the outside panel, their dead load is included with the dead load of the panel.

4. Determine Wheel Load Fraction for live load distribution.

Longitudinal glulam panels are designed as individual members to resist applied loads. In the direction of the deck span, no longitudinal distribution of wheel loads is assumed, and wheel loads act as concentrated loads. The portion of the wheel line laterally distributed to each panel is based on the WLF. For live load moment, vertical shear, and deflection, the WLF is based on the panel width and span in feet and is specified separately for bridges designed for one traffic lane, and bridges designed for two or more traffic lanes (AASHTO 3.25.3.1):

For bridges designed for one traffic lane, WLF is computed by

$$\text{WLF} = \frac{W_p}{4.25 + L/28} \quad \text{or} \quad \frac{W_p}{5.50}, \text{ whichever is greater} \quad (8\text{-}4)$$

where   WLF = the portion of the maximum force or deflection produced by one wheel line that is supported by one deck panel,

   W_p = panel width (ft), and

   L = length of span for simple-span decks and the length of the shortest span for continuous-span decks, measured center to center of the bearings (ft).

For bridges designed for two or more traffic lanes, WF is computed by

$$\text{WLF} = \frac{W_p}{3.75 + L/28} \quad \text{or} \quad \frac{W_p}{5.00}, \text{ whichever is greater} \quad (8\text{-}5)$$

5. Determine dead load and live load moment.

Compute the maximum panel dead load moment based on the deck dead loads previously determined. Compute live load moment by multiplying the maximum moment for one wheel line of the design vehicle by the WLF:

$$M_{LL} = M_{m}(\text{WLF}) \quad (8\text{-}6)$$

where M_{LL} = live load moment applied to one panel (in-lb), and

   M_{m} = maximum moment produced by one wheel line of the design vehicle (in-lb).

Maximum simple-span moments for standard AASHTO vehicles are given in Table 16-8. For multiple-span continuous bridges, maximum moments are computed for the controlling truck or lane load by analyzing the deck as a continuous beam.

6. Compute bending stress and select a deck combination symbol.

Compute deck bending stress by dividing the sum of the maximum live load and dead load bending moments by the panel section modulus ($f_b = M/S_y$). Based on the magnitude of the stress, select a panel combination symbol from Table 2 of *AITC 117-Design*, which provides the required bending capacity. As with transverse glulam decks, the most common combination symbols for longitudinal decks are No. 2 for western species (F_{bx} = 1,800 lb/in^2) and No. 47 for Southern Pine (F_{bx} = 1,750 lb/in^2). Applied bending stress, f_b must not be greater than the allowable bending stress, F_b' as computed by

$$F_b' = F_{by}C_FC_M \tag{8-7}$$

where C_F = size factor for panels less than 12 inches thick and

C_M = wet-use factor for glulam = 0.80.

t (in.)	C_F
5 or 5-1/8	1.10
6-3/4	1.07
8-1/2 or 8-3/4	1.04
10-1/2 or 10-3/4	1.01

Allowable bending stress may be increased by a factor of 1.33 for over-loads in AASHTO Load Group IB.

If $f_b \leq F_b'$, the initial deck thickness and combination symbol are satisfactory in bending. When f_b is significantly lower than F_b', a thinner deck or lower-grade combination symbol may be more economical, however, no changes should be made in the panel combination symbol or thickness until after live load deflection is checked.

If $f_b > F_b'$, the deck is insufficient in bending and the deck thickness or grade must be increased. If deck thickness or width is changed, the design sequence must be repeated.

7. Check live load deflection.

Live load deflection is resisted by the full moment of inertia, I_y, of the panel section. The deflection applied to each panel is the maximum deflection produced by the one wheel line of the design vehicle times the WLF (AASHTO 3.25.3.3), as computed by

$$\Delta_{LL} = \Delta_{WL}(\text{WLF}) \tag{8-8}$$

where Δ_{LL} = live load panel deflection (in.), and

Δ_{WL} = maximum live load deflection produced by one wheel line of the design vehicle (in.)

Deck live load deflection is computed by standard methods of elastic analysis, with the glulam modulus of elasticity *(E)* adjusted for wet-use conditions. Deflection coefficients for standard AASHTO loads on simple spans are given in Table 16-8.

Requirements for live load deflection in longitudinal glulam decks are not included in AASHTO specifications, and the acceptable deflection limit is left to designer judgment. It is recommended that maximum panel deflection not exceed $L/360$. Because continuity from panel to panel is provided only at stiffener-beam locations, relative panel displacements do occur at locations between these beams. At this time, there is no accurate method for predicting the interpanel displacements between stiffener beams; however, with a maximum panel live load deflection of $L/360$, ISU studies indicate that the interpanel displacement will not exceed approximately 0.10 inch in most applications (see stiffener-beam design later in this section). As discussed in Chapter 7, the 0.10-inch limit on relative panel displacement is considered the maximum allowable for acceptable asphalt wearing surface performance. A further reduction in deflection is desirable to reduce the potential for minor asphalt cracks at the panel joints, or when the bridge includes a pedestrian walkway.

8. Check horizontal shear.

Horizontal shear is normally not a controlling factor in longitudinal deck design because of the relatively large panel area. It is checked based on the magnitude of the maximum vertical shear occurring at the same locations used for beams (Chapters 5 and 7). Dead load vertical shear is computed at a distance from the support equal to the deck thickness, t, neglecting all loads within the distance t from the supports, using

$$V_{DL} = w_{DL}\left(\frac{L}{2} - t\right) \tag{8-9}$$

where V_{DL} = dead load vertical shear at a distance t from the support (lb), and

w_{DL} = uniform panel dead load (lb/ft).

Live load vertical shear is based on the maximum vertical shear occurring at a distance from the support equal to three times the deck thickness ($3t$) or the span quarter point ($L/4$), whichever is less. The live load shear applied to each panel is equal to the maximum shear produced by one wheel line of the design vehicle times the WLF for the panel, as computed by

$$V_{LL} = V_{wL}(\text{WLF}) \tag{8-10}$$

where V_{LL} = live load vertical shear (lb), and

V_{wL} = maximum vertical shear produced by one wheel line of the design vehicle at the lesser distance of $3t$ or $L/4$ from the support (lb).

Horizontal shear stress is assumed to be resisted by the total area of the panel cross section. Applied stress must not be greater than the allowable shear stress for the deck combination symbol, as given by

$$f_v = \frac{1.5V}{A} \leq F_v' = F_{v_j} C_M \qquad (8\text{-}11)$$

where $\qquad V = V_{DL} + V_{LL}$ (lb),

$\qquad A = $ panel cross-sectional area (in^2), and

$\qquad C_M = $ wet-use factor for shear $= 0.875$.

When $f_v > F_v'$ the only options are to increase the deck thickness or panel width. In both cases the design procedure must be repeated.

9. Determine stiffener spacing and configuration.

Transverse stiffener beams are placed across the deck width to distribute loads and deflections among the individual panels (Figure 8-4). As previously discussed, current design criteria for stiffener beams are empirical and are based on analytical and experimental data collected during the ISU studies. A more formal design procedure is currently being developed. In practice, stiffener beams are often used for guardrail post attachment, and therefore, stiffener spacing, strength, and connections may be dictated by more restrictive railing requirements (Chapter 10).

Figure 8-4. - Transverse glulam stiffener beam attached to the underside of a longitudinal glulam deck bridge (photo courtesy of Dave Nordenson, USDA Forest Service).

Stiffener beams typically consist of horizontally laminated glulam beams or shallow steel shapes (Figure 8-5). AASHTO specifications require that a stiffener beam be placed at midspan for all deck spans, and at intermediate spacings not to exceed 10 feet (AASHTO 3.25.3.4). A more restrictive intermediate stiffener-beam spacing of 8 feet is recommended by the AITC, which will be used in this chapter.[4] Stiffener design consists of sizing the beam so that the stiffness factor, EI, of the member is not less than 80,000 k-in^2; however, this is an approximate value that should not be significantly exceeded. Experimental and analytical tests at ISU have shown that the connection may be overstressed if the stiffness factor is very large, on the order of twice the minimum value. Load distribution between panels is more effectively improved by decreasing stiffener beam spacing, rather than by increasing the beam size substantially above the required minimum.

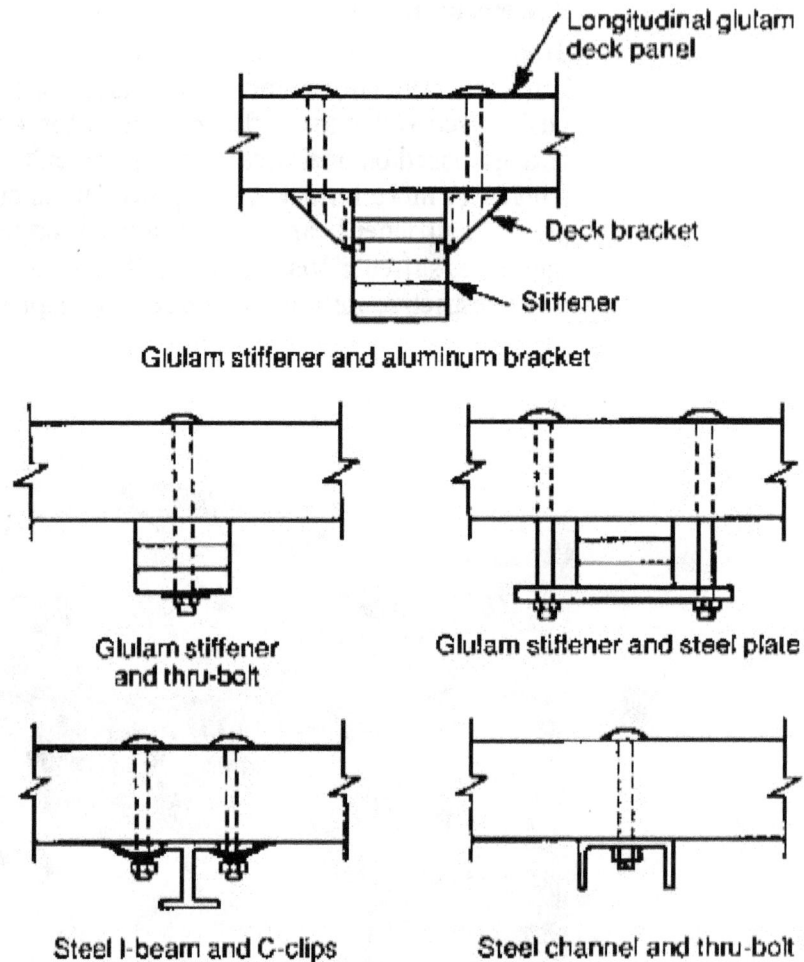

Figure 8-5. - Types of transverse stiffener-beam configurations for longitudinal glulam deck panels.

Connections between the stiffener beam and the deck panels are placed approximately 6 inches from each panel edge (Figure 8-6). The type of connection depends on the stiffener-beam material and configuration. Through-bolting is used for glulam beams and steel channels. Deck brackets or steel plates are also used for glulam beams, and C-clips are used for steel I-beams. A minimum bolt diameter of 3/4 inch is recommended for single through-bolt connections while a minimum 5/8-inch diameter bolt is used for bracket connections. The type of connection is left to designer judgment since all connector types shown in Figure 8-5 were modeled in the ISU study. However, experimental results at ISU indicate that the through-bolt type connections provide more favorable load distribution in the panels and reduce the potential for localized stress conditions in the region of the connection to the stiffener beams. They are also more effective in reducing interpanel displacements that occur between stiffener-beam locations.

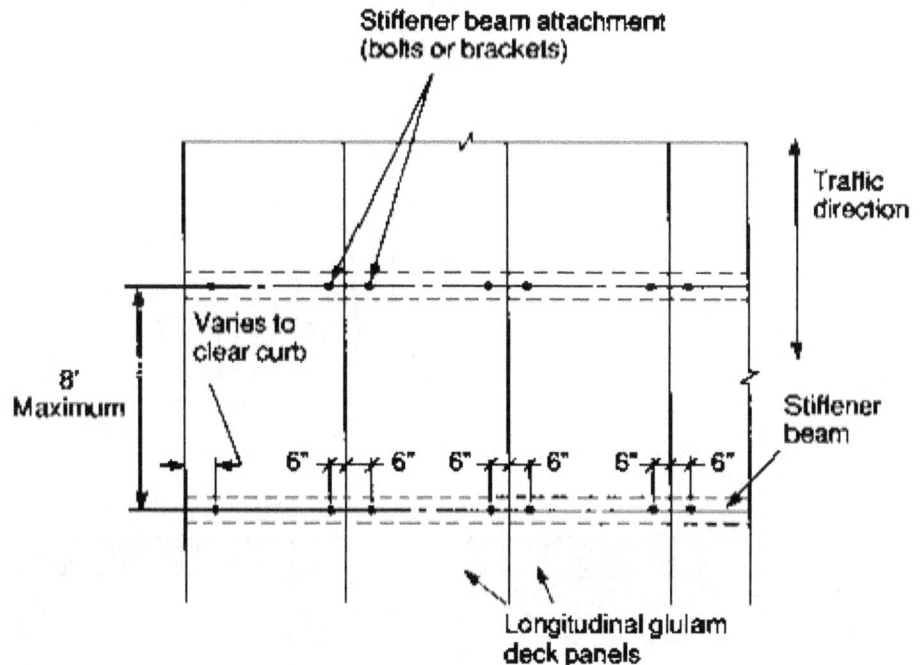

Figure 8-6. - Stiffener-beam attachment for longitudinal glulam decks.

10. Determine bearing configuration and check bearing stress.

Bearings are designed to resist the vertical and lateral forces in the same manner previously discussed for glulam beams. For longitudinal deck bridges however, the required bearing length is normally controlled by considerations for bearing configuration, rather than stress in compression perpendicular to grain. From a practical standpoint, a bearing length of 10 to 12 inches is recommended for stability and deck attachment.

Because of the long, continuous width associated with deck bridges, bearing attachments are normally made through the deck to the supporting cap or sill, or from the deck underside. For short-span crossings, a side attachment using steel angles may also be feasible. Two common configurations are shown in Figure 8-7.

Galvanized dome head bolts placed
through deck and cap

Longitudinal glulam deck

Substructure cap or sill

Dome head bolt

Longitudinal
glulam deck

Neoprene bushing

Steel angle or
welded steel plates

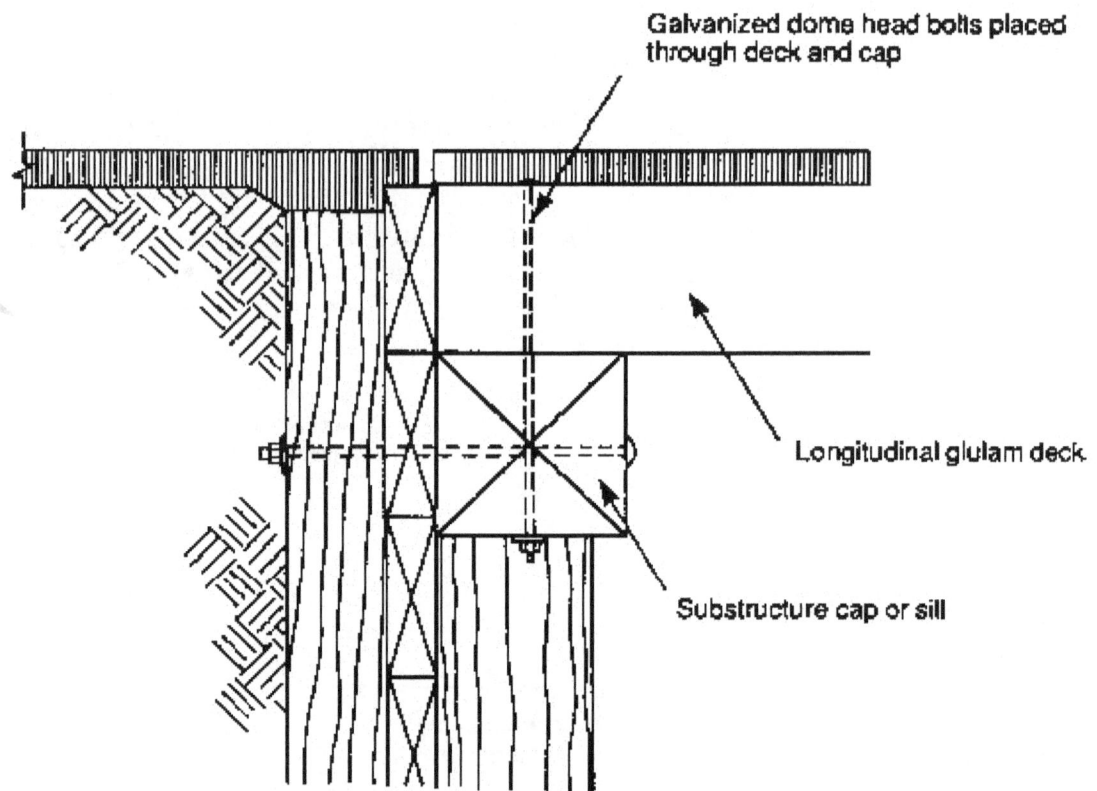

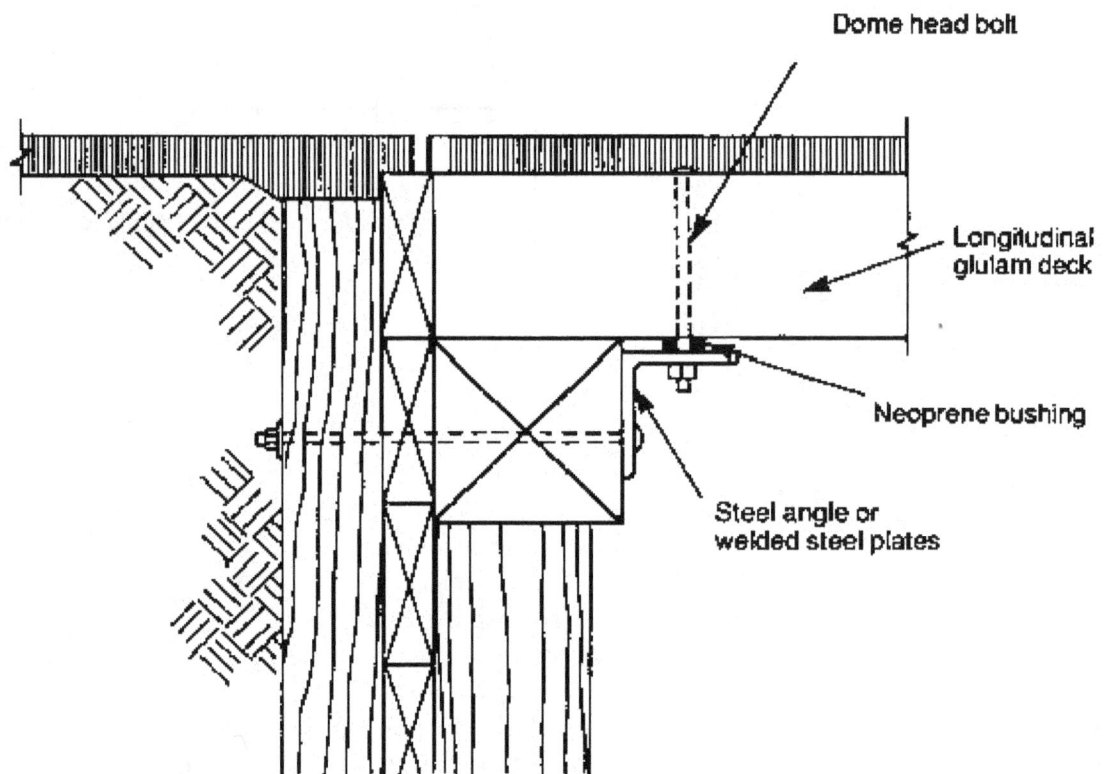

Figure 8-7. - Typical bearing configurations for longitudinal glulam decks.

Based on the bearing configuration, dead load reactions are computed by conventional methods using the unit dead load of the panel. Live load reactions for single- and multiple-lane bridges are based on the following WLF for reactions (AASHTO 3.25.3.2):

$$\text{WLF} = \frac{W_p}{4} \text{ , but not less than 1.0} \tag{8-12}$$

The live load reaction distributed to each panel is the maximum reaction of the design vehicle times the WLF given by Equation 8-12:

$$R_{LL} = R_{mL}(\text{WLF}) \tag{8-13}$$

where R_{LL} = live load reaction distributed to each deck panel (lb), and

             R_{mL} = maximum reaction produced by one wheel line of the design vehicle (lb).

Applied stress in compression perpendicular to grain at reactions must not be greater than the allowable stress in compression perpendicular to grain for the panel combination symbol:

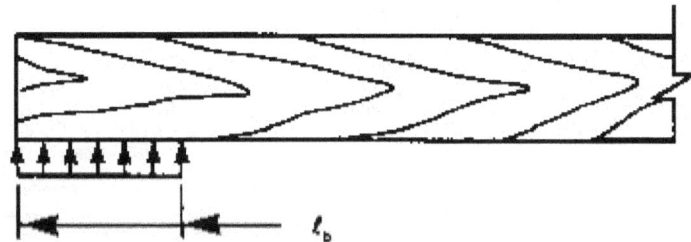

$$f_{cL} = \frac{R_{DL} + R_{LL}}{w_p \ell_b} \leq F_{cL}' = F_{cL} C_M \tag{8-14}$$

where ℓ_b is the length of panel bearing in inches.

Example 8-1 - Longitudinal glulam deck bridge; two-lane HS 20-44 loading

An existing bridge on a city street is to be removed and replaced with a longitudinal glulam deck bridge. The bridge spans 20 feet center-to-center of bearings and supports two lanes of AASHTO HS 20-44 loading over a roadway width of 26 feet. Design this bridge, assuming the following:

1. Vehicular railing with a dead load of 55 lb/ft per side is attached to transverse stiffener beams.

2. The rail face extends inward approximately 6 inches from the outside deck edge.

3. The deck will be paved with 3 inches of asphalt pavement.

4. Live load deflection must be limited to $L/400$.

5. Glulam is visually graded western species.

Solution

Define Deck Geometric Requirements and Design Loads

With a roadway width of 28 feet, and railing that projects 6 inches inward from each deck edge, a bridge width of 29 feet is required. Design loading will be one HS 20-44 wheel line in AASHTO Load Group I.

Estimate Panel Thickness and Width and Compute Section Properties

An initial panel thickness of 10-3/4 inches is selected from Table 8-1. Panel width must be 42 to 54 inches in 1-1/2 inch increments (lamination thickness). The selected configuration will be two outside panels, 51 inches wide, and five interior panels, 49-1/2 inches wide, for a total deck width of 29 feet 1-1/2 inches:

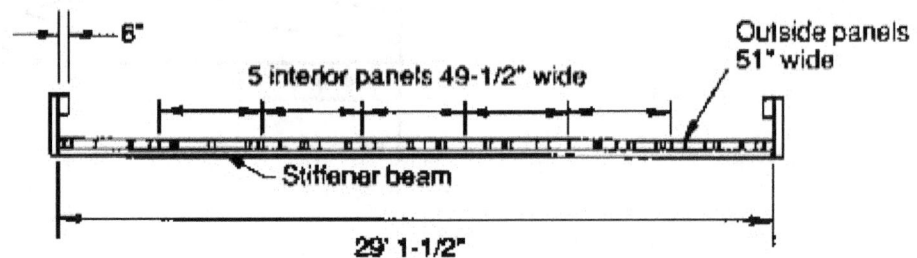

Section properties are computed for the smaller 49.5-inch panel width:

$$t = 10.75 \text{ in.}$$

$$w_p = 49.5 \text{ in.}$$

$$A = t(w_p) = 10.75(49.5) = 532.13 \text{ in}^2$$

$$S_y = \frac{w_p(t^2)}{6} = \frac{49.5(10.75)^2}{6} = 953.39 \text{ in}^3$$

$$I_y = \frac{w_p(t^3)}{12} = \frac{49.5(10.75)^3}{12} = 5,124.47 \text{ in}^4$$

Compute Panel Dead Load

From Table 8-2, the dead load of the 49.5inch wide panel with a 3-inch asphalt wearing surface is 339.5 lb/ft. Railing dead load is distributed equally over the deck width. For a total railing load of $2(55) = 110$ lb/ft, the load supported by each panel is

$$\frac{110 \text{ lb/ft}}{7 \text{ panels}} = 15.7 \text{ lb/ft}$$

An additional estimated dead load of 8 lb/ft will also be applied to each panel for the stiffener beams and associated attachment hardware.

$$w_{DL}/\text{panel} = 339.5 + 15.7 + 8 = 363.2 \text{ lb/ft}$$

Determine Wheel Load Fraction for Live Load Distribution

By Equation 8-5 for a two-lane bridge,

$$\text{WLF} = \frac{W_p}{3.75 + L/28} \quad \text{or} \quad \frac{W_p}{5.00}, \text{ whichever is greater}$$

$$W_p = \frac{w_p}{12 \text{ in/ft}} = \frac{49.5}{12} = 4.13 \text{ ft}$$

$$\frac{W_p}{3.75 + L/28} = \frac{4.13}{3.75 + (20/28)} = 0.93 \text{ WL/panel}$$

$$\frac{W_p}{5.00} = \frac{4.13}{5.00} = 0.83 \text{ WL/panel}$$

Therefore, WLF = 0.93WL/panel.

Determine Dead Load and Live Load Moment

Dead load moment is computed by assuming each panel is a simply supported beam:

$$M_{DL} = \frac{w_{DL}L^2}{8} = \frac{363.2(20)^2}{8} = 18,160 \text{ ft-lb}$$

Live load moment is the product of the WLF and the moment produced by one wheel line of the design vehicle. From Table 16-8, the maximum moment from one wheel line of HS 20-44 loading is 80,000 ft-lb:

$$M_{LL} = 0.93 \text{ WL/panel}(80,000 \text{ ft-lb}) = 74,400 \text{ ft-lb}$$

$$M = M_{DL} + M_{LL} = 18,160 + 74,400 = 92,560 \text{ ft-lb}$$

Compute Bending Stress and Select a Deck Combination Symbol

$$f_b = \frac{M}{S_y} = \frac{92{,}560 \text{ ft-lb } (12 \text{ in/ft})}{953.39} = 1{,}165 \text{ lb/in}^2$$

From *AITC 117-Design*, combination symbol No. 2 is selected with the following tabulated values:

$F_{by} = 1{,}800 \text{ lb/in}^2$ $\qquad$ $C_M = 0.80$

$F_{vy} = 145 \text{ lb/in}^2$ $\qquad$ $C_M = 0.875$

$F_{c\perp} = 560 \text{ lb/in}^2$ $\qquad$ $C_M = 0.53$

$E = 1{,}700{,}000 \text{ lb/in}^2$ $\qquad$ $C_M = 0.833$

By Equation 8-7,

$$F_b' = F_{by} C_r C_M = 1{,}800(1.01)(0.80) = 1{,}454 \text{ lb/in}^2$$

$f_b = 1{,}165 \text{ lb/in}^2 < F_b' = 1{,}454 \text{ lb/in}^2$ so the combination symbol is satisfactory for bending. The combination symbol could be reduced to No. 1 ($F_{by} = 1{,}450 \text{ lb/in}^2$) and still be acceptable in bending; however, it is anticipated that deflection criteria will not be met at the lower E value of $1{,}500{,}000 \text{ lb/in}^2$. Live load deflection will be checked before changing the combination symbol.

Check Live Load Deflection

The deflection coefficient for one wheel line of HS 20-44 loading on a 20-foot span is obtained from Table 16-8:

$$\Delta_{WL} = \frac{4.61 \times 10^9}{E' I_y}$$

$$E' = EC_M = 1{,}700{,}000(0.833) = 1{,}416{,}100 \text{ lb/in}^2$$

$$\Delta_{WL} = \frac{4.61 \times 10^9}{1{,}416{,}100(5{,}124.47)} = 0.64 \text{ in.}$$

Deck deflection is computed by Equation 8-8:

$$\Delta_{LL} = \Delta_{WL}(\text{WLF}) = (0.64 \text{ in.})(0.93) = 0.60 \text{ in.} = L/400$$

Live load deflection equals the maximum allowable deflection of $L/400$. The combination symbol No. 2 panel is retained since any reduction in E will result in excessive deflection.

Check Horizontal Shear

Dead load vertical shear is computed at a distance t from the support by Equation 8-9:

$$V_{DL} = w_{DL} = \left(\frac{L}{2} - t\right) = 363.2 \text{ lb/ft} \left(\frac{20}{2} - \frac{10.75}{12 \text{ in/ft}}\right) = 3{,}307 \text{ lb}$$

Live load vertical shear is computed at the lesser distance of $3t$ or $L/4$ from the support:

$$3t = \frac{3(10.75)}{12 \text{ in/ft}} = 2.69 \text{ ft} \qquad \frac{L}{4} = \frac{20}{4} = 5 \text{ ft}$$

Maximum vertical shear 2.69 feet from the support is computed for one HS 20-44 wheel line:

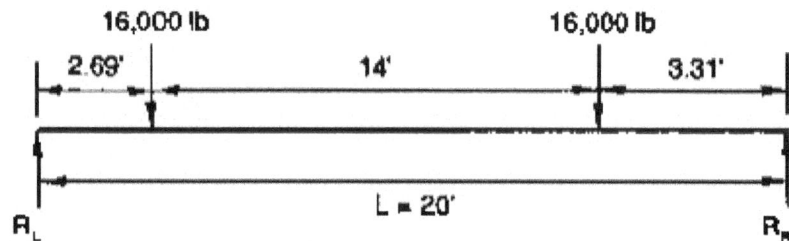

$$V_{WL} = R_L = \frac{(16{,}000 \text{ lb})(3.31 \text{ ft} + 17.31 \text{ ft})}{20 \text{ ft}} = 16{,}496 \text{ lb}$$

By Equation 8-10,

$$V_{LL} = V_{WL}(\text{WLF}) = 16{,}496(0.93) = 15{,}341 \text{ lb}$$

Stress in horizontal shear is computed by Equation 8-11:

$$V = V_{DL} + V_{LL} = 3{,}307 + 15{,}341 = 18{,}648 \text{ lb}$$

$$f_v = \frac{1.5V}{A} = \frac{1.5(18{,}648)}{532.13} = 53 \text{ lb/in}^2$$

$$F_v' = F_v C_M = 145 \text{ lb/in}^2(0.875) = 127 \text{ lb/in}^2$$

$F_v' = 127 \text{ lb/in}^2 > f_v = 53 \text{ lb/in}^2$, so shear is satisfactory.

Determine Stiffener Spacing and Configuration

Maximum spacing for stiffener beams is 8 feet. For this bridge, stiffener beams will be placed at the span third points for a spacing of 6 feet 8 inches:

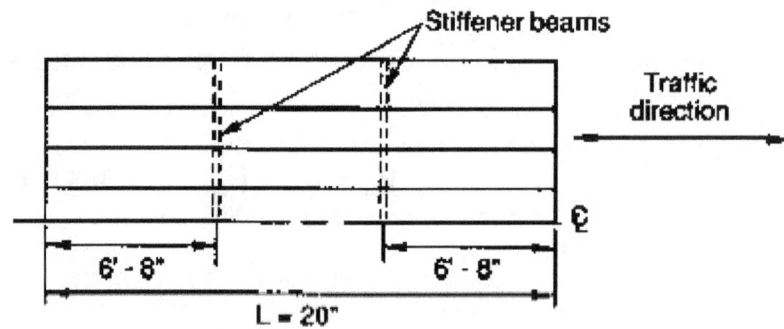

The size and stiffness of the stiffener beam must be sufficient to provide a minimum EI value of 80,000 k-in^2. Selecting a combination symbol No. 2 glulam stiffener, 6-3/4 inches wide and 4-1/2 inches deep (dry-use conditions may be used for glulam stiffener beams if they are protected by a watertight deck):

$$E' = EC_M = 1,700,000(1.0) = 1,700,000 \text{ lb/in}^2$$

$$I = \frac{bd^3}{12} = \frac{6.75(4.5)^3}{12} = 51.26 \text{ in}^4$$

$$E' I = \frac{1,700,000 \text{ lb/in}^2}{1,000 \text{ lb/k}} \left(51.26 \text{ in}^4\right) = 87,142 \text{ k-in}^2$$

87,142 k-in^2 > 80,000 k-in^2, so 6-3/4 by 4-1/2-inch stiffener beams are satisfactory. The beams will be attached to the deck with 3/4-inch-diameter bolts located 6 inches from the panel edge (Figure 8-6).

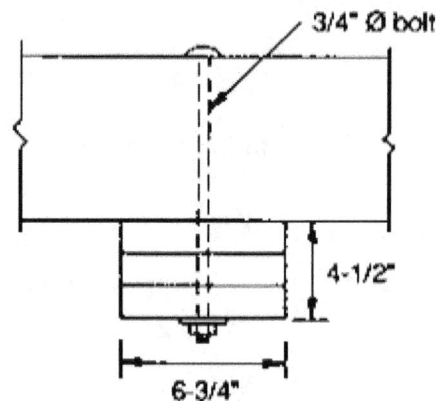

Checking the stiffener beam dead load,

$$DL/\text{stiffener} = \frac{(6.75 \text{ in.})(4.5 \text{ in.})(50 \text{ lb/ft}^3)(29 \text{ ft})}{144 \text{ in}^2/\text{ft}^2} = 306 \text{ lb}$$

$$\text{Stiffener} = w_{DL} / \text{panel} = \frac{(306\,\text{lb})(2\,\text{stiffeners})}{(20\,\text{ft})(7\,\text{panels})} = 4.4\ \text{lb/ft}$$

4.4 lb/ft is less than the 8 lb/ft assumed, but revision of panel dead load is not required or warranted.

Determine Bearing Configuration and Check Bearing Stress

The length of bearing required for longitudinal glulam deck bridges is generally dictated by requirements for deck attachment to the substructure. In this case, it is assumed that attachment will be by through bolting to a 12-inch by 12-inch sill. For a bearing length, ℓ_b, of 12 inches:

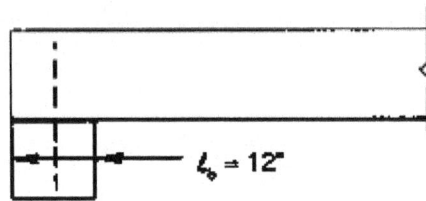

Dead load reactions are determined by assuming the panel acts as a simple beam between supports. For an out-out panel length of 21 feet,

$$R_{DL} = \frac{(363.2\ \text{lb/ft})(21\ \text{ft})}{2} = 3{,}814\ \text{lb}$$

Live load reactions are computed by multiplying the maximum reaction for one wheel line times the wheel load fraction for reactions (Equation 8-12):

$$\text{WLF} = \frac{W_p}{4} \geq 1.0 = \frac{4.13\ \text{ft}}{4} = 1.03\ \text{WL/panel}$$

From Table 16-8, the maximum reaction for one wheel line of an HS 20-44 vehicle is 20,800 pounds. By Equation 8-13,

$$R_{LL} = R_{wL}(\text{WLF}) = (20{,}800\ \text{lb})(1.03) = 21{,}424\ \text{lb}$$

For a length of bearing (ℓ_b) of 12 inches,

$$f_{c\perp} = \frac{R_{DL} + R_{LL}}{w_p(\ell_b)} = \frac{3{,}814 + 21{,}424}{49.5(12)} = 42\ \text{lb/in}^2$$

$$F_{c\perp}' = F_{c\perp}(C_M) = 560\ \text{lb/in}^2(0.53) = 297\ \text{lb/in}^2$$

$f_{c\perp} = 42\ \text{lb/in}^2 < F_{c\perp}' = 297\ \text{lb/in}^2$, so a bearing length of 12 inches is satisfactory. The out-to-out length of the panels will be 21 feet.

Summary

The bridge will consist of seven 10-3/4-inch thick glulam panels, 21 feet long, manufactured to *AITC 117--Design* combination symbol No. 2. The five interior panels are 49-1/2 inches wide and the two outside panels are 51 inches wide. Stiffener beams are 6-3/4-inch by 4-1/2-inch combination symbol No. 2 glulam, placed at the span third points. Stresses and deflection are as follows:

$$f_b = 1,165 \ lb/in^2$$

$$F_b' = 1,454 \ lb/in^2$$

$$\Delta_{LL} = 0.60 \ in. = L/400$$

$$f_v = 53 \ lb/in^2$$

$$F_v' = 127 \ lb/in^2$$

$$f_{c\perp} = 42 \ lb/in^2$$

$$F_{c\perp}' = 297 \ lb/in^2$$

Example 8-2- Longitudinal glulam deck bridge; single-lane with overload

---

A longitudinal glulam deck bridge with a 14-foot roadway width is to be constructed on a forest road. The bridge will span 15 feet center to center of bearings and support AASHTO HS 20-44 loading with an occasional U80 overload. Design this bridge, assuming the following:

1. Rough-sawn 12-inch by 12-inch curbs are provided along the roadway edges.

2. The deck will be provided with a 4-inch full-sawn lumber wearing surface.

3. Live load deflection for HS 20-44 loads must be limited to *L/360*.

4. Glulam is visually graded Southern Pine.

Solution
Define Deck Geometric Requirements and Design Loads
For a roadway width of 14 feet with 12-inch curbs, an out-to-out bridge width of 16 feet is required. Design loading will be an HS 20-44 wheel line in AASHTO Load Group I and a U80 wheel line in AASHTO Load Group IB (33 percent stress increase permitted for occasional overloads).

Estimate Panel Thickness and Width and Compute Section Properties

An initial panel thickness of 8-1/2 inches is estimated from Table 8-1. Panel width will be 48-1/8 inches, rounded to 48 inches for design calculations:

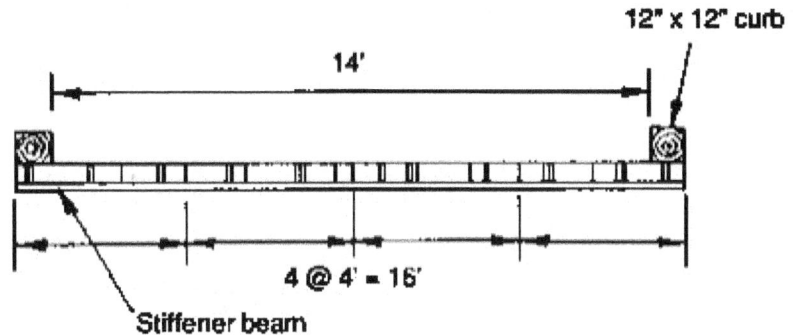

Section properties are as follows:

$$t = 8.5 \text{ in.}$$

$$w_p = 48 \text{ in.}$$

$$A = t(w_p) = 8.5(48) = 408 \text{ in}^2$$

$$S_y = \frac{w_p(t^2)}{6} = \frac{48(8.5)^2}{6} = 578 \text{ in}^3$$

$$I_y = \frac{w_p(t^3)}{12} = \frac{48(8.5)^3}{12} = 2,456.5 \text{ in}^4$$

Compute Panel Dead Load

For an 8-1/2-inch deck and 4-inch lumber wearing surface, dead load is computed over the 48-inch panel width:

$$w_{DL} = \frac{(48 \text{ in.})(8.5 \text{ in.} + 4 \text{ in.})}{144 \text{ in}^2/\text{ft}^2}\left(50 \text{ lb/ft}^3\right) = 208.3 \text{ lb/ft}$$

Curb dead load is assumed to be distributed equally across the deck width. For a total curb load of 2(50 lb/ft) = 100 lb/ft, the load supported by each panel is

$$\frac{100 \text{ lb/ft}}{4 \text{ panels}} = 25 \text{ lb/ft}$$

With one stiffener beam on a 15 foot span, the dead load of the stiffener beam and attachment hardware will be negligible.

Total w_{DL} per panel = 208.3 + 25 = 233.3 lb/ft.

Determine Wheel Load Fraction for Live Load Distribution

By Equation 8-4 for a one-lane bridge,

$$WLF = \frac{W_p}{4.25 + L/28} \quad \text{or} \quad \frac{W_p}{5.50} \quad \text{whichever is greater}$$

$$W_p = \frac{w_p}{12 \text{ in/ft}} = \frac{48}{12} = 4 \text{ ft}$$

$$\frac{W_p}{4.25 + L/28} = \frac{4}{4.25 + (15/28)} = 0.84 \text{ WL/panel}$$

$$\frac{W_p}{5.50} = \frac{4}{5.50} = 0.73 \text{ WL/panel}$$

WLF = 0.84 WL/panel will be used.

Determine Dead Load and Live Load Moment

$$M_{DL} = \frac{wL^2}{8} = \frac{233.3(15)^2}{8} = 6,562 \text{ ft-lb}$$

From Table 16-8 for a 15-foot span, the maximum moment for one wheel line is 60,000 ft-lb for HS 20-44 loading and 100,250 ft-lb for U80 loading.

$$\text{HS 20-44 } M_{LL} = (0.84 \text{WL/panel})(60,000) = 50,400 \text{ ft-lb}$$

$$\text{U80 } M_{LL} = (0.84 \text{WL/panel})(100,250) = 84,210 \text{ ft-lb}$$

Compute Bending Stress and Select a Deck Combination Symbol

The deck will be designed for the U80 load, then checked for the HS 20-44 load.

$$M = M_{DL} + \text{U80 } M_{LL} = 6,562 + 84,210 = 90,772 \text{ ft-lb}$$

$$f_b = \frac{M}{S_y} = \frac{(90,772 \text{ ft-lb})(12 \text{ in/ft})}{578 \text{ in}^3} = 1,885 \text{ lb/in}^2$$

From *AITC 117-Design*, combination symbol No. 48 is selected with the following tabulated values:

$F_{by} = 2,000 \text{ lb/in}^2$ $\qquad C_{df} = 0.80$

$F_{vy} = 175 \text{ lb/in}^2$ $\qquad C_{df} = 0.875$

$F_{c\perp} = 650 \text{ lb/in}^2$ $\qquad C_{df} = 0.53$

$E = 1,700,000 \text{ lb/in}^2$ $\qquad C_{df} = 0.833$

Allowable bending stress is computed by Equation 8-7 with a 33-percent increase for group IB loading:

$$F_b' = F_{by}C_PC_M(1.33) = 2,000(1.04)(0.80)(1.33) = 2,213 \text{ lb/in}^2$$

$F_b' = 2,213 \text{ lb/in}^2 > f_b = 1,885 \text{ lb/in}^2$, so the combination symbol is satisfactory in bending for U80 loading.

Check HS 20-44 loading:

$$M = M_{DL} + M_{LL} = 6,562 + 50,400 = 56,962 \text{ ft-lb}$$

$$f_b = \frac{M}{S_y} = \frac{(56,962 \text{ ft-lb})(12 \text{ in./ft})}{578 \text{ in}^3} = 1,183 \text{ lb/in}^2$$

$$F_b' = F_{by}C_FC_M = 2,000(1.04)(0.80) = 1,664 \text{ lb/in}^2$$

$F_b' = 1,664 \text{ lb/in}^2 > f_b = 1,183 \text{ lb/in}^2$, so HS 20-44 loading is also satisfactory.

The combination symbol and deck thickness are acceptable in bending, but the applied stress is considerably lower than the allowable stress. The panel combination symbol could be lowered to a No. 47 ($F_b = 1,750 \text{ lb/in}^2$), but the E value would be reduced to 1,400,000 lb/in². Deflection will be checked before any changes are made.

Check Live Load Deflection
The deflection coefficient for one wheel line of HS 20-44 loading on a 15-foot span is obtained from Table 16-8:

$$\Delta_{WL} = \frac{1.94 \times 10^9}{E I_y}$$

$$E' = EC_M = 1,700,000(0.833) = 1,416,100 \text{ lb/in}^2$$

$$\Delta_{WL} = \frac{1.94 \times 10^9}{1,416,100(2,456.50)} = 0.56 \text{ in.}$$

Deck deflection is computed by Equation 8-8:

$$\Delta_{LL} = \Delta_{WL}(WLF) = (0.56 \text{ in.})(0.84) = 0.47 \text{ in.} = L/383$$

$L/383 < L/360$, so the deck deflection is acceptable with $E = 1,700,000$ lb/in².

For a panel combination symbol No. 47:

$$E' = EC_M = 1,400,000(0.833) = 1,166,200 \text{ lb/in}^2$$

$$\Delta_{WL} = \frac{1.94 \times 10^9}{1,166,200(2,456.50)} = 0.68 \text{ in.}$$

$$\Delta_{LL} = \Delta_{WL}(\text{WLF}) = (0.68 \text{ in.})(0.84) = 0.57 \text{ in.} = L/316$$

The deck deflection for combination symbol No. 47 exceeds the allowable. Combination symbol No. 48 will be retained.

Check Horizontal Shear

Dead load vertical shear is computed at a distance t from the support by Equation 8-9:

$$V_{DL} = w_{DL} = \left(\frac{L}{2} - t\right) = (233.3 \text{ lb/ft})\left(\frac{15}{2} - \frac{8.5}{12 \text{ in/ft}}\right) = 1,585 \text{ lb}$$

Live load vertical shear is computed at the lesser of $3t$ or $L/4$ from the support:

$$3t = \frac{3(8.5)}{12 \text{ in/ft}} = 2.13 \text{ ft} \qquad\qquad \frac{L}{4} = \frac{15}{4} = 3.75 \text{ ft}$$

For U80 loading,

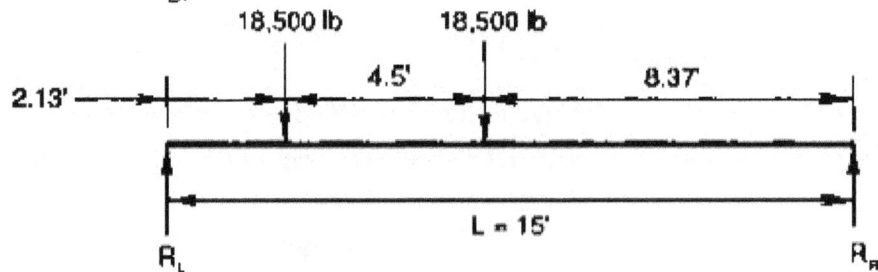

$$V_{WL} = R_L = \frac{(18,500 \text{ lb})(8.37 \text{ ft} + 12.87 \text{ ft})}{15 \text{ ft}} = 26,196 \text{ lb}$$

$$V_{LL} = V_{WL}(\text{WLF}) = 26,196(0.84) = 22,005 \text{ lb}$$

$$V = V_{DL} + V_{LL} = 1,585 + 22,005 = 23,590 \text{ lb}$$

$$f_v = \frac{1.5V}{A} = \frac{1.5(23,590)}{408} = 87 \text{ lb/in}^2$$

$$F_v' = F_v(C_M)(1.33) = (145 \text{ lb/in}^2)(0.875)(1.33) = 169 \text{ lb/in}^2$$

$F_v' = 169 \text{ lb/in}^2 > f_v = 87 \text{ lb/in}^2$, so shear is satisfactory for the U80.

For HS 20-44 loading,

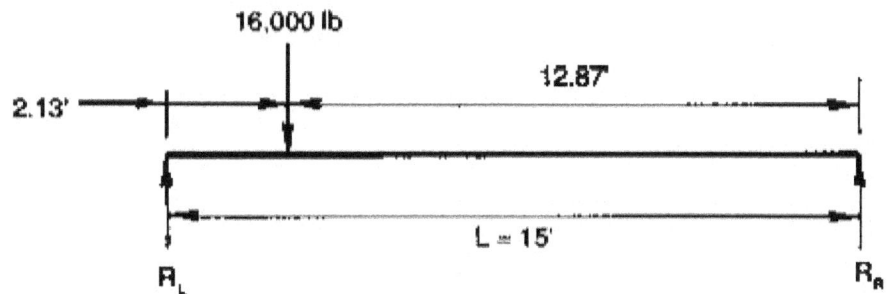

$$V_{WL} = R_L = \frac{(16,000 \text{ lb})(12.87 \text{ ft})}{15 \text{ ft}} = 13,728 \text{ lb}$$

$$V_{LL} = V_{WL}(\text{WLF}) = 13,728(0.84) = 11,532 \text{ lb}$$

$$V = V_{DL} + V_{LL} = 1,585 + 11,532 = 13,117 \text{ lb}$$

$$f_v = \frac{1.5V}{A} = \frac{1.5(13,117)}{408} = 48 \text{ lb/in}^2$$

$$F_v' = F_{vy}C_M = (145 \text{ lb/in}^2)(0.875) = 127 \text{ lb/in}^2$$

$F_v' = 127 \text{ lb/in}^2 > f_v = 48 \text{ lb/in}^2$, so shear is also satisfactory for the HS 20-44.

Determine Stiffener Spacing and Configuration

A stiffener beam will be placed at the span centerline for a spacing of 7.5 feet. The size, configuration, and calculations for the stiffener are the same as shown in Example 8-1 (combination symbol No. 48 has the same E' value as a combination symbol No. 2).

Determine Bearing Configuration and Check Bearing Stress

The bearing for this bridge will use the steel angle configuration (Figure 8-7) with a length of bearing, ℓ_b, of 10 inches. Panel length will be 15 feet 10 inches. The dead load reaction is computed as follows:

$$R_{DL} = \frac{(233.3 \text{ lb/ft})(15.83 \text{ ft})}{2} = 1,847 \text{ lb}$$

Live load reactions are computed by Equation 8-12:

$$\text{WLF} = \frac{W_P}{4} > 1.0 = \frac{4 \text{ ft}}{4} = 1.0 \text{ WL/panel}$$

The maximum live load reaction will be controlled by the heavier U80 vehicle, without the 33-percent increase for AASHTO Load Group IB (allowable stress increases for overloads are generally not applied to $F_{\perp}$). From Table 16-8, the maximum reaction for one wheel line of a U80 vehicle on a 15-foot span is 31,450 pounds. By Equation 8-13,

$$R_{LL} = R_{wL}(WLF) = (31{,}450 \text{ lb})(1.0) = 31{,}450 \text{ lb}$$

For $\ell_b = 10$ inches,

$$f_{c\perp} = \frac{R_{DL} + R_{LL}}{w_p(\ell_b)} = \frac{1{,}847 \text{ lb} + 31{,}450 \text{ lb}}{(48 \text{ in.})(10 \text{ in.})} = 69 \text{ lb/in}^2$$

$$F_{c\perp}' = F_{c\perp}(C_M) = (650 \text{ lb/in}^2)(0.53) = 345 \text{ lb/in}^2$$

$F_{c\perp}' = 345 \text{ lb/in}^2 > f_{c\perp} = 69 \text{ lb/in}^2$, so a bearing length of 10 inches is satisfactory.

Summary

The bridge will consist of four 8-1/2-inch-thick glulam panels, 48-1/8 inches wide, and 15 feet 10 inches long, manufactured to *AITC 117--Design* combination symbol No. 48. A 6-3/4-inch by 4-1/2-inch combination symbol No. 48 stiffener beam will be placed at the span center. Stresses and deflection are as follows:

HS 20-44 loading		U80 overload	
f_b	$= 1{,}183 \text{ lb/in}^2$	f_b	$= 1{,}885 \text{ lb/in}^2$
F_b'	$= 1{,}664 \text{ lb/in}^2$	F_b'	$= 2{,}213 \text{ lb/in}^2$
Δ_{LL}	$= 0.47 \text{ in.} = L/383$	Δ_{LL}	$= \text{N/A}$
f_v	$= 48 \text{ lb/in}^2$	f_v	$= 87 \text{ lb/in}^2$
F_v'	$= 127 \text{ lb/in}^2$	F_v'	$= 169 \text{ lb/in}^2$
$f_{c\perp}$	$= < \text{U80}$	$f_{c\perp}$	$= 69 \text{ lb/in}^2$
$F_{c\perp}'$	$= 345 \text{ lb/in}^2$	$F_{c\perp}'$	$= 345 \text{ lb/in}^2$

8.4 LONGITUDINAL NAIL-LAMINATED LUMBER DECK BRIDGES

Longitudinal nail-laminated deck bridges consist of a series of lumber laminations that are placed on edge and nailed together on their wide faces. They may be constructed either as continuous decks or as panelized decks (Figure 8-8). In continuous decks, each lamination is nailed to the adjacent lamination, making the deck continuous across the bridge width. For panelized decks, laminations are prefabricated into a series of panels that are placed longitudinally between supports and interconnected with transverse stiffener beams. Provisions for panelized decks without distributor beams are also contained in AASHTO, but such decks are not commonly used and are not included in this chapter. Laminations for both continuous and panelized configurations must be one piece over the span length (no butt joints). The bridge clear span is therefore limited by the available length of lumber. Longer crossings are made with a series of simple spans with joints between successive spans over intermediate supports (Figure 8-9).

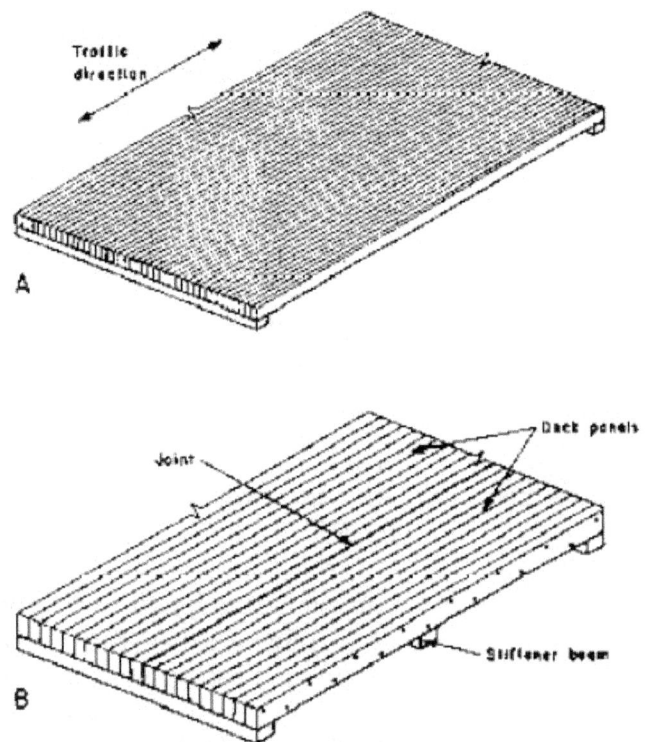

Figure 8-8. - Longitudinal nail-laminated lumber decks are constructed as continuous decks and as panelized decks. (A) In continuous decks, laminations are progressively nailed to adjacent laminations to form a continuous deck across the structure width. (B) For panelized decks, lumber is nail-laminated into panels that are interconnected with transverse stiffener beam(s).

Figure 8-9. - Multiple-span longitudinal nail-laminated lumber deck bridge consisting of a series of simple spans (photo courtesy of Wheeler Consolidated, Inc.).

Longitudinal nail-laminated decks are constructed from lumber laminations that are 2 to 4 inches thick, and 5 inches or wider, in the Joist and Plank size classification.[1][2] Both continuous and panelized configurations can be constructed from any lumber size provided it is a minimum of 6 inches in nominal depth (AASHTO 3.25.2.2). From a practical standpoint, however, continuous decks are normally constructed of 2-inch nominal material that is 6 to 12 inches wide to facilitate field nailing and handling. Panelized systems commonly use 4-inch nominal material that is 10 to 16 inches wide, which is more economical and practical for shop fabrication.

CONTINUOUS NAIL-LAMINATED LUMBER BRIDGES

Continuous nail-laminated lumber bridges are practical for simple spans up to approximately 19 feet for HS 20-44 and H 20-44 loads and 21 feet for HS 15-44 and H 15-44 loads. Load distribution and continuity across the bridge are provided by the nails that are placed through two and one-half laminations, in the same pattern used for transverse nail-laminated decks (Figure 8-10). Transverse stiffener beams are not required. The performance of longitudinal nail-laminated bridges is similar in many respects to transverse nail-laminated decks and depends primarily on the effectiveness of the nails in transferring loads between adjacent laminations. Field experience has shown that many nail-laminated decks demonstrate a tendency to loosen or delaminate from cyclic loading and moisture content changes in the laminations. This subsequently leads to reduced load distribution and deterioration of asphalt wearing surfaces. In longitudinal deck bridges, the potential for delamination is normally higher than for transverse configurations because the deck spans and associated deflections are generally larger. Performance can be improved by limiting live load deflections and using edge-grain lumber for laminations, but these measures may not be totally effective in eliminating deck loosening.

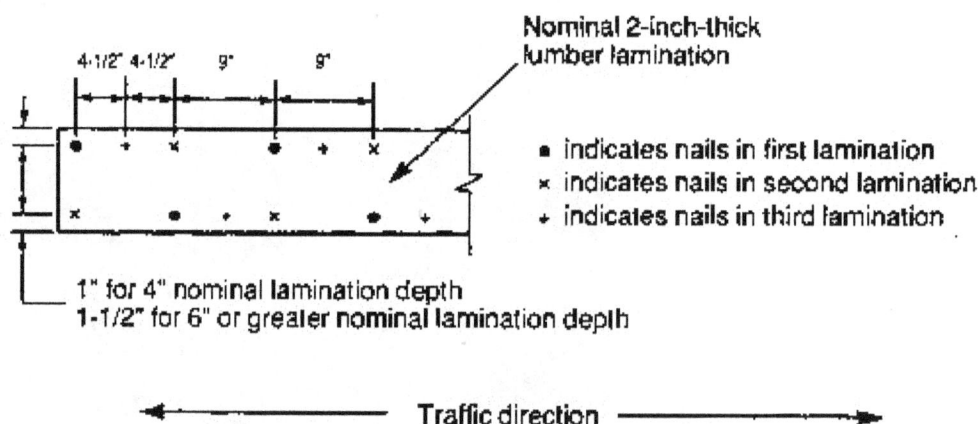

Figure 8-10. - Nailing pattern for continuous longitudinal nail-laminated lumber decks constructed of nominal 2-inch-thick sawn lumber.

When properly designed, longitudinal nail-laminated deck bridges are generally suitable for low-volume local or rural roads that are not required to carry heavy highway loads. They are not recommended for primary or secondary road systems, or crossings that require an asphalt wearing surface.

Design Procedures

Design procedures for longitudinal continuous nail-laminated bridges are similar to those for transverse nail-laminated decks discussed in Chapter 7. For longitudinal decks, however, the span is measured center to center of bearings and different criteria are used for live load distribution (AASHTO 3.25.2). In the longitudinal direction, wheel loads are assumed to act as point loads. In the transverse direction, wheel loads are distributed over a wheel load distribution width, D_w equal to the tire width plus twice the deck thickness (Figure 8-11), as computed by

$$D_w = b_t + 2t \qquad\qquad (8\text{-}15)$$

where
D_w = wheel load distribution width transverse to the deck span (in.),

b_t = truck tire width perpendicular to traffic = $\sqrt{0.025P}$ (in.),

P = wheel load (lb), and

t = deck thickness (in.).

The effective deck section defined by the deck thickness, t, and wheel-load distribution width, D_w is designed as a beam to resist the bending, deflection, shear, and reactions produced by one wheel line of the design

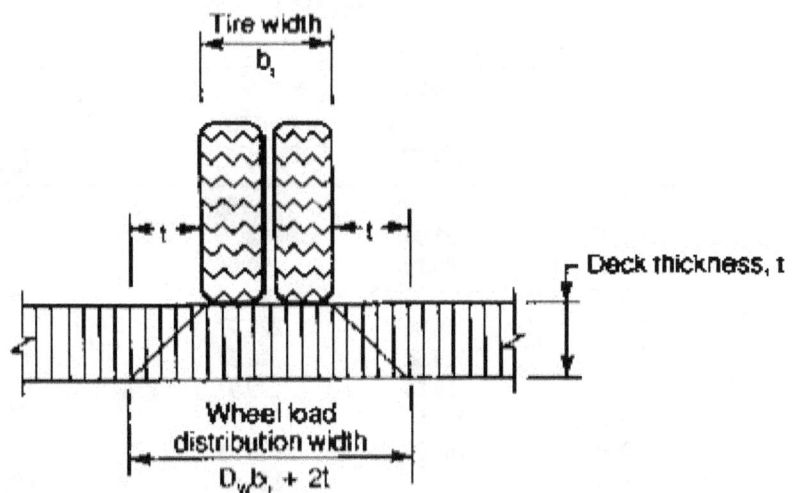

Figure 8-11. - Wheel load distribution width for continuous longitudinal nail-laminated lumber decks.

vehicle. It is generally most convenient to start with a selected species and grade of lumber, size the deck thickness based on deflection, and then check bending and shear. Because of the susceptibility of the deck to loosening or delamination, a maximum live load deflection of $L/500$ is recommended. Effective deck section properties and typical dead loads are given in Tables 8-3 and 8-4, respectively.

Table 8-3. - Effective deck section properties for continuous longitudinal nail-laminated decks and longitudinal nail-laminated deck panels with adequate shear transfer between panels.

t (in.)	H 15-44 and HS 15-44 12,000-pound wheel load				H 20-44 and HS 20-44 16,000-pound wheel load			
	D_w (in.)	A (in²)	S (in³)	I (in⁴)	D_w (in.)	A (in²)	S (in³)	I (in⁴)
5-1/2	28.32	155.76	142.78	392.65	31.00	170.50	156.29	429.80
6	29.32	175.92	175.92	527.76	32.00	192.00	192.00	576.00
7-1/4	31.82	230.70	278.76	1,010.49	34.50	250.13	302.23	1,095.60
8	33.32	266.56	355.41	1,421.65	36.00	288.00	384.00	1,536.00
9-1/4	35.82	331.34	510.81	2,362.49	38.50	356.13	549.03	2,539.25
10	37.32	373.20	622.00	3,110.00	40.00	400.00	666.67	3,333.33
11-1/4	39.82	447.98	839.95	4,724.74	42.50	478.13	896.48	5,042.72
12	41.32	495.84	991.68	5,950.08	44.00	528.00	1,056.00	6,336.00
13-1/4	43.82	580.62	1,282.19	8,494.52	46.50	616.13	1,360.61	9,014.04
14	45.32	634.48	1,480.45	10,363.17	48.00	672.00	1,568.00	10,976.00
15-1/4	47.82	729.26	1,853.52	14,133.11	50.50	770.13	1,957.40	14,925.18
16	49.32	789.12	2,104.32	16,834.56	52.00	832.00	2,218.67	17,749.33

Table 8-4. - Deck dead load for the wheel distribution width (D_w) in lb/ft of deck span for longitudinal continuous nail-laminated decks and longitudinal nail-laminated deck panels with adequate shear transfer between panels.

t (in.)	H 15-44 and HS 15-44 12,000-pound wheel load			H 20-44 and HS 20-44 16,000-pound wheel load		
	D_w (in.)	Deck only (lb/ft)	Deck plus 3 in. of asphalt (lb/ft)	D_w (in.)	Deck only (lb/ft)	Deck plus 3 in. of asphalt (lb/ft)
5-1/2	28.32	54.08	142.58	31.00	59.20	156.08
6	29.32	61.08	152.71	32.00	66.66	166.67
7-1/4	31.82	80.10	179.54	34.50	86.84	194.65
8	33.32	92.55	196.67	36.00	99.99	212.50
9-1/4	35.82	115.04	226.97	38.50	123.65	243.97
10	37.32	129.58	246.21	40.00	138.88	263.88
11-1/4	39.82	155.54	279.97	42.50	166.01	298.83
12	41.32	172.16	301.28	44.00	183.32	320.82
13-1/4	43.82	201.59	338.53	46.50	213.92	359.23
14	45.32	220.29	361.95	48.00	233.32	383.32
15-1/4	47.82	253.20	402.66	50.50	267.39	425.20
16	49.32	273.98	428.11	52.00	288.87	451.40

A two-lane, 24-foot-wide bridge is to be constructed on a low-volume county road. The bridge spans 19 feet center-to-center of bearings and supports two lanes of AASHTO HS 15-44 loading. Design this bridge as a continuous nail-laminated deck, assuming the following:

1. The deck is covered with a full-width wearing surface of dressed (S4S) 4-inch by 12-inch planks.

2. A modified vehicular railing system will be provided with the rail face extending 8 inches inward from the deck edges. Dead load of the railing is 70 lb/ft per side.

3. Bearing at each end is on a 12-inch by 12-inch timber pile cap.

4. Live load deflection must be limited to $L/500$.

5. Laminations are dressed 2-inch nominal visually graded Southern Pine.

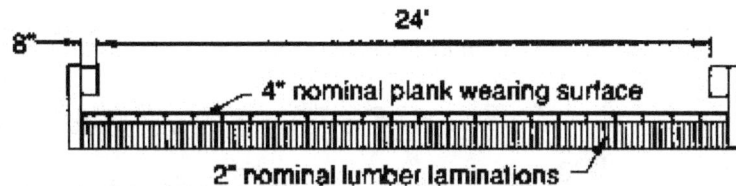

Solution

It is anticipated that the design will be controlled by the maximum live load deflection requirement of $L/500$. A species of lumber will be selected and the deck initially will be designed based on deflection, then checked for bending and shear.

Define Deck Geometric Requirements and Design Loads

For a 24-foot roadway width, and railing that projects 8 inches inward from each deck edge, the total bridge width of 25 feet 4 inches (25.33 feet) is required (203 nominal 1-1/2-inch-thick lumber laminations). Design loading will be one wheel line of an HS 15-44 vehicle in AASHTO Load Group I. Because this bridge is designed for HS 15-44 loads, which are less than H 20-44 loads, the design must also be checked in AASHTO Load Group IA using a 100-percent increase in live load forces and a 50-percent increase in allowable stresses (Chapter 6). This requirement does not apply to live load deflection.

Select a Species and Grade of Lamination

From NDS Table 4A, No. 1 visually graded Southern Pine is selected in the J&P size classification from the table "surfaced dry, used at 19%

maximum m.c." Per NDS footnotes, stresses for this grade when the moisture content will exceed 19 percent are taken from the table "surfaced green, used any condition", and further adjustment by C_M is not required. Tabulated values are as follows:

$$F_b = 1,350 \text{ lb/in}^2 \text{ (repetitive member uses)}$$

$$F_v = 85 \text{ lb/in}^2$$

$$F_{c\perp} = 375 \text{ lb/in}^2$$

$$E = 1,500,000 \text{ lb/in}^2$$

Determine Deck Thickness Based on Live Load Deflection

A deflection of $L/500$ on a 19-foot span is equivalent to 0.46 inches From Table 16-8, the deflection coefficient for an HS 15-44 vehicle on a 19-foot span is 2.96×10^9 lb-in^3. Equating the allowable deflection to the deflection coefficient for one wheel line,

$$0.46 \text{ in.} = \frac{2.96 \times 10^9}{E' I}$$

In this case,

$$E' = EC_M = 1,500,000(1.0) = 1,500,000 \text{ lb/in}^2$$

Rearranging terms, the deflection equation is solved for the required moment of inertia of the effective deck section:

$$I = \frac{2.96 \times 10^9}{E'(0.46)} = \frac{2.96 \times 10^9}{1,500,000(0.46)} = 4,289.86 \text{ in}^4$$

For a minimum $I = 4,289.86$ in^4, an 11-1/4-inch (12-inch nominal) deep lamination is selected from Table 8-3. Effective deck section properties from that table are as follows:

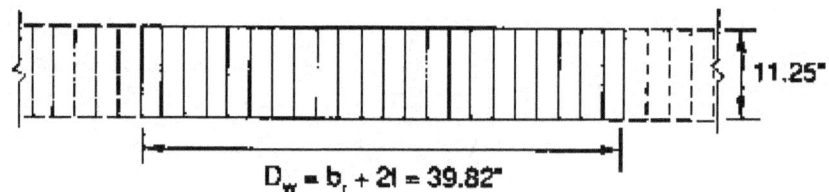

$$D_w = b_t + 2t = 39.82"$$

11.25"

$$D_w = 39.82 \text{ in.}$$

$$A = 447.98 \text{ in}^2$$

$$S = 839.95 \text{ in}^3$$

$$I = 4,724.74 \text{ in}^4$$

The actual live load deflection is computed:

$$\Delta_{LL} = \frac{2.96 \times 10^9}{1,500,000(4,724.74)} = 0.42 \text{ in.} = L/543$$

L/543 < L/500, so dressed 2-inch by 12-inch No. 1 Southern Pine laminations are acceptable for live load deflection.

Compute Deck Dead Load

The dead load of an 11.25-inch deck and 3.5-inch wearing surface are computed over the effective wheel load distribution width of 39.82 inches:

$$\text{Deck } w_{DL} = \frac{(39.82 \text{ in.})(11.25 \text{ in.} + 3.5 \text{ in.})}{144 \text{ in}^2/\text{ft}^2} \left(50 \text{ lb/ft}^3 \right) = 203.94 \text{ lb/ft}$$

Dead load of the railing system is uniformly distributed across the deck width:

$$\text{Rail } w_{DL} = \frac{39.82 \text{ in.}}{12 \text{ in/ft}} = \left(\frac{70 \text{ lb/ft}(2)}{25.33 \text{ ft}} \right) = 18.3 \text{ lb/ft}$$

$$\text{Total } w_{DL} = 203.9 + 18.3 = 222.2 \text{ lb/ft}$$

Compute Applied Moments and Bending Stress

Dead load moment is computed by assuming the effective deck section is a simply supported beam:

$$M_{DL} = \frac{w_{DL}L^2}{8} = \frac{222.2(19)^2}{8} = 10,027 \text{ ft-lb}$$

Live load moment is the maximum moment for one wheel line of an HS 15-44 vehicle obtained from Table 16-8:

$$M_{LL} = 57,000 \text{ ft-lb}$$

$$M = M_{DL} + M_{LL} = 10,027 + 57,000 = 67,027 \text{ ft-lb}$$

Bending stress is computed for the effective deck section:

$$F_b' = F_b C_u C_r = 1,350(1.0)(1.0) = 1,350 \text{ lb/in}^2$$

$$f_b = \frac{M}{S} = \frac{67,027(12 \text{ in/ft })}{839.95} = 958 \text{ lb/in}^2$$

$F_b' = 1,350 \text{ lb/in}^2 > f_b = 958 \text{ lb/in}^2$, so the deck is satisfactory in bending in AASHTO Load Group I. Deflection obviously controls design as indicated by the considerable difference between f_b and F_b'.

Bending is next checked for AASHTO Load Group IA loading, using a 100-percent increase in live load moment and a 50-percent increase in allowable bending stress:

$$M = M_{DL} + 2(M_{LL}) = 10,027 + (2)57,000 = 124,027 \text{ ft-lb}$$

$$F_b' = 1.5(1,350) = 2,025 \text{ lb/in}^2$$

$$f_b = \frac{M}{S} = \frac{127,027(12 \text{ in/ft})}{839.95} = 1,772 \text{ lb/in}^2$$

$F_b' = 2,025 \text{ lb/in}^2 > f_b = 1,772 \text{ lb/in}^2$, so the deck is satisfactory in bending in AASHTO Load Group IA.

Check Horizontal Shear

Dead load vertical shear is computed at a distance t from the support, neglecting loads that act within a distance t from the supports:

$$V_{DL} = w_{DL}\left(\frac{L}{2} - t\right) = (222.2 \text{ lb/ft})\left(\frac{19}{2} - \frac{11.25}{12 \text{ in/ft}}\right) = 1,903 \text{ lb}$$

Live load vertical shear is computed at the lesser of $3t$ or $L/4$ from the support:

$$3t = \frac{3(11.25)}{12 \text{ in/ft}} = 2.81 \text{ ft} \qquad \frac{L}{4} = \frac{19}{4} = 4.75 \text{ ft}$$

The maximum vertical shear 2.81 feet from the support is computed for one wheel line of an HS 15-44 vehicle:

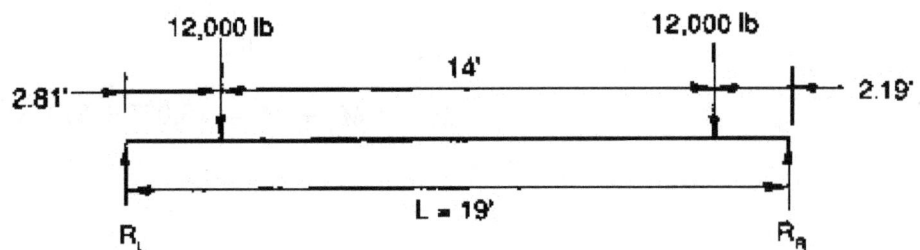

$$V_{LL} = R_L = \frac{(12,000 \text{ lb})(2.19 \text{ ft} + 16.19 \text{ ft})}{19 \text{ ft}} = 11,608 \text{ lb}$$

$$V = V_{DL} + V_{LL} = 1,903 + 11,608 = 13,511 \text{ lb}$$

$$f_v = \frac{1.5V}{A} = \frac{1.5(13,511)}{447.98} \approx 45 \text{ lb/in}^2$$

$$F_v' = F_v C_M \text{ (shear stress modification factor)}$$

Using a 2.0 shear stress modification factor (Table 7-17) for nail-laminated lumber treated with oil-type preservatives,

$$F_v' = F_v C_M (2.0) = 85(1.0)(2.0) = 170 \text{ lb/in}^2$$

$F_v' = 170$ lb/in$^2 > f_v = 45$ lb/in^2, so shear is satisfactory. By examination, shear is also acceptable for AASHTO Load Group IA.

Determine Bearing Configuration and Check Bearing Stress

Bearings for this bridge will involve nailing the laminations to a 12-inch by 12-inch pile cap. Dead load reaction is computed as follows, based on a 20-foot bridge length:

$$R_{DL} = \frac{(222.2 \text{ lb/ft})(20 \text{ ft})}{2} = 2,222 \text{ lb}$$

The maximum live load reaction for one wheel line of an HS 15-44 is obtained from Table 16-8:

$$R_{LL} = 15,160 \text{ lb}$$

For $\ell_b = 12$ inches,

$$f_{c\perp} = \frac{R_{DL} + R_{LL}}{D_w(\ell_b)} = \frac{2,222 + 15,160}{39.8(12)} = 36 \text{ lb/in}^2$$

$$F_{c\perp}' = F_{c\perp}(C_M) = (375 \text{ lb/in}^2)(1.0) = 375 \text{ lb/in}^2$$

$F_{c\perp}' = 375$ lb/in$^2 > f_{c\perp} = 36$ lb/in^2, so a bearing length of 12 inches is sufficient. The out-to-out length of the lumber laminations will be 20 feet.

Determine Nail Size and Pattern

Nails must be of sufficient length to penetrate two and one-half laminations. For an actual lamination thickness of 1-1/2 inches, 20d (4-inch long) nails will be used in the pattern shown in Figure 8-10.

Summary

The bridge will consist of 203 nominal 2-inch by 12-inch lumber laminations, 20 feet long. Lumber will be No. 1 or better Southern Pine that is surfaced dry. Stresses and deflection are as follows:

$$f_b = 958 \text{ lb/in}^2$$
$$F_b' = 1{,}350 \text{ lb/in}^2$$
$$\Delta_{LL} = 0.42 \text{ in.} = L/543$$
$$f_v = 45 \text{ lb/in}^2$$
$$F_v' = 170 \text{ lb/in}^2$$
$$f_{c\perp} = 36 \text{ lb/in}^2$$
$$F_{c\perp}' = 375 \text{ lb/in}^2$$

PANELIZED NAIL-LAMINATED LUMBER BRIDGES

Panelized nail-laminated decks are practical for simple spans up to approximately 34 feet for HS 20-44 and H 20-44 loads and 38 feet for HS 15-44 and H 15-44 loads (Figure 8-12). Load distribution within the panels is provided by spikes placed through the laminations, while load transfer between panels is provided by stiffener beams. Some designs also use a lapped joint between panels to further improve load distribution and continuity between panels (Figure 8-13). Panels for longitudinal nail-laminated bridges are prefabricated before shipment to the construction site and are of approximately equal width, but normally not greater than 7-1/2 feet wide for transportation and erection considerations. Laminations

Figure 8-12. - Panelized longitudinal nail-laminated deck bridges. (A) During construction. (photos courtesy of Wheeler Consolidated, Inc.).

are spiked together with galvanized 5/16- or 3/8-inch-diameter spikes that are of sufficient length to penetrate four laminations. The placement pattern uses two basic spike patterns involving pairs of adjacent laminations that alternate over the panel width (Figure 8-14). To prevent splitting and reduce potential deterioration, spike lead holes are drilled in the laminations before pressure treatment with preservatives.

Figure 8-12. - Panelized longitudinal nail-laminated deck bridges (continued). (B) Typical multiple-span bridge configuration (photos courtesy of Wheeler Consolidated, Inc.).

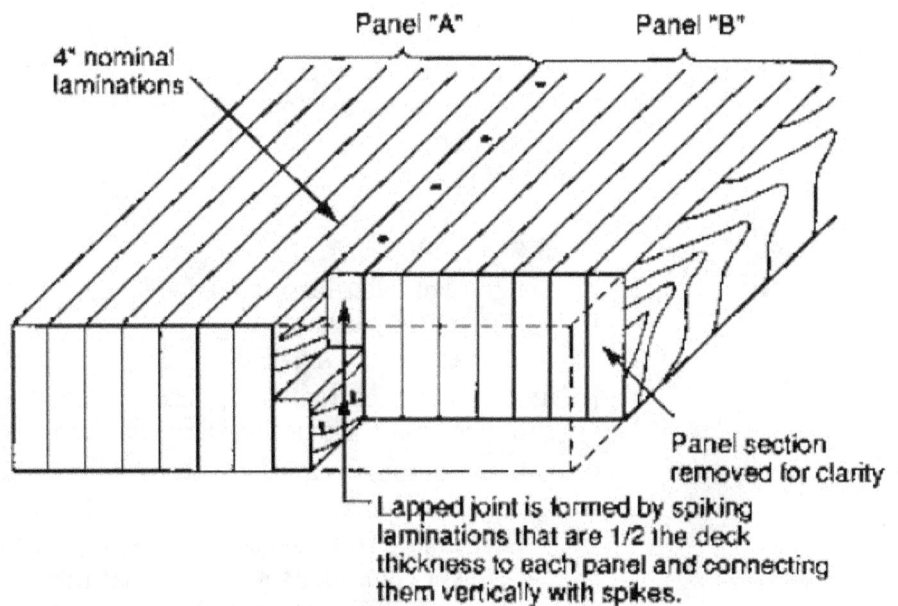

Figure 8-13. - Overlap joint configuration for longitudinal nail-laminated lumber deck panels

8-39

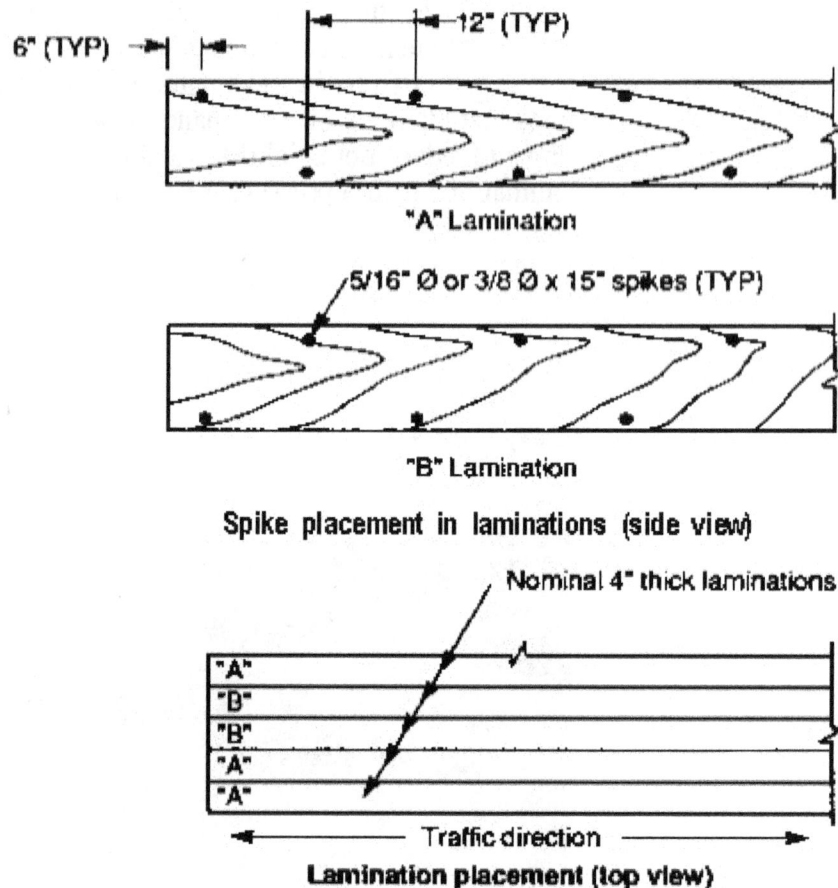

Figure 8-14. - Spike placement for longitudinal nail-laminated deck panels constructed of nominal 4-inch-thick lumber laminations.

Because of the larger laminations and increased spike size and length, performance of longitudinal nail-laminated panels is improved over conventional continuous nail-laminated decks. They are commonly used on secondary and local road systems and are capable of supporting repetitive highway loads.

Design Procedures

Longitudinal nail-laminated panels are designed using the same basic procedures and live load distribution as continuous longitudinal nail-laminated decks (AASHTO 3.25.2). With panelized decks, however, the live load distribution width cannot exceed the panel width. Transverse stiffener beams are designed for the same requirements used for glulam, with a minimum required stiffness factor, EI, of 80,000 k-in^2. One stiffener is placed at the bridge center, with subsequent stiffeners at intervals not greater than 8 feet. Because of the improved performance of panelized decks over continuous decks, a maximum live load deflection of $L/360$ is recommended. Effective deck section properties and typical dead loads for panelized decks are given in Tables 8-3 and 8-4.

8-40

A two-lane, 24-foot-wide bridge is to be constructed on a secondary state
road. The bridge spans 31 feet center to center of bearings and supports
two lanes of AASHTO HS 20-44 loading. Design this bridge as a panelized nail-laminated deck, assuming the following:

1. The deck is covered with a 3-inch asphalt wearing surface.

2. Vehicular railing is provided with the rail face extending 6 inches
    inward from the deck edges. Dead load of the railing is 75 lb/ft
    per side.

3. Bearing at each end will be on a 12-inch by 12-inch timber pile
    cap.

4. Live load deflection must not exceed $L/360$.

5. Laminations are 4-inch nominal S4S visually graded Douglas Fir-
    Larch.

Solution

The design sequence for this panelized bridge will follow the same procedures used for the continuous nail-laminated deck in Example 8-3, but will
include stiffener beam design similar to that used for longitudinal glulam
decks.

Define Deck Geometric Requirements and Design Loads

For a 24-foot roadway width, and railing that projects 6 inches inward
from each deck edge, the total bridge width of 25 feet is required. Based
on an actual lamination thickness of 3-1/2 inches, four panels will be used:
two panels of 21 laminations (6 feet 1-1/2 inches wide) and two panels of
22 laminations (6 feet 5 inches wide). The bridge width out-to-out will be
25 feet 1 inch (25.08 feet). Design loading will be one wheel line of an
HS 20-44 vehicle in AASHTO Load Group I.

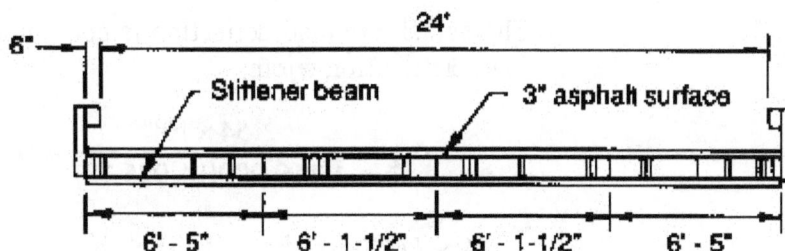

Select a Species and Grade of Lamination

From NDS Table 4A, No. 1 visually graded Douglas Fir-Larch is selected in the J&P size classification. Tabulated values are as follows:

$$F_b = 1{,}750 \text{ lb/in}^2 \text{ (repetitive member uses)} \qquad C_M = 0.86$$

$$F_v = 95 \text{ lb/in}^2 \qquad\qquad\qquad\qquad\qquad C_M = 0.97$$

$$F_{c\perp} = 625 \text{ lb/in}^2 \qquad\qquad\qquad\qquad\quad C_M = 0.67$$

$$E = 1{,}800{,}000 \text{ lb/in}^2 \qquad\qquad\qquad\quad C_M = 0.97$$

Determine Deck Thickness Based on Live Load Deflection

A deflection of $L/360$ on a 31-foot span is equivalent to 1.03 inch. From Table 16-8, the deflection coefficient for an HS 20-44 vehicle on a 31-foot span is 2.54×10^{10} lb-in^3:

$$1.03 \text{ in.} = \frac{2.54 \times 10^{10}}{E' I}$$

$$E' = E C_M = 1{,}800{,}000(0.97) = 1{,}746{,}000 \text{ lb/in}^2$$

Rearranging terms,

$$I = \frac{2.54 \times 10^{10}}{E'(1.03)} = \frac{2.54 \times 10^{10}}{1{,}746{,}000(1.03)} = 14{,}123.82 \text{ in}^4$$

For a minimum $I = 14{,}123.82$ in^4, a 15-1/4-inch-deep (16-in. nominal) lamination is selected from Table 8-3. Effective deck section properties from that table are as follows:

$D_w = 50.50$ in.

$A = 770.13$ in^2

$S = 1{,}957.40$ in^3

$I = 14{,}925.18$ in^4

The actual live load deflection is computed based on the 50.50-inch wheel load distribution width:

$$\Delta_{LL} = \frac{2.54 \times 10^{10}}{1{,}746{,}000(14{,}925.18)} = 0.97 = L/384$$

$L/384 < L/360$, so dressed 4-inch by 16-inch No. 1 Douglas Fir-Larch laminations are acceptable for live load deflection.

Compute Deck Dead Load

From Table 8-4, the dead load of a 15.25-inch deck and 3-inch asphalt wearing surface over the wheel distribution width of 50.50 inches is 425.2 lb/ft. An additional dead load of 8 lb/ft will be added for the stiffener beams and attachment hardware. Dead load of the railing system is assumed to be uniformly distributed across the deck width:

$$\text{Rail } w_{DL} = \frac{50.50 \text{ in.}}{12 \text{ in/ft}} = \left(\frac{(75 \text{ lb/ft})(2)}{25.08 \text{ ft}} \right) = 25.2 \text{ lb/ft}$$

$$\text{Total } w_{DL} = 425.2 + 8 + 25.2 = 458.4 \text{ lb/ft}$$

Compute Applied Moments and Bending Stress

$$M_{DL} = \frac{w_{DL}L^2}{8} = \frac{458.4(31)^2}{8} = 55,065 \text{ ft-lb}$$

Maximum moment for one wheel line of an HS 20-44 vehicle on a 31-foot span is obtained from Table 16-8:

$$M_{LL} = 148,650 \text{ ft-lb}$$

$$M = M_{DL} + M_{LL} = 55,065 + 148,650 = 203,715 \text{ ft-lb}$$

Bending stress is computed for the effective deck section:

$$F_b' = F_b C_M C_r = 1,750 \ (0.86)(1.0) = 1,505 \text{ lb/in}^2$$

$$f_b = \frac{M}{S} = \frac{203,715 \, (12 \text{ in/ft})}{1,957.40} = 1,249 \text{ lb/in}^2$$

$F_b' = 1,505 \text{ lb/in}^2 > f_b = 1,249 \text{ lb/in}^2$, so the deck is satisfactory in bending.

Check Horizontal Shear

$$V_{DL} = w_{DL}\left(\frac{L}{2} - t\right) = 458.4 \text{ lb/ft} \left(\frac{31}{2} - \frac{15.25}{12 \text{ in/ft}}\right) = 6,523 \text{ lb}$$

Live load vertical shear is computed at the lesser of $3t$ or $L/4$ from the support:

$$3t = \frac{3\,(15.25)}{12 \text{ in/ft}} = 3.81 \text{ ft} \qquad\qquad \frac{L}{4} = \frac{31}{4} = 7.75 \text{ ft}$$

The maximum vertical shear 3.81 feet from the support is computed for one HS 20-44 wheel line:

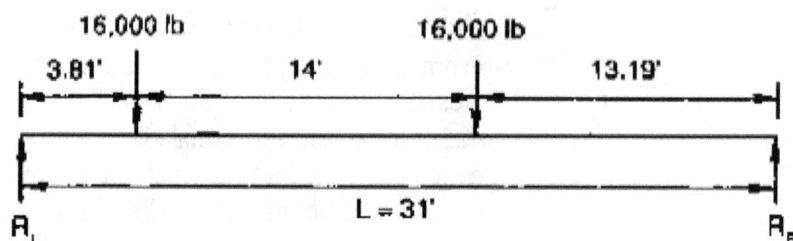

$$V_{LL} = R_L = \frac{(16,000 \text{ lb})(13.19\text{ft} + 27.19 \text{ ft})}{31 \text{ ft}} = 20,841 \text{ lb}$$

$$V = V_{DL} + V_{LL} = 6,523 + 20,841 = 27,364 \text{ lb}$$

$$f_v = \frac{1.5V}{A} = \frac{1.5(27,364)}{770.13} = 53 \text{ lb/in}^2$$

$$F_v' = F_v C_M (\text{shear stress modification factor})$$

Using a 2.0 shear stress modification factor (Table 7-17) for nail-laminated lumber treated with oil-type preservatives:

$$F_v = F_v C_M (2.0) = 95(0.97)(2.0) = 184 \text{ lb/in}^2$$

$F_v' = 184 \text{ lb/in}^2 > f_v = 53 \text{ lb/in}^2$, so shear is satisfactory.

Determine Bearing Configuration and Check Bearing Stress

For bearing on a 12-inch pile cap, the bridge length will be 32 feet:

$$R_{DL} = \frac{(458 \text{ lb/ft})(32 \text{ ft})}{2} = 7,333 \text{ lb}$$

The maximum live load reaction for one wheel line of an HS 20-44 is obtained from Table 16-8:

$$R_{LL} = 25,160 \text{ lb}$$

For $\ell_b = 12$ inches,

$$f_{c\perp} = \frac{R_{DL} + R_{LL}}{D_w(\ell_b)} = \frac{7,333 + 25,160}{50.50(12)} = 54 \text{ lb/in}^2$$

$$F_{c\perp}' = F_{c\perp}(C_M) = (625 \text{ lb/in}^2)(0.67) = 419 \text{ lb/in}^2$$

$F_{c\perp}' = 419 \text{ lb/in}^2 > f_{c\perp} = 54 \text{ lb/in}^2$, so a bearing length of 12 inches is acceptable. The out-to-out length of the lumber laminations will be 32 feet.

Determine Spike Size and Pattern

Spikes must be of sufficient length to penetrate four laminations. For an actual lamination thickness of 3-1/2 inches, 3/8-inch-diameter by 15-inch-long spikes will be used in the pattern shown in Figure 8-14.

Determine Stiffener Spacing and Configuration

Design requirements for stiffener beams on panelized nail-laminated decks are the same as those for longitudinal glulam decks. For this bridge, stiffener beams will be placed at the span quarter points for a spacing of 7.75 feet:

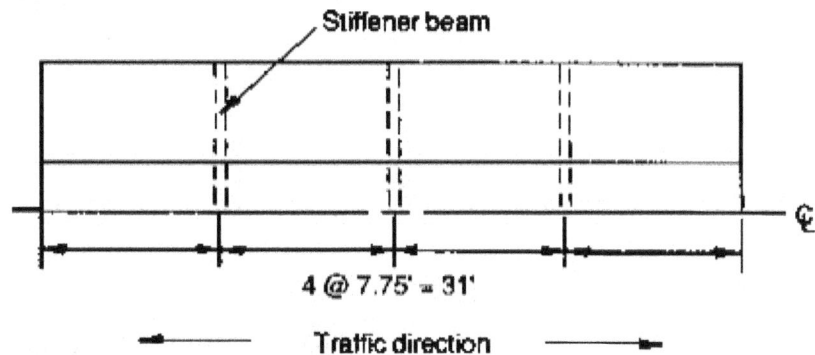

The size and stiffness of an individual stiffener beam must be sufficient to provide a minimum EI value of 80,000 k-in². A glulam stiffener will be used because of the improved dimensional stability of glulam compared to sawn timber. Selecting a combination symbol No. 2 stiffener, 5-1/8 inches wide and 6 inches deep:

$$E' = EC_M = 1,700,000(0.833) = 1,416,100 \text{ lb/in}^2$$

$$I = \frac{bd^3}{12} = \frac{5.125\,(6.0)^3}{12} = 92.25 \text{ in}^4$$

$$E'I = \frac{1,416,100 \text{ lb/in}^2}{1,000 \text{ lb/k}}\,\left(92.25 \text{ in}^4\right) = 130,635 \text{ k-in}^2$$

130,635 k-in² > 80,000 k-in², so 5-1/8-inch by 6-inch stiffener beams are satisfactory. Stiffener attachment will be with 3/4-inch-diameter bolts as described in Example 8-1.

Checking the stiffener beam dead load per panel,

$$DL/stiffener = \frac{(5.125 \text{ in.})(6 \text{ in.})\left(50 \text{ lb/ft}^3\right)(25.08 \text{ ft})}{144 \text{ in}^2/\text{ft}^2} = 267.8 \text{ lb}$$

$$\text{Stiffener} = w_{DL}/\text{panel} = \frac{(267.8 \text{ lb})(3 \text{ stiffeners})}{31(4 \text{ panels})} = 6.5 \text{ lb/ft}$$

6.5 lb/ft is less than the 8 lb/ft assumed, so no dead load revision is required.

Summary

The bridge will consist of four nail-laminated panels constructed of S4S 4-inch by 16-inch lumber, 32 feet long. The two outside panels will be 6 feet 5 inches wide (22 laminations) and the two interior panels will be 6 feet 1-1/2 inches wide (21 laminations). Lumber will be No. 1 or better

Douglas Fir-Larch in the J&P size classification. Stresses and deflection are as follows:

$$f_b = 1,249 \text{ lb/in}^2$$

$$F_b' = 1,505 \text{ lb/in}^2$$

$$\Delta_{LL} = 0.97 \text{ in.} = L/384$$

$$f_v = 53 \text{ lb/in}^2$$

$$F_v' = 184 \text{ lb/in}^2$$

$$f_{c\perp} = 54 \text{ lb/in}^2$$

$$F_{c\perp}' = 419 \text{ lb/in}^2$$

Stiffener beams will consist of three 5-1/8-inch-wide by 6-inch-deep by 25-feet-1-inch-long glulam beams, manufactured to combination symbol No. 2.

8.5 LONGITUDINAL DECKS ON TRANSVERSE FLOORBEAMS

One of the primary applications of longitudinal timber decks has been on transverse floorbeams. Floorbeams are transverse beams that either support a longitudinal deck directly or support longitudinal stringers, which in turn support a transverse deck (Figure 8-15). They are used primarily in truss and arch superstructures, and on beam superstructures where the beam spacing exceeds the economical span for transverse deck configurations. Longitudinal decks with floorbeams are used for new structures, but they have also demonstrated distinct advantages in the rehabilitation of existing structures, predominantly as a replacement for deteriorated concrete decks. Not only can a concrete deck be economically replaced with timber, but the lighter dead loads and improved live load distribution frequently result in an increased capacity for existing structures.[9] In many cases, dead load is further reduced when existing stringers are removed and the timber replacement deck is placed directly on the floorbeams (Figure 8-16). Longitudinal timber decks have been used in many cases to restore structurally deficient bridges to full capacity for modern highway loads (Chapter 15).

FLOORBEAM DESIGN

Floorbeams are designed to support the deck dead load and vehicle live loads over the tributary deck span. Their design follows the same basic beam design procedures discussed in Chapters 5 and 7; however, AASHTO gives specific live load distribution criteria for transverse floorbeams (AASHTO 3.23.3). In both the transverse and longitudinal directions, no wheel load distribution is assumed and the wheel loads are

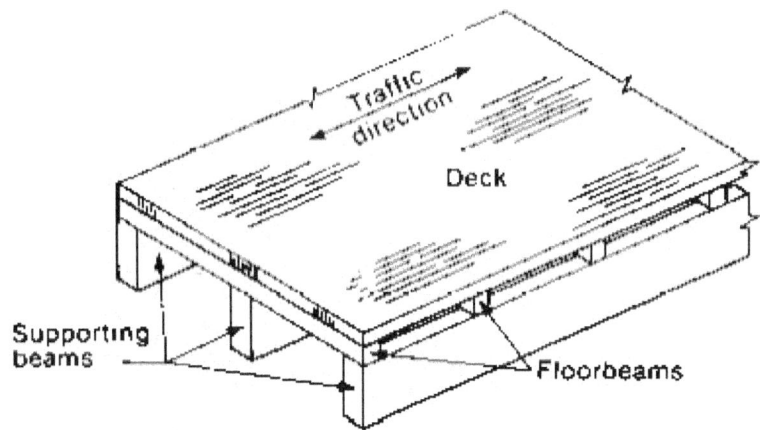

Longitudinal deck supported on transverse floorbeams

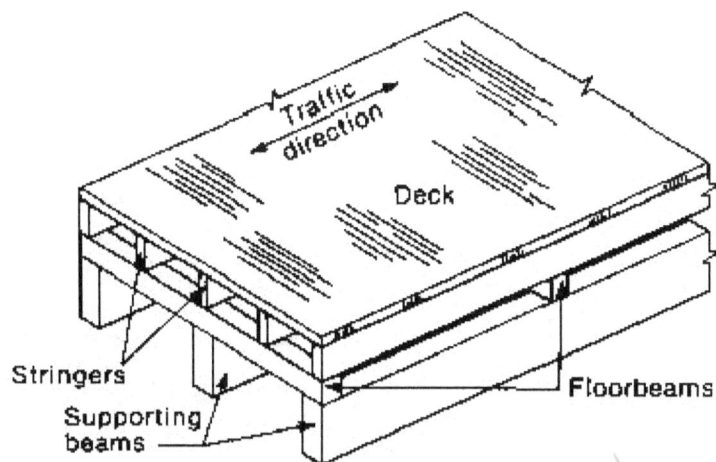

**Transverse deck supported on longitudinal stringers,
supported on transverse floorbeams**

Figure 8-15. - Timber bridge floorbeam configurations.

assumed to act as concentrated loads (Figure 8-17). When the deck is
supported directly on the floorbeams, the portion of the wheel loads
longitudinally distributed to each beam depends on the deck type and the
center-to-center floorbeam spacing. For beam spacings of approximately
4-1/2 to 5-1/2 feet, depending on the deck type and thickness, the fraction
of the wheel load applied to each floorbeam is determined from empirical
equations given in AASHTO (Table 8-5). For greater floorbeam spacings,
the load on each beam is the reaction of the wheel loads, assuming the
deck acts as a simple span between floorbeams. It should be noted that the
AASHTO empirical equations in Table 8-5 are based on the ability of the
deck to distribute loads longitudinally among adjacent floorbeams. For
floorbeams at bridge ends, longitudinal distribution is limited because
there is no adjacent beam on the approach roadway. End floorbeams
should therefore be designed for the reaction of the wheel lines, assuming
the deck acts as a simple span between beams.

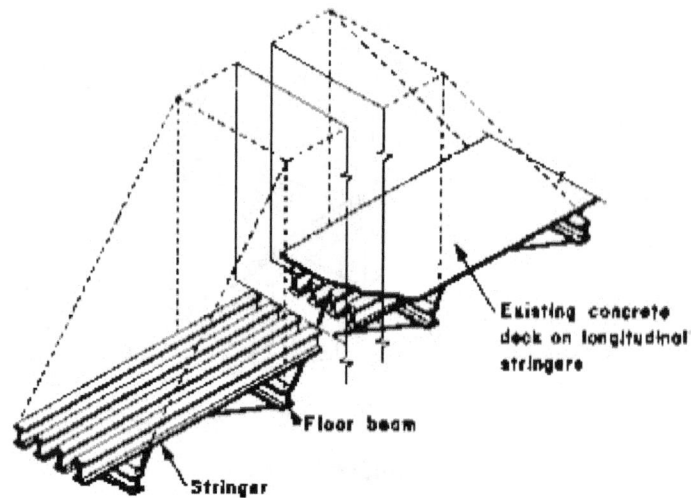

Existing structure with deteriorated concrete deck

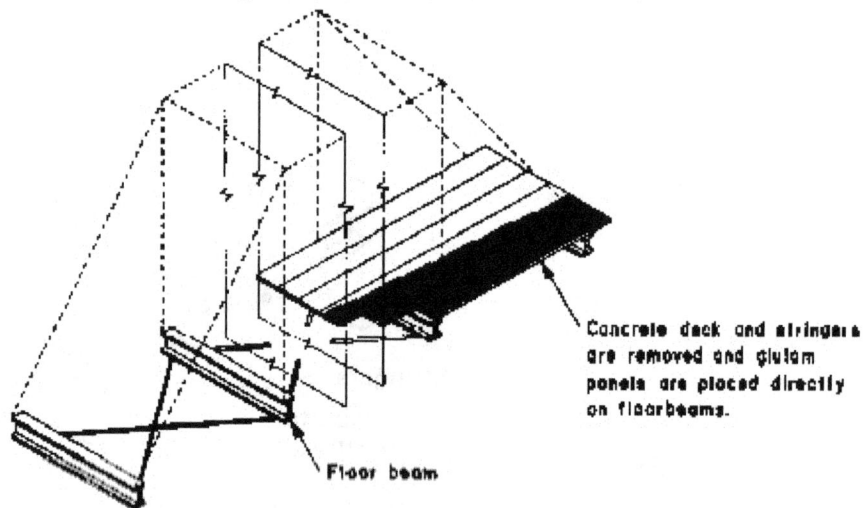

Deck replacement with longitudinal glulam panels

Figure 8-16 - Typical truss rehabilitation with longitudinal glulam deck panels.

Table 8-5. - Distribution of wheel loads to transverse floorbeams.

Type of deck	Fraction of wheel load to each floor beam[a]
Nail-laminated lumber or glulam, 4 in. in nominal thickness.	$S/4.5$
Nail-laminated lumber or glulam, 6 in. or more in nominal thickness.	$S/5$[b]

[a] S = floorbeam spacing in feet.

[b] If S exceeds the denominator, the load shall be the reaction of the wheel loads assuming the flooring between beams acts as a simple span.

From AASHTO[1] Table 3.23.3.1; © 1983. Used by permission.

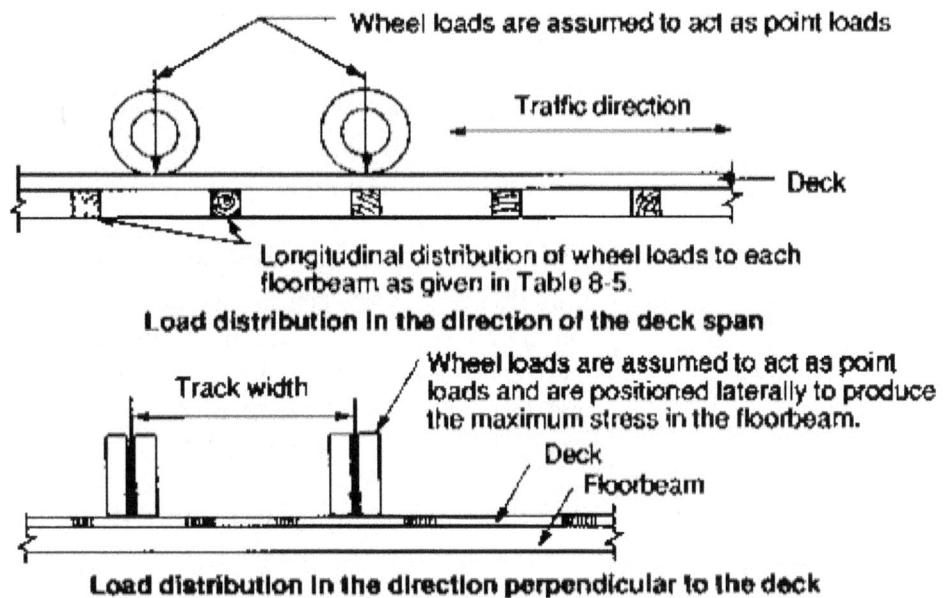

Figure 8-17. - Wheel load distribution to transverse floorbeams that directly support a longitudinal timber deck.

Example 8-5 - Transverse glulam floorbeam design

A beam bridge carries two lanes of AASHTO HS 20-44 loading on a 26-foot roadway width. The beam system consists of five 10-1/2-inch-wide glulam beams, spaced 6 feet on center. The deck is a series of longitudinal glulam panels that are supported by transverse glulam floorbeams, spaced 7 feet on center. Design the floorbeams for this structure, assuming the following:

1. The deck is 8-1/2 inches thick and is provided with a 3-inch asphalt wearing surface.

2. Floorbeams are visually graded Southern Pine glulam and are provided with continuous lateral support from the deck. Floorbeam attachment to the supporting beams is adequate to prevent sliding or overturning of the floorbeams.

3. The deck is watertight and protects floorbeams from exposure to weathering. With the exception of compression perpendicular to grain, dry condition stresses may be used for design.

4. Floorbeam live load deflection must not exceed $L/500$.

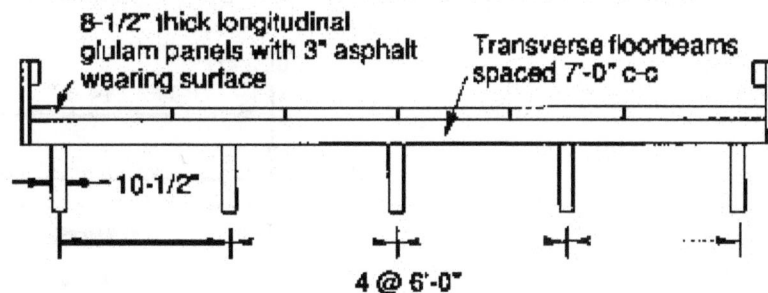

8-49

Solution

The design procedure for the floorbeams will follow the same basic procedures used for the glulam beams. Because of the short span, design will initially be based on horizontal shear, then checked for bending, deflection and bearing.

Define Basic Configuration and Design Criteria

The floorbeams are continuous over five supports. Analysis will be based on the conservative assumption that floorbeams act as simple spans between supports, using 80 percent of the simple span moment and deflection to account for span continuity. An alternative would be to use continuous beam analysis with yielding supports.

The floorbeam span, L, is the center-to-center distance between supporting beams:

$$L = 72 \text{ in.}$$

The design loading is two traffic lanes of HS 20-44 loading.

Select a Beam Combination Symbol

Floorbeams are subject to both positive and negative bending moments, and a balanced Southern Pine combination symbol, 24F-V5, is selected from *AITC 117--Design*. Tabulated values are as follows:

$$F_{bx} = 2,400 \text{ lb/in}^2$$

$$F_{c\perp x} = 650 \text{ lb/in}^2$$

$$F_{vx} = 200 \text{ lb/in}^2$$

$$E_x = 1,700,000 \text{ lb/in}^2$$

Compute Longitudinal Wheel Load Distribution

The glulam deck is supported directly by floorbeams, so longitudinal wheel load distribution is obtained from Table 8-5. From that table, the floorbeam spacing of 7 feet exceeds the denominator value of 5.0 for an 8-1/2-inch glulam deck. Longitudinal wheel load distribution is therefore computed by assuming the deck acts as a simple span between floorbeams:

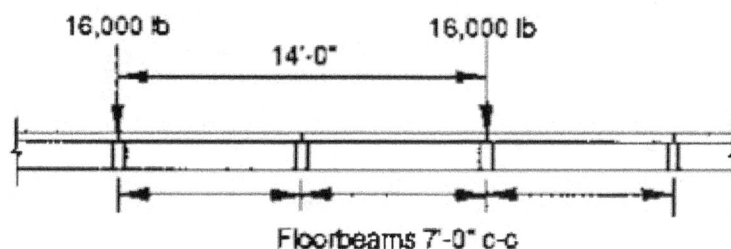

Floorbeams 7'-0" c-c

For the minimum 14-foot axle spacing, maximum longitudinal distribution is one axle load per floorbeam.

Determine Deck Dead Load and Dead Load Moment

Each floorbeam supports a tributary deck span of 7 feet. Assuming that the deck acts as a simple span between floorbeams, dead load of the deck and wearing surface is computed in lb/ft of floorbeam span:

Deck w_{DL}

$$= (7 \text{ ft}) \left[\frac{(8.5 \text{ in.})(50 \text{ lb/ft}^3) + (3 \text{ in.})(150 \text{ lb/ft}^3)}{12 \text{ in/ft}} \right] = 510.4 \text{ lb/ft}$$

Determine Floorbeam Size Based on Horizontal Shear

Using the simple-span beam analogy, maximum deck dead load vertical shear is computed at a distance d from the supports, neglecting loads that occur within a distance d. Estimating a floorbeam depth, $d = 18$ inches, deck dead load shear is computed by Equation 7-6:

$$\textbf{Deck } V_{DL} = w_{DL} \left(\frac{L}{2} - d \right) = 510.4 \left(\frac{6}{2} - 1.5 \right) = 765.6 \text{ lb}$$

Live load vertical shear is computed at the lesser distance from the support of $3d$ or $L/4$:

$$3d = 3(1.5 \text{ ft}) = 4.5 \text{ ft} \qquad \frac{L}{4} = \frac{6 \text{ ft}}{4} = 1.5 \text{ ft}$$

$L/4 = 1.5$ feet controls and the two traffic lanes (4 wheel lines) are positioned laterally to produce the maximum live load shear at that location. In this case, wheel loads from adjacent lanes can both be on the center spans:

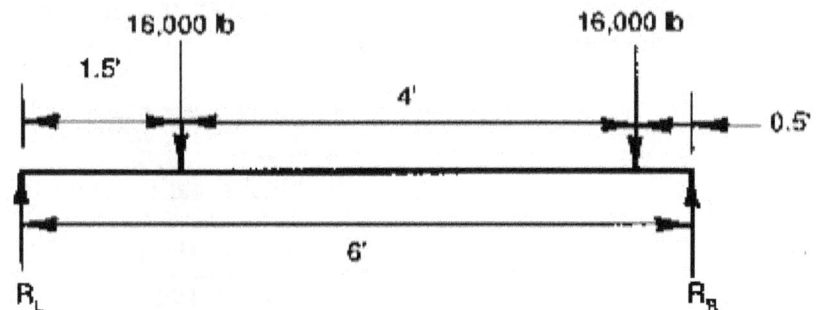

$$V_{LL} = R_L = \frac{16,000(4.5 \text{ ft} + 0.5 \text{ ft})}{6 \text{ ft}} = 13,333 \text{ lb}$$

$$\textbf{Deck } V_{DL} + V_{LL} = 765.6 + 13,333 = 14,099 \text{ lb}$$

8-51

Assuming $F_v' = f_v$, the minimum required floorbeam area is computed using a modified form of Equation 7-7:

$$F_v' = F_{vx}(C_M) = (200)(1.0) = 200 \text{ lb/in}^2$$

$$A = \frac{1.5V}{F_v'} = \frac{1.5(14,099)}{200} = 105.7 \text{ in}^2$$

From Table 16-4, two glulam beam sizes are feasible: 6-3/4 inches by 17-7/8 inches; or 8-1/2 inches by 13-3/4 inches. The 6-3/4-inch by 17-7/8-inch size is selected because it will provide a greater moment of inertia (I) for increased stiffness. Floorbeam properties are as follows:

$$A = 120.7 \text{ in}^2$$

$$S_x = 359.5 \text{ in}^3$$

$$C_r = 0.96$$

$$I_x = 3,212.6 \text{ in}^4$$

Beam weight = 41.9 lb/ft

Beam dead load shear is computed at a distance d from the support. A rounded floorbeam depth of 18 inches is used and revision of previous deck dead load calculations is not required.

$$\textbf{Beam } V_{DL} = w_{DL}\left(\frac{L}{2} - d\right) = 41.90\left(\frac{6}{2} - 1.5\right) = 62.9 \text{ lb}$$

$$V_{DL} = \text{Beam } V_{DL} + \text{Deck } V_{DL} = 62.9 + 765.6 = 829 \text{ lb}$$

Live load vertical shear is controlled by the $L/4$ distance and revision is not required.

$$V = V_{DL} + V_{LL} = 829 + 13,333 = 14,162 \text{ lb}$$

Stress in horizontal shear is computed by Equation 7-7:

$$f_v = \frac{1.5V}{A} = \frac{1.5(14,162)}{120.7} = 176 \text{ lb/in}^2$$

$f_v = 176 \text{ lb/in}^2 < F_v' = 200 \text{ lb/in}^2$, so a 6-3/4-inch by 17-7/8-inch floorbeam is satisfactory in horizontal shear.

Check Bending

For a deck dead load of 510.4 lb/ft and floorbeam dead load of 41.9 lb/ft, dead load moment is computed by Equation 7-2:

$$M_{DL} = \frac{w_{DL}L^2}{8} = \frac{(510.4 + 41.9)(6)^2}{8} = 2,485 \text{ ft-lb}$$

Live load moment is determined by positioning the wheel loads laterally to produce the maximum moment in the floorbeam. For a 6-foot floorbeam span, maximum moment is produced with one wheel load centered on a span:

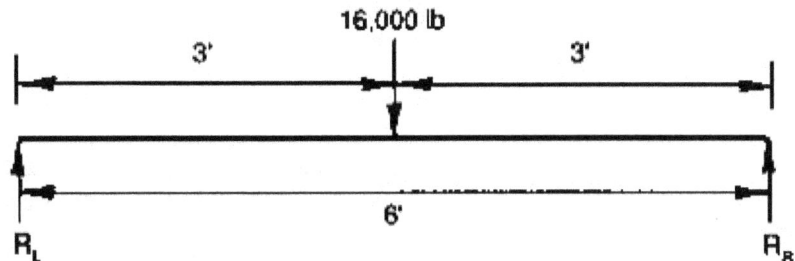

$$M_{LL} = \frac{PL}{4} = \frac{(16,000 \text{ lb})(6 \text{ ft})}{4} = 24,000 \text{ ft-lb}$$

Allowable bending stress is computed using the beam size factor, C_F. Consideration of lateral stability is not required because the floorbeams are continuously supported by the deck:

$$F_b' = F_{bx}C_M C_F = (2,400 \text{ lb/in}^2)(1.0)(0.96) = 2,304 \text{ lb/in}^2$$

$$M = M_{DL} + M_{LL} = 2,485 + 24,000 = 26,485 \text{ ft-lb}$$

$$f_b = \frac{M}{S_x} = \frac{26,485(12 \text{ in/ft})}{359.5} = 884 \text{ lb/in}^2$$

$f_b = 884$ lb/in^2 is substantially less than $F_b' = 2,304$ lb/in^2, indicating that beam size is controlled by horizontal shear. By examining the various visually graded Southern Pine combination symbols in *AITC 117-- Design*, it is seen that F_{vx} for most combinations is 200 lb/in^2 although F_{bx} varies from 1,600 lb/in^2 to 2,400 lb/in^2. In this application, a new balanced combination symbol 16F-V5 is selected with the following section properties:

$$F_{bx} = 1,600 \text{ lb/in}^2$$

$$F_{c\perp x} = 560 \text{ lb/in}^2$$

$$F_{vx} = 200 \text{ lb/in}^2$$

$$E_x = 1,400,000 \text{ lb/in}^2$$

Allowable bending stress is recomputed for the revised combination symbol:

$$F_b' = (1,600 \ lb/in^2)(0.96) = 1,536 \ lb/in^2 > f_b = 884 \ lb/in^2$$

Check Live Load Deflection

Live load deflection is computed with a wheel load centered on a floorbeam span. Using 80 percent of the simple span deflection to account for span continuity,

$$E' = E_x(C_M) = 1,400,000(1.00) = 1,400,000 \ lb/in^2$$

$$\Delta_{LL} = 0.80 \frac{PL^3}{48E'I} = 0.80 \frac{16,000(72)^3}{48(1,400,000)(3,212.6)} = 0.03 \ in.$$

0.03 in. $= L/2,400 < L/500$, so live load deflection is acceptable.

Check Bearing Stress

Bearing stress between the floorbeam and the longitudinal supporting beam is checked for a bearing area, A, equal to the floorbeam width times beam width, D_w:

$$A = (6.75 \ in.)(10.50 \ in.) = 70.88 \ in^2$$

Dead load and live load reactions are computed by assuming that the deck acts as a simple span between floorbeams. From bending calculations, floorbeams support a dead load of 552.3 lb/ft of deck width. The dead load reaction is computed based on a tributary deck width equal to the spacing of the supporting beams,

$$R_{DL} = (552.3 \ lb/ft)(6ft) = 3,314 \ lb$$

Maximum live load reaction occurs with one wheel load over the beam,

$$R_{LL} = 16,000 \ lb$$

$$f_{c\perp} = \frac{R_{DL} + R_{LL}}{A} = \frac{3,314 + 16,000}{70.88} = 272 \ lb/in^2$$

$$F_{c\perp}' = F_{c\perp}C_M = 560(0.53) = 297 \ lb/in^2$$

$f_{c\perp} = 272 \ lb/in^2 < F_{c\perp}' = 297 \ lb/in^2$, so bearing is satisfactory.

Summary
Floorbeams will be 6-3/4-inch by 17-7/8-inch visually graded Southern Pine glulam combination symbol No. 16F-V5. Stresses, and deflection are as follows:

$$f_b = 884 \text{ lb/in}^2$$
$$F_b' = 1,536 \text{ lb/in}^2$$
$$\Delta_{LL} = 0.03 \text{ in.} = L/2,400$$
$$f_v = 176 \text{ lb/in}^2$$
$$F_v' = 200 \text{ lb/in}^2$$
$$f_{c\perp} = 272 \text{ lb/in}^2$$
$$F_{c\perp}' = 297 \text{ lb/in}^2$$

Floorbeam and Deck Attachment

The attachment between floorbeams and the supporting beams or other components depends primarily on the beam materials. Several common attachments for timber and steel beams are shown in Figure 8-18. In each case, the attachment must sufficiently resist all applied vertical and transverse loads and meet minimum connector design requirements discussed in Chapter 5. Deck attachment to floorbeams uses the same connections previously discussed for transverse deck configurations. Bolted brackets or clips are recommended because they compensate for minor construction tolerances and do not require field drilling or cutting.

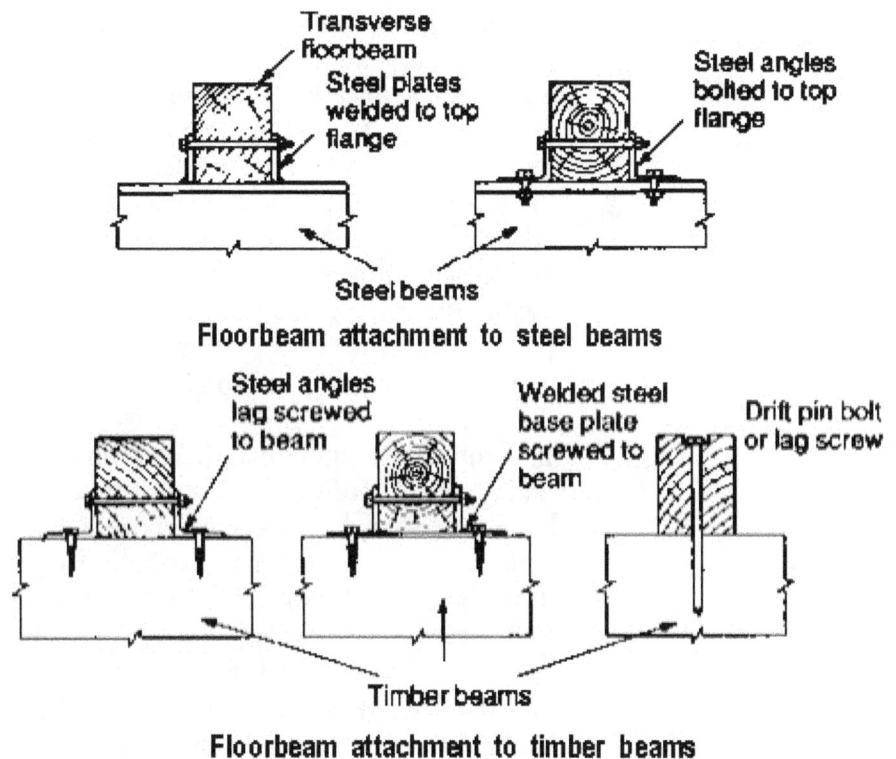

Figure 8-18. - Typical floorbeam attachment details.

DECK DESIGN

The design of glulam or nail-laminated longitudinal decks on transverse floorbeams is basically the same as the design of longitudinal deck bridges. The primary difference is that floorbeam spans are generally less than those typically encountered in superstructure design. Because the longitudinal deck functions specifically as a deck or floor in floorbeam applications, rather than the primary support for the bridge, the design criteria are also slightly different. When used on floorbeams, the deck span, s, is the clear distance between the floorbeams plus one-half the width of one beam, but not greater than the clear span plus the floor thickness (AASHTO 3.25.2.3). In addition, the assumptions used in deck analysis may vary among applications. In continuous multiple-span longitudinal deck bridges, the deck is normally analyzed as a continuous beam. On floorbeams, spans are usually substantially less, and AASHTO permits the deck to be designed as a series of simple spans. If the deck is continuous over more than two spans, the maximum positive moment and deflection from the design truck load are assumed to be 80 percent of those computed for a simple span (AASHTO 3.25.4). This simple span assumption may be adequate for most longitudinal decks on floorbeams, but for long deck spans or unusual configurations the designer should analyze the deck as if it were a continuous member, rather than a series of simple spans.

Longitudinal Glulam Decks

Glulam is normally the preferred material for longitudinal decks over floorbeams because of its higher strength, improved performance, and longer panel lengths compared to sawn lumber (Figure 8-19). In longitudinal deck applications, glulam panels may be used with transverse stiffener beams or as noninterconnected panels without stiffener beams. When stiffener beams are used, the deck is designed in the same manner as was the longitudinal deck bridge based on the ISU studies previously discussed. However, for those design criteria to be applicable, a transverse stiffener beam must be provided between floorbeams to provide lateral continuity and load distribution among the panels. Therefore, the glulam panel stiffener-beam configuration is most practical for long deck spans of approximately 8 feet or more.

In addition to longitudinal decks with stiffener beams, a noninterconnected glulam panel configuration is also used on floorbeam spacings of approximately 8 feet or less. Noninterconnected glulam panels function independently, and there is no load distribution among adjacent panels. In the direction of the deck span, wheel loads are assumed to act as point loads. In the transverse direction, the wheel loads are laterally distributed to the panel over a wheel load distribution width, D_w, equal to the tire width, b_t, plus the deck thickness (Figure 8-20). The deck is then designed as a beam, assuming that the deck section of thickness t and width D_w resists the forces produced by one wheel line of the design vehicle. Many of the design limitations and maximum spans for longitudinal noninterconnected panels closely parallel those for transverse glulam panels discussed in Chapter 7. Because there is no load sharing among panels, a maximum panel deflection of approximately 0.10 inch is recommended.

Figure 8-19. - Longitudinal glulam deck on transverse glulam floorbeams.

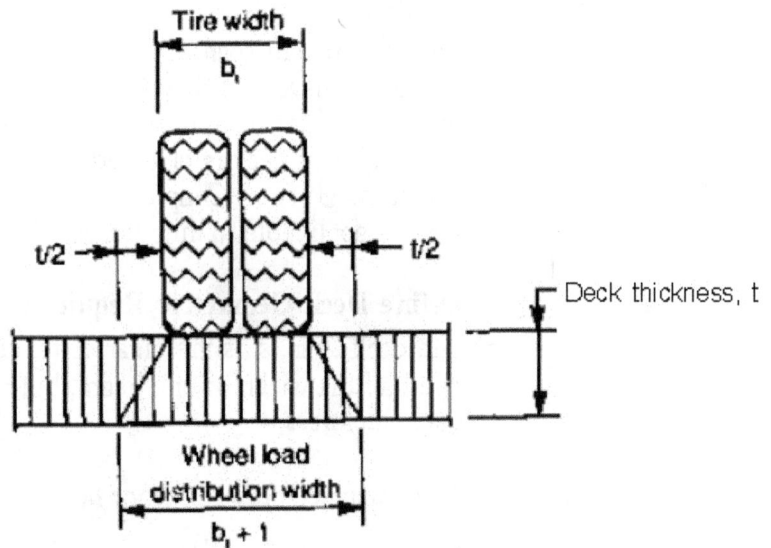

Figure 8-20. - Wheel load distribution width for longitudinal noninterconnected glulam decks.

Example 8-6 - Longitudinal glulam deck on transverse floorbeams

A steel bridge carries two traffic lanes of AASHTO HS 20-44 loading on a roadway width of 24 feet. Rehabilitation of the structure will involve replacement of the existing concrete deck with a longitudinal glulam deck. The new deck will be placed on 10-inch-wide transverse steel floorbeams that are spaced 6 feet on center. Design a glulam deck for this bridge, assuming the following:

1. The deck will be provided with a 3-inch asphalt wearing surface.

2. The dead load of the railing system is carried by the steel floorbeams.

3. Live load deflection must be limited to 0.10 inch.

4. Glulam deck panels are manufactured from visually graded western species.

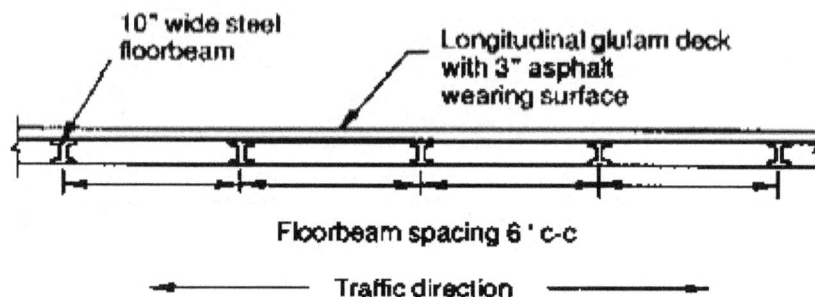

Floorbeam spacing 6 ' c-c

Traffic direction

Solution

For a floorbeam span of 6 feet, noninterconnected glulam deck panels without transverse stiffener beams will be used. The panels will initially be designed for bending, then checked for deflection, shear, and bearing. Although this deck is oriented longitudinally, many of the design aids and equations given in Chapter 7 for transverse noninterconnected decks will also be applicable to this design.

Define Deck Geometric Requirements and Design Loads

The deck span, s, is the clear distance between supporting floorbeams plus one-half the width of one beam, but not greater than the clear span plus the deck thickness, t.

Clear distance between floor beams = 72 in. − 10 in. = 62 in.

$$s = 62 \text{ in.} + \frac{10}{2} \text{ in.} = 67 \text{ in.} \le 62 + t$$

Six 4-foot-wide panels are selected for the 24-foot roadway width. Design loading will be one HS 20-44 wheel line in AASHTO Load Group I.

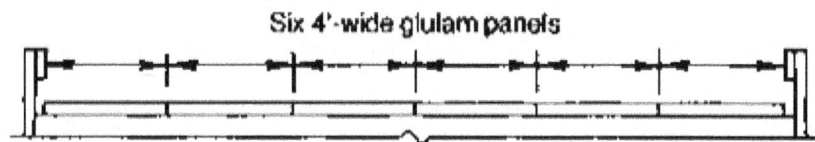

Six 4'-wide glulam panels

Estimate Panel Thickness

Based on a similar span for a transverse noninterconnected glulam deck (Chapter 7), an initial panel thickness of 6-3/4 inches is selected. For this thickness, $s = 67$ inches will control.

Determine Wheel Load Distribution Widths and Effective Deck Section Properties

In the direction of the deck span, wheel loads are assumed to act as point loads. In the direction perpendicular to the deck span, wheel loads are distributed over a width, D_w, equal to the tire width, b_t, plus the deck thickness. For an HS 20-44, 16,000-pound wheel load,

$$b_t = \sqrt{0.025P} = \sqrt{0.025(16,000)} = 20 \text{ in.}$$

$$D_w = b_t + t = 20 + 6.75 = 26.75 \text{ in.}$$

Effective deck section properties are computed:

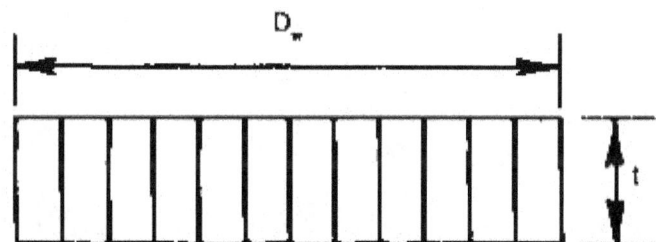

$t = 6.75$ in.

$D_w = 26.75$ in.

$A = t(D_w) = 6.75(26.75) = 180.56 \text{ in}^2$

$$S_y = \frac{D_w(t^2)}{6} = \frac{26.75(6.75)^2}{6} = 203.13 \text{ in}^3$$

$$I_y = \frac{D_w(t^3)}{12} = \frac{26.75(6.75)^3}{12} = 685.57 \text{ in}^4$$

8-59

Compute Panel Dead Load and Dead Load Moment

For a 6-3/4-inch deck with a 3-inch asphalt wearing surface, dead load is computed for the 26.75-inch distribution width:

$$w_{DL} = \frac{26.75\left[(6.75 \text{ in.})\left(50 \text{ lb/ft}^3\right) + (3 \text{ in.})\left(150 \text{ lb/ft}^3\right)\right]}{1,728 \text{ in}^3/\text{ft}^3} = 12.2 \text{ lb/in}$$

Dead load moment is computed by assuming the deck acts as a simple span between floorbeams:

$$M_{DL} = \frac{w_{DL}s^2}{8} = \frac{12.2(67)^2}{8} = 6,846 \text{ ft-lb}$$

Compute Live Load Moment

The maximum live load moment occurs with the 16,000-pound wheel load centered on the deck span:

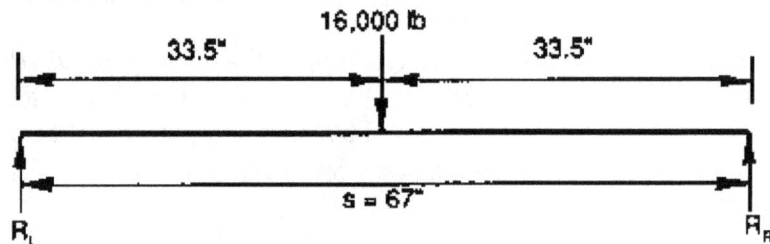

$$M_{LL} = \frac{Ps}{4} = \frac{16,000(67)}{4} = 268,000 \text{ in-lb}$$

Compute Bending Stress and Select a Deck Combination Symbol

The deck is continuous over more than two spans, so the maximum bending moment is 80 percent of that computed for a simple span to account for span continuity:

$$M = M_{DL} + M_{LL} = 6,846 + 268,000 = 274,846 \text{ in-lb}$$

$$f_b = \frac{0.80M}{S_y} = \frac{0.80(274,846)}{203.13} = 1,082 \text{ lb/in}^2$$

From *AITC 117--Design*, combination symbol No. 1 is selected with the following tabulated values:

$F_{by} = 1,450$ lb/in²	$C_M = 0.80$
$F_{vy} = 145$ lb/in²	$C_M = 0.875$
$F_{c\perp} = 560$ lb/in²	$C_M = 0.53$
$E = 1,500,000$ lb/in²	$C_M = 0.833$

Allowable bending stress is computed by Equation 8-7:

$$F_b' = F_b C_r C_M = 1{,}450(1.07)(0.80) = 1{,}241 \text{ lb/in}^2$$

$F_b' = 1{,}241 \text{ lb/in}^2 > f_b = 1{,}082 \text{ lb/in}^2$, so the combination symbol and deck thickness are satisfactory in bending.

Check Live Load Deflection

As with moment, the wheel load is positioned at the span centerline for maximum deflection. Deflection is computed by standard engineering methods using 80 percent of the simple span deflection to account for span continuity:

$$E' = EC_M = 1{,}500{,}000(0.833) = 1{,}249{,}500 \text{ lb/in}^2$$

$$\Delta_{LL} = 0.80 \frac{Ps^3}{48\,E'I_y} = 0.80\,\frac{16{,}000(67)^3}{48\,(1{,}249{,}500)(685.57)} = 0.09 \text{ in.}$$

0.09 inch < 0.10 inch, so deck deflection is acceptable.

Check Horizontal Shear

Dead load vertical shear is computed by at a distance t from the supports by Equation 8-9:

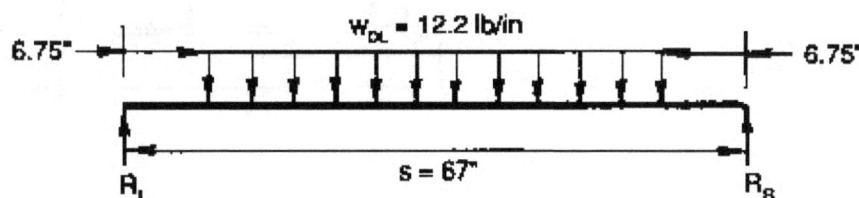

$$V_{DL} = R_L = w_{DL}\left(\frac{s}{2}-t\right) = 12.2\left(\frac{67}{2}-6.75\right) = 326.4 \text{ lb}$$

Live load vertical shear is computed at the lesser distance of $3t$ or $s/4$ from the support:

$$3t = 3\,(6.75) = 20.25 \text{ in.} \qquad\qquad \frac{s}{4} = \frac{67}{4} = 16.75 \text{ in.}$$

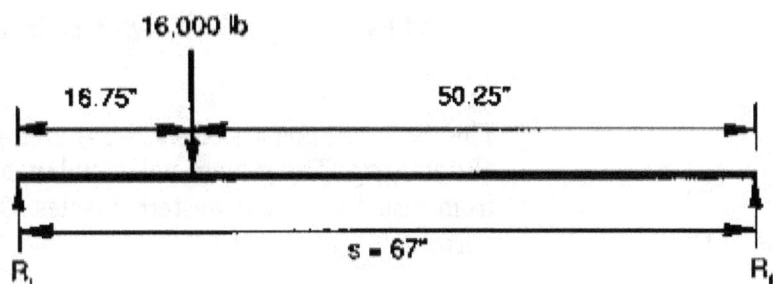

$$V_{LL} = R_L = \frac{16,000(50.25)}{67} = 12,000 \text{ lb}$$

$$V = V_{DL} + V_{LL} = 326.4 + 12,000 = 12,326 \text{ lb}$$

$$f_v = \frac{1.5V}{A} = \frac{1.5(12,326)}{180.56} = 102 \text{ lb/in}^2$$

$$Fv' = F_v(C_M) = (145 \text{ lb/in}^2)(0.875) = 127 \text{ lb/in}^2$$

$f_v = 102 \text{ lb/in}^2 < F_v' = 127 \text{ lb/in}^2$, so shear is satisfactory.

Check Bearing Stress

Bearing stress between the deck and floorbeam is checked for a bearing area, A, equal to the floorbeam width times the wheel load distribution width, D_w:

$$A = (10 \text{ in.})(26.75 \text{ in.}) = 267.5 \text{ in}^2$$

Dead load and live load reactions are computed by assuming that the deck acts as a simple span between floorbeams:

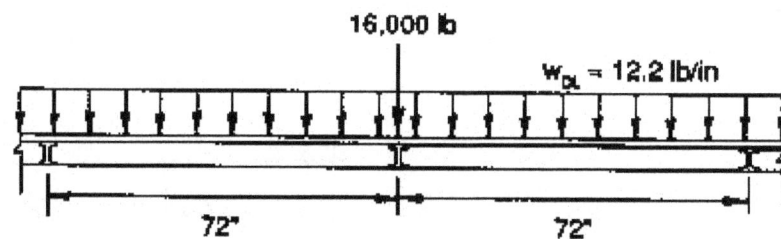

$$R_{DL} = s(w_{DL}) = (72 \text{ in.})(12.2 \text{ lb/in}) = 878.4 \text{ lb}$$

$$R_{LL} = 16,000 \text{ lb}$$

$$f_{c\perp} = \frac{R_{DL} + R_{LL}}{A} = \frac{878.4 + 16,000}{267.5} = 63 \text{ lb/in}^2$$

$$F_{c\perp}' = F_{c\perp} C_M = 560(0.53) = 297 \text{ lb/in}^2$$

$f_{c\perp} = 63 \text{ lb/in}^2 < F_{c\perp}' = 297 \text{ lb/in}^2$, so bearing is satisfactory.

Summary

The deck will consist of six 6-3/4-inch glulam panels, 48 inches wide and 68 feet long. The glulam will combination symbol No. 1, manufactured from visually graded western species. Stresses and deflection are as follows:

$$f_b = 1,082 \text{ lb/in}^2$$
$$F_b' = 1,241 \text{ lb/in}^2$$
$$\Delta_{LL} = 0.09 \text{ in.}$$
$$f_v = 102 \text{ lb/in}^2$$
$$F_v' = 127 \text{ lb/in}^2$$
$$f_{c\perp} = 63 \text{ lb/in}^2$$
$$F_{c\perp}' = 297 \text{ lb/in}^2$$

Longitudinal Nail-Laminated Lumber Decks

Longitudinal nail-laminated decks over floorbeams are designed for the effective span, s, using the same design procedures as longitudinal bridges. The continuous configuration is most practical for spans up to approximately 10 feet where live load deflection can be limited to $s/500$. For longer spans, the panelized configuration is used with a stiffener beam placed at center span between floorbeams and at maximum intervals of 8 feet for longer spans.

Example 8-7 - Longitudinal continuous nail-laminated lumber deck on transverse floorbeams

---

A single-lane bridge on a private road will be redecked with a longitudinal continuous nail-laminated lumber deck. The deck is supported by 6-inch-wide transverse floorbeams, spaced 4 feet on-center, and must carry an HS 15-44 truck. Design the deck for this bridge, assuming the following:

1. The deck will be provided with a lumber wearing surface consisting of 3-inch thick rough-sawn planks.

2. Live load deflection must be limited to $s/500$.

3. Lumber laminations will be S4S visually graded Southern Pine.

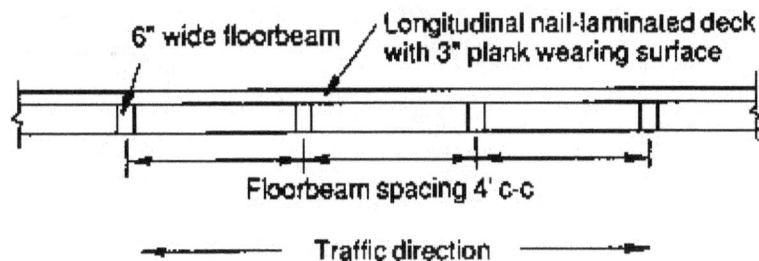

Longitudinal nail-laminated deck with 3" plank wearing surface

6" wide floorbeam

Floorbeam spacing 4' c-c

Traffic direction

Solution

Given the relatively short span, it is anticipated that AASHTO requirements for a minimum nominal deck thickness of 6 inches will control. It is also suspected that horizontal shear will be the controlling stress. The deck initially will be designed for shear, then checked for bending, deflection, and bearing.

Define Deck Geometric Requirements and Design Loads

The deck span, s, is the clear distance between supporting floorbeams plus one-half the width of one beam, but not greater than the clear span plus the deck thickness, t.

Clear distance between floorbeams = 48 in. - 6 in. = 42 in.

$$s = 42 \text{ in.} + \frac{6}{2} \text{ in.} = 45 \leq 42 + t$$

$s = 45$ inches will control the effective deck span.

Design loading will be one HS 15-44 wheel load (12,000 pounds) in AASHTO Load Group I.

Select a Species and Grade of Lamination

From NDS Table 4A, No. 2 visually graded Southern Pine is selected in the J&P size classification from the table labeled "surfaced dry, used at 19% maximum m.c." Per NDS footnotes, wet-use values are obtained from the table "surfaced green, used any condition." Further adjustment by C_M is not required.

$F_b = $ 1,100 lb/in^2 (repetitive member uses)

$F_v = $ 85 lb/in^2

$F_{c\perp} = $ 375 lb/in^2

$E = $ 1,400,000 lb/in^2

Estimate Deck Thickness and Compute Section Properties

For this short span, the minimum deck thickness of 5-1/2 inches (6 inches nominal) is selected. In the direction perpendicular to the deck span, the wheel load is distributed over a deck width, D_w equal to the tire width plus twice the deck thickness. From Table 8-3,

$t = 5.5$ in.

$D_w = 28.32$ in.

$A = 155.76$ in^2

$$S = 142.78 \text{ in}^3$$

$$I = 392.65 \text{ in}^4$$

Check Horizontal Shear

From Table 8-4, the dead load of the 5-1/2-inch lumber deck and 3-inch plank wearing surface over a distribution width $D_w = 28.32$ inches is 142.6 lb/ft, or 11.9 lb/in. Dead load vertical shear is computed at a distance t from the support by Equation 8-9:

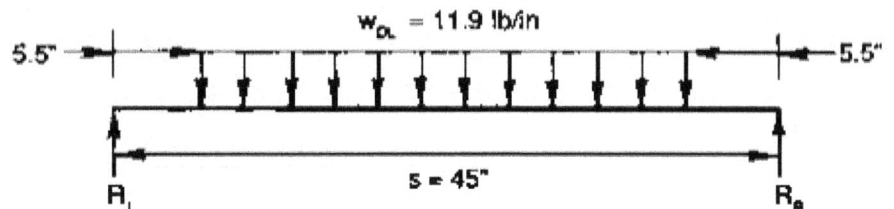

$$V_{DL} = R_L = w_{DL}\left(\frac{L}{2} - t\right) = (11.9 \text{ lb/in})\left(\frac{45}{2} - 5.5\right) = 202.3 \text{ lb}$$

Live load vertical shear is computed at the lesser of $3t$ or $s/4$ from the support:

$$3t = 3(5.5) = 16.50 \text{ in.} \qquad \frac{s}{4} = \frac{45}{4} = 11.25 \text{ in.}$$

$s/4 = 11.25$ inches controls:

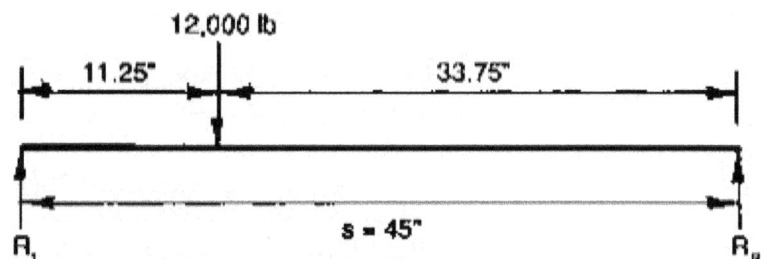

$$V_{LL} = R_L = \frac{12,000(33.75)}{45} = 9,000 \text{ lb}$$

$$V = V_{DL} + V_{LL} = 202.3 + 9,000 = 9,202.3 \text{ lb}$$

$$f_v = \frac{1.5V}{A} = \frac{1.5(9,202.3)}{155.76 \text{ in}^2} = 89 \text{ lb/in}^2$$

8-65

$$F_v' = F_v C_M \text{(shear stress modification factor)}$$

Using a shear stress modification factor of 2.0 for nail-laminated lumber treated with oil-type preservatives (Table 7-17),

$$F_v' = 85(1.0)(2.0) = 170 \text{ lb/in}^2$$

$f_v = 89 \text{ lb/in}^2 < F_v' = 170 \text{ lb/in}^2$, so the deck is satisfactory for horizontal shear.

Check Bending

Dead load moment is computed by assuming that the effective deck section is a simply supported beam:

$$M_{DL} = \frac{w_{DL} L^2}{8} = \frac{11.9(45)^2}{8} = 3,012 \text{ in-lb}$$

Live load moment is computed with the wheel load centered on the deck span:

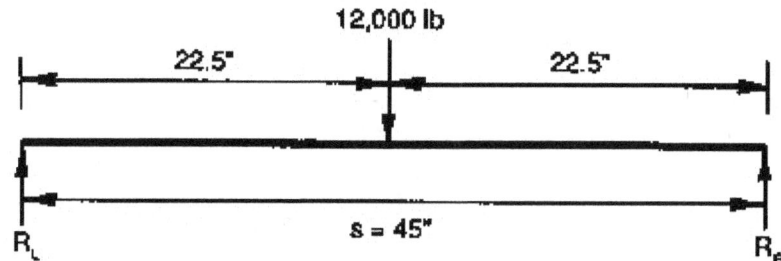

$$M_{LL} = \frac{Ps}{4} = \frac{16,000(45)}{4} = 135,000 \text{ in-lb}$$

The applied moment is 80 percent of the simple span moment to account for deck continuity:

$$M = 0.80(M_{DL} + M_{LL}) = 0.80(3,012 + 135,000) = 110,410 \text{ in-lb}$$

Bending stress is computed for the effective deck section:

$$F_b' = F_b C_M C_r = 1,100(1.0)(1.0) = 1,100 \text{ lb/in}^2$$

$$f_b = \frac{M}{S} = \frac{110,410}{142.78} = 773 \text{ lb/in}^2$$

$f_b = 773 \text{ lb/in}^2 < F_b' = 1,100 \text{ lb/in}^2$, so the deck is satisfactory in bending.

Check Live Load Deflection

Maximum live load deflection is produced with the wheel load centered on the deck span:

$$\Delta_{LL} = \frac{Ps^3}{48\,E'I}$$

$$E' = EC_M = 1,400,000(1.0) = 1,400,000 \text{ lb/in}^2$$

Using 80 percent of the simple span deflection to account for span continuity,

$$\Delta_{LL} = 0.80 \ \frac{12,000(45)^3}{48(1,400,000)(392.65)} = 0.03 \text{ in.} = s/1,500$$

$s/1,500 < s/500$, so deflection is acceptable.

Check Bearing Stress

Bearing stress between the deck and floorbeam is checked for a bearing area, A, equal to the floorbeam width times the wheel load distribution width, D_w:

$$A = (6 \text{ in.})(28.32 \text{ in.}) = 169.92 \text{ in}^2$$

Dead load and live load reactions are computed by assuming that the deck acts as a simple span between floor-beams:

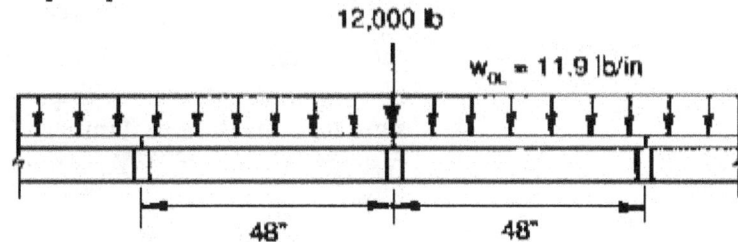

$$R_{DL} = (48 \text{ in.})(w_{DL}) = (48 \text{ in.})(11.9 \text{ lb/in}) = 571.2 \text{ lb}$$

$$R_{LL} = 12,000 \text{ lb}$$

$$f_{c\perp} = \frac{R_{DL} + R_{LL}}{A} = \frac{571.2 + 12,000}{169.92} = 74 \text{ lb/in}^2$$

$$F_{c\perp}' = F_{c\perp}(C_M) = 375(1.0) = 375 \text{ lb/in}^2$$

$f_{c\perp} = 74 \text{ lb/in}^2 < F_{c\perp}' = 375 \text{ lb/in}^2$, so bearing is satisfactory.

Summary

The deck will consist of S4S Southern Pine laminations that are visually graded No. 2 or better. The laminations will be nailed in the pattern shown in Figure 8-10 using 20d nails. Stresses and deflection are as follows:

$$f_b = 773 \text{ lb/in}^2$$

$$F_b' = 1,100 \text{ lb/in}^2$$

$$\Delta_{LL} = 0.03 \text{ in.} = s/1,500$$

$$f_v = 89 \text{ lb/in}^2$$

$$F_v' = 170 \text{ lb/in}^2$$

$$f_{c\perp} = 74 \text{ lb/in}^2$$

$$F_{c\perp}' = 375 \text{ lb/in}^2$$

8.6 SELECTED REFERENCES

1. American Association of State Highway and Transportation Officials. 1983. Standard specifications for highway bridges. 13th ed. Washington, DC: American Association of State Highway and Transportation Officials. 394 p.

2. American Association of State Highway and Transportation Officials. 1983. Manual for maintenance inspection of bridges. Washington, DC: American Association of State Highway and Transportation Officials. 50 p.

3. American Institute of Timber Construction. 1987. Design standard specifications for structural glued laminated timber of softwood species. AITC 117-87-Design. Englewood, CO: American Institute of Timber Construction. 28 p.

4. American Institute of Timber Construction. 1988. Glulam bridge systems. Vancover, WA: American Institute of Timber Construction. 33 p.

5. American Institute of Timber Construction. 1985. Timber construction manual. 3d ed. New York: John Wiley and Sons, Inc. 836 p.

6. Commonwealth of Pennsylvania, Department of Transportation. 1984. Standard plans for low cost bridges. Series BLC-540, timber spans. Pub. No. 130. [Pittsburgh, PA]: Commonwealth of Pennsylvania, Department of Transportation. 28 p.

7. Erickson, E.C.O.; Romstad, K.M. 1965. Distribution of wheel loads on timber bridges. Res. Pap. FPL 44. Madison, WI: U.S. Department of Agriculture, Forest Service, Forest Products Laboratory. 62 p.

8. Gutkowski, R.M.; Williamson, T.G. 1983. Timber bridges: state-of-the-art. Journal of Structural Engineering. 109(9): 2175-2191.

9. Klaiber, F.W.; Dunker, K.F.; Wipf, T.J.; Sanders, W.W., Jr. 1987. Methods of strengthening existing highway bridges. National Cooperative Highway Research Program Rpt. 293. Washington, DC: National Research Council, Transportation Research Board. 114 p.

10. Mielke, K.F. 1977. Experimental project for glued-laminated timber deck panels on highway bridges. Juneau, AK: State of Alaska, Department of Highways. [50 p.].

11. National Forest Products Association. 1986. Design values for wood construction. A supplement to the national design specification for wood construction. Washington, DC: National Forest Products Association. 34 p.

12. National Forest Products Association. 1986. National design specification for wood construction. Washington, DC: National Forest Products Association. 87 p.

13. Ou, F.L. 1986. An overview of timber bridges. In: Trans. Res. Rec. 1053. Washington, DC: National Academy of Sciences, National Research Council, Transportation Research Board: 1-12.

14. Sanders, W.W., Jr.; Klaiber, F.W.; Wipf, T.J. 1985. Load distribution in glued laminated longitudinal timber deck highway bridges. Ames, IA: Iowa State University, Engineering Research Institute. 47 p.

15. Sanders, W.W. 1984. Distribution of wheel loads on highway bridges. National Cooperative Highway Research Program, Synthesis of Highway Practice No. 3. Washington, DC: National Academy of Sciences, National Research Council, Transportation Research Board. 22 p.

16. Sprinkel, M.M. 1985. Prefabricated bridge elements and systems. National Cooperative Highway Research Program, Synthesis of Highway Practice No. 119. Washington, DC: National Academy of Sciences, National Research Council, Transportation Research Board. 75 p.

17. Stacey, W.A. 1935. The design of laminated timber bridge floors. Wood Preserving News 13(4): 44-46, 55-56.

18. State of Wisconsin, Department of Transportation. 1979. Bridge manual. Madison, WI: State of Wisconsin, Department of Transportation. [350 p.].

19. State of Wisconsin, Department of Transportation. 1985. Standard bridge plans. Madison, WI: State of Wisconsin, Department of Transportation. [50 p.].

20. Stone, M.F. [1975]. New concepts for short span panelized bridge design of glulam timber. Tacoma, WA: Weyerhaeuser Co. 8 p.

21. U.S. Department of Transportation, Federal Highway Administration. 1979. Standard plans for highway bridges. Timber Bridges. Washington, DC: U.S. Department of Transportation, Federal Highway Administration. Vol. 3. 19 p.

22. Weyerhaeuser Company. [1980]. Digest of five research studies made on longitudinal glued laminated wood bridge decks. St. Paul, MN: Weyerhaeuser Co. 19 p.

23. Weyerhaeuser Company. 1976. Weyerhaeuser bridge deck bracket. SL-495. Tacoma, WA: Weyerhaeuser Co. 4 p.

24. Weyerhaeuser Company. 1980. Weyerhaeuser glulam wood bridge systems. Tacoma, WA: Weyerhaeuser Co. 114 p.

25. Weyerhaeuser Company. 1975. Weyerhaeuser panelized bridge system for secondary roadways, highways & footbridges. SL-1318. Tacoma, WA: Weyerhaeuser Co. 4 p.

26. Wheeler Consolidated, Inc. 1986. Timber bridge design. St. Louis Park, MN: Wheeler Consolidated, Inc. 42 p.

27. Wipf, T.J.; Funke, R.W.; Klaiber, F.W.; Sanders, W.W., Jr. 1986. Experimental and analytical load distribution behavior of glued laminated timber deck bridges. In: Proceedings of the 2d International Conference on Short and Medium Span Bridges: 63-77. Vol. 2.

28. Wipf, T.J.; Klaiber, F.W.; Sanders, W.W. 1986. Load distribution criteria for glued-laminated longitudinal timber deck highway bridges. In: Trans. Res. Rec. 1053. Washington, DC: National Academy of Sciences, National Research Council, Transportation Research Board: 31-40.

CHAPTER 9

DESIGN OF LONGITUDINAL STRESS-LAMINATED DECK SUPERSTRUCTURES

9.1 INTRODUCTION

Longitudinal stress-laminated deck superstructures consist of a series of lumber laminations that are placed edgewise between supports and are compressed transversely with high-strength prestressing elements (Figure 9-1). The bridges are similar in configuration to glulam or nail-laminated longitudinal decks previously discussed; however, with stress-laminated decks the load transfer between laminations is developed totally by compression and friction between the laminations, rather than by glue or mechanical fasteners. This friction is created by transverse compression applied to the deck using the same type of high-strength steel-stressing elements that are commonly used for prestressed concrete. These elements, which have historically been high-strength steel rods, are placed at regular intervals through prebored holes in the wide faces of the laminations and are stressed in tension using a hydraulic jack. In a typical stress-laminated lumber deck, each rod may have from 80,000 to 100,000 pounds of tension that is transferred into the deck to develop compression between the laminations. The total force from all prestressing elements on a 32-foot-long bridge, for example, may be as high as 1 million pounds. That 1 million pounds compresses the laminations so tightly that the deck behaves like one large, solid plate of wood (Figure 9-2).

Stress-laminating is the newest development in timber bridge construction and offers many advantages over conventional nail-laminated lumber systems. Deck superstructures can be prefabricated locally into panels, or into complete units, that are shipped to the project site and lifted into place. Once installed, the deck acts as a continuous slab without transverse or longitudinal joints that adversely affect wearing surface performance. In addition, the stress-laminated lumber deck will not delaminate over time, which is a problem associated with nail-laminated lumber construction. Another advantage of stress-laminated decks is the length of lumber required for the laminations. Because load transfer between the laminations is developed from friction, all laminations do not have to be continuous (one piece) over the bridge length. Discontinuous laminations using butt joints are permitted within certain limitations. This provides advan-

---

This chapter was coauthored by Michael A. Ritter and Michael G. Oliva, Ph. D., Assistant Professor, Department of Civil and Environmental Engineering, University of Wisconsin, Madison.

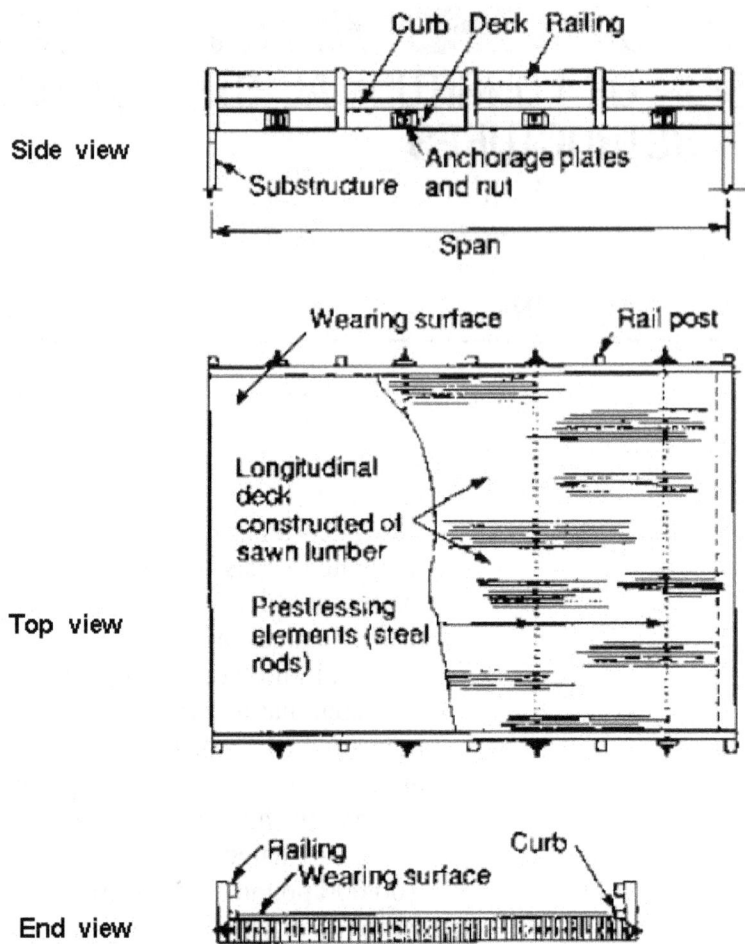

Figure 9-1. - Typical configuration for a longitudinal stress-laminated lumber deck bridge.

Figure 9-2. — Stress-laminated deck bridge built over Iron River on the Cheguamegon National Forest in 1988.

tages over conventional nail-laminated systems because shorter lumber can be used. It also allows longer spans to be cambered to offset dead load deflection.

The concept of stress-laminated lumber was originally developed in Ontario, Canada, in the mid-1970's. Design procedures and specifications were subsequently included in the *Ontario Highway Bridge Design Code* (OHBDC) in 1979.[10,11] Although numerous stress-laminated lumber superstructures have been built in the United States, design provisions are not included in the AASHTO bridge specifications,[1] but are currently being proposed. This chapter presents a brief history of developmental work completed in Ontario and in the United States relative to stress-laminated deck performance and design. The basic characteristics for longitudinal stress-laminated lumber decks are presented and followed by suggested design procedures and examples.

9.2 DEVELOPMENT OF STRESS-LAMINATED BRIDGE SYSTEMS

Stress-laminated lumber has been used as a method of bridge construction for more than a decade. Its inception and development are the result of pioneering efforts in Ontario. Further research and development has occurred in the United States. This section presents a brief summary of the development of stress-laminated lumber bridge systems, including an overview of recent developments in stress-laminating technology and their application to new bridge systems.

DEVELOPMENT IN ONTARIO

Stress-laminating was first used for timber bridges in Ontario in 1976. At that time, the Ontario Ministry of Transportation (MTO, formerly the Ontario Ministry of Transportation and Communication, MTC) was interested in developing new methods for rehabilitating deteriorated nail-laminated lumber bridge decks. Many such decks in Ontario were separating or delaminating under repeated heavy highway loading. Although the static strength and condition of the laminations was good, load distribution between laminations was severely reduced, and the delamination was causing asphalt wearing surfaces to crack and separate from the deck. It appeared to MTO engineers that structural integrity and continuity could be reestablished in the decks by using prestressing techniques to preload and recompress the wood laminae.

In 1976, a pilot project was carried out in Ontario on the Hebert Creek Bridge.[8] This bridge, a longitudinal nail-laminated lumber deck, was in an advanced state of delamination and was scheduled for replacement in 1977. The bridge had an overall length of 55 feet, with the longest span between bents being 20 feet. Steel prestressing rods were placed above and below the existing deck and were tensioned to recompress the deck (Figure 9-3). While stressing the rods, it was found that the compression caused the bridge width to decrease and additional laminations had to be added to maintain the original roadway width. After stressing was complete, MTO load tested the bridge to assess the results.[27] The effects of the stress-laminating were rather dramatic and substantially increased the bridge load-carrying capacity. The rehabilitation method proved so successful that the scheduled bridge replacement was cancelled (more detailed information on the Hebert Creek Bridge and other rehabilitation projects completed in Ontario is presented in Case History 15.5 in Chapter 15).

Figure 9-3. - Prestressing rod configuration of the type used on the Hebert Creek Bridge. Rods were placed above and below the existing deck and were anchored to steel plates along the deck edges.

Although the first application of stress laminating in Ontario involved the rehabilitation of an existing bridge, the method offered a variety of possibilities for the construction of new bridges. A long series of development studies was undertaken by MTO and Queen's University to provide an understanding of the fundamentals of stress laminating and to identify possible problems and associated design implications. Among the investigations were tests to determine (1) the friction force developed between the laminations and its dependence on the level of compressive prestress, (2) the mechanism and magnitude of deck bending and deformation, (3) time-related prestress losses, and (4) effective plate stiffness properties of the stress-laminated system. In addition, analytic models were developed to predict deck behavior.

Load testing of small decks in the laboratory of Queen's University proved that stress-laminated lumber decks behave like an orthotropic plate, with different stiffness in the directions parallel to the laminations and perpendicular to the laminations.[6] The stiffness parallel to the laminations was found to depend on the lamination depth and the modulus of elasticity parallel to the wood grain. The transverse system stiffness across the laminations, perpendicular to grain direction, was found to be substantially lower and was expressed as a fraction of the longitudinal stiffness. In comparison to a similar longitudinal glulam deck, the stress-laminated deck showed slightly less transverse stiffness, probably from minor variations in lamination thickness or warp, which reduces interlaminar contact. Thus, a stress-laminated lumber deck is slightly less efficient than a continuous glulam deck of the same size.

Based on research work conducted by MTO and Queen's University,[5,6,7] as well as successful rehabilitation projects in Ontario, a design procedure for stress-laminated decks was developed and included in the 1979 edition of the OHBDC. Subsequently, the system has been successfully used on numerous bridge rehabilitation and new construction projects in Ontario.

DEVELOPMENT IN THE UNITED STATES

Research and development on stress-laminated bridges has been completed in the United States at the University of Wisconsin, Madison (UW) in cooperation with the USDA Forest Service, Forest Products Laboratory (FPL). The focus of this research centered on expanding work done in Ontario and at Queen's University to develop a design procedure for use in the United States. Extensive evaluation and testing were conducted over a 3-year period starting in 1985. During the research, two full-size stress-laminated bridge decks were constructed and tested in the structures laboratory at UW.[9,15,16,17] The first deck was constructed of heavy timber laminations using nominal 4-inch-wide by 16-inch-deep lumber (Figure 9-4). The second deck was built from dimension lumber using nominal 2-inch-wide by 12-inch-deep laminations. Both decks were tested extensively under simulated truck loads for spans up to 48 feet for the heavy timber laminations, and spans up to 24 feet for the dimension-lumber laminations.

The results of the UW/FPL research confirmed many of the Ontario findings and exhibited good correlation with previous truck load tests. The results also were verified for nominal 4-inch-thick laminations, which had not been previously tested. In addition, UW/FPL research investigated new areas of stress-laminated deck behavior, including (1) the effects of lamination butt joints on wheel-load distribution and deck stiffness, (2) the mechanism of stress transfer into the deck and related edge effects on wheel-load distribution, (3) the effects of transverse bending on the required level of compressive prestress, and (4) requirements for anchorage of prestressing rods (which resulted in a new anchorage design without the

Figure 9-4. - Full-scale experimental stress-laminated deck in the structures laboratory at the University of Wisconsin.

steel channel bulkhead traditionally used in Ontario). Additional research is continuing at FPL and at other universities in the United States to substantiate the performance of prototype stress-laminated deck bridges. Cooperative work between West Virginia University and FPL is currently in progress to develop design procedures and performance characteristics for stress-laminated decks constructed from native hardwood species. A long-term moisture study also is being conducted by FPL to determine the effects of moisture variations in the laminations on the level of compressive prestress.

To date, nearly 20 stress-laminated decks have been built in the United States. Many of these bridges are being periodically monitored and load-tested to assess field performance and verify design criteria (Figure 9-5). In 1989-1990, approximately 60 new stress-laminated bridges will be constructed through the USDA Forest Service Timber Bridge Initiative; Approximately half of these bridges will be built in West Virginia under the supervision of West Virginia University. The remainder will be distributed across more than 20 states. Data obtained from monitoring these bridges will provide a great deal of information on stress-laminated deck performance in a wide range of environmental conditions.

Figure 9-5. - Load test of the Zuni Creek stress-laminated deck on the Idaho Panhandle National Forests.

NEW STRESS-LAMINATED SYSTEMS

The stress-laminated bridge investigations previously described have involved the use of longitudinal sawn lumber laminations with transverse prestressing (some projects in Ontario have used transverse lumber laminations with longitudinal prestressing). Although these designs have proved successful in short-span applications, the moment of inertia of the deck is limited by the available depth of lumber laminations, which is generally 16 inches nominal. Like other longitudinal deck systems constructed of glulam or nail-laminated lumber, span capabilities of longitudinal stress-laminated decks are normally controlled by stiffness (deflection), rather than stress. The need for longer spans has focused attention on developing new designs for stress-laminated timber bridges that provide additional stiffness. Although these new systems are in a developmental stage at this time, and no design criteria or procedures are available, design criteria should be forthcoming.

Work has recently been completed at UW/FPL on a new bridge system using parallel-chord trusses that are stress-laminated together.[9.19] By using parallel-chord trusses in place of the sawn lumber laminations, a deeper, stiffer system was obtained using the same volume of lumber. A full-size, 52-foot span, stress-laminated parallel-chord bridge was built and tested in the UW structures laboratory under various simulated loading conditions. Individual truss laminations consisted of 4-inch-wide by 6-inch-deep top and bottom chords, separated by 4-inch-wide by 12-inch-deep discontinuous web members (Figure 9-6). The connection between the top and bottom chord and the web was made with steel-drive spikes that were placed through the chords, into the web. The stress-laminated trusses produced a significant increase in bridge stiffness compared to sawn lumber laminations, yet exhibited many of the same characteristics previously observed for longitudinal decks. After laboratory testing, a prototype stress-laminated parallel-chord truss bridge was built by the Forest Service on the Hiawatha National Forest in late 1987. This structure has been load tested on two occasions and is being continuously monitored.

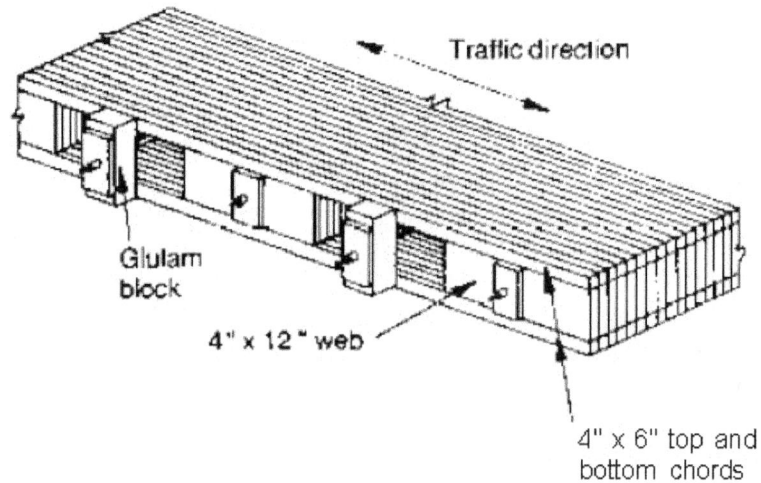

Figure 9-6. - (Top) Drawing of a stress-laminated parallel-chord truss. (Bottom) Prototype stress-laminated parallel-chord bridge built on the Hiawatha National Forest (shown during construction).

In addition to the parallel-chord truss work done by UW/FPL, new applications of stress-laminating are being investigated in Ontario and at several universities in the United States. Ontario is investigating the development of a stress-laminated cellular or box girder-type of bridge, the advantages of which have already been recognized in steel and concrete bridge construction. If the Ontario development work is successful, the cellular stress-laminated wood system may be very competitive with other systems for longer-span applications. West Virginia University also has performed laboratory tests and has constructed a prototype bridge using a T or ribbed cross section. In this design, deep, laminated veneer lumber (LVL) beams are stress-laminated to a relatively thin, sawn lumber deck (Figure 9-7). The ribs formed by the deeper LVL laminations contribute substantially to the longitudinal bridge stiffness, making longer spans possible. A similar system using glulam rather than LVL beams also is feasible and is being developed. Other cooperative work between West Virginia University and FPL is aimed at developing stress-laminated box girder systems and new methods for adapting stress-laminated decks to other bridge superstructures constructed of glulam, steel, or concrete. In addition, Pennsylvania State University is developing a stress-laminated wood-steel composite bridge system that is intended to increase bridge stiffness and reduce long-term deflection.

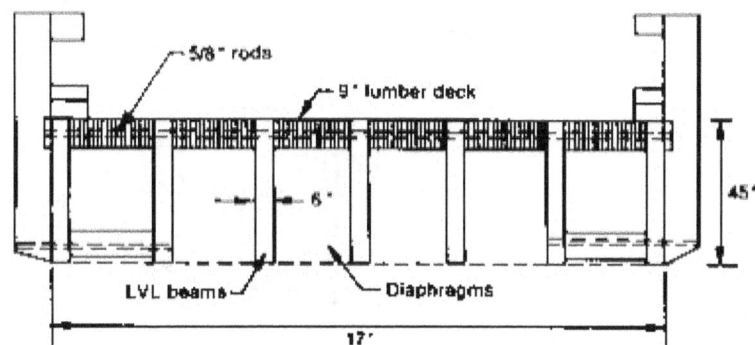

Figure 9-7. - Drawing of a stress-laminated T-bridge cross section.

9.3 CHARACTERISTICS OF LONGITUDINAL STRESS-LAMINATED LUMBER DECKS

As previously discussed, stress-laminating creates a large plate of wood that is held together by compressive forces applied through the prestressing elements. When subjected to vehicle loading, a stress-laminated bridge deck acts as an orthotropic plate with different properties in the longitudinal and transverse directions. When a wheel load is placed at any point on the deck, the entire deck deflects downward (except at locations over the supports), resulting in displacements in both the longitudinal and transverse directions. Because of this behavior, bending moments are also developed in the longitudinal and transverse directions. The magnitude of these moments depends primarily on five variables: (1) load magnitude, (2) deck span, (3) deck width, (4) longitudinal deck stiffness, and (5) transverse deck stiffness. The longitudinal bending moment produces bending stress and deflection that controls the required deck thickness. The transverse moment, which also produces bending stress and deflection, dictates the amount of compressive prestress that must be applied between the laminations.

When a truck wheel load is placed over the deck laminations, two primary actions occur that deteriorate the platelike behavior of the deck (Figure 9-8). The first action results from transverse bending, which produces a tendency for opening between the laminations on the deck underside. The second type of action is from transverse shear, which creates a tendency for the laminations to slip vertically. In both cases, the actions will not occur if the deck has a sufficient level of compressive prestress between the laminations. In the case of transverse bending, the compressive stress directly offsets the tension effects on the deck underside. For shear, vertical slip is prevented by friction between the laminations resulting from the compressive prestress. Maintaining an adequate level of prestress is the single most important aspect of stress-laminated bridge construction.

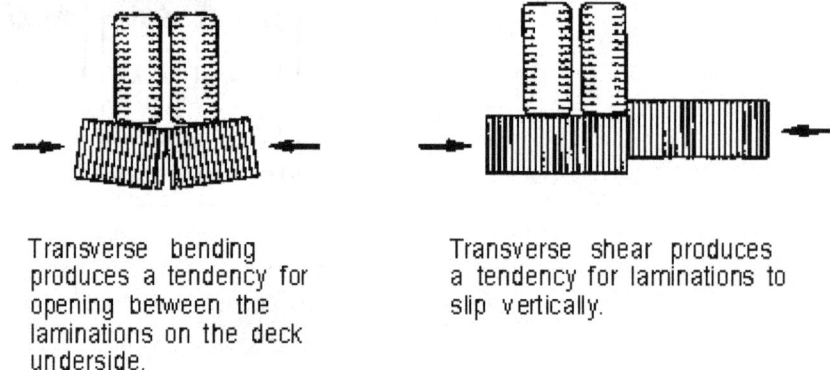

Transverse bending produces a tendency for opening between the laminations on the deck underside.

Transverse shear produces a tendency for laminations to slip vertically.

Figure 9-8. - Actions that tend to reduce the platelike behavior of a longitudinal stress-laminated bridge deck.

Stress-laminating is a relatively new concept for bridge construction in the United States. Although there have been numerous stress-laminated bridges built in this country, information about system characteristics and design requirements are not as widely available as they are for other, more conventional timber bridge systems. Many aspects of longitudinal stress laminated deck design are similar to those for other longitudinal deck systems, but several characteristics are unique to stress laminating. The most important of these characteristics are related to the lumber laminations, prestress elements and anchorages, time-related stress loss, and construction methodology.

LUMBER LAMINATIONS

The lumber laminations of a stress-laminated bridge provide the required strength and stiffness for bridge performance and serviceability. Of particular interest are characteristics related to material requirements, load sharing, and lamination joints.

Material Requirements

Longitudinal stress-laminated bridges are constructed from visually graded or MSR lumber in the Joists and Planks size classification (nominally 2 to 4 inches thick, 5 inches and wider). Although decks could theoretically be constructed from any lumber thickness, the 2- to 4-inch thickness range has proven most efficient and economical. The laminations may be dressed, rough-sawn or full-sawn; however, rough-sawn and full-sawn material must be surfaced to a uniform thickness to ensure even bearing between the laminations. To date, most bridges constructed in the United States have used rough-sawn 4-inch nominal lumber that is surfaced on one side (S1S) to provide a uniform thickness.

Stress-laminated bridge decks can generally be built from any lumber species provided it meets design requirements for strength and stiffness and is treatable with preservatives. At this time, however, the number of suitable species is somewhat limited because parameters for stress-laminated deck design have not been established for all species. Research has been completed for Douglas Fir-Larch, Hem-Fir (North), Red Pine and Eastern White Pine.[5,6,9] Research for other species is currently in progress and will be available in the near future. For all species, the lumber laminations used for stress-laminated decks should be treated with oil-type preservatives (Chapter 4). As previously discussed for other bridge types, the oil-type preservatives provide a protective barrier that helps reduce wood moisture content variations and associated dimensional changes. This is especially important for stress-laminated construction because dimensional changes in the laminations can affect the level of compressive prestress in the bridge.

Load Sharing

When lumber laminations are stressed together, the strength-reducing characteristics of the individual laminations are dispersed throughout the cross section in the same manner previously discussed for glulam (Chapter 3). Like glulam, the bending strength of stress-laminated lumber is substantially greater than a comparable sawn lumber member of the same size. Research conducted in Canada showed that stress laminating increases usable bending strength by 50.8 to 82.5 percent, depending on the grade and species of lamination.[26,29] For stress-laminated bridges, the

OHBDC currently allows a bending stress increase of 50 percent for lumber of mixed grades No. 1 and No. 2, and 30 percent for lumber graded Select Structural.

Lamination Joints

As previously discussed, load transfer between laminations in a stress-laminated bridge is accomplished by friction between the laminations induced by the high level of compressive prestress. Because this friction is sufficient to prevent movement between the laminations, it can be used as a means of longitudinally splicing the laminations. Thus, the laminations for a stress-laminated bridge deck need not be continuous over the bridge span and can be provided with longitudinal butt joints. When butt joints are used, the OHBDC requires that not more than one butt joint occur in any four adjacent laminations within a 4-foot distance, measured along the bridge span (Figure 9-9).

The ability to use butt joints in stress-laminated decks provides an advantage over conventional nail-laminated construction because shorter laminations can be used, resulting in reduced costs and improved availability. However, research at UW has shown that butt joints reduce longitudinal stiffness, and therefore must be compensated for in design. In addition, the discontinuity at the joint reduces the effective deck section available to resist bending stress. The effects of butt joints are discussed further in the design procedures given later in this chapter.

PRESTRESSING SYSTEMS

The prestressing system is perhaps the most important part of a stress-laminated bridge because it holds the bridge together and develops the necessary friction between the laminations. The system generally consists of two parts: the prestressing elements and the anchorages. The prestressing elements are placed transverse to the bridge span and are stressed in tension. The anchorages hold the prestressing elements along the deck edges where the tension is transferred into the lumber lamina-

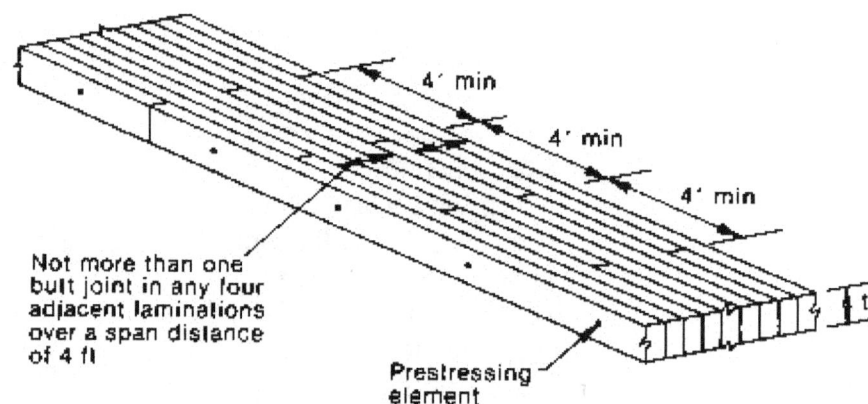

Figure 9-9. - Minimum requirements for butt joints in longitudinal stress-laminated bridge decks.

tions. The function of the prestressing system is to develop the required uniform compressive force between the laminations. Research at UW has shown that the compressive prestress is localized at the anchorage, but becomes uniformly distributed at interior locations, away from the anchorage.

Prestressing Elements

Prestressing elements for stress-laminated bridge decks must be carefully selected for their strength and corrosion-resistance properties. All stress-laminated bridges constructed to date have used high-strength threaded steel rods that conform to ASTM A 722, Uncoated High-Strength Steel Bar for Prestressing Concrete.[4] These rods have a minimum ultimate stress in axial tension of 150,000 lb/in^2 and are available in diameters ranging from 5/8 inch to 1-3/8 inch. Steel prestressing strand has not been used but is being investigated and may prove to be an alternate material in the future. The potential advantages of strand include its higher strength (270,000 lb/in^2 ultimate tensile stress) and lower cost. A disadvantage with strand is that the anchor chuck damages the strand so that it cannot be restressed. As a result, the strand must be replaced each time the bridge is restressed.

Because steel prestressing elements are under high stress, and are particularly susceptible to corrosion, it is essential that special corrosion protection be provided. Existing applications have predominantly used galvanizing to protect the rods and this method should be used until alternative techniques are proven. Galvanizing is generally provided by the rod manufacturer using processes that avoid embrittlement or strength loss in the high-strength steel. Epoxy coatings, similar to those used for concrete-reinforcing steel, are being evaluated and have been used with good results in several applications. In addition, some bridge rehabilitation applications in Ontario have successfully used plastic (PVC) pipes that are placed over the rods and are filled with grease (see Case History 15.5 in Chapter 15).

Anchorages

The anchorages for prestressing elements must transfer the required stress to the lumber laminations without causing wood crushing in the outside laminations. Additionally, they must be capable of developing the full capacity of prestressing elements. Anchorage systems for steel prestressing rods have traditionally used steel plates or shapes. The rod is placed through the steel components and anchored with a nut. The nuts match the coarse thread pattern on the rods and are made from high-strength steel by the rod manufacturer. Standard nuts are not compatible with high-strength stressing rods.

Two types of anchorages are used for longitudinal stress-laminated decks; one for deck rehabilitation where rods are placed externally, over and under the lumber laminations, and one for new deck construction where rods are placed internally, through holes in the laminations. For bridge

9-13

rehabilitation, the external channel bulkhead anchorage was developed in Ontario and employs a continuous steel channel along the deck edges (Figure 9-10 A). The rods extend beyond the channel and are attached with nuts to rectangular steel anchorage plates. For new construction, two anchorage configurations are currently used: the channel bulkhead configuration and the bearing plate configuration. The channel bulkhead configuration was developed in Ontario and is currently a design requirement in the OHBDC (Figure 9-10 B). It is similar to the external channel bulkhead used for deck rehabilitation, but the rods extend through the center of the channel and attach to rectangular steel bearing plates along the channel web. Although considered necessary for bridge rehabilitation, research at UW showed that the steel channel contributed little to load transfer or bridge performance for new construction. A new anchorage employing a large, rectangular steel bearing plate and a smaller, outside anchorage plate was developed by UW/FPL (Figure 9-10 C). Most of the stress-laminated deck bridges constructed in the United States have substituted this steel-plate configuration for the continuous channel. It should be noted, however, that although the steel channel is not considered necessary from a structural standpoint, in certain circumstances it may be desirable to cover the outside laminations. In past applications, the steel plates used without channels have caused some local wood crushing and created an indentation in the outside laminations. The channel effectively covers these areas so they are not visible.

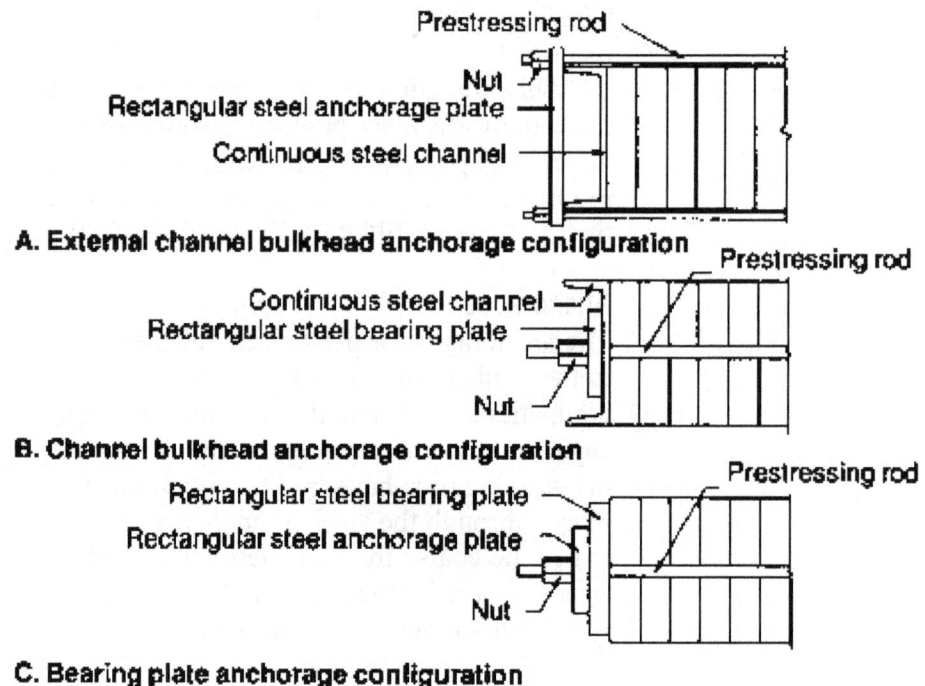

A. External channel bulkhead anchorage configuration

B. Channel bulkhead anchorage configuration

C. Bearing plate anchorage configuration

Figure 9-10. - Types of anchorages for steel prestressing rods.

9-14

TIME-RELATED STRESS LOSS

A sufficient level of uniform, compressive prestress must be maintained between the lumber laminations in order for a stress-laminated bridge to perform properly. With time, the initial level of prestress placed in the deck at installation will be affected primarily by two factors: creep in the wood and variations in wood moisture content. The Ontario research proved that when a constant compressive force is applied to wood over time, the wood slowly compresses or creeps. This occurs because the wood cells gradually change shape and become permanently compressed. Thus, when the deck laminations are compressed by the prestressing force, they slowly become narrower. Unfortunately, the level of prestress decreases when this occurs. Work in Ontario found that this loss of compression from creep increased when the cross-sectional area of the steel prestressing components increased. To reduce this loss effect, it was found necessary to use high-strength steel rods to carry the large prestressing force with a minimum cross-sectional area of steel. In addition, design limits were placed on the ratio of the wood area to the steel area (discussed in the next section on design).

Although creep is a natural wood characteristic that adversely affects the compressive prestress level, the research done in Ontario has developed a method of effectively controlling this phenomenon. Specifically, the amount of creep in a stress-laminated deck was found to be directly related to the number of times the deck is stressed (Figure 9-11). If a deck is stressed only once during construction, 80 percent or more of the initial compression may be lost to creep. If the deck is restressed within a relatively short period, the subsequent stress loss is less. If the deck is restressed a second time within a specified time period, the total compression loss over time can be limited to a maximum of 60 percent. Research at UW/FPL found that a stress-laminated deck would perform acceptably at a compressive prestress level as low as 24 lb/in². Because this is many

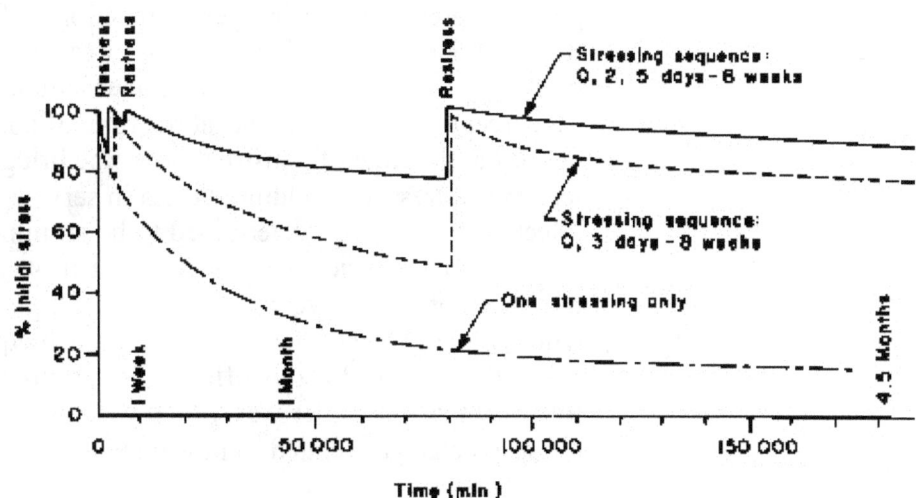

Figure 9-11. - Effects of restressing on time-related stress loss (from Csagoly and Taylor).[*]

9-15

times lower than the strength of the wood in compression perpendicular to grain, the level of compressive prestress placed in a bridge during the initial stressing operations is increased to compensate for subsequent creep losses over the life of the structure (the actual prestress level depends on a number of factors discussed later in the design procedures). Thus, a subsequent stress loss from creep of 60 percent over the life of the bridge will still leave the minimum prestress level required for acceptable performance, plus an additional margin for safety. To maintain the minimum prestress level, the following stressing sequence is used:

1. The deck is initially assembled and stressed to the design level required for the structure.

2. The deck is restressed to the full level approximately 1 week after the initial stressing.

3. Final stressing is completed 4 to 6 weeks after the second stressing.

When this stressing sequence is followed, the maximum expected loss in prestress from creep will be limited to approximately 60 percent of the initial level (40 percent of the initial stress level will be maintained). It may be desirable, however, to periodically recheck stress levels over the life of the structure as part of a preventative maintenance program.

In addition to stress loss from creep, the prestress level in stress-laminated lumber decks can be affected by variations in the moisture content of the lumber laminations. As discussed in Chapter 3, wood below the fiber saturation point (approximately 30 percent moisture content) shrinks when moisture is lost and expands when moisture is gained. The effects of these moisture changes can result in a loss or gain in prestress. Research on moisture-related stress changes has involved a few laboratory tests and periodic monitoring of bridges installed in different environmental conditions. Although some changes in prestress have been observed, they have been relatively minor when the lumber laminations were dry (less than 19 percent moisture content) at the time of construction. When lumber is not dry at the time of construction, some bridges have shown an increased loss in prestress as the lumber dries in service. At this time, moisture effects have not been determined to be an important consideration for stress-laminated bridges when dry lumber is used. When lumber with a moisture content above 19 percent is used, periodic restressing may be required until the lumber laminations reach equilibrium moisture content. An evaluation of moisture effects for various lumber species and preservative treatments is under way which will provide insight into the potential for stress changes related to moisture.

Several characteristics related to the construction of stress-laminated decks are unique compared with other bridge systems. Although many of the general principles of timber bridge construction apply to stress-laminated construction, unique methodology is involved in the areas of bridge assembly, camber, and stressing. A brief description of these topics is presented below. A more complete description of the construction of a stress-laminated lumber deck is presented in case histories given Chapter 15.

Bridge Assembly

Stress-laminated bridges can be assembled using three different methods. Two methods involve on-site assembly, while the third involves preassembly at a fabrication facility. The first on-site method involves assembly over the abutments or intermediate supports. Using this method, the laminations are sequentially placed and aligned, and the prestressing rods are inserted and stressed. This method is generally acceptable when the laminations span the full distance between supports and there are no butt joints. When laminations with butt joints are used, scaffolding or other temporary supports must be used to support the laminations until the bridge is stressed. As a result, this method is seldom practical when butt joints are used.

Another option for on-site assembly is to assemble the bridge at a staging area adjacent to the crossing, then lift the entire deck into place. This method offers some advantage over the previous method because the laminations can be supported by the ground rather than by scaffolding. A disadvantage, however, is that a crane or other equipment is required to lift the bridge into place. For both on-site assembly methods, all stressing must be accomplished in the field. After initial construction, two additional trips must be made to the site to complete the required stressing sequence.

In many applications, the preferable method of bridge assembly involves prefabrication at a manufacturing or fabrication facility. With this method, the bridge is fabricated in a series of stressed panels that are normally 7 to 10 feet wide, depending on transportation restrictions and lifting capacity at the site. The panels are shipped to the bridge site, lifted into place, and stressed together to form a continuous deck. To join the panels, the stress in alternate opposing rods is released and the anchorage plates on the joint edge of the released rods are removed (Figure 9-12). The released rods are then inserted into a special coupler on the opposite stressed rod, and the two panels are stressed together (see Case History 15.9 in Chapter 15). Most stress-laminated lumber bridges constructed in the United States have utilized prefabricated panels. The method has been most economical and requires a minimum time for field erection. Another advantage of using the prefabricated panel method is that the restressing sequence can be completed at the fabrication facility. Repeated trips to the bridge site for restressing are not required.

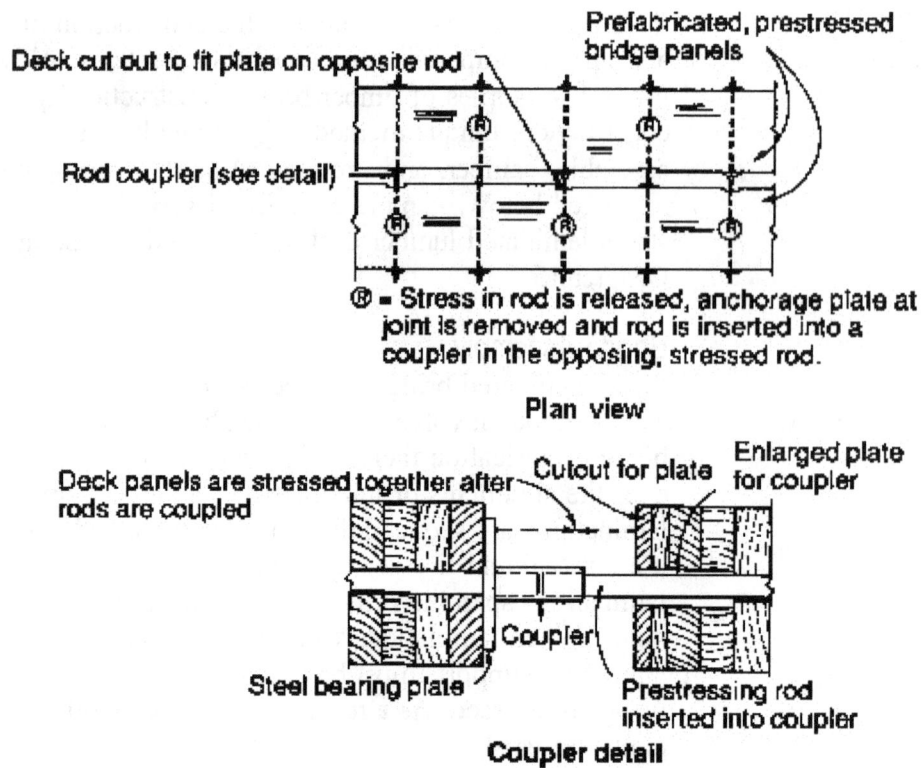

Plan view

Coupler detail

Figure 9-12. - Method of joining two prefabricated, prestressed longitudinal stress-laminated bridge panels.

Camber

Camber is an upward curvature that is placed in a bridge to offset vertical dead load deflection. Stress-laminated decks are unique among timber decks because when butt joints are used, the deck can be cambered (lumber decks without butt joints cannot be cambered). Cambering is accomplished by slightly offsetting the laminations at butt joints before stressing. When the deck is prefabricated or assembled on the ground, this is done with sleeper blocks that are placed under the laminations (Figure 9-13). If the bridge is assembled on scaffolding, the same effect is achieved by

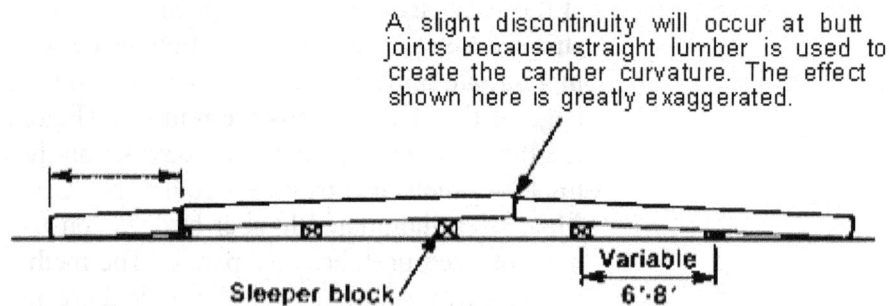

A slight discontinuity will occur at butt joints because straight lumber is used to create the camber curvature. The effect shown here is greatly exaggerated.

Figure 9-13. - Method of cambering longitudinal stress-laminated bridge joints with butt joints.

varying scaffolding height. After the desired amount of curvature is built into the deck, it is stressed together and the camber is locked in. Because stress-laminated decks use straight lumber, cambering causes slight discontinuities at the butt joints. However, these discontinuities are normally of little consequence. The amount of centerspan camber recommended for stress-laminated decks is a minimum of two times, and preferably three times, the deck dead load deflection.

Stressing

Stress-laminated lumber decks are stressed together with a hydraulic jack that applies tension to the prestressing rod by pulling the rod away from steel anchorage plates. After the tension is applied, the nut is tightened against the anchorage plate and the tension remains in the rod when jack pressure is released. Two types of jacks have been used for stress-laminated decks, both of which are hollow-core jacks (the prestressing rod is inserted through the jack body)(Figure 9-14). The first type uses a built-in ratchet to tighten the nut after stress has been applied. The second type involves a standard hollow-core jack used with a prefabricated steel chair. The rachet-type jacks are available from rod manufacturers and are simple and convenient to operate; however, they are expensive to purchase or rent. The hollow core and steel chair arrangement is much less expensive, but the nut must be tightened with a wrench rather than a built-in ratchet.

The method used for stressing a stress-laminated lumber deck depends on the number of jacks that are available. In Ontario, bridges have generally been stressed using a series of up to 24 jacks. Although it is expensive to purchase or rent a large number of jacks, this method is most convenient because the entire deck can be stressed in one operation. In the United States, most stress-laminated lumber decks have used a single jack that is sequentially used for each rod. When using the single-jack method, jacking starts at the first rod on one end of the bridge and progresses to the last rod on the opposite end. After all rods are stressed the first time, three or more additional passes are necessary to restress each rod to the required level. This restressing is necessary because the initial stress in one rod squeezes the laminations together and reduces the stress in adjacent rods. In most cases, the proper uniform stress is achieved by making four passes along the deck.

Figure 9-14. - Types of hollow core jacks used for stress-laminated bridges. (A) With a built-in ratchet. (B) With a steel chair assembly.

9.4 DESIGN OF LONGITUDINAL STRESS-LAMINATED LUMBER DECKS

The design of longitudinal stress-laminated lumber decks is controlled by four basic design constraints. The first and most obvious constraint is ensuring safety by limiting the material to allowable stresses that provide an acceptable factor of safety. The second constraint involves maintaining sufficient stiffness within the deck to avoid long term sagging and unacceptable live load deflection. The third constraint requires that the necessary minimum uniform level of compressive prestress be maintained to keep the bridge laminated together over the design life. Finally, the stress induced by the prestress compression must be within acceptable limits to avoid wood damage.

This section presents sequential design procedures and examples for longitudinal stress-laminated lumber decks. As previously discussed, design provisions for stress-laminated lumber are not included in current AASHTO bridge design specifications, although they are currently being developed. The design procedures presented here are from a preliminary AASHTO proposal based on laboratory and field research conducted by UW/FPL. In addition, provisions from the *Ontario Highway Bridge Design Code* are included, based on research completed by Queen's University and MTO. Other design procedures currently being developed at West Virginia University and at MTO will be considered by the appropriate AASHTO committees when research is completed. The basis for the procedures proposed by West Virginia University, including equations for deflection and bending moment, are presented in the West Virginia University Civil Engineering Report *Wheel Load Distributions on Highway Bridges.*[20]

DESIGN CRITERIA AND DEFINITIONS

General design requirements related to stress-laminated deck design are summarized below. Additional criteria related to specific component design are addressed in more detail in the design procedures and examples that follow.

Deck Configuration
The following limitations apply to stress-laminated lumber decks designed using this procedure:

1. The deck is constructed of sawn lumber laminations that are placed edgewise between supports and are transversely stressed together.

2. Deck width is constant.

3. Deck thickness is constant and is not less than 8 inches nominal.

4. The deck is a rectangle in plan, or is skewed less than 20 degrees.

5. End or intermediate supports are continuous across the deck width.

6. Butt joints are permitted in the laminations provided no more than one butt joint occurs in any four adjacent laminations within a span distance of 4 feet.

Loads

Loads are based on AASHTO loading requirements discussed in Chapter 6. Design procedures and examples are limited to AASHTO Load Group I and IB, where design is routinely controlled by a combination of structure dead load and vehicle live load. As with other timber bridge types, allowable design stresses may be increased by 33 percent for overloads.[2] AASHTO special provisions for H 20-44 and HS 20-44 wheel loads (Chapter 6) do not apply to longitudinal stress-laminated decks.

Lumber Laminations

Design procedures are valid for sawn lumber laminations of Douglas Fir-Larch, Hem-Fir (North), Red Pine, or Eastern White Pine. Behavior and performance data on other species are not currently available but are being developed. Conditions of use are based on a normal duration of load and wet-use conditions, without adjustments for temperature and fire-retardant treatments. *All wood components are assumed to be pressure treated with an oil-type preservative prior to fabrication.*

Tabulated values for lumber are taken from the 1986 edition of the NDS.[12] To account for load-sharing characteristics of the stress-laminated system, the tabulated bending stress for single-member use is increased 30 percent for lumber graded Select Structural, and 50 percent for lumber graded No. 1 or No. 2. These increases are based on research conducted in Canada (discussed in Section 9.3) and are somewhat less than load-sharing increases currently allowed in the United States for glulam.

Prestressing System

Prestressing elements are high-strength steel rods conforming to ASTM A722.[4] The rods are placed through the laminations and are attached to anchorages with high-strength nuts (refer to OHBDC for design requirements related to rod configurations placed above and below, rather than through, the laminations). Design procedures are included for both the steel plate anchorage and the channel bulkhead anchorage. Either system may be used at the prerogative of the designer. All prestressing components and metal hardware are galvanized or otherwise provided with acceptable corrosion protection.

Live Load Deflection

AASHTO specifications do not include design criteria or guidelines for live load deflection in timber bridges. The recommendations given in this section are based on field experience and common design practice, and are consistent with recommendations previously given for other timber bridge types. Although it is highly recommended that these maximum-deflection guidelines be followed for best performance, deflection criteria should be based on specific design circumstances and are left to designer judgment.

DESIGN PROCEDURES

The design of longitudinal stress-laminated decks is basically a two-part process involving design of the lumber laminations followed by design of the prestressing system. Lamination design is based on a wheel load distribution width similar to that used for longitudinal nail-laminated decks. Using this approach, the deck is assumed to act as a beam and is designed for bending, deflection, and compression at the supports. Horizontal shear is not a controlling factor in stress-laminated deck design, and need not be considered. Design of the prestressing system is based on the deck configuration and the magnitude of the transverse moment and shear. For both the deck and the prestressing system, design procedures use graphs that are based on variable relationships developed by analytic modeling, verified by full-scale structure performance.

The basic design procedures for longitudinal stress-laminated lumber decks are outlined in the following steps. The sequence of the procedures assumes that the deck thickness is initially based on bending, then checked for deflection. In many applications, deflection will control; however, the acceptable level of deflection may vary for different design applications. The order of the procedures may be rearranged as necessary.

1. Define deck geometric requirements and design loads.

   a. Define geometric requirements for bridge span, width, and the number of design traffic lanes. The effective deck span, L, is the distance measured center to center of supports. Deck width is the roadway width plus additional width required for curb and railing systems.

   b. Identify design vehicles (including overloads), other applicable loads, and AASHTO load combinations discussed in Chapter 6. Also note design requirements for live load deflection and other site-specific requirements for geometry or loading.

2. Select a species and grade of lamination and compute allowable design values.

Stress-laminated decks are normally constructed from lumber in the Joists and Planks size classification (2 to 4 inches thick, 5 inches and wider). Grades for visually graded lumber are generally No. 2 or better for nominal 2-inch material and No. 1 or better for nominal 4-inch material. Select a species and grade of lumber from the NDS Table 4A (Douglas Fir-Larch, Hem-Fir (North), Red Pine, or Eastern White Pine) and compute allowable design values for bending (F_b'), modulus of elasticity (E'), and compression perpendicular to grain $(F_{c\perp}')$ by Equations 9-1, 9-2, and 9-3, respectively. Tabulated single-member bending stress given in the NDS Table 4A is increased by the load-sharing factor, C_{LS}.

$$F_b' = F_b C_M C_{LS} \tag{9-1}$$

$$E' = E C_M \tag{9-2}$$

$$F_{c\perp}' = F_{c\perp} C_M \tag{9-3}$$

where C_M = moisture content factor from Table 5-7, and

C_{LS} = load sharing factor (1.30 for lumber graded Select Structural, 1.50 for lumber graded No. 1 or No. 2).

3. Determine preliminary lamination layout.

The design of stress-laminated lumber decks depends on the configuration of the laminations and the frequency of butt joints. Butt joints reduce the required length of lamination but create discontinuities in the deck. As a result, longitudinal deck stiffness is decreased, which improves load distribution. However, the discontinuities caused by the butt joints decrease the deck section and reduce load capacity. The decision to use butt joints, and their relative frequency, depends on the availability and relative economics of lumber sizes and must be evaluated on a site-specific basis.

Determine the preliminary lamination layout including the length of laminations and the frequency and location of butt joints. Not more than one butt joint may occur in any four adjacent laminations over a span distance of 4 feet (Figure 9-9).

4. Compute the transverse moduli for the stress-laminated system.

In addition to material design values, stress-laminated deck design must consider the transverse bending modulus, E_{TS}, and the transverse shear modulus, G_{TS} of the stress-laminated system. These values are derived from research data on stress-laminated deck behavior and depend on the species of lumber lamination and the level of prestress (the minimum prestress level required for acceptable deck performance is used). They are based on overall system behavior and should not be confused with the clear wood values discussed in Chapter 3.

At this time, values of E_{TS} and G_{TS} derived by testing are limited to the Douglas Fir-Larch, Hem-Fir (North), Red Pine, or Eastern White Pine laminations. Design values for these species are computed by

$$E_{TS} = 0.013 \, E' \tag{9-4}$$

$$G_{TS} = 0.03 \, E' \tag{9-5}$$

Research is currently in progress to determine E_{TS} and G_{TS} for other softwoods and hardwoods and values should be available in the near future.

5. Compute maximum vehicle live load moment.

Compute the maximum moment for one wheel line of the design vehicle. Maximum simple-span moments for standard AASHTO vehicles and selected overloads are given in Table 16.8 of Chapter 16. For multiple-span continuous bridges, maximum moments are computed for the controlling truck or lane load by analyzing the deck as a continuous beam.

6. Compute wheel load distribution width.

Stress-laminated decks are designed as a beam, assuming that one wheel line of the design vehicle is distributed over a wheel load distribution width, D_w. The value of D_w is based on orthotropic plate behavior and is slightly larger for decks with butt joints because of the lower longitudinal stiffness caused by the joints. The effect of butt joints on load distribution depends on butt-joint frequency and is expressed by a butt joint factor, C_B given in Table 9-1.

Determine D_w from Figure 9-15 using values of α and θ computed by

$$\alpha = \frac{2G_{TS}}{\sqrt{E'(C_B)(E_{TS})}} \tag{9-6}$$

$$\theta = \frac{b}{2L}\left[\frac{E'(C_B)}{E_{TS}}\right]^{0.25} \tag{9-7}$$

Table 9-1. - Butt-joint factor, C_B for longitudinal stress-laminated lumber bridges.

Butt joint frequency	C_B
1 in 4	0.80
1 in 5	0.85
1 in 6	0.88
1 in 7	0.90
1 in 8	0.93
1 in 9	0.93
1 in 10	0.94
No butt joints	1.00

Number of butt joints in number of adjacent laminations, measured within a distance of 4 feet along the bridge span (1 in 4 indicates that one butt joint occurs in 4 adjacent laminations as shown in Figure 9-9).

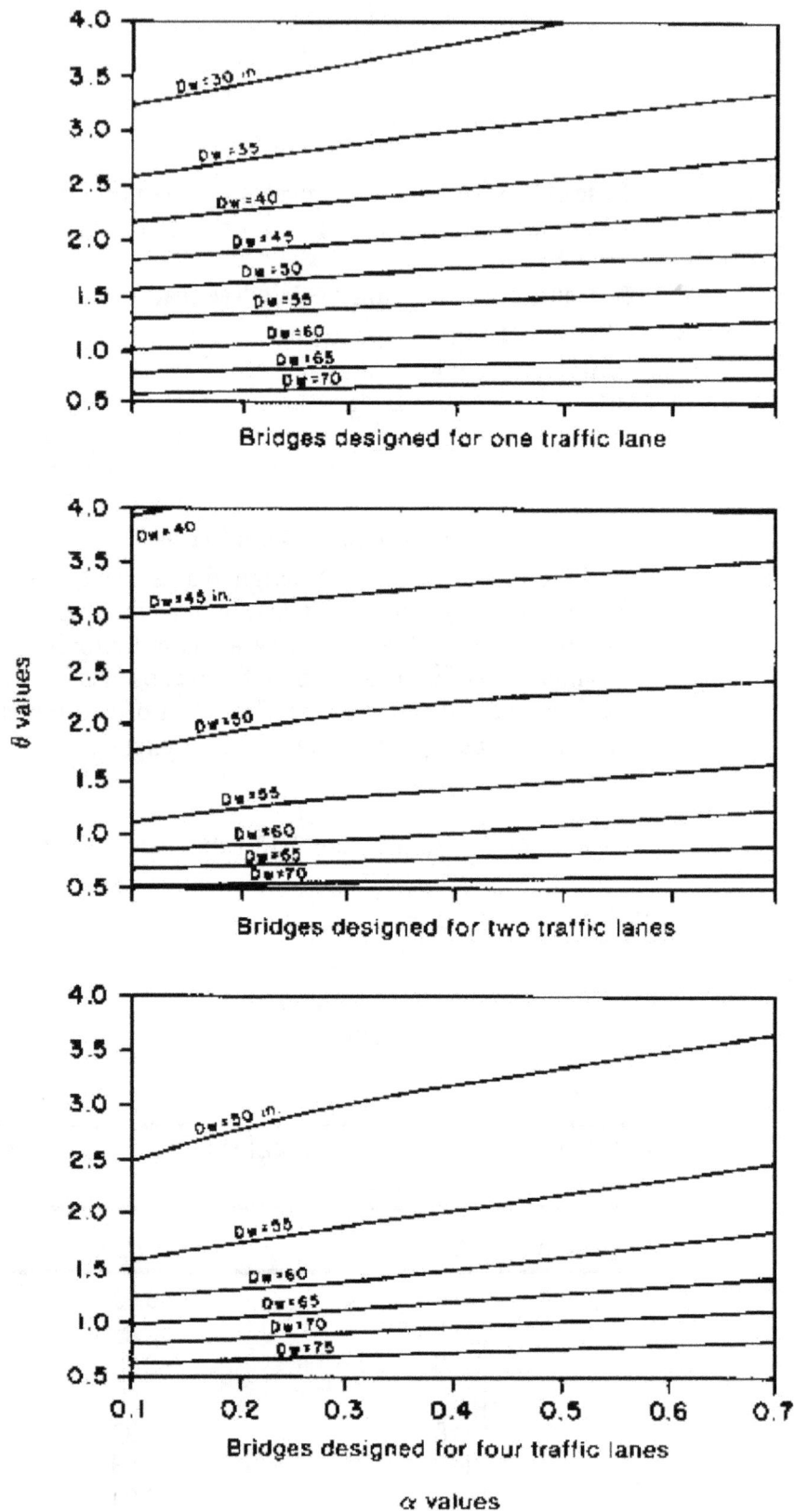

Figure 9-15. - Graphs for determining the wheel load distribution width (D_w) for longitudinal stress-laminated bridge decks.

where G_{TS} = transverse shear modulus of the stress-laminated system (lb/in^2),

E' = allowable modulus of elasticity for the lumber laminations (lb/in^2),

C_B = butt joint factor from Table 9-1,

E_{TS} = transverse modulus of elasticity for the stress-laminated system (lb/in^2),

b = deck width measured between the outside deck edges (ft), and

L = deck span measured center to center of bearings (ft).

The distribution width, D_w, must not be greater than the bridge width divided by the total number of wheel lines, assuming two wheel lines per design traffic lane.

7. Estimate deck thickness and compute effective deck-section properties.

Deck thickness must be estimated for initial calculations. Approximate deck thickness span relationships that may be used for estimating deck thickness are shown in Table 9-2.

Select an initial deck thickness, t, and compute section properties of the effective deck section using Equations 9-8 and 9-9 below (note that D_w is adjusted by C_B). When the hole diameter in laminations for prestressing rods is less than or equal to 20 percent of the deck thickness, holes may be ignored when computing section properties. When the hole diameter exceeds 20 percent of the deck thickness, the hole area must be deducted from the effective deck section.

Table 9-2. - Approximate maximum spans for longitudinal stress-laminated deck bridges for purposes of estimating deck thickness.

Deck thickness (in.)	Approximate maximum span (ft)	
	Single-lane bridges	Double-lane bridges
7-1/4	14	14
8	17	17
9-1/4	22	21
10	24	23
11-1/4	25	24
12	27	26
13-1/4	31	29
14	33	31
15-1/4	37	35
16	39	37

Spans listed in this table are based on HS 20-44 loading and No. 1 Douglas Fir-Larch laminations, and are typically limited by a live load deflection of L/360. Longer spans may be possible with an increased deflection, subject to designer judgment. For lumber species with lower tabulated bending stress or modulus of elasticity, indicated maximum spans should be reduced accordingly.

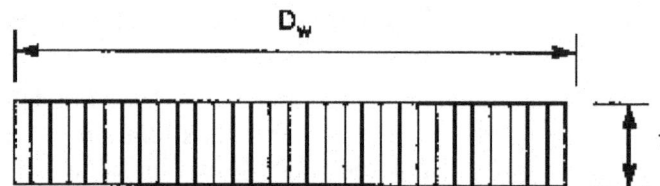

$$S = \text{effective deck section modulus} \left(\text{in}^3\right) = \frac{D_w (C_B)(t)^2}{6} \qquad (9\text{-}8)$$

$$I = \text{effective deck moment of inertia} \left(\text{in}^4\right) = \frac{D_w (C_B)(t)^3}{12} \qquad (9\text{-}9)$$

where t is the deck thickness (in.)

8. Compute deck dead load and dead load moment.

Compute the uniform dead load, DL, of the deck and wearing surface in pounds per square foot using the unit material weights given in Chapter 6. Typical values of DL for decks with asphalt or timber wearing surfaces are given in Table 9-3. From this, determine the uniform dead load acting over D_w per foot of deck span, w_{DL}. When the deck is provided with curbs, railings, or other attached components, the dead load of these components is assumed to be uniformly distributed across the entire deck width and is added to w_{DL}.

Dead load moment for simple-span decks with uniform loads is computed by Equation 9-10:

Table 9-3. -Typical dead load unit weights for stress-laminated lumber bridge decks.

Deck thickness (in.)	Dead load (lb/ft²)		
	Deck only	Deck with 3-inch asphalt surface	Deck with 3-inch lumber surface
7-1/4	30.2	67.7	42.7
8	33.3	70.8	45.8
9-1/4	38.5	76.0	51.0
10	41.7	79.2	54.2
11-1/4	46.9	84.4	59.4
12	50.0	87.5	62.5
13-1/4	55.2	92.7	67.7
14	58.3	95.8	70.8
15-1/4	63.5	101.0	76.0
16	66.7	104.2	79.2

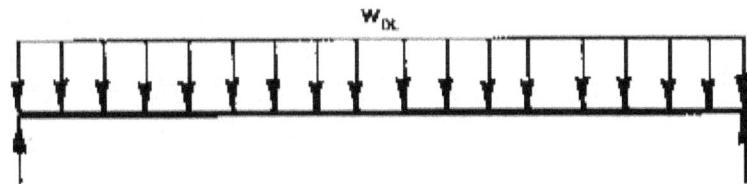

$$M_{DL} = \frac{w_{DL} L^2}{8} \tag{9-10}$$

where M_{DL} = maximum dead load moment (ft-lb),

w_{DL} = uniform dead load over the wheel load distribution width, D_w per foot of deck span (lb/ft), and

L = bridge span length (ft).

9. Compute bending stress.

Deck bending stress is computed by dividing the sum of the maximum live load and dead load bending moments by the effective deck section modulus, as computed by

$$f_b = \frac{M}{S} \tag{9-11}$$

where $M = M_{DL} + M_{LL}$, the sum of the maximum dead load moment and the maximum live load moment from one wheel line of the design vehicle (in-lb), and

S = effective deck section modulus from Equation 9-8 (in³).

The applied bending stress must not exceed the allowable bending stress for the selected species and grade of lumber lamination, as computed by

$$f_b \leq F_b'$$ (9-12)

The allowable bending stress may be increased by a factor of 1.33 for overloads in AASHTO Load Group IB.

If $f_b \leq F_b'$, the deck is sufficient in bending. If f_b is substantially less than F_b', a thinner deck or lower-grade material may be more economical; however, no changes in deck thickness or grade should be made until after live load deflection is checked.

If $f_b > F_b'$, the deck is insufficient in bending and the initial deck thickness or lumber grade (tabulated bending stress) must be increased. In either case, the design sequence must be repeated.

10. Check live load deflection.

Live load deflection is computed by standard methods of elastic analysis for one wheel line of the design vehicle. Because deflection is a serviceability design criterion, an acceptable method without safety factors is desired. Because the orthotropic behavior of the deck results in a wider distribution width for deflection than for bending, the deck moment of inertia used to calculate live load deflection should be taken as 1.33 times the effective deck moment of inertia computed by Equation 9-9. Deflection coefficients for standard AASHTO loads and selected overloads on simple spans are given in Table 16-8. The computed live load deflection must not exceed the allowable deflection established for the structure. If deflection exceeds an acceptable level, the deck thickness or modulus of elasticity must be increased and the design sequence repeated.

Recommended live load deflection criteria for timber bridges is not specified by AASHTO, and the maximum permissible deflection is left to designer judgment. The maximum live load deflection recommended for a stress-laminated deck with asphalt wearing surface is $L/360$. If the structure is provided with a pedestrian walkway, a further reduction in live load deflection is recommended to avoid dynamic effects and the human perception of motion. Acceptance of deflection values exceeding $L/360$ is at the designers discretion and should be related to the relative magnitude of the deflection and its effect on the overall bridge performance.

11. Compute dead load deflection and camber.

For longitudinal stress-laminated lumber decks with butt joints, it is recommended that the bridge be cambered to offset sagging caused by long-term creep. The amount of camber depends on the initial dead load deflection resulting from the uniform dead load acting over a deck width, D_w. For a simple-span deck, dead load deflection is computed by Equation 9-13:

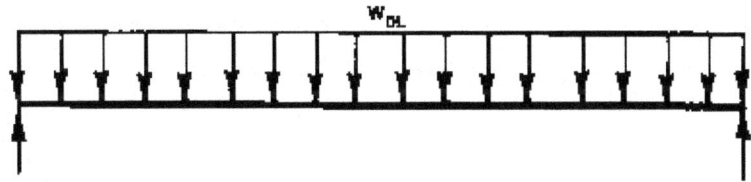

$$\Delta_{DL} = \frac{5 w_{DL} L^4}{384 E' I}$$

(9-13)

where Δ_{DL} = dead load deflection (in.),

w_{a} = uniform dead load over the wheel load distribution width, D_{w} per inch of deck span (lb/in),

L = deck span (in.), and

I = effective deck moment of inertia from Equation 9-9 (in^4).

The amount of camber placed in the deck at the time of stressing should be a minimum of two times, and preferably three times, the computed deck dead load deflection.

12. Determine the required prestress level.

The level of compressive prestress between the laminations must be sufficient to offset flexural tension stress caused by transverse moment and slip caused by transverse shear. For stress-laminated deck design, the level of uniform prestress must be determined for two conditions; in service and at installation. The prestress level in service represents the minimum compressive prestress required for adequate deck performance, assuming all time-related stress loss has occurred. The prestress level at installation is the amount of prestress that must be introduced into the deck at the time of stressing.

Required compressive prestress levels depend on the magnitude of transverse bending and transverse shear from applied loads. Values for both forces are determined from curves based on orthotropic deck behavior in response to applied wheel loads. The magnitude of transverse bending moment, M_p is obtained from Figure 9-16 using the values of α and θ computed by Equations 9-6 and 9-7. Transverse shear, V_p is determined from Figure 9-17 using the parameter β, defined by

$$\beta = \pi \left(\frac{b}{L} \right) \sqrt{\frac{E'(C_B)}{2 G_{TS}}}$$

(9-14)

Values of M, and V, obtained from Figures 9-16 and 9-17 are based on an HS 20-44 truck with a 16,000-pound wheel load. When other wheel loads

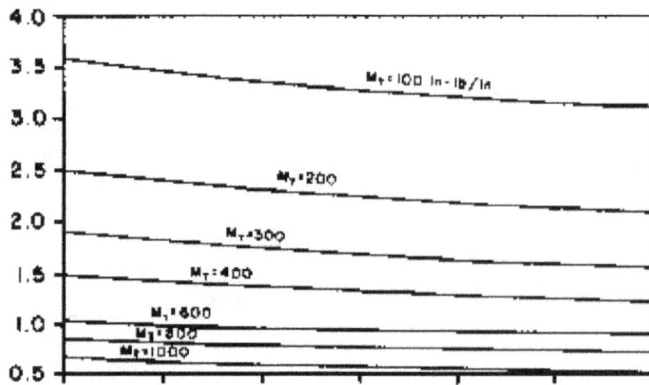

Bridge designed for one traffic lane

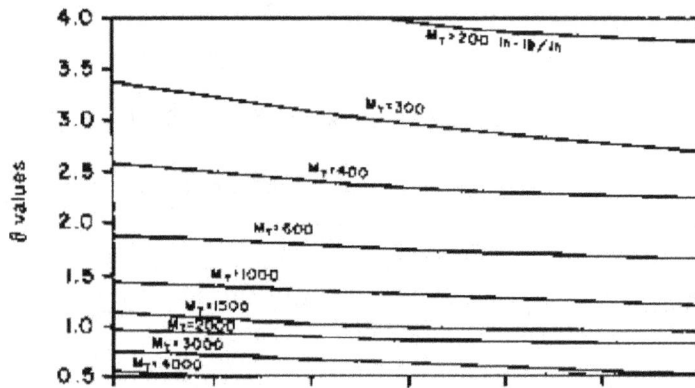

Bridge designed for two traffic lanes

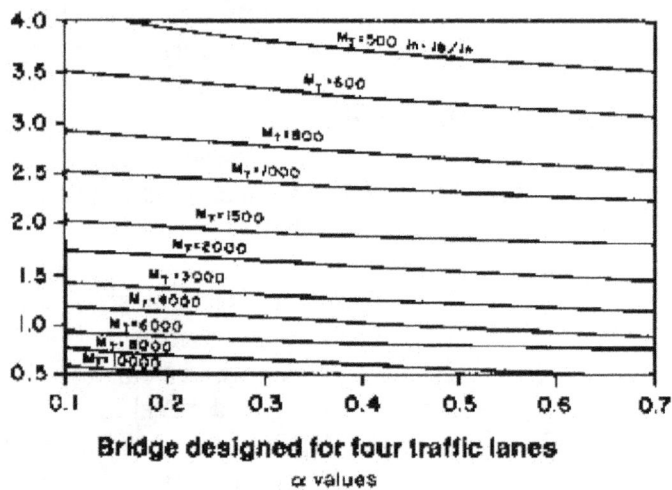

Bridge designed for four traffic lanes

α values

Graphs are based on a HS 20-44 vehicle with maximum wheel load of 16,000-lb. For other wheel loads, multiply the graph value of M_t by the ratio of design wheel load to a 16,000-lb wheel load.

Figure 9-16. - Graphs for determining the magnitude of transverse bending (M_T) for longitudinal stress-laminated bridge decks.

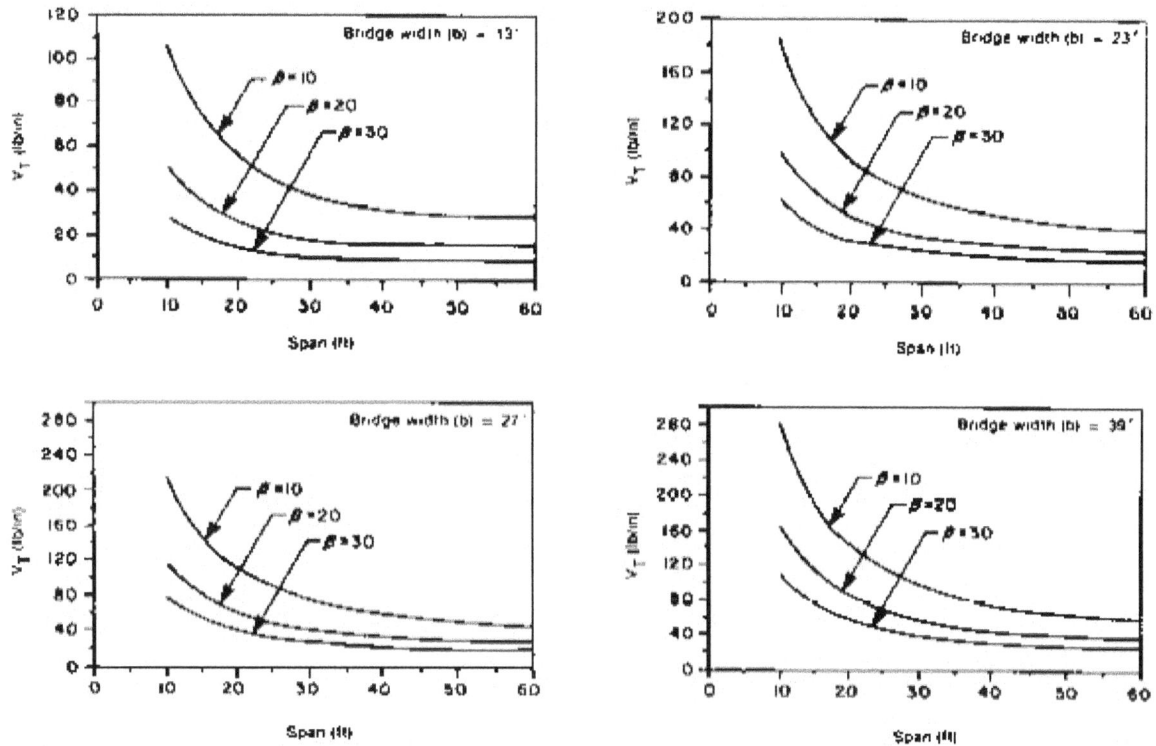

Graphs are based on a HS 20-44 vehicle with a maximum wheel load of 16,000 lb. For other wheel loads, multiply the graph value of V_T by the ratio of design wheel load to a 16,000-lb wheel load. Use interpolation and/or extrapolation for intermediate bridge widths and β-values.

Figure 9-17. - Graphs for determining the magnitude of transverse shear (V_T) for longitudinal stress-laminated bridge decks.

are used, values must by multiplied by the ratio of the design wheel load to the HS 20-44 wheel load.

The minimum level of uniform compressive prestress in service, N, is the largest value obtained from the following equations, but not less than 40 lb/in², as computed by

$$N = \frac{6M_T}{t^2} \qquad \text{or} \qquad N = \frac{1.5V_T}{t(\mu)} \text{ , whichever is greater} \qquad (9\text{-}15)$$

$$N \geq 40 \text{ lb/in}^2 \qquad (9\text{-}16)$$

where N = minimum uniform compressive prestress in service (lb/in²),

t = deck thickness (in.),

M_T = magnitude of transverse bending from applied wheel loads (in-lb/in),

9-33

V_r = magnitude of transverse shear from applied wheel loads (lb/in), and

μ = coefficient of friction (0.35 for surfaced (S4S) lumber, 0.45 for rough-sawn lumber or lumber that is surfaced on one side (S1S)).

Over the bridge life, time-related creep losses are assumed to reduce the level of compressive prestress to 40 percent of the initial level at installation 60-percent stress loss). This assumption is based on research and field performance for softwood laminations that are properly treated with oil-type preservatives and are installed at a moisture content of 19 percent or less (there has been no research or experience with hardwood species or waterborne treatments; however, research in these areas is in progress). To compensate for the gradual 60-percent stress loss, the level of uniform prestress at the time of installation, N_i must be greater than or equal to 2.5 times the minimum required prestress level in-service, as computed by

$$N_i \geq 2.5N \tag{9-17}$$

where N_i is the level of uniform compressive prestress required at the time of installation (lb/in^2).

13. Determine spacing and size of prestressing rods and the required prestressing force.

Prestressing rods for stress-laminated decks are threaded high-strength steel conforming to ASTM A 722, Uncoated High-Strength Steel Bar for Prestressing Concrete. The rods are 5/8-inch, 1-inch or 1-1/4-inch diameter with properties shown in Table 9-4. The specified minimum ultimate tensile stress of the prestressing rods, f_{pu} is 150,000 lb/in^2. The maximum allowable tensile stress, at or after anchorage, cannot exceed 70 percent of ultimate tensile strength (105,000 lb/in^2). During jacking, the maximum short-term tensile stress cannot exceed 80 percent of the ultimate tensile strength (120,000 lb/in^2). These values should be further reduced by any strength reductions recommended by the rod manufacturer.

The spacing of the prestressing rods, S_p must be sufficient to induce the required uniform compressive prestress in areas adjacent to vehicle wheel loads. As previously discussed, compressive prestress is not uniform at the deck edge but becomes uniform at some interior distance. Rod spacing therefore depends on the wheel load placement in relation to the deck edge. Using the same requirements previously discussed for other timber deck systems, the center of the wheel load is placed 1 foot from the nearest face of the curb or rail.

Table 9-4. - Properties of steel prestressing rods used for stress-laminated lumber bridge decks.

Rod diameter (in.)	Rod area, A_s (in²)	Maximum allowable tensile load[a] (lb)	
		At or after anchorage[b]	During jacking (short-term)[c]
5/8	0.28	29,400	33,600
1	0.85	89,250	102,000
1-1/4	1.25	131,250	150,000

[a] For rods conforming to ASTM A 722 with a specified minimum ultimate tensile strength, f_{pu}, of 150,000 lb/in².

[b] $0.70 f_{pu} A_s$

[c] $0.80 f_{pu} A_s$

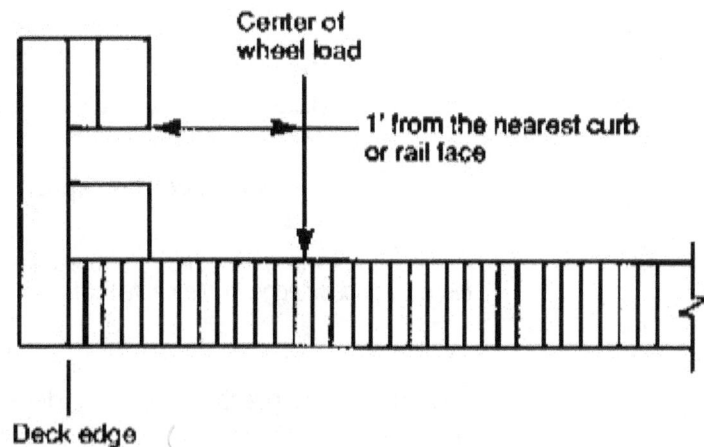

The maximum spacing of prestressing rods is obtained using the curve in Figure 9-18, based on the distance from the outside deck edge to the center of the wheel load. The spacing of the first rod from the deck end is generally equal to one-half the center-to-center spacing.

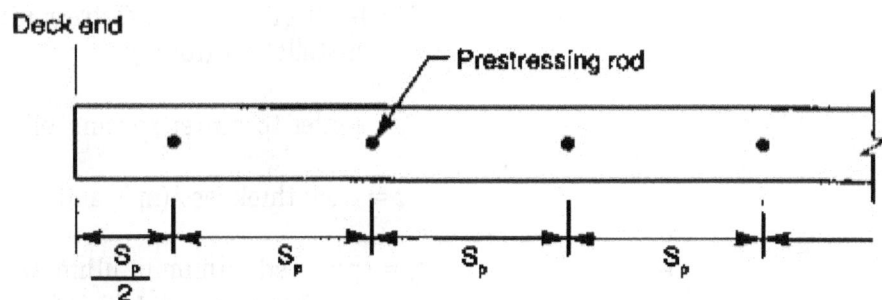

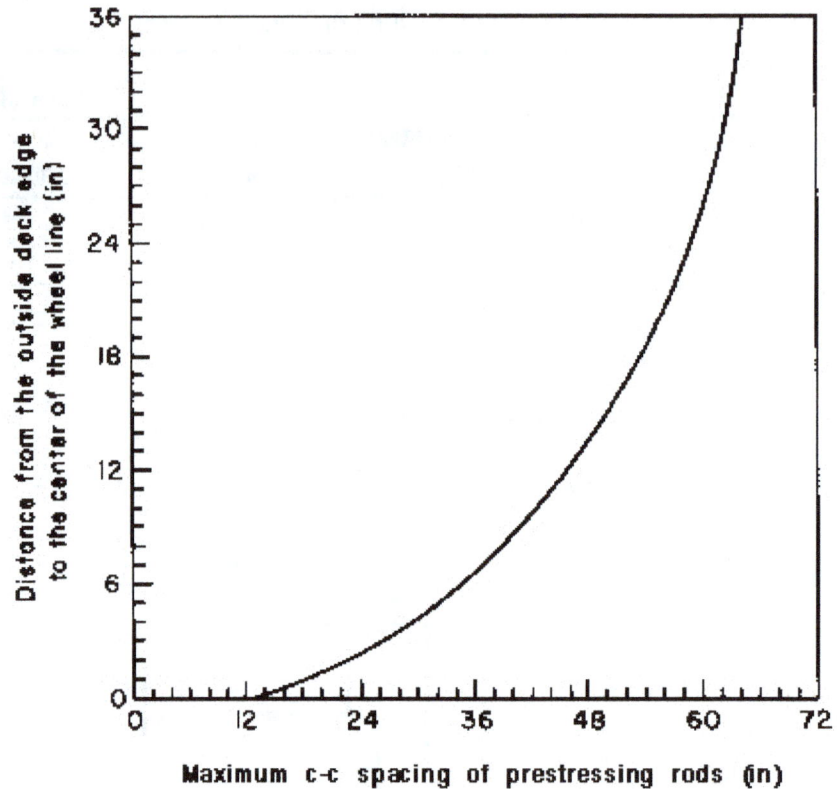

Figure 9-18. - Maximum spacing of prestressing rods as a function of the distance from the outside deck edge to the center of the vehicle wheel line.

The size of the prestressing rods depends on the required level of compressive prestress at installation, N_i and the rod spacing, S_p. In addition, rod area must be limited so that the ratio of the steel area to the wood area is less than or equal to 0.0016, as computed by

$$\frac{N_i(S_p)(t)}{0.70(f_{pu})} \leq A_s \qquad \text{and} \qquad \frac{A_s}{S_p(t)} < 0.0016 \qquad (9\text{-}18)$$

where A_s = cross-sectional area of the steel prestressing rod (in^2),

N_i = level of uniform prestress required at the time of installation (lb/in^2),

S_p = center-to-center spacing of the prestressing rods (in.),

t = deck thickness (in.), and

f_{pu} = specified minimum ultimate tensile stress for the prestressing rod, 150,000 lb/in^2.

Select a rod spacing and diameter that satisfy maximum spacing requirements and the steel area requirements of Equation 9-18. Rod spacing should also consider possible conflicts with other structural components, such as guardrail posts. Approximate spacing requirements for various rod diameters and deck thicknesses are given in Table 9-5.

The prestressing force required in each prestressing rod, F_{pa}, is computed by

$$F_{pa} = N_i(S_p)(t) \tag{9-19}$$

14. Design the anchorage system.

The anchorage system for prestressing rods must securely hold the rods and effectively transfer the prestressing force to the lumber laminations. In addition, the anchorage must be of sufficient size to prevent excessive wood crushing in the outside laminations. The two anchorage configurations used are the bearing-plate configuration developed at UW/FPL and the channel bulkhead configuration developed in Ontario and included in the OHBDC. With the exception of the high-strength steel rods and nuts, components for both systems are normally fabricated of galvanized steel (ASTM A 36) or weathering steel (ASTM A 588).

As previously discussed, the bearing-plate anchorage configuration may result in some localized wood crushing in the vicinity of the bearing plates that may not be acceptable in all cases. The channel bulkhead configuration covers the outside laminations with a steel channel and any wood crushing

Table 9-5. - Approximate spacing requirements for prestressing rods used for stress-laminated lumber decks.

t (in.)	Rod spacing (in.)					
	5/8-in. Ø rods		1-in. Ø rods		1-1/4 in. Ø rods	
	Max.	Min.	Max.	Min.	Max.	Min.
7-1/4	41	24	—	—	—	—
8	37	22	—	—	—	—
9-1/4	32	19	—	—	—	—
10	29	18	89	53	—	—
11-1/4	26	16	79	47	—	—
12	25	15	74	44	—	—
13-1/4	—	—	67	40	99	59
14	—	—	64	38	94	56
15-1/4	—	—	59	35	86	51
16	—	—	56	33	82	49

Maximum rod spacing is based on a uniform compressive prestress level of 100 lb/in².
Minimum rod spacing is based on a maximum wood/steel ratio of 0.0016.

is not visible; however, the channel bulkhead is more costly. Design procedures for both configurations are presented below. The choice of the most appropriate system is left to designer judgment based on specific project requirements.

Bearing-Plate Anchorage Configuration

The bearing-plate anchorage consists of an inner-steel bearing plate, an outer-steel anchorage plate and a high-strength steel nut (Figure 9-19). Design of this anchorage primarily involves determining the length, width, and thickness of the inner bearing plate. The outer anchorage plate is available from the rod manufacturer and is normally standardized (by manufacturer) based on the prestressing rod diameter (Table 9-6).

The area of the bearing plate must be sufficient to limit compressive stress under the plate to the allowable compression perpendicular to grain for the lumber laminations, as computed by

$$A_P \geq \frac{F_{ps}}{F_{c\perp}}.$$

(9-20)

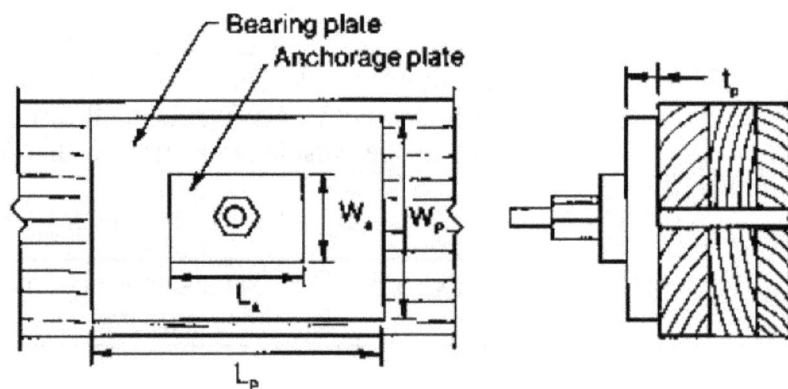

Figure 9-19. - Bearing-plate anchorage configuration.

Table 9-6. - Typical sizes for prestressing-rod anchorage plates.

Prestressing rod diameter (in.)	Anchorage plate dimensions (in.) width (W_A) x length (L_A) x thickness (t_A)	
	Typical plate size	Alternate plate size
5/8	2 x 5 x 1	3 x 3 x 0.75
1	4 x 6.5 x 1.25	4 x 7 x 1
1-1/4	5 x 8 x 1.5	5 x 8 x 1.25

Plate sizes may vary and should be verified with the rod manufacturer. Other sizes may be specified by the designer to meet specific design requirements.

where A_p = bearing plate area (in²),

F_{pt} = rod prestressing force, from Equation 9-19 (lb), and

$F_{c\perp}'$ = allowable stress in compression perpendicular to grain for the lumber laminations (lb/in²).

In addition, the ratio of the bearing plate length to width must not be less than 1.0, nor greater than 2.0, as computed by

$$1.0 \geq \frac{L_P}{W_P} \geq 2.0$$

(9-21)

where L_P = bearing-plate length (in.), and

W_P = bearing-plate width (in.).

Determine an acceptable bearing plate size based on the requirements of Equations 9-20 and 9-21 and compute the lamination bearing stress in compression perpendicular to grain by

$$f_{c\perp} = \frac{F_{pt}}{A_P}$$

(9-22)

where $f_{c\perp}$ is the applied bearing stress in compression perpendicular to grain (lb/in²).

Based on the bearing-plate area and bearing stress, select a bearing-plate thickness that satisfies:

$$t_P = \sqrt{\frac{3(f_{c\perp})(k^2)}{F_b}}$$

(9-23)

where

$$k = \frac{W_P - W_A}{2} \quad \text{or} \quad k = \frac{L_P - L_A}{2}, \text{ whichever is greater} \quad (9\text{-}24)$$

t_p = bearing plate thickness (in.),

F_b = 0.55F_y = allowable bending stress for the steel plate (lb/in²),

F_y = specified minimum yield point for the steel plate (lb/in²), from AASHTO Table 10.2A (36,000 lb/in² for A36 steel and 50,000 lb/in² for A588 steel),

W_A = anchor-plate width (in.), and

L_A = anchor-plate length (in.).

If an acceptable bearing-plate size that limits compression perpendicular to grain to an allowable value cannot be achieved, or if the plate thickness is excessive, rod spacing must be decreased, and the anchorage design must be repeated.

Channel Bulkhead Anchorage Configuration

The channel bulkhead anchorage consists of a continuous steel channel, a steel bearing plate, and a high-strength steel nut (Figure 9-20). For this anchorage configuration, design involves sizing both the steel channel and the bearing plate. The design provisions given here are based on current requirements of OHBDC.

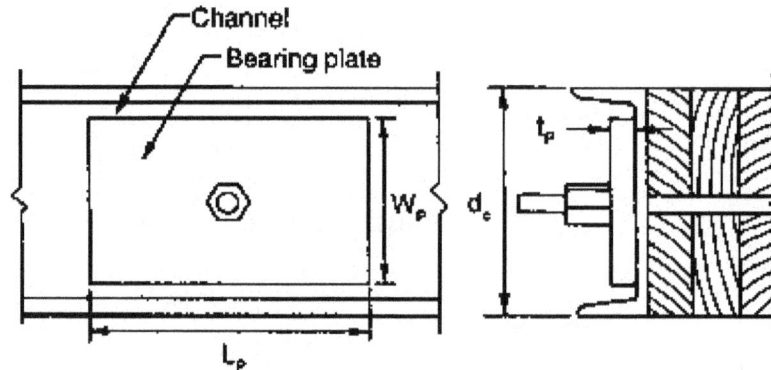

Figure 9-20. - Channel bulkhead anchorage configuration.

The steel channel for the bulkhead configuration is continuous along the bridge span but may be discontinuous over supports. Channel depth is based on deck thickness and must be within 85 and 100 percent of the lamination depth, as computed by

$$0.85t \leq d_c \leq t \qquad (9\text{-}25)$$

where t = deck thickness (in.), and

d_c = depth of steel channel (in.).

Section properties for steel channels should also meet minimum requirements given in Table 9-7. Select a channel size based on the requirements of Equation 9-25 and Table 9-7 and compute an initial bearing-plate length using

$$L_p \geq \frac{F_{ps}}{d_c F_{c\perp}'} - 2(t_w) \qquad (9\text{-}26)$$

where: L_p = bearing-plate length (in.),

F_{ps} = rod prestressing force, from Equation 9-19 (lb),

9-40

Table 9-7. - Minimum section properties for steel channel bulkheads used for stress-laminated lumber decks.

Nominal lamination depth, t (in.)	Minimum channel moment of inertia[a] (in⁴)	Minimum channel web thickness (in.)
8	1.3	0.38
10	2.4	0.43
12	3.3	0.43
14	5.1	0.51
16	9.2	0.52

[a] Moment of inertia about the minor axis.

$F_{c\perp}'$ = allowable stress in compression perpendicular to grain for the lumber laminations (lb/in²), and

t_w = steel channel web thickness (in.).

Select the bearing-plate width and thickness based on (bearing-plate width must also permit the plate to fit between the tapered flanges of the channel)

$$1.0 \leq \frac{L_P}{W_P} \leq 2.0 \qquad (9-27)$$

$$t_P \geq \frac{L_P}{12} \qquad (9-28)$$

where W_P = bearing-plate width (in.), and

t_P = bearing-plate thickness (in.).

The bearing area of the channel bulkhead must be sufficient to limit the compressive stress at the anchorage to the allowable compressive stress perpendicular to grain for the lumber laminations. The effective bearing area, A_E is based on a length equal to the bearing-plate length plus twice the channel thickness, and a width equal to the channel depth, as computed by

$$A_E = d_c(L_P + 2t_w) \qquad (9-29)$$

where A_E is the effective bearing area in in².

The bearing stress in compression perpendicular to grain is computed by

$$f_{c\perp} = \frac{F_{pu}}{A_E} \qquad (9-30)$$

This value must not exceed the allowable compression perpendicular to grain for the lumber laminations computed by

$$f_{c\perp} \leq F_{c\perp}'$$

(9-31)

If $f_{c\perp} > F_{c\perp}$, the size of the bearing plate or steel channel must be increased or the rod spacing must be decreased. In either case, the anchorage design must be repeated.

15. Determine the support configuration and check bearing stress.

Support attachments for longitudinal stress-laminated decks must be designed to resist the vertical and lateral forces transmitted from the superstructure to the substructure. As with other longitudinal deck superstructures, the required bearing length is normally controlled by considerations for bearing configuration, rather than stress in compression perpendicular to grain. From a practical standpoint, a bearing length of 10 to 12 inches is recommended for stress-laminated decks. Bearing attachments are normally made through the deck to the supporting cap or sill, or from the deck underside, using the same details previously discussed for longitudinal glulam decks (Figure 8-7).

Stress in compression perpendicular to grain at the bearing is checked for a deck width equal to the wheel load distribution width, D_w, using

$$f_{c\perp} = \frac{R_{DL} + R_{LL}}{D_w(\ell_b)}$$

(9-32)

where
R_{DL} = dead load reaction for a deck width D_w based on the out-out bridge length (lb),

R_{LL} = maximum reaction produced by one wheel line of the design vehicle, from Table 16-8 (lb), and

ℓ_b = bearing length (in.).

Stress in compression perpendicular to grain must not exceed the allowable stress for the species and grade of lumber lamination, as computed by

$$f_{c\perp} \leq F_{c\perp}'$$

(9-33)

An existing bridge on a county road will be replaced with a longitudinal stress-laminated lumber deck bridge. The bridge spans 37 feet center-to-center of bearings and carries two lanes of AASHTO HS 20-44 on a roadway width of 24 feet. Support for the structure is provided by existing pile abutments with 12-inch-wide caps. Design this bridge, assuming the following:

1. The bridge will include 12-inch by 12-inch timber curbs and vehicular railing with a combined dead load of 85 lb/ft, per side.

2. The deck will be paved with 3 inches of asphalt pavement.

3. Live load deflection must be limited to *L/360*.

4. Lumber laminations are full-sawn, surfaced one side (S1S) Douglas Fir-Larch.

5. The deck will have butt joints at the minimum spacing.

6. A bearing plate anchorage configuration will be used.

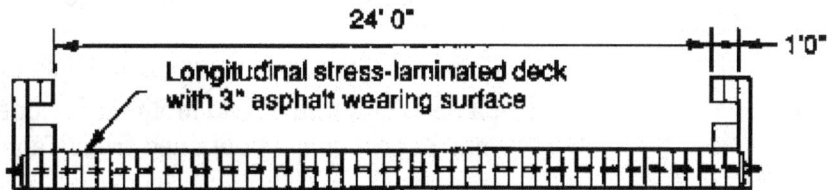

Solution
Define Deck Geometric Requirements and Design Loads

The bridge supports two traffic lanes over a span of 37 feet. With a roadway width of 24 feet, and 12-inch-wide curbs on each side, a bridge width of 26 feet is required. Design loading will be one HS 20-44 wheel line in AASHTO Load Group I.

Select a Species and Grade of Lamination and Compute Allowable Design Values

From the NDS Table 4A, Douglas Fir-Larch that is visually graded No. 1 or better in the J&P size classification is selected. Tabulated values are as follows:

$F_b = 1,500 \ lb/in^2$ (single-member use)

$E = 1,800,000 \ lb/in^2$

$F_{c\perp} = 625 \ lb/in^2$

Allowable design values are computed using the applicable moisture content factor (C_M) from Table 5-7. The tabulated bending stress for single-member use is increased by the load sharing factor, $C_{LS}= 1.50$:

$$F_b' = F_b C_M C_{LS} = 1,500(0.86)(1.50) = 1,935 \text{ lb/in}^2$$

$$E' = E C_M = 1,800,000(0.97) = 1,746,000 \text{ lb/in}^2$$

$$F_{c\perp}' = F_{c\perp} C_M = 625(0.67) = 419 \text{ lb/in}^2$$

Determine the Preliminary Lamination Layout

The minimum butt joint spacing is assumed. Not more than one butt joint will occur in any four adjacent laminations within a span distance of 4 feet.

Compute the Transverse Moduli for the Stress-Laminated System

Values of the transverse bending modulus (E_{rs}) and transverse shear modulus (G_{rs}) are computed by Equations 9-4 and 9-5:

$$E_{TS}= 0.013E' = 0.013(1,746,000) = 22,698 \text{ lb/in}^2$$

$$G_{TS}= 0.03E' = 0.03(1,746,000) = 52,380 \text{ lb/in}^2$$

Compute Maximum Live Load Moment

The maximum live load moment for one wheel line of an HS 20-44 truck on a 37-foot span is obtained from Table 16-8:

$$M_{LL}= 198,300 \text{ ft-lb}$$

Compute Wheel Load Distribution Width

Values of α and θ are computed using Equations 9-6 and 9-7, respectively. Assuming one butt joint in every 4 adjacent laminations, a butt joint factor $C_B= 0.80$ is obtained from Table 9-1.

$$\alpha = \frac{2G_{TS}}{\sqrt{E'(C_B)E_{TS}}} = \frac{2(52,380)}{\sqrt{1,746,000(0.80)(22,698)}} = 0.59$$

$$\theta = \frac{b}{2L}\left[\frac{E'(C_B)}{E_{TS}}\right]^{0.25} = \frac{26}{2(37)}\left[\frac{1,746,000(0.80)}{22,698}\right]^{0.25} = 0.98$$

The distribution width, D_w is obtained from Figure 9-15 using the curves for bridges with two traffic lanes:

$$D_w = 63 \text{ in.}$$

9-44

Estimate Deck Thickness and Compute Effective Section Properties

From Table 9-2, an initial nominal deck thickness of 16 inches is selected. Effective deck section properties are computed by Equations 9-8 and 9-9, assuming that holes for the prestressing rods are less than 20 percent of the deck thickness:

$$S = \frac{D_w (C_s)(t)^2}{6} = \frac{63(0.80)(16)^2}{6} = 2,150 \text{ in}^3$$

$$I = \frac{D_w (C_s)(t)^3}{12} = \frac{63(0.80)(16)^3}{12} = 17,203 \text{ in}^4$$

Compute Deck Dead Load and Dead Load Moment

From Table 9-3, the dead load of the 16-inch deck with a 3-inch asphalt wearing surface is 104.2 lb/ft². The 85lb/ft dead load for the curb and railing is increased by an estimated 10 lb/ft for the prestressing system, and is assumed to be uniformly distributed across the deck width:

$$DL = 104.2 \text{ lb/ft}^2 + \frac{2(85 \text{ lb/ft} + 10 \text{ lb/ft})}{26 \text{ ft}} = 111.5 \text{ lb/ft}^2$$

For the distribution width of 63 in.,

$$w_{DL} = \frac{63 \text{ in.}}{12 \text{ in/ft}}(111.5 \text{ lb/ft}^2) = 585.4 \text{ lb/ft}$$

Maximum dead load moment is computed by Equation 9-10:

$$M_{DL} = \frac{w_{DL}L^2}{8} = \frac{585.4(37)^2}{8} = 100,117 \text{ ft-lb}$$

Compute Bending Stress

Bending stress is computed by Equation 9-11:

$$f_b = \frac{M}{S} = \frac{(198,300 + 100,177)(12 \text{ in/ft})}{2,150 \text{ in}^3} = 1,666 \text{ lb/in}^2$$

$f_b = 1,666 \text{ lb/in}^2 < F_b' = 1,935 \text{ lb/in}^2$, so bending stress is acceptable. Because of the large difference between f_b and F_b', it may be possible to reduce deck thickness, but no changes will be made until after deflection is checked.

Check Live Load Deflection

From Table 16-8, the deflection coefficient for one wheel line of an HS 20-44 truck on a 37-foot simple span is 4.74 x 10¹⁰ lb-in³. Live load deflection is computed using 133 percent of the effective deck moment of inertia:

$$\Delta_{LL} = \frac{4.74 \times 10^{10}}{E'(1.33)(I)} = \frac{4.74 \times 10^{10}}{1,746,000(1.33)(17,203)} = 1.19 \text{ in.} = L/373$$

$L/373 < L/360$, so live load deflection is satisfactory. The deflection is close to the allowable level, so a reduction in deck thickness is not feasible.

Compute Dead Load Deflection and Camber

Dead load deflection is computed by Equation 9-13 for $w_{DL} = 585.4$ lb/ft:

$$\Delta_{DL} = \frac{5w_{DL}L^4}{384E'I} = \frac{5(585.4)[37(12 \text{ in/ft})]^4}{(12 \text{ in/ft})(384)(1,746,000)(17,203)} = 0.82 \text{ in.}$$

The deck will be cambered a minimum of 2.5 inches, which is approximately 3 times the computed dead load deflection.

Determine the Required Prestress Level

Using the previously computed values of α **and** θ, M_r is obtained for a two-lane bridge from Figure 9-16:

$$M_r = 1,500 \text{ in-lb/in}$$

The variable β is computed by Equation 9-14:

$$\beta = \pi\left(\frac{b}{L}\right)\sqrt{\frac{E'(C_b)}{2G_{TS}}} = 3.14\left(\frac{26}{37}\right)\sqrt{\frac{1,746,000(0.80)}{2(52,380)}} = 8.06$$

By interpolation and extrapolation of Figure 9-17,

$$V_r = 80 \text{ lb/in}$$

The minimum required level of compressive prestress in service, N, is the largest value computed by Equation 9-15, but not less than 40 lb/in^2:

$$N = \frac{6M_r}{t^2} = \frac{6(1,500)}{(16)^2} = 35.2 \text{ lb/in}^2$$

Based on transverse shear,

$$N = \frac{1.5V_r}{t(\mu)} = \frac{1.5(80)}{16(0.45)} = 16.7 \text{ lb/in}^2$$

Both values are less than the minimum 40 lb/in^2, so $N = 40$ lb/in^2 will control. Based on this value, the required level of uniform prestress at installation, N_i, is computed by Equation 9-17:

$$N_i = 2.5N = 2.5(40) = 100 \text{ lb/in}^2$$

Determine Spacing and Size of Prestressing Rods and the Required Prestressing Force

Positioning the wheel line 1 foot from the curb face places the center of the wheel line 2 feet from the deck edge:

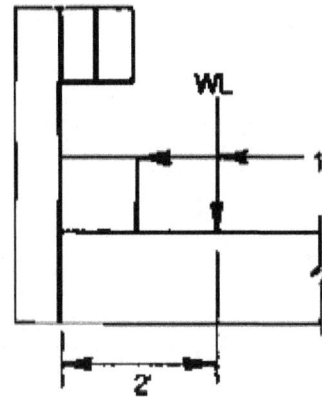

Using the curve in Figure 9-18, the maximum spacing of prestressing rods is approximately 58 inches.

From Table 9-5, 1-inch-diameter ASTM A722 rods are selected. For the 16-inch deck thickness, rods must be spaced between 33 and 56 inches on-center. For a bridge length of 38 feet (37-foot span on 1-foot-wide sills), a spacing of 48 inches will be used, with end rods 12 inches from the deck end:

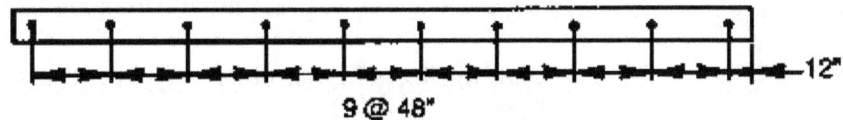

9 @ 48"

From Table 9-4 for a 1-inch-diameter rod, $A_s = 0.85$ in^2. The minimum required rod area and the steel/wood ratio are checked by Equation 9-18:

$$\frac{N_i S_p t}{0.70 f_{pu}} = \frac{100(48)(16)}{0.70(150,000)} = 0.73 \text{ in}^2 < 0.85 \text{ in}^2$$

$$\frac{A_s}{S_p(t)} = \frac{0.85}{48(16)} = 0.0011 < 0.0016$$

The prestressing force required in each rod, F_{ps}, is computed by Equation 9-19:

$$F_{ps} = N_i(S_p)(t) = 100(48)(16) = 76,800 \text{ lb}$$

Design Anchorage System

Using the bearing plate anchorage configuration illustrated in Figure 9-19, the minimum bearing plate area is computed by Equation 9-20:

$$A_P = \frac{F_{ps}}{F_{c\perp}{}'} = \frac{76{,}800}{419} \approx 183 \text{ in}^2$$

For the 16-inch-thick deck, a plate depth, W_p, of 14 inches is selected. The minimum required plate length is computed by dividing the plate area by the plate width:

$$L_p = \frac{A_P}{W_P} = \frac{183}{14} = 13.1 \text{ in.}$$

A 14-inch length will be used, and

$W_p = 14$ in.

$L_p = 14$ in.

$A_p = (14 \text{ in.})(14 \text{ in.}) = 196 \text{ in}^2$

The ratio of the bearing plate length to width is checked by Equation 9-21:

$$\frac{L_P}{W_P} = \frac{14}{14} = 1.0$$

$1.0 \le 1.0 \le 2.0$, so plate dimensions are satisfactory.

Bearing stress in compression perpendicular to grain is computed by Equation 9-22:

$$f_{c\perp} = \frac{F_{ps}}{A_P} = \frac{76{,}800}{196} = 392 \text{ lb/in}^2$$

$f_{c\perp} = 392$ lb/in² $< F_{c\perp}{}' = 419$ lb/in², so bearing stress is acceptable.

Dimensions for the steel anchorage plate are obtained from Table 9-6 and k values are computed by Equation 9-24:

$W_A = 4$ in.

$L_A = 6.5$ in.

$t_A = 1.25$ in.

$$k = \frac{W_P - W_A}{2} = \frac{14 - 4}{2} = 5.00 \text{ in.}$$

or

$$k = \frac{L_P - L_A}{2} = \frac{14 - 6.5}{2} = 3.75 \text{ in.}$$

The largest k value of 5.00 controls and the required bearing plate thickness for an A36 steel plate is computed by Equation 9-23:

$$t_P = \sqrt{\frac{3(f_{c\perp})(k^2)}{F_b}} = \sqrt{\frac{3(392)(5)^2}{0.55(36,000)}} = 1.2 \text{ in.}$$

A minimum plate thickness of 1.25 inches will be used.

Determine the Support Configuration and Check Bearing Stress

Superstructure support is provided by 12-inch-wide pile caps on existing abutments. The bridge will be anchored to the caps with bolts placed through the deck and cap:

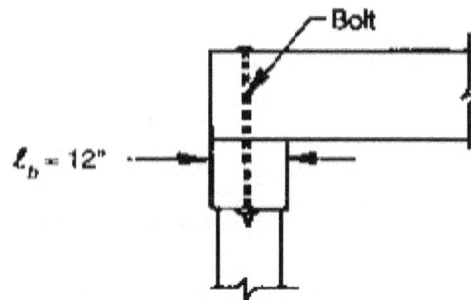

Bearing stress is checked for the bearing length, ℓ_b, of 12 inches. From Table 16-8, the maximum reaction for one wheel line of an HS 20-44 truck on a 37-foot span is 26,920 pounds. The dead load reaction is computed using the bridge length of 38 feet:

$$R_{DL} = \frac{w_{DL}(38 \text{ ft})}{2} = \frac{(585.4 \text{ lb/ft})(38 \text{ ft})}{2} = 11,123 \text{ lb}$$

Bearing stress in compression perpendicular to grain is computed by Equation 9-32:

$$f_{c\perp} = \frac{R_{DL} + R_{LL}}{D_W(\ell_b)} = \frac{11,123 + 26,920}{63(12)} = 50.3 \text{ lb/in}^2$$

$f_{c\perp} = 50.3 \text{ lb/in}^2 < F_{c\perp} = 419 \text{ lb/in}^2$, so the bearing configuration is satisfactory.

Summary

The replacement bridge will consist of a longitudinal stress-laminated lumber deck, 38 feet long, with a span of 37 feet center-to-center of bearings. The bridge will be 26 feet wide and carry two lanes of AASHTO HS 20-44 loading on a roadway width of 24 feet. The lumber laminations will be S1S full-sawn 4-inch by 16-inch Douglas Fir-Larch, visually graded No. 1 or better. The stressing system will consist of galvanized 1-inch-diameter high-strength steel rods conforming to ASTM A722. The rods will be spaced 48 inches on center with end rods 12 inches from the deck end. The rod anchorage system will consist of a 14-inch by 14-inch by 1.25-inch bearing plate and a 4-inch by 6.5-inch by 1.25-inch anchorage plate, manufactured of galvanized A36 steel.

Stresses, deflections, prestressing force and camber are as follows:

$$f_b = 1,666 \text{ lb/in}^2$$

$$F_b' = 1,935 \text{ lb/in}^2$$

$$\Delta_{LL} = 1.19 \text{ in.} = L/373$$

$$\Delta_{DL} = 0.82 \text{ in.}$$

$$\text{Camber} = 2.5 \text{ in.}$$

$$N = 40 \text{ lb/in}^2$$

$$N_i = 100 \text{ lb/in}^2$$

$$F_{ps} = 76,800 \text{ lb}$$

$$f_{c\perp} \text{ at anchorage} = 392 \text{ lb/in}^2$$

$$f_{c\perp} \text{ at bearings} = 50.3 \text{ lb/in}^2$$

$$F_{c\perp}' = 419 \text{ lb/in}^2$$

Example 9.2 - Channel bulkhead anchorage for longitudinal stress-laminated lumber decks

---

Design a channel bulkhead anchorage for the bridge of Example 9-1. The following values apply:

$$F_{ps} = 76,800 \text{ lb}$$

$$F_{c\perp}' = 419 \text{ lb/in}^2$$

Solution

The channel bulkhead configuration is illustrated in Figure 9-20. Design will involve selecting a channel size and a bearing plate size, then checking bearing stress on the lumber laminations.

Determine Channel Size

By Equation 9-25, the channel depth must be within 85 to 100 percent of the deck thickness:

$$0.85(t) = 0.85(16) = 13.4 \text{ in.}$$

$$13.4 \text{ in.} \leq d_c \leq 16 \text{ in.}$$

From Table 9-7, minimum channel section properties for a 16-inch deck are as follows:

$$I = 9.2 \text{ in}^4$$

$$t_w = 0.52 \text{ in.}$$

From the *Steel Construction Manual,* a C15x40 channel is selected with the following properties:

$$d_c = 15 \text{ in.}$$

$$I = 9.23 \text{ in}^4$$

$$t_w = 0.52 \text{ in.}$$

Determine Bearing Plate Size

The minimum bearing plate length is computed by Equation 9-26:

$$L_p \geq \frac{F_{pz}}{d_c F_{cl}'} - 2(t_w) = \frac{76,800}{15(419)} - 2(0.52) = 11.2 \text{ in.}$$

An initial plate size of 12 inches by 10 inches is selected and the length/width ratio is checked by Equation 9-27:

$$L_p = 12 \text{in.}$$

$$W_p = 10 \text{ in.}$$

$$\frac{L_p}{W_p} = \frac{12}{10} = 1.2$$

The ratio is between 1.0 and 2.0 and is acceptable.

Minimum plate thickness is computed by Equation 9-28:

$$t_p \geq \frac{L_p}{12} = \frac{12}{12} = 1 \text{ in.}$$

A plate thickness of 1 inch will be used.

Check Bearing Stress

The effective bearing area of the channel bulkhead is computed by Equation 9-29:

$$A_g = d_c(L_p + 2t_w) = 15[12 + (2)(0.52)] = 196 \text{ in}^2$$

Bearing stress is computed by Equation 9-30:

$$f_{c\perp} = \frac{F_{pt}}{A_g} = \frac{76,800}{196} = 392 \text{ lb/in}^2$$

$f_{c\perp} = 392$ lb/in² $< F_{c\perp}' = 419$ lb/in², so the channel and plate sizes are acceptable.

Summary

The anchorage will consist of a C15x40 steel channel with 12-inch by 10-inch by 1-inch bearing plates.

Example 9-3 - Longitudinal stress-laminated lumber deck; single lane, HS 25-44 loading

---

A single-lane stress-laminated lumber bridge will be built on a remote logging road where the design speed is 5 mph. The bridge will span 22 feet center-to-center of bearings and carry one lane of AASHTO HS 25-44 loading a roadway width of 14 feet. Bridge ends are supported on abutments with a 12-inch length of bearing. Design this bridge, assuming the following:

1. The bridge will include 12-inch by 12-inch timber curbs and a 3-inch thick lumber wearing surface.

2. Because of the low design speed, live load deflection is not a consideration.

3. Lumber laminations are surfaced Red Pine.

4. Butt joints are not required.

5. A bearing plate anchorage configuration will be used.

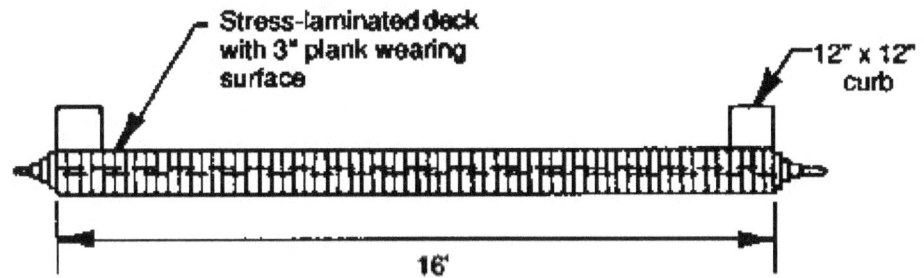

Stress-laminated deck with 3" plank wearing surface

12" x 12" curb

16'

Solution

Define Deck Geometric Requirements and Design Loads

The bridge supports one traffic lane over a 22-foot span. With a roadway width of 14 feet and 12-inch-wide curbs, a bridge width of 16 feet is required. Design loading will be one HS 25-44 wheel line in AASHTO Load Group I.

Select a Species and Grade of Lamination and Compute Allowable Design Values

From the NDS Table 4A, No. 1 Red Pine visually graded to NLGA rules is selected. Tabulated values are as follows:

$F_b = 1,000$ lb/in^2 (single-member use)

$E = 1,300,000$ lb/in^2

$F_{c\perp} = 440$ lb/in^2

Allowable design values are computed using the applicable moisture content factor (C_M) from Table 5-7:

$$F_b' = F_b C_M C_{LS} = 1,000(0.86)(1.50) = 1,290 \text{ lb/in}^2$$

$$E' = E C_M = 1,300,000(0.97) = 1,261,000 \text{ lb/in}^2$$

$$F_{c\perp}' = F_{c\perp} C_M = 440(0.67) = 294 \text{ lb/in}^2$$

Determine the Preliminary Lamination Layout

Lumber laminations will be continuous over the bridge span. Butt joints are not required.

Compute the Transverse Moduli for the Stress-Laminated System

Values of the transverse bending modulus (E_{TS}) and transverse shear modulus (G_{TS}) are computed by Equations 9-4 and 9-5:

$$E_{TS} = 0.013E' = 0.013(1,261,000) = 16,393 \text{ lb/in}^2$$

$$G_{TS} = 0.03E' = 0.03(1,261,000) = 37,830 \text{ lb/in}^2$$

Compute Maximum Live Load Moment

The maximum live load moment for one wheel line of an HS 25-44 truck on a 22-foot simple span is obtained from Table 16-8:

$$M_{LL} = 110,000 \text{ ft-lb}$$

Compute Wheel Load Distribution Width

Values of α **and** θ are computed using Equations 9-6 and 9-7, respectively. From Table 9-1, $C_s = 1.0$:

$$\alpha = \frac{2G_{TS}}{\sqrt{E'(C_s)(E_{TS})}} = \frac{2(37,830)}{\sqrt{1,261,000(1.0)(16,393)}} = 0.53$$

$$\theta = \frac{b}{2L}\left[\frac{E'(C_s)}{E_{TS}}\right]^{0.25} = \frac{16}{2(22)}\left[\frac{1,261,000(1.0)}{16,393}\right]^{0.25} = 1.08$$

The distribution width, D_w is obtained from Figure 9-15 using the curves for bridges with one traffic lane:

$$D_w = 62 \text{ in.}$$

Estimate Deck Thickness and Compute Effective Section Properties

An initial deck thickness of 11-1/4 inches (12 inches nominal) is selected from Table 9-2. Although the table is based on HS 20-44 loading, for this span it should be reasonably accurate for HS 25-44 loads. Effective deck section properties are computed by Equations 9-8 and 9-9:

$$S = \frac{D_w(C_s)(t)^2}{6} = \frac{62(1.0)(11.25)^2}{6} = 1,308 \text{ in}^3$$

$$I = \frac{D_w(C_s)(t)^3}{12} = \frac{62(1.0)(11.25)^3}{12} = 7,356 \text{ in}^4$$

Compute Deck Dead Load and Dead Load Moment

From Table 9-3, the dead load of an 11.25inch deck with a 3-inch lumber wearing surface is 59.4 lb/ft^2 Based on a unit weight for wood of 50 lb/ft^3, curb dead load is 50 lb/ft. The curb dead load is increased by an estimated 10 lb/ft for the prestressing system, and is assumed to be uniformly distributed across the deck width:

$$DL = 59.4 \text{ lb/ft}^2 + \frac{2(50 \text{ lb/ft} + 10 \text{ lb/ft})}{16 \text{ ft}} = 66.9 \text{ lb/ft}^2$$

For the distribution width of 62 in.,

$$w_{DL} = \frac{62 \text{ in.}}{12 \text{ in/ft}}\left(66.9 \text{ lb/ft}^2\right) = 345.7 \text{ lb/ft}$$

$$M_{DL} = \frac{w_{DL}L^2}{8} = \frac{345.7(22)^2}{8} = 20,915 \text{ ft-lb}$$

Compute Bending Stress

Bending stress is computed by Equation 9-11:

$$f_b = \frac{M}{S} = \frac{(110,000 + 20,915)(12 \text{ in/ft})}{1,308} = 1,201 \text{ lb/in}^2$$

$f_b = 1,201 \text{ lb/in}^2 < F_b' = 1,290 \text{ lb/in}^2$, so bending stress is acceptable.

Check Live Load Deflection

Although live load deflection is not a controlling consideration for design, it will be computed for reference. From Table 16-8, the deflection coefficient for one wheel line of an HS 25-44 truck on a 22-foot simple span is $7.99 \times 10^9 \text{lb-in}^2$. Live load deflection is computed using 133 percent of the effective deck moment of inertia:

$$\Delta_{LL} = \frac{7.99 \times 10^9}{E'(1.33)(I)} = \frac{7.99 \times 10^9}{1,261,000(1.33)(7,356)} = 0.65 \text{ in.}$$

Determine the Required Prestress Level

Using the previously computed values of α and θ, M_r is obtained for HS 20-44 loading on a two-lane bridge from Figure 9-16:

HS 20-44 $M_r = 480$ in-lb/in

Because this design is for HS 25-44 loading, the value of M_r from Figure 9-16 must be multiplied by the ratio of the design wheel load (20,000 pounds from Example 6-1) to the HS 20-44 wheel load (16,000 pounds):

$$\frac{20,000 \text{ lb}}{16,000 \text{ lb}} = 1.25$$

$M_r = 1.25 \ (480 \text{ in-lb/in}) = 600 \text{ in-lb/in}$

The variable β is computed by Equation 9-14:

$$\beta = \pi\left(\frac{b}{L}\right)\sqrt{\frac{E'(C_b)}{2G_{TS}}} = 3.14\left(\frac{16}{22}\right)\sqrt{\frac{1,261,000(1.0)}{2(37,830)}} = 9.32$$

By interpolation and extrapolation of Figure 9-17,

HS 20-44 $V_r = 60$ lb/in

For HS 25-44 loading,

$$V_r = 1.25(60 \text{ lb/in}) = 75 \text{ lb/in}.$$

The minimum level of compressive prestress is computed is computed by Equation 9-15. Based on transverse bending,

$$N = \frac{6M_T}{t^2} = \frac{6(600)}{(11.25)^2} = 28.4 \text{ lb/in}^2$$

Based on transverse shear,

$$N = \frac{1.5V_T}{t\mu} = \frac{1.5(75)}{11.25(0.35)} = 28.6 \text{ lb/in}^2$$

Both values are less than the minimum 40 lb/in², so $N = 40$ lb/in² will control. By Equation 9-17,

$$N_i = 2.5N = 2.5(40) = 100 \text{ lb/in}^2$$

Determine Spacing and Size of Prestressing Rods and the Required Prestressing Force

From Table 9-5 for an 11.25-inch deck, two rod diameters are feasible; 5/8-inch-diameter rods at a spacing of 16 to 26 inches, or 1-inch-diameter rods at a spacing of 47 to 79 inches. From Figure 9-18, maximum rod spacing is limited to approximately 58 inches.

It is anticipated that 1-inch-diameter rods at the minimum 47-inch spacing will require an excessive bearing plate size. Therefore, 5/8-inch-diameter rods will be used. For a bridge length of 23 feet (22- foot span on 1-foot-wide sills), rods will be spaced 24 inches on-center with the end rods spaced at 12 inches and 18 inches:

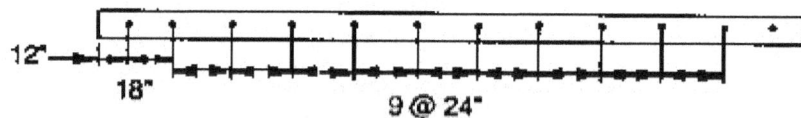

From Table 9-4 for a 5/8-inch-diameter rod, $A_s = 0.28$ in². The minimum required rod area and the steel/wood ratio are checked by Equation 9-18:

$$\frac{N_i S_p t}{0.70 f_{pu}} = \frac{100(24)(11.25)}{0.70(150,000)} = 0.26 \text{ in}^2 < 0.28 \text{ in}^2$$

$$\frac{A_S}{S_p(t)} = \frac{0.28}{24(11.25)} = 0.0010 < 0.0016$$

The prestressing force required in each rod, F_{pe} is computed by Equation 9-19:

$$F_{pe} = N_i(S_p)(t) = 100(24)(11.25) = 27,000 \text{ lb}$$

Design Anchorage System

The minimum bearing plate area is computed by Equation 9-20:

$$A_p = \frac{F_{pe}}{F_{c\perp}{}'} = \frac{27,000}{294} = 92 \text{ in}^2$$

For the 11.25-inch-thick deck, a plate depth, W_p of 10 inches is chosen. The minimum required plate length is computed by dividing the plate area by the plate width:

$$L_p = \frac{A_p}{W_p} = \frac{92}{10} = 9.2 \text{ in.}$$

A 10-inch-square plate will be used, and

$W_p = 10$ in.

$L_p = 10$ in.

$A_p = 100$ in

For the square plate, the ratio of the bearing plate length to width is acceptable by Equation 9-21, and bearing stress in compression perpendicular to grain is computed by Equation 9-22:

$$f_{c\perp} = \frac{F_{pe}}{A_p} = \frac{27,000}{100} = 270 \text{ lb/in}^2$$

$f_{c\perp} = 270$ lb/in^2 < $F_{c\perp}{}'$ = 294 lb/in^2, so bearing stress is acceptable.

From Table 9-6, an anchorage plate size of 3 inches by 3 inches by 0.75 inch is selected and k values are computed by Equation 9-24:

$W_A = 3$ in.

$L_A = 3$ in.

$t_A = 0.75$ in.

$$k = \frac{W_P - W_A}{2} = \frac{L_P - L_A}{2} = \frac{10 - 3}{2} = 3.5 \text{ in.}$$

The required bearing plate thickness for an A36 steel plate is computed by Equation 9-23:

$$t_P = \sqrt{\frac{3\left(f_{c\perp}\right)\left(k^2\right)}{F_b}} = \sqrt{\frac{3\,(270)(3.5)^2}{0.55(36{,}000)}} = 0.71 \text{ in.}$$

A plate thickness of 0.75 inch will be used.

Determine the Support Configuration and Check Bearing Stress

Superstructure support is provided by a bearing length, ℓ_b, of 12 inches. From Table 16-8, the maximum reaction for one wheel line of an HS 25-44 truck on a 22-foot span is 27,270 pounds. The dead load reaction is computed using the bridge length of 23 feet:

$$R_{DL} = \frac{w_{DL}(23 \text{ ft})}{2} = \frac{\left(345.7 \text{ lb/ft}\right)(23 \text{ ft})}{2} = 3{,}976 \text{ lb}$$

Bearing stress in compression perpendicular to gram is computed by Equation 9-32:

$$f_{c\perp} = \frac{R_{DL} + R_{LL}}{D_W(\ell_b)} = \frac{3{,}976 + 27{,}270}{62(12)} = 42 \text{ lb/in}^2$$

$f_{c\perp} = 42 \text{ lb/in}^2 < F_{c\perp}{}' = 294 \text{ lb/in}^2$, so the bearing configuration is satisfactory.

Summary

The bridge will consist of a longitudinal stress-laminated lumber deck, 23 feet long, with a span of 22 feet center to center of bearings. The bridge will be 16 feet wide and carry one lane of AASHTO HS 25-44 loading on a roadway width of 14 feet. The lumber laminations will be S4S 2-inch by 12-inch Red Pine, visually graded No. 1 or better to NLGA rules. The stressing system will consist of galvanized 5/8-inch-diameter high-strength steel rods conforming to ASTM A722, spaced 24 inches on-center. The rod anchorage system will consist of a 10-inch by 10-inch by 0.75-inch bearing plate and a 3-inch by 3-inch by 0.75-inch anchorage plate, manufactured of galvanized A36 steel.

Stresses, deflections, prestressing force and camber are as follows:

$$f_b = 1{,}201 \text{ lb/in}^2$$

$$F_b{}' = 1{,}290 \text{ lb/in}^2$$

$$\Delta_{LL} = 0.65 \text{ in.}$$

$$N = 40 \text{ lb/in}^2$$

$$N_i = 100 \text{ lb/in}^2$$

$$F_{ps} = 27{,}000 \text{ lb}$$

$$f_{c\perp} \text{ at anchorage} = 270 \text{ lb/in}^2$$

$$f_{c\perp} \text{ at bearings} = 42 \text{ lb/in}^2$$

$$F_{c\perp}' = 294 \text{ lb/in}^2$$

9.5 SELECTED REFERENCES

1. American Association of State Highway and Transportation Officials. 1983. Standard specifications for highway bridges. 13th ed. Washington, DC: American Association of State Highway and Transportation Officials. 394 p.

2. American Association of State Highway and Transportation Officials. 1983. Manual for maintenance inspection of bridges. Washington, DC: American Association of State Highway and Transportation Officials. 50 p.

3. American Institute of Steel Construction. 1980. Manual of steel construction. Chicago, IL: American Institute of Steel Construction. [1000 p.]

4. American Society for Testing and Materials. [Current edition]. Standard specification for uncoated high-strength steel bar for prestressing concrete. ASTM A 722. Philadelphia, PA: ASTM.

5. Batchelor, B.; Van Dalen, K.; Hachbom, A.; Lee, E. 1979. Structural characteristics of transversely prestressed laminated wood bridge decks. Report 8305. Kingston, ON, Can.: Queen's University. 127 p.

6. Batchelor, B.; Van Dalen, K.; Morrison, T.; Taylor, R.J. 1981. Structural characteristics of red pine and hem-fir in prestressed laminated wood bridge decks. Report 23122. Kingston, ON, Can.: Queen's University. 165 p.

7. Batchelor, B.; Van Dalen, K.; Taylor, R.J. 1982. Use of oil-borne creosoted hem-fir in prestressed laminated wood bridge decks. Report 23142. Kingston, ON, Can.: Queen's University. 60 p.

8. Csagoly, P.F.; Taylor, R.J. 1979. A development program for wood highway bridges. 79-SRR-7. Downsview, ON, Can.: Ministry of Transportation and Communications. 57 p.

9. Dimakis, A.G.; Oliva, M.G. 1987. Behavior of post-tensioned wood bridge decks: full-scale testing, analytic correlation. Rpt. 87-2. Madison, WI: University of Wisconsin, Department of Civil and Environmental Engineering. 76 p.

10. Ministry of Transportation and Communications. 1983. Ontario highway bridge design code. Downsview, ON, Can.: Ministry of Transportation and Communications. 357 p.

11. Ministry of Transportation and Communications. 1983. Ontario highway bridge design code commentary. Downsview, ON, Can.: Ministry of Transportation and Communications. 279 p.

12. National Forest Products Association. 1986. Design values for wood construction. A supplement to the national design specification for wood construction. Washington, DC: National Forest Products Association. 34 p.

13. National Forest Products Association. 1986. National design specification for wood construction. Washington, DC: National Forest Products Association. 87 p.

14. Nowak, A.S.; Taylor, R.J. 1986. Ultimate strength of timber deck bridges. In: Trans. Res. Rec. 1053. Washington, DC: National Academy of Sciences, National Research Council, Transportation Research Board: 26-30.

15. Oliva, M.G.; Dimakis, A. 1986. Behavior of a prestressed timber highway bridge. Madison, WI: University of Wisconsin, Department of Civil and Environmental Engineering. 32 p.

16. Oliva, M.G.; Dimakis, A. 1988. Behavior of stress-laminated timber highway bridges. Journal of Structural Engineering 114(8): 1850-1869.

17. Oliva, M.G.; Dimakis, A.G.; Tuomi, R.L. 1985. Interim report: behavior of stressed-wood deck bridges. Report 85-1/A. Madison, WI: University of Wisconsin, College of Engineering, Structures and Materials Test Laboratory. 40 p.

18. Oliva, M.G.; Tuomi, R.L.; Dimakis, A.G. 1986. New ideas for timber bridges. In: Trans. Res. Rec. 1053. Washington, DC: National Academy of Sciences, National Research Council, Transportation Research Board: 59-64.

19. Oliva, M.G.; Lyang, J. 1988. Behavior of parallel chord trusses for stress-lam wood bridges. Rpt. 87-3. Madison, WI: University of Wisconsin, Department of Civil and Environmental Engineering. 39 p.

20. Raju, P.R.; GangaRao, V.S.H. 1989. Wheel load distributions on highway bridges. CFC Rpt. No. 89-100. Morgantown, WV: West Virginia University, College of Engineering, Constructed Facilities Center. 93 p.

21. Taylor, R.J. 1983. Appendix to SRR-83-02. Downsview, ON, Can.: Ministry Of Transportation and Communications. 108 p.

22. Taylor, R.J. 1983. Design of prestressed wood bridges using the Ontario highway bridge design code. SRR-83-03. Downsview, ON, Can.: Ministry Of Transportation and Communications. 30 p.

23. Taylor, R.J. 1984. Design of wood bridges using the Ontario highway bridge design code. SRR-83-02 revised. Downsview, ON, Can.: Ministry Of Transportation and Communications. 24 p.

24. Taylor, R.J. 1984. Prestressed wood applications. Paper presented at Western Area Association of State Highway and Transportation Officials meeting. Rapid City, SD. 12 p.

25. Taylor, R.J. 1983. Wood bridge calibration study for the Ontario highway bridge design code. SRR-83-04. Downsview, ON, Can.: Ministry Of Transportation and Communications. 37 p.

26. Taylor, R.J.; Batchelor, B.; Van Dalen, K. 1983. Prestressed wood bridges. SRR-83-01. Downsview, ON, Can.: Ministry of Transportation and Communications. 15 p.

27. Taylor, R.J.; Csagoly, P.F. 1979. Transverse post-tensioning of longitudinally laminated timber bridge decks. Downsview, ON, Can.: Ministry of Transportation and Communications. 16 p.

28. Taylor, R.J.; Walsh, H. 1984. A prototype prestressed wood bridge. SRR-83-07. Downsview, ON, Can.: Ministry Of Transportation and Communications. 75 p.

29. Sexsmith, R.G.; Boyle, P.D.; Rovner, B.; Abbot, R.A. 1979. Load sharing in vertically laminated post-tensioned bridge decking. Tech. Rpt. No. 6. Vancover, BC, Can.: Forintek Canada Corp., Western Forest Products Laboratory. 18 p.

CHAPTER 10

RAIL SYSTEMS FOR TIMBER DECKS

10.1 INTRODUCTION

Railing is provided on bridges for the protection of vehicles and pedestrians that use the structure. It is normally placed along bridge sides to prevent users from going off the edge, but railing is also used to separate vehicle from pedestrian traffic and to protect exposed structural components. The four basic types of bridge railing are vehicular, pedestrian, bicycle, and combination railing (Figure 10-1). Vehicular railing is placed along roadway edges to safely contain and redirect impacting vehicles. Pedestrian and bicycle railings are installed on the outside edge of sidewalks intended for foot or bicycle traffic. Combination railing is a combination of vehicular and pedestrian or bicycle railing placed primarily to separate vehicle traffic from pedestrian or bicycle traffic.

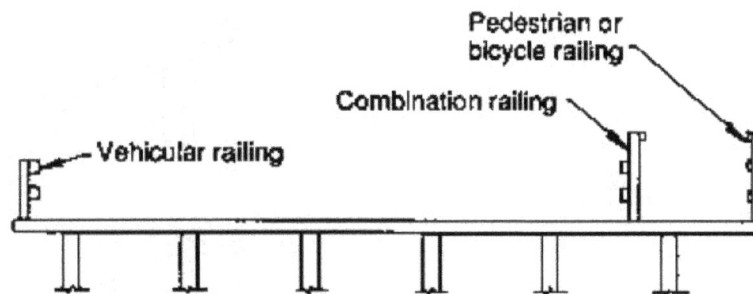

Figure 10-1. - Types of timber bridge railing.

All types of bridge railing must be strong enough to contain the intended traffic, be resistant to damage, be economical in construction and maintenance, and have a pleasing functional appearance. Specific design requirements for railing geometry and loads are given in AASHTO.[3] These requirements represent the minimum criteria for railing design, but allow the designer moderate flexibility in determining the most appropriate configuration and materials for a specific structure. This chapter discusses AASHTO railing requirements, including design considerations and recommended criteria for timber decks.

10.2 VEHICULAR RAILING

The purpose of vehicular railing is to safely restrain an impacting vehicle. In addition, consideration must be given to the protection of the occupants in the vehicles, the protection of other vehicles or pedestrians near the collision, the effects of railing impact on the structure, and the railing appearance. Although each of these considerations may be addressed somewhat independently, they all interact to determine the performance of the railing system.

DESIGN REQUIREMENTS
Vehicular railing systems for timber bridges normally consist of horizontal rails mounted on vertical posts, solid timber parapets, or a combination of the two (Figure 10-2). The design requirements for these systems are given in AASHTO as geometric requirements for railing height, spacing and alignment, and static load requirements for rails, posts, and parapets. Although actual loads are dynamic in nature, the use of static loading simplifies design and has been used by AASHTO since 1964. Materials for vehicular railing may be timber, metal, or concrete; however, metal materials must have a minimum 10-percent tested elongation (AASHTO 2.7.1.1.2). Any railing configuration may be used provided it complies with the minimum criteria stated in AASHTO or has been verified by full-scale crash testing.

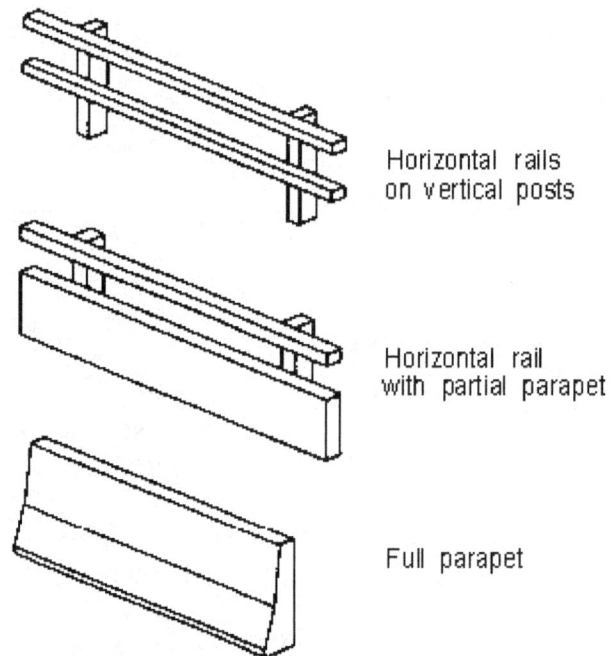

Horizontal rails
on vertical posts

Horizontal rail
with partial parapet

Full parapet

Figure 10-2. - Typical configurations for vehicular railing used on timber bridges.

Current AASHTO railing requirements (through 1987 interim) are independent of the service level or type of structure and are based on static load design criteria. The same requirements apply to all bridges from single-lane bridges on dirt roads to multiple-span structures on interstate highways. These criteria have been under criticism for several years on the premise that they represent a compromise approach that does not accurately reflect loading and safety requirements for all bridges. For heavily traveled highways, the static load criteria may be insufficient, while use of the same criteria on low-volume rural roads could result in overly conservative designs. There have been several proposals for a service-level approach to railing design that would vary requirements for structures based on the functional classification of the roadway, bridge geometry, and the type, speed, and volume of traffic.[9] There is also a movement to eliminate static load requirements and require full-scale crash tests of all vehicular railing systems (Figure 10-3). Although AASHTO does not currently require full-scale crash tests for railing acceptance, guide specifications for railing crash testing are being prepared by AASHTO and will be available in the near future. It is expected that full-scale crash testing will eventually be required for all vehicular railing systems. Current design requirements, based on AASHTO geometric requirements and static load criteria, are discussed in the following paragraphs.

Geometric Requirements

Vehicular railing must be positioned to safely contain an impacting vehicle without allowing it to pass over, under, or through the rail elements. In addition, it must be free of features that may catch on the vehicle or cause it to overturn or decelerate too rapidly. To ensure a minimum level of safety and uniformity for vehicular railing, the following minimum geometric requirements are given in AASHTO (Figure 10-4).

1. **Reference Surface.** Vertical requirements for railing height and spacing are measured relative to a roadway reference surface defined as the top of the roadway surface, the top of the future overlay if resurfacing of the roadway is anticipated, or the top of the curb when the curb projects more than 9 inches beyond the traffic face of the railing (AASHTO 2.7.1.2.1). When the reference surface is a future overlay, minimum heights are measured from the overlay elevation while maximum heights are measured from the original roadway elevation.

2. **Railing Height.** The height of vehicular railing shall not be less than 2 feet 3 inches above the reference surface (AASHTO 2.7.1.2.2). The height of parapets designed with sloping traffic faces intended to allow vehicles to ride up them under low-angle contacts shall be at least 2 feet 8 inches above the reference surface.

3. **Railing Placement.** The maximum clear opening below the bottom rail shall not exceed 17 inches. The maximum clear opening between succeeding rails shall not exceed 15 inches (AASHTO 2.7.1.2.4). The lower rail element should consist of a rail centered 15 to 20 inches above the reference surface, or a parapet projecting a minimum of 18 inches above the reference surface (AASHTO 2.7.1.2.3).

10-3

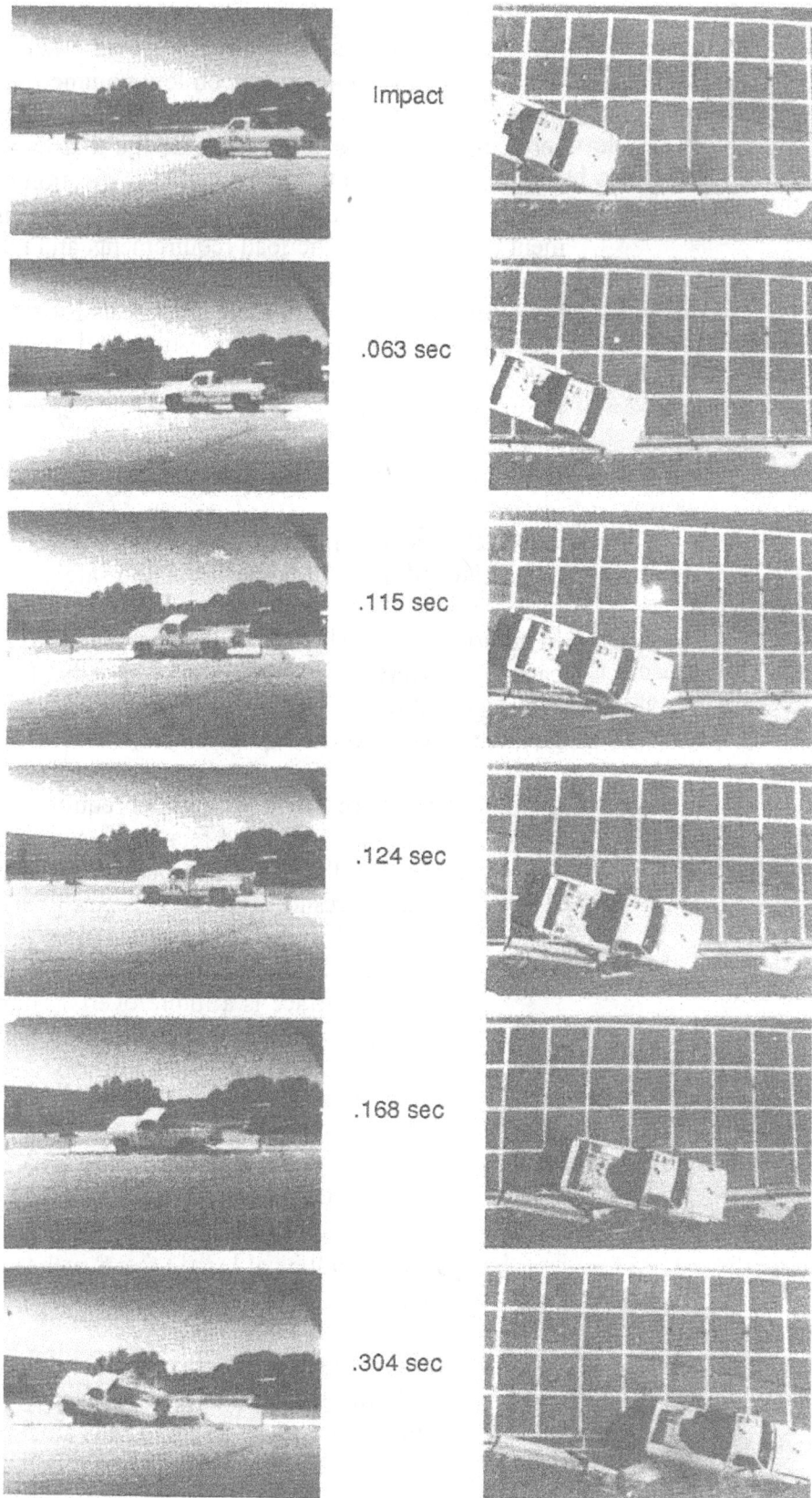

Impact

.063 sec

.115 sec

.124 sec

.168 sec

.304 sec

Figure 10-3. - Partial sequence of a full-scale crash test of vehicular railing (photos courtesy of Dr. Edward Post, University of Nebraska at Lincoln).

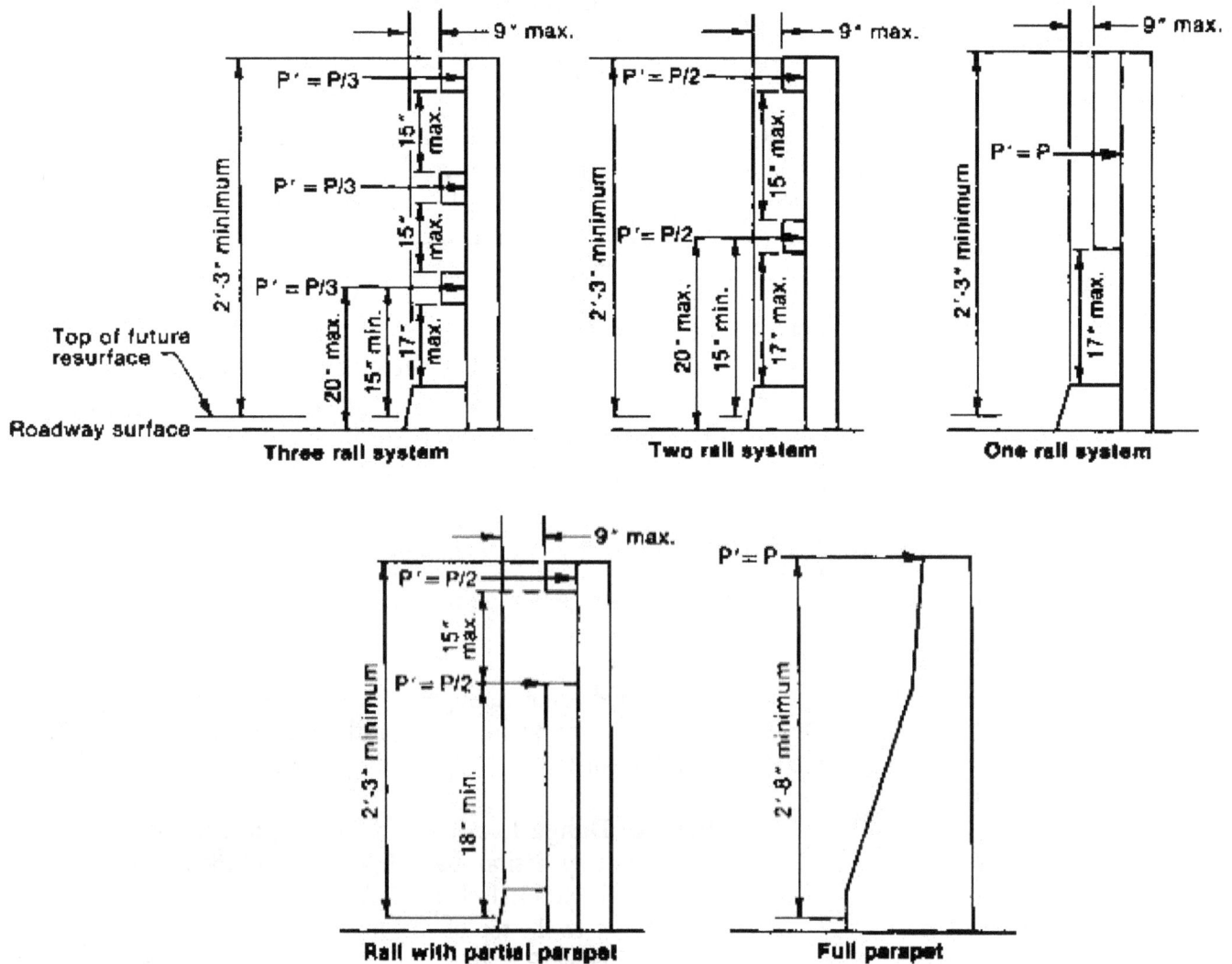

Three rail system

Two rail system

One rail system

Rail with partial parapet

Full parapet

Notes:

1. Rail and post shapes are illustrative only. Any material or combination of materials may be used in any configuration provided minimum AASHTO requirements are met.
2. Refer to AASHTO for illustrations of other railing configurations and for design requirements when the curb projects more than 9 inches from the traffic face of railing.
3. Additional post and rail loading requirements are illustrated in Figures 10-5 and 10-6.

Figure 10-4. - AASHTO requirements for vehicular railing geometry and outward transverse static loads when there is no curb or the curb projects 9 inches or less from the traffic face of railing (adapted from AASHTO Figure 2.7.4B). © 1983. Used by permission.

4. **Vertical Alignment.** The traffic face of all rails must be within 1 inch of a vertical plane through the traffic face of the rail closest to traffic (AASHTO 2.7.1.2.5).

In addition to the above requirements, vehicular railing should provide a smooth, continuous traffic face with posts set back from the rail face. Protrusions or depressions at rail joints are acceptable provided their thickness or depth is no greater than the wall thickness of the rail members or 3/8 inch, whichever is less (AASHTO 2.7.1.1.4).

Loading Requirements

AASHTO specifications state that the primary purpose of vehicular railing is to contain the average vehicle using the structure. Although the average vehicle is not defined in the specifications, it is generally considered to be a full-size domestic passenger car weighing approximately 4,500 pounds. The static design loads are intended to safely contain the design vehicle at an impact angle of approximately 25 degrees at a speed of 60 miles per hour (mph). Railing configurations that have been successfully tested by full-scale impact tests are exempt from these static load requirements (AASHTO 2.7.1.3.7).

Design loads for vehicular railing are based on a minimum highway design load that is distributed to post, rail, and parapet elements. Requirements for load magnitude and distribution are as follows:

1. **Highway Design Load.** The basic design load for posts and rails is the highway design load, P. The magnitude of P depends on the height of the top rail element above the reference surface. When the distance to the top of the upper rail is less than or equal to 2 feet 9 inches, $P = 10,000$ pounds (AASHTO Figure 2.7.4B). When the height of the top rail exceeds 2 feet 9 inches, P equals 10,000 pounds times the adjustment factor, C, as computed by

$$C = 1 + \left(\frac{h - 33}{18} \right) \quad \text{and} \quad C \geq 1.0 \quad (10\text{-}1)$$

where h is the height of the top of the top rail element above the reference surface, in inches.

2. **Post Loads.** The highway design load, P, is distributed to each rail post as an outward transverse load. The distribution of P along the post height depends on the number and position of rail elements. When the railing configuration complies with minimum AASHTO geometric requirements, P is distributed equally at the center of each rail, and the distributed outward transverse post load, P' equals P, $P/2$, or $P/3$, depending on the railing

configuration (Figure 10-4). Rails with a traffic face more than 1 inch behind the vertical plane through the face of the rail closest to traffic, or centered less than 15 inches above the roadway reference surface, are not considered as traffic rails for distributing P (AASHTO 2.7.1.3.2). However, they may be used in determining the maximum vertical clear opening, provided they are designed for a transverse loading equal to that applied to an adjacent traffic rail or $P/2$, whichever is less (see the following discussions on rail loads).

In addition to the outward transverse loads, rail posts must also be designed to resist longitudinal loads and inward transverse loads (Figure 10-5). A longitudinal post load equal to $P'/2$ is applied simultaneously with the outward transverse load and is divided among not more than four posts in a continuous rail length (AASHTO 2.7.1.3.3). Posts must be designed to resist an independently applied inward transverse load equal to $P'/4$.

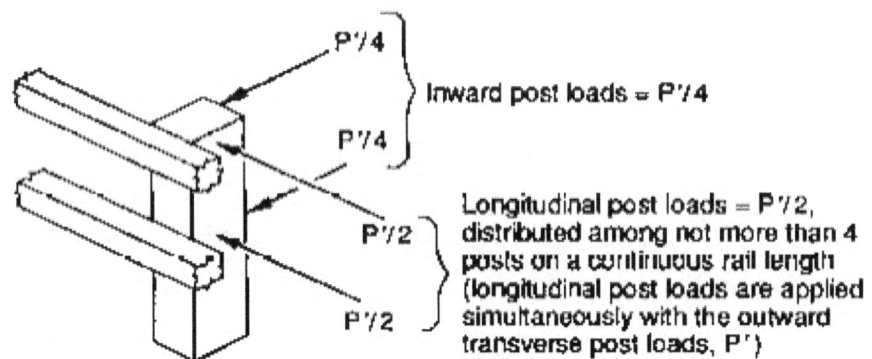

P' = the outward transverse post load applied to the post at each rail location (P, P/2, or P/3 from Figure 10-4)

Figure 10-5. - AASHTO requirements for longitudinal and inward transverse post loads, illustrated for a two-rail system.

3. **Rail Loads.** Rails are designed for a moment from an outward transverse load applied at the center of the panel and at the posts, equal to $P'L/6$ where L is the post spacing and P' is the portion of the outward transverse post load $(P, P/2,$ or $P/3)$ applied to the post at each rail location (AASHTO 2.7.1.3.5). The rail attachment to the post must be designed to resist a vertical load, applied alternately upward or downward, equal to $P'/4$ (AASHTO 2.7.1.3.4). The rail attachment must be designed to resist an inward transverse load equal to $P'/4$ (Figure 10-6).

10-7

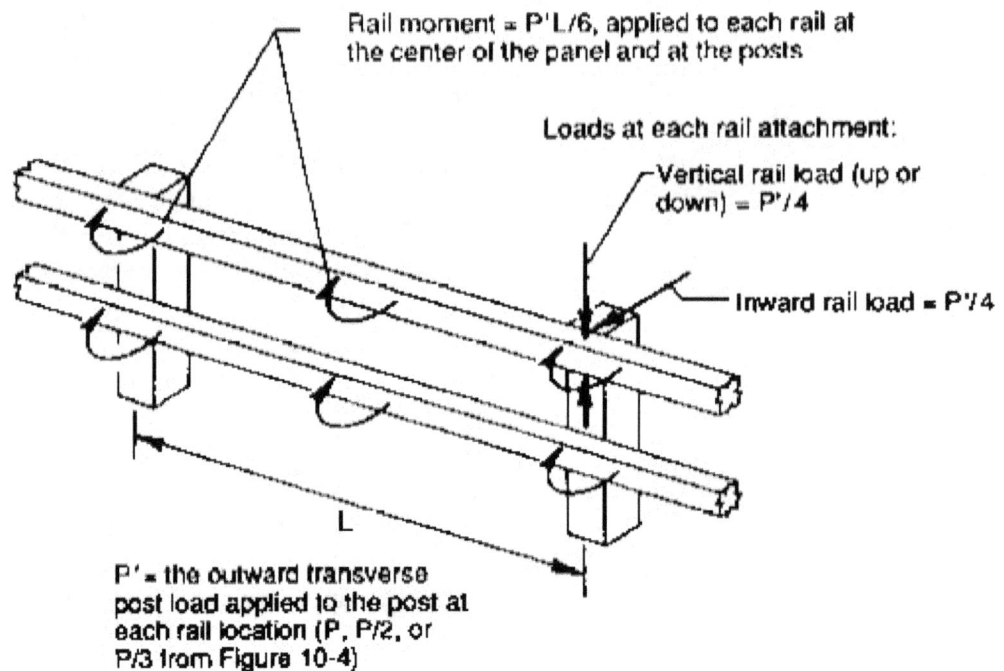

Rail moment = P'L/6, applied to each rail at
the center of the panel and at the posts

Loads at each rail attachment:

Vertical rail load (up or
down) = P'/4

Inward rail load = P'/4

L

P' = the outward transverse
post load applied to the post at
each rail location (P, P/2, or
P/3 from Figure 10-4)

Figure 10-6. - AASHTO requirements for rail loading, illustrated for a two-rail system.

4. **Parapet Loads.** The highway design load, P, is applied as an
outward transverse load along the top of parapets. The load is
assumed to act at any location along the parapet and is distributed
over a longitudinal length of 5 feet (AASHTO 2.7.1.3.6).

Although AASHTO requires that all vehicular railing be designed for a
minimum highway design load of $P = 10,000$ pounds, some agencies
have reduced this loading for certain types of bridges. For example, the
USDA Forest Service uses 50 percent of the AASHTO loading, or
$P = 5,000$ pounds, for all single-lane low-volume bridges with a design
speed less than 45 mph and a probable vehicle-railing impact angle less
than or equal to 15 degrees. Many counties also follow reduced AASHTO
loading criteria on similar low-volume roads. It is expected that AASHTO
will eventually recognize a service level design approach that will allow
lower railing loads for certain types of bridges.

DESIGN GUIDELINES

Within the railing design requirements given in AASHTO, many railing
configurations can be used on timber bridges. The system that is most
appropriate for a specific bridge depends on factors such as the deck
configuration and material as well as the economy and availability of
railing materials. Some of the design considerations for rail elements,
posts, and parapets are discussed in the following paragraphs.

10-8

Rails

Railing selection depends on the post spacing and the aesthetic qualities desired for the structure. Because AASHTO rail loads are directly related to the post spacing, the required rail load increases as post spacing increases. In most cases, the choice for configuration is between a one-rail or a two-rail system. When one-rail designs are used, the rail must resist all applied loads and be deep enough to meet AASHTO geometric requirements for the clear opening below the rail. If a curb is not provided, the rail must be a minimum of 10 inches deep, assuming minimum rail height of 2 feet 3 inches. This depth can be decreased when a curb reduces the clear opening below the rail. Although not as common as single-rail systems, two-rail designs are widely used on timber bridges. Two-rail systems are generally more expensive than one-rail systems, but loads are equally distributed to each rail element, reducing the individual rail loads to 50 percent of that required for a single rail. In addition, the load distribution to two rails reduces the reaction at the post attachment, which is normally the most critical railing design consideration on timber decks.

The three types of vehicular rails most commonly used on timber bridges are timber, semirigid steel, and rigid steel (Figure 10-7). Each of these railing types is discussed in subsequent paragraphs, and approximate maximum post spacings for various configurations are shown in Table 10-1.

1. **Timber Rails.** Timber rails constructed of sawn lumber or glulam are widely used on timber bridges because of their good energy-absorbing properties and the pleasing appearance of wood. Lumber rails for one-rail configurations are generally 4 to 6 inches thick and 10 to 12 inches deep. For two-rail lumber configurations, 6- by 8-inch members are typically used with the 8-inch dimension horizontal. Glulam rails are normally 10-3/4 inches deep for single rails and 6-3/4 inches deep for double rails. Glulam rails are preferable to sawn lumber in most applications because they can be manufactured in longer lengths (up to the bridge length) and provide better dimensional stability in service.

2. **Semirigid Steel Rails.** Semirigid steel rails are cold-formed standard sections including the W-beam or Thrie-beam (Table 10-2). These rails must conform to the requirements of AASHTO M 180 and are fabricated in standard 12-foot-6-inch and 25-foot-0-inch sections. The rails are available in two thicknesses: Class A, which is 0.105 inch thick (12 gage), and Class B, which is 0.135 inch thick (10 gage). Sections are available in the following four types, depending on the surface finish of the rail: Type 1, zinc coated, 1.80 oz/ft^2; Type 2, zinc coated, 3.60 oz/ft^2; Type 3, uncoated, to be painted; and Type 4, corrosion-resistant steel (weathering steel).

Figure 10-7. - Types of vehicular rails commonly used on timber bridges. (A) Glulam beams. (B) Sawn lumber beam.

Figure 10-7. - Types of vehicular rails commonly used on timber bridges (continued). (C) Semirigid steel W-beam. (D) Rigid-steel structural tubes.

Table 10-1. - Approximate maximum post spacing for vehicular railing designed to full AASHTO static load criteria.

Material	Rails (number)	Rail description[a]	Approximate maximum post spacing (ft)
Glulam[b]	1	6 in. x 10-3/4 in. (vertical laminations)	7.2
	1	5-1/8 in. x 12 in. (horizontal laminations)	6.9
	1	6-3/4 in. x 12 in. (horizontal laminations)	11.6
	2	6 in. x 6-3/4 in. (vertical laminations)	9.1
Sawn lumber[c]	1	4 in. x 10 in. Grade Dense No.1	2.7
	1	4 in. x 12 in. Grade No.1	2.7
	1	4 in. x 12 in. Grade Dense No.1	3.2
	1	6 in. x 10 in. Grade No.1	5.3
	1	6 in. x 10 in. Grade Dense No.1	6.1
	1	6 in. x 12 in. Grade No.1	6.6
	1	6 in. x 12 in. Grade Dense No.1	7.6
	2	6 in. x 8 in. Grade No.1	8.4
	2	6 in. x 8 in. Grade Dense No. 1	9.7
Flexible steel[d]	1	10 gage W-beam	2.6
	1	12 gage W-beam	2.1
	1	10 gage W-beam, double layer	5.2
	1	12 gage W-beam, double layer	4.1
	1	10 gage Thrie-beam	4.3
	1	12 gage Thrie-beam	3.2
	1	10 gage Thrie-beam, double layer	8.6
	1	12 gage Thrie-beam, double layer	6.5
	1	10 gage W-beam, back to back	8.7
	1	12 gage W-beam, back to back	6.8
	1	10 gage Thrie-beam, back to back	13.9
	1	12 gage Thrie-beam, back to back	10.7
Semirigid steel[e]	2	4 in. x 4 in. x 1/4 in. structural tube	8.9
	1	5 in. x 5 in. x 3/8 in. structural tube	9.8
	2	6 in. x 3 in. x 3/8 in. structural tube	11.2
	2	6 in. x 3 in. x 1/4 in. structural tube	12.9

[a] When dimensions are given, the first dimension denotes the rail width (horizontal) dimension.

[b] Combination 2 Western species adjusted by C_M, C_D, and C_F, when applicable. F_b = 1800 lb/in^2 for horizontal laminations and 1,700 lb/in^2 for vertical laminations per AITC 117—Design.[4]

[c] Based on dressed lumber sizes for the Douglas-fir grades shown. 4-inch lumber widths are in the Joists and Planks size classification. 6-in. lumber is in the Beams and Stringers size classification graded to WWPA rules. Tabulated bending stresses from the NDS[20] were adjusted by C_M, C_D, and flatwise use factors. when applicable.

[d] F_y = 50,000 lb/in^2 per AASHTO M180; F_b = 0.60F_y per AASHTO 2.7.4.2.[3]

[e] F_y = 36,000 lb/in^2 per ASTM A501; F_b = 0.60F_y per AASHTO 2.7.4.2.[3]

Loads are based on full AASHTO railing requirements for the materials specified. Values are for estimating purposes only and should be verified for specific design applications.

Table 10-2. - Section properties of W-beam and Thrie-beam guardrail.

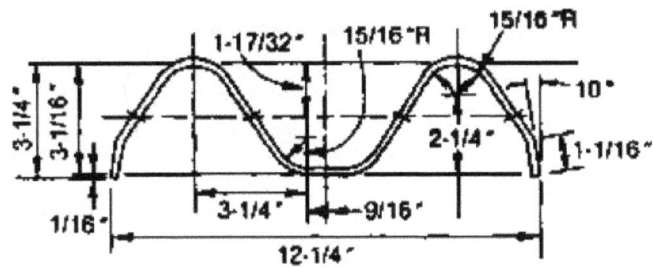

W - beam

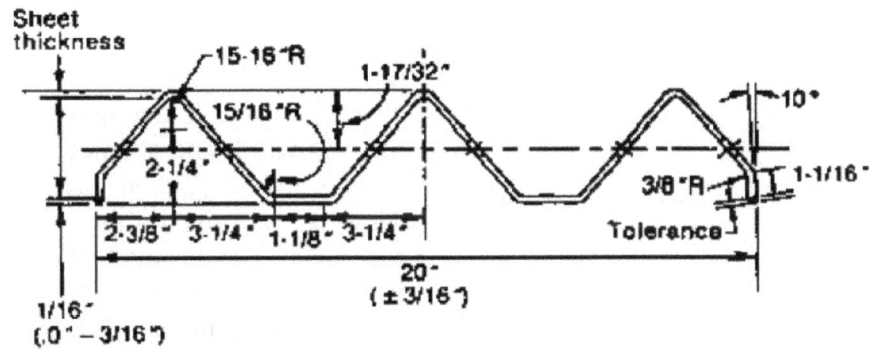

Thrie - beam

Uncoated thickness (in.(gage))	Area (in²)	Moment of inertia (in⁴)	Section modulus (in³)
W-beam			
0.1046 (12)	1.987	2.296	1.364
0.1345 (10)	2.555	2.919	1.732
Thrie-beam			
0.1046 (12)	3.140	3.590	2.150
0.1345 (10)	4.020	4.790	2.870

Bolt holes are not considered in section properties.

From *Handbook of Steel Drainage and Highway Construction Products.*[6] Used by Permission.

The W-beam and Thrie-beam sections are primarily used as highway guardrail and median barriers. Because of the low moment of inertia of the sections, their use as bridge railing is generally restricted to single-lane bridges where the design loading is 50 percent of that required by AASHTO. The strength and span capabilities of semirigid railing can be increased by doubling the rail elements (nesting one section inside the other), or by placing two elements back to back. Additional strength is achieved by backing the sections with steel pipes, channels, or timber members. Because these rail elements are quite flexible, the rail should be blocked away from the post 6 to 8 inches to prevent impacting vehicles from catching the post.

3. **Rigid Steel Rails.** Rigid steel rails are structural steel shapes adapted for use as bridge railing. They are normally rectangular or round steel tubes that are used in both one- and two-rail configurations. Although any steel shape can be used provided it meets strength and geometric size requirements, the most practical and economical designs are from standardized shapes specifically adapted for bridge railing. These typically consist of tubular steel sections or box beams that are available in a variety of sizes.[6,7] Rigid steel rails provide the highest stiffness to prevent vehicles from snagging the posts on impact. They may be attached directly to the post or to offset blocks that project the traffic face away from the post.

For all types of railing, two considerations that must be addressed are rail splices and the transition from bridge railing to roadway approach railing. Splices are important because they give continuity and strength to the overall rail system. With the exception of glulam rails, which can be fabricated in one piece for the bridge length, all types of rails must normally be spliced on the bridge. AASHTO loading criteria require that rails meet strength requirements at the post and at the center of the span, so that the strength of the splice will be sufficient to develop the full strength of the rail. For lumber and glulam rails, splices are normally made at the posts using steel angle or plate splices to transfer applied bending and tension. For W-beam and Thrie-beam sections, splicing is accomplished by bolting sections in prefabricated slots that are normally 6 feet 3 inches on center. For steel tubes, splices are made with smaller tube sections that are inserted inside the rail and bolted in place. Splices for steel rails serve not only to facilitate transportation and construction but also to provide a mechanism for expansion and contraction from temperature changes.

When designing bridge railing, careful attention must be given to the treatment of the railing at the bridge ends. Exposed rail ends, posts, and sharp changes in the geometry of the rail present a significant hazard to vehicles and must be avoided. The transition between the bridge and the approach roadway is generally accomplished by continuing the bridge railing a distance along the roadway or by transitioning the bridge railing to approach roadway railing (Figure 10-8). In both cases, the transition must be smooth and of sufficient strength to protect the traffic from direct collision with the bridge-rail ends.

Figure 10-8. - Standard rail transition between a glulam bridge rail and a steel W-beam approach rail.

Posts

Rail posts for timber bridges consist of timber or steel posts attached to the deck edge, or steel posts welded to base plates bolted to the top deck surface (Figure 10-9). Timber posts are either sawn or glulam members 8 to 12 inches wide and 10 to 12 inches deep. Steel posts are WF 6 x 20 or WF 6 x 25 sections fabricated from galvanized steel (ASTM A 36) or weathering steel (ASTM A 588). For edge-mounted posts, configurations vary for decks with and without curbs. When curbs are provided, posts are generally bolted through the curb at their midsection, with the lower end connected to brackets attached to supporting beams. When curbs are absent, posts are attached with steel brackets that bolt around or over the deck edge. The top mount configuration uses a steel base plate that bolts through the deck, most commonly a transverse-laminated deck.

Static load requirements for posts in AASHTO are the same regardless of post spacing. Hypothetically, posts spaced 1 foot apart are designed for the same loads as posts spaced 10 feet apart. In practice, the most common post spacing on timber decks is between 5 and 8 feet. Economically, materials and installation costs for posts increase as spacing decreases, while the required strength and cost of rails decrease as post spacing

10-15

decreases. Post spacing on glulam deck panels should consider the economics of panel fabrication and should be related to panel width. Placing a post on every other or every third panel in a repeating sequence allows standardized panel fabrication that can reduce costs and construction time.

One of the primary considerations in post design is the load transfer mechanism from the post to supporting components of the superstructure. When improperly designed, rail impact can cause substantial damage to the structure, requiring extensive and costly repairs (Figure 10-10). Longitudinal glulam and nail-laminated lumber decks require special attention because rail forces produce bending at the deck attachment, which in turn introduces tension perpendicular to the wide faces of the laminations. Because wood is weak in tension perpendicular to grain, these loads can cause longitudinal glulam decks to separate or break when railing loads are applied. On longitudinal nail-laminated decks, the same effects can cause the deck to separate between laminations. When a post is attached to a longitudinal beam, the outward post load can also produce loading against the weak axis of the beam. This can cause torque, tension perpendicular to grain, or lateral displacement of the beam. These effects can damage large members such as glulam beams, but their effects are much more pronounced in smaller beams, particularly sawn lumber. When attaching rail components to longitudinal beams, the beams must be sufficiently braced to distribute loads to adjacent members of the superstructure and to prevent adverse loading conditions on the members.

Figure 10-9. - Typical vehicular railing configurations used on timber bridges. (A) Steel posts welded to base plates that are bolted through a glulam deck.

Figure 10-9. - Typical vehicular railing configurations used on timber bridges (continued).
(B) Lumber posts attached to steel plates that are bolted through a glulam deck.
(C) Lumber posts bolted to a lumber curb with braces attached to a glulam beam.

Figure 10-9. - Typical vehicular railing configurations used on timber bridges (continued). (D) Glulam posts and rail with a partial parapet bolted to a glulam deck (photo courtesy of LamFab Wood Structures, inc.). (E) Lumberposts bolted to the lumber curb on a longitudinal nail-laminated lumber deck (photo courtesy of Wheeler Consolidated, Inc.).

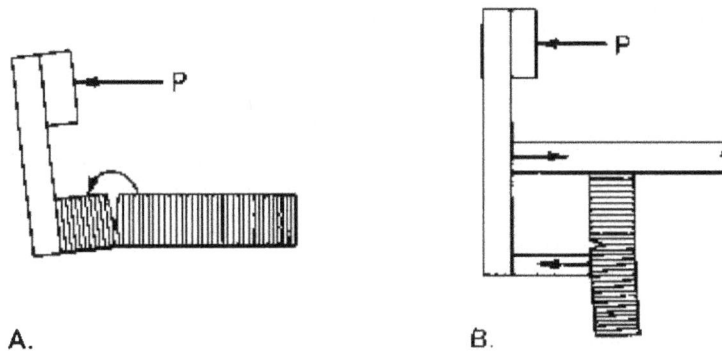

Figure 10-10. - Two types of potential timber bridge damage resulting from rail impact loads. (A) Separation in a longitudinal glulam or nail-laminated lumber deck from tension perpendicular to the laminations. (B) Beam damage resulting from forces transferred by a post brace.

Parapets

Parapets are solid barrier walls that are designed to resist vehicle impact loads and safely redirect vehicles without causing significant damage to the structure or injury to the vehicle passengers. The most widely used type of parapet is the New Jersey-style barrier fabricated from reinforced concrete. Although these concrete barriers can be used on timber bridges they are generally impractical because of their high dead load. The same configuration can be fabricated from glulam when a barrier-type containment is desired (Figure 10-11). These barriers bolt to the bridge deck and must be evaluated in terms of deck effect in the same manner previously discussed for post configurations.

Figure 10-11. - Glulam parapets (photo courtesy of the Weyerhaeuser Co.).

Pedestrian or bicycle railing is provided along the outside edge of sidewalks when vehicle and pedestrian traffic is separated by vehicular or combination railing. Pedestrian railing is used when the walkway is limited to foot traffic, while bicycle railing is used for bicycle traffic or a mix of bicycle and pedestrian traffic. Both railing types are designed for pedestrian loads and are not intended to resist vehicle impact. If a vehicle barrier is not provided between pedestrian and vehicle traffic, sidewalk railing should be combination railing discussed later in this chapter.

DESIGN REQUIREMENTS

Design requirements for pedestrian and bicycle railing are based on minimum geometric and static load criteria given in AASHTO. Railing components should be proportioned commensurate with the type and volume of anticipated traffic with consideration given to safety and appearance. As with vehicular railing, any configuration or combination of materials is permissible provided minimum AASHTO requirements for geometry and loading are met. In cases where the structure will carry equestrians or other specialized traffic, more restrictive design requirements may be appropriate based on designer judgment. Requirements in AASHTO for rail geometry and loads are discussed below and shown in Figure 10-12.

Geometric Requirements

Geometric requirements are the same for pedestrian and bicycle railing, with the exception of minimum rail height, which must be higher for bicycles. In both cases, the railing should provide a safe barrier to prevent adults and children from falling through. The system should also be designed to be difficult or impossible to crawl over or under. Minimum AASHTO requirements for railing geometry are as follows.

1. **Rail Height.** The minimum height of the railing measured from the top of the walkway surface to the top of the top rail is 3 feet 6 inches for pedestrian railing (AASHTO 2.7.3.2.1) and 4 feet 6 inches for bicycle railing (AASHTO 2.7.2.2.1).

2. **Rail Spacing.** Within a vertical band bordered by the walkway surface and a horizontal line 3 feet 6 inches above the surface for pedestrian railing, and 4 feet 6 inches above the surface for bicycle railing, the maximum clear vertical opening between horizontal rail elements is 15 inches (AASHTO 2.7.1.2.4 and 2.7.2.2.2). Vertical elements of the railing assembly shall have a maximum clear spacing of 8 inches within this band. If the railing uses both horizontal and vertical elements, the spacing requirements apply to one or the other, but not both.

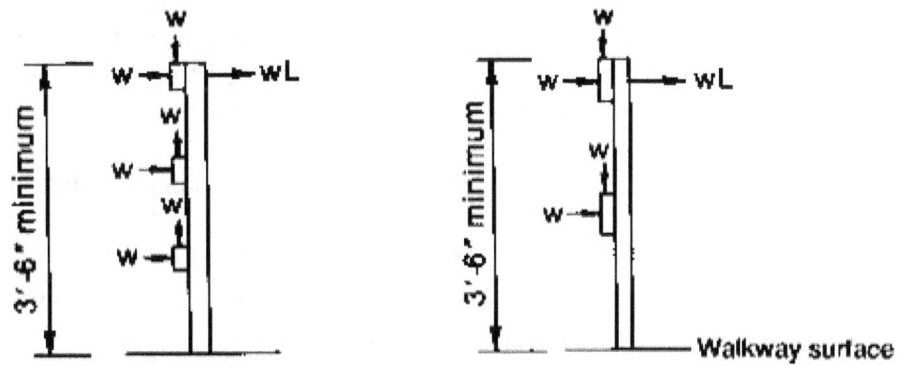

Pedestrian railing

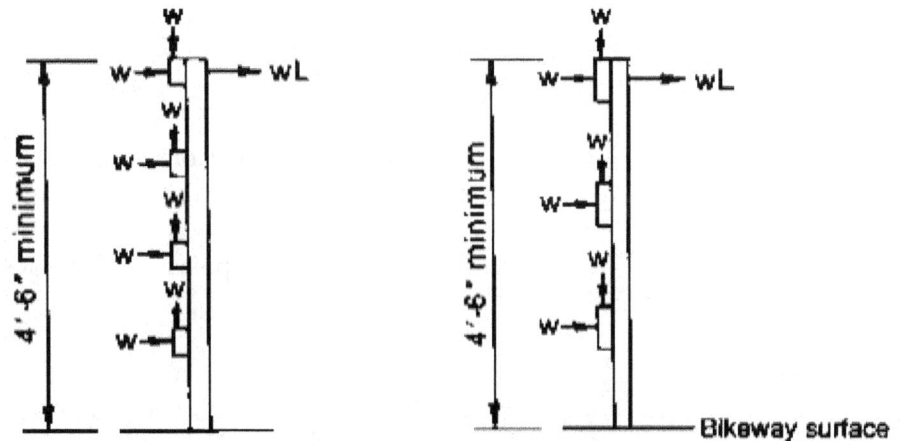

Bicycle railing

Notes:

1. Loadings shown to the left of the post are applied to the rails. Loads shown to the right of the post are applied to the post.

2. $w = 50$ lb/ft; $L =$ post spacing in feet.

3. The maximum clear opening between rails, or between the lower rail and the walkway or bikeway surface, is 15 inches.

4. Rail and post shapes are illustrative only. Any material or combination of materials may be used in any configuration provided minimum AASHTO requirements are met.

5. Refer to AASHTO for illustrations of other railing configurations.

Figure 10-12 - AASHTO requirements for pedestrian and bicycle railing geometry and static loads (adapted from AASHTO Figure 2.7.4A); © 1983. Used by permission.

Loading Requirements

Load requirements for pedestrian and bicycle railing are based on a uniformly distributed load acting on rail elements. Unlike vehicular railing, post loads are directly related to post spacing. Minimum requirements for rail and post loads are as follows:

1. **Rail Loads. The** minimum design loading for each pedestrian and bicycle rail element is $w = 50$ lb/ft, applied simultaneously in the transverse and vertical directions (AASHTO 2.7.3.2.2 and 2.7.2.2.3). When rails are located more than 5 feet above the walkway for pedestrian railing, or 4 feet 6 inches above the walkway for bicycle railing, AASHTO loading is not required and loads are left to designer judgment (AASHTO 2.7.3.2.2 and 2.7.2.2.4).

2. **Post Loads.** Posts are designed for an outward transverse load wL, where L is the post spacing and $w = 50$ lb/ft, as described above (AASHTO 2.7.3.2.3 and 2.7.2.2.5). The load is applied to the post at the center of gravity of the upper rail member, but not more than 5 feet above the walkway for pedestrian railing or 4 feet 6 inches above the walkway for bicycle railing (more severe loading for higher posts is left to designer judgment).

DESIGN GUIDELINES

The most common pedestrian and bicycle railing configurations for timber bridges use horizontal rails on vertical posts, or vertical pickets on longitudinal rails (Figure 10-13). Rails are generally 3-1/8-inch glulam, nominal 2-inch dimension lumber, or steel tubes. Posts are 5-1/8-inch glulam, nominal 4-inch or 6-inch dimension lumber, or steel tubes. Posts are mounted to the sidewalk or supporting beam sides, or to the deck top with a base plate, in the same manner discussed for vehicular railing (Section 10.2). Although lower in magnitude, the structural effects of loads produced by pedestrian or bicycle railing also must be given the same attention discussed for vehicular railing. Of particular concern are the forces produced at the post attachment, where bending and shear can introduce tension perpendicular to grain in supporting members. On beam-type structures, transverse loads can also produce torsion in the beams, and the resulting stresses must be evaluated.

Pedestrian and bicycle railing differs significantly from traffic railing in one very important aspect: it is subject to human contact. The railing should be free of both chemical and physical hazards. Railing components should not be treated with oil-type preservatives that may cause skin irritations. Rather, surfaces should be treated with waterborne preservatives that are dried after treatment to prevent checking and warping (Chapter 4). Where aesthetic considerations are important, treated surfaces may be stained or painted to the desired color. Timber surfaces and edges should be planed and may be sanded smooth so that the potential for abrasion and

Figure 10-13. - Typical pedestrian/bicycle railing configurations used on timber bridges. (A) Glulam rails mounted on glulam posts. (B) Lumber posts with horizontal rails and vertical pickets.

Figure 10-13. - Typical pedestrian/bicycle railing configurations used on timber bridges (continued). (C) Lumber rails on lumber posts that are bolted to a glulam beam. (D) Lumber rails on lumberposts that are bolted to a lumber beam (photo courtesy of Wheeler Consolidated, Inc.).

splintering is reduced. Hardware should be countersunk, with threaded bolt ends and nuts placed on the side opposite the sidewalk. When steel components are used, all edges and weldments should be ground smooth so that sharp edges and weld points are eliminated.

10.4 COMBINATION RAILING

Combination railing is a multipurpose railing designed to perform the dual functions of vehicular railing and pedestrian or bicycle railing. It is used to separate sidewalks and bikeways from adjacent vehicle traffic, or is used along the outside edge of sidewalks when vehicle and sidewalk traffic are not separated by railing (Figure 10-14). AASHTO specifications require combination railing between the sidewalk and roadway for bridges on urban expressways (AASHTO 2.7). On other structures, the separation can be made with vehicular railing or combination railing; however, combination railing is recommended on bridges with an anticipated high volume of pedestrian or bicycle traffic to provide added protection for users.

DESIGN REQUIREMENTS

Combination railing must be designed to function safely for two types of users. The lower traffic portion of the railing must meet the requirements specified for vehicular railing, while the upper portion must comply with the requirements for pedestrian or bicycle railing, including minimum rail height. The loading and geometric requirements previously given for

Figure 10-14. - Combination traffic and pedestrian railing placed along the outside edge of a timber bridge.

vehicular, pedestrian, and bicycle railing also apply to the respective portions of combination railing, with the following exceptions:

1. The maximum vertical clear opening between the lowest rail and the reference surface is 15 inches rather than the 17 inches specified for vehicular railing (AASHTO 2.7.1.2.4).

2. Handrail members of combination railings are designed for a moment at the center of the panel and at the posts of $0.1\,wL^2$, where $w = 50$ lb/ft and L is the post spacing in feet (AASHTO 2.7.1.3.5).

Minimum AASHTO requirements for combination railing geometry and outward transverse post loads arc illustrated in Figure 10-15.

DESIGN GUIDELINES

The most significant design consideration for combination railings used between a roadway and walkway/bikeway is the attachment of the posts to the deck or supporting components. On glulam, stress-laminated lumber, and transverse nail-laminated lumber decks, the most convenient and practical approach is generally to use steel posts that are welded to base

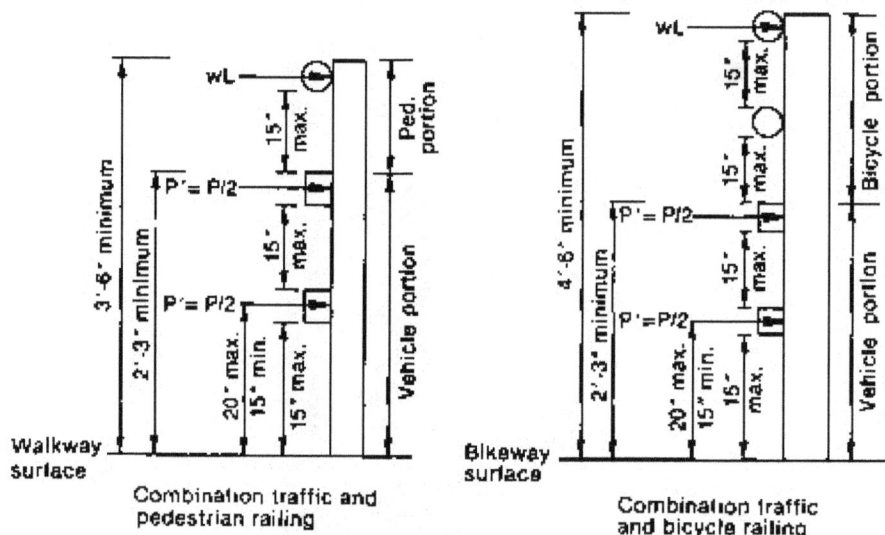

Notes:

1. w = 50 lb/ft; L = post spacing in feet.
2. Rail and post shapes are illustrative only. Any material or combination of materials may be used in any configuration provided minimum AASHTO requirements are met.
3. Refer to AASHTO for illustrations of other railing configurations.

Figure 10-15. - AASHTO requirements for combination railing geometry and outward transverse static loads (adapted from AASHTO Figure 2.7.4B); © 1983. Used by permission.

plates and bolted through the deck in the same manner previously discussed for vehicular railing. An alternate approach, and one that can be adapted to other deck types, is to carry the post through a cutout in the deck and attach it directly to the supporting beam (Figure 10-16). When this is done, the beam capacity must be sufficient to resist potential railing loads and the torsion they create. In addition, attachments of this type require that transverse bracing between the beams be of sufficient strength and spacing to adequately distribute loads applied through the posts.

Figure 10-16. - Combination railing posts attached to glulam beams (arrow) through a cutout in the glulam deck (photo courtesy of Western Wood Structures, Inc.).

10.5 SELECTED REFERENCES

1. American Association of State Highway and Transportation Officials. 1984. A policy on geometric design of highways and streets. Washington, DC: American Association of State Highway and Transportation Officials. 1087 p.
2. American Association of State Highway and Transportation Officials. 1982. AASHTO materials: pt. 1, specifications. Washington, DC: American Association of State Highway and Transportation Officials. 1094 p.
3. American Association of State Highway and Transportation Officials. 1983. Standard specifications for highway bridges. 13th ed. Washington, DC: American Association of State Highway and Transportation Officials. 394 p.

4. American Institute of Timber Construction. 1987. Design standard specifications for structural glued laminated timber of softwood species. AITC 117-87-Design. Englewood, CO: American Institute of Timber Construction. 28 p.

5. American Institute of Timber Construction. 1985. Timber construction manual. 3d ed. New York: John Wiley and Sons, Inc. 836 p.

6. American Iron and Steel Institute. 1983. Handbook of steel drainage and highway construction products. 3d ed. Washington, DC: American Iron and Steel Institute. 414 p.

7. American Road and Transportation Builders Association [and others]. 1978. A guide to standardized highway barrier rail hardware. Tech. Bull. No. 268-B. Washington, DC: American Road and Transportation Builders Association. 224 p.

8. American Society for Testing and Materials. [current edition]. Specifications for hot-formed welded and seamless carbon steel structural tubing. ASTM A 501. Philadelphia, PA: ASTM.

9. Bronstad, M.E.; Michie, J.D. 1981. Multiple-service-level highway bridge railing selection procedures. NCHRP Report No. 239. Washington, DC: National Research Council, Transportation Research Board. 155 p.

10. Bruesch, L.D. 1975. Traffic railing for bridges on low-volume roads. Paper presented at the 1975 Northwest Bridge Engineers' Seminar; 1975 September 16-18; Boise, ID. Washington, DC: U.S. Department of Agriculture, Forest Service Division of Engineering. 7 p.

11. Buffalo Specialty Products, Inc. [1974]. Guide rail. York, PA: Buffalo Specialty Products, Inc. 15 p.

12. Buth, E. 1984. Safer bridge railings. Summary report. Report No. FHWA/RD-82/072. McLean, VA: U.S. Department of Transportation, Federal Highway Administration, Turner-Fairbank Highway Research Center. Vol. 1. 154 p.

13. Canadian Institute of Timber Construction. 1970. Modem timber bridges, some standards and details. 3d ed. Ottawa, Can.: Canadian Institute of Timber Construction. 48 p.

14. Commonwealth of Pennsylvania, Department of Transportation. 1984. Standard plans for low cost bridges. Series BLC-540, timber spans. Pub. No. 130. [Pittsburgh, PA]: Commonwealth of Pennsylvania, Department of Transportation. 28 p.

15. Elliott, A.L. 1968. Bridges. Pt. 1. Steel and concrete bridges. In: Gaylord, E.H., Jr.; Gaylord, C.N. eds. Structural engineering handbook. New York: McGraw Hill. Chapter 18.

16. Engineering News Record. 1987. Guardrail absorbs impact. Engineering News Record 218(22): 15.

17. Hale, C.Y. 1977. Static load tests of Weyerhaeuser bridge rail systems. Rep. No. RDR 045-1609-1. Tacoma, WA: Weyerhaeuser Co. 29 p.

18. Kozak, J.J.; Leppmann, J.F. 1976. Bridge engineering. In: Merrit, F.S., ed. Standard handbook for civil engineers. New York: McGraw-Hill Co. Chapter 17.

19. Michie, J.D.; Bronstad, M.E. 1976. Upgrading safety performance in retrofitting traffic railing systems. Report No. FHWA-RD-77-40. Washington, DC: U.S. Department of Transportation, Federal Highway Administration, Offices of Research and Development. 129 p.

20. National Forest Products Association. 1986. National design specification for wood construction. Washington, DC: National Forest Products Association. 87 p.

21. Olson, R.M.; Ivey, D.L.; Post, E.R.; Gunderson, R.H. [and others]. 1974. Bridge rail design [factors, trends, and guidelines]. NCHRP Report No. 149. Washington, DC: National Research Council, Transportation Research Board. 49 p.

22. Olson, R.M.; Post, E.R.; McFarland, W.F. 1970. Tentative service requirements for bridge rail systems. NCHRP Report No. 86. Washington, DC: National Academy of Sciences-National Academy of Engineering, National Research Council, Highway Research Board. 62 p.

23. State of Wisconsin, Department of Transportation. 1985. Standard bridge plans. Madison, WI: State of Wisconsin, Department of Transportation. [50 p.].

24. U.S. Department of Agriculture, Forest Service, Northern Region. 1985. Bridge design manual. Missoula, MT: U.S. Department of Agriculture, Forest Service, Northern Region. 299 p.

25. U.S. Department of Transportation, Federal Highway Administration. 1979. Standard plans for highway bridges. Timber bridges. Washington, DC: U.S. Department of Transportation, Federal Highway Administration. Vol. 3. 19 p.

26. Weyerhaeuser Company. 1980. Weyerhaeuser glulam wood bridge systems. Tacoma, WA: Weyerhaeuser Co. 114 p.

27. Wheeler Consolidated, Inc. 1986. Timber bridge design. St. Louis Park, MN: Wheeler Consolidated, Inc. 42 p.

CHAPTER 11

WEARING SURFACES FOR TIMBER DECKS

11.1 INTRODUCTION

A wearing surface is a layer placed on the bridge deck to form the roadway surface. It is the only portion of the bridge in direct contact with vehicle traffic. On timber bridges, a wearing surface is one of the most important components of the superstructure and serves two primary purposes. First, it provides a safe, smooth surface for vehicle traffic and improves the poor skid resistance of treated timber decks. Second, the wearing surface protects the deck from the abrasion and physical action of vehicle traffic. Without this protection, timber decks can wear rapidly, resulting in accelerated deterioration and reduced structural capacity.

Wearing surfaces vary in material and configuration and are classified as full or partial depending on the extent of deck coverage (Figure 11-1). A full wearing surface covers the entire bridge deck and is constructed of asphalt pavement, asphalt chip seal, lumber planks, or aggregate. A partial surface covers two longitudinal strips for vehicle tracking and is constructed from lumber planks or steel plates. Full surfaces are used on most bridges while partial surfaces are limited to single-lane, low-volume bridges only. This chapter discusses the performance considerations and design requirements for several full and partial wearing surfaces commonly used on timber decks.

11.2 DESIGN CONSIDERATIONS

Selection and design of a wearing surface depend on the weight, volume, and speed of traffic, as well as construction and maintenance costs. The objective is to provide the safest, most economical surface that meets use and performance requirements for the structure. Asphalt pavement or chip seals are normally the only acceptable surfaces for highway bridges and other bridges on paved roads. When bridges are located on local or low-volume roads, however, a wearing surface constructed of other materials may meet design objectives at a lower cost.

A wearing surface must interact with other bridge components for overall structure performance. In many cases, design considerations for the wearing surface are interrelated with those of the deck and other members of the structure and must be considered concurrently. Some of the general design and performance considerations for wearing surfaces are discussed below.

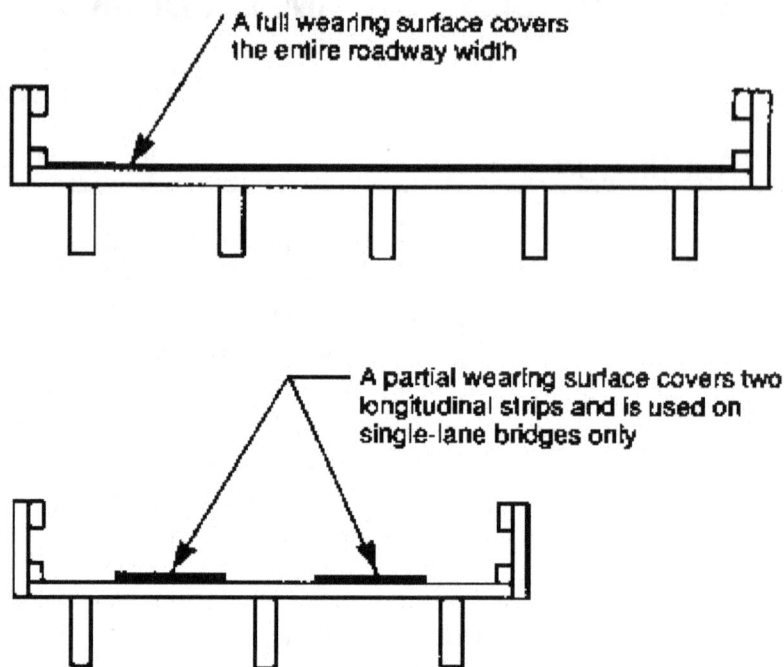

A full wearing surface covers the entire roadway width

A partial wearing surface covers two longitudinal strips and is used on single-lane bridges only

Figure 11-1. - General wearing surface configurations.

STRUCTURAL INTEGRITY

The wearing surface is a sacrificial component; that is, it is intended to wear away over a period of time. Thus, its performance and integrity cannot be ensured for the life of the structure. The wearing surface is not considered as a structural element for the purposes of load capacity or distribution; however, it must be designed to transmit vehicle loads to the bridge deck. In addition to vehicle live load, the surface may be subjected to longitudinal and transverse loads from vehicle braking, wind, and centrifugal force (Chapter 6). The strength of the wearing surface and the connection or bond between it and the deck must be sufficient to transmit these loads.

USER SAFETY

The wearing surface is the only portion of the structure that directly contacts passing vehicles. As a result, it is one of the most important components for user safety. Although many factors influence safety, perhaps the single most important factor is skid resistance. Asphalt pavement or chip seals provide the best skid resistance. The relative skid resistance of other materials, such as timber and steel, depends on the age and condition of the surface but is considerably less than that of asphalt. Skid resistance is related to deck drainage, regardless of the wearing surface material. When water collects on the deck surface, vehicles may hydroplane and become uncontrollable. Wearing surfaces must be free-draining and provide a level of skid resistance commensurate with the type

11-2

and speed of traffic. Lumber and steel wearing surfaces are not recommended when design speeds exceed approximately 30 miles per hour (mph) because of the poor skid resistance of these surfaces, particularly when wet.

In addition to skid resistance, the configuration of the wearing surface influences safety. Partial surfaces cover only a portion of the deck width, delineating the intended roadway for vehicle tracking. The lane width presented to the driver is restricted to two relatively narrow strips, and safe clearance is implied for any vehicle position on the strips. On single-lane bridges, one vehicle uses the structure at a time, and a partial surface may be acceptable. On multiple-lane bridges, however, lateral clearance is restricted when partial surfaces are used, and the potential for collision is greater. In addition, some partial wearing surfaces are elevated above the bridge deck. If a light vehicle rides off the surface, the change in elevation can cause a loss of vehicle control. As a result of these considerations, partial wearing surfaces should be restricted to single-lane bridges. When partial wearing surfaces are used on bridges intended for passenger vehicles, the thickness of the surface should not be more than 2 inches to reduce the probability of a vehicle riding off the surface and losing control.

DECK PROTECTION

One of the primary functions of a wearing surface is to protect the bridge deck. The surface material and thickness should be based on the expected weight and density of traffic. A thicker or more abrasion-resistant surface is required for heavy truck traffic or tire chain use. Partial wearing surfaces offer the least protection and frequently result in deck wear from vehicle off-tracking (Figure 11-2). In addition to protection from vehicle damage, the wearing surface should protect the deck from moisture and weathering effects. The best wearing surface is watertight and shields the deck and supporting members from direct exposure to the elements. Full-width asphalt or chip-seal surfaces drain water and protect the deck from moisture. Lumber, steel, and aggregate surfaces tend to trap moisture and increase susceptibility to decay.

ECONOMICS

The relative economy of wearing surfaces should be evaluated in terms of initial construction cost, the design life of the surface, and estimated costs for maintenance and replacement over the life of the structure. Wearing surface design life depends on the material and configuration of the surface as well as the weight and density of traffic. Surface life is difficult to estimate for the general case and should be based on site-specific information for projected traffic. Relative approximations of service life are given in the following sections of this chapter.

Figure 11-2. - Severe abrasion on a nail-laminated lumber deck caused by vehicles off-tracking the partial steel plate wearing surface (photo courtesy of Sakee Poulakidas, USDA Forest Service).

When evaluating wearing surfaces, simple construction cost comparisons do not give an accurate indication of total economy and can be misleading. Although the initial cost for some surfaces may be higher than others, savings in future maintenance and replacement expenses over the life of the structure can more than offset the additional cost. Maintenance for plank surfaces generally involves complete replacement, while minor crack repair for asphalt surfaces may significantly extend service life without replacement. In general, maintenance costs are higher for partial surfaces that are elevated above the deck because they trap water and debris and require cleaning at regular intervals. Maintenance costs are also high for surfaces that are bolted to the deck and require access to the deck underside for tightening or replacement.

The type of wearing surface may have an effect on the service life of the deck. When field drilling for fasteners such as spikes, lag screws, or bolts is required for deck attachment, the preservative envelope of the deck is broken. This may lead to accelerated deck decay or deterioration, especially when the replacement interval of the wearing surface is frequent. Although effects on deck life are difficult to predict, the potential should be considered.

RIDEABILITY

The rideability or user comfort provided by the wearing surface should be considered in design. In most applications, rideability is evaluated for light passenger vehicles and is related to the traffic speed. The roadway should be as smooth as possible without abrupt changes in texture or elevation.

11-4

The riding quality of the wearing surface should equal or surpass that of the adjacent approach roadways.

DEAD LOAD

The dead load of wearing surfaces can vary significantly for the same surface thickness. For example, a 3-inch asphalt surface weighs approximately three times more than a lumber surface of the same thickness. Although this weight difference generally has little influence on the design of new structures, it may be an important consideration in the rehabilitation of existing bridges.

11.3 ASPHALT PAVEMENT

An asphalt pavement wearing surface consists of a layer of bituminous concrete that is spread and compacted on the bridge deck to produce a smooth, well-consolidated surface (Figure 11-3). It is perhaps the most desirable of all wearing surfaces because it effectively protects the entire deck from traffic abrasion and moisture and provides a smooth, skid-resistant surface. It is the only surface compatible with high-speed paved highways. The service life of an asphalt wearing surface depends not only on the weight and volume of traffic but also on the type of deck, local environmental conditions, and the preparation, design, and application of the asphalt pavement. When properly applied and maintained, asphalt wearing surfaces can provide good service for periods of 15 years or more.

Figure 11-3. - Asphalt pavement wearing surface on a timber bridge deck.

Although the overall performance of asphalt wearing surfaces on timber decks has been good, there have been cases where the surfacing has cracked or separated while in service. The suitability of asphalt on timber decks is primarily a matter of deck compatibility. For asphalt concrete to perform properly, deck deflection under vehicle loads must be limited to prevent pavement cracking or disintegration. Decks constructed of glulam, stress-laminated lumber, and nail-laminated lumber are suitable for paving provided deflections are limited to reasonable levels, as discussed in Chapters 7, 8, and 9. Plank decks should not be paved because plank deflection and movement from moisture variations are difficult to control within acceptable limits. If cracks do appear in paved decks, they can be filled with an asphalt-sand mixture or commercial crack fillers with no significant economic or performance loss. The best solution, however, is to prevent or reduce the incidence of cracking through proper deck design.

This section discusses some of the considerations related to asphalt paving on timber bridge decks. Discussions on asphalt manufacture and the design of asphalt pavements are beyond the scope of coverage, and readers are referred to references listed at the end of this chapter.[6,7,9]

MATERIALS

Asphalt pavement consists of a combination of well-graded, high-quality aggregate that is uniformly mixed and coated with an asphalt binder. Three types of asphalt binders are used: asphalt cements, cutbacks, and emulsions. Asphalt cements are undiluted refined asphalt, while cutbacks are asphalt cement dissolved in petroleum solvents. Emulsions consist of asphalt in an emulsified solution with water. The use of cutbacks has declined in recent years because of increased petroleum costs and environmental considerations related to solvent evaporation. They are slowly being superseded by emulsions, which contain little or no solvent and can be used for many of the same purposes as cutbacks.

The most common asphalt pavement for bridge applications is hot-asphalt plant mix (hot-mix). Hot-mix is manufactured at a central batching plant where aggregate and asphalt cement are heated to 250 to 325 °F before mixing. While the paving mixture is still hot, it is shipped to the construction site and placed. As an alternative to hot-mix, cold-mix pavements are used on bridges with light to medium traffic. Cold-mixes are manufactured with asphalt cutbacks or emulsions and are transported, spread, and compacted at ambient temperatures. They offer advantages in outlying areas where transportation of hot-mix pavements is impractical.

Asphalt paving mixtures are produced from a wide range of mix designs involving aggregate combinations and variations in the amount and grade of asphalt used. Dense-graded mixtures are used exclusively for timber bridges because they provide a dense, water-resistant surface over the deck. Open-graded mixtures provide no moisture protection and are not recommended. Specifications and mix designs suitable for timber bridges are normally maintained by state and federal agencies with responsibilities for road paving and maintenance. In most states, it is practical to use one of the standard mixes normally available from asphalt mix suppliers in the state.

SURFACE PREPARATION
Surface preparation of the bridge deck is perhaps the most important step in asphalt paving. As discussed in Chapter 4, bridge components are treated with oil-type preservatives because of the added moisture protection oil-type preservatives afford. Some of the same qualities that provide this added protection affect the physical properties and bonding capabilities of asphalt pavements. When the deck surface contains excess preservatives, the asphalt cannot bond properly to the deck and will eventually soften and disintegrate, or separate from the deck surface. Problems of this type can be eliminated when the deck is properly prepared.

Planning for asphalt pavement starts during the design process when the specifications are prepared. In an effort to provide as clean a surface as possible, treating specifications should require treatment by an empty-cell process, followed by an expansion bath or steaming. Depending on the treater, material treated in this manner will generally be free of excessive surface deposits of preservative or solvent. The level of free preservatives may be further reduced by specifying one of the new clean creosote treatments mentioned in Chapter 4. It may be beneficial to discuss treatment alternatives with a local treater or national treating organization to determine the best treatment based on local availability.

After treating, most material will continue to exude preservative or solvent volatiles, and time must be allowed for excess material within the wood to evaporate. Unless the preservatives stabilize, a satisfactory bond will not be achieved between the deck surface and the asphalt. The rate at which these volatiles leave the wood depends on the type of preservative and temperature. Preservatives in heavy-oil solvents leave the surface at a slower rate than light-oil solvents, but the rate for both increases as temperatures rise. When practical, treated timber decks should not be paved for 30 to 45 days after the material has been treated. In the interim, deck material can be stored where air can circulate freely around all surfaces, or be installed with a blotter material (discussed in the following paragraphs) and paved at a future date.

When decks must be placed with free surface preservatives, or before all residuals have evaporated, application of a surface blotter before paving can greatly improve asphalt bonding. A blotter mixture of dust and 10 to 20 percent crushed material passing the No. 8 sieve, spread at a rate of 10 to 15 lb/yd^2, is recommended. The blotter is spread on the deck and immediately rolled with a rubber-tired roller. After the excess preservative has been absorbed (approximately 1 week), the blotter is removed by

11-7

brooming and additional blotter applied if necessary. The effectiveness of surface blotters in bridge paving is noted by Bruesch and Pelzner as follows:[11]

> Two recently completed timber bridge projects serve to demonstrate the problem. Briefly, on both projects significant quantities of free, oil-borne preservatives were in evidence on the surface of installed deck panels. The free preservatives may have been on the panel surfaces when timber came out of the treating cylinder, or may have bled to the surface prior to installation. In one case, the asphalt surfacing, placed directly over the free preservative, was softened and easily removed by lateral forces. In the other case, a mixture of sand and fines was used to blot up the free preservatives prior to application of the asphalt surfacing; that surfacing appears to be adequately bonded and is functioning properly. These field experiences and technical advice from the concerned industries lead us to recommend use of a blotter to neutralize the free preservative.

PLACEMENT AND
CONFIGURATION

After the deck is free of excessive preservatives, the surface is thoroughly cleaned of all dirt and other debris and a tack coat is applied. The tack coat is a thin layer of asphalt that serves to glue the asphalt pavement to the deck surface. On timber decks, it is normally a slow-setting asphalt emulsion that is diluted 50 percent by volume with water and sprayed on the deck at an application rate of 0.05 to 0.15 gal/yd^2. After emulsion tack coats are sprayed, they must be allowed to break or set before pavement is placed (breaking is the separation of the asphalt cement from the water).

Asphalt pavement is applied to the deck to a compacted thickness of 2 to 3 inches using standard paving procedures and equipment. For drainage purposes, pavement may be sloped or crowned to a minimum compacted edge thickness of 1-1/2 inches (Figure 11-4). Recommended transverse crown is 1/2 inch per traffic lane or 1/2 inch total, whichever is greater. Retainer strips are normally installed along curbs or railings to form a neat edge and prevent the pavement from filling drainage openings. These strips can be constructed using galvanized steel angles or treated dimension lumber that is cut to the required pavement thickness and connected to the deck with lag screws (Figure 11-5). For drainage purposes, it is important that the top of the strip not be higher than the adjacent pavement.

When a bridge is located on dirt or gravel roads, service life and performance of the wearing surface can be significantly increased if road approaches are paved a minimum of 75 feet beyond the bridge ends. This reduces the amount of gravel and other debris tracked onto the deck and eliminates the potholes that commonly form at the bridge ends.

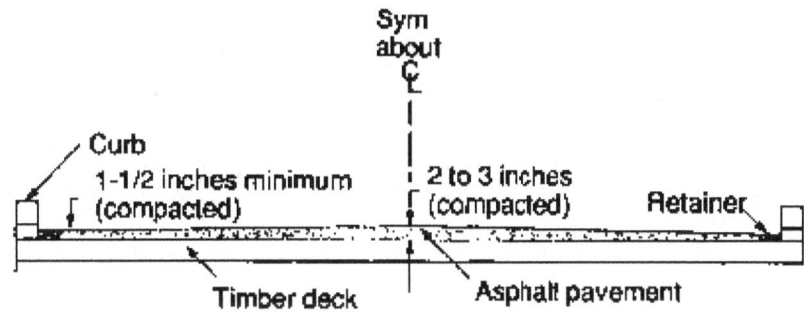

Figure 11-4. - Typical asphalt pavement wearing surface cross section.

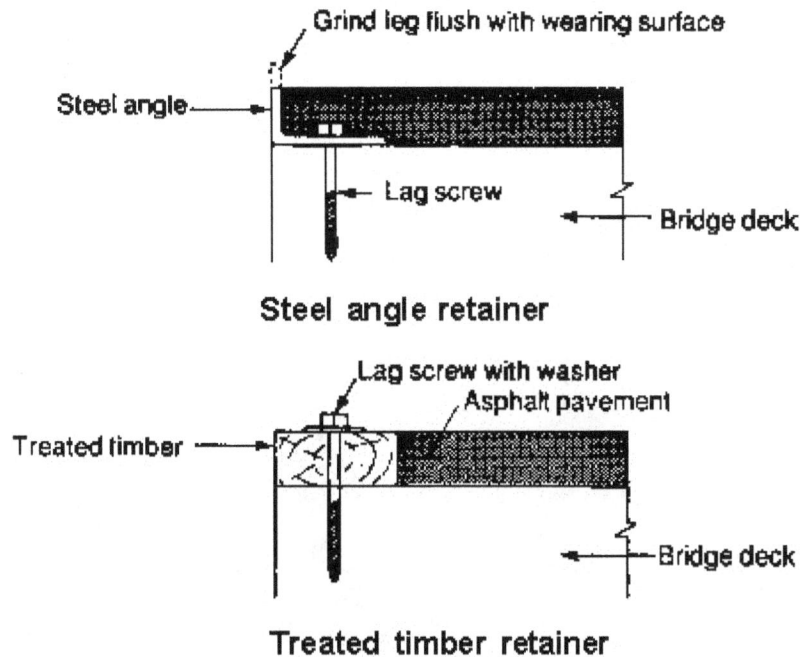

Steel angle retainer

Treated timber retainer

Figure 11-5. - Types of retainer strips for asphalt pavement wearing surfaces.

GEOTEXTILE FABRICS

Geotextile fabrics are synthetic engineering fabrics that were originally developed to provide additional stability and load distribution in numerous geotechnical (soils) and hydraulic applications. Specialized paving fabrics have been used for several years to improve pavement performance and longevity. When properly placed between the bridge deck and asphalt pavement, geotextile fabrics can improve the bond between the asphalt and the deck surface, provide increased moisture resistance of the surface, and reduce or eliminate pavement cracking at glulam panel joints.

Geotextile fabrics for bridge paving are available in two types: plain and asphalt impregnated. Plain fabrics consist of a nonwoven geotextile fabric only and are commonly available in rolls 12 feet wide. Impregnated

11-9

fabrics have a layer of rubberized asphalt bonded to one side and are normally available in 12- and 36-inch widths. The impregnated fabrics are most commonly used on timber decks where heat from the asphalt causes the rubberized asphalt layer to bond to the deck. This provides improved adhesion and an impermeable barrier to moisture.

Paving with geotextile fabrics involves the same deck surface preparation previously discussed. After the deck is free of excess preservative and debris, the fabrics can be placed. A tack coat is necessary before placing plain fabrics but is not required for impregnated fabrics. The fabric is rolled on the deck with an overlap between adjacent strips of 2 to 3 inches. On transverse glulam decks, the narrow-width impregnated fabrics also can be placed transverse over panel joints only (Figure 11-6). After the fabric is rolled in place, a tack coat between the fabric and asphalt concrete layer is required for both plain and impregnated fabrics. This generally consists of an asphalt emulsion spread to achieve a residual asphalt layer of 0.10 to 0.15 gal/yd^2 (this may vary among fabric brands and should be verified with the manufacturer). Pavement is then applied to the surface in the usual manner. A sequence of photos showing a deck-paving project using impregnated fabric is given in Figure 11-7.

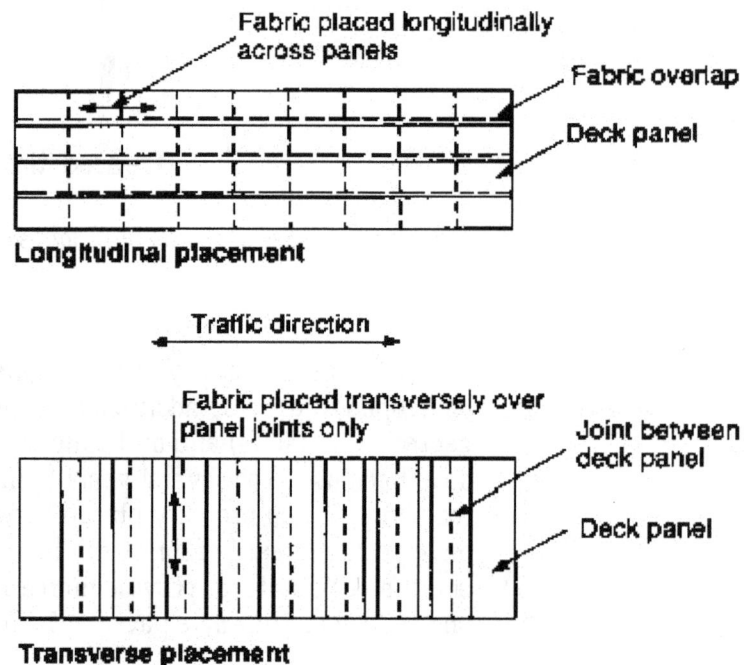

Figure 11-6. - Placement of impregnated geotextile fabric on transverse glulam deck panels.

Figure 11-7. - Asphalt paving sequence on a glulam bridge deck using impregnated geotextile fabric. (A) Geotextile fabric is rolled longitudinally over the cleaned deck (note that the backing paper on the asphalt side of the fabric is removed as the fabric is rolled). (B) Completed fabric placement.

Figure 11-7. - Asphalt paving sequence on a glulam bridge deck using impregnated geotextile fabric (continued). (C) Hand-application of a tack coat to the fabric. (D) Hot-mix asphalt is spread for compaction by rollers.

Figure 11-7. - Asphalt paving sequence on a glulam bridge deck using impregnated geotextile fabric (continued). (E) A corner of the fabric is pulled back after application of the hot asphalt, showing the bond between the rubberized asphalt on the fabric and the bridge deck. (F) The completed wearing surface (photos courtesy of Ron Vierra, USDA Forest Service).

An asphalt chip seal consists of a sprayed application of liquid asphalt covered with a layer of selected aggregate (Figure 11-8). It is not considered a pavement but an asphalt surface treatment that seals the deck surface and protects it from the abrasive effects of traffic. Chip seals have been used with great success on timber bridge decks and have provided service lives of 15 years or more, depending on traffic conditions. They are well suited for most timber bridge applications and provide a smooth, even surface that is compatible with paved roadways. The thinner chip seal surface normally provides added flexibility that is less susceptible to cracking than the more rigid asphalt-concrete pavements.

Figure 11-8. - Asphalt chip seal wearing surface on a timber bridge deck.

MATERIALS

Materials for chip seals consist of the asphalt binder and the aggregate. Rapid-setting emulsified asphalts and soft grades of cutbacks are usually best suited for chip seals. Application rates vary with the type of binder and aggregate but are normally in the range of 0.20 to 0.35 gal/yd² for emulsions and 0.15 to 0.25 gal/yd² for cutbacks. Aggregates are normally 3/8- or 1/2-inch angular material that is as uniformly graded as economically practical. Most hard aggregates such as gravel, crushed stone, or slag can be successfully used if they are clean. If aggregates are dirty or covered with dust, the coating forms a film that prevents asphalt-aggregate adhesion. Aggregate spread rates depend on the size and quality of aggregate and range from approximately 20 to 25 lb/yd² for 3/8-inch material to 25 to 30 lb/yd² for 1/2-inch material. As with asphalt pavement, the designer should check with local state or county road agencies to determine the best asphalt-aggregate combination and application rates for the local area. Additional information is given in references listed at the end of this chapter.[37]

PLACEMENT AND CONFIGURATION

Chip seals can be applied as a single treatment or as a multiple treatment. Single treatments consist of one layer of asphalt and one layer of aggregate. Multiple treatments are built by adding additional layers of asphalt and progressively smaller-size aggregate. For bridge applications, a double treatment approximately 3/4 inch thick provides much better performance than a single treatment. Thicker surfaces can be built by increasing aggregate size.

Surface preparation for asphalt chip seals is the same as previously discussed for asphalt pavement. Unless the surface is clean and free of excess preservatives, the asphalt will not adhere to the deck. After the deck is cleaned, the asphalt binder is applied to the bridge by an asphalt distributor. The distributor is a tank truck equipped with a heater, pump, and spray-bar assembly that uniformly sprays the asphalt over the deck surface (Figure 11-9). The spray bar is extended from the rear of the truck to cover a width in one pass of 6 to 30 feet, depending on the capacity of the pump. During the spraying process, it is important that the pump and spray bar nozzles be properly calibrated and adjusted to deliver a uniform, even layer of asphalt at the required rate.

Figure 11-9. - Asphalt binder for a chip seal being applied to a timber bridge deck by an asphalt distributor truck (photo courtesy of Paul Cole, USDA Forest Service).

11-15

Aggregate is applied over the asphalt using a spreader. Spreaders range from simple vane spreaders or mechanical spreaders to highly efficient self-propelled machines. Vane spreaders attach to the dump truck tailgate and fan out slightly more than the truck width. The application rate is controlled by the feed gate opening and the speed of the truck as it backs up. Mechanical spreaders are hoppers on wheels that connect to the truck tailgate. Although the application rate also depends on truck speed, mechanical spreaders provide a more controlled, even aggregate spread across the lane than vane spreaders. The most suitable spreaders are self-propelled models (Figure 11-10). The aggregate truck hitches to the rear of the spreader, dumps aggregate into a receiving hopper, and is pulled by the spreader. Aggregate from the hopper is moved by conveyer to the front of the spreader where it is evenly distributed by a spread roller. For all types of spreaders, a check of the aggregate application rate can be made by laying 1 yd^2 of cloth or building paper on the ground and weighing the amount of aggregate distributed after the spreader passes.

Figure 11-10. - Crushed aggregate chips are applied over an asphalt binder by a self-propelled aggregate spreader (photo courtesy of Paul Cole, USDA Forest Service).

Immediately following chip application, it is important that the surface be compacted to properly seat the aggregate in the asphalt binder. A towed or self-propelled rubber-tire roller is recommended for use on chip seals because the tires force the aggregate firmly into the asphalt without crushing (Figure 11-11). Steel-wheel rollers bridge over smaller particles or depressions in the surface and may crush the aggregate. After the layer is compacted, the asphalt is allowed to set so that the aggregate is tightly bonded. The layer may then be brushed or broomed with motorized equipment to remove excessive chips, and the second treatment is applied using the same procedures.

Figure 11-1. - Rubber-tire roller of the type used for compacting asphalt chip seal wearing surfaces.

GEOTEXTILE FABRICS

Geotextile fabrics previously discussed for asphalt pavement can be used with asphalt chip seals. Plain fabrics are recommended at this time because the use of impregnated fabrics with chip seals is still in the developmental stage, and results are not yet conclusive. When fabrics are used, the rate of asphalt application must be increased to saturate the fabric layer. This increase is generally 0.10 to 0.15 gal/yd^2 residual asphalt, but should be verified with the fabric manufacturer.

11.5 LUMBER SURFACE

Lumber wearing surfaces consist of a series of lumber planks placed edge to edge across the deck width (Figure 11-12). They are frequently used on single- and multiple-lane bridges and are compatible with all types of timber decks. Lumber surfaces are probably the most economical full surface to construct and maintain on bridges located on low-speed, unpaved roads. Service life is typically 5 to 12 years depending on the weight and volume of traffic and plank thickness. When gravel or other abrasive material is tracked on the surface, service life is significantly decreased.

11-17

Figure 11-12. - Typical lumber wearing surface.

MATERIALS

Lumber surfaces are constructed of planks that are 10 to 12 inches wide and a minimum of 8 feet long. Random-length, rough-sawn planks are commonly used and are field cut to required length. It is desirable to leave the wide faces of the planks unplaned to provide additional surface texture. Plank edges are rough sawn, or are edge planed (S2E) to provide consistent plank widths. Plank thickness depends on traffic weight and density. Guidelines for thickness based on vehicle weight are given below; however, the wearing surface should not be thicker than the bridge deck.

Vehicle weight (tons)	Recommended plank thickness (inches)
≤ 20	2
≤ 50	3
> 50	4

Selection of wood species for planks should be based on the considerations of wearability and dimensional stability. Both of these properties are directly related to species density (Chapter 3). As density increases, planks wear better but are more susceptible to dimensional changes and deformation because of moisture content changes. Species such as Douglas Fir-Larch, Hemlock, and Spruce provide good wearability with acceptable dimensional stability. Regardless of species, wearability and dimensional stability are increased when edge-grain planks are used (Figure 11-13). Flat-grain planks wear faster and may cup or twist because of moisture

11-18

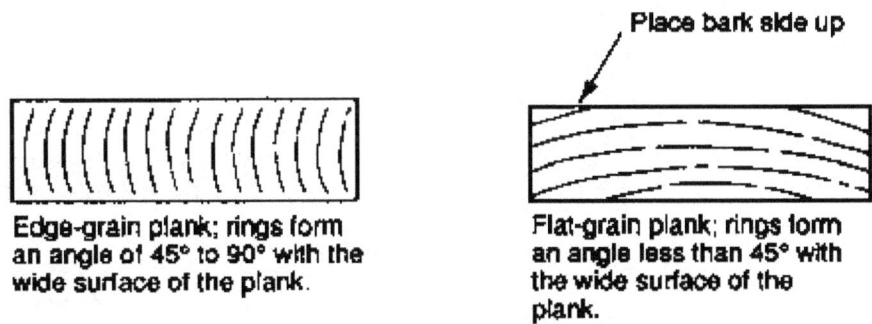

Edge-grain plank; rings form an angle of 45° to 90° with the wide surface of the plank.

Place bark side up

Flat-grain plank; rings form an angle less than 45° with the wide surface of the plank.

Figure 11-13. - Edge-grain and flat-grain plank orientations for lumber wearing surfaces.

changes. When edge-grain material is not available, flat-grain planks should be used with the bark side up (heart side down).

Under low traffic volumes or light vehicle loads, planks may decay before they wear out, especially in the areas not contacted by traffic. Under these conditions it may be economically beneficial to treat planks with preservatives to extend their service life. When planks are treated, waterborne preservatives should be used (Chapter 4). Oil-type preservatives reduce skid resistance and may create a vehicle safety hazard.

CONFIGURATION

Wearing surface planks are typically oriented in the direction of traffic. A transverse or diagonal orientation may be used but planks wear faster when traffic is across the grain. The most economical arrangement is an alternating repetition of longitudinal planks with odd lengths at the bridge ends (Figure 11-14). End joints in adjacent planks are staggered by a minimum of 3 feet. When seasoned planks are used, a gap of approximately 1/4 inch is left between edge joints to allow for expansion. Tight edge joints are used for unseasoned (green) planks.

The configuration of a lumber surface at bridge ends should minimize the effects of vehicle impact on planks, especially on dirt or gravel roads where potholes develop at bridge approaches. Beveling of plank ends on bridge approaches reduces vehicle impact forces and improves wearing surface performance and longevity (Figure 11-15).

DECK ATTACHMENT

Performance of a lumber wearing surface depends on the plank attachment to the bridge deck. The connection must keep the planks firmly attached, minimize deck damage, and permit easy removal for plank replacement. The two fasteners most commonly used are spikes and lag screws. Bolts are not economical and require access to the deck underside for installation and removal. Whenever possible, deck fasteners and hardware should be recessed below the roadway surface. This reduces tire damage and protects fasteners from road maintenance vehicles such as snow plows and

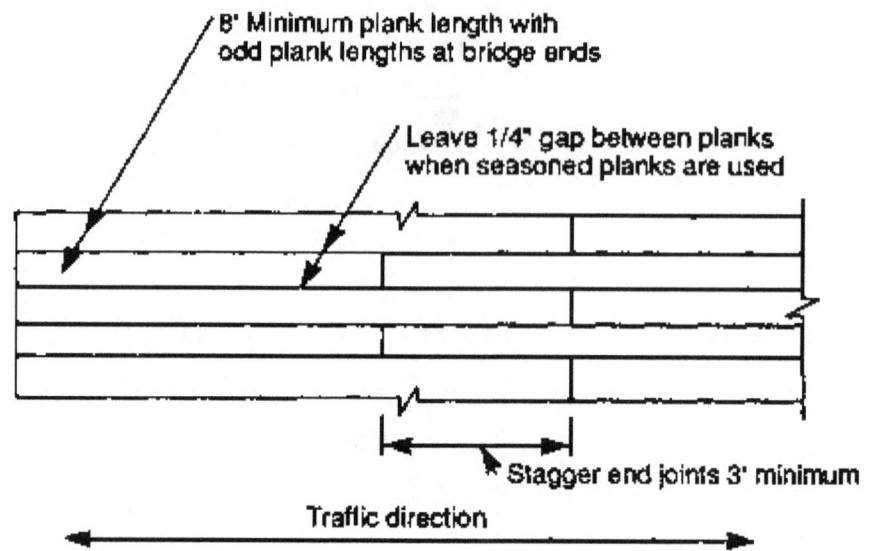

8' Minimum plank length with odd plank lengths at bridge ends

Leave 1/4" gap between planks when seasoned planks are used

Stagger end joints 3' minimum

Traffic direction

Figure 11-14. - Typical plank layout for lumber wearing surfaces.

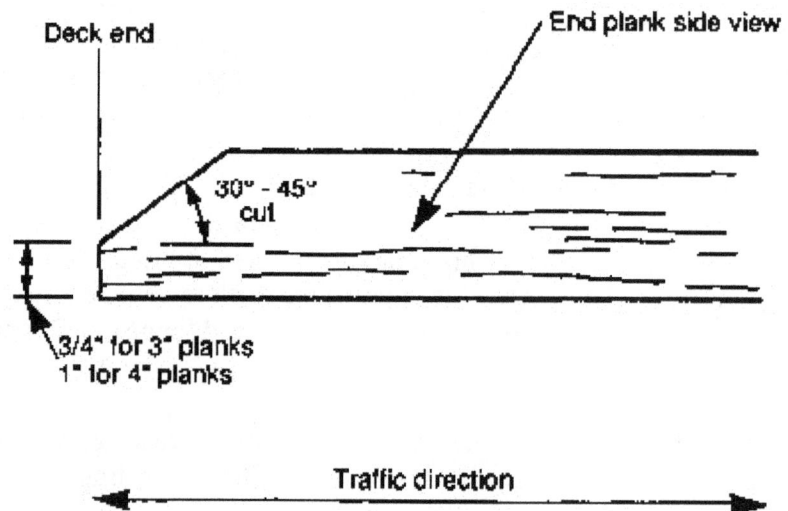

Deck end

End plank side view

30° - 45° cut

3/4" for 3" planks
1" for 4" planks

Traffic direction

Figure 11-15. - Beveled end-plank configuration to minimize vehicle impact at bridge ends.

motor graders. The recessed hole does provide a trap for dirt, water, and other material, but this has little or no effect on the deck or wearing surface. These depressions can be sealed with mastic compound or caulked if considered necessary by the designer.

Field placement of fasteners such as spikes or lag screws requires penetration of the preservative envelope of the deck, providing access for organisms that decay untreated timber. Decay susceptibility in decks is especially significant because the deck has a high exposure to moisture and debris accumulation. To minimize decay and prevent splitting, all fasteners should be placed in lead holes that are prebored and field treated with liquid wood preservative. When fasteners are permanently removed, as when planks are replaced, holes are re-treated with preservatives and tightly plugged with treated wood dowels (Figure 11-16). Protection of timber members from decay is critical to the longevity of the deck and cannot be overemphasized. Failure to properly install and replace fasteners can result in accelerated decay, which reduces deck service life.

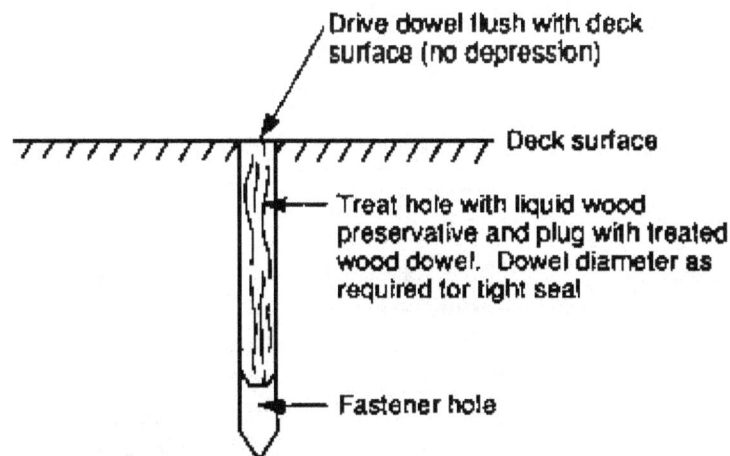

Figure 11-16. - Treated dowel plug for wearing surface fastener holes.

Spikes

Spikes are the most common fastener for lumber wearing surfaces because they are inexpensive and simple to install. One disadvantage of using spikes is their tendency to loosen from moisture loss in the planks or from structure vibrations. Safety hazards can result when planks move or spikes project above the wearing surface. These problems can be minimized by proper spike placement and maintenance.

Spikes for wearing surfaces should be annularly (ring shanked) or helically (spiral) threaded (Figure 11-17). Common steel spikes with a smooth finish are not recommended because they loosen under repeated loading. A minimum spike diameter of 1/4 inch is recommended for planks 3 inches thick or less. When planks are more than 3 inches thick, 5/16- or

11-21

3/8-inch diameter spikes should be used, depending on the weight and volume of traffic. Spike length should be approximately twice the plank thickness, but not greater than the combined depth of the wearing surface (minus countersink depth) and deck. Spikes should be galvanized, especially when de-icing salts may be applied to the deck. Although corrosion protection may not be warranted in all areas, additional cost for galvanizing is low and the zinc finish provides additional resistance to withdrawal.

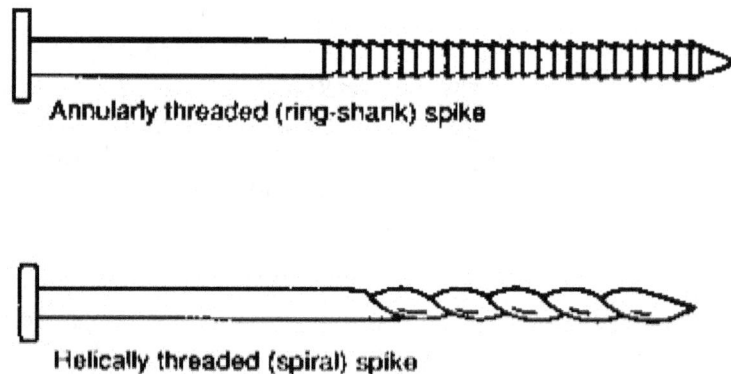

Figure 11-17. - Types of spikes used for attaching lumber wearing surfaces.

The recommended attachment pattern for spiked planks is shown in Figure 11-18. Two spikes are placed at each end with single spikes at an intermediate spacing of approximately 2 feet, staggered to alternate sides. Spikes are placed a minimum of 2 inches from plank edges and 4 inches from ends, and are normally countersunk below the roadway surface. Resistance to withdrawal is increased when spikes are driven at an angle of 10 to 20 degrees in the plank direction (Figure 11-19 A). All spikes are driven in prebored holes that are approximately 75 percent of the spike diameter (Chapter 5). Deformed-shank spike diameters may vary between manufacturers and should be verified by the designer before specifying prebore diameters.

Lag Screws
Lag screws are threaded fasteners that are inserted by turning rather than by driving. Although they cost more than spikes, lag screws are stronger and less susceptible to loosening from moisture changes or vibration. Lag screws provide some benefit over spikes because they can be reused when planks are replaced. In some cases, the same lead hole is used, reducing the number of new holes required in the deck.

Attachment patterns for lag screws are the same as those used for spikes (Figure 11-18) but they are inserted vertically with a round steel cut washer under the head (Figure 11-19 B). A minimum lag screw diameter

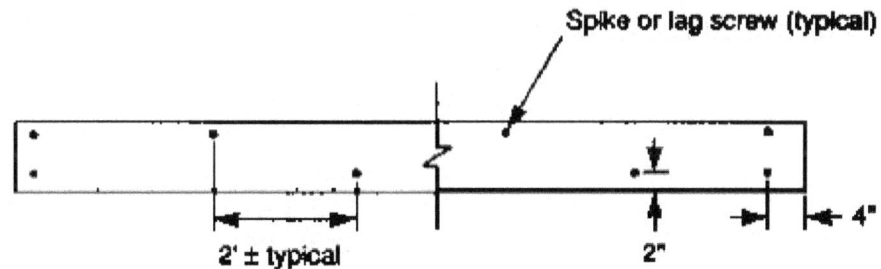

Figure 11-18. - Lumber wearing surface plank attachment pattern using spikes or lag screws.

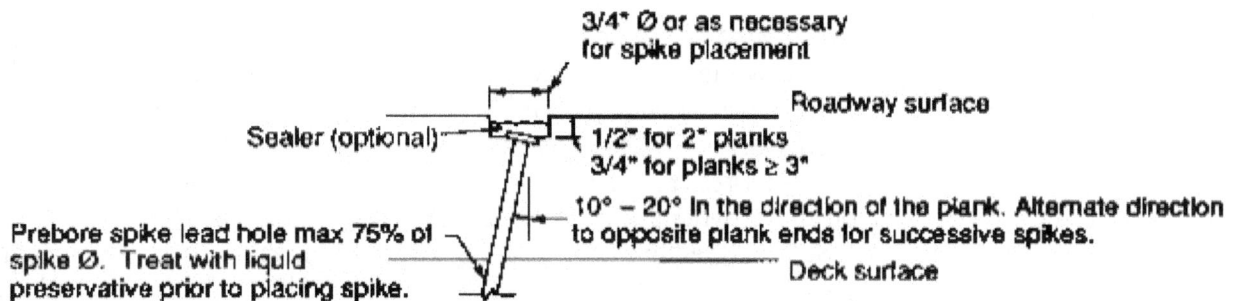

A. Spike attachment.

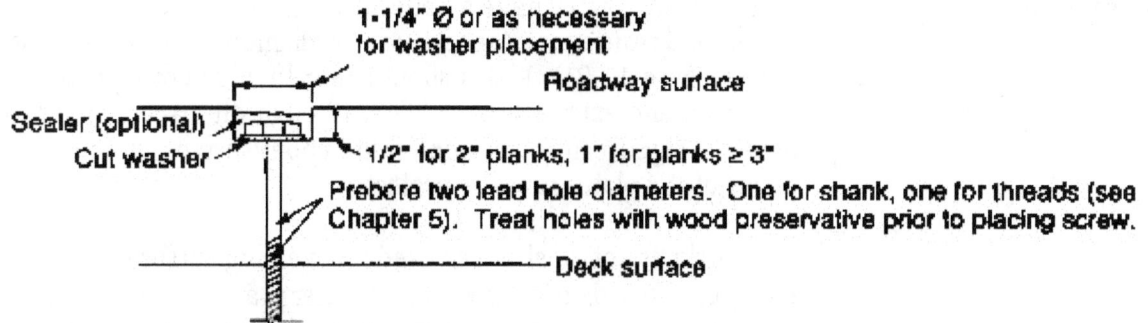

B. Lag screw attachment.

Figure 11-19. - Spike and lag screw attachment details for lumber wearing surfaces.

of 3/8 inch is used for planks 3 inches thick or less (smaller diameters tend to break from twisting before they are completely inserted). When planks are more than 3 inches thick, 7/16- or 1/2-inch-diameter screws are used. Lag screws should be long enough to penetrate the deck approximately 8 diameters for Douglas Fir-Larch or Southern Pine and 10 to 11 diameters for other species. All lag screws and cut washers should be galvanized for corrosion protection.

11-23

Lead holes for lag screws are prebored with two diameters: one diameter for the upper shank portion, and a smaller diameter for the threaded length. The lead hole for the shank portion is 1/16 inch greater in diameter than the lag screw shank. The lead hole for the threaded length varies from 40 to 75 percent of the screw shank diameter depending on deck species. For Douglas Fir-Larch and Southern Pine, a hole diameter of 60 to 75 percent of the shank diameter is used. Prebore diameters for other species are given in Chapter 5.

11.6 STEEL RUNNING PLATES

Steel running plates consist of a series of steel plates placed in two strips, oriented symmetrically about the bridge centerline (Figure 11-20). They provide a partial wearing surface over the portion of the deck intended for vehicle tracking. The center and outside portions of the deck are not protected. Steel plates are used on low-volume, single-lane bridges and typically provide a service life of 25 years or more. They are resistant to abrasion and require little maintenance other than periodic attachment tightening. A disadvantage of steel running plates is their poor skid resistance, especially when wet or frosty. For this reason, use of steel running plates should be limited to low-speed applications.

MATERIALS

Steel running plates have a patterned surface to provide texture and additional skid resistance. A checkered or diamond pattern is most commonly used (Figure 11-21). Plates should be galvanized or painted to control corrosion and extend service life. Although the friction of vehicle tires prevents significant corrosion on the upper surface, the underside and edges of the plates must be protected.

The thickness of steel plates used for wearing surfaces is influenced by strength rather than wearability. Thicker plates are more capable of transmitting loads and resisting buckling or deformation from heavy trucks. Recommended plate thicknesses based on vehicle weight are as follows:

Vehicle weight (tons)	Recommended plate thickness (inches)
< 50	3/16
< 75	1/4
> 75	5/16 or 3/8

Figure 11-20. – Steel running plate wearing surface on a timber bridge deck.

Figure 11-21. – Typical checkered surface pattern on steel running plates.

CONFIGURATION

The configuration of steel plates must be adequate to protect the deck over the expected range in vehicle track widths. These widths vary from less than 5 feet for compact cars to 7 feet or more for off-highway trucks. The inside spacing between plates is commonly 2 to 4 feet, with plate width from 2 to 4 feet, depending on vehicle track widths (Figure 11-22). When approach roadways are curved, additional plate width should be provided to protect the deck from vehicle off-tracking.

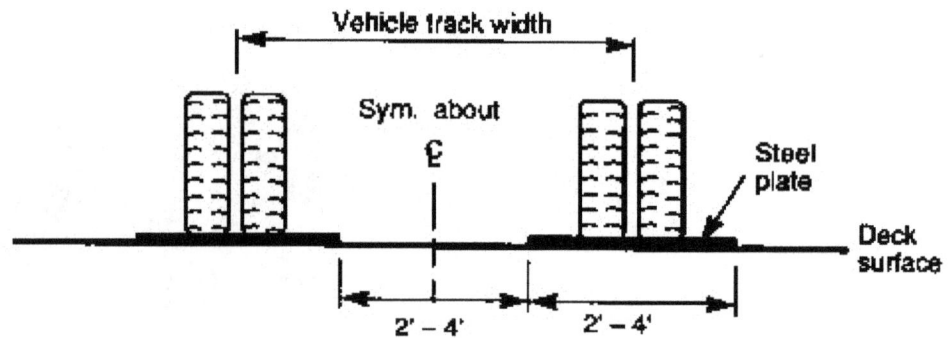

Figure 11-22. - Typical steel plate wearing surface cross section.

Individual plates should be no less than 8 feet long in the longitudinal direction. For short-span bridges, plates may be welded at butt joints to form a continuous surface. However, if welding is done on the deck, precautions must be taken to avoid deck damage during the welding process. On longer spans, continuous plate length should be limited to approximately 12 feet and a 1/4-inch gap left at butt joints to allow for thermal expansion of the steel.

DECK ATTACHMENT

Steel plates are attached to the bridge deck with 1/2-inch-diameter bolts or lag screws (Figure 11-23). Bolts should be provided with malleable iron or steel cut washers and self-locking nuts on the deck underside. Lag screws should be the same length as those recommended for lumber surfaces. Fasteners for steel plates cannot be recessed below the roadway and fastener heads should be smooth to avoid tire damage. Bolts are preferable to lag screws because they provide a more positive connection, although they must be tightened from the deck underside. All fasteners should be galvanized and placed in prebored holes treated with a liquid wood preservative.

The attachment configuration for steel plates is the same for bolts and lag screws (Figure 11-24). Plate ends are attached with three fasteners: one at the plate center and one on each edge. Intermediate fasteners are placed along plate edges at 1-1/2- to 2-1/2-foot intervals. The distance from the center of the fastener to the plate end or edge should be 1-1/2 to 2 inches.

Fastener holes in the steel plates are commonly 1/16 to 1/8 inch larger than the fastener diameter, but may be slotted or oversized to allow for construction tolerances or plate expansion. Whenever possible, plate holes should be located on the flat unpatterned portion of the plate. When slotted or oversized holes are used, a steel cut washer should be placed under the fastener head (washers are not required when dome head bolts are used).

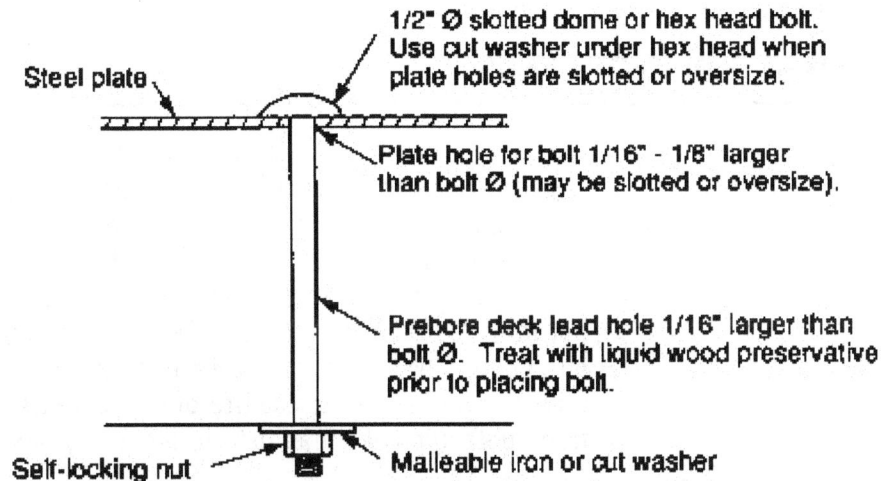

Steel plate

1/2" Ø slotted dome or hex head bolt.
Use cut washer under hex head when
plate holes are slotted or oversize.

Plate hole for bolt 1/16" - 1/8" larger
than bolt Ø (may be slotted or oversize).

Prebore deck lead hole 1/16" larger than
bolt Ø. Treat with liquid wood preservative
prior to placing bolt.

Self-locking nut

Malleable iron or cut washer

A. Bolted attachment

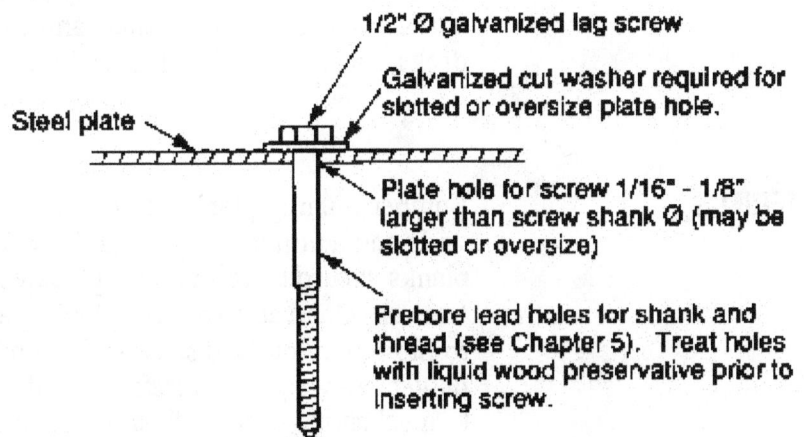

1/2" Ø galvanized lag screw

Galvanized cut washer required for
slotted or oversize plate hole.

Steel plate

Plate hole for screw 1/16" - 1/8"
larger than screw shank Ø (may be
slotted or oversize)

Prebore lead holes for shank and
thread (see Chapter 5). Treat holes
with liquid wood preservative prior to
inserting screw.

B. Lag screw attachment

Figure 11-23. - Bolt and lag screw attachment details for steel plate wearing surfaces.

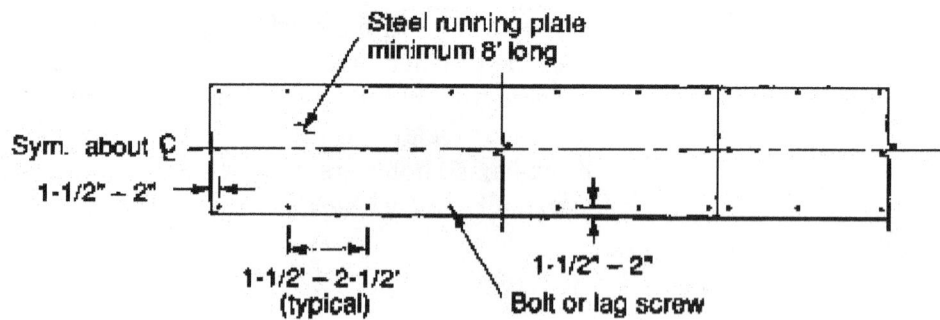

Figure 11-24. - Steel plate wearing surface attachment pattern using bolts or lag screws.

11.7 LUMBER RUNNING PLANKS

Lumber running planks are a series of sawn lumber planks placed edge to edge to form two longitudinal surfaces (Figure 11-25). They are similar to steel running plates in that they provide a wearing surface over the portion of the deck intended only for vehicle tracking. Lumber running planks are used on single-lane, low-speed, rural bridges and on special-purpose roads not intended for public traffic. Service life depends on the traffic weight and volume as well as the surface thickness. Under light loads, planks typically provide a service life of 4 to 8 years. When subjected to heavy truck traffic, planks may deteriorate in 2 years or less.

There are two notable disadvantages with lumber running planks. First, the difference in elevation between the wearing surface and the deck can be a safety hazard when vehicles track off the surface. This hazard is most serious for light passenger traffic or when thick planks are used. Second, the opening between the running planks serves as a trap for debris, requires increased maintenance, and can cause water to pond on watertight decks, creating a safety hazard to motorists and increasing the potential for deck decay.

MATERIALS

Lumber running planks are constructed of planks that are 10 to 12 inches wide and a minimum of 8 feet long. Planed edges are not required, and planks should be left in a rough-sawn condition for enhanced vehicle traction. Considerations for plank species and grain orientation are similar to those previously discussed for lumber surfaces. Plank thickness is based on vehicle weight and traffic density. Running planks are more susceptible to mechanical damage than are comparable lumber surfaces because of vehicle off-tracking on the outside plank edges. Planks for bridges that carry heavy vehicles must be thicker. Recommended plank thicknesses based on vehicle weight are given below; however, running planks on bridges intended for public traffic should not be more than 2 inches thick because cars may lose control if they leave the plank surface.

11-28

Figure 11-25. - Timber bridge deck with lumber running planks.

Vehicle weight (tons)	Recommended plank thickness (inches)
< 50	3
< 100	4
> 100	5

Running planks will normally wear out or deteriorate from mechanical damage before they decay. Treatment with preservatives is required only when low traffic volumes or light loads will result in reduced abrasion and mechanical damage. Under these conditions, biological attack may become important and planks may be treated with waterborne preservatives for extended life.

CONFIGURATION

The transverse configuration of lumber running planks is based on anticipated vehicle track widths discussed for steel running plates (Section 11.6). In addition, surface spacing and width should be based on consideration of elevation differences between the deck and wearing surface. Additional width should be provided as necessary to reduce the potential for vehicle off-tracking, especially when passenger vehicles use the structure or when approach roadways are curved. For passenger vehicles, a maximum spacing of 2 feet between surfaces is recommended (Figure 11-26). Surface widths vary for different track widths, but are commonly four planks wide (approximately 4 feet when nominal 12-inch planks are used).

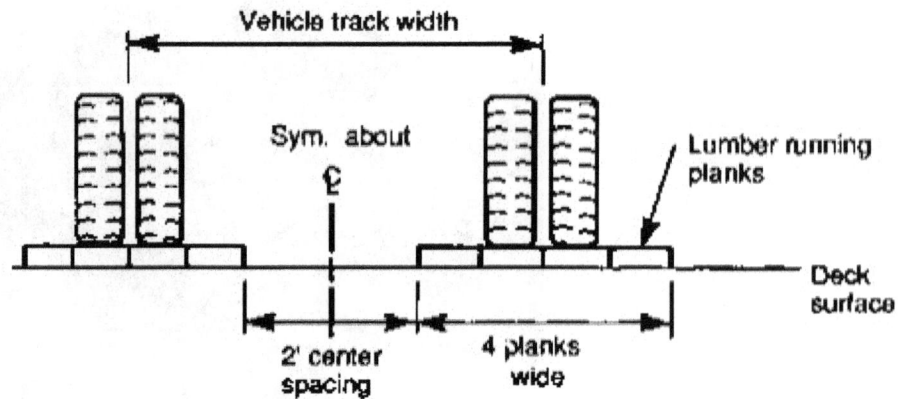

Figure 11-26. - Typical lumber running plank wearing surface cross section.

In the longitudinal direction, lumber running planks are similar to full lumber surfaces. The most economical configuration is an alternating repetition of plank lengths with odd lengths at the bridge ends (Figure 11-27). End joints in adjacent planks should be staggered a minimum of 3 feet, and plank ends on bridge approaches should be beveled to minimize vehicle impact.

DECK ATTACHMENT

Lumber running planks are attached to the deck with spikes or lag screws, as discussed for lumber surfaces. A bolted attachment configuration that employs threaded rods and steel angle brackets can also be used (Figure 11-28). Using this bolted configuration, deck attachment holes in glulam panel decks can be bored before preservative treatment of the panels. Thus, running planks can be installed and replaced without boring

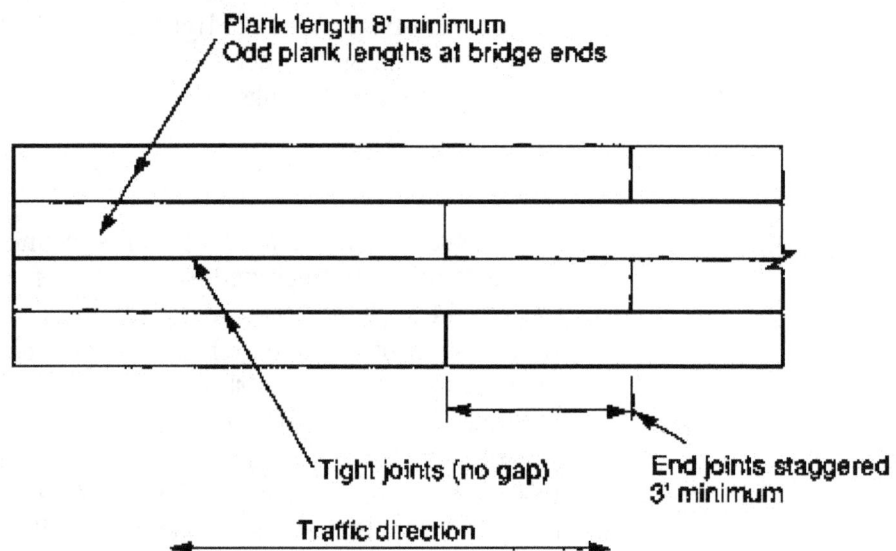

Figure 11-27. - Typical plank layout for a lumber running plank wearing surface.

11-30

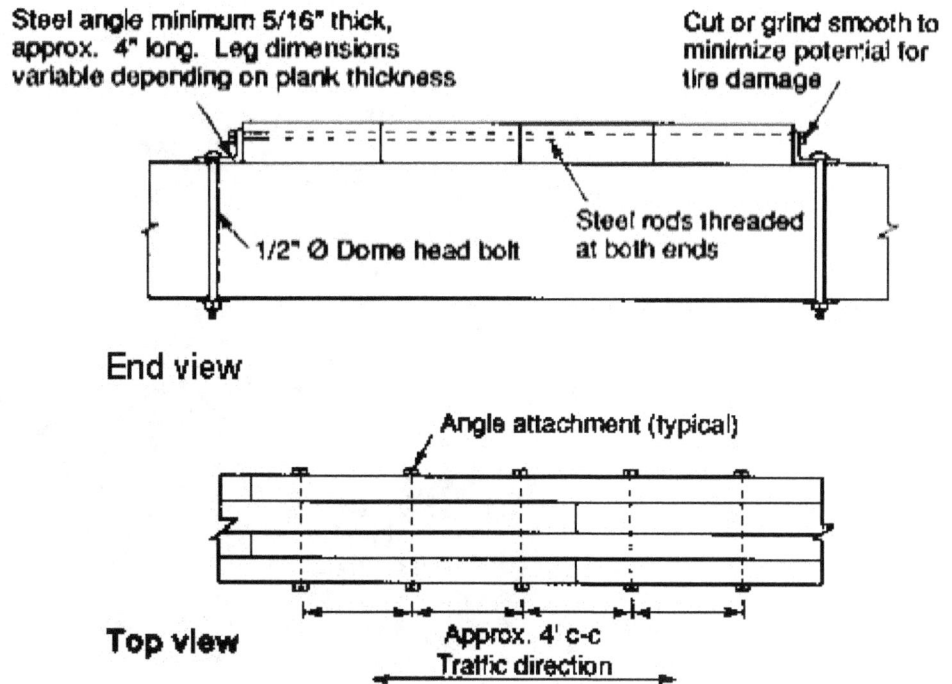

Steel angle minimum 5/16" thick, approx. 4" long. Leg dimensions variable depending on plank thickness

Cut or grind smooth to minimize potential for tire damage

1/2" Ø Dome head bolt

Steel rods threaded at both ends

End view

Angle attachment (typical)

Top view

Approx. 4' c-c
Traffic direction

Figure 11-28. - Attachment detail for lumber running planks using threaded rods and steel angle brackets.

additional deck holes. However, plank installation and replacement using this attachment configuration is more difficult, compared to conventional attachment with spikes or lag screws.

11.8 AGGREGATE SURFACE

Aggregate wearing surfaces consist of a layer of crushed rock or other material placed across the bridge deck (Figure 11-29). These surfaces are inexpensive, easy to construct and maintain, and blend into the surrounding landscape. Aggregate surfaces are used primarily on native log stringer bridges or other temporary structures on low-volume, unpaved roads. They are not commonly used on lumber or glulam decks because they are heavy, hold moisture, and can cause severe abrasion to the bridge deck when the surface thickness is reduced by traffic.

MATERIALS

Aggregate surfaces are constructed of any material that provides a good traffic surface and drains well. Materials that are frequently used include gravel, crushed rock, pit-run, shot rock, coarse sand, and coarse mineral soil. The material should provide a good running surface and resist decomposition from moisture and repeated vehicle loading.

Figure 11-29. - Aggregate wearing surface on a timber bridge deck.

CONFIGURATION

The depth of an aggregate surface must be sufficient to prevent abrasion and protect the deck during maintenance operations such as grading and snow removal. Minimum recommended depths are 4 inches for light vehicles and 6 inches for heavy truck traffic. In either case, depth should not be less than three times the diameter of the largest material in the surface. Where considerations for stream siltation are important, aggregate surfaces are placed on geotextile fabric (Figure 11-30).

11.9 SELECTED REFERENCES

1. American Association of State Highway and Transportation Officials. 1983. Standard specifications for highway bridges. 13th ed. Washington, DC: American Association of State Highway and Transportation Officials 394 p.

2. American Association of State Highway and Transportation Officials. 1986. AASHTO guide for design of pavement structures. Washington, DC: American Association of State Highway and Transportation Officials. [350 p.].

3. American Institute of Timber Construction. 1985. Timber construction manual. 3d ed. New York: John Wiley and Sons, Inc. 836 p.

4. Anderson, L.O.; Heebink, T.B.; Oviatt, A.E. 1971. Construction guide for exposed wood decks. Portland, OR: U.S. Department of Agriculture, Forest Service, Pacific Northwest Forest and Range Experiment Station. 78 p.

Figure 11-30. - Geotextile fabric placed across a log stringer bridge before placement of an aggregate wearing surface (photo courtesy of Neil Newlun, USDA Forest Service).

5. Asphalt Institute. 1979. A basic asphalt emulsion manual. Man. Series No. 4. College Park, MD: The Asphalt Institute. 262 p.

6. Asphalt Institute. 1983. A simplified method for the design of asphalt overlays for light to medium traffic pavements. Inf. Series No. 139 (IS-139). College Park, MD: The Asphalt Institute. 4 p.

7. Asphalt Institute. 1983. Asphalt technology and construction practices, instructor's guide. Educ. series No. 1 (ES-1). College Park, MD: The Asphalt Institute. [300 p.].

8. Asphalt Institute. 1981. Design techniques to minimize low-temperature asphalt pavement transverse cracking. Res. Rep. No. 81-1. College Park, MD: The Asphalt Institute. 75 p.

9. Asphalt Institute. 1973. Model specifications for small paving jobs. Constr. Leafl. No. 2 (CL-2). College Park, MD: The Asphalt Institute. 4p.

10. Barnhart, J.E. 1986. Ohio's experiences with treated timber for bridge construction. In: Trans. Res. Rec. 1053. Washington, DC: National Academy of Sciences, National Research Council, Transportation Research Board: 56-58.

11. Bruesch, L.; Pelzner, A. 1976. Surfacing treated decks with bituminous materials. U.S. Department of Agriculture, Forest Service. Engineering Field Notes. 84: 1-3.

12. Canadian Institute of Timber Construction. 1970. Modern timber bridges, some standards and details. 3d ed. Ottawa, Can.: Canadian Institute of Timber Construction. 48 p.

13. Faurot, R.A. 1984. Use of geotextiles as bridge paving underlayment. U.S. Department of Agriculture, Forest Service. Engineering Field Notes. 16: 15-17.

14. Nagy, M.M.; Trebett, J.T.; Wellburn, G.V. 1980. Log bridge construction handbook. Vancouver, Can.: Forest Engineering Research Institute of Canada. 421 p.

15. National Forest Products Association. 1986. National design specification for wood construction. Washington, DC: National Forest Products Association. 87 p.

16. Sprinkel, M.M. 1978. Glulam timber deck bridges. VHTRC 79-R26. Charlottesville, VA: Virginia Highway and Transportation Research Council. 33 p.

17. US. Department of Transportation, Federal Highway Administration. 1979. Standard plans for highway bridges. Timber bridges. Washington, DC: U.S. Department of Transportation, Federal Highway Administration. Vol. 3. 19 p.

18. Weyerhaeuser Company. 1980. Weyerhaeuser glulam wood bridge systems. Tacoma, WA: Weyerhaeuser Co. 114 p.

CHAPTER 12

TIMBER BRIDGE FABRICATION AND CONSTRUCTION

12.1 INTRODUCTION

The performance and serviceability of any bridge depend on the accuracy and quality of fabrication and construction. When correct procedures are followed, the bridge can be economically built and can provide many years of service. When improper or negligent practices dominate, both the economics and long-term serviceability of the bridge will be adversely affected. Timber bridges are especially suited to economical fabrication and construction because they can be completely prefabricated at a shop facility and shipped to the project site for assembly. Components are lightweight compared to those using other bridge materials and can be quickly installed without highly skilled labor or specialized equipment.

This chapter addresses proper techniques and procedures for timber bridge fabrication and construction. Topics include the preparation of engineering drawings, bridge fabrication, handling, transportation, storage, and construction. Discussions are general in nature and are applicable to most timber bridge types. Because construction specifications and administrative procedures vary for different projects and jurisdictions, details related to these two areas are not included.

12.2 ENGINEERING DRAWINGS

Successful bridge fabrication and construction depend on the accuracy and completeness of the engineering drawings. Two types of drawings are normally used: design drawings and shop drawings. Design drawings show the structure configuration and provide information necessary for field assembly. Shop drawings provide more detailed information for the fabrication of individual components. Design drawings are prepared by the organization responsible for the design of the structure. The same organization may also prepare shop drawings, or the fabricator may prepare them from the design drawings. In some cases, the design drawings are completed in sufficient detail to serve as both design drawings and shop drawings.

This chapter was coauthored by Michael A. Ritter and Charles B. Schmokel, Bridge Systems Manager, Western Wood Structures, Inc.

Engineering drawings are usually the only means of communicating design and fabrication information to the material fabricator and construction crew. They must be complete, legible, and accurate, and must contain all necessary information, including material specifications and material lists. Individual components and assembly details should be laid out clearly with all dimensions, hole sizes, and assembly locations accurately shown. When laying out the drawings it is desirable to assign mark numbers to individual timber members. These numbers can be placed on the components during fabrication where they will help identify material during erection. Some common symbols and abbreviations used for detailing timber structures are shown in Table 12-1.

The drawings should include material specifications that are referenced to standard specifications discussed in previous chapters (AITC, AWA, AASHTO, or ASTM) and should include specific information related to timber grades, surfacing, preservative treatments, steel and hardware grades, and corrosion protection. Material information should be summarized in a materials list that includes the required number, size, and weight for all components and hardware. Such lists are important because they often serve as the basis for competitive bidding, transportation estimates, and checklists of material quantities delivered to the project site. In addition, drawings should include any special assembly instructions or requirements for transportation, handling, or storage. Complete, accurate drawings increase the likelihood that correct materials and quantities will arrive at the jobsite. An example of a good-quality engineering drawing for a timber bridge is shown in Figure 12-1.

Drawing preparation is an integral part of the design process. As such, the attention given to detailing can have a substantial effect on both the economy and long-term performance of the structure. When preparing drawings, consideration should be given to material selection, ease of assembly, fabrication and erection tolerances, and details that affect bridge performance. Some of the important points related to detailing and specifications discussed in previous chapters are reiterated as follows:

1. Use standard material sizes and grades for glulam and sawn lumber (Chapter 3).

2. Use timber species that are readily treatable with preservatives (Chapter 4).

3. Specify appropriate wood preservatives for the intended application (Chapter 4). Oil-type preservatives, such as creosote, pentachlorophenol, or copper naphthenate in heavy oil, provide the best protection for bridge components. When members are subject to human contact, waterborne preservatives or oil-type preservatives in light petroleum solvents should be used.

Table 12-1. - Typical detailing symbols and abbreviations for timber.

Detailing symbols[a]

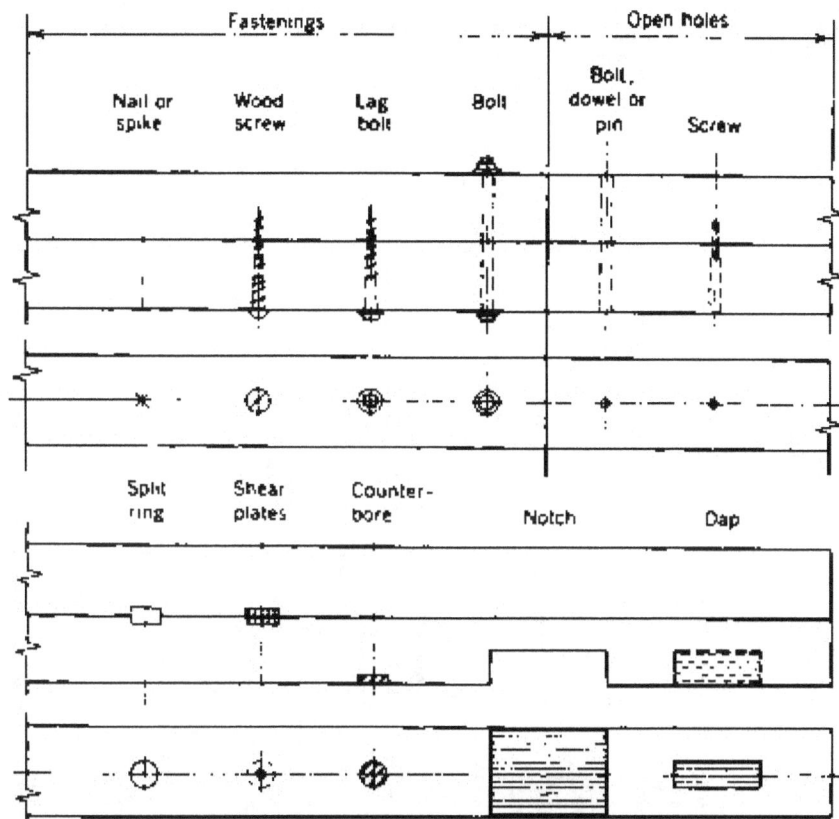

Detailing abbreviations[b]

Angle(s) (metal section)	∠, ∠s	Malleable iron washer	M.I.
Both sides	B.S.	Mark(ed)	MK
Carriage bolt	C.B.	Mold loft	ML
Centerline	₵	Near side	N.S.
Center-to-center	c-c	Ogee (cast) washer	O.G.
Chamfer	chfr	On centers	o.c.
Channel(s) (metal section)	[, [s	Plate	₽
Counterbore	cbr	Radius	R, rad
Countersink	csk	Steel section	S
Cut washer	C.W.	Shear plate	Sh. Pl.
Diameter	φ	Split ring	S.R.
Each side	E.S.	Threaded rod	thrd
Far side	F.S.	Turnbuckle	tbkl
Lag bolt	L.B.	Steel section	W
Lamination	lam	Wrought washer	W.W.
Machine bolt	M.B.		

[a] Symbols are intended for large-scale details only. For all symbols, the sizes and quantities required must be indicated on the drawings. If sizes are mixed, or if these symbols do not provide a clear explanation of the connection, a detail showing the hardware arrangement should be included.

[b] Additional abbreviations are given in Chapter 16.

From *Timber Construction Manual;*[2] American Institute of Timber Construction; ©1974, John Wiley & Sons, Inc. Reprinted by permission.

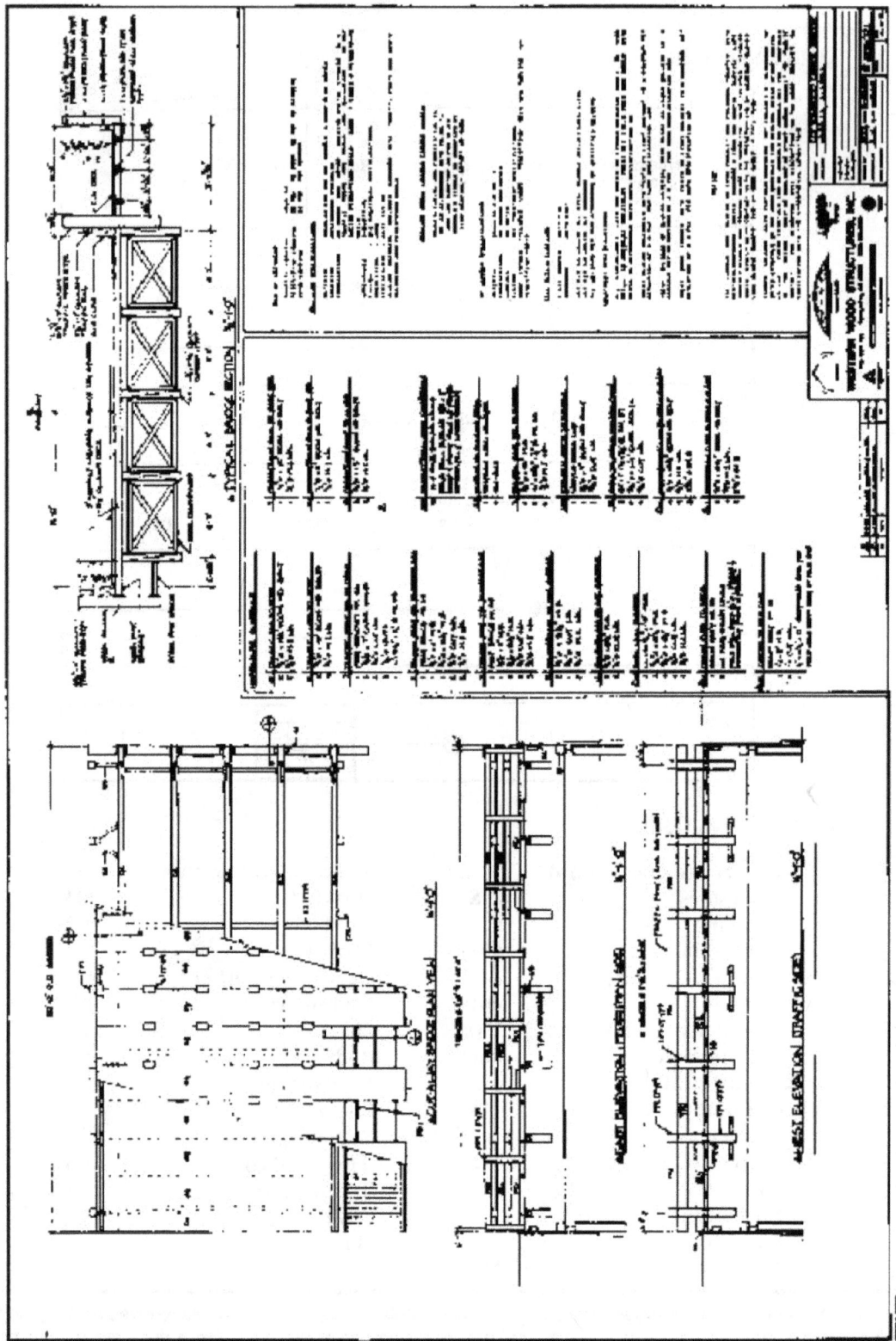

Figure 12-1.—Example of an engineering drawing for a timber bridge (photo courtesy of Western Wood Structures, Inc.).

12-4

4. Detail members so that fabrication can be completed before pressure treatment with preservatives.

5. Use standardized members in a repetitious arrangement, especially for glulam deck panels.

6. Avoid details that trap water, debris, or other material.

7. Use standard connection details whenever practical. Typical connection details are given in *AITC 104 - Typical Construction Details.*[2]

As a final step in drawing preparation, it is important that all work be independently checked for completeness and accuracy before putting an order into production. Do not depend on the fabricator or contractor to check the accuracy of dimensions, quantities, or specifications. A few hours of checking in the office can save thousands of dollars in field expenses.

Familiarity of the design engineer and draftsperson with material availability, cost, and common fabrication and construction practices can greatly improve economy and ease of construction. It is beneficial for design personnel to visit fabrication facilities and construction sites to observe procedures. This is also a good opportunity to discuss processes with fabrication and construction personnel and solicit comments on methods for improving the fabrication and installation of future timber bridges.

12.3 BRIDGE FABRICATION

Accurate fabrication is essential for the quick installation of a timber bridge. It is more economical to accomplish as much work as possible at the fabrication plant since costs are normally lower there than in the field. Also, plant equipment is generally faster and more accurate (Figure 12-2).

Glulam and dressed lumber is initially manufactured to the dimensions discussed in Chapter 3. The expected tolerance for the holes and cuts made during fabrication is approximately 1/16 inch. In some cases, tolerances may be slightly greater depending on the type of component and the condition of the timber at the time of fabrication. A member that is precisely fabricated at a shop facility may undergo slight dimensional changes because of variations in moisture content during treatment, transportation, and storage. Therefore, minor dimensional changes may occur before the material reaches the construction site.

Figure 12-2.- Shop fabrication of timber bridge components, such as these nail-laminated lumber deck panels, is more accurate and more cost effective than field fabrication (photo courtesy of Wheeler Consolidated, Inc.).

Most glulam and sawn lumber manufacturers provide fabrication services such as trimming, drilling, counter boring, notching, tapering, and in some cases, incising. Some glulam manufacturers also have layout areas where they can pattern fabrication templates for more accurate fabrication of multiple members (Figure 12-3). Many treating plants offer fabrication services comparable to those of material suppliers and manufacturers, including incising. A number of businesses deal strictly in timber fabrication. These operations have fabrication capabilities similar to those of sawn lumber manufacturers or treating plants and can usually cut, bore, and incise timber components, as well as package units ready for pressure treatment with preservatives. This results in reduced handling requirements at the treating plant, and can lower treating costs. Some manufacturers and fabricators will also preassemble bridge components to ensure proper fit; however, this service can be relatively expensive.

Most timber bridges require fabricated steel and fastening hardware. Fabrication procedures and tolerances for steel are just as critical as those for wood, and a reputable steel fabricator should be used to ensure that correct fabrication procedures are used. Steel bearing shoes, hangers, or saddles should be manufactured approximately 1/4 inch larger than the member over which they will fit. Holes should be 1/16 inch oversized, or slotted where provisions for movement or field adjustment are required. Weldments or other protrusions that may conflict with other bridge components should be ground smooth. Bolts, nuts, washers, and other

Figure 12-3. - Template production area at a glulam manufacturing plant. Plywood templates shown in the background are used to ensure consistent, accurate, and complete fabrication of multiple glulam members.

hardware can be purchased from hardware manufacturers and suppliers. Most hardware required for bridge construction is standard and often readily available. When special hardware is required, additional time should be allowed for manufacturing. All steel components should be hot-dip galvanized or painted for corrosion protection and longevity.

Several businesses currently sell complete timber bridge packages. These packages, which may include structural design at the option of the purchaser, provide all fabricated bridge materials including treated sawn lumber or glulam, steel components, and hardware. The materials are packaged for a specific project and shipped to the construction site in bundles and containers, ready for construction. In many cases, packages of this type are the most economical source of bridge material.

12.4 TRANSPORTATION, HANDLING, AND STORAGE

Timber is a naturally durable material that can withstand moderate abuse without damage. However, it is necessary that reasonable care be exercised in transportation, handling, and storage to ensure that good quality is maintained.

12-7

TRANSPORTATION

Bridge materials can be transported from the plant to the jobsite by trucks, rail cars, or barges. Highway trucks are the most common method of transportation and are capable of hauling between 45,000 and 60,000 pounds (maxi-trucks). It may be necessary to obtain special equipment to haul long lengths (over 50 feet), wide loads (over 8 feet), and members with a large amount of curvature. Long material may also require pilot cars or steering trailers. In most States, there are curfews, permits, or other regulatory laws that must be considered. Some States have length limitations that may not permit hauling long material. In addition, roads to the project site may have limited vertical clearance and could have hairpin turns, which require special equipment or handling.

Many manufacturing and treating plants have rail sidings, but most job-sites do not. This means that material must be taken off the rail cars at some point and transported by truck to final destinations. Rail cars have width, length, and other restrictions, including minimum weight limits that may increase costs. Some water-locked sites may only be accessible by barge. The same types of length, width, and weight restrictions that apply to rail transport may apply to barges and should be considered.

Timber components are normally stacked in units for easy loading and transportation. It is common and accepted practice to bundle a number of pieces and band them together with steel straps. When steel straps are used, corner guards must be placed to prevent damage to the wood (Figure 12-4). It is advisable to place a piece of nominal 2-inch lumber across the bundle, under the band, to protect the material and provide access for lifting when bundles are stacked. In most cases, members are handled and transported flat. Short members can be transported on a flatbed truck, while loads over 75 feet may require a log truck or other specialized vehicle (Figure 12-5). If more than one member is being trucked at a time, they should be strapped together with nylon binders.

With competitive prices and the ability to haul materials directly to the jobsite, truck delivery is normally most economical. Bridge members can be hauled hundreds or even thousands of miles to a jobsite and still be competitive. For example, it is not uncommon for glulam bridges manu-factured in Oregon to win competitive bids for projects in Virginia or Florida. Therefore, transportation distance should not be a limiting factor in soliciting project bids.

HANDLING

Treated sawn lumber and glulam must be handled with reasonable care to avoid breaking the material or the preservative treatment envelope. Minor scuffs usually do not affect the end use of the material unless they cause an appearance problem. More severe damage, such as cuts or breaks in the tension laminations of glulam beams, can have adverse effects on struc-tural capacity. It is recommended that timber members be handled with nylon slings to prevent damage. The sling is placed around the member

Figure 12-4. - (A) Properly placed corner guards prevent wood damage from steel straps (photo courtesy of Western Wood Structures, Inc.). (B) Wood damage resulting from steel straps when corner guards are not used.

Figure 12-5. - Truck transportation of timber bridge components. (A) Short members, such as these nail-laminated lumber deck panels, can be transported on a flatbed truck (photo courtesy of Wheeler Consolidated, Inc.). (B) Long glulam beams require specialized vehicles such as this log truck and dolly (photo courtesy of Western Wood Structures, Inc.)

with the loop at a corner (choke position) so the member rides vertically
(Figure 12-6). Chains or cables are not recommended because they can cut
into the wood surface. If they are the only rigging available, steel corner
protectors must be used to protect the wood members.

Figure 12-6. - Proper placement of a nylon sling on a glulam beam, with the loop in the
sling at the corner of the beam.

Because of their relatively light weight, lifting timber components can be
done with a variety of equipment, depending on what is available near the
project site. Cranes are usually the most desirable, but forklifts, front-end
loaders, backhoes, or other equipment can be used, depending on the size
and type of component. Short glulam members can be picked up and
moved in the flat position while longer beams must be tipped and lifted on
edge (Figure 12-7). When glulam deck panels or beams are lifted flatwise,
they should not be lifted by the edges parallel to the wide face of the
laminations. This can induce high bending stress perpendicular to grain
and may cause structural damage. Members of this type should be lifted in
a vertical position, with the laminations horizontal (supports placed across
the wide face of the lamination) or with fabricated steel C-shaped brackets
that fit over the member ends (Figure 12-8).

Figure 12-7. - A large glulam bridge beam is tipped from its horizontal storage position before being lifted into place on edge (photo courtesy of Tim Chittenden, USDA Forest Service).

Handling preservative-treated timber is generally not hazardous to construction workers. However, a few common-sense procedures should be followed. Workers should use chemically impervious gloves and wear long-sleeved shirts and long pants when working with treated materials. Eye protection (goggles and face masks) should be used when sawing or machining treated lumber. After handling treated wood, workers should wash exposed skin areas carefully before eating, drinking, or using tobacco products. By law, all shipments of pressure-treated wood must be accompanied by an EPA Consumer Information Sheet (copies are included in Chapter 16). All workers should read and understand these sheets before construction begins.

STORAGE

For short- or long-term storage, timber should be neatly stacked in dry, level areas that are clear of plant growth and debris (Figure 12-9). The bottom layer of material should be approximately 8 inches above ground level and be supported on spacer blocks placed 10 to 15 feet apart, de-

12-12

Figure 12-8. - Lifting a glulam deck panel from steel C-brackets that fit over the panel ends.

pending on the material. If sagging is evident, additional supports should be added. Layers in the stack are added on 2-inch nominal sawn-lumber spacers (stickers) that extend across the full width of the stack. The stickers separate the layers to allow free air circulation and provide access for lifting equipment. It is important that all stickers be aligned vertically and be spaced at regular intervals. Otherwise, stacked members may be subjected to bending stress and might twist or warp during extended storage.

When properly stacked, it is normally not necessary to cover timber that has been treated with oil-type preservatives. Free air circulation is all that is required. If dried sawn lumber treated with waterborne preservatives is stored, a cover may be desirable for protection during inclement weather conditions, depending on the anticipated length of storage. When covers are necessary, impervious membranes such as polyethylene film should not be left in place during dry weather because they trap moisture that evaporates from the ground or from the timber members.

12.5 BRIDGE PRECONSTRUCTION

Before the arrival of bridge materials, a thorough job of preconstruction engineering at the bridge site can save time and money during construction. The first step is to review all drawings and specifications to understand the sequence of construction and any special handling or equipment that might be required. If there are questions, the bridge designer or

12-13

Figure 12-9. - Glulam deck panels stacked for storage.

supplier should be contacted for clarification. Personnel and equipment should not be kept idle while drawings are being interpreted. After drawings are reviewed, the bridge substructure should be inspected for correct placement. Sills must be spaced the correct distance apart and at the correct elevations shown on the drawings. Holes for the bridge bearing anchor bolts must be in their correct positions, both longitudinally and transversely. The substructures should be measured corner to corner to verify squareness (Figure 12-10). Many of the problems that develop in timber bridge construction can be eliminated by doing this preliminary review and inspection before actual receipt of materials for construction of the bridge superstructure.

To efficiently construct a timber bridge, proper lifting equipment and tools must be available at the jobsite. A crane is usually most practical for large components such as glulam beams or large, prefabricated bridge sections (Figure 12-11), while other types of equipment such as forklifts, front-end loaders, or backhoes can be used for smaller components. When determining required equipment capacity, the weights of bridge components are normally on the drawings or can be calculated from member dimensions. If this is not possible, weights can be obtained from the bridge designer or supplier. When possible, lifting equipment should be provided with two nylon slings that are long enough to be used in the choke position and strong enough to lift at least half the weight of the largest member. Deck-lifting brackets should also be available when glulam panels are a part of the project.

12-14

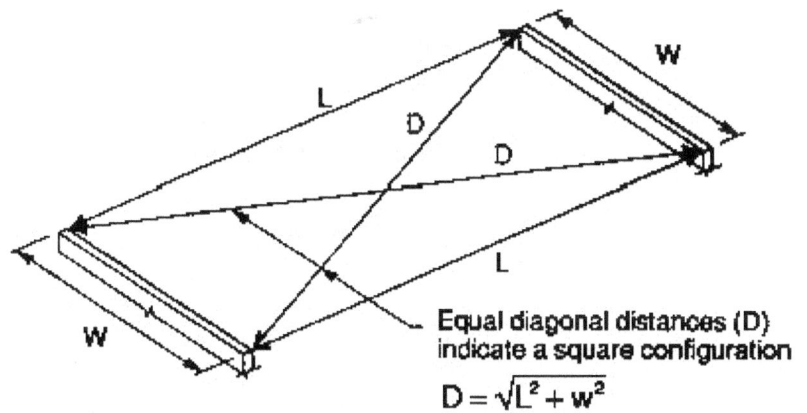

Figure 12-10. — Verifying squareness and alignment of substructures by measurement of diagonals.

Equal diagonal distances (D) indicate a square configuration

$$D = \sqrt{L^2 + w^2}$$

Figure 12-11. - One-half of a stress-laminated parallel-chord bridge is lifted into place by a crane.

It is helpful in bridge construction to have a power source or generator for electric tools. Even though bridges are usually totally prefabricated, field construction and adjustment tools such as drills, reamers, and power saws should be available. Cutting torches and welders may also be required, in unusual cases, to modify steel members. A supply of wood preservative (which may be shipped with the bridge) and galvanizing paint should be

12-15

available for field touchup during construction. Other tools, including impact wrenches, sockets, pry bars, come-a-longs, sledge hammers, and spud wrenches are helpful. For most types of timber bridges, access to the bridge underside is necessary for placement of transverse bracing and fasteners. A tall ladder may be sufficient for small structures, but for most bridges a scaffold system is required. Scaffolding should be movable from the top of the bridge using staffpower or available equipment (Figure 12-12).

Before material delivery, survey the site to make sure there is adequate access for delivery trucks and equipment. Most trucking companies allow an hour of free unloading time and then charge an hourly standby fee, so prompt truck unloading is important. Locate adequate material storage sites before unloading begins. When selecting storage locations, consider lifting equipment access to various stacks.

When bridge materials are delivered to the site, immediately make an initial visual inspection and inventory of materials. Any items that are obviously damaged or missing should be noted on the bill of lading accompanying the shipment. If the damages appear serious, notify the supplier at once. Next, all material should be carefully sorted to make sure the proper sizes and quantities are present. Verify dimensions of all fabricated components as soon as possible after delivery. It is better to find incorrect sizes or fabrication errors when the material arrives than to wait until construction has started.

12.6 BRIDGE ASSEMBLY

The methods and techniques of bridge assembly differ slightly among bridge types and materials. This section discusses bridge assembly for glulam beam bridges with transverse glulam deck panels. These general procedures also apply to longitudinal deck bridges or to bridges with sawn lumber beams or decking. More specific assembly procedures are presented in case histories in Chapter 15.

For beam-type glulam bridges, assembly is normally started with one of the outside beams. The beam is lifted upright from the storage stack using two nylon slings in the choke positions, placed approximately at the beam one-third points. When available, a spreader beam is normally used for long members. When the beam is upright, shop drawings should be checked to ensure that the beam has the correct mark number and is standing with the top mark up. In some designs, beam fabrication may not be symmetrical so it is important that the member is properly orientated. At this point, it is desirable to attach some of the steel components, such as bearing shoes, steel cross-frames, and railing brackets, to the beams. It is easier and safer to attach these components while the beam is on the ground and readily accessible. After steel components are in place, the

Figure 12-12. - Typical scaffolding configurations for accessing the underside of a glulam beam bridge. (A) Along the deck overhang (photo courtesy of Tim Chittenden, USDA Forest Service). (B) Between beams (photo courtesy of Western Wood Structures, Inc.).

beam is lifted into position (Figure 12-13). Ropes should be tied at each end of the beam so personnel can keep it aligned as it is moved. The beam is positioned on the substructure, and anchor bolts are placed and finger-tightened. Nuts should not be tightened until the bridge is completely assembled to allow for adjustment during the course of construction. After the first beam is in place, the remainder of the glulam beams are correctly orientated and lifted into place in the same manner (Figure 12-14).

Figure 12-13. - Glulam beam is lifted into position with steel components in place. Note the use of a spreader beam and nylon straps for lifting (photo courtesy of Western Wood Structures, Inc.).

After all beams are in place, deck panels can be placed. Again, it is impor-tant to check the mark numbers on the deck panels against those on the shop drawings to ensure proper placement and sequence. If the panels are not interchangeable, or are not symmetrical, panel size and fabrication should be visually verified against the shop drawings. Panels should also be checked to be sure they are not upside down; however, some deck panel layouts require that the last panel at the bridge ends be turned upside down to match bolt spacing. Deck panels can be picked up and set easily with a backhoe, forklift, or crane using deck-lifting brackets (Figure 12-15). As panels are placed, a mastic sealer is usually applied to the panel interface (Chapter 7). To aid in panel placement, it is beneficial to tack a piece of colored flagging at the center of each panel to assist in visually aligning the panels on the beams. Once the panels are in place they can be easily adjusted with a pry bar. When all deck panels are on the bridge, the crane (if used) may be discharged and other, less expensive equipment may be used to finish the bridge construction.

12-18

Figure 12-14. - Glulam beams for this bridge are placed from left to right, starting with the outside beam. As the beams are placed in position, bolts are inserted through bearings and steel cross-frames (photo courtesy of Western Wood Structures, Inc.).

Figure 12-15. - Glulam deck panels are sequentially lifted into place with a backhoe, using C-brackets over the panel ends.

If load-transferring devices, such as dowels, are used between the deck panels, they are installed progressively as the panels are placed. For dowels, the first deck panel should be placed in position and attached to supporting beams. This panel then serves as a starting point for installing the balance of the deck panels. Dowels are partially inserted into the stationary deck panel, the next panel is moved into position, and the dowels are inserted into the corresponding dowel holes. The deck panels are then either pulled together with come-a-longs or pushed together with equipment or jacks (Figure 12-16). For a complete description of the installation procedures for a doweled glulam deck, refer to *Erection Procedure for Glued Laminated Timber Bridge Decks With Dowel Connector.*

If the deck panels are attached to beams with brackets or clips, the bolt holes in the panels must match the routed grooves in the beams. The deck connection bolts are placed through the deck panels and the brackets or clips are loosely attached to the beams. Next, the curbs are set into place and fastened to the deck with connecting bolts. If some of the holes do not line up, a spud wrench, pry bar, or sledgehammer (with a softening device) may be used to slightly adjust the deck or curb so holes align (Figure 12-17). From time to time, some components may not fit perfectly because of minor misfabrication, and some adjustments may be required. Twisting or tapping a bolt may help solve the problem. It may be necessary in some cases to ream or enlarge holes so that the bolt will slide through. Before reaming, the engineer should determine if enlarging the

Figure 12-16. - Doweled glulam deck panels are pulled together with a come-a-long. Note the steel C-brackets that are placed over the panel edges to prevent damage during jacking.

Figure 12-17. - Workers use a pry bar to align the curb for deck attachment.

hole will alter the strength of the connection. This type of reaming will usually not expose untreated wood, and field treatment will not be necessary; however, if new holes are drilled or members are cut, field treating should be done in accordance with procedures discussed later in this chapter.

Once the curbs are in place, rail posts and railing are installed using the same procedures as those used in bolt alignment. Alignment and appearance are important on the curbs and railings, and the system must be level and straight. Once this is done, all bolts on the bridge can be tightened. The quickest way to tighten nuts is to use an air or electric impact wrench and sockets. A torque ratchet is desirable to ensure that bolts are tightened to approximately 50 ft-lb of torque. This can be noted visually when the washers begin to pull into the treated wood. It is necessary to go under the bridge to tighten most types of deck-attachment hardware, and a ladder or scaffolding will be required for access.

After the bridge is assembled, and all connections are tightened, the substructure backwalls can be placed and the approach roadway can be backfilled. Backfill should be placed from both sides at approximately the same rate to prevent the bridge from being pushed out of line by the uneven backfill loads. Once backfilling is complete, and the roadway approaches are in place, the wearing surface is placed over the bridge deck.

Most contractors are surprised at how quickly a glulam bridge can be erected if the bridge shop drawings are accurate, if the bridge is properly fabricated, and if the preconstruction techniques and reviews are followed. Cases have been documented where 60-foot glulam beam bridges have been completely assembled in 60 work hours. Once a crew has gained experience, the bridge construction time and cost can be reduced even further, thus making a glulam bridge one of the fastest, easiest, and least expensive bridges to install.

12.7 FIELD TREATMENTS

Occasionally, treated timber bridges may be damaged or require field modification during installation. This can expose untreated wood that must be field treated to protect the member from future decay and deterioration. Field treatment procedures are outlined in AWPA Standard M4-84,[4] which requires that all cuts, holes, and injuries to treated wood be protected by brushing, spraying, dipping, or soaking in an approved preservative. Field application is not nearly as effective as pressure treating, so field fabrication and field treatment should be kept at a minimum.

Most timber bridges are pressure treated with such oil-type preservatives as creosote, creosote solutions, pentachlorophenol, and occasionally, waterborne salts. In the past, these treating solutions could be purchased over the counter from a number of sources and then applied by construction crews as necessary, but because of recent EPA rulings, a state applicator's license is required for the purchase and application of most wood preservatives. Even with an applicator's license, it can be very difficult to locate and obtain common wood preservative solutions. The most widely available and approved field treatment solution is copper naphthenate in an oil solvent. This product is available over the counter and does not require an applicator's license. AWPA Standard M4 states that copper naphthenate solutions may be used to field treat wood that was originally treated with creosote, creosote solutions, pentachlorophenol, or waterborne preservatives. The preservative solution is prepared with a solvent conforming to AWPA Standard P9 and must have a minimum concentration of 2 percent copper metal. When available to licensed applicators, other wood preservatives can be used in accordance with the guidelines in AWPA Standard M4.

Preservatives for field treating are usually applied by brushing, dipping, or squirting (Figure 12-18). For each method, the surface of the wood must be saturated with the preservative to provide adequate protection. Even small openings in the preservative can provide an avenue for decay entry. In order to adequately protect wood, all wood preservatives must be toxic to intended targets such as fungi and insects. Workers applying field treatments must wear protective clothing, gloves, and eye protection.

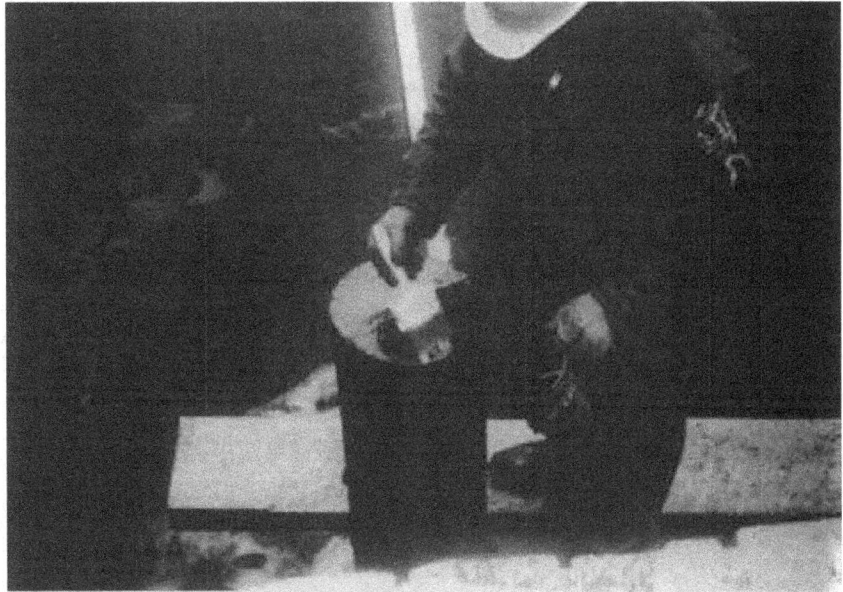

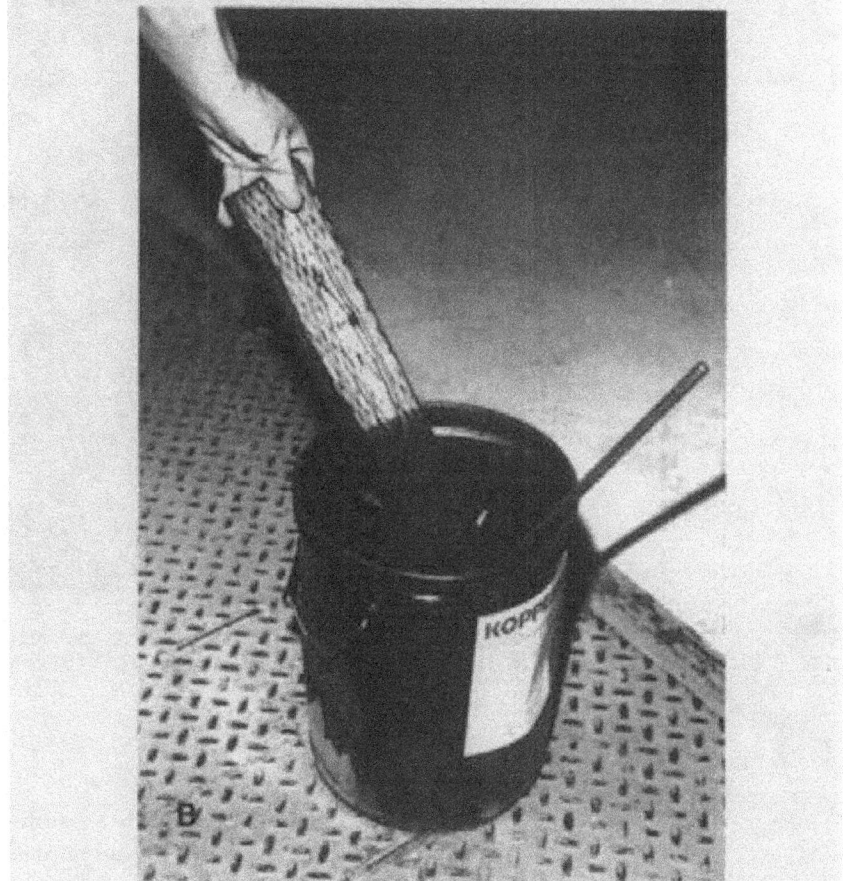

Figure 12-18. – Commonly used methods of field treating timber members. (A) Brushing. (B) Dipping.

Figure 12-18. – Commonly used methods of field treating timber members (continued). (C) Squirting into a horizontal hole. (D) Squirting into a vertical hole.

Methods of applying preservatives for field treating depend on the type and orientation of the area to be treated. When members can be moved, the best method of field treating is dipping or soaking. The area with exposed untreated wood is immersed in the preservative solution for 3 to 5 minutes, or longer. This completely saturates the wood surface and allows some preservative absorption into the wood. Unfortunately, most field treatments must be made when members are a part of the structure, and then soaking or dipping is impractical.

When treating in-place members with field cuts, abrasions, or breaks in the wood surface, the preservative is normally brushed over the surface. For horizontal surfaces, the area can be saturated with preservative solution, with time allowed for the preservative to soak into the wood. On vertical surfaces, excess preservative will run off and the amount that can be applied in one application will be limited. In such a case, three or four successive brushings must be applied with adequate time allowed between each brushing for the preservative to soak in.

Through-holes, whether horizontal or vertical, are more difficult to treat than exposed cuts because access to the untreated wood is limited. It is generally necessary to squirt or spray preservative to one end and catch the excess preservative coming out the other end. Holes can also be treated by plugging one hole end, treating the other end, and then reversing the procedure. When the plug is removed, the excess preservative must be collected in a container to avoid spillage.

Bore holes that do not go through the member, such as those for lag screws and spikes, are field treated by filling the hole with preservative. This is done with an oil can or plastic squeeze bottle that allows a controlled amount of the liquid to be inserted directly into the hole. After the preservative is applied, time must be allowed for the preservative to soak into the wood before fasteners are placed. All preservative will not be absorbed into the wood, and fasteners must be placed with caution to prevent the preservative from being squirted out of the hole. It is beneficial to place a rag around the fastener to contain any preservative that may be forced out as the fastener is driven or screwed.

12.8 INSPECTION AND CERTIFICATION

Owners and specifiers of timber bridges are often concerned as to whether quality requirements for material, treatments, and construction methods are being met or exceeded. This is especially true in bid situations where the lowest bid must be accepted. Quality control and material compliance can be ensured in a number of ways. Many large organizations, such as government agencies, utilities, and railroads, maintain their own inspection personnel. These inspectors visit manufacturing or fabrication facilities and conduct quality control inspections and tests to ensure specification compliance. For most purchasers, however, maintaining a full-time inspection staff is impractical. An alternative is to hire an independent third-party inspection agency. There are numerous agencies of this type located across the United States that have many years of experience as well as good reputations. Several specialize in wood products, including sawn lumber, glulam, and preservative treatments. These inspection firms charge a fee for their services, which varies with the amount and type of inspection required. In some cases, material manufacturers and suppliers will charge a fee for third-party inspection to compensate for the extra handling and the potential for material rejection that inspectors may cause.

For most timber bridge projects, acceptable quality control is achieved by industry material-certification programs discussed in previous chapters. Certificates of conformance issued through such programs provide written documentation that the material was manufactured in accordance with the applicable specifications and standards established by that organization. Examples of these programs include a grading-stamp certification program for sawn lumber administered by various grading rules agencies (Chapter 3), a glulam certification program administered by the American Institute of Timber Construction (Chapter 3), and a preservative treatment certification program administered by the American Wood Preservers Bureau (Chapter 4). Under these programs, participating manufacturers, treaters, and inspection agencies are routinely checked for quality control and compliance by the administering association. If they comply, the producer is authorized to use quality stamps and/or issue certifications of material conformance (Figure 12-19). If they are found deficient, corrective action must be taken immediately or the producer will lose the quality certification. There is a small charge for these association certification programs, which is normally absorbed by the manufacturer or treater and included in material prices.

In addition to material certification by industry associations, quality certification may be indicated by mill certificates. Mill certificates are material certifications issued by individual manufacturers or suppliers, rather than industry quality control associations. Thus, their validity is usually based solely on the word of the manufacturer. Many reputable firms issue mill certificates based on extensive in-house testing and quality control programs. Other firms may have few or no quality control programs. Before accepting mill certificates as proof of material compliance, it is a good idea to verify the reputation of the manufacturer and check on the extent and depth of its quality control and testing programs.

CERTIFICATE OF ![AITC] CONFORMANCE

THE UNDERSIGNED, COMPANY A AND COMPANY B, hereby make the following certifications:

(1) Both Companies A and B certify that all the products identified below and on the attached sheets Nos. _____ were manufactured, in part, in the plant of Company A at _____ and, in part, in the plant of Company B at _____

(2) Company A certifies that the said products so made by it were manufactured in conformance with the applicable provisions of American National Standard ANSI/AITC A190.1–1983, Structural Glued Laminated Timber, at its said plant, which plant operates under a quality control system approved by the Inspection Bureau of the AMERICAN INSTITUTE OF TIMBER CONSTRUCTION and is inspected periodically by such Bureau; and

(3) Company B certifies that the said products made by it were manufactured in conformance with the applicable provisions of American National Standard ANSI/AITC A190.1–1983, Structural Glued Laminated Timber, at its said plant, which plant operates under a quality control system approved by the Inspection Bureau of the AMERICAN INSTITUTE OF TIMBER CONSTRUCTION and is inspected periodically by such Bureau.

The manufacture of these members complies with the manufacturing and fabricating provisions of Chapter 25 of the Uniform Building Code. ☐ Not applicable.

JOB NAME _____

JOB LOCATION _____

CUSTOMER'S ORDER NO _____ DATE _____ MFGR'S ORDER NO _____

_____

_____

SIGNATURE _____ COMPANY A _____

TITLE _____ ADDRESS _____ DATE _____

CUSTOMER'S ORDER NO _____ DATE _____ MFGR'S ORDER NO _____

_____

_____

SIGNATURE _____ COMPANY B _____

TITLE _____ ADDRESS _____ DATE _____

AITC HEREBY CERTIFIES that Companies A and B, at their respective plants, are licensed by the AMERICAN INSTITUTE OF TIMBER CONSTRUCTION to use the AITC Collective Mark in respect of products which comply with the applicable provisions of said Standard, that the adequacy of the quality control system in effect at each said plant is periodically inspected and verified by the Inspection Bureau of the AMERICAN INSTITUTE OF TIMBER CONSTRUCTION, and that, in the judgment of AITC, said Companies A and B are capable of complying with the applicable manufacturing and testing provisions of said Standard in respect of products manufactured at said plants. Conformance with the Standard with respect to any specific or particular product is the sole responsibility of said manufacturer. AITC's guarantee hereunder being that said Companies A and B are qualified to produce products meeting the said Standard and that each of said plants is periodically inspected and verified by the AITC Inspection Bureau.

SEAL 1952

AITC FORM 18CC

AITC *Certificate No.* 00021 **C**

AMERICAN INSTITUTE OF TIMBER CONSTRUCTION

© 1983 AMERICAN INSTITUTE OF TIMBER CONSTRUCTION

Figure 12-19. – Certificate of material conformance for glulam issued through the AITC Quality Control Program (photo courtesy of the American Institute of Timber Construction).

12.9 SELECTED REFERENCES

1. American Association of State Highway Officials. 1973. Construction manual for highway bridges and incidental structures. Washington, DC: American Association of State Highway Officials. 66 p.

2. American Institute of Timber Construction. 1985. Timber construction manual. 3d ed. New York: John Wiley and Sons, Inc. 836 p.

3. American Institute of Timber Construction. 1984. Typical construction details. AITC 104-84. Englewood, CO: American Institute of Timber Construction. 32 p.

4. American Wood-Preservers' Association. 1987. Book of standards. Stevensville, MD: American Wood-Preservers' Association. [240 p.]

5. Canadian Wood Council. 1985. Canadian wood construction. Storage and handling. CWC datafile CW-1. Ottawa, Can.: Canadian Wood Council. 4 p.

6. Morrell, J. J.; Helsing, G.G.; Graham, R.D. 1984. Marine wood maintenance manual: a guide for proper use of Douglas fir in marine exposure. Res. Bull. 48. Corvallis, OR: Oregon State University, Forest Research Laboratory. 62 p.

7. Muchmore, F.W. 1986. Designing timber bridges for long life. GPO 693-017. Missoula, MT: U.S. Department of Agriculture, Forest Service, Northern Region. 15 p.

8. Nagy, M.M.; Trebett, J.T.; Wellburn, G.V. 1980. Log bridge construction handbook. Vancouver, Can.: Forest Engineering Research Institute of Canada. 421 p.

9. Tuomi, R.L. 1976. Erection procedure for glued-laminated timber bridge decks with dowel connectors. Res. Pap. FPL 263. Madison, WI: U.S. Department of Agriculture, Forest Service, Forest Products Laboratory. 15 p.

10. U.S. Department of Agriculture, Forest Service, Northern Region. 1985. Bridge design manual. Missoula, MT: U.S. Department of Agriculture, Forest Service, Northern Region. 299 p.

11. U.S. Department of Agriculture, Forest Service. 1985. Forest Service specifications for construction of bridges & other major drainage structures. EM-7720-100B. Washington, DC: U.S. Department of Agriculture, Forest Service. 241 p.

12. Western Wood Products Association. 1973. Storage of lumber. WWPA tech. guide TG-5. Portland, OR: Western Wood Products Association. 1 p.

13. Weyerhaeuser Company. 1980. Weyerhaeuser glulam wood bridge systems. Tacoma, WA: Weyerhaeuser Co. 114 p.

CHAPTER 13

BRIDGE INSPECTION FOR DECAY AND OTHER DETERIORATION

13.1 INTRODUCTION

Wood is an amazing combination of polymers that exhibits both strength and durability as a structural material. Nevertheless, from the time it is formed in the tree, wood is subject to deterioration by a variety of agents. Damage ranges from relatively minor discolorations caused by fungi or chemicals to more serious decay and insect attack. Wood degradation is beneficial in the ecosystem, returning carbon and other elements to the soil and air, but it becomes detrimental when the deteriorating material is part of a bridge or other structure. Wood outperforms most other materials when used in a properly designed and maintained structure; however, when used in adverse environments, it must be protected to ensure adequate performance. Although the use of pressure-treated wood has significantly extended the life of timber, decay is still the primary cause of bridge deterioration.

The decision to establish a management program for timber bridges is a difficult one that often comes after the user has experienced losses because of previous poor management. Like any investment, a timber bridge must be inspected and maintained on a regular basis to maximize the investment. Yet, most users simply install the structure and walk away, hoping that all will be well. If it is not, they blame the material, when in fact, poor design, poor construction practices, and poor management were probably major factors in the decline. Over the life of a timber bridge, deterioration can be minimized by alert inspectors who identify and record information on structure condition and performance. With such information, timely maintenance operations can be undertaken to correct situations that could otherwise lead to extensive repair or even replacement.

Timber bridge inspectors have the difficult task of accurately assessing the condition of an existing structure. They must understand the biotic and physical factors associated with wood deterioration as well as the relative rate at which these processes occur in a given environment. Timber inspection is a learned process that requires some knowledge of wood pathology, wood technology, and timber engineering. This chapter covers the fundamentals of timber bridge inspection for decay and deterioration; it identifies the agents of deterioration and outlines inspection methods.

This chapter was co-authored by Michael A. Ritter and Jeffrey J. Morrell, Ph.D., Associate Professor, Department of Forest Products, Oregon State University.

Additional information on more general aspects of bridge inspection is available in references listed at the end of this chapter.[1,52,54]

13.2 AGENTS OF WOOD DETERIORATION

Wood deterioration is a process that adversely alters wood properties. In broad terms, it can be attributed to two primary causes: biotic (living) agents and physical (nonliving) agents. In most cases, wood deterioration is a continuum, whereon the degrading actions from one or more agents alter wood properties to the degree required for other agents to attack. The inspector's familiarity with the agents of deterioration is one of the most important aids in effective bridge inspection. With this knowledge, inspection can be approached with a thorough understanding of the processes involved in deterioration and the factors that favor or inhibit its development.

BIOTIC AGENTS OF DETERIORATION

Wood is remarkably resistant to biological deterioration but a number of organisms have evolved with the ability to utilize wood in a manner that alters its properties. Organisms that attack wood include bacteria, fungi, insects, and marine borers. Some of these organisms use the wood as a food source, while others use it for shelter.

Biotic Requirements
Biotic agents require certain conditions for survival. These requirements include moisture, available oxygen, suitable temperatures, and an adequate source of food, which is generally the wood. Although the degree of dependency on these requirements varies among different organisms, each must be present for deterioration to occur. When any one is removed, the wood is safe from biotic attack.

Moisture
Although many wood users speak of *dry rot*, the term is misleading since wood must contain water for most biological attacks to occur. Wood moisture content is a major determinant of the types of organisms present and the rate at which they degrade the wood. Generally, wood below the fiber saturation point will not decay, although some specialized fungi and insects can attack wood at much lower moisture levels. While keeping wood dry makes sense, it is sometimes difficult to implement, particularly in exposed timber bridges.

Moisture in wood serves several purposes in the deterioration process. For fungi and insects, it is required for many metabolic processes. For fungi, it also provides a diffusion medium for enzymes that degrade the wood structure. When water enters wood, the microstructure swells until the

13-2

fiber saturation point is reached (about 30 percent wood moisture content). At this point, free water collects in the wood cell cavities, and many fungi can begin to degrade the wood. The swelling associated with water is believed to make the cellulose more accessible to fungal enzymes, enhancing the rate of decay. Additionally, repeated wetting and drying or continuous exposure to moisture can result in leaching of toxic heartwood extractives and some preservatives, reducing decay resistance.

Oxygen

With the exception of anaerobic bacteria, all organisms require oxygen for respiration. While depriving them of oxygen may seem a logical decay control strategy, it is generally impractical in bridge applications since most fungi can survive at very low oxygen levels. An exception is piling that is totally submerged or placed below the water table. In marine environments, piling may be wrapped in plastic or concrete so that marine borers are unable to exchange nutrients and oxygen with the surrounding seawater. In many cases, untreated piling in fresh water will decay to the water line, but remain sound underwater where oxygen is absent.

Temperature

Most organisms thrive in an optimum temperature range of 70 to 85 °F; however, they are capable of surviving over a considerably wider range. At temperatures below 32 °F, the metabolism of most organisms slows, or they produce resistant survival structures to carry them through the unfavorable period. As temperatures rise above freezing, they once again begin to attack wood, but activity slows rapidly as the temperature approaches 90 °F. At temperatures above 90 °F, the growth of most organisms declines, although some extremely tolerant species continue to thrive up to 104 °F. Most organisms succumb at prolonged exposure above this level, and it is generally accepted that 75 minutes of exposure to 150 °F will eliminate all decay fungi established in wood.[9]

In the context of timber bridges, temperature is not controllable, but the inspector should realize that decay will be much more serious in warm environments where the rate of biological activity is higher. This factor has been used, in combination with rainfall, to develop a climate index that expresses temperature and rainfall for an area to formulate a decay hazard index.[46] Although this index cannot account for small variations in regional weather patterns, it does provide a relative guide to decay hazard.

Food

Most biotic agents that attack wood use it as a food source. When wood is treated with preservatives, the food source is poisoned, and infestation can occur only where the preservative treatment envelope is inadequate, or has been broken. If the exposed wood is from a naturally durable species it will initially have some degree of resistance to attack, but this resistance will be reduced rapidly by weathering and leaching. Maintaining an effective preservative treatment is essential for preventing biotic attack.

Bacteria

Bacteria are small, single-cell protists that are among the most common organisms on earth. They recently have been shown to be important colonizers of untreated wood in very wet environments, causing increased permeability and softening of the wood surface. Bacterial decay is normally an extremely slow process, but can become serious in situations where untreated wood is submerged for long periods. Many bacteria are also capable of degrading preservatives and may modify treated wood in such a way that it becomes more susceptible to less chemically tolerant organisms.[13] Although significant strength loss may develop in untreated wood that remains saturated for very long periods, bacterial decay does not appear to be a significant hazard to the pressure-treated timber typically used for bridge construction.

Fungi

Fungi are simple, plantlike organisms that break down and utilize wood material as a food source. They move through the wood as a network of microscopic, threadlike hyphae that grows through the pits or directly penetrates the wood cell wall (Figure 13-1). As the hyphae elongate, they secrete enzymes that degrade cellulose, hemicellulose, or lignin and absorb the degraded material to complete the digestion process. Once the fungus obtains a sufficient amount of energy from the wood, it produces a sexual or asexual fruiting body to distribute reproductive spores that can invade other wood. Fruiting bodies vary from single-cell spores produced at the end of the hyphae to elaborate perennial fruiting bodies that produce millions of spores (Figure 13-2). These spores are so widely spread by wind, insects, and other means that they can be found on most exposed surfaces. As a result, all wood structures are subject to fungal attack when moisture and other requirements conducive to fungal growth are present.

Although wood decay has been noted throughout recorded history, it was not until 1878 that R. Hartig accurately described the relationship between fungal hyphae and wood decay.[20] Even today, we continue to discover new species and intriguing relationships among the organisms that colonize wood. Although there are hundreds of fungal species, the fungi that attack wood can be divided into three types: mold fungi, stain fungi, and decay fungi. These fungi are similar in many ways, but differ substantially in their effects on timber structures.

Mold and Stain Fungi

Mold and stain fungi colonize wood soon after it is cut and continue to grow as long as the moisture content remains high (above approximately 25 percent for softwoods). The primary effect of these fungi is to stain or discolor the wood (Figure 13-3). They are considered nondecay fungi and are of practical consequence primarily where wood is produced for its aesthetic qualities. Mold fungi infect the wood surface, causing blemishes that can generally be removed by brushing or planing, but stain fungi cause serious concerns because they penetrate deeper and discolor the

13-4

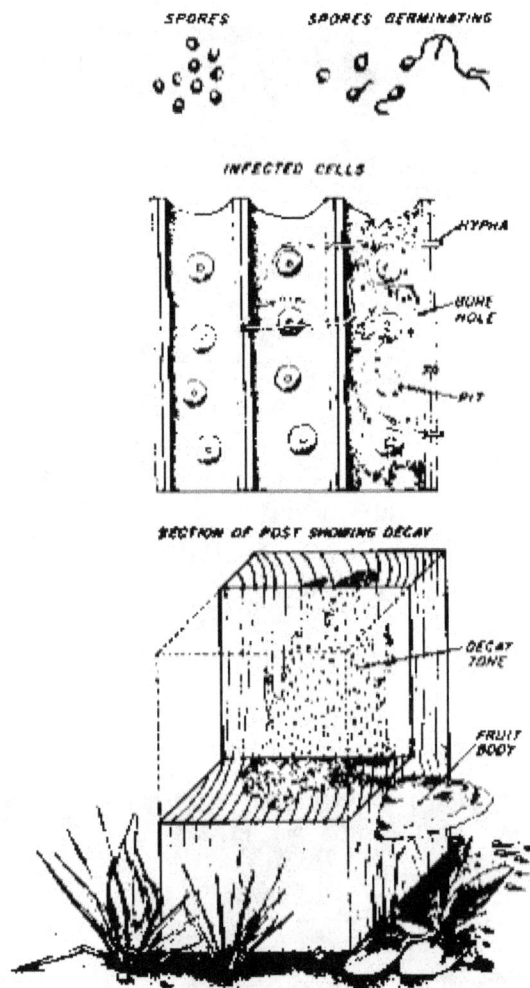

Figure 13-1. - The decay cycle (top to bottom). Fungi begin as minute spores that germinate and grow through the wood. Once enough energy has been obtained, the fungus produces a fruiting body and releases spores that spread and infect other wood.

wood. Under optimum conditions, some stain fungi may also continue to degrade wood, causing decreased toughness and increased permeability; consequently, stained wood is generally rejected during grading for structural uses.

Mold and stain fungi use the contents of the wood cell for food, and do not degrade the cell wall. They do not adversely affect strength, but their presence can indicate conditions favorable for more serious decay fungi. The continued growth of some mold and stain fungi may cause a slow detoxification of natural wood toxins or surface preservatives that can lead to accelerated attack by decay fungi. Since most species attack sapwood, they are more of a problem on thick-sapwood species such as Southern Pine.

Figure 13-2. - A typical fungal fruiting body. Such growths vary considerably in size, color, and shape among species of fungi.

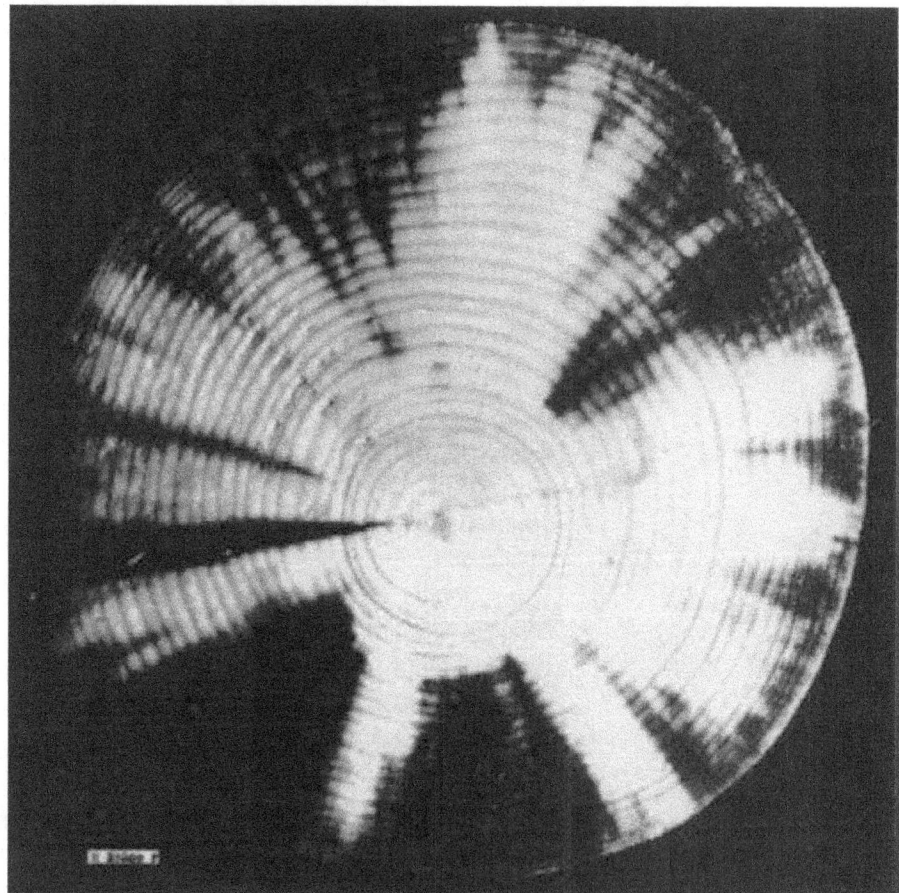

Figure 13-3. - Log cross section showing discoloration caused by stain fungi.

Decay Fungi

Decay in timber bridges is normally caused by decay fungi. These fungi are grouped into three broad classes based on the manner in which they attack wood and the appearance of the decayed material. The three types of decay fungi are brown rot fungi, white rot fungi, and soft rot fungi.

Brown rot fungi, as the name implies, give decayed wood a brownish color. In advanced stages, brown rotted wood is brittle and has numerous cross checks, similar in appearance to the face of a heavily charred timber (Figure 13-4). In the 1700's, scientists examining brown rotted wood stated that the wood had combusted, and it was not until the latter 1800's that fungi were associated with this damage. The brown rots primarily attack the cellulose and hemicellulose fractions of the wood cell wall and modify the residual lignin, causing weight losses of nearly 70 percent. Because cellulose provides the primary strength to the cell wall, the brown rot fungi cause substantial strength losses at the very early stages of decay. At this point, the wood appears sound and the fungus may have removed only 1 to 5 percent of the wood weight, but some strength properties may be reduced by as much as 60 percent.[36]

Figure 13-4. - Wood infected with brown rot fungi in an advanced stage. The decayed wood has a darkened color with a cracked, brittle surface that resembles charred wood.

Of the three types of decay fungi, brown rots are among the most serious because of their pattern of attack. Enzymes produced by these fungi migrate or diffuse far from the point where the fungal hyphae are growing. As a result, strength losses in wood may extend a substantial distance from locations where the decay can be visibly detected.

White rot fungi produce decay that resembles normal wood in appearance, but may be whitish or light tan in color with dark streaks. In the advanced stages of decay, infected wood has a distinctively soft texture, and individual fibers can be peeled from the wood (Figure 13-5). The white rots differ from brown rots in that they attack all three components of the wood cell wall, causing weight loss of up to 97 percent. In most cases, the associated strength loss is approximately comparable to weight loss. The enzymes produced by white rot fungi normally remain close to the growing hyphae, and the effects of infestation are not as noticeable at the early decay stages.

Figure 13-5. - Wood infected with white rot fungi. The decayed wood is abnormally light colored with dark streaks (arrows).

Soft rot fungi are a more recently recognized group that generally confine their attack to the outer wood shell (Figure 13-6). They typically attack wood subjected to continuous wetting or changing moisture conditions, and may occur in low-oxygen environments that inhibit conventional decay fungi. Most soft rot fungi require the addition of exogenous nutrients to cause substantial attack. These nutrients are often inadvertently provided by fertilizers in agricultural soils, pulp waste in cooling towers, and other miscellaneous nutrient sources. Although they may be encountered in some situations, soft rot fungi normally are not associated with significant strength loss in bridge components.

Figure 13-6. - Soft rot decay in a timber pole. Note the shallow depth of decay.

For descriptive purposes, the degree of decay in wood can be classified into three stages: incipient, intermediate, and advanced. Incipient decay occurs at the advancing margin or newest part of the infection, where the damage is difficult to detect because there are no visible signs of attack. Significant changes in wood properties can occur in the incipient stages. As decay enters the intermediate stage, the wood becomes softened, discolored, and retains little, if any, strength. In the advanced stages of decay, wood retains virtually no strength, decay pockets or voids are formed, or the wood is literally dissolved. Detecting decay in the initial or incipient stage is the most difficult, but also the most important, part of bridge inspection. At this point, decay can be most effectively controlled to prevent more severe damage to the structure.

Insects

Insects are among the most common organisms on earth, and it is not surprising that a number of species have developed the ability to use wood for shelter or food. Of the 26 insect orders, 6 cause wood damage. Termites (Isoptera), beetles (Coleoptera), and bees, wasps, and ants (Hymenoptera) are the primary causes of most insect-related deterioration. Insect attack is generally apparent from tunnels or cavities in the wood, which often contain wood powder or frass (insect feces). Powder posting, a pile of wood powder or frass on the outside of the wood, is another sign of attack. In addition to removing portions of the wood structure, insects may also carry stain and decay fungi that further deteriorate wood. One insect even carries a fungus that causes hard pines to wilt.

13-9

Termites

Over 2,000 species of termites are distributed in areas where the average annual temperature is 50 °F or higher. In some cases, termites extend their range into cooler climates by living in heated humanmade structures. They attack most wood species, but the heartwood of a few species, such as juniper and southern cypress, exhibits some resistance to attack. Termites are social insects, organized into a series of castes that perform specific functions. The colony's leader is a queen whose sole purpose is to lay eggs. The queen is protected by soldiers and nurtured and fed by workers, who also build the nest and cause wood damage (Figure 13-7). Like all creatures, termites have certain requirements, including wood at a high moisture content, a suitable food source (wood), a high carbon dioxide level, and oxygen. Termite colonies range in size from several hundred to a million or more members.

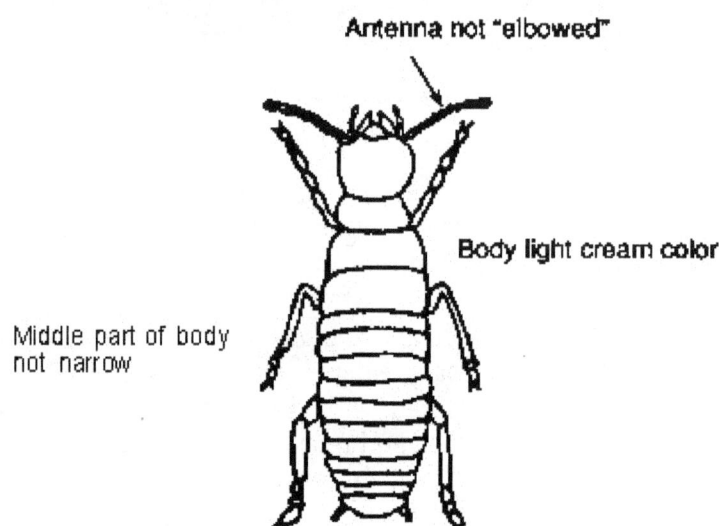

Figure 13-7. - Drawing of a termite worker showing general anatomical features.

Termites that attack wood are separated into five families, three of which are found within the continental United States. The species most associated with wood damage are the subterranean, dampwood, and drywood termites.

Subterranean Termites

Subterranean termites (Rhinotermitidae) attack virtually any available wood, but they need a moisture source and typically nest in the ground. They have developed the ability to attack wood aboveground by constructing earthen tubes that protect them from light and carry moisture to the wood. In the United States, subterranean termites are common throughout the southeast and extend northward into less temperate climates (Figure 13-8). Wood damaged by subterranean termites has numerous tunnels through the springwood, but there are no exit holes to the surface that indicate the termite's presence. Often, a sharp tap on the wood surface will reveal that only a thin veneer of wood remains. Subterranean termite

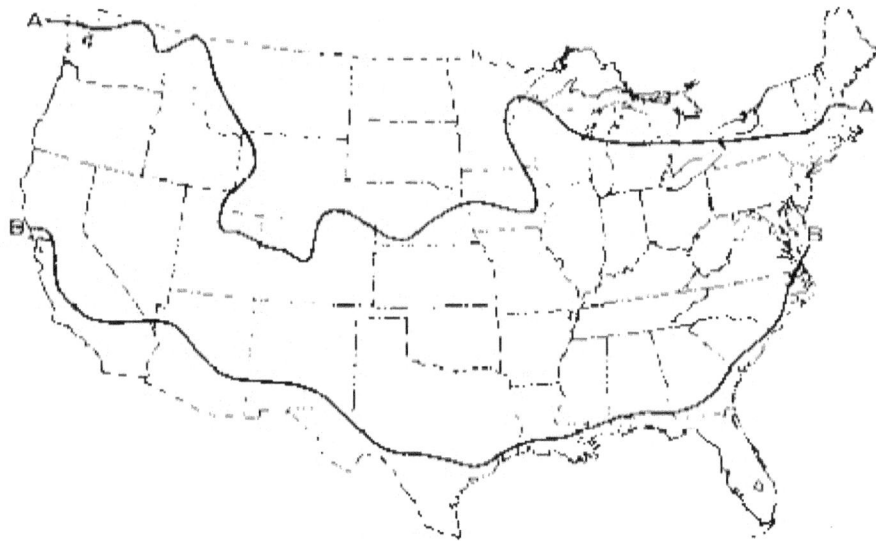

Figure 13-8. - (A) The northern limit of recorded damage done by subterranean termites in the United States. (B) The northern limit of damage done by drywood termites.

tunnels are filled with a mixture of frass and debris and have a dirty appearance (Figure 13-9). The economic impact of these insects in the United States has been conservatively estimated at $1.5 billion per year.[12]

A variety of subterranean termites known as Formosan termites *(Coptotermes formosanus)* recently has moved into several Southeastern States. The presence of this species is cause for concern because of its ability to attack preservative-treated wood, the large size of its colonies, and its habit of occasionally nesting in moist wood not in ground contact. Fortunately, the Formosan termite has been found only at some ports of entry along the southern portions of the United States; however, their capabilities are cause for concern throughout the warmer Southern States.

Dampwood Termites

Dampwood termites are common to the Pacific Northwest, although one group is found in the more arid Southwest. The most common dampwood species is found along the Pacific coast from northern California to British Columbia. Like the subterranean termites, dampwood species need wood that is very wet, and their attack is often associated with decay. These insects are a problem in freshly cut lumber, utility poles, and any untreated wood in ground contact. Tunnels made by dampwood termites are fairly large, but like the subterranean species, they tend to avoid the harder summerwood. The tunnels often contain small amounts of pelletlike frass, but the wood looks somewhat cleaner than that attacked by the subterranean species. Dampwood termite attack can be prevented or arrested by removing the moisture source or by using preservative-treated wood in situations requiring ground contact.

13-11

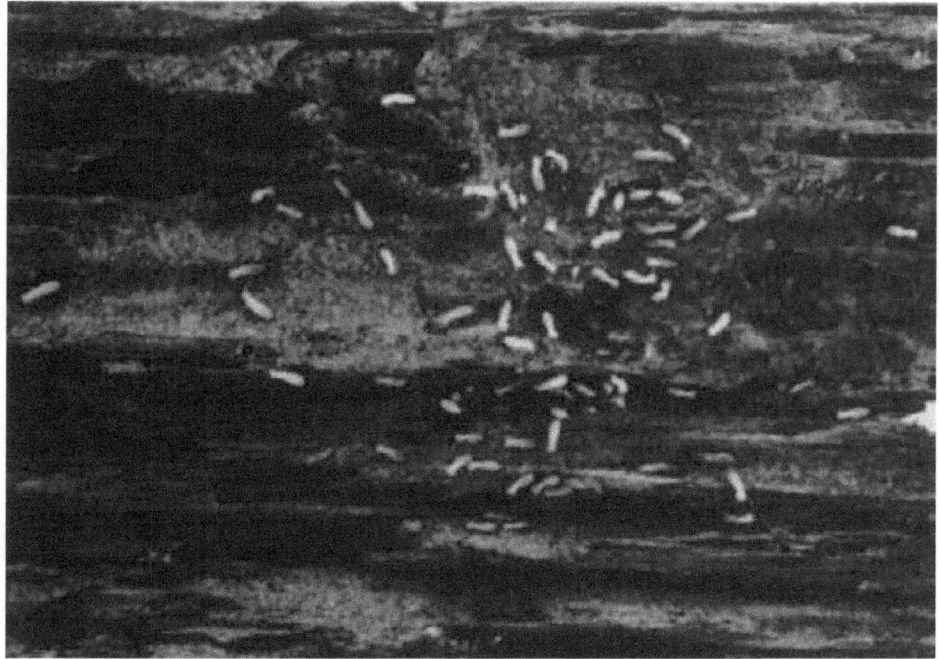

Figure 13-9. - Subterranean termites and the wood damage they cause. Note the frass and debris accumulations in the tunnels (photo courtesy of USDA Forest Service, Forest Sciences Laboratory, Gulfport, Mississippi).

Drywood Termites

Drywood termites (Kalotermitidae) differ from subterranean and dampwood termites in their ability to attack wood that is extremely dry (5 to 6-percent moisture content). As a result, drywood termites attack wood not in ground contact and away from visible moisture sources. Wood damaged by these insects has large, smooth tunnels that are free of either frass or debris. In addition, there is no variation in attack between springwood and summerwood. Drywood termites will frequently clean out their nests by chewing holes to the surface and kicking out debris, which collects below the infested wood. Although these holes are resealed, the presence of debris below a kick hole is a good sign of attack. In general, clusters of infestations are found in one geographic area, and prevention poses some difficulty. Should an infestation occur, the use of structural fumigation has been reported to be effective. Fortunately, the drywood termite is confined to a relatively small geographic region.

Beetles

Beetles (Coleoptera) represent the largest order of insects and contain nine families that cause substantial damage to wood (Table 13-1). Many beetles in these families attack only living trees or freshly cut timber, but they will be briefly discussed because their damage may be encountered during inspection and can be confused with active deterioration.

13-12

Table 13-1. - Families of wood-attacking beetles.

Family	Common name	Wood damage
Anobiidae	Powder post beetle	Powder posting
Bostricidae	Powder post beetle	Powder posting
Brentidae	Timber worm	Pin holes
Buprestidae	Flat-headed borer	Grub holes
Cerambycidae	Round headed borer	Grub holes
Lyctidae	Powder post beetle	Powder posting
Lymexylidae	Timber worm	Pin holes
Platypodidae	Ambrosia beetles	Pin holes & stain
Scolytidae	Ambrosia and bark beetles	Pin holes & stain

Powder Post Beetles

The powder post beetles are insects whose larvae attack wood, leaving behind a series of small tunnels packed with powderlike frass (Figure 13-10). The three families of powder post beetles are the Anobiidae, the Bostrichidae, and the Lyctidae. These insects cause serious damage to seasoned wood and are a particular problem in museums, where wooden artifacts may go unobserved for long periods. In the field, the Anobiidae and Bostrichidae attack moist wood in dead branches but will also attack untreated construction timbers. The damage is worsened by emerging adults reinfesting the same piece of wood. The Lyctidae, or true powder post beetles, are found on hardwoods throughout the world and attack wood at moisture contents above 8 percent. As the larvae of these beetles tunnel, they push frass out of the wood. This frass collects beneath

Figure 13-10. - Emergence holes in wood damaged by powder post beetles. The beetle larvae tunnel through the wood, without discoloring it, and leave behind a flourlike frass.

13-13

the affected wood and is a good sign of powder post infestation. The use of preservative treatments or sealing of the wood surface will prevent Lyctidae infestation. However, powder post beetle attack can become a problem where untreated wood is used in older existing bridges.

Brentidae and Lymexylidae

The Brentidae, or primitive weevils, and the Lymexylidae, or ship timber beetles, attack freshly cut hardwood logs. The larvae of these beetles make extensive galleries in the wood and cause considerable reduction in lumber quality. The effects of the Brentidae and Lymexylidae can be minimized by removing woody debris that may serve as breeding areas, by ponding logs before processing, or by debarking logs as soon as possible. Neither species is capable of surviving in the seasoned wood once the bark has been removed, although the damage cannot be eliminated. In general, damage caused by these beetles is mainly cosmetic and should not adversely affect strength.

Scolytidae

Scolytidae attack freshly cut timber while the bark remains intact, producing pinholes and providing an avenue of entry for stain fungi. As a result, the wood is aesthetically ruined, and its value decreases. Most Scolytids are confined to the wood cambial layer, and damage is relatively minor; however, some species, such as the Ambrosia beetle, penetrate to greater depths and carry stain fungi deep into the wood interior (Figure 13-11). Adult beetles bore into the wood to lay their eggs and deposit a small amount of fungal material with each egg. The fungus grows into the wood

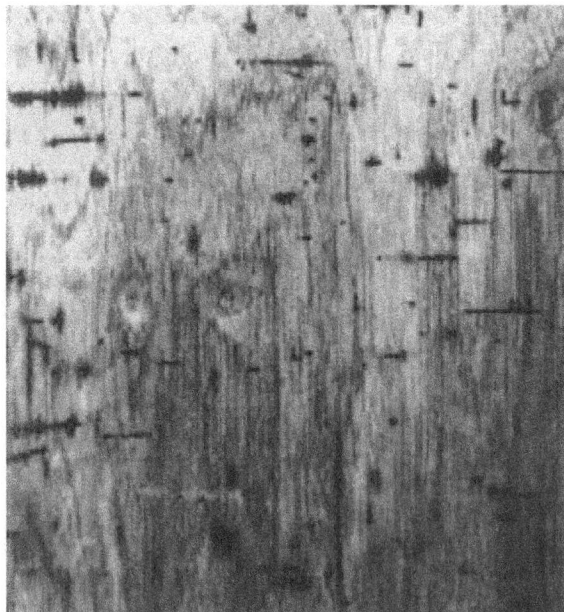

Figure 13-11. - Damage by Ambrosia beetles in green wood. The galleries are free of residue and the surrounding wood is darkly stained.

13-14

structure and the larvae consume the wood to obtain the fungal nutrition. Ambrosia beetles are found throughout the United States and their control is difficult. Although log ponding is an effective preventive measure, surfaces exposed to the air can be reinfested. Prompt bark removal appears to be the most practical solution for limiting damage by this beetle, but this removal permits more rapid entry by stain and decay fungi unless the wood is rapidly processed and dried.

Buprestidae

The Buprestidae, also called flat-headed or metallic wood borers, are almost entirely dependent on trees to complete their life cycle. They cause significant damage by attacking living trees, leaving damage that may be evident in lumber or other wood products. Buprestids lay their eggs on the surface of bark or in tree wounds, and hatching larvae burrow into the wood to varying depths. Over the course of their 1- to 3-year life cycles, the larvae tunnel extensively in the wood, leaving galleries tightly packed with frass. The mature larvae pupates, and the adult chews its way out through a D-shaped exit hole. In addition to the species that attack living trees, one species, the Golden Buprestid *(Buprestis aurulenta),* is capable of attacking Douglas-fir in service. The Golden Buprestid causes serious damage to utility poles, where its attack is often associated with extensive decay (Figure 13-12). The Golden Buprestid larvae are extremely resistant to dry conditions and have been reported to live in seasoned wood for over 50 years.

Figure 13-12. - Golden Buprestid next to a surface entrance hole. These insects tunnel through the wood of western species and are often associated with internal decay.

13-15

Long Horned Beetles

Long horned beetles (Cerambycidae) include a number of wood degraders that generally have antennae longer than their bodies. They attack wood in all conditions, depending on the species, and cause substantial damage. Some, like the sugar maple borer and poplar borer, attack only living trees, eventually killing them and reducing the value of the wood. Other species attack freshly cut pine, rapidly degrading the wood. One interesting attacker of green wood is the ponderous borer, whose larvae attack Douglas-fir and ponderosa pine, producing tunnels nearly 1 inch in diameter. Although this larva can complete its development in the sawn timber, it will not reinfest the seasoned wood.

In addition to the long horned beetles that attack living or freshly harvested trees, several species cause damage to wood in service. The telephone pole borer was once a common inhabitant of untreated utility poles and was associated with extensive internal decay; however, the use of preservative treated wood has decreased the incidence of this species. Another species, the old house borer, is one of the most destructive wood borers and prefers dry coniferous wood. The old house borer has been reported to cause extensive damage to structural timber along the coastal southeastern United States, but does not cause serious problems elsewhere. Generally, infestations by these beetles can be prevented by using preservative-treated wood.

Ants, Bees, and Wasps

Ants, bees, and wasps are collectively included in the order Hymenoptera. Several members of this order can attack wood, but discussions here are limited to carpenter ants and carpenter bees because these two groups attack wood in service.

Carpenter Ants

Carpenter ants (Formicidae) differ from the insects previously discussed because they use wood for shelter rather than for food. They are social insects with a complex organization revolving around a queen. To sustain the colony and rear their young, carpenter ant workers must forage great distances from the nest to obtain food, which can consist of insect secretions, insects, and sugary food sources. As the colony grows from the original queen to its eventual 100,000 members, the workers gradually enlarge their nest, causing serious internal wood damage. Many colonies seem to prefer wood that is above the fiber saturation point and are often associated with internal decay. Wood damaged by carpenter ants is characterized by the presence of clean, frass-free tunnels that are largely confined to the softer earlywood, and extend parallel to and across the grain (Figure 13-13). As the workers attack the wood, they remove large amounts of fibrous frass that collect at the base of the piece under attack and provide a readily identifiable sign of infestation. Carpenter ants are often confused with termites but there are several easy methods for distinguishing attack by these two species (Table 13-2).

13-16

Figure 13-13. - Wood damaged by carpenter ants. The tunnels are generally clear of debris and extend parallel to, and across, the grain.

Table 13-2. - Differences between termites and carpenter ants.

Termite   Carpenter ant

Characteristic	Termites	Carpenter ants
Body segments	Equal size, no constrictions	Variable size, with constrictions
Mature workers	Cream color, rarely seen outside nest	Dark colored, often seen outside nest
Wings	2 pairs of equal sized wings	2 pairs of unequal sized wings
Wood damage	Tunnels contain frass	Tunnels free of frass
Food source	Digested wood	Sugar, other insects

13-17

Carpenter Bees

Like carpenter ants, carpenter bees (*Xylocopa* sp.) use wood only for shelter and for rearing their young. In this process, they tunnel along the grain of coniferous wood, creating 5- to 18-inch long by 0.3- to 0.5-inch wide galleries (Figure 13-14). Carpenter bees look remarkably similar to bumble bees but differ slightly in coloration. They are not common, but when infestation does occur, damage can be serious. The adults of this species tunnel into the wood and lay their eggs in individual cells that are provisioned with food for the growing larvae. The adults emerge and can reinfest the wood. These insects have also been found attacking wood treated with inorganic arsenicals at aboveground retentions.

Marine Borers

When timber substructures are located in salt or brackish waters, severe damage may occur from attack by marine borers. The marine borers that cause wood damage in the United States are classified into three groups based upon their morphology and pattern of wood attack (Figure 13-15): pholads, shipworms, and *Limnoria*. Collectively, these organisms cause over $250 million in damage each year,[25] but their damage is often overlooked because it usually occurs in isolated areas over relatively long time periods. More spectacular short-term losses, such as the $25 million loss in San Francisco Bay during the 1920's, have highlighted the importance of these organisms in marine environments and stimulated interest in their control.[27]

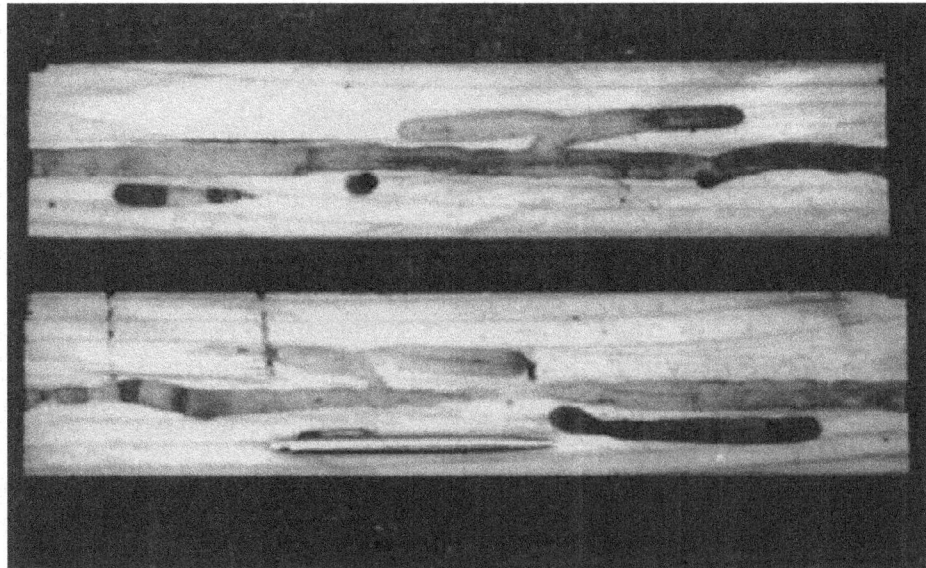

Figure 13-14 - Carpenter bee damage in wood. The bees bore long tunnels along the grain to lay their eggs (photo courtesy of USDA Forest Service, Forest Sciences Laboratory, Gulfport, Mississippi).

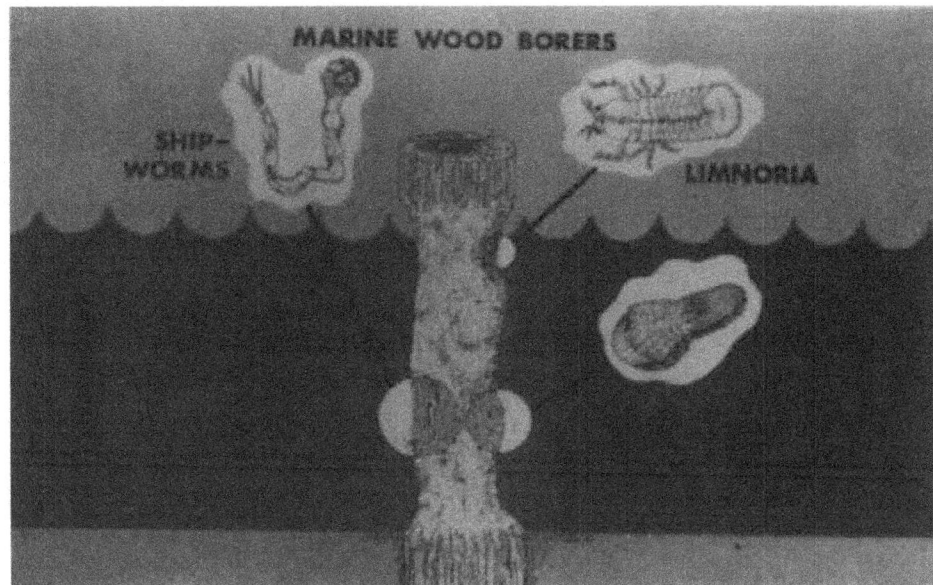

Figure 13-15. - Marine borers that cause wood damage in U.S. waters.

Pholads

Pholads are clamlike mollusks that burrow into wood and filter food from the surrounding water. They begin life as tiny free-swimming larvae that eventually settle onto a suitable wood surface and become permanently established in the wood. Pholads grow to be approximately 2.5 inches long and leave an entry hole in the wood surface about 0.25 inches in diameter. As pholads burrow into wood, the surface eventually weakens and tends to break away under wave action. Internal damage is generally identifiable by characteristic pear-shaped borings (Figure 13-16). Eventually, the wood area decreases to the point where it fails. Although pholads do not pose a problem along the continental United States, one species, *Martesia striata,* causes extensive damage to wood in more tropical marine environments. Attack can be prevented by the use of creosoted wood; however, other wood-degrading organisms in the tropical environment are resistant to creosote so dual treatment with both creosote and a waterborne inorganic arsenical is required. In temperate waters rock burrowing pholads also cause damage to concrete structures.

Shipworms

Shipworms are long, wormlike mollusks that cause interior damage to wood while leaving only a small hole on the surface as evidence of their attack. Like pholads, shipworms begin life as small, free-swimming larvae, then settle down to begin their sedentary, wood-inhabiting life. In the 1700's, ship captains exploited this portion of the life cycle by sailing their infested wooden ships upriver into fresh water where the trapped shipworms would succumb to the lack of salinity.

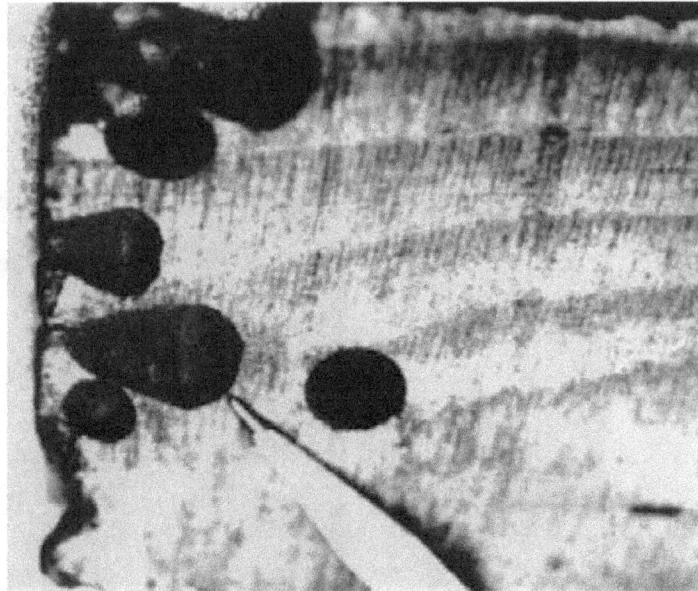

Figure 13-16. - Internal wood damage caused by pholads. These borers generally burrow near the wood surface and are characterized by pear-shaped borings.

Two shipworm species, *Teredo navalis* and *Bankia setacea*, are commonly encountered along the United States continental coasts. These species differ in their morphology, with the former growing to be 3.5 to 7 inches long and 0.5 inch in diameter, and the latter growing to be 59 to 71 inches long and 0.8 inch in diameter. Generally, *T. navalis* has a greater tolerance to low salinity and can survive far upstream in many estuaries, while *B. setacea* is more temperature resistant and is found in more northerly harbors.

As shipworms become established in wood, two hard, clamlike shells near the tops of their heads begin to rasp away at the wood, leaving tunnels with a characteristic white coating (Figure 13-17). The shipworm gradually enlarges the tunnel within the wood, but the initial hole it entered rarely enlarges beyond 0.06 inch in diameter. From the safety of their wood burrow, shipworms extend a pair of feathery siphons into surrounding water. These siphons function in the exchange of nutrients, oxygen, and waste products. At any sign of danger, the siphons are retracted and the surface hole is covered by a hardened pallet that protects the organism from attack. The protection of the pallet also allows the shipworm to survive in wood out of water for 7 to 10 days. The small size of the surface hole and the presence of the pallet make visual detection of internal shipworm attack unreliable, but recent advances in acoustic detection have improved the prospects for detecting infestations before substantial damage occurs.

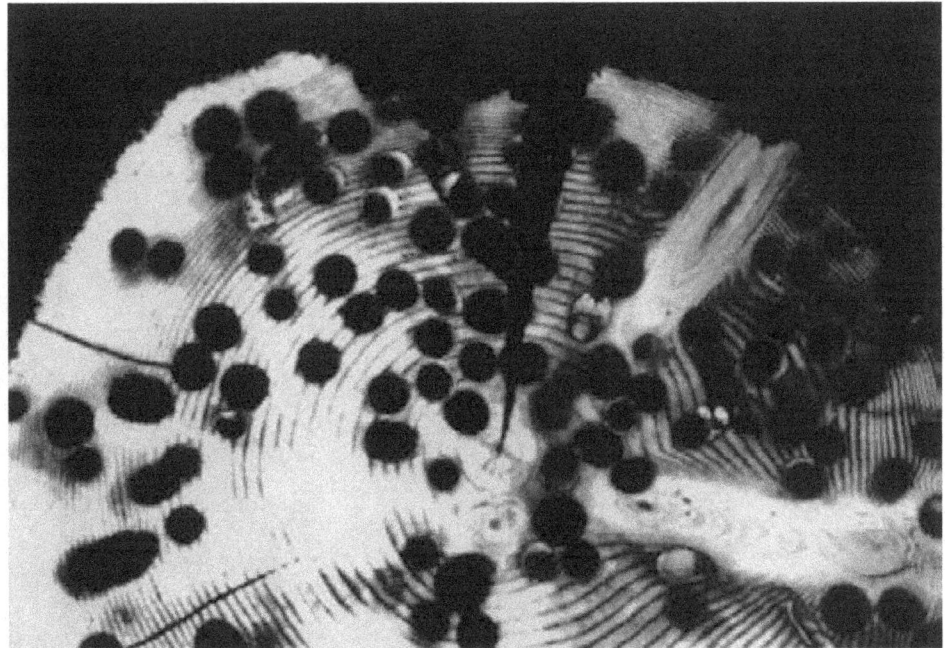

Figure 13-17 - Internal wood damage caused by ship worms. Tunnels extend throughout the cross section and are usually covered with white calcium deposits.

Limnoria

Limnoria or gribbles are mobile crustaceans that differ from shipworms and pholads in their ability to move from one piece of wood to another during their life cycle. There are 20 species of *Limnoria* that attack wood in marine waters, but only 3 cause major damage in the United States. Two of these species are capable of attacking only untreated wood, but the other species, *L. tripunctata,* attacks creosote-treated wood in waters south of San Francisco Bay on the west coast and all along the east coast of the United States. Specimens of this species have been removed from creosoted wood and the preservative could literally be squeezed from their bodies, yet they continued to attack the wood. This remarkable resistance has both fascinated and stymied scientists, who have yet to develop a plausible explanation for this phenomenon.

Limnoria damage wood by burrowing small-diameter (0.12 inch) tunnels near the wood surface. Although the damage is minimal, continued removal of the weakened wood by wave action exposes new wood to attack. Eventually, the member area is reduced to the point where the structure fails or must be replaced. A classic sign of *Limnoria* attack is the hourglass shape that severely attacked piling take about the tidal zone (Figure 13-18); however, attack can and does extend to the mudline if oxygen and salinity conditions are suitable.

Other Marine Borers

A relatively new concern for wood users in semitropical waters is *Sphaeroma terebrans,* a mobile crustacean native to the Florida mangrove

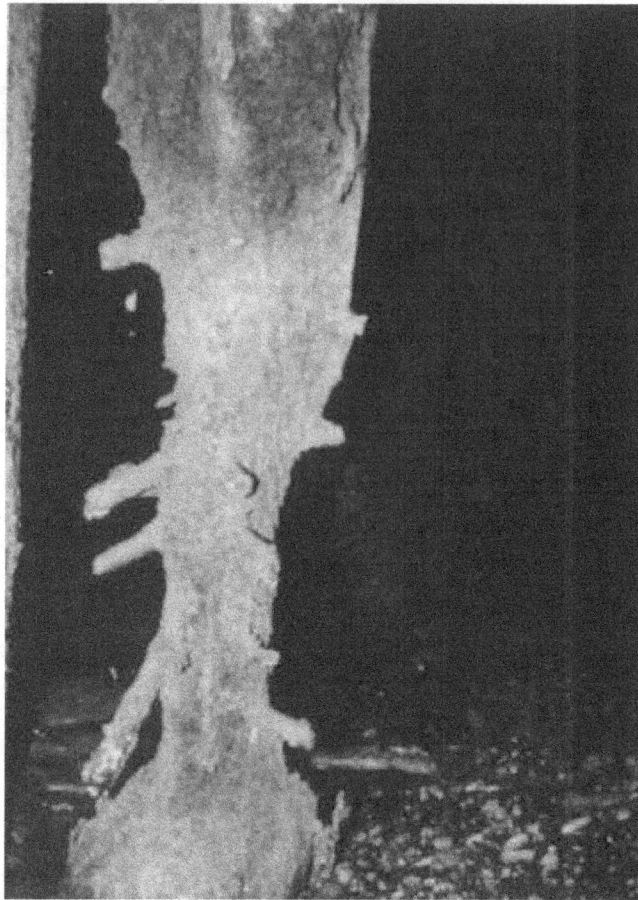

Figure 13-18. - Limnoria damage to a timber pile, evidenced by the characteristic hourglass shape in the tidalzone.

swamps. This species exhibits greater tolerance to copper-containing wood preservatives and may become an important factor in Florida and other warm-water regions.

PHYSICAL AGENTS OF DETERIORATION

Although wood deterioration is traditionally viewed as a biological process, wood can also be degraded by physical agents. These agents are generally slow acting, but can become quite serious in specific locations. Physical agents include mechanical abrasion or impact, ultraviolet light, metal corrosion by-products, and strong acids or bases. Damage by physical agents can be mistaken for biotic attack, but the lack of visible signs of fungi, insects, or marine borers, plus the general appearance of the wood, can alert the inspector to the nature of the damage. Although destructive in their own right, physical agents can also damage the preservative treatment, exposing untreated wood to attack by biotic agents.

Mechanical Damage

Mechanical damage is probably the most significant physical agent of timber bridge deterioration. It is caused by a number of factors and varies considerably in its effects on the structure. Most commonly, mechanical damage is from vehicle abrasion, which produces worn or marred surfaces and reduces the effective wood section. Obvious examples of this damage occur in the bridge deck area where abrasion produces degradation of wearing surfaces and wheel guards. More severe mechanical damage may be caused by long-term exposure to vehicle overloads, foundation settlements, and debris or ice floes in the stream channel (Figure 13-19).

Ultraviolet Light Degradation

Some of the most visible wood deterioration results from the action of the ultraviolet portion of sunlight, which chemically degrades the lignin near the wood surface (Figure 13-20). Ultraviolet degradation typically causes light woods to darken and dark woods to lighten, but this damage penetrates only a short distance below the surface." The damaged wood is slightly weaker, but the shallow depth of the damage has little influence on strength except where continued removal of damaged wood eventually reduces the member dimensions.

Corrosion

Wood degradation from metal corrosion is frequently overlooked as a cause of bridge deterioration. This type of degradation can be significant in some situations, particularly in marine environments where saltwater galvanic cells form and accelerate degradation." Corrosion begins when

Figure 13-19. - Severe mechanical damage to a glulam bridge caused by debris flow during high stream levels.

13-23

Figure 13-20. - Ultraviolet light degradation of the end grain of a guardrail post. Note the minor surface erosion of earlywood between the latewood (growth rings).

moisture in the wood reacts with iron in a fastener to release ferric ions that in turn deteriorate the wood cell wall. As corrosion progresses, the fastener becomes an electrolytic cell with an acidic end (anode) and an alkaline end (cathode). Although the conditions at the cathode are not severe, the acidity at the anode causes cellulose hydrolysis and severely reduces wood strength in the affected zone. Wood attacked in this fashion is often dark and appears soft (Figure 13-21). In many wood species, discoloration also occurs where iron contacts the heartwood.

In addition to the deterioration caused by corrosion, the high moisture conditions associated with this damage can initially favor the development of fungal decay. As corrosion progresses, the toxicity of the metal ions and the low pH in the wood eventually eliminate fungi from the affected zone, although decay may continue at some distance away from the fastener. The effect of wood metal corrosion can be limited by using galvanized or noniron fasteners.

Chemical Degradation

In isolated cases, the presence of strong acids or bases can cause substantial damage to wood. Strong bases attack the hemicellulose and lignin, leaving the wood a bleached white color. Strong acids attack the cellulose and hemicellulose, causing weight and strength losses. Wood damaged by acid is dark in color and its appearance is similar to that of wood damaged by fire. Strong chemicals will normally not contact a timber bridge unless accidental spills occur.

13-24

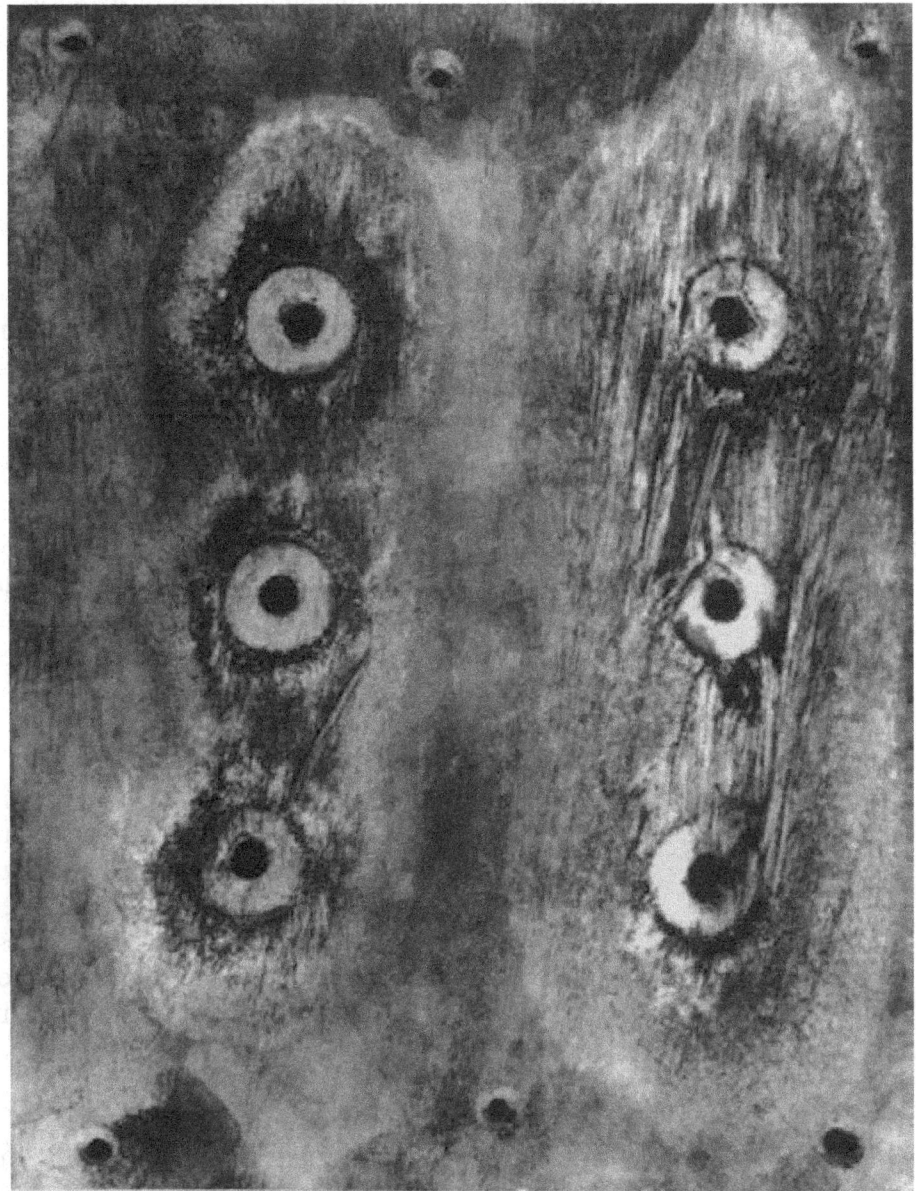

Figure 13-21. - Wood damage around bolt holes caused by corrosion of the metal fasteners.

13.3 METHODS FOR DETECTING DETERIORATION

Until this point, discussions have been fairly specific about the effects that various organisms have on wood. Unfortunately, our ability to detect wood deterioration has lagged far behind our knowledge of deterioration mechanisms. As a result, the inspection process varies widely among regions, although the tools of the trade are fairly standard. There is no magic box that will accurately determine the condition of a given structure, but a number of tools used in combination can give a reasonable estimate of the amount and degree of wood deterioration present.

Methods for detecting deterioration in bridges are divided into two categories: those for exterior deterioration and those for interior deterioration. In both cases, specific methods or tools are appropriate for certain types of damage, and their usefulness varies depending on the type of structure. Although a variety of inspection methods may be employed, in practice the inspector uses only a few tools. The methods or tools are often dictated by budget, previous experience, and the types of problems that are encountered. No equipment can replace a well-trained inspector who has a broad knowledge of wood systems.

METHODS FOR DETECTING EXTERIOR DETERIORATION

Exterior deterioration is the easiest to detect because it is often readily accessible to the inspector. The ease of detection depends on the severity of damage and the method of inspection. The four methods or tools most commonly used include visual inspection, probing, the pick test, and the Pilodyn. When areas of exterior deterioration are located by these methods, further investigation by other methods is required in order to confirm and define the extent of damage.

Visual Inspection

The simplest method for locating deterioration is visual inspection. The inspector observes the structure for signs of actual or potential deterioration, noting areas for further investigation. Visual inspection requires strong light and is suitable for detecting intermediate or advanced surface decay. It will not detect decay in the early stages, when control is most effective, and should never be the sole method employed. Some of the more common visual signs of deterioration include the following (Figure 13-22):

Fruiting bodies provide positive indication of fungal attack, but do not indicate the amount or extent of decay. Some fungi produce fruiting bodies after small amounts of decay have occurred, while others develop only after decay is extensive. Because fruiting bodies are not common on bridges, they almost certainly indicate serious decay problems when they are present.

Sunken faces or localized surface depressions can indicate underlying decay. Decay voids or pockets may develop close to the surface of the member, leaving a thin, depressed layer of intact, or partially intact, wood at the surface.

Staining or discoloration indicates that members have been subjected to water and potentially high moisture contents suitable for decay. Rust stains from connection hardware are also a good indication of wetting.

Figure 13-22. – Visual signs of potential deterioration. (A) Fruiting bodies. (B) Sunken faces (shown with the thin surface layer removed).

Figure 13-22. - Visual signs of potential deterioration (continued). (C) Water staining. (D) Insect activity (powder posting).

Figure 13-22. - Visual signs of potential deterioration (continued). (E) Plant growth.

Insect activity is visually characterized by holes, frass, powder posting, or other signs previously discussed. The presence of insect activity may also indicate the presence of decay.

Plant or **moss growth** in splits, cracks, or soil accumulations on the structure indicate that adjacent wood has been at a relatively high moisture content suitable for decay for a sustained period of time.

Probing

Probing with a moderately pointed tool, such as an awl or knife, locates decay near the wood surface by revealing excessive softness or a lack of resistance to probe penetration. Although probing is a simple inspection method, experience is required to interpret results. Care must be taken to differentiate between decay and water-softened wood that may be sound but somewhat softer than dry wood. It is also sometimes difficult to assess damage in soft-textured woods such as western redcedar.

Pick Test

The pick test is one of the simplest, yet most widely used, methods for detecting surface decay. A pointed pick, awl, or screwdriver is driven a short distance into the wood and used to pry out a sliver (Figure 13-23). The wood break is examined to determine if the break is brash (decayed) or splintered (sound). Sound wood has a fibrous structure and splinters when broken across the grain. Decayed wood breaks abruptly across the grain or crumbles into small pieces. Several studies indicate that the pick test is reasonably reliable for detecting surface decay. The only drawback to this method is having to remove a large sliver of wood for each test.

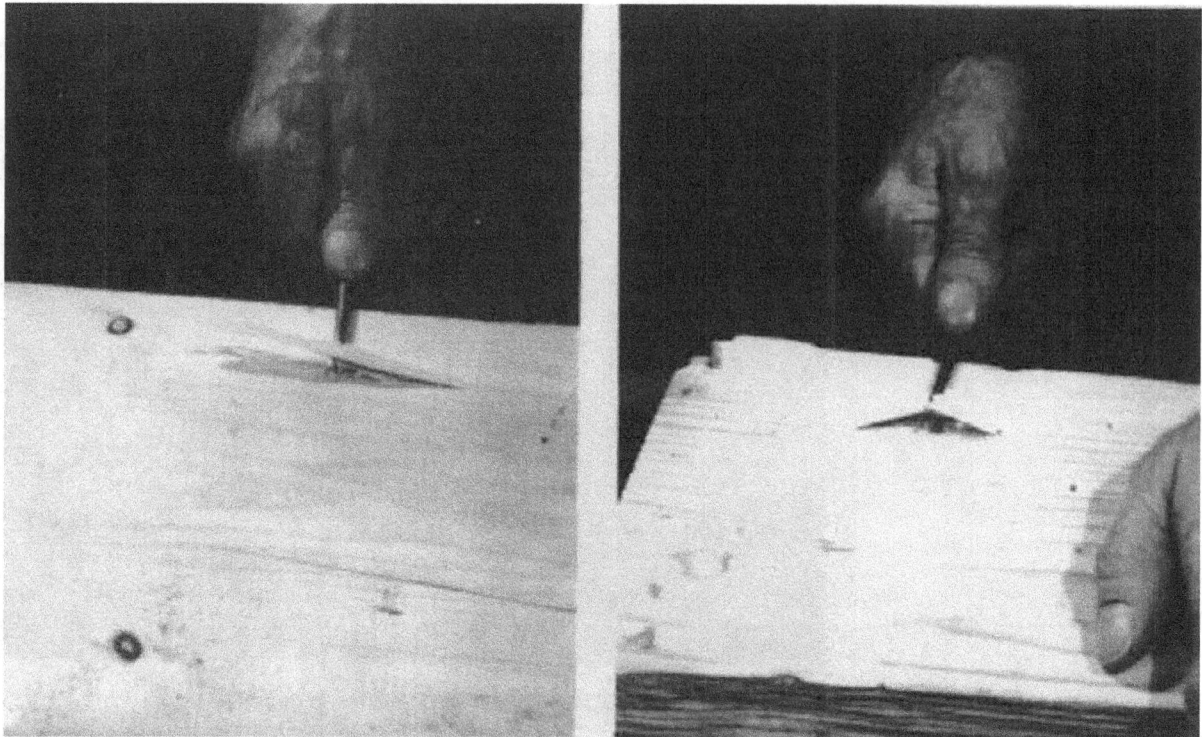

Figure 13-23. - The pick test for detecting earlywood decay. (Left) Sound wood pries out as long slivers. (Right) Decayed wood breaks abruptly across the grain without splintering.

Pilodyn

Like the pick test, the Pilodyn is also used to detect surface damage. The Pilodyn is a spring-loaded pin device that drives a hardened steel pin into the wood (Figure 13-24). The depth of pin penetration is used as a measure of the degree of decay. The Pilodyn is used extensively in Europe, where soft rot attack is more prevalent. It is also used to measure the specific gravity of wood for tree improvement programs. Where surface damage is suspected, the Pilodyn can produce an accurate assessment, provided corrections are incorporated for moisture content and the wood species tested.[48]

METHODS FOR DETECTING INTERIOR DETERIORATION

Unlike exterior deterioration, interior deterioration is difficult to locate because there may be no visible signs of its presence. Numerous methods and tools have been developed to evaluate internal damage that range in complexity from sounding the surface with a hammer to sophisticated sonic or radiographic evaluation. In addition, such tools as moisture meters are used to help the inspector identify areas where conditions are suitable for development of internal decay.

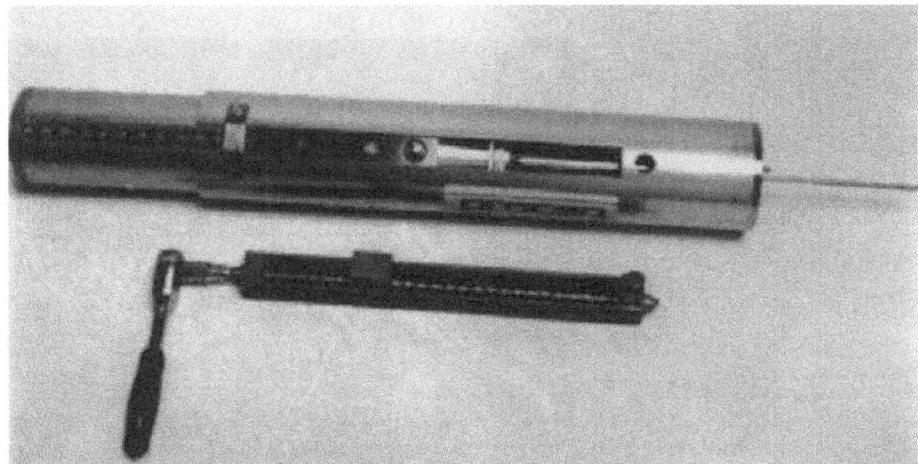

Figure 13-24. - The Pilodyn uses a spring-loaded pin that is forced into the wood surface. The depth of pin penetration provides a measure of wood condition.

Sounding

Sounding the wood surface by striking it with a hammer or other object is one of the oldest and most commonly used inspection methods for detecting interior deterioration (Figure 13-25). Based on the tonal quality of the ensuing sounds, a trained inspector can interpret dull or hollow sounds that may indicate the presence of large interior voids or decay. Although sounding is widely used, it is often difficult to interpret because factors other than decay can contribute to variations in sound quality. In addition, sounding provides only a partial picture of the extent of decay present and will not detect wood in the incipient or intermediate stages of decay. Nevertheless, sounding still has its place in inspection and can quickly identify seriously decayed structures. When suspected decay is encountered, it must be verified by other methods such as boring or coring.

Moisture Meters

As wood decays, certain electrolytes are released from the wood structure and electrical properties of the material are altered. Based on this phenomenon, several tools can be used for detecting decay hazard by changes in electrical properties. One of the simpler tools is the resistance type moisture meter. This unit uses two metal probes (pins) driven into the wood to measure electrical resistance, and thus, moisture content (Figure 13-26). Moisture meters must be corrected for temperature and are most accurate at wood moisture contents between 12 and 22 percent. Pins are available in various lengths for determining moisture content at depths up to 3 inches.

Although it does not detect decay, the moisture meter will help identify wood at high moisture content and is recommended to initially check suspected areas of potential decay. Moisture contents higher than 30 percent indicate conditions suitable for decay development unless the wood in the immediate area is treated with preservatives and no breaks

13-31

Figure 13-25. - A decay pocket near the wood surface is detected by sounding with a hammer.

are occurring in the treatment envelope. If inspection is conducted after an unusually long period of dry weather, lower moisture levels in the range of 20 to 25 percent should be used as an indication of potentially hazardous conditions. Information on the use and limitations of moisture meters is more thoroughly discussed elsewhere.[29]

Shigometer

The Shigometer, a device that has been compared to the moisture meter, uses a pulsed current to measure changes in electrical conductivity associated with decay (Figure 13-27). A small hole is drilled into the wood, and a twisted wire probe connected to a meter is inserted into the hole. As the probe encounters zones of decreased resistance, the meter reading drops. Zones of large meter declines (50 to 75 percent of that indicated for sound wood) are then bored or drilled to determine the nature of the defect. The Shigometer has performed very well in detecting decay in living trees, but wood in service is normally too dry to permit the use of this instrument. Nevertheless, several studies show that the Shigometer is a reasonable method for detecting decay if it is used under proper conditions by trained operators who understand its operation and interpretation.[30]

Drilling and Coring

Drilling and coring are the most common methods for detecting internal deterioration in bridges.[34] Both techniques are used to detect the presence of voids and to determine the thickness of the residual shell when voids are present. Drilling and coring are similar in many respects and will be discussed together.

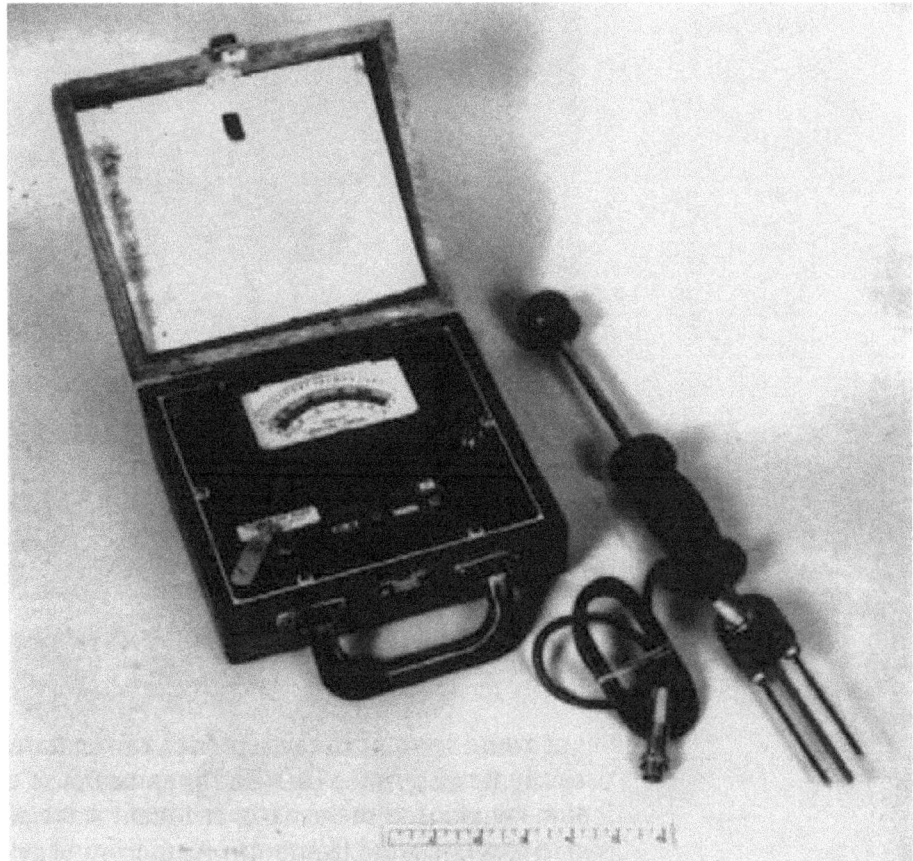

Figure 13-26. - The resistance-type moisture meter uses two steel pins that are driven into the wood to measure moisture content (the middle probe between the pins is a depth indicator). This device can help determine whether the wood moisture content is suitable for decay organisms.

Drilling is usually done with an electric power drill or hand-crank drill equipped with a 3/8- to 3/4-inch-diameter bit. Power drilling is faster, but hand drilling allows the inspector a better feel and may be more beneficial in detecting pockets of deterioration. Generally, the inspector drills into the structure, noting zones where the drilling becomes easier (torque releases), and observes the drill shavings for evidence of decay (Figure 13-28). The presence of common wood defects such as knots, resin pockets, and abnormal grain must be anticipated while drilling and must not be confused with decay. If decay is detected, the inspection hole can also be used to add remedial treatments to the wood.

Coring with increment borers also provides information on the presence of decay pockets and other voids, and coring produces a solid wood core that can be carefully examined for evidence of decay (Figure 13-29). Where appropriate, the core also can be used to obtain an accurate measure of the depth of preservative penetration and retention. Where structures are not

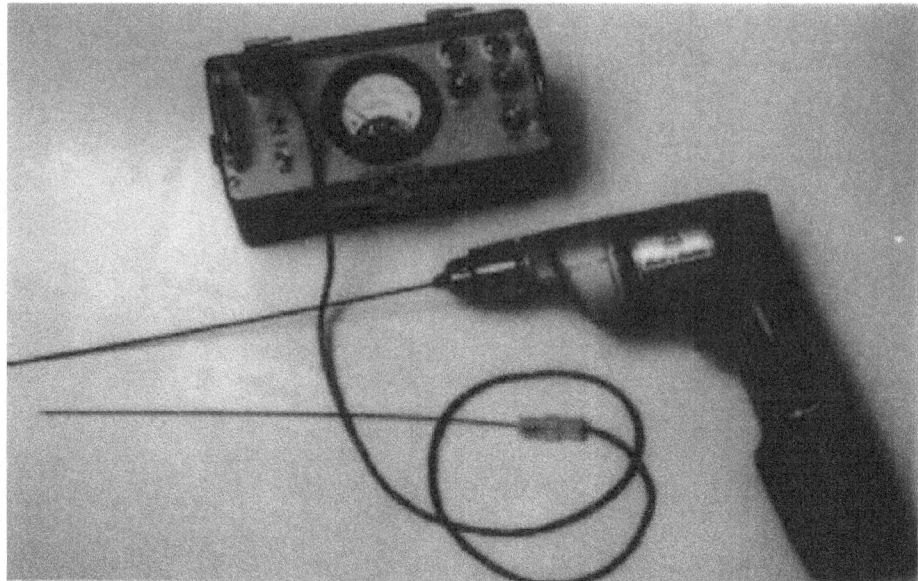

Figure 13-27. - A Shigometer and the drill and bit used to bore holes for insertion of the wire probe.

yet showing signs of decay, cores can be cultured to detect the presence of decay fungi (Figure 13-30). The presence of such fungi usually indicates that the wood is in the early or incipient stage of decay and should be remedially treated (Chapter 14). Culturing provides a simple method for assessing the potential decay hazard and many laboratories provide routine culturing services.[39] Because of the wide variety of fungi near the surface, culturing is not practical for assessing the hazard of external decay.

Drilling and coring are generally used to confirm suspected areas of decay identified by the use of moisture meters or other methods. When decay is detected, drilling and coring are also used to further define the decay's extent and limits. Inspectors may find drilling best for initial inspection until some evidence of decay is found. When decay is detected, coring may be preferred for defining the limits of the infection and extracting samples for further examination and analysis. It is important to use sharp tools for both drilling and coring and the inspector should always carry extra bits or increment borers. Dull tools tend to crush or break wood fibers and cause excessive core or shaving breakage that may be confused with decay.

Shell-Depth Indicator

A tool that is useful when drilling or coring is the shell-depth indicator. This tool is a metal bar, notched at the end and inscribed in inches, that is inserted into the inspection hole and pulled back along the hole sides (Figure 13-31). As it moves along the wood, the hook will catch on the edges of voids. In this way the inspector can note the depth of the solid shell, which can be used to estimate residual wood strength.

13-34

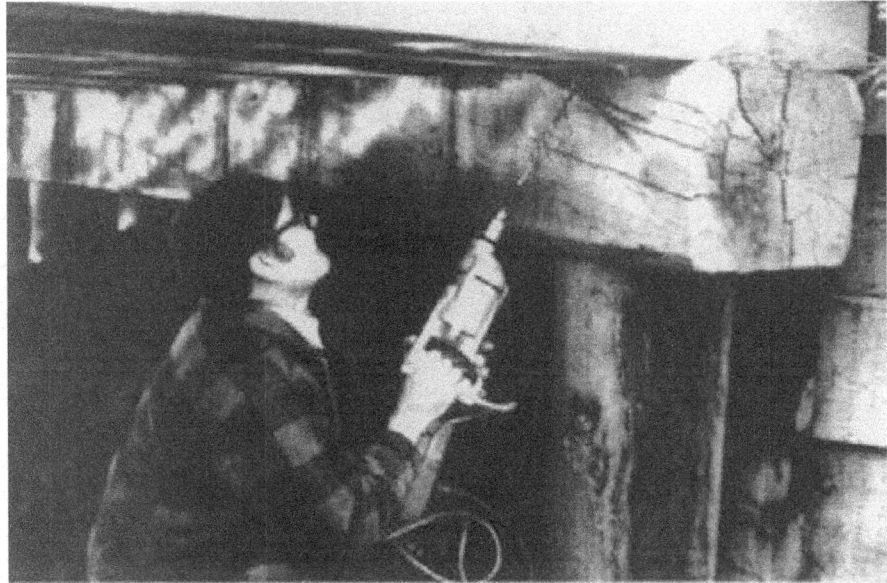

Figure 13-28. - Drilling the underside of a timber bridge beam to detect internal voids. The inspector feels and listens for torque release as the drill bit enters the wood, and examines shavings for evidence of decay.

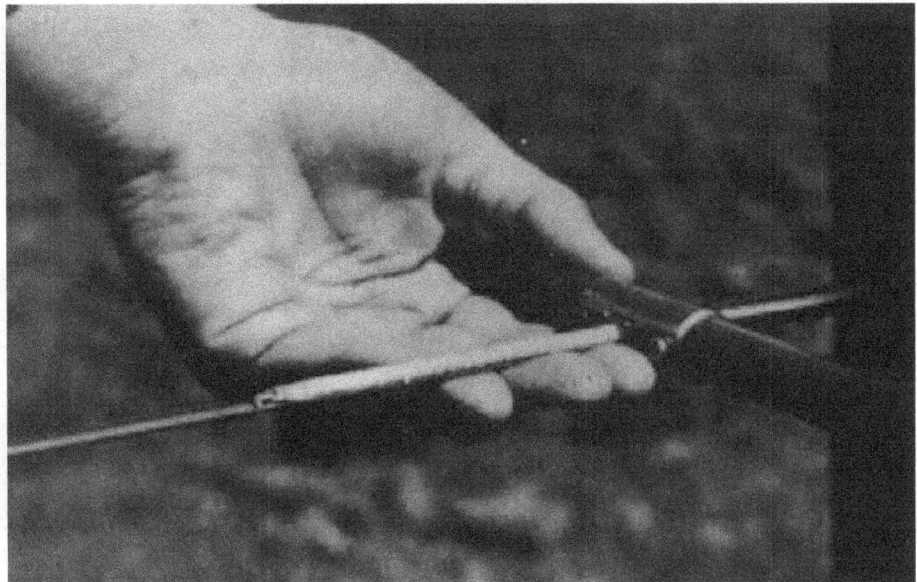

Figure 13-29. - Solid wood core removed with an increment borer. Such cores can be examined to determine the location and extent of decay.

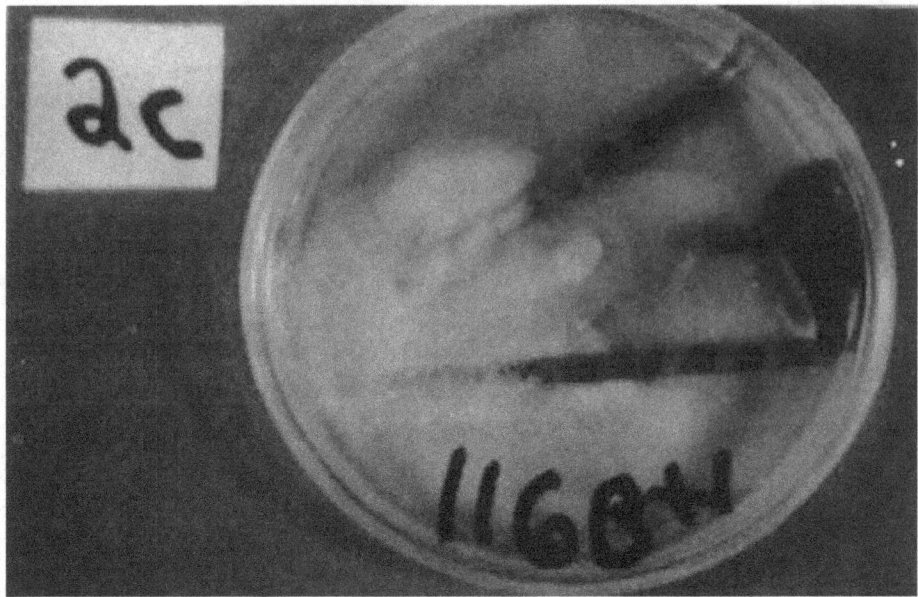

Figure 13-30. - Culturing increment cores to determine the presence of decay fungi. This process can detect decay before visible damage occurs and provides a method of assessing future risk.

Figure 13-31. - Use of a shell depth indicator, illustrated with a portion of the member removed. The tool is inserted into an inspection hole and moved along the hole sides to feel for decay voids.

Sonic Evaluation

Sophisticated sonic tests for evaluating wood condition have been developed in recent years. Several of these methods, including sonic wave velocity, acoustic emission, and stress wave analysis have been investigated. The simplest of the sonic techniques uses an instrument to measure the velocity changes of a sound wave moving across the wood (Figure 13-32). The earliest versions of these tools were used with mixed results on utility poles. More recent efforts have concentrated on measuring how the sonic wave is altered by wood defects. The altered sonic wave or fingerprint can be used to determine the exact size and nature of a defect. Several sonic methods are nearing commercialization and offer a significant advancement in decay detection capabilities; however, where defects are detected, other methods must still be used to determine the cause.

X Rays and Tomography Scanners

X rays were once commonly used for detecting internal voids in wood? As the x rays pass through the wood, the presence of knots or other defects alters the density of the resulting radiograph (Figure 13-33). X-ray technology has advanced considerably since the first field units were developed; however, the high cost of equipment, along with the safety factors associated with the use of ionizing radiation and the need for expert interpretation of results, have largely eliminated its use in wood. Despite these problems, x rays are particularly useful for detecting insect and marine borer infestations in wood.

Figure 13-32. - A sonic inspection device for detecting internal defects in wood.

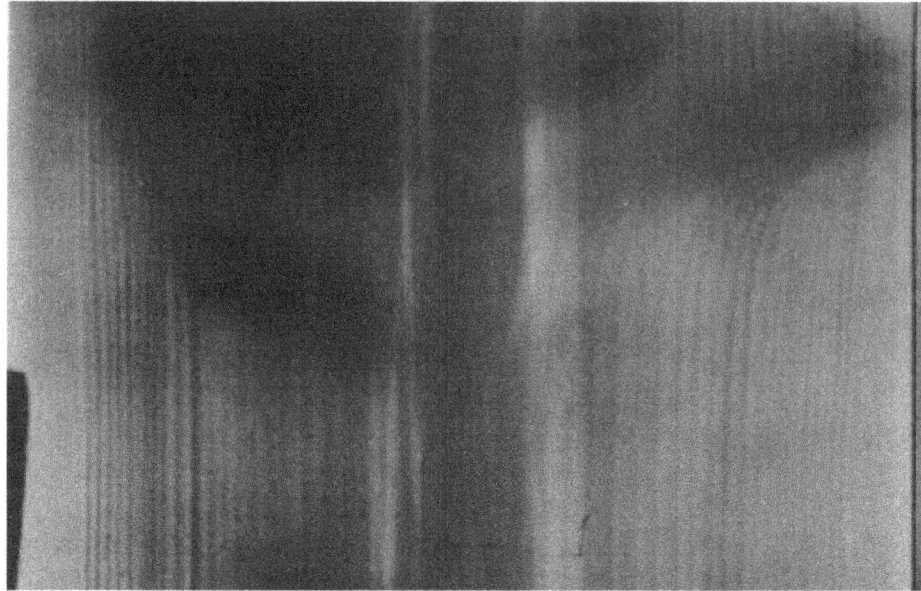

Figure 13-33. - X-ray radiograph of a timber member. X rays can be used to detect internal wood defects, but are particularly useful for locating insect or marine borer damage.

Recently, several European universities have developed computer-aided tomography scanners for wood poles. The scanners move up or down a pole and provide an image of internal wood conditions. Prototypes of these devices are in the early stages of development, and further refinements are necessary to speed up the process of data evaluation.

POSTINSPECTION TREATMENT

Several inspection methods involve techniques that destroy or remove a portion of the wood. Splinters, probe holes, and borings may become avenues for decay entry if not properly treated at the conclusion of the inspection. All surface damage should be treated with liquid or paste wood preservative (Chapter 14). For bore holes, liquid wood preservative should be squirted into the hole, which then should be plugged with a preservative-treated dowel slightly larger in diameter than the inspection hole (Figure 13-34). Treatment with creosote or copper naphthenate is generally sufficient for most bridge inspections, but other treatments should be used for additional protection in areas of marine borer hazard. When wood is subject to attack by *Limnoria*, surfaces and plugs should be treated with waterborne salts. In areas where pholads may attack, treatment with both creosote and waterborne salts is advisable. Failure to follow these procedures may result in accelerated decay development or deterioration in the structure.

Figure 13-34.- After treating an inspection hole in a bridge deck with liquid wood preservative it is plugged with a treated wood dowel (photo courtesy of Frank Muchmore, USDA Forest Service).

13.4 INSPECTION PROCEDURES

Inspection procedures for timber bridges depend on such variables as the age and type of bridge and the environment in which the bridge is located. Therefore, detailed recommendations for specific procedures are somewhat impractical. In general, the inspector must thoroughly examine the bridge for decay and other deterioration and record findings in sufficient detail for an engineering appraisal. The specific procedures and methods, however, will vary substantially from bridge to bridge.

Bridge inspection can be divided into three major steps: preinspection evaluation, field inspection, and preparation of reports and records. Although the specific procedures in each step vary among bridges, the basic process is the same. Discussions in this section are intended to provide the inspector with an understanding of the general characteristics of deterioration and the concepts related to inspection procedures. With this understanding, specific inspection procedures can be developed that are best suited to a particular structure.

PREINSPECTION EVALUATION

The potential for deterioration in a timber bridge depends on its environment. A preliminary assessment of hazard potential will reduce the need to speculate on potential causes and effects and better prepare the inspector to formulate methods of inspection. From an environmental viewpoint, decay potential varies considerably among localities, and local experience is the best information source.

Preinspection evaluation involves an office review of information before field inspection. The purpose of the evaluation is to learn as much as possible about the history of the bridge to better prepare the inspector for the field work. During the evaluation, the inspector should make a thorough study of historical records, reports, and other available information. It is also beneficial to discuss factors related to the bridge with people who are familiar with its location and history. A little effort spent on preinspection evaluation will help the inspector anticipate potential problems and make field inspection more effective.

The previous inspection reports are one of the best sources of bridge information. These reports provide the most current information on bridge condition and familiarize the inspector with the types and locations of previous damage. In addition, the original bridge construction drawings and documents are good sources of information. As-built drawings are most informative, but when they are not available design drawings may be used. The drawings provide information about the dimensions, species, and grade of material used as well as the type and retentions of preservative treatments. Other construction documents such as contract specifications, inspection records, material certifications, and shipping invoices are also good sources of information.

When local information is not available, the general potential for fungal attack can be correlated geographically based on variations in average rainfall and temperature. The Southeastern region of the United States, with abundant rainfall and moderate temperatures, represents an area with high decay potential. The Northwest Pacific coast area is also in this category because of the unusually high annual rainfall. Bridges in areas having less than approximately 25 inches per year of rainfall or abnormally short growing seasons have reduced potential for decay. Maps are available that depict insect and decay hazards based on climatic conditions in broad regions (Figure 13-35); however, local conditions within these regions may vary considerably.

FIELD INSPECTION

Field inspection is the physical examination of a bridge for evidence of deterioration. Variations in bridge configurations and exposure conditions make this a complex task. It is therefore necessary for the inspector to be well acquainted with the agents of deterioration, the areas conducive to decay, and the fundamentals of component inspection. With this knowledge as a guide, the inspector is better prepared to identify and locate deterioration and accurately define its extent.

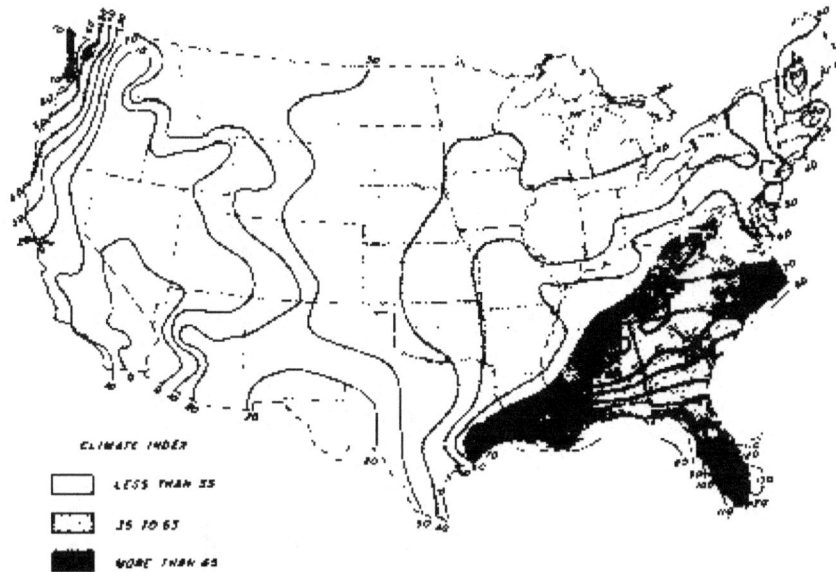

Figure 13-35. - Climate index map for decay hazard. The higher numbers indicate a higher decay hazard.

Areas Susceptible to Decay

Wood decay can occur only when proper conditions prevail for fungal growth. Although timber bridges differ in many respects, there are several common areas where decay is most likely to occur. These areas involve situations where the wood moisture is high and where breaks in the preservative envelope (or insufficient preservative penetration) provide an entry point for decay organisms. Signs of high moisture content and sites around fasteners, checks, or mechanical damage should be considered areas of high decay potential (Figure 13-36).

The moisture content of bridge components is not uniform, and substantial variations occur within and between members. End-grain surfaces absorb water much quicker than do side-grain surfaces (Figure 13-37). With other conditions equal, permeability in the longitudinal direction (parallel to grain) is 50 to 100 times greater than in the transverse direction (perpendicular to grain). Decay development is most affected by the moisture content of the wood in the immediate vicinity of the infection. Therefore, a member may remain generally dry and uninfected along most of its length but be severely decayed in localized areas where untreated wood is exposed and water is continuously or intermittently trapped. Bridge moisture conditions are also subject to seasonal variations and may be altered by maintenance operations or changes in drainage patterns. Wood that appears thoroughly dry may have been exposed to high moisture contents in the past and could be seriously decayed. The inspector must be alert for any visual or intuitive indications of wetting. Visual signs may appear as

13-41

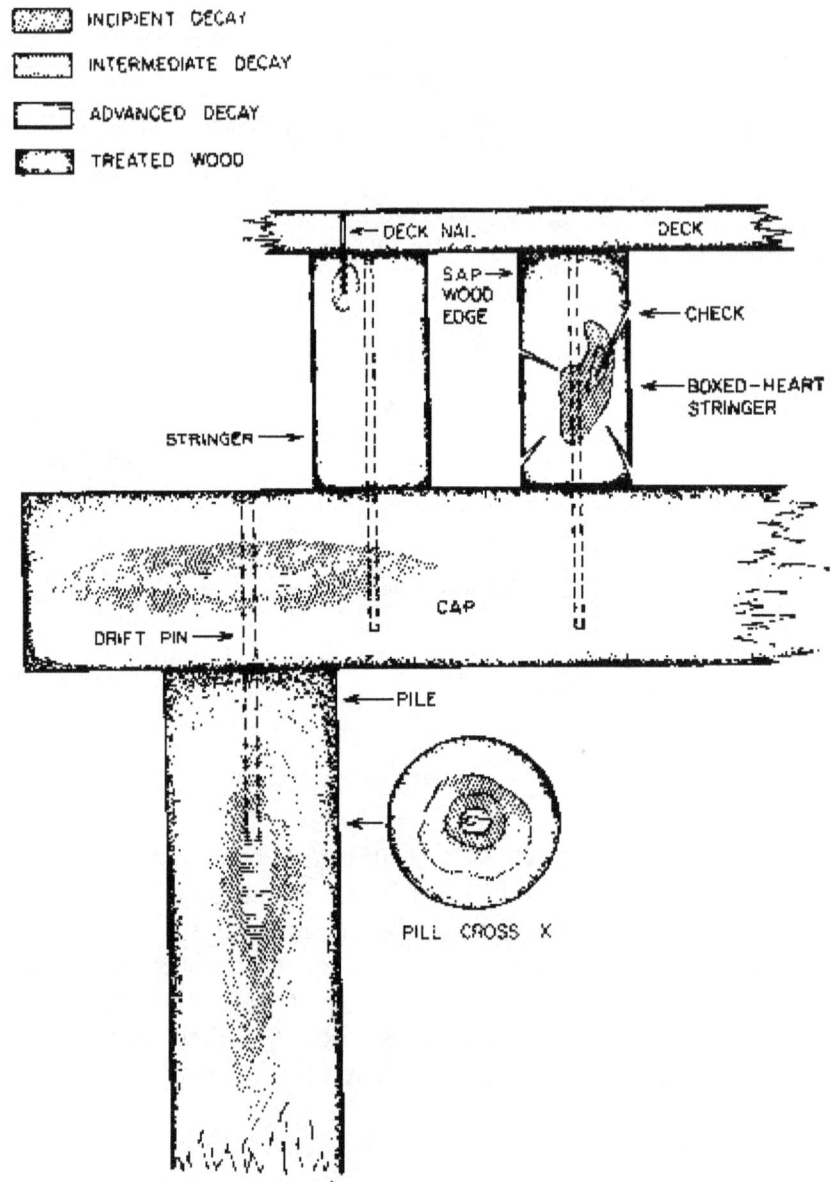

Figure 13-36. - Diagram depicting potential decay locations in a timber bridge.

watermarks, staining, or light mud stains. Intuitive signs include any horizontal surfaces, contact areas, depressions, or other features that may trap water and therefore indicate potentially high moisture exposures.

As discussed, the potential for bridge decay is highest where untreated wood is exposed. This condition occurs most often in the vicinity of seasoning checks, fasteners, and areas of mechanical damage. Conditions for deterioration are enhanced at these locations because moisture enters cracks or other crevices where air circulation and drying are inhibited. Seasoning checks commonly develop in large lumber members, and, to a

13-42

Figure 13-37. - Decay in the end grain of a timber rail post (photo courtesy of Duane Yager, USDA Forest Service).

lesser extent, in glulam. Although the size of the check influences the area of exposed untreated material, very small openings are still sufficient to allow entry of decay organisms (Figure 13-38). Holes for bolts, nails, or other hardware can trap water, which will be absorbed deep into the wood end grain by capillary action. Decay susceptibility at connections is higher because fasteners may be placed in field-bored holes that are not adequately treated with preservatives (Figure 13-39). Mechanical damage from improper handling, overloads, vehicle abrasion, and support settlements also breaks the preservative barrier and provides an entry point for decay organisms. In addition to increasing the decay hazard, mechanical damage may also affect structural capacity, depending on the decay's nature, location, and extent (Figure 13-40).

Component Inspection
Component inspection involves the systematic examination of individual bridge members. When deterioration is found, its location and extent must be defined and noted so that the load-carrying capacity of the structure can be determined by engineering analysis. At some locations, deterioration may have no significant effect on member strength. In other locations, any deterioration will reduce capacity. In both cases, the inspector must accurately locate, define, and record all deterioration, notwithstanding its perceived effects on structural capacity.

Because of the large number of structural components and the variety of locations where conditions for decay development exist in a bridge, the degree of accuracy for assessing the extent of deterioration depends on the judgment of the inspector. Regardless of bridge size, no inspection can

13-43

Figure 13-38. – Cross section of a timber curb, exposed by sawing, reveals interior decay resulting from seasoning checks in the upper surface.

Figure 13-39. – Decay in timber members around field-bored fastener holes.

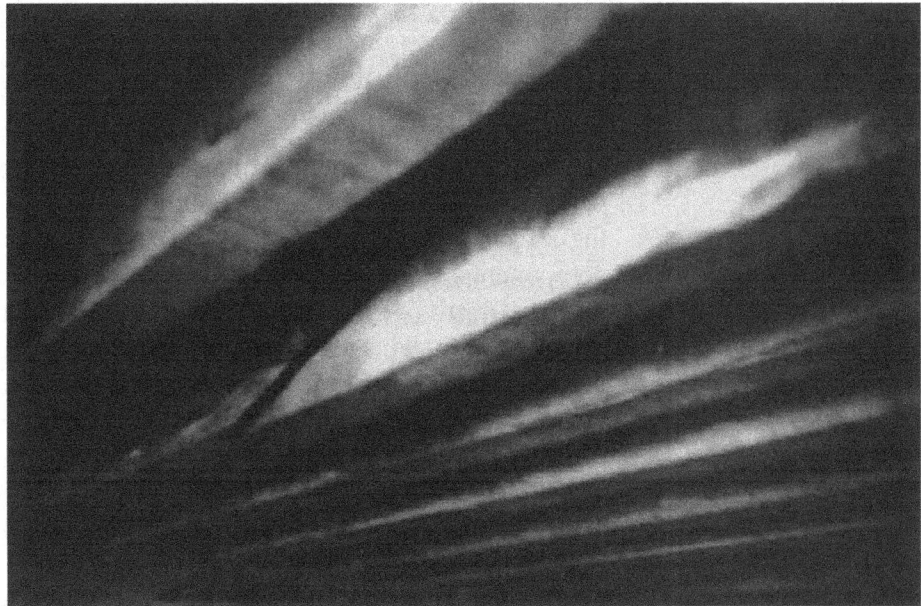

Figure 13-40. - Large crack in a sawn lumber bridge beam caused by vehicle overloads (photo courtesy of Duane Yager, USDA Forest Service).

reasonably or economically examine every bridge component. Rather, the inspector must base the degree of inspection on information from the preinspection evaluation and knowledge of bridge deterioration and its causes, signs, and probable locations. For example, it may not be practical to examine the area around each fastener when deck members are attached with penetrating fasteners in each beam. Instead, the inspector should select the most probable areas of deterioration for evaluation. If deterioration is found, its extent is determined and additional inspections are made at other locations. If no deterioration is found in high-hazard zones, it is unlikely that other areas are affected.

One of the most important aspects of component inspection is the sequence and coordination of inspection efforts. To ensure that all critical areas are covered, a systematic, well-defined plan must be developed. When more than one inspector is involved, the responsibilities of each must be clearly defined to avoid either missing areas or excessive duplication. The preferred inspection sequence generally follows the sequence of construction. After initially surveying the structure, the inspector begins with the lower substructure members and progresses upward to the top of the superstructure. Following this sequence, the inspector can observe the behavior of members under load before their actual inspection.

Initial Survey

The best way to begin a bridge inspection is to take a brief walk across and around the structure, observing general features and looking for obvious signs of deterioration or distress. Particular attention should be given to changes in the longitudinal or transverse deck elevation that may indicate foundation movement, deck swelling, or other adverse conditions. The rail and curb elements should also be checked for position and alignment. Slanted posts or separated rails may indicate deck swelling or superstructure movement. This is also a good time to observe drainage patterns on approach roadways and obstructions to deck drains, as well as the effectiveness of the deck and wearing surface in protecting underlying components. General observations of this type can alert the inspector to potentially adverse situations requiring more detailed examination later in the inspection. This inspection also can provide an opportunity to prepare initial sketches of the structure and to define the directions and other features used in recording inspection findings.

Substructure Inspection

The substructure is the portion of a bridge that is probably most susceptible to deterioration. Soil-contacting members such as posts, piling, abutments, and wing walls are exposed in varying degrees to nearly constant wetting, resulting in wood moisture contents suitable for decay. Surrounding soil frequently contains large numbers of fungal spores and woody plant material in which decay fungi can live and spread to infect bridge members. Substructure decay potential is also greater because of the high incidence of field fabrication (cutting and drilling) and the large number of penetrating fasteners.

Initial inspection of the substructure should begin with a visual examination of abutments for signs of deterioration, mechanical damage, and settlement. The most probable locations for decay are in the vicinity of the ground line, at connections between the cap and column, and at framing connections for bracing, tie rods, and backwall or wingwall planks. Starting at the base of the abutment, soil should be removed around a representative number of members in order to inspect for indications of decay or insect attack. When soil is very wet or covered by water, decay is generally limited to areas close to ground level because the lack of oxygen below the surface limits the growth of most fungi. As soil moisture content decreases, conditions below ground become more favorable, and decay may occur at depths of 2 feet or more in moderately dry soils. Surface decay and insect damage can be revealed by visual observation and probing. When evidence of decay is found, its extent is further defined by drilling or coring (Figure 13-41). Detecting internal decay is generally accomplished by using a combination of sounding and drilling or coring. Because sounding will reveal only serious internal defects, it should never be the only method used.

Figure 13-41. - Hand drilling at the base of a timber pile.

From below the ground line, inspection should proceed upward, with particular attention given to connections, seasoning checks, and mechanical damage. Timber backwalls, wingwalls, and incidental bracing should also be examined for breakage or bulging from earth pressure. Exposed end grain on pile or post tops should also be inspected for decay. Many tops are intentionally cut at an angle in the belief that water will run off. Instead, angled cuts expose more untreated end grain, increasing the decay potential. When tops are provided with protective sheet-metal caps, the condition of the cap should be checked for holes or tears in the surface. Damaged caps allow water to enter through the break and penetrate end grain, creating ideal conditions for internal decay (Figure 13-42).

Above the supporting piles or posts, the cap supporting the superstructure provides a horizontal surface that traps debris and water runoff from the deck. Connections into the cap and horizontal checks that trap water and debris are critical zones. The connection between the cap and column is especially important because many connections are made with drift pins or bolts that extend deep into the column end grain. Water from the cap flows into these connections and can result in substantial internal decay with little evidence of exterior damage (Figure 13-43). The inspector should also check for crushed zones at bearing points along the cap that trap water and damage the treated wood shell. Crushing can also indicate overloads or load redistribution from settlement and should be further investigated in other components of the structure.

13-47

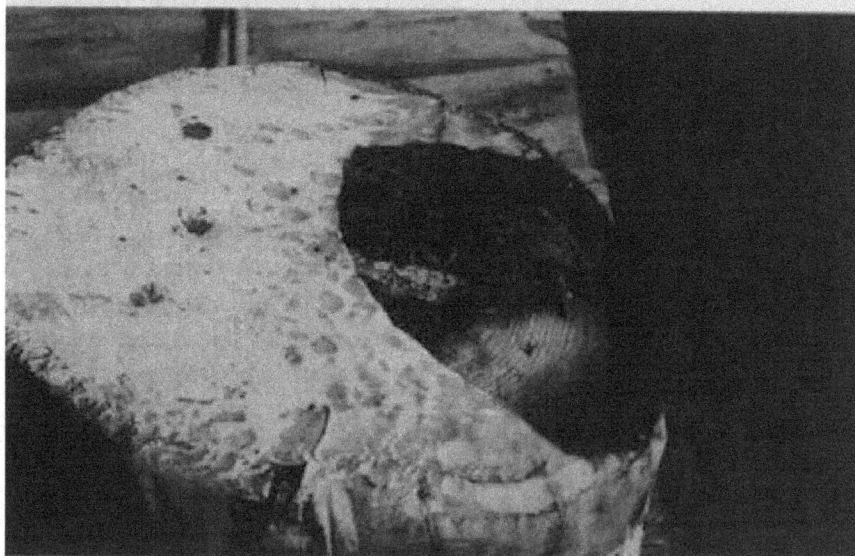

Figure 13-42. - (Top) Damaged metal pile caps allow water to enter, but restrict air circulation and drying. (Bottom) Pile decay is exposed when the damaged cap is removed.

13-48

Figure 13-43. – Internal decay in a timber pile where it was drift-pinned to the cap. Before the breakage of the outside shell, caused by cap removal, the pile showed little exterior sign of the interior decay.

Portions of the substructure containing piers or bents use the same basic inspection criteria for the same potential problem areas as abutments. If these structures are in water, however, inspection is much more difficult because access is limited. In water locations, members are also more susceptible to mechanical damage from floating debris and ice. In shallow water, inspectors can wear hip-waders to examine exposed members, whereas in deeper water a small boat or float is required. When inspection below the water level is necessary, the service of a diver is required. Underwater inspections require a high degree of skill and must be well coordinated to accurately identify and record deficiencies.[B]

For substructures located in seawater, low tides present the best opportunity to inspect for marine borer damage. Low-tide inspection is best suited for detecting *Limnoria*, which attack the external faces of members. A scraper and probe can be used to remove fouling organisms from the pile surface and thus permit better examination around bolt holes and adjoining wood members. Damage signs include an hourglass shape of piles in the tidal zone, bore holes; a general softening of wood in the attack areas; and loose bolts and bracing. Intertidal inspection is less effective for detecting damage by shipworms because they leave only a very small entrance hole on the wood surface, making visual detection difficult. Inspection methods using sonic instruments represent the best method for evaluating shipworm damage.

In areas where marine borer attack is suspected, an assessment of the hazard potential can be made by immersing sacrificial blocks of untreated wood at various depths around the substructure. These blocks are then removed periodically and examined for evidence of borer attack. Do not depend on the collection of driftwood to evaluate marine borer hazard because there is no way of knowing whether the wood came from sites outside the immediate area. Exposing wood samples can accurately assess marine borer hazard while providing a means for continually monitoring the long-term hazard.

Superstructure Inspection

After completing the substructure inspection, the inspector moves to the underside of the superstructure. It is best to thoroughly inspect all components from the bridge underside before moving to the roadway, since critical components are obscured by the wearing surface and deck. Superstructure inspection is generally hindered because access to the center portions of the underside is difficult or impossible without specialized equipment. When areas cannot be reached with ladders, a vehicle equipped with a mechanical arm or snooper may be required in order to adequately inspect the structure. Because ladders and other inspection equipment must be moved frequently to provide access to elevated areas, it is advisable that the inspection be performed by zones rather than by components. For the purposes of clarity, the following discussions are ordered by component.

Although most elements of the superstructure are out of ground contact, decay potential can be high in areas where water passes from the deck and collects at member interfaces, connections, checks, and crevices where air circulation and drying are inhibited. In many cases, this decay occurs with little or no surface evidence, although the member may be severely decayed inside. As a result, the inspector must be alert for conditions conducive to decay and must investigate areas where these conditions are likely to occur. As previously discussed, a moisture meter is a good tool for locating moisture conditions favorable to decay development (Figure 13-44). At least one boring should be made in areas of high moisture content where decay potential is considered highest. If decay is detected, additional borings should be taken to define its area, degree, and extent. If no decay is detected, but preservative penetration is shallow or moisture content is above 30 percent, it is desirable to remove a core for culturing to determine whether decay fungi are present.

The highest potential for decay in beams occurs at the deck-beam interface and attachment points, framing connections to other members, bearings, and seasoning checks. The deck-beam interface is one of the most frequent decay areas because water passing through the deck is trapped and enters fastener holes at the beam top. The hazard is highest when decks are attached with nails or lag screws that penetrate the top surface of the beam

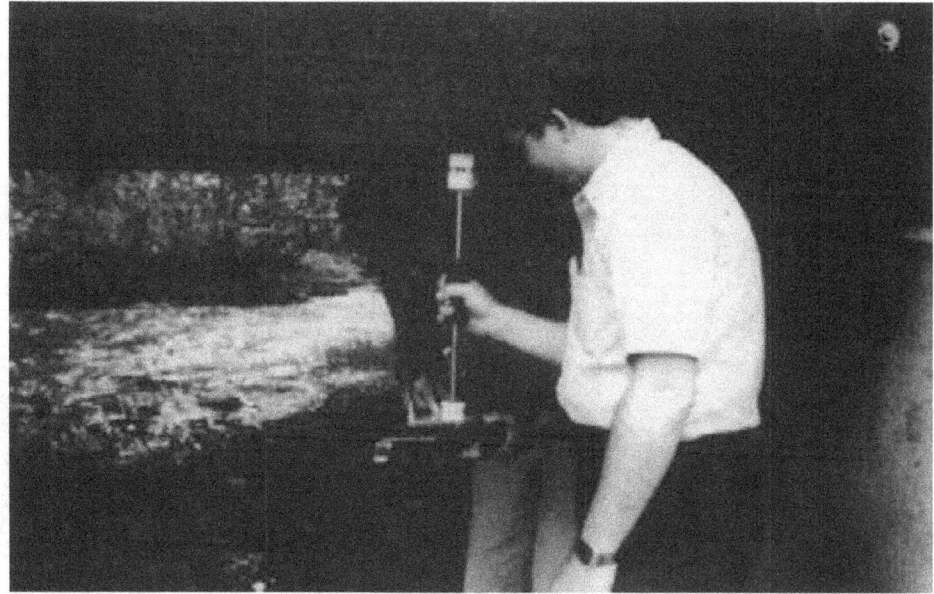

Figure 13-44. - The moisture content of a timber beam is measured with a resistance-type moisture meter.

(Figure 13-45). Glulam deck panels with bolted brackets do not involve attachments that penetrate the beam; thus, there is no significant increase in decay potential. On the deck underside, the inspector should be alert for signs of water movement and the presence of moisture at joint interfaces. Although stains are generally visible when water has passed through the deck, asphalt wearing surfaces tend to filter runoff, and visible signs are more difficult to detect. If significant decay is found along beam tops, it is advisable to remove deck sections to further examine beam condition.

In addition to the deck-beam interface and attachments, beam decay may develop in checks or delaminations, especially in the areas where end grain is exposed. Large checks or delaminations are not common in glulam and may be an indication of more severe structural problems. Bearings that trap water or show signs of beam crushing, and fasteners for transverse bracing or diaphragms are other potential decay locations. Sagging, splintering, or excessive deflections under load may also indicate mechanical damage or possible advanced decay. In some situations, surface decay may be present on a beam side or bottom that does not appear to be in an environment conducive to decay (Figure 13-46). Decay in such locations can occur in sawn lumber beams because of incomplete preservative penetration of heartwood.

Concurrent with beam inspection, the deck underside should be examined for signs of deterioration and conditions conducive to decay. Signs to observe include abnormal deflections and loose joints or fasteners, both of which may result from decay. Nail-laminated decks are frequently

13-51

Figure 13-45. - Severe decay in the tops of sawn lumber beams where the deck was attached to the beams with spikes.

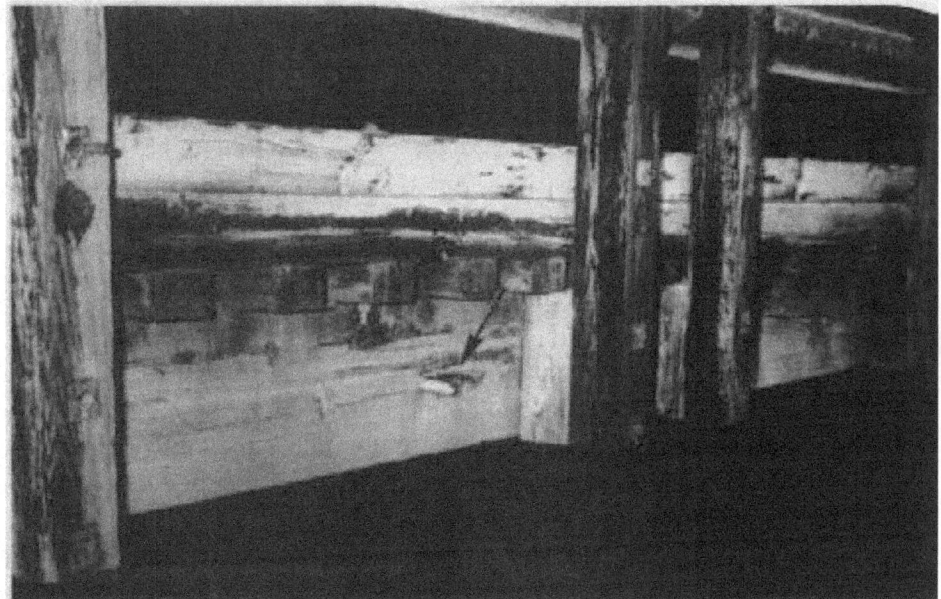

Figure 13-46. - Surface decay on the side of a sawn lumber beam (arrow). Decay in such locations is usually the result of poor preservative penetration of the heartwood.

delaminated by dynamic loading. Although delamination may not adversely affect strength, it does create voids between laminations, allowing water to flow on supporting beams and other components. Susceptibility to internal deck decay is highest with nail-laminated lumber or plank decks because they are interconnected and/or attached with nails or spikes (Figure 13-47). All fabrication for glulam panels is generally done before preservative treatment and the decay potential is lower unless panels are attached with spikes, lag screws, or other fasteners placed after the deck is treated.

Figure 13-47. - Decay on the underside of a spike-attached lumber plank deck at the deck-beam interface (arrow).

When inspection of the bridge underside is complete, efforts are next directed to the roadway portion of the deck. The upper deck is subject to wear and abrasion from traffic, and the horizontal surface facilitates water and debris accumulation. The highest decay potential occurs at fasteners or zones of mechanical damage and is influenced by the degree of protection provided by the wearing surface. A partial wearing surface affords the least deck protection because the gap between the running surfaces traps debris and moisture. On watertight glulam or stress-laminated timber decks, standing water may accumulate between running planks and remain for long periods. Moisture is also trapped under steel plate or full-plank surfaces where penetrating fasteners are normally placed after deck treatment. Asphalt wearing surfaces do not use mechanical fasteners, but moisture can accumulate at the deck interface when the surface is cracked or otherwise broken from excessive deflection.

The moisture content of timber decks generally averages 20 percent, but may frequently be much higher.[35] The inspector should carefully check exposed deck surfaces for moisture content and other conditions conducive to decay. When deck moisture contents are high, it is advisable to remove a number of cores from sites near the fasteners and other high-hazard locations. If necessary, portions of the wearing surface should be removed to assess deck condition. If evidence of substantial deterioration is found, the entire wearing surface should be removed to thoroughly inspect the deck.

Timber rails and curbs (wheel guards) are some of the most exposed elements of the bridge superstructure, yet are often ignored in bridge inspection. Although they are not critical for support of the structure, they are important for user safety and should be thoroughly inspected. Rails and curbs are susceptible to weathering, seasoning checks, and vehicle impact or abrasion. Rails and curbs are commonly the last components installed during the construction process and their installation presents an increased potential for field cutting and boring to meet alignment requirements. The inspector should pay particular attention to fasteners and areas that trap water and debris. One very probable decay situation occurs when approach railposts are embedded in concrete (Figure 13-48).

Figure 13-48. - Decay in a timber railpost embedded in concrete at the abutment. Concrete spalling was caused when water trapped in the post cavity was subjected to freeze-thaw cycles.

REPORTS AND RECORDS While detecting decay or other wood damage is the major goal of bridge inspections, it is important to ensure that all pertinent inspection information is accurately recorded. The report prepared by the inspector provides the only means of communicating information about the structure and serves to

1. identify conditions that may limit the capacity of the structure or otherwise make it unsafe for public travel,

2. develop a chronological record of structural condition and provide the information necessary to complete a structural analysis when conditions change,

3. provide a basis for identifying current and future maintenance needs through the detection of early structural defects or deficiencies, and

4. provide a reference source for future inspections and comparative analysis.

When properly completed, the bridge inspection report is an important document and plays a critical role in ensuring the safety of users and in allocating funds for maintenance and replacement. In addition, it is a legal record that may be an important part of any future litigation. Although specific report formats vary among different jurisdictions and structure types, all must be well organized, clear, and concise. Each report should include a title page; drawings or sketches of the structure, labeling all components; a condition assessment of the structure, by component; a narrative summary of inspection findings; and recommendations for maintenance and corrective action. For large or complex structures, a notebook format is most appropriate. For smaller or less complex structures, standard inspection forms are more practical and convenient. In either case, a complete inspection report should be prepared for each bridge inspection, regardless of the purpose or depth of the inspection. Although no changes may be evident during the inspection, and the condition seems relatively unimportant, accurate documentation of the inspection can be valuable in the future.

A good inspection report documents detected deterioration and notes any details of the structure that deviate from the as-built drawings. During the course of the inspection, these deficiencies should be noted as they are found in order to avoid loss of detail. The inspector should be as objective as possible, recording what is seen and measured. For timber bridges, it is critical that all decay and its location be accurately and completely described. This must include both the location of the deterioration in specific components and the longitudinal and transverse dimensions of the decayed wood. It is also beneficial for correction to note the probable source of water and its pathway to the decay site. Additionally, the report should

13-55

note any indication of member weakness or failure, including evidence of excessive deflections, crushing, buckling, cracking, collapse, abnormal looseness of joints, or member displacement at joints. Further investigations should be recommended whenever they are considered necessary, either because the inspector does not have sufficient training or because more sophisticated equipment is required.

Sketches, drawings, and photographs are invaluable for illustrating inspection results and should be used freely to locate, identify, and clarify the condition of the bridge components. Drawings and sketches should define the location and extent of deterioration in sufficient detail and accuracy so that other inspectors or maintenance personnel can easily locate the area in question. When available, as-built drawings or drawings from previous inspection reports can be copied and used for this purpose. Photographs are also very useful for showing structure condition and areas of deterioration. As a minimum, two photos should be included with each inspection report: one of the roadway view looking down the bridge and one of a side elevation. Additional photos showing defects or other important features should also be included when the inspector believes they will be helpful.

Each inspection report should include a summary of inspection findings and the recommendations of the inspector. The summary should outline the general condition of the structure and significant deficiencies encountered during the inspection. It may also include information and recommendations that the inspector believes are necessary to emphasize important inspection findings, including estimates of the materials and work hours required to perform the repairs and maintenance activities.

An example of a good timber bridge inspection report using a standard report format is shown in Figure 13-49. Additional information on inspection reports, including sample formats, is given in references listed at the end of this chapter.[152]

TIMBER BRIDGE INSPECTION REPORT

ROUTE NO. _463_ MILE POST _05.7_ NAME _Timber Bridge_

FEATURE CROSSED _Big Creek_ STRUCTURE TYPE _Single Span Timber Br._

DWG. NO. _None_ FOREST _Flatfoot_ DISTRICT _Lokeside_

T _23N_, R _20W_, SEC. _12_ YEAR BUILT _1952_ DESIGN LOADING _Unk._ SKEW _0°_

BRIDGE LENGTH _36'-6_ WIDTH _14'-1_ DATE INSP. _8/20/87_ NEXT INSP. _8/89_

INSPECTION TEAM NAMES AND FIRM _G. Pabel, R.O. Engineering_ _B. Miller_

S.O. Engineering

BRIDGE COMPONENT CONDITION RATING: COMPOSITE RATING: _8_

DECK: _8_ SUPERSTRUCTURE: _7_ SUBSTRUCTURE: _8_

STREAM CHANNEL: _8_ APPROACHES: _7_

CONDITION RATING DEFINITION:

N	Not applicable
9	New condition
8	Good condition; no repairs needed
7	Generally good condition; potential exists for minor maintenance
6	Fair condition; potential exists for major maintenance
5	Generally fair condition; potential exists for minor rehabilitation
4	Marginal condition; potential exists for major rehabilitation
3	Poor condition; repair or rehabilitation required immediately
2	Critical condition; the need for repair or rehabilitation is urgent. Close the facility until the repair is complete.
1	Critical condition; close the facility. Conduct a study to determine the feasibility for repair.
0	Critical condition; facility is closed and is beyond repair.

BRIDGE APPRAISAL RATING: COMPOSITE APPRAISAL: _6_

DECK GEOMETRY: _6_ STRUCTURAL: _7_ CLEARANCES: _6_

LOAD CAPACITY: _6_ WATERWAY: _6_ APPROACH ALIGNMENT: _6_

APPRAISAL RATING DEFINITIONS:

N	Not applicable
9	Condition superior to present desirable criteria
8	Condition equal to present desirable criteria
7	Condition better than present minimum criteria
6	Condition equal to present minimum criteria
5	Condition somewhat better than minimum adequacy to tolerate being left in place as is
4	Condition meeting minimum tolerable limits to be left in place as it
3	Basically intolerable condition, requiring high priority of repair
2	Basically intolerable condition, requiring high priority for replacement
1	Immediate repair necessary to put back in service
0	Immediate replacement necessary to put back in service

RI - FS - 7700-4 (7/87)

Figure 13-49. - Timber bridge inspection report using a standard report format (courtesy of Duane Yager, USDA Forest Service). See following pages.

BRIDGE NAME AND NUMBER ___Timber Bridge No. 463-05.7___ DATE __8/20/87__

Approaches — DRAINAGE, CONDITION, ETC. _Washboard @ N. end causes high impact_
___on the bridge. Otherwise fair condition___

Load Capacity and Other Signs — LEGIBILITY, VISIBILITY, ETC. _Signs in good condition. Object_
markers are Black and White. Two are missing (H-1L)

Waterway

1. ESTIMATED VELOCITY OF STREAM: __3 fps 8/20/87__

2. CHANNEL STABILITY (SCOUR, DEPOSITS, ETC.) DIKES AND BANK PROTECTION, OBSTRUCTIONS (ABOVE AND BELOW SITE).
   BACKWATER FROM FLOODING.

   COMMENTS: _Good condition. Small riprap has been placed @ both_
   abutments. Stream now flowing down the center of the structure.
   Channel appears to be stable now.

Abutments — UNDERMINING OR SETTLEMENT, DRIFT OR ICE DAMAGE, DECAY.

   COMMENTS: _N. upstream post sounded hollow. 8 inch core wet but no decay._
   Hollow sound is a shake. Abutment in good condition.

Superstructure

A. CURB, RUNNING PLANK, DECK, RAILING (DECAY, LOOSENESS, ETC.)

   COMMENTS: _Treated timber running plank new. Curbs are split_
   and starting to rot. Core #1 upstream between post #2&3 - 2"Good 1"Rot. 3"good.
   Core #2 downstream between post #4&5 - 2"Good, 3½Rot. 2½good. Deck Good condition

B. STRINGERS

   1. DECAY AT BEARING _None - Good Condition_

   2. DECAY BETWEEN DECK AND STRINGER (THREE PLACES ON BRIDGE, BORE AND LOCATE ON PLAN SKETCH) _____

Core #1.	3" R. Plank Good,	5⅛"Deck Good,	½ Rot Stringer 4"Good Stringer #1
Core #2	3" " " "	5⅛ " "	4½" Good Stringer #4
Core #3	3" " " "	5½ " "	6" " " #7

Condition Rating Of Each Member Or Element

NA =   — NOT APPLICABLE.

NOB =   — APPLICABLE, BUT NOT OBSERVED (Give reason unless obvious.)

G = GOOD   — ELEMENT IN NEW OR GOOD CONDITION WITH NO REPAIRS NECESSARY.

F = FAIR   — ELEMENT IS STILL PERFORMING THE FUNCTION FOR WHICH IT WAS INTENDED BUT MAY NEED
                MAINTENANCE.

P = POOR   — ELEMENT STILL PERFORMING THE FUNCTION FOR WHICH IT WAS INTENDED BUT IS IN NEED OF REPAIRS.

C = CRITICAL — ELEMENT IS NOT PERFORMING THE FUNCTION FOR WHICH IT WAS INTENDED.

R1 - FS - 7700-4 (7/87)

Figure 13-49. - (continued).

BRIDGE NAME & NO. _Timber Bridge No. 463-05.7_   DATE _8/20/87_

B. STRINGERS (CONTINUED)

3. DECAY OF STRINGERS: BORE TO A DEPTH OF AT LEAST HALF THE DIAMETER OF THE STRINGER. AFTER THE HOLES HAVE BEEN BORED, THEY SHALL BE FILLED WITH A 5 PERCENT SOLUTION OF PENTACHLOROPHENOL AND PLUGGED WITH A ROUND WOOD STOCK SOAKED IN THE PENTACHLOROPHENOL SOLUTION.

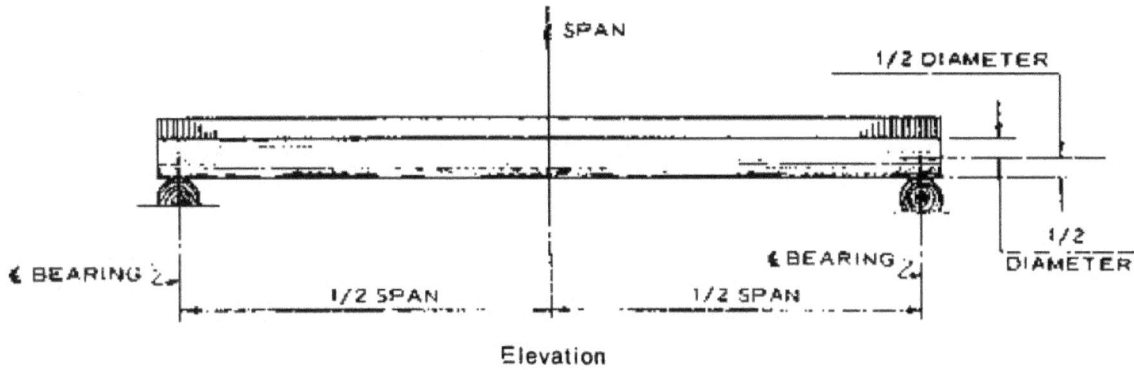

Elevation

Condition of Stringer*

STRINGER NO.	₵ BEARING @ ABUTMENT NO. 1	₵ SPAN	₵ BEARING @ ABUTMENT NO. 2
1	#1 Top Stringer 87 ½" Rot 4" Good		
2			#6 Side Stringer 87 4" Good
3			
4		#2 Top Stringer 87 2½" Good	
5	#5 Side Stringer 87 4½" Good		
6		#4 Side Stringer 87 5" Good	
7			#3 Top Stringer 87 6" Good
8			
9			

*DESCRIBE THE LOCATION OF THE DETERIORATED PORTION OF THE STRINGER. EXAMPLE: CENTER 3", OUTSIDE 2", NO DECAY, ETC.

Comments: _____

_____

_____

_____

_____

_____

RI - FS - 7700-4 (7/R7)

Figure 13-49. - (continued).

13-59

BRIDGE NAME (STREAM) __Big Creek_____ BRIDGE NO. __463-05.7_____

Dimensions

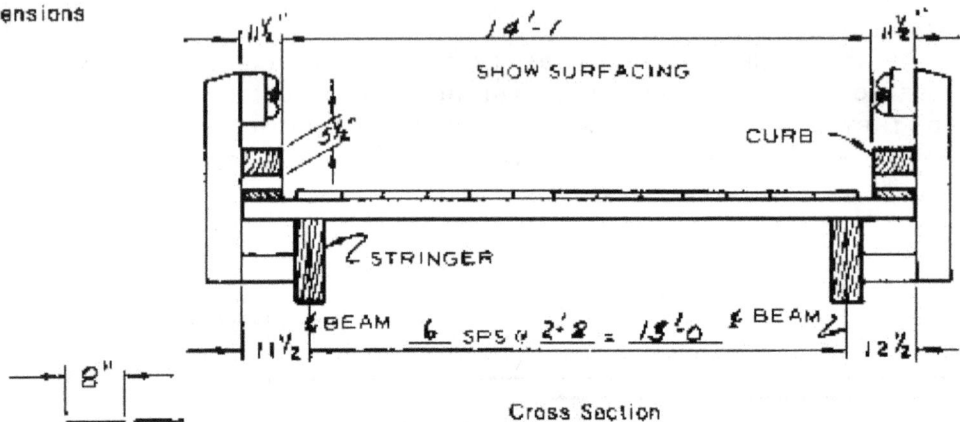

Cross Section

Rail Post

WIDTH __2½"__

DEPTH __9½"__

LENGTH __5'-6½__

SPACING __6'-3__

Deck

DIMENSION __5⅜" x 1½"__

TYPE

PLANK _____

NAIL LAMINATED __Treated__

GLUE LAMINATED _____

Stringer

SOLID SAWN (SPECIE) __D.Fir_____
            (TREATED) (UNTREATED)

LAMINATED _____
            (TREATED) (UNTREATED)

Guardrails

BRIDGE GUARDRAIL

   TIMBER (SIZE) _____

   STEEL FLEXBEAM __Double - 12 gauge__

   OTHER _____

   NONE _____

APPROACH GUARDRAIL

   FLARED END __Cable Anchored__

   "BURIED END" SECTION _____

   OTHER _____

   NONE _____

Wearing Surface

STEEL PLATE RUNNING PLANK (SIZE) _____

TIMBER RUNNING PLANK (SIZE) __3 x 12"_____
                        (TREATED) (UNTREATED)

GRAVEL (DEPTH) _____

ASPHALT _____

NONE _____

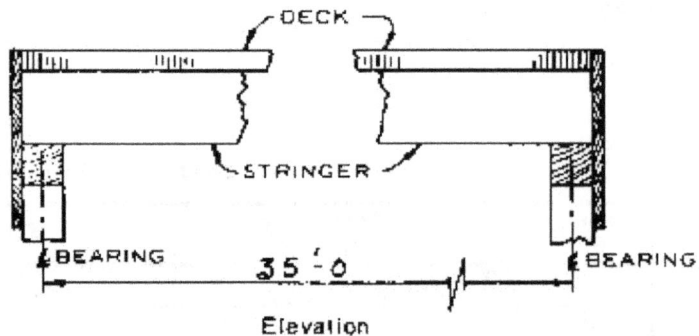

Elevation

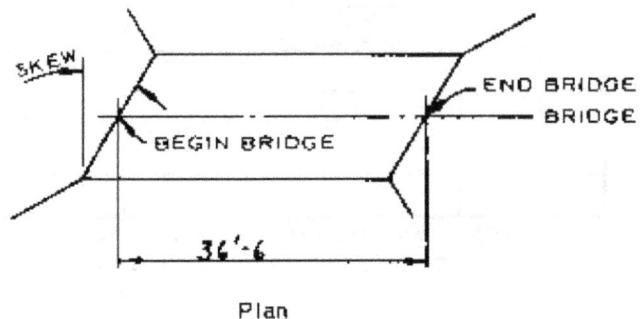

Plan

Skew

(LEFT OR RIGHT AHEAD)

0° __✓__ 15° ____ 30° ____ 45° ____

Clear Height

DISTANCE BETWEEN BOTTOM OF STRINGERS
AND STREAMBED __5'-3"_____

Figure 13-49. - (continued).

R1 - FS - 7700-4 (7/87)

BRIDGE NAME AND NUMBER _Timber Bridge No 463-05.7_ **INSPECTION DATE** _8/20/87_

<u>Sketches</u> (if changed from last inspection)

Coring No. 0

Big Creek

missing object Member

Str #7

missing object Member

← N

← ② Photo

PLAN (Show north arrow)

3%

4'

1-3

Riprap Added

0%

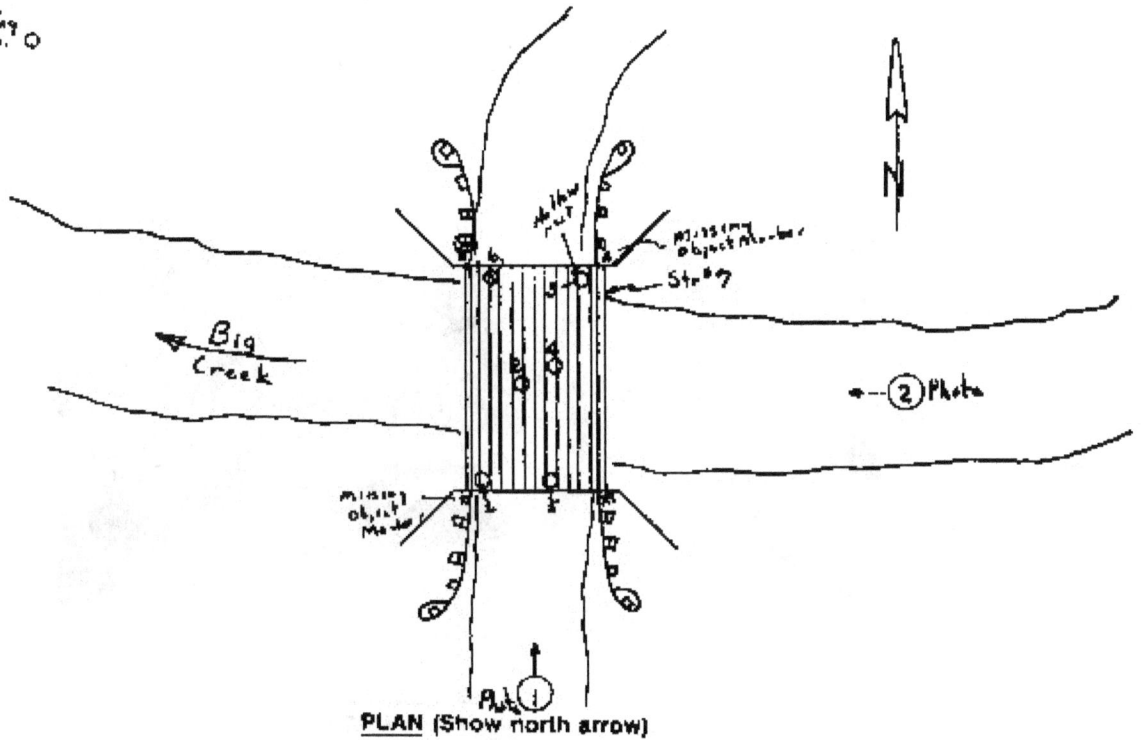

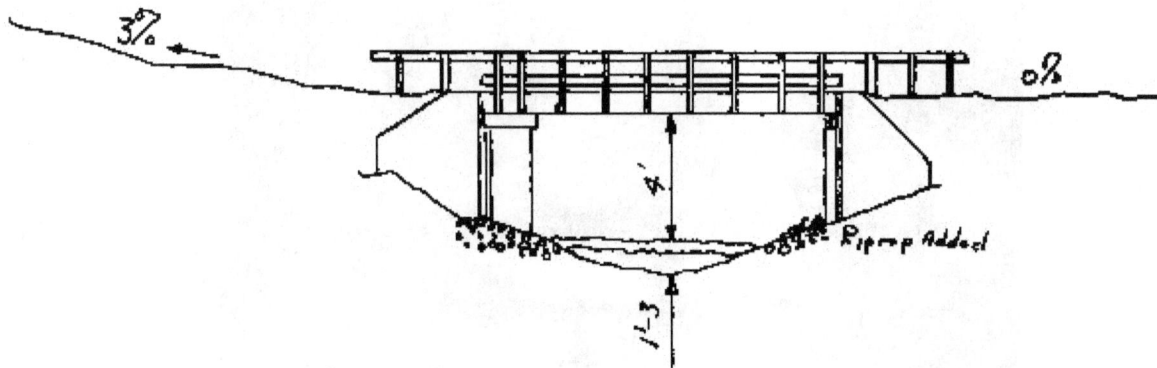

<u>ELEVATION LOOKING</u> _Downstream_
(Direction)

Figure 13-49. - (continued).

R1 - FS - 7700-4 (7/87)

BRIDGE NAME AND NUMBER _Timber Bridge No. 463 05.7_ INSPECTION DATE _8/20/87_

Pictures (Elevation, Approach Views, Views and Others of Significance)

1. APPROACH LOOKING _North_
(Direction)

2. PROFILE LOOKING _Downstream_
(Up/Down Stream)

R1 - FS - 7700-6 (7/87)

Figure 13-49. - (continued).

13.5 STRENGTH LOSS FROM DECAY

Bridge members infected with decay fungi experience progressive strength loss as the fungi develop and degrade the wood structure. The degree of strength reduction depends on the area of the infection and the stage of decay development, whether advanced, intermediate, or incipient. In the advanced or intermediate stages, wood deterioration has progressed to the point where no strength remains in infected areas. At this stage, suitable detection methods can be used by the inspector to accurately define the affected areas with some degree of certainty. At the incipient or early stages of development, detection is much more difficult and the effect of strength loss varies among types of fungi.

Little information exists on assessing strength loss at the incipient stages of decay, but several researchers have correlated strength to weight loss in small wood samples. These investigations found that strength loss associated with some brown rot fungi can be as high as 50 to 70 percent when the weight is reduced by only 3 percent or less.[25,30] These findings are especially significant for timber bridges because (1) most bridge decay is from brown rot rather than white rot fungi, (2) incipient brown rot decay, with its minimal weight loss, is difficult to detect, and (3) the effects of brown rot fungi usually extend a substantial distance away from areas where decay is visible.

Although the strength effects for white rot fungi may be less than those for brown rot, differentiating between the two is not possible in the field. Thus, all decay should be assumed to be significant. In light of the large strength losses associated with early brown rot development, it is recommended that no strength value be assigned to wood showing evidence of decay in any stage of development. Although this approach may result in a slightly conservative evaluation in some instances, it is the only safe approach for assessing strength, given the large number of variables involved. Although numerous cores may be taken to define the decayed area, the possibility remains that the entire area of infection will not have been sampled. Additionally, decay will continue to further reduce strength unless immediate maintenance actions are undertaken to arrest its growth.

13.6 SELECTED REFERENCES

1. American Association of State Highway and Transportation Officials. 1983. Manual for maintenance inspection of bridges. Washington, DC: American Association of State Highway and Transportation Officials. 50 p.

2. American Association of State Highway and Transportation Officials. 1976. AASHTO manual for bridge maintenance. Washington, DC: American Association of State Highway and Transportation Officials. 251 p.

3. American Institute of Timber Construction. 1986. Checking in glued laminated timber. AITC Tech. Note No. 11. Englewood, CO: American Institute of Timber Construction. 1 p.

4. American Society of Civil Engineers. 1982. Evaluation, maintenance, and upgrading of wood structures. Freas, A., ed. New York: American Society of Civil Engineers. 428 p.

5. American Society of Civil Engineers. 1986. Evaluation and upgrading of wood structures: case studies. New York: American Society of Civil Engineers. 111 p.

6. Baker, A.J. 1974. Degradation of wood by products of metal corrosion. Res. Pap. FPL 229. Madison, WI: U.S. Department of Agriculture, Forest Service, Forest Products Laboratory. 6 p.

7. Better Roads. 1987. Better ways to inspect bridges. Better Roads. 57(11): 24-25.

8. Better Roads. 1987. How to document underwater inspections. Better Roads. 57(11): p. 22.

9. Chidester, M.S. 1937. Temperatures necessary to kill fungi in wood. In: Proceedings, American Wood Preserver's Association 33: 316-324.

10. Daniel, G.; Nilsson, T. 1986. Ultrastructural observations on wood degrading erosion bacteria. IRG:WP:1283. Stockholm: International Research Group on Wood Preservation.

11. Duncan, C.G. 1960. Wood attacking capacities and physiology of soft rot fungi. Report no. 2173. Madison, WI: U.S. Department of Agriculture, Forest Service, Forest Products Laboratory.

12. Ebeling, W. 1968. Termites: identification, biology, and control of termites attacking buildings. Extension Service Manual 38. California Agricultural Experiment Station.

13. Ellwood, E.L.; Eklund, B.A. 1959. Bacterial attack of pine logs in storage. Forest Products Journal 9: 283-292.

14. Eslyn, W.E.; Clark, J.W. 1979. Wood bridges-decay inspection and control. Agric. Handb. 557. Washington, DC: U.S. Department of Agriculture, Forest Service. 32 p.

15. Eslyn, W.E. 1976. Wood preservative degradation by marine bacteria. In: Proceedings of the 3rd International Biodeterioration Symposium; 1976. London: Applied Sciences Publishers, Ltd.

16. Eslyn, W.E.; Clark, J.W. 1975. Appraising deterioration of submerged piling. Tech. article. Sup. 3 to materials and organisms. Madison, WI: U.S. Department of Agriculture, Forest Service, Forest Products Laboratory. 44-52 pp.

17. Feist, W. 1983. Weathering and protection of wood. In: Proceedings American Wood Preserver's Association; 1983; 79: 195-205.

18. Gower, L.E. 1979. Maintenance and inspection of logging bridges. White Rock, BC, Can.: Big Wheel Publications Ltd. 46 p.

19. Gower, L.E. 1986. Remaining glulam bridges should be inspected carefully. Logging and Sawmilling Journal [Can.] 17(8): 42-43.

20. Graham, R.D. 1973. History of wood preservation. Wood preservation and its prevention by preservative treatments. New York: Syracuse University Press: 1-30.

21. Graham, R.D.; Helsing, G.G. 1979. Wood pole maintenance manual: inspection and supplemental treatment of Douglas-fir and western redcedar poles. Res. Bull. 24. Corvallis, OR: Oregon State University, Forest Research Laboratory.

22. Graham, R.D.; Wilson, M.M.; Oteng-Amoaka, A. 1976. Wood-metal corrosion: an annotated survey. Res. Bull. 21. Corvallis, OR: Oregon State University, Forest Research Laboratory.

23. Greaves, H. 1976. An illustrated comment on the soft rot problem in Australia and Papua New Guinea. Holzforschung 31: 71-79.

24. Greaves, H. 1971. The bacterial factor in wood decay. Wood Science and Technology. 51(1): 6-16.

25. Hartley, C. 1958. Evaluations of wood decay in experimental work. Rep. No. 2119. Madison, WI: U.S. Department of Agriculture, Forest Service, Forest Products Laboratory. 53 p.

26. Henningsson, B.; Nilsson, T. 1976. Microbiological, microscopic, and chemical studies of some salt treated utility poles installed in Sweden in the years 1941-1946. Swedish Wood Preservation Institute Report E-117.

27. Hill, C.L.; Kofoid, C.A. 1927. Marine borers and their relation to marine construction on the Pacific coast: final report of the San Francisco Bay piling committee. Berkeley, CA: University of California Press.

28. Hurlbut, B.B. 1978. Basic evaluation of the structural adequacy of existing timber bridges. Washington, DC: National Academy of Sciences, National Research Council, Transportation Research Board: 6-9.

29. James, W.L. 1975. Electric moisture meters for wood. Gen. Tech. Rep. FPL 6. Madison, WI: U.S. Department of Agriculture, Forest Service, Forest Products Laboratory. 28 p.

30. Kennedy, R.W. 1958. Strength retention in wood decayed to small weight losses. Forest Products Journal 10(8): 308-314.

31. LaQue, F.L. 1975. Marine corrosion: causes and prevention. New York: John Wiley and Sons. 332 p.

32. Lew, J.D.; Wilcox, W.W. 1981. The role of selected deuteromycetes in the soft-rot of wood treated with pentachlorophenol. Wood and Fiber 13(4): 252-264.

33. Lindgren, R.M. 1952. Permeability of southern pine as affected by mold growth and other fungus infection. In: Proceedings American Wood Preserver's Association; 1952; 48: 158-174.

34. Maeglin, R.R. 1979. Increment cores-how to collect, handle, and use them. Gen. Tech. Rep. FPL 25. Madison, WI: U.S. Department of Agriculture, Forest Service, Forest Products Laboratory. 19 p.

35. McCutcheon, W.J.; Gutkowski, R.M.; Moody, R.C. 1986. Performance and rehabilitation of timber bridges. Trans. Res. Rec. 1053. Washington, DC: National Academy of Sciences, National Research Council, Transportation Research Board: 65-69.

36. McDonald, K.A. 1978. Lumber defect detection by ultrasonics. Res. Pap. FPL 311. Madison, WI: U.S. Department of Agriculture, Forest Service, Forest Products Laboratory. 21 p.

37. McDonald, K.A.; Cox, R.G.; Bulgrin, E.H. 1969. Locating lumber defects by ultrasonics. Res. Pap. FPL 120. Madison, WI: U.S. Department of Agriculture, Forest Service, Forest Products Laboratory. 12 p.

38. McGee, D. 1975. The timber bridge inspection program in Washington State. Portland, OR: U.S. Department of Transportation, Federal Highway Administration, Region 10. 52 p.

39. Morrell, J. J.; Helsing, G.G.; Graham, R.D. 1984. Marine wood maintenance manual: a guide for proper use of Douglas fir in marine exposure. Res. Bull. 48. Corvallis, OR: Oregon State University, Forest Research Laboratory. 62 p.

40. Mothershead, J.S.; Stacey, S.S. 1965. Applicability of radiography to inspection of wood products. In: Proceedings 2nd Symposium on Non-Destructive Testing of Wood; 1965; Spokane, WA.

41. Muchmore, F.W. 1984. Techniques to bring new life to timber bridges. Journal of Structural Engineering 110(8): 1832-1846.

42. Naval Facilities Engineering Command. 1985. Inspection of wood beams and trusses. NAVFAC MO-111.1. Alexandria, VA: Naval Facilities Command. 56 p.

43. Nilsson, T. 1973. Studies on wood degradation and cellulolytic activity of microfungi. Stockholm, Sweden: Studia Forestalia Suecica Nr 104.

44. Organisation for Economic Co-operation and Development. 1976. Road Research Group. Bridge inspection. Paris, France: Organisation for Economic Co-operation and Development, Road Research Group. 133 p.

45. Park, S.H. 1980. Bridge inspection and structural analysis. Trenton, NJ: S.H. Park. 312 p.

46. Scheffer, T.C. 1971. A climate index for estimating potential for decay in wood structures above ground. Forest Products Journal 21(10): 25-31.

47. Scheffer, T.C. 1986. O_2 requirements for growth and survival of wood-decaying and sapwood-staining fungi. Canadian Journal of Botany 64: 1957-1963.

48. Smith, S.M.; Morrell, J.J. 1986. Correcting Pilodyn measurement of Douglas-fir for different moisture levels. Forest Products Journal 36(1): 45-46.

49. Tabak, H.H.; Cook, W.B. 1968. The effects of gaseous environments on the growth and metabolism of fungi. Botanical Review 34: 126-252.

50. U.S. Department of Agriculture. 1987. Wood handbook: wood as an engineering material. Agric. Handb. No. 72. Madison, WI: U.S. Department of Agriculture, Forest Service, Forest Products Laboratory. 466 p.

51. U.S Department of Transportation, Federal Highway Administration. 1979. Recording and coding guide for the structural inventory and appraisal of the nation's bridges. Washington, DC: U.S. Department of Transportation, Federal Highway Administration. 50 p.

52. U.S. Department of Transportation, Federal Highway Administration. 1979. Bridge inspector's training manual. Washington, DC: U.S Department of Transportation, Federal Highway Administration. 246 p.

53. United States Navy. 1965. Marine biological operational handbook: inspection, repair, and preservation of waterfront structures. NAVDOCKS MO-311. Washington, D.C.: U.S. Department of Defense, Bureau of Yards and Docks.

54. White, K.R.; Minor, J.; Derocher, K.N.; Heins, C.P., Jr. 1981. Bridge maintenance inspection and evaluation. New York: Marcel Dekker, Inc. 257 p.

55. White, K.; Minor, J. 1978. The New Mexico bridge inspection program. In: Bridge engineering. Trans. Res. Rec. 664. Washington, DC: National Academy of Sciences, National Research Council, Transportation Research Board: 7-13. Vol. 1.

56. Wilcox, W.W. 1978. Review of literature on the effects of early stages of decay on wood strength. Wood and Fiber 9(4): 252-257.

57. Wilcox, W.W. 1983. Sensitivity of the "pick test" for field detection of early wood decay. Forest Products Journal 33(2): 29-30.

58. Zabel, R.A.; Wang, C.J.K.; Terracina, F.C. 1980. The fungal associates, detection, and fumigant control of decay in treated southern pine poles. EPRI-EL 2768. Palo Alto, CA: Electric Power Research Institute.

CHAPTER 14

MAINTENANCE AND REHABILITATION OF TIMBER BRIDGES

14.1 INTRODUCTION

Wood is one of the most durable bridge materials, but over extended periods it may be subject to deterioration from decay, insect attack, or mechanical damage. Timber bridges must be periodically maintained or rehabilitated in order to keep them in a condition that will give optimum performance and service life. Effective bridge maintenance programs improve public safety, extend the service life of the structure, and reduce the frequency and cost of repairs. The objective is not only to repair existing deficiencies, but also to take corrective measures to prevent or reduce future problems. When tied to a competent bridge inspection program, regular maintenance represents the most cost-effective approach for achieving long service life from existing structures. Unfortunately, maintenance is often neglected until critical problems develop that require major restoration or replacement of the structure. In times of declining budgets, the first program reduced as a money-saving measure is often maintenance, when, in fact, reduced maintenance substantially increases long-term costs.

In general terms, bridge maintenance includes those activities necessary to preserve the utility of a bridge and ensure the safety of road users. In practice, all maintenance is either preventative or remedial. Maintenance activities are divided into categories that vary in definition and scope among different agencies. In this chapter, timber bridge maintenance is divided into the three following categories:

1. **Preventative maintenance** involves keeping the structure in a good state of repair to reduce future problems. At this stage, decay or other deterioration has not started, but the conditions or potential are present.

2. **Early remedial maintenance** is performed when decay or other deterioration is present but does not affect the capacity or performance of the bridge in normal service. At this stage, more severe structural damage is imminent unless corrective action is taken.

3. **Major maintenance** involves immediate corrective measures that restore a bridge to its original capacity and condition. Deterioration has progressed to the point where major structural components have experienced moderate to severe strength loss and repair or replacement is mandatory to maintain load-carrying capacity.

Bridge rehabilitation is another form of restoration performed on bridges that are functionally or structurally obsolete. Rehabilitation is similar to maintenance in some ways because it involves many of the same methods and techniques; however, rehabilitation is performed to improve the geometric or load-carrying capacity of an existing bridge, rather than to restore the original capacity. Rehabilitation is most commonly performed on older bridges that were built to lesser geometric or loading standards than those required for today's modern traffic.

This chapter discusses several maintenance and rehabilitation practices and methods that are commonly used for timber bridges. Because deficiencies develop from a variety of causes, it is impractical to address each type of potential problem. Rather, preventative and remedial methods are discussed that can be adapted to the specific circumstances of the structure. These methods include moisture control, in-place preservative treatment, mechanical repair, epoxy repair, and component replacement. Applications of these techniques to actual projects are given in case histories presented in Chapter 15. For additional guidelines and information related to bridge maintenance in general, consult the references listed at the end of this chapter. [1,3,28,37]

14.2 MOISTURE CONTROL

Moisture control is the simplest, most economical method of reducing the hazard of decay in timber bridges. It can be used as an effective and practical maintenance technique to extend the service life of many existing bridges. When exposure to wetting is reduced, members can dry to moisture contents below that required to support most fungal and insect growth (approximately 25 percent). Moisture control was the only method used for protecting many covered bridges constructed of untreated timber, some of which have provided service lives of 100 years or more (Figure 14-1). Although modern timber bridges are protected with preservative treatments, decay can still occur in areas where the preservative layer is shallow or broken. This damage is the major cause of deterioration in timber bridges.

Moisture control involves a common sense approach of identifying areas with visible wetting or high moisture contents, locating the source of water, and taking corrective action to eliminate the source. For example, drainage patterns on approach roadways can be rerouted to channel water away from the bridge rather than onto the deck. Cleaning dirt and debris from the deck surface, drains, and other horizontal components also reduces moisture trapping and improves air circulation (Figure 14-2). One of the most effective approaches to moisture control is restricting or preventing water passage through the deck. Decks that are impervious to moisture penetration will protect critical structural members and

Figure 14-1. - Many covered bridges constructed of untreated timber, such as this one in New Hampshire, have lasted more than 100 years because they were protected from moisture.

substantially reduce the potential for decay. Glulam or stress-laminated decks afford the best protection because they can be placed to form a watertight surface. Leaks between glulam panels or at butt joints in stress-laminated decks can be resealed using bituminous roofing cement.

The deck wearing surface also plays an important role in moisture protection. Wearing surfaces constructed of lumber planks or steel plates provide little protection and often trap moisture under the planks or plates. Lumber running planks are a particular problem because they inhibit drainage on watertight decks and often cause water ponding on the deck surface. When ponding occurs, the only practical option for its removal is to install tubes through the deck to drain water down and away from the deck, rather than onto the deck underside and supporting members (Figure 14-3).

On glulam, stress-laminated, and some nail-laminated decks, the addition of an asphalt wearing surface provides a moisture barrier that protects not only supporting members but also the deck. The effectiveness of the surface protection is increased when the asphalt is placed on geotextile

14-3

Figure 14-2. - Dirt and debris on the deck surface can trap moisture and lead to premature deterioration. Material such as this should be removed periodically as part of a good preventative maintenance program.

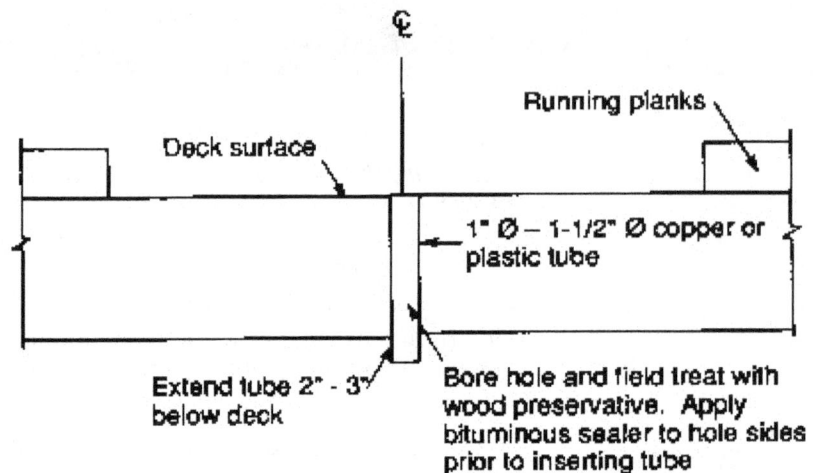

Figure 14-3. - Detail of drain tube for removing trapped water between lumber running planks.

fabric (Chapter 11). All glulam and stress-laminated decks are normally suitable for asphalt surfaces; however, use of asphalt surfaces on nail-laminated decks may be limited by the condition of the deck. Nail-laminated decks commonly show varying degrees of looseness after 5 to 10 years of service under heavy loading. Paving these decks is futile because the separation and movement of laminations will cause the pavement to crack and disintegrate. The best approach to waterproofing a loose nail-laminated deck is to apply stressing to restore deck integrity

14-4

(discussed later in this chapter), followed by application of an asphalt wearing surface. When this is not practical, deck replacement is usually the only other option.

On bridges with asphalt surfaces, breaks in the surface may develop in service from deck deflections, improper bonding, or poor construction practices. Deficiencies of this type should be repaired as soon as possible to prevent more serious deterioration. Cracking may result from a number of causes but is typically caused by differential deck deflections at panel joints or at bridge ends. Cracks of this type should be thoroughly cleaned with a stiff brush and compressed air, then filled with emulsion slurry or liquid asphalt mixed with sand (Figure 14-4). If pavement is broken or missing, surrounding pavement must be removed to the point where it is sound and tightly bonded to the deck, and a patch must be applied. For best results, the repair area should be cut in a square or rectangular shape with vertical sides, be thoroughly cleaned, and be patched with a dense grade of asphalt pavement.

14.3 IN-PLACE PRESERVATIVE TREATMENT

In-place treating involves the application of preservative chemicals to prevent or arrest decay in existing structures. Two types of treatment are commonly used: surface treatments and fumigants. Surface treatments are applied to prevent infection of exposed wood, whereas fumigants are used to treat internal decay. In-place treating can provide a safe, effective, and economical method for extending the service life of timber bridges. Most of the techniques and treatments were developed for use on railroads or utility poles, for which they have been used effectively for many years. A large number of timber bridges have been treated in-place, extending service life by as much as 20 years or more (see case histories in Chapter 15).

SURFACE TREATMENTS

Surface treatments are applied to existing bridge members to protect newly exposed, untreated wood from decay or to supplement the initial treatment some years after installation. This type of treatment is most effective when applied before decay begins and is commonly used for treating checks, splits, delaminations, mechanical damage, or areas that were field-fabricated during construction. The ease of application and effectiveness of surface treatments as toxic barriers make them useful in preventive maintenance; however, the shallow penetration limits their effectiveness against established internal decay.

Surface treating uses the same basic procedures discussed for field treatment (Chapter 12). Conventional liquid wood preservatives are applied by

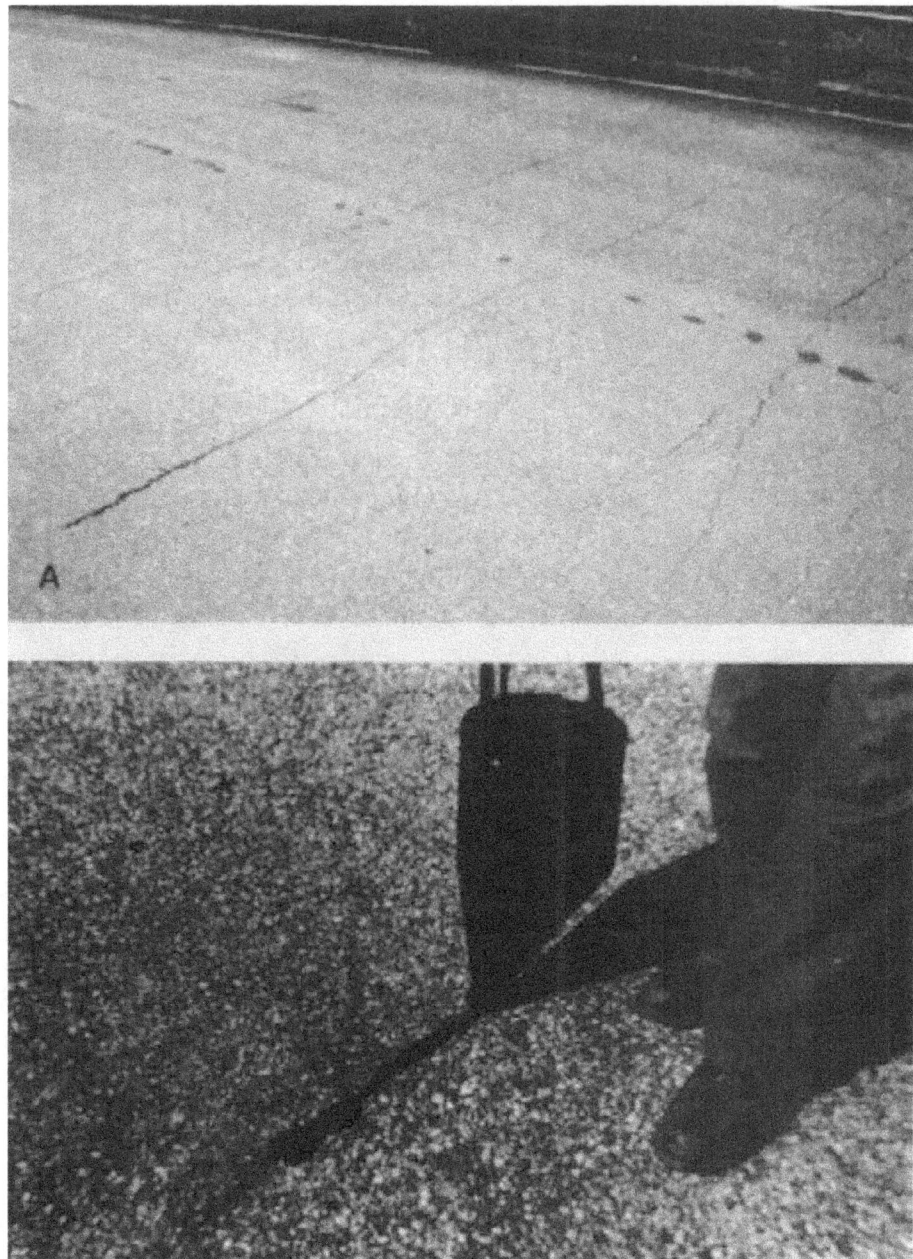

Figure 14-4. - (A) Minor cracking in an asphalt wearing surface from differential deck deflections. (B) Sealing the cracks with an asphalt emulsion slurry.

brushing, squirting, or spray-flooding the wood surface (Figure 14-5). Creosote heated to 150 to 200 °F is probably the most commonly used preservative, but penta and copper naphthenate are also used. The wood surface should be thoroughly saturated with preservative so that all cracks and crevices are treated; however, care must be exercised to prevent excessive amounts from spilling or running off the surface and contaminating water or soil.

14-6

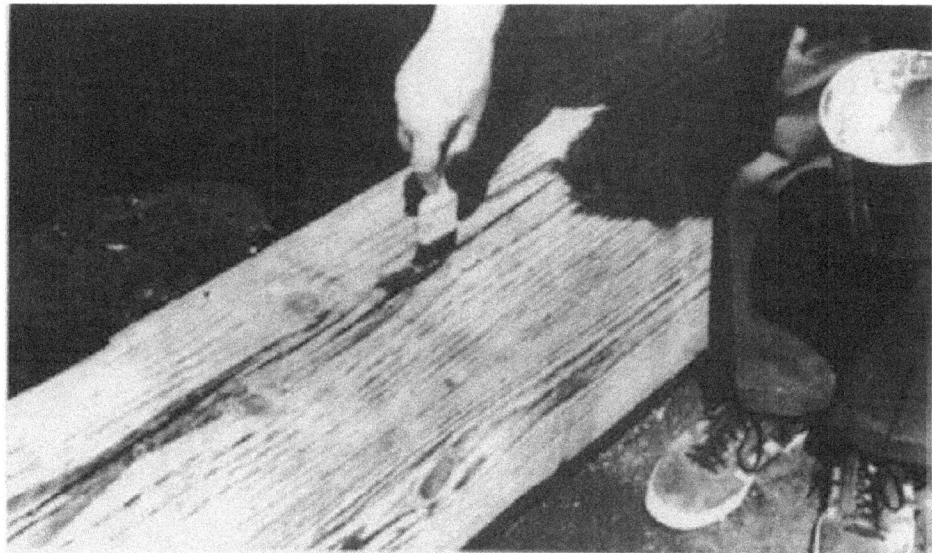

Figure 14-5. - Liquid wood preservative is applied to a check in a timber curb by brushing.

In addition to preservative liquids, some preservative compounds are available in semisolid greases or pastes. These preservatives, which generally use sodium fluoride, creosote, or pentachlorophenol as the primary preservative chemical, are useful for treating vertical surfaces or openings. Their primary advantage is that larger quantities of the toxic chemical can be locally applied in heavy coatings that adhere to the wood. Preservative adsorption over an extended period of time can produce deeper penetration than single surface applications of liquid treatments. Semisolid preservatives are commonly used at the groundline of posts, poles, and piling, where they are brushed on the surface from several inches above the groundline to 18 to 24 inches below the groundline (Figure 14-6). After the preservative is applied, the treated portion is wrapped with polyethylene, or other impervious material, to exclude moisture and prevent leaching of the treatment into the surrounding soil.

The effectiveness of surface treatments depends on the thoroughness of application, wood species, size, and moisture content at the time of treatment. Wet wood absorbs less preservative than does dry wood. This factor is significant in timber bridges because many areas requiring treatment are subject to wetting. Tests indicate that improved treatment of wet wood was obtained by using preservatives at double the normal 3- to 5-percent concentration.[17] Although field tests show that surface treatments in aboveground locations can prevent decay infections for up to 20 years or more,[33] it is recommended that treatments used for bridge applications be systematically reapplied at intervals of 3 to 5 years to ensure adequate protection from decay.

14-7

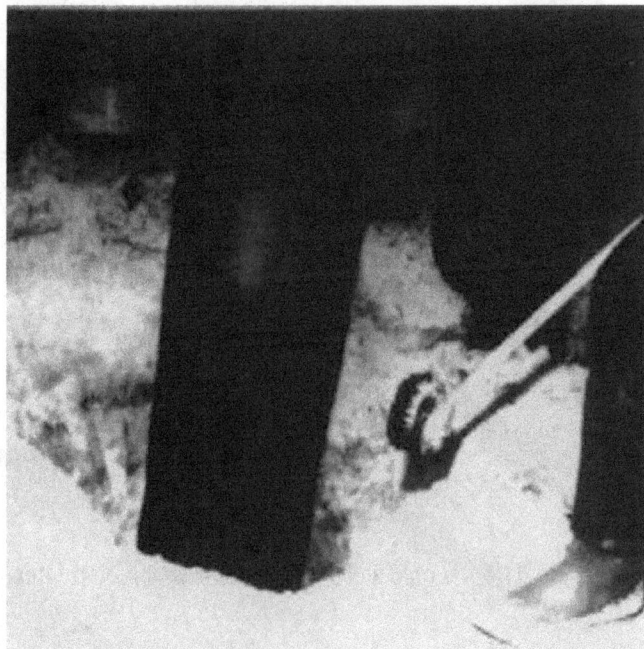

Figure 14-6. - Paste wood preservative is applied to a timber pile around the groundline. Note the wrapping material at the upper end of the treated section (photo courtesy of Osmose Wood Preserving, Inc.).

FUMIGANTS

Fumigants are specialized preservative chemicals in liquid or solid form that are placed in prebored holes to arrest internal decay. Over a period of time, the fumigants volatilize into toxic gases that move through the wood, eliminating decay fungi and insects. Fumigants can diffuse in the direction of the wood grain for 8 feet or more from the point of application in vertical members, such as poles. In horizontal members, the distance of movement is approximately 2 to 4 feet from the point of application. The three chemicals most commonly used as liquid fumigants are Vapam (33-percent sodium N-methyldithiocarbamate), Vorlex (20-percent methylisothiocyanate, 80-percent chlorinated C₃ hydrocarbons), and chloropicrin (trichloro-nitromethane). Solid fumigants are available in capsules of methylisothiocyanate (MIT), which is the active ingredient of Vapam and Vorlex. Solid fumigants provide increased safety, reduce the risk of environmental contamination, and permit fumigant use in previously restricted applications.

To be most effective, fumigants must be applied to sound wood. When applied in very porous wood or close to surfaces, some of the fumigant is lost by diffusion to the atmosphere. Before applying fumigants, the condition of the member should be carefully assessed to identify the optimal boring pattern that avoids fasteners, seasoning checks, badly decayed wood, and other openings to the atmosphere. In vertical members such as piles, holes should be bored at a steep downward angle toward the center of the member to avoid crossing seasoning checks (Figure 14-7). It is best to begin by boring almost perpendicular to the member, then quickly raising the drill to a 45 to 60-degree angle once the bit catches in the wood. For horizontal members, holes are bored in pairs straight down to within 1-1/2 to 2 inches from the bottom side. If large seasoning checks are present in horizontal members, holes should be bored on each side of the check to more completely protect the timber (Figure 14-8). The amount of chemical and the size and number of treatment holes depends on the member size and orientation. Table 14-1 gives some examples of the number and size of holes and fumigant dosages required to treat vertical piling. For horizontal members, pairs of holes should not be more than 4 feet apart Additional information and recommended dosages for fumigants may be obtained from the chemical manufacturers.

When solid fumigants are used, they are inserted directly into the prebored holes. Liquid fumigants are applied using commercial equipment but can also be applied from 1-pint polyethylene squeeze bottles (Figure 14-9).[27] When using polyethylene, it is helpful to replace the plastic cap with a reusable cap fastened to a 1-foot length of plastic or rubber tubing. After adding the required dosage of fumigant, the original cap is replaced so the remaining liquid stays in the bottle, and the fumigant is returned to its original container (liquid fumigants should not be stored in plastic bottles for long periods because they can cause the plastic to become brittle and crack). If leaks are observed while applying liquid fumigants, it is

Figure 14-7. - Treating holes for fumigants in vertical members are bored at a steep downward angle (photo courtesy of Jeff Morrell, Oregon State University).

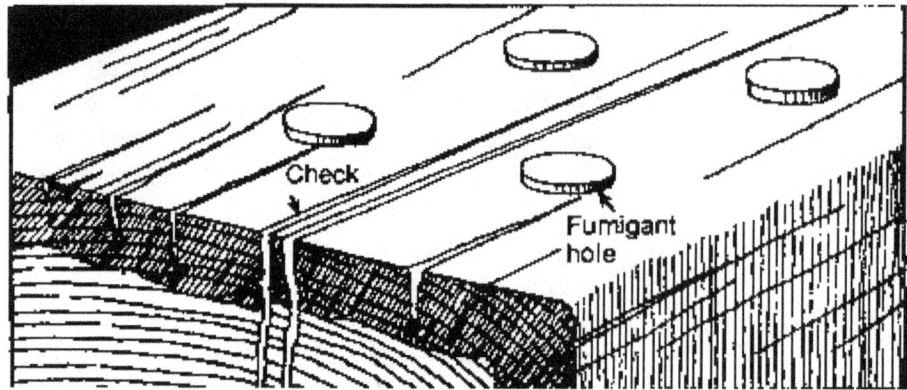

Figure 14-8. - Treating holes for fumigants in horizontal members should be placed on both sides of checks or splits, and be bored to within 1-1/2 to 2 inches of the bottom of the member.

Figure 14-9. - Application of liquid fumigants. (A) Liquid fumigants applied with commercial equipment (photo courtesy of Osmose Wood Preserving, Inc.).

Table 14-1. - Number and size of holes and dosage of fumigant required for piles.

Hole dimensions (in.)		Fumigant dosage (pints per in. of hole)	Numbers of holes for piles of various circumferences (and dosages)		
Diameter	Length[a]		< 32 in. (3/4 pint)	32 – 45 in. (1 pint)	> 45 in. (2 pints)
5/8	15	0.010	6	—	—
	18	0.010	5	—	—
3/4	15	0.015	4	6	—
	18	0.015	—	5	—
	21	0.015	—	4	—
	24	0.015	—	3	6
7/8	21	0.024	—	3	5
	24	0.024	—	—	4

[a] Effective length of treating hole is 3 inches less to allow for a 3-inch treated plug. From Morrell and others.[27]

Figure 14-9. - Application of liquid fumigants (continued). (B) Liquid fumigants applied from a polyethylene squeeze bottle (photo courtesy of Jeff Morrell, Oregon State University).

14-11

important to stop filling, to plug the hole, and to bore another hole into sound wood. Immediately after placing the chemicals, the hole is plugged with a tight-fitting, treated-wood dowel driven slowly to avoid splitting the wood. For liquid fumigants, sufficient room (1.5 to 2 inches) must be left in the treating hole so the plug can be driven without squirting the chemical.

Fumigants will eventually diffuse out of the wood, allowing decay fungi to recolonize. In properly treated solid wood, Vorlex and chloropicrin will remain effective for 10 to 15 years, while Vapam is somewhat less effective (Figure 14-10). These periods will be reduced when the wood has many fastener holes, splits, checks, or end grain where the chemical can diffuse to the atmosphere. Retreatment can be made at periodic intervals in the same holes used for the initial treatment. The old plug is drilled or pulled, new fumigant is added, and the hole is replugged with a new, treated dowel. Until retreatment cycles are better defined, it is recommended that a 10-year treatment cycle be used with a regular inspection program at 5-year intervals. When inspections indicate the presence of active decay, the protective effects of the fumigant have declined below a toxic threshold, and retreatment is required. It is important to keep accurate records of all in-place treating, including the date and location of the application, the type of chemical, and the dose (such records are required in some States). It may also be beneficial to place a metal tag on the

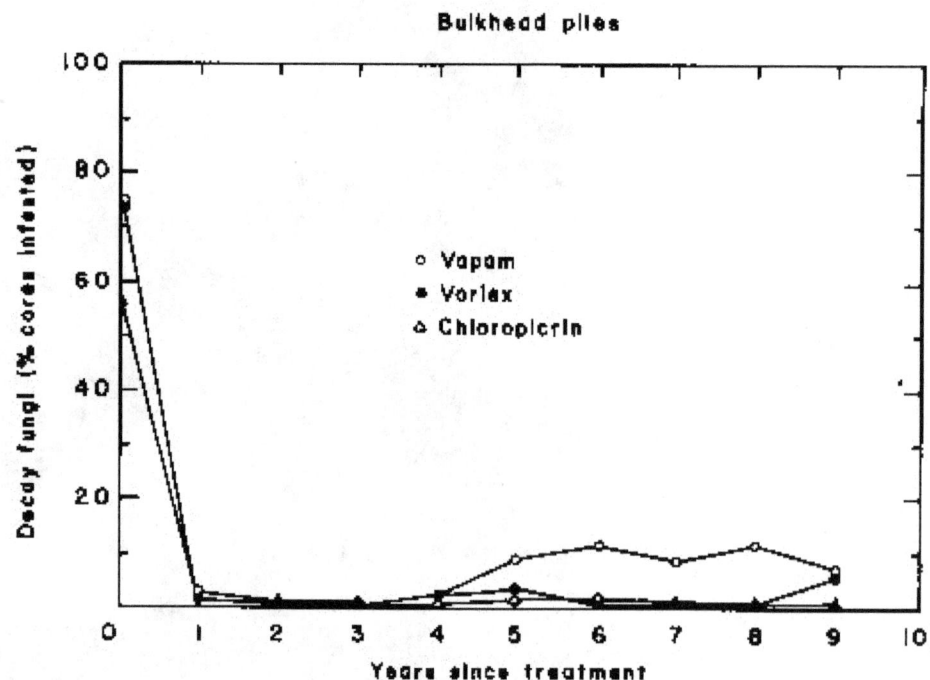

Figure 14-10. - Annual changes in the population of decay fungi isolated from creosoted Douglas-fir piles treated with various fumigants. Each value on the curve represents 60 cores from each of 12 piles. From Morrell and others. ''

14-12

member noting treatment information; however, these tags may be stolen or vandalized and should not be the sole means of recording treatment information.

PRECAUTIONS FOR IN-PLACE TREATING

As with other preservatives and pesticides, wood preservatives and fumigants for in-place treating are toxic to humans and must be used in accordance with State and Federal laws. When properly applied, the treatments pose no environmental or health hazard; however, the potential for environmental damage can be higher in some field locations because of variable conditions and the proximity to streams and other water sources. In-place treatments must be applied only by trained and licensed personnel who fully understand their use and the required safeguards. In addition to the precautions for wood preservatives discussed in Chapter 12, fumigant applicators should also have a gas mask with the appropriate filter available for emergency use. If fumigant vapors are detected by their strong odor or eye irritation, all personnel should move upwind from the treating area and allow vapors to clear. When any form of in-place treatment is used, the procedures, precautions, and contingency for accidental spillage or injury should be well planned before beginning treatment.

In general, in-place treating by local maintenance crews is limited by the scope of the treatment required. For routine maintenance, the amount of treating required is normally minor, and local crews can be used when properly trained and licensed personnel are available. For larger projects involving many members or an entire structure, it is advisable to contract the project to specialists in the field. There are companies that have provided in-place treating services for many years with excellent safety records and results. When selecting a contractor, previous experience and performance histories should be carefully evaluated to ensure that the contractor is qualified to perform the required treating.

14.4 MECHANICAL REPAIR

Mechanical methods of repair use steel fasteners and additional wood or steel components to strengthen or reinforce members. The three methods of mechanical repair discussed in this section are member augmentation, clamping and stitching, and stress laminating.

MEMBER AUGMENTATION

Member augmentation involves the addition of material to reinforce or strengthen existing members. The additional pieces, commonly wood or steel plates attached with bolts, serve to increase the effective section and thus load capacity. The two most widely used methods of member augmentation are splicing and scabbing. Although the distinction between the two is rather vague, splicing generally applies to a defined

14-13

location where load transfer is restored at a break, split, or other defect (Figure 14-11 A). Scabbing is more frequently associated with strengthening members where existing capacity is insufficient and may involve adding reinforcing pieces over a substantial portion or even over the entire member length (Figure 14-11 B). In both cases, a thorough structural analysis is required to ensure the capacity of the repair and to verify stress distribution in the members. Situations that introduce eccentric loads or tension perpendicular to grain must be avoided. When using splices, it is recommended that the defective member be cut entirely through to more equally distribute loads to splice plates.[3]

In addition to wood or steel augmentation methods, reinforced concrete can be used to strengthen deteriorated timber piling sections (Figure 14-12). Using this procedure, the pile is wrapped with a jacket-type form of fiber-reinforced plastic or fabric that fits the pile like a sleeve. Reinforcing steel is placed around the pile, and the sleeve is filled with concrete. The reinforced concrete increases pile strength and prevents further deterioration, but the pile size is increased and specialized equipment is required for construction.[15,16,20]

CLAMPING AND STITCHING

A typical problem associated with timber members is the development of longitudinal splits. These splits commonly develop in sawn lumber as the member seasons and checks in place. To a lesser degree, splits may also develop in glulam if delamination occurs at the glue lines, although this problem has become very rare with the introduction of waterproof adhesives. In both sawn and glulam members, splits can also develop from overloads or poor design details that introduce tension perpendicular to grain at connections. When splitting is detected it must be determined whether the splits are the result of normal seasoning or the result of a more serious structural problem. Several references are available that provide a good overview of the potential structural effects of splitting in timber members.[3,25]

Clamping and stitching are maintenance operations that use fasteners and steel assemblies to arrest cracks, splits, or delaminations in timber members. These methods are most commonly used for buildings, but also apply to some bridge components, particularly trusses or other structures with a high number of small members or fastened connections. The objective is not to close the split or check, but rather to prevent its further development by drawing the two parts together. Clamping uses bolts with steel-plate assemblies, while stitching uses bolts or lag screws through the member (Figure 14-13). Although both methods have been used effectively, clamping with bolts and steel plates is generally preferable because the section of the member is not reduced. Aside from fastener design requirements discussed in Chapter 5, there are no specific design criteria for clamping and stitching, and the configuration, number, and size of fasteners must be

Top view

Side view

A. Splicing

Top view

Side view

B. Scabbing

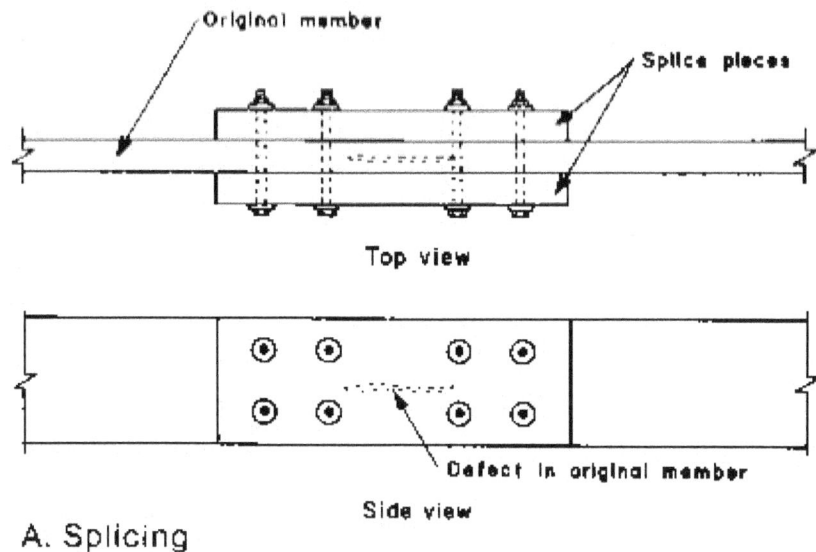

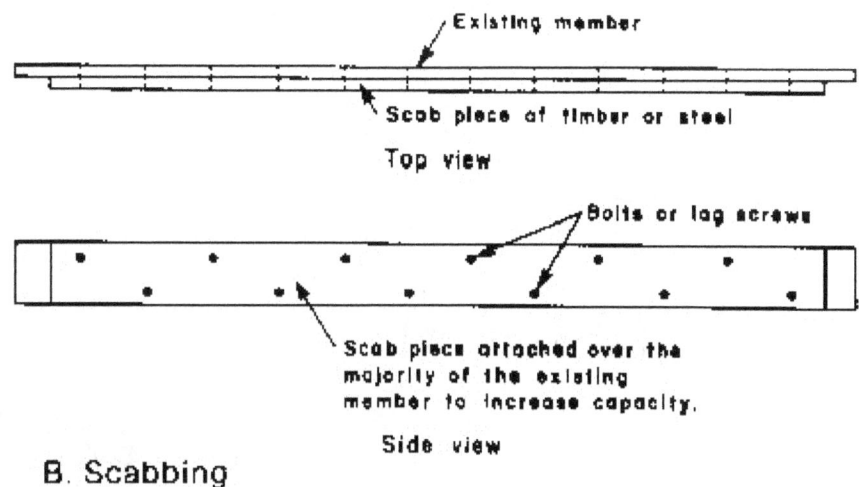

Figure 14-11. - Splicing and scabbing methods of member augmentation.

based on designer judgment on a case by case basis. The following guidelines for stitching are recommended by Ketchum, May, and Hanrahan:[D]

When used at the end of a piece, stitch bolts should be placed between 2 inches and 3 inches from the end. Small 3/8 or 1/2 inch diameter bolts are suggested. Ordinarily, when bored at a critical stress section of a member, the area of the cross-section removed by the hole for the stitch bolt should not exceed the cross-sectional area occupied by the maximum knot permitted in the structural grade. In drawing up the stitch bolts they should be tightened only to the point where the bolts begin to take tension. No attempt should be made to close a split or check as this may extend the split on the other side of the joint. In servicing structures, stitch bolts should be tightened as well as other bolts.

14-15

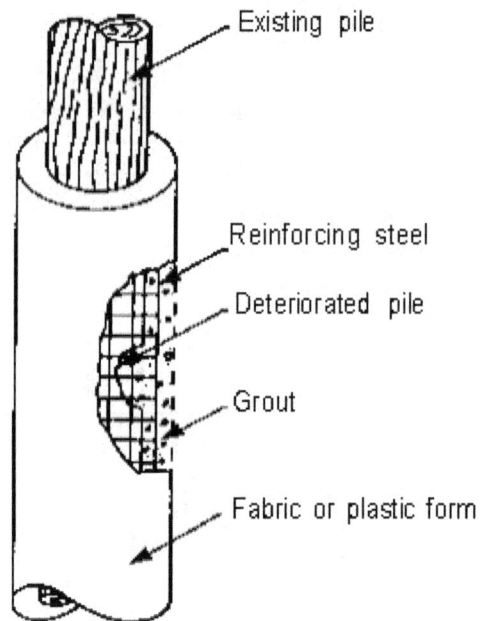

Figure 14-12. - Reinforced concrete jacket for pile augmentation.

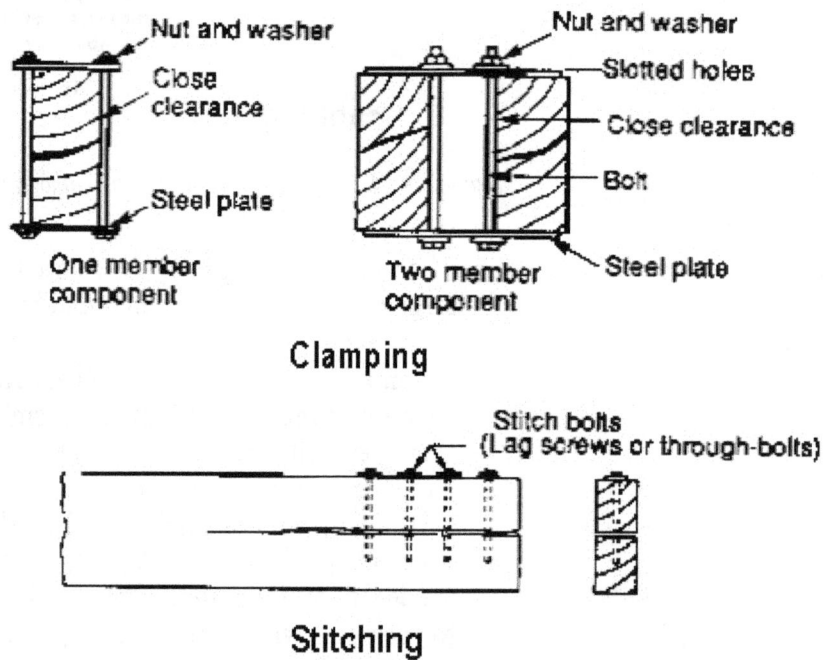

Clamping

Stitching

Figure 14-13. - Typical configurations for clamping and stitching timber members.

STRESS LAMINATING

Stress laminating is probably the most effective method for the mechanical repair of existing nail-laminated decks. Such decks frequently separate and delaminate from repeated loading, causing breakup of asphalt wearing surfaces, water penetration through the deck, and a loss in live load distribution width. In these cases, the static strength and condition of the deck is generally maintained, but its serviceability and ability to distribute loads between individual laminations is greatly reduced. In this situation, the laminations no longer act together to distribute loads, and local failures occur. This condition also increases the rate of deterioration, eventually leading to failures that require complete deck replacement.

The system for stress laminating existing nail-laminated decks was originally developed in 1976 by the Ministry of Transportation and Communications in Ontario, Canada. Since that time, it has been successfully used on a number of bridges to restore the integrity of the existing decks.[318,36] Using this approach, which uses the same design criteria discussed in Chapter 9, the laminations are stressed with a series of high-strength steel rods applied transverse to the length of the laminations. The stress squeezes the laminations together and greatly increases the load distribution characteristics of the deck. Additionally, the stress seals the deck as the laminations are pressed together, providing a watertight surface.

Stress laminating for existing decks differs in configuration from new construction in that stressing rods are positioned on the outside of the laminations, rather than in holes through the laminations (Figure 14-14). This allows the stressing operation to take place without removing the deck and without costly fabrication operations, while traffic is still using the bridge. It is usually necessary to add laminations to the deck before stressing because the rod force squeezes laminations together, reducing the deck width 10 inches or more, depending on the original width. Stress laminating provides a good long-term solution for repairing existing nail-laminated decks to increase load capacity and substantially extend the service life of the structure. More specific information on stress laminating existing nail-laminated decks is presented in a case history in Chapter 15.

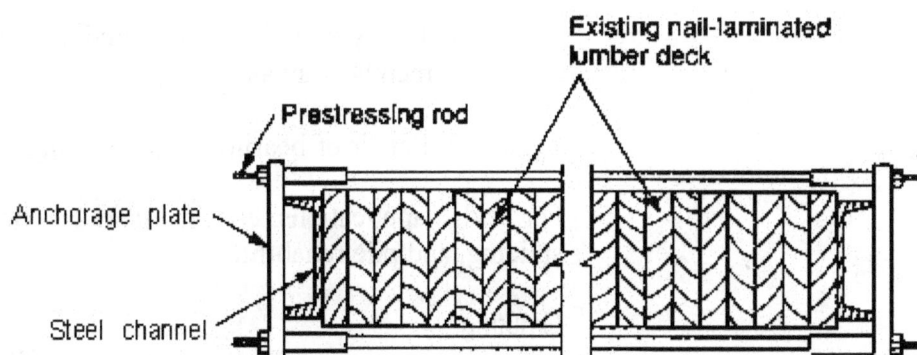

Figure 14-14. - Typical rod and anchorage configuration for stressing existing nail-laminated lumber decks.

Epoxies consist of basic resins and resin-hardening agents that are blended together in a liquid or gel (putty) form. When mixed, the epoxy compounds harden to form a solid, durable material that provides a high degree of adhesion to most clean surfaces. Epoxies were originally developed by the paint and aircraft industries in the 1950's and have been used extensively to repair cracks in concrete since the 1960's. The first reported study on epoxy use for timber repair was presented by Avent[8] in 1976. Since that time there has been a considerable research effort to develop design criteria and to evaluate the effectiveness of epoxy repairs in timber members. Although there are currently no codes or specifications with design criteria or allowable stresses, epoxy repair techniques have been successfully used on timber bridges (some since the early 1960's). The information presented in this section is based on referenced research publications and successful field applications.

TYPES OF EPOXY REPAIRS

Epoxy is used for timber repair as a bonding agent (adhesive) and/or grout (filler) in both structural and semistructural repairs. It is commonly injected under pressure but is also manually applied as a gel or putty. Epoxy is most effective when used as a bonding agent to provide shear resistance between members for structural repairs in dry locations. For semistructural repairs, it is used to fill voids or repair bearing surfaces. Avent[6] describes six basic types of epoxy repairs for structural (Type A) and semistructural (Type B) repairs, as follows:

Type A-1. Epoxy injection of cracked and split members at truss joints.

Type A-2. Epoxy injection and reinforcement of decayed wood.

Type A-3. Splicing and epoxy injection of broken members.

Type A-4. Epoxy injection of delaminated beams.

Type B-1. Epoxy injection of longitudinal cracks and splits in truss members away from joints.

Type B-2. Repair of bearing surfaces using epoxy gel.

For bridge applications, epoxy repairs can be categorized as grouting, splicing, and pile rehabilitation.

Grouting

As a grouting material, epoxy is used for filling checks, splits, delaminations, insect damage, and decay voids. The epoxy seals the affected area, preventing water and other debris from entering. It can also restore the bond between separated sections, increase shear capacity, and reduce further splitting. In building applications, epoxy has been successfully used in structural repairs to fill splits in truss connections.[2,3,8,9] It has also been used in conjunction with reinforcing rods to replace severely decayed portions of existing members.[1] In bridge applications, its use as a grout has been limited primarily to semistructural or cosmetic repairs involving surface damage or internal insect damage. For surface repairs, voids or other defects are filled with epoxy gel (Figure 14-15). For internal repairs involving splits or insect damage, liquid epoxy must be injected to the inside of the member to fill the void.

Figure 14-15. - Epoxy gel surface repair of a timber pile (photo courtesy of Osmose Wood Preserving, Inc.).

Splicing

Splicing repairs involve the addition of splice pieces that are lapped over the split or deteriorated members and are epoxied in place. In this type of repair, epoxy is used as an adhesive to bond the splices in place. While other types of adhesives are available for wood, epoxies are preferable for field repairs because of their high strength and rapid cure rate. Epoxy splicing has been used mostly in buildings and is not a common type of repair in bridge applications at this time. However, one method of splicing that has been used to a limited degree involves the reconstruction of glulam. In this method, damaged or decayed laminations are cut from the glulam member and replaced with new laminations that are epoxied in place. The laminations in the replacement section are lapped over existing laminations a sufficient distance to develop the required shear strength at the epoxied joint. There is evidence that variations in the moisture content of timber members can in time cause a significant reduction in the bonding strength of epoxy. Therefore, splicing repairs in members exposed to weathering or significant fluctuations in moisture content are not recommended. Also, epoxy splicing should not be used on material treated with oil-type preservatives because of poor bonding between the wood and the epoxy.

Pile Rehabilitation

Pile rehabilitation employs epoxy (using grouting and splicing) for the repair of timber piles loaded primarily in axial compression. The two methods of pile rehabilitation most commonly used are pile posting and pile restoration. In pile posting, the damaged section of pile is completely removed and a new section of similar cross section is installed in its place (Figure 14-16). The new section is positioned with a 1/8- to 1/4-inch gap at the top and bottom and is wedged tightly against the existing pile cutoffs. Following placement of the new section, holes are bored at a steep downward angle above each joint, spaced approximately 90 degrees apart. Steel pins are then driven through the holes to mechanically join the two sections. The sides of the joints are next sealed with epoxy gel, plastic film, or tape, and epoxy is injected into the joints, filling the voids and

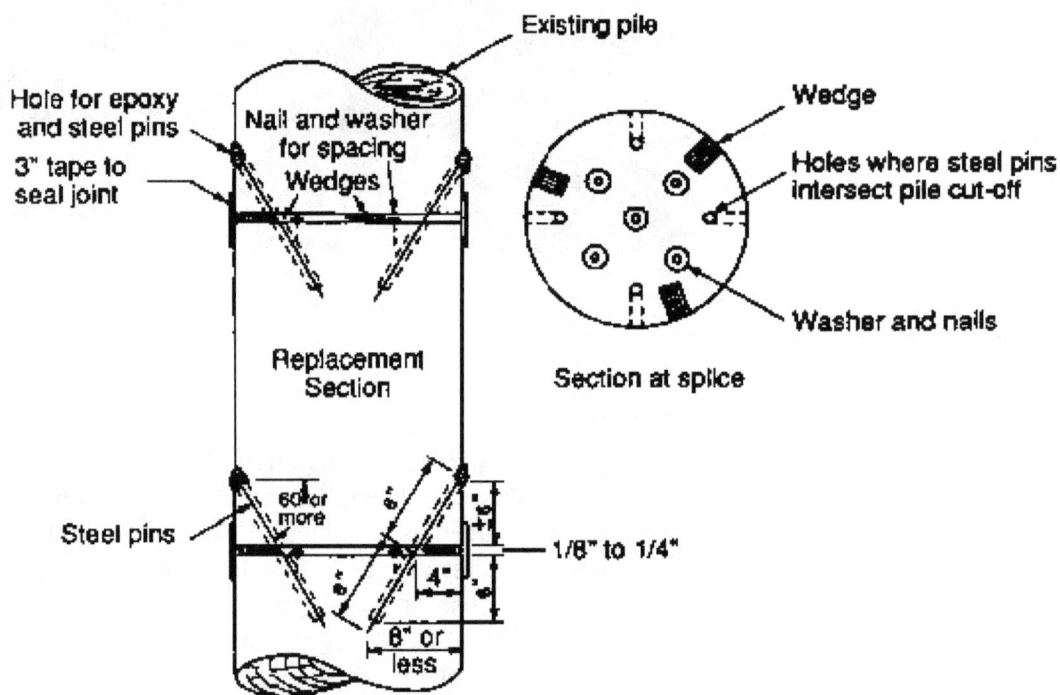

Figure 14-16. - Schematic diagram of pile posting.

14-20

bonding the old and new pile sections. This type of repair has proven to be an economical method of substructure repair that effectively restores the compressive strength of deteriorated members. Additional information on pile posting can be found in case histories presented in Chapter 15.

Pile restoration involves the removal and replacement of a vertical wedge-shaped section of piling rather than the entire cross section. This type of repair has been successfully used on piling where localized deterioration occurs in an otherwise sound section. Using this method, a wedge-shaped section is removed from the existing pile by cutting and chiseling (Figure 14-17). A matching replacement section is fabricated from new treated material. The replacement section is fitted to match the removed section, but is slightly smaller in size. After the replacement is fabricated, the contact surfaces of both old and new sections are covered with epoxy gel applied with a putty knife. The new section is placed in position, and metal bands are installed around the section to hold it in place while the epoxy cures. Pile restoration is more expensive than posting and is normally used only when posting is impractical because of limited access.

Figure 14-17. - Pile repair using pile restoration techniques. (A) The deteriorated pile area is removed as a wedge-shaped section. (photos courtesy of Osmose Wood Preserving, Inc.).

Figure 14-17. - Pile repair using pile restoration techniques (continued). (B) A replacement section is cut, and epoxy gel is applied to the contact surfaces. (C) The replacement section is placed and banded to the existing pile (photos courtesy of Osmose Wood Preserving, Inc.).

The procedures for the use of epoxy vary with the type and extent of repair. The basic procedures for epoxy injection can be summarized in four steps: member preparation, port setting and joint sealing, epoxy injection, and finishing.[6] For manual, nonpressure application, port setting and joint sealing are not required. As with all types of repairs, a structural evaluation and analysis of existing components must be made to determine load capacity before and after repair. The cause of the problem should also be identified and corrective measures taken to prevent its recurrence.

Member Preparation

The degree of member preparation required for epoxy repair varies with the type of repair and the wood condition. When the defect or weakness in the original member is the result of decay, actions must be taken to remove the damaged wood, arrest the infections, and prevent renewed damage. If areas to be repaired show early signs of decay, in-place treatment may be sufficient to arrest decay, provided sufficient strength remains in the member. When visible decay is present, the most thorough approach is to remove the infected section. For such cases, the following guidelines are given by Clark and Eslyn:[17]

> The undetectable extensions of the infecting fungi may reach 6 to 12 inches in the grain direction beyond the apparent limits of the decay. A safe rule in removing decayed parts of members is to include the visible decay plus an additional 2 feet of the adjacent wood in the grain direction.

In addition to removing or treating decay areas, the moisture source to the infected member should be identified and eliminated, if possible. When moist wood (greater that 20 percent moisture content) is found, the member should be dried before repairs are made. Although there are epoxies that will bond to moist wood, the presence of moisture levels greater than 20 percent may provide suitable conditions for continued fungal growth and continued deterioration.

As a final preparation step for all epoxy repairs, surfaces must be thoroughly cleaned of all dirt and debris so that a good bond can be achieved between the wood and the epoxy. Areas should be free of excess oil preservatives, which may affect the bond. Although there have been no studies on the bonding strength of epoxies to wood treated with oil-type preservatives, successful piling repairs (compressive loading) have been made on existing members treated with creosote that have been in place for a number of years. Splicing or shear-type repairs are not recommended on surfaces treated with oil-type preservatives because of the questionable bonding to the member surfaces.

Port Setting and Joint Sealing

When epoxy is applied by pressure injection, the repair area must be provided with injection ports and completely sealed before epoxy placement. The injection ports are holes bored into the joint area that permit epoxy injection into interior portions of the repair, vent displaced air as epoxy fills the void, and provide a visual means of observing epoxy distribution. These ports are generally 1/4 to 3/8 inch in diameter and are topped with a small copper or plastic tube that projects from the wood surface. The number and location of ports varies depending on the size and configuration of the repair area. The minimum number of ports is two, one for the injection and one as an escape for displaced air. For most types of repairs, additional ports are added to ensure epoxy penetration to all areas of the joint.

After injection ports are set, areas of the joint must be completely sealed (with the exception of injection port openings). Incomplete sealing allows epoxy to seep from the repair area, wasting material and creating voids in the epoxy that reduce its effectiveness. Methods of joint sealage vary depending on the configuration of the members being repaired. For most repairs, openings can be sealed with an epoxy gel, provided the gel viscosity is sufficiently low to span the distance of the opening. Another common method for sealing piling and other exposed, smooth locations is to staple plastic wraps or tape to the outside of the member (Figure 14-18). With porous wood, it may be beneficial to seal the outside surface with thick epoxy paint to fill hairline cracks and other small openings. These

Figure 14-18. - A joint for a posting-type epoxy repair is sealed with plastic wrap stapled to the members. Small wood strips are then nailed across the plastic to provide an additional seal (photo courtesy of Osmose Wood Preserving Inc.).

openings will allow epoxy to escape even though they may not be evident during visual inspection.

Epoxy Application

Epoxy is applied using manual nonpressure methods or pressure injection, depending on the type of repair. Nonpressure methods are usually limited to exposed surface applications. The two epoxy components are thoroughly mixed in a bowl or other container and are applied with a knife or brush. Surface repairs on angled or vertical surfaces may require a plastic wrap or special tape to keep the epoxy in position as it cures. For pressure injection, the epoxy is applied through one injection port at each joint. As the epoxy fills the voids in the joint, venting ports begin to leak an even flow of epoxy and are progressively sealed. Injection is accomplished using either a caulking gun and tubes of epoxy that are mixed manually before application (Figure 14-19) or an automatic injection gun that mixes the epoxy components in the nozzle. For both techniques, the injection pressure must be sufficient to completely fill the void without breaking joint seals. A maximum injection pressure of 40 lb/in^2 is recommended.[6]

Finishing

The time required for epoxy to cure to its full strength varies among brands of epoxy and the curing temperature. Most epoxies set in a few hours, but complete curing can take several days. After final curing, the epoxy surface can be finished to meet aesthetic requirements of the site, including removal of projecting injection ports, sanding, and painting of the epoxy surface.

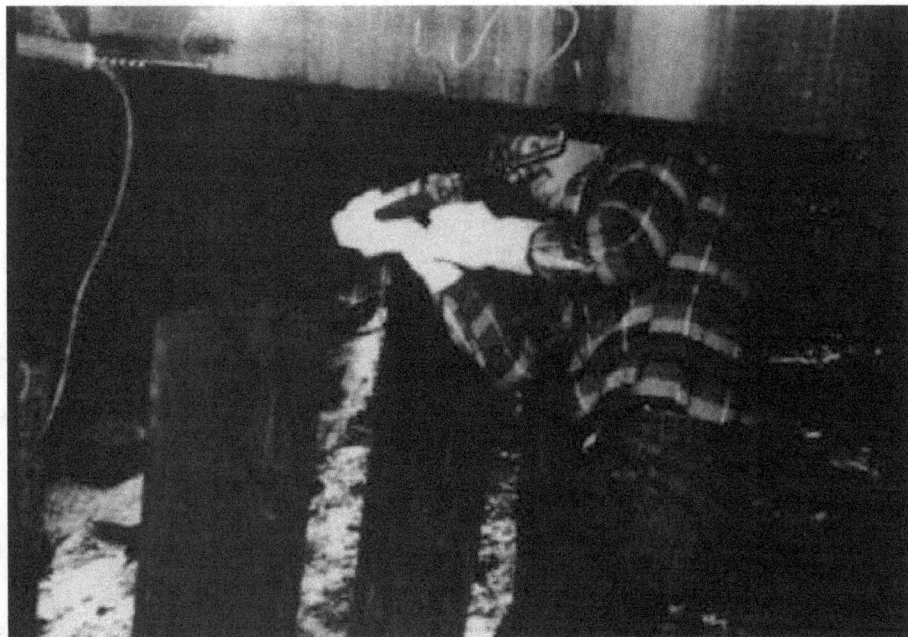

Figure 14-19. - Epoxy is manually injected between a timber pile and cap using a caulking gun (photo courtesy of Osmose Wood Preserving, Inc.).

14-25

QUALITY CONTROL FOR EPOXY REPAIRS

A key factor in epoxy effectiveness is the level of quality control provided during the repair process. Although little has been published on this subject, the following guidelines on quality control are given by Avent:[4]

In many cases laboratory testing is not possible for wood repair in contrast to concrete repair where test cylinders can be taken. For example, lack of quality control can result in serious problems for epoxy repaired members. Many epoxies are very sensitive to mix proportions. The standard injection equipment consists of two positive-displacement pumps driven by a single motor geared to obtain the proper mix. The two epoxy components are mixed at the nozzle; thus a fairly continuous flow prevents hardening of the epoxy in the nozzle. However, crimped lines, malfunctioning pumps, or line blockages can sometimes occur. In severe cases the epoxy will not harden at all, but in other cases the problem may result in soft spots within the joints. Frequent collecting of small samples in containers will verify if the epoxy is hardening as expected, and this is routinely done by contractors on an hourly basis. The detection of weak but hardened material is much more difficult. One method is to inject shear block specimens at the beginning of operations and after the repair of every fifth member. A shear specimen [see Figure 14-20] is cut into four shear blocks after curing and each is tested in single shear. The failure stress level should be approximately equal to the ultimate shear strength of the *wood*. This level of shear strength indicates a high-quality bond.

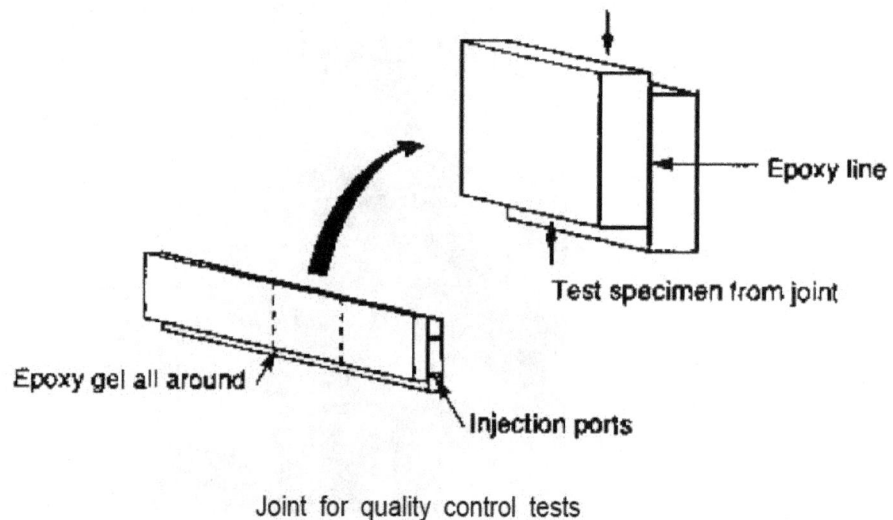

Epoxy line

Test specimen from joint

Epoxy gel all around

Injection ports

Joint for quality control tests

Figure 14-20. - Typical shear block specimen for evaluating the strength of an epoxied joint.

14-26

Another quality control problem is that of determining epoxy penetration into voids. Special sampling techniques are currently in the development process, but none have proven completely satisfactory as yet. This problem is often heightened because there are two types of repair: structural and non-structural. Non-structural repairs are associated with sealing in applications such as waterproofing, crack sealing to prevent contamination, and cosmetic repairs. Many contractors are familiar only with this type. The approach to non-structural repairs is to inject from port to port without undue concern for complete penetration. Often air voids become trapped by such an approach. The key to successful structural repair is to fill all voids. To ensure complete penetration, it is best to inject from only one port while letting others serve as vents. The successive bleeding and capping of these ports gives a high degree of confidence in the amount of penetration. An average repair often involves at least 12 ports and many have considerably more. However, without close supervision of the injection operation, a contractor may revert to his usual approach for non-structural repairs, especially since the different goals of these types of repair are usually not appreciated. Close supervision thus becomes the primary method of quality control.

14.6 COMPONENT REPLACEMENT

There are situations where a lack of maintenance or other causes leads to deterioration so severe that replacement of the member is the only economically viable alternative. In these cases, the structure must be temporarily supported (when required), the old member removed, and a new one installed in its place. Before replacing members, the cause of deterioration in the original member must be determined and corrected. If the problem is structural, an increased capacity for the replacement may be warranted. If decay is the source of deterioration, corrective measures should be taken to exclude moisture from newly installed members. Whenever a member is replaced, it is advisable to thoroughly inspect all adjacent and contacting components for decay that may not have been apparent when the member was in place. Confirmed or suspected areas of decay should be treated in place before the new member is installed. Remember that failure of the original member resulted from a specific cause that could also cause premature failure or high maintenance costs for the replacement.

On some structures it may be impractical to replace a member because of difficulties with removing the old member or positioning a new member in its place. An alternative solution is to add a sister member that is structurally capable of resisting the loads previously applied to the original member. The use of sister members is most applicable when damage occurs from overloads or other mechanical damage (Figure 14-21). When existing members are decayed, appropriate steps must be taken to eradicate the infection and prevent its spread to the new component. The decayed portions of the member should be removed and the remaining portions treated in place. Again, the source of moisture that provided the suitable decay conditions must also be eliminated.

Figure 14-21. - A sister member in a glulam beam superstructure. The outside beam, which was damaged by a vehicle overload, could not be easily replaced. The sister member was added along the outside of the beam to restore the capacity to the structure.

14.7 SELECTED REFERENCES

1. American Association of State Highway and Transportation Officials. 1976. AASHTO manual for bridge maintenance. Washington, DC: American Association of State Highway and Transportation Officials. 251 p.

2. American Society of Civil Engineers. 1986. Evaluation and upgrading of wood structures: case studies. New York: American Society of Civil Engineers. 111 p.

3. American Society of Civil Engineers. 1982. Evaluation, maintenance, and upgrading of wood structures. Freas, A., ed. New York: American Society of Civil Engineers. 428 p.

4. Avent, R.R. 1985. Decay, weathering and epoxy repair of timber. Journal of the Structural Division, ASCE 111(2): 328-342.

5. Avent, R.R. 1986. Design criteria for epoxy repair of timber structures. Journal of Structural Engineering 112(2): 222-240.

6. Avent, R.R. 1985. Factors affecting strength of epoxy-repaired timber. Journal of the Structural Division, ASCE 112(2): 207-221.

7. Avent, R.R. 1986. Repair of timber bridge piling by posting and epoxy grouting. In: Trans. Res. Rec. 1053. Washington, DC: National Academy of Sciences, National Research Council, Transportation Research Board: 70-79.

8. Avent, R.R.; Emkin, L.Z.; Howard, R.H.; Chapman, C.L. 1976. Epoxy-repaired bolted timber connections. Journal of the Structural Division, ASCE 102(4): 821-838.

9. Avent, R.R.; Emkin, L.Z.; Sanders, P.H. 1978. Behavior of epoxy-repaired full-scale timber trusses. Journal of the Structural Division, ASCE 104(6): 933-951.

10. Avent, R.R.; Sanders, P.H.; Chapman, C.L. 1982. Space-age adhesives. The Military Engineer 74(477): 20-22.

11. Avent, R.R.; Sanders, P.H.; Emkin, L.Z. 1978. Epoxy repair of timber structures comes of age. Construction [Adhesive Engineering Co.] 13(2): 6.

12. Avent, R.R.; Sanders, P.H.; Emkin, L.Z. 1979. Structural repair of heavy timber with epoxy. Forest Products Journal 29(3): 15-18.

13. Avent, R.R; Issa, R.R.A.; Baylot, J.T. 1982. Weathering effects of epoxy-repaired timber structures. In: Structural Uses of wood in adverse environments. New York: Van Nostrand Reinhold Co.: 208-218.

14. Better Roads. 1980. Bridge pilings can be protected; FPR jackets stop deterioration. Better Roads May: 20-25.

15. Better Roads. 1973. For these timber piles, life begins at 58 years. Railway Track and Structures September: 27-29.

16. Better Roads. 1980. Gribbles' attack collapses bridge...but polyester sleeves thwart them. Better Roads April: 10-14.

17. Clark, J.W.; Eslyn, W.E. 1977. Decay in wood bridges: inspection and preventive & remedial maintenance. Madison, WI: U.S. Department of Agriculture, Forest Service, Forest Products Laboratory. 51 p.

18. Csagoly, P.F.; Taylor, R.J. 1979. A development program for wood highway bridges. 79-SRR-7. Downsview, ON, Can.: Ministry of Transportation and Communications. 57 p.

19. Eslyn, W.E.; Highley, T.L. 1985. Efficacy of various fumigants in the eradication of decay fungi implanted in Douglas-fir timbers. Phytopathology 75(5): 588-592.

20. Gerke, R.C. 1969. New process restores timber piles. Ocean Industry May: 92-93.

21. Heising, G.D. 1979. Controlling wood deterioration in waterfront structures. Sea Technology 20(6): 20-21.

22. Highley, T.L. 1980. In-place treatment for control of decay in waterfront structures. Forest Products Journal 30(9): 49-50.

23. Highley, T.L.; Eslyn, W.E. 1982. Using fumigants to control interior decay in waterfront timbers. Forest Products Journal 32(2): 32-34.

24. Hurlbut, B.B. 1978. Basic evaluation of the structural adequacy of existing timber bridges. Washington, DC: National Academy of Sciences, National Research Council, Transportation Research Board: 6-9.

25. Ketchum, V.T.; May, T.K.; Hanrahan, F.J. 1944. Are timber checks and cracks serious? Engineering News Record July 27: 90-93.

26. McCutcheon, W.J.; Gutkowski, R.M.; Moody, R.C. Performance and rehabilitation of timber bridges. Trans. Res. Rec. 1053. Washington, DC: National Academy of Sciences, National Research Council, Transportation Research Board: 65-69.

27. Morrell, J. J.; Helsing, G.G.; Graham, R.D. 1984. Marine wood maintenance manual: a guide for proper use of Douglas fir in marine exposure. Res. Bull. 48. Corvallis, OR: Oregon State University, Forest Research Laboratory. 62 p.

28. Muchmore, F.W. 1983. Timber bridge maintenance, rehabilitation, and replacement. GPO 693-015. Missoula, MT: U.S. Department of Agriculture, Forest Service, Northern Region. 31 p.

29. Railway Track and Structures. 1973. Grout-filled timber piles. Railway Track and Structures February: 28-29.

30. Railway Track and Structures. 1980. In-place decay treatment "supports" timber trestle. Railway Track and Structures June: 38.

31. Railway Track and Structures. 1977. Second-generation in-place treatment for these ICG timber trestles. Railway Track and Structures January: 28-30.

32. Railway Track and Structures. 1971. Wood piles wrapped to keep out marine borers. Railway Track and Structures. April: 1.

33. Sanders, P.H.; Emkin, L.Z.; Avent, R.R. 1978. Epoxy repair of timber roof trusses. Journal of the Construction Division, ASCE 104(3): 309-321.

34. Scales, M. 1964. Epoxy based structural adhesives. Adhesives Age 7 (November): 22-24.

35. Scheffer, T.C.; Eslyn, W.E. 1982. Twenty-year test of on-site preservative treatments to control decay in exterior wood of buildings. Material u. Organismen 17(3): 181-198.

36. Taylor, R.J.; Csagoly, P.F. 1979. Transverse post-tensioning of longitudinally laminated timber bridge decks. Downsview, ON, Can.: Ministry of Transportation and Communications. 16 p.

37. White, K.R.; Minor, J.; Derocher, K.N.; Heins, C.P., Jr. 198 1. Bridge maintenance inspection and evaluation. New York: Marcel Dekker, Inc. 257 p.

38. Williams, J.R. 1965. Redrive timber piling or treat in place-why not a program that utilizes both? Progressive Railroading July: 3.

39. Williams, J.R.; Norton, K.J. 1976. Decay in timber trestles: what is its rate of growth? Railway Track and Structures April: 26-29.

CHAPTER 15

BRIDGE MAINTENANCE, REHABILITATION, AND REPLACEMENT: CASE HISTORIES

Over time, bridges may become structurally or functionally deficient. Structurally, the deficiency can result from deterioration, damage, or increased load requirements in excess of the design capacity. Hydraulically, the original waterway opening under the bridge may become inadequate as a result of changing drainage patterns in the watershed or because the hydraulic parameters on which the original design was based are inadequate. Bridges may also become functionally deficient when the roadway width, vertical clearance, or geometry are inadequate for current traffic requirements.

In most cases, structural deficiencies that develop are corrected by preventative or routine maintenance. If such maintenance is continually neglected, major maintenance may be required to restore the bridge to its original capacity. When hydraulic or geometric deficiencies are encountered, bridge rehabilitation can improve the conditions. If the bridge is severely deficient structurally, hydraulically, or geometrically, complete replacement may be the only option.

Previous chapters discuss the methods of timber bridge design, maintenance, and rehabilitation. In this chapter, case histories illustrate how these methods have been applied. These case histories include the following:

Case History 15.1 - In-Place Preservative Treatment of Deteriorating Timber Bridges
Case History 15.2 - Extending Bridge Life: In-Place Treatment of a Timber Bridge
Case History 15.3 - Pepin County Bridge Widening
Case History 15.4 - Union County Covered Bridge Rehabilitation
Case History 15.5 - Rehabilitation of Nail-Laminated Timber Decks by Transverse Stressing
Case History 15.6 - Sauk County Bridge Redecking
Case History 15.7 - Bruneau River Bridge Rehabilitation
Case History 15.8 - Uinta Canyon Canal Bridge Replacement
Case History 15.9 - Cook County Bridge Replacement

CASE HISTORY 15.1-
IN-PLACE PRESERVATIVE TREATMENT OF DETERIORATING TIMBER BRIDGES

Contributed by Edgar E. Hedgecock, Civil Engineer, USDA Forest Service, Francis Marion and Sumter National Forests

In 1982, the decision was made to arrest internal deterioration and surface decay by undertaking in-place preservative treatment of the existing 391-foot-long Bay Creek Bridge located on the Apalachicola National Forest in Florida. The in-place treatment also included the replacement of deteriorated sections of structurally deficient timber piling.

The National Forests in Florida contracted with Osmose Wood Preserving, Inc., to treat the Bay Creek Bridge. The work was completed over a 9-day period in the fall of 1982. The cost for the in-place treatment was slightly more than $28,000 ($72/lin ft). Replacing the bridge would have cost approximately $450,000.

All chemicals used in treating the Bay Creek Bridge were formulated to eradicate existing decay fungi. In addition, the treatment is intended to retard any new fungus infection for 12 to 15 years. The following wood-preserving chemicals, produced by Osmose Wood Preserving, Inc., of Buffalo, New York, were used in treating the bridge:

Tie-Gard. Tie-Gard cartridges were placed in predrilled holes at the groundline or waterline of all the bridge piles. The cartridges are solidified preservatives consisting of 37.5-percent sodium fluoride, 37.5-percent potassium bifluoride, 19-percent sodium dichromate, 5-percent 2,4 dinitro-phenol, and 1-percent inert material. The ingredients of the cartridges become active when exposed to moisture.

Timber Fume. Vials of Timber Fume were placed in holes between the Tie-Gard cartridges and the tops of the piles. This chemical is a highly poisonous liquid-fumigant solution that volatilizes and diffuses into the wood to arrest internal wood decay. The solution consists of 99-percent chloropicrin and 1-percent inert ingredients.

Osmose 24-12 Solution. Osmose 24-12 solution was injected under pressure into the pile caps and into the top area of piles. This liquid wood-preservative solution prevents wood mold and decay fungi. The solution is composed of 4.48-percent pentachlorophenol, 0.52-percent other chloro-phenols, 5.16-percent aromatic petroleum solvent, 46.52-percent aliphatic petroleum solvent, and 43.32-percent inert ingredients.

Osmoplastic-F. Osmoplastic-F was injected into predrilled holes in all bridge stringers. This chemical is a paste wood preservative that kills existing decay fungi and inhibits new fungi growth. The preservative is

20-percent sodium fluoride, 8.9-percent pentachlorophenol, 1.1-percent other chlorophenols, 15-percent creosote, and 55-percent inert ingredients. The contractor had years of experience in treating wood railroad bridges and utility poles but had not treated a bridge like the Bay Creek unit. Their standard operating plan, with only a few minor changes, followed this sequence:

1. Drill inspection holes
2. Inspect for internal decay
3. Establish a treatment pattern for piles, caps, and stringers
4. Develop environmental protection measures
5. Replace sections ("posting") of piles
6. Treat piling
7. Treat caps
8. Treat stringers
9. Remove and dispose of all protective material and waste

Each of the procedures involved specific steps to ensure maximum efficiency in the treatment. Brief summaries of each step follow:

1. **Drill inspection holes.** The contractor made a preliminary inspection of the bridge to assess requirements for treating and component repair. All members were sounded with a hammer, and borings were made at locations of suspected decay in piles, caps, and stringers. The results of this inspection were used to develop a cost estimate on which the contract price was based.

As part of the contract, the contractor drilled a predetermined pattern of 3/8-inch diameter inspection holes into piles, caps, and stringers (Figure 15-1). The patterned holes permitted inspection of critical areas near the groundline and at the pile-cap-stringer connections. The contractor modified a standard railroad bridge inspection pattern to better fit the condition and configuration of the Bay Creek Bridge.

Figure 15-1. - Typical investigative borehole pattern.

2. Inspect for internal decay. The condition of each bridge member was determined using the inspection holes and a special metal probe. The contractor estimated the location and extent of sound wood and deteriorated wood in each member, and the figures for shell and void were marked beside each inspection hole.

3. Establish treatment pattern. A treatment pattern was established based on the inspection data and surface-water conditions at the site. A variety of chemicals were selected because of the high water table and the variation in size of the structural members.

4. Develop environmental protection measures. A special effort was made to prevent any pollution or contamination of water. Highly toxic chemicals were used, and extra care was required during treatment near the water. Plastic draping was placed around and under pile caps to contain any spillage or leakage. In addition, a metal funnel-shaped collar was attached to the base of each pile to collect any chemicals that might run down the pile.

A two-person crew performed all treatment operations. One crew member applied the chemical treatment, while the other acted as a guard to locate and control spills or leakage.

5. Post rejected piles. The contractor has methods and equipment to completely replace deteriorated piling, matching the replacement pile to the alignment and batter of the original pile. Further, the contractor can replace only a section of deteriorated piling above the water- or groundline using pile posting techniques. On the Bay Creek Bridge, two piles were rejected because of decay, and sections of each were replaced by posting from just above the groundline to the pile cap (Figures 15-2 and 15-3). The creosote-treated replacement sections were connected to the existing piles with eight fluted-steel pins, 3/8 inch in diameter by 16 inches long, and were welded in place with epoxy resin (Figure 15-4).

Figure 15-2. - Section of deteriorated pile removed from the structure.

Normally, the contractor has a separate crew do the pile posting; however, because this bridge was a relatively small project the treatment crew performed the piling repairs.

6. Treat piling. In-place treatment of the piling was complicated by the high, fluctuating water level of Bay Creek. To achieve maximum effectiveness in the treatment, the contractor used a combination of three products: Osmose 24-12, Timber-Fume, and Tie-Gard. The following procedure allowed optimum treatment of wood with the least risk of chemical spillage or leakage:

Figure 15-3. - Cutting a replacement pile section using a specially designed pile cutter.

Figure 15-4. - Completed pile posting with protective plastic in place.

15-6

a. In addition to the inspection borings, extra holes were drilled in the piling at the groundline. Tie-Gard preservative cartridges were inserted into the holes, which were then plugged with treated wood dowels (Figures 15-5 and 15-6). The cartridges dissolve over a period of time and the preservative diffuses into the pile by capillary action.

Figure 15-5. - Drilling holes for solid preservative cartridges.

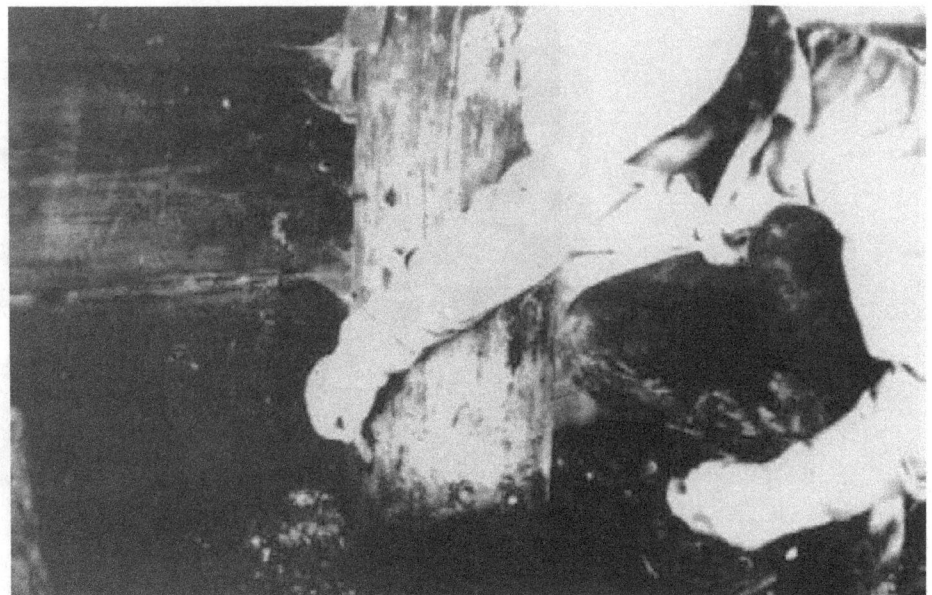

Figure 15-6 - Insertion of solid preservative cartridges into a timber pile. Note the use of rubber safety gloves.

b. Vials of Timber-Fume fumigant were inserted into holes drilled in the midsection of all piles (Figure 15-7). This chemical becomes a gas and diffuses into the pile. Timber-Fume is not effective when water is present. It is a highly toxic substance so extra care was used in preparing and placing the vials (Figure 15-8).

Figure 15-7. - Fumigant vial in wood-plugged hole, with location tab visible.

Figure 15-8. - Preparing fumigant vials. Note the required safety mask and gloves.

c. For extra protection against decay, Osmose 24-12 preservative was injected into the inspection holes at the top area of each pile (Figure 15-9).

7. Treat caps. Osmose 24-12 preservative was injected under pressure into all inspection holes in the pile caps. A treated wood plug was driven into each hole to seal the chemical in the wood after injection. Plastic draping was used to catch any chemical that seeped out of cracks during injection (Figures 15-10 and 15-11).

In the injection process, a relief valve and catch bucket were used to suppress back-pressure spills. The two-person crew worked well in preventing spills (Figure 15-12).

8. Treat stringers. Stringer treatment caused the contractor some problems. The narrow 3-inch stringer width, and the high water condition at the site, prevented use of liquid wood preservatives. The contractor chose to use Osmoplastic paste preservative, which was injected into the stringers with a grease gun. However, the high sodium fluoride content of Osmoplastic caused the rubber seals in the grease gun to disintegrate, and forced the contractor to switch to Osmoplastic-F, which contains ingredients that are less active. With this compound, the contractor was able to maintain pressure and properly inject the compound into the inspection holes.

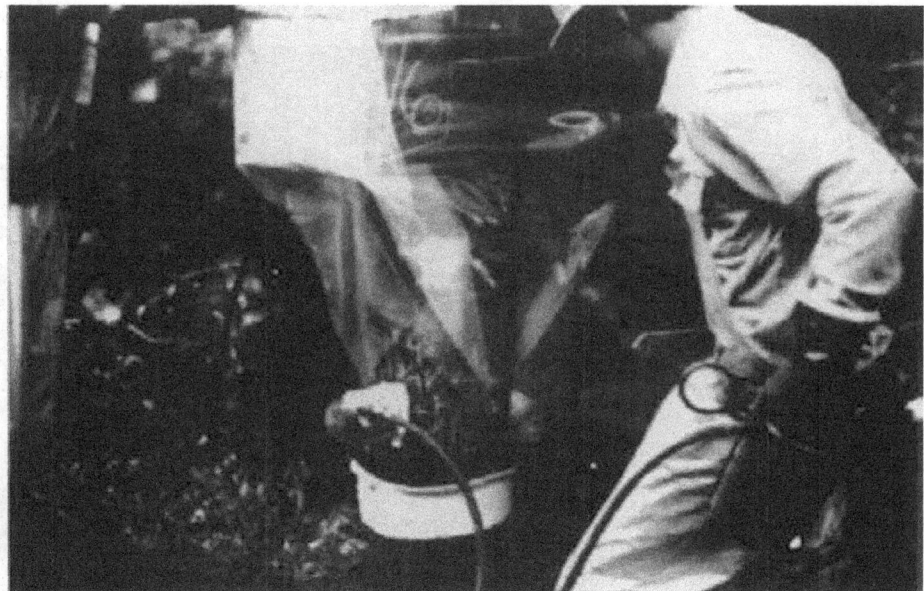

Figure 15-9. - Injection of liquid preservative solution into a timber pile.

Figure 15-10. - Typical individual cap and pile draping and funnel for protection against leakage of the liquid preservative.

Figure 15-11. - Bridge draping for spillage protection, in place and ready for preservative injection.

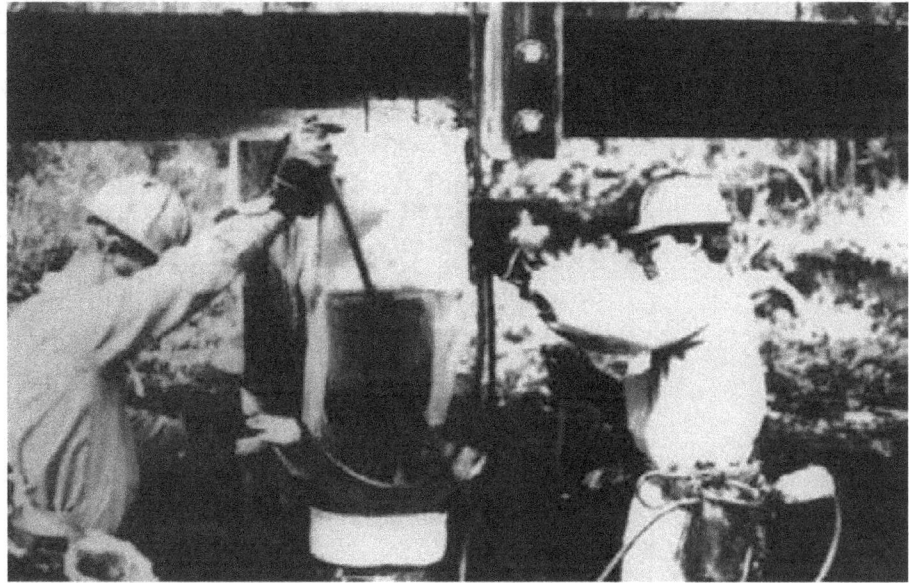

Figure 15-12. - Two-person crew injects liquid wood preservative solution into piles and caps. Note the safety backflow valve on the worker's belt.

9. Cleanup. The contractor waited 2 to 3 hours after treatment before removing protective plastic draping. All plastic, waste rags, and containers were carefully rolled and placed in plastic trash bags for removal from the site.

Throughout the project, the Forest Service monitored water quality at the Bay Creek site. Water samples were taken before treatment commenced to establish a typical quality level. During the project, samples were taken upstream and downstream from the bridge and were sent to the Florida Department of Environmental Regulation for testing. Analysis of the samples indicated no detectable levels of the treatment chemicals. Further, there was no noticeable variation in water quality between the samples taken before treatment and those taken after treatment.

This was the first contract awarded by the Forest Service in the Southern Region for this particular service. If the Forest Service's experience with this method is as satisfactory as that of the railroads, we will be able to save considerable maintenance and reconstruction funding in the future.

CASE HISTORY 15.2-
EXTENDING BRIDGE LIFE: IN-PLACE TREATMENT OF A TIMBER BRIDGE

Contributed by Bill Grabner, Regional Bridge Engineer, USDA Forest Service, Pacific Northwest Region

The Sullivan Lake Outlet Bridge, located on the Colville National Forest in northeastern Washington, is a single-lane, 10-span timber trestle that is 191 feet long. Originally constructed in 1935, the bridge consists of a series of sawn lumber stringer spans supported on timber pile bents. Inspections of the structure completed in 1979 and 1981 indicated that extensive decay was present in several of the timber piles and pile caps. Because the bridge was not located on a heavily used road and was subjected primarily to light administrative and recreation traffic, no funds were available for bridge replacement or for major bridge rehabilitation work within the foreseeable future.

In the summer of 1981, it was learned that a Midwest-based company had been engaged in the in-place treatment of utility poles and timber railroad structures for a number of years. This company had not previously done any in-place timber bridge treatment for public agencies; however, negotiations were held between the Forest Service and this company, and a contract was awarded. The bridge rehabilitation work subsequently took place in November 1981.

To begin the work, a thorough inspection of the entire 191-foot-structure was made by an experienced bridge inspection crew employed by the contractor. The inspection procedure consisted of drilling 2,126 holes, each 3/8 inch in diameter, in areas where decay was most likely to occur. When decay was located, its extent was determined by the use of a wire probe and additional drilling, as needed. This allowed the inspector to completely define the void or deteriorated region in each pile and cap. The results of the inspection were noted in a detailed report for each of the eleven pile bents and ten stinger spans. Based on the inspection report and the recommendations of the contractor, the decision as to which members were to be treated and which were to be replaced was made by the Forest Service.

Four piles were repaired by removing badly deteriorated pile sections, varying in length from 5 to 15 feet, and replacing them with new sections of treated pile (pile posting). The replacement sections were secured into position using 16-inch steel pins and epoxy resin. One cap was replaced by the contractor using a unique jacking method to lift and support the superstructure while installing a new treated-timber cap. This work was accomplished with little traffic interruption.

The result of the work was to restore the structure to full capacity and extend the usable life by an estimated 10 to 15 years. This was achieved by replacing piles and caps that were inadequate to support loads, and by destroying all fungi and preventing their reintroduction into the timber by the continuing presence of fumigant. This in-place treatment work can be repeated once the fumigant level has reached nontoxic levels, with a further increase in structure life. Fumigant toxicity level can be determined by assay methods.

Total cost of the treatment, including replacement of piles and cap, was $31,140. Comparing this to an estimated bridge replacement cost of $250,000 indicates that this pilot project proved to be very cost effective. Further use of in-place treatment for timber structures is contemplated in the Pacific Northwest Region of the Forest Service as a method of reducing replacement costs by extending timber bridge life. Various phases of this project are depicted in Figures 15-13 through 15-17.

Figure 15-13. - View of the Sullivan Lake Outlet Bridge as in-place treatment was commencing. Note the canvas slings placed under the bridge to prevent water contamination should accidental spillage of treatment chemicals occur.

Figure 15-14. - Liquid fumigants being applied to the timber abutment.

Figure 15-15 - Underside view of an intermediate bent cap after treatment was completed. Holes drilled for treatment application have been plugged with treated-wood dowels.

Figure 15-16. - Timber pile after treatment of the soil contact area. The treated portion of the pile was wrapped in waterproof paper to prevent potential soil contamination.

Figure 15-17. - Pile bents after replacement of deteriorated sections by pile posting.

Contributed by Ken Johnson, Civil Engineer, Wheeler Consolidated, Inc.

Pepin County is a small rural county in western Wisconsin. Many miles of rural highways with a large number of bridges were built in the county between 1950 and 1970. These facilities were constructed according to the standards existing at that time. The factor used to project average daily traffic for determining the level of design underestimated the actual growth in traffic. The growth in traffic above projected levels made many bridges functionally deficient by current standards.

This condition adversely affected the Pepin County Highway Department in two ways. First, the bridges posed a continuing hazard to the traveling public because they were too narrow to adequately handle the traffic volume. Second, the sections of highway with functionally deficient bridges did not qualify for Federal or State funding for resurfacing work.

This case study covers three bridges, all located on the same section of a county highway. The three bridges were similar in construction, crossed the same river, and had been constructed at the same time. The bridges were single, simple-span concrete structures. The span for each was a cast-in-place concrete slab. The curb and railing were also concrete but were not monolithic with the deck slab. Abutments were concrete, with load-bearing piles in the back walls and wing walls. The back wall extended from the outside edges of the deck about 2 feet on each side. The wing walls flared from the back wall at a 45 degree angle. The bridges were structurally and hydraulically adequate but were deficient in roadway width. The vertical alignment of the bridges and highway was less than desirable, as each bridge was placed at the low point of a very short sag vertical curve. Horizontal alignment was satisfactory.

The county explored all the options that could be applied. The "do nothing" option was immediately ruled out, as any continuation of the present conditions was unacceptable. Complete replacement of the structures, the most costly solution, was considered. The two replacement alternatives, building a culvert or a bridge at each location, would require the construction of a temporary bypass at two of the locations. The third option, widening the existing structures, offered the best solution.

The most economical method of widening the bridges was then addressed by the county. The proposal to match the original type of construction, cast-in-place concrete, was analyzed from both engineering and cost of construction perspectives. A concrete deck supported on steel stringers was considered. Another option investigated was the use of a prefabricated treated-timber deck. This option was less costly than any of the others considered and was selected by Pepin County.

The first step was to measure the existing bridges to produce accurate as-built plans. These plans were then used by the County Engineer to design the rehabilitation project. The deck was designed as a longitudinal lumber deck, mechanically laminated with dowel-like spikes. The basic concept for this type of widening is to make use of the extra width of the abutment back wall, including a portion of the wing walls, to support the additional deck width.

After the plans and proposals were completed, a contract was awarded for the actual construction work. This was a furnish and install type of contract.

The contractor began the project by removing the concrete curbs and railings. A construction joint between the curb and the deck slab facilitated removal (most bridges of this type have a construction joint that facilitates this type of rehabilitation).

A bearing area for the new panels was constructed on the portions of the abutments and wing walls that were cut down. The elevation of the bearing areas was established so that the top elevation of the timber deck panels would be the same as the top of the concrete slab. The wing walls were raised to retain the additional fill needed to widen the grade.

Holes for anchor bolts were drilled into the outside edge of the existing concrete bridge deck. Cinch-type anchors and galvanized machine bolts were used to attach the first timber plank, measuring 4 inches thick by 14 inches wide, to the existing bridge. A splice plank was next attached to the this first plank. The splice plank was one-half the depth of the deck panels to create a ship-lap joint between the first deck lamination and the remainder of the additional new deck section.

A sequence of photos describing the project is presented in Figures 15-18 through 15-26.

Figure 15-18. - Typical concrete bridge on Pepin County Trunk Highway Z before rehabilitation. The bridges were adequate hydraulically and structurally, but did not meet minimum standards for roadway width.

Figure 15-19. - The first step in the rehabilitation was removal of the concrete curb and railing. Next, a concrete seat was poured on the wing walls to support the additional deck width. This bridge also had 2 feet of overburden removed as part of the contract.

Figure 15-20. - A 4-inch-thick by 14-inch-wide creosote-treated Douglas-fir plank was attached to the edge of the existing concrete deck with galvanized machine bolts and cinch type anchors. The ends of this plank rest on the concrete placed on the wing walls. A 4-inch-thick by 7-inch-wide splice plank was spiked to the bottom half of the plank to form a ship-lap-type joint with the timber deck panel.

Figure 15-21. - Anchor bolts are placed in the concrete wingwalls that match predrilled holes in the timber deck panels.

Figure 15-22. - Rail posts are attached to the deck panels before the panels are placed on the abutments.

Figure 15-23. - A prefabricated deck panel is lifted onto the abutments. Each panel was prefabricated to fit each bridge.

Figure 15-24. - A panel is lowered over the anchor bolts.

Figure 15-25. - A panel in final position. The ship-lap joint between the deck panel and first plank are interconnected with 5/8-inch-diameter by 13-inch-long drive spikes placed vertically through the joint.

Figure 15-26. - The completed project with widened roadway, ready for paving.

CASE HISTORY 15.4-
UNION COUNTY COVERED BRIDGE REHABILITATION

Contributed by Jeff Stauch, Civil Engineer, Union County, Ohio

During the 1800's, more than 3,500 wooden covered bridges were built in Ohio. Many different types and designs made up the once-abundant population of Ohio's covered bridges, of which only about 145 remain. Some of the remaining structures must be completely replaced, others are being moved to local fairgrounds or parks to be used as pedestrian crossings, and in some cases new bridges are being built alongside the old to divert all traffic away from the existing structures. But the ideal preservation practice involves rehabilitation of the existing bridge, leaving it in place with the ability to carry modern loads, to remain a part of the local transportation system.

In Union County, Ohio, located in the central part of the State, five covered bridges remain, four of which are an integral part of the county road system. The County Engineer and Commissioners have recognized the importance of preserving these structures. The decision was made to upgrade each of these four bridges, rehabilitating one every year with Union County forces.

The first candidate was chosen based on its low traffic volume (dead-end road) and generally poor condition throughout the bridge. This truss, spanning 95 feet, had a noticeable twist caused primarily by nearly broken

15-22

lower chords at opposite corners. The ends of some diagonals and lower chords were decayed and crushed from years of termite attack and general deterioration. Two steel piers were placed under the bridge in the 1950's, along with other various supports added in attempts to keep the bridge standing. An accurate analysis of the bridge was nearly impossible because of the unique design, the poor condition of the truss, and all the supports installed over the years. The bridge had a posted load limit of only 3 tons.

Various design options were considered, many of which would have worked well. Most centered around a bridge-within-a-bridge concept, where the existing floor system would be removed and replaced by a system that would remain independent of the wooden truss. This concept was especially attractive to us because of the uncertainty of the live-load capabilities of the old truss. Armed with this central idea, other more specific design parameters were formulated, including the following:

1. The waterway adequacy must not be constricted by the improvement.
2. Bridge capacity must be increased to handle a two-axle, 18-ton fire truck.
3. Timber will be used in the improvement for aesthetic compatibility.
4. The new system will help support the truss against further sag and twist, and straighten the truss.
5. All construction will be performed by Union County crews.
6. One of the piers cannot be relied on because of possible foundation problems.
7. Original appearance must be maintained as much as possible.
8. The project must meet economic criteria; 50 other structures in the county are load reduced, and also need attention.

A final design solution was selected based on a great deal of discussion, preliminary design calculations and sketches, and help from Ashtabula County (Ohio) Engineer John Smolen, whose covered bridge rehabilitation and construction programs are known nationwide.

Two large glulam girders (10-3/4 inches wide by 42 inches deep) were set inside the bridge at roadway elevation, and transverse glulam floorbeams (5-1/8 inches wide by 14-1/4 inches deep) were hung from the girders with steel rods. The glulam members, fabricated from Southern Pine, were pressure treated with pentachlorophenol in a heavy oil. A longitudinal, nail-laminated lumber floor was then placed on the floorbeams. Because only one existing pier could be used, two unequal simple spans were necessary; one 60 feet long and the other 34 feet long. The steel hanger assembly for floorbeams consisted of 3/4-inch-diameter threaded steel rods (ASTM A108) and 3-1/2-inch by 3/8-inch steel angles (ASTM A36). The floorbeams were placed 30 inches on-center and were extended beyond the girders underneath the lower truss chords to help straighten the chords by drawing up the beams with the threaded rods. This configuration did not help the trusses as much as expected because their condition was worse than originally thought. However, it was expected that the floorbeams would lend a great deal of support.

Eventually, the decision was made to repair all four truss corners, especially the lower chords, with new poplar timbers. Once this was done the truss squared up nicely.

The design for loading, in excess of AASHTO H 15-44 loading, was based on the current AASHTO, NDS, and AITC specifications. The floorbeam spacing (30 inches on-center) was a result of the AASHTO wheel-load distribution guidelines.

In retrospect, the project was a success. A covered bridge was saved and left in service. Some historians and covered bridge purists may argue the methods used, or question the authenticity or aesthetic value that remains, but there is probably no perfect or absolutely correct way to improve these bridges' deficiencies and still preserve them. Too many factors are involved to ideally address each problem area of the bridge. It tends to become a give-and-take exercise. Various phases of this project are depicted in Figures 15-27 through 15-37.

Figure 15-27. - Covered bridge on Winget Road before rehabilitation work began. Note the excessive rack and twist of the truss.

Figure 15-28. – Another view of the bridge before rehabilitation. One objective of the project was to maintain the bridge's appearance.

Figure 15-29. – Support for the roadway was obtained by installing two pressure-treated glulam girders inside the existing bridge, one along each side of the roadway.

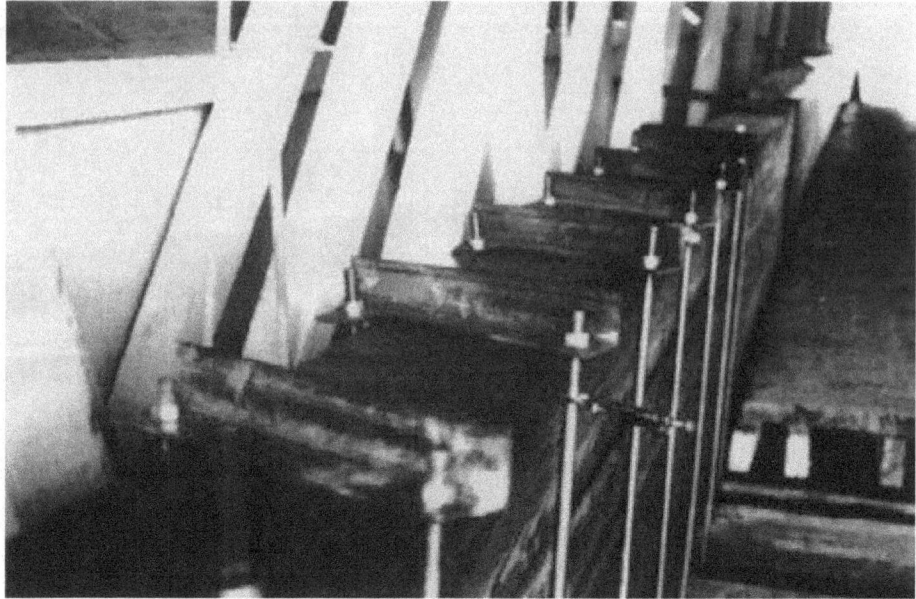

Figure 15-30. - Transverse glulam floorbeams were suspended from the tops of the girders by steel-rod hangers.

Figure 15-31. - Floorbeams are installed on the hangers from the bridge underside.

Figure 15-32. - Severe deterioration at each corner of the bridge was one cause of the poor truss alignment. New floorbeams were extended beyond the girders to help support each truss.

Figure 15-33. - Broken lower chords at two of the four corners prompted the replacement of some existing truss members.

Figure 15-34. – Completed repair of one corner of the bridge, using poplar timbers. Note the closer spacing of the endmost floorbeams to account for the reduced floorbeam load distribution adjacent to the abutment.

Figure 15-35. – A view inside the bridge showing the glulam girders, floorbeams, and longitudinal nail-laminated lumber floor. The existing floor, still in place at the far end of the roadway, was removed in sections to facilitate construction.

Figure 15-36. - The final result of the bridge-within-a-bridge concept. Four kneebraces to the roof system, which had been removed over the years, were reinstalled for stability and appearance. Note the improved truss alignment.

Figure 15-37. - The completed rehabilitation results in a significantly increased load capacity (18 tons), while maintaining the historic bridge's aesthetic appeal.

Planning for a second restoration is under way for a shorter (63-foot span) structure. Several improvements have been incorporated into the design, both aesthetic and structural. The hanger system will be totally hidden, connecting the floor system with the girders through holes bored along the centerline of the girders. In addition, a panelized transverse glulam deck with a plank wearing surface will replace the floorbeam and nail-laminated lumber floor system used before. The combination of 10-3/4-inch wide by 48-inch deep glulam girders (single span) and an 8-3/4-inch-thick glulam deck will permit AASHTO H 15-44 loadings. Use of the glulam deck will also increase the vertical clearance within the structure, without decreasing the waterway opening.

CASE HISTORY 15.5-
REHABILITATION OF NAIL-LAMINATED TIMBER DECKS BY TRANSVERSE STRESSING

Contributed by Raymond J. Taylor, Associate Research Engineer, Ministry of Transportation, Ontario, Canada

The concept of transverse prestressing was developed in 1976 as a method of rehabilitating deteriorated nailed-laminated wood bridge decks. Since that time, it has been developed as a totally new form of wood bridge deck design through considerable research and development. The current design specifications in the Ontario Highway Bridge Design Code (OHBDC) cover both the rehabilitation of old nail-laminated decks and the design of new ones. Since the first bridge was rehabilitated in 1976, this method of repair has been applied nearly a dozen times. This brief summary describes the field operations involved in the rehabilitation of nailed-laminated timber decks in Ontario.

Figure 15-38 displays the common problem associated with the delamination of nailed-laminated timber bridge decks. Their inability to maintain an unbroken asphalt wearing surface makes them a constant maintenance problem. To correct the problem, the deck is squeezed back together by applying pressure perpendicular to the laminations. This creates adequate friction between the lamination surfaces to reinstate load sharing and prevent breakup of the wearing surface.

Figure 15-38. - Breakup of asphalt pavement because of delamination of the nail-laminated timber deck.

The typical detail of the prestressing system as specified in the OHBDC is shown in Figure 15-39. The arrangement consists of a pair of high-strength bars attached to steel anchorage plates. This acts like a large flexible clamping system. The anchorage plates bear against steel channel bulkheads, which run the full length of the deck, and help to distribute the high prestressing forces, preventing local crushing of the wood.

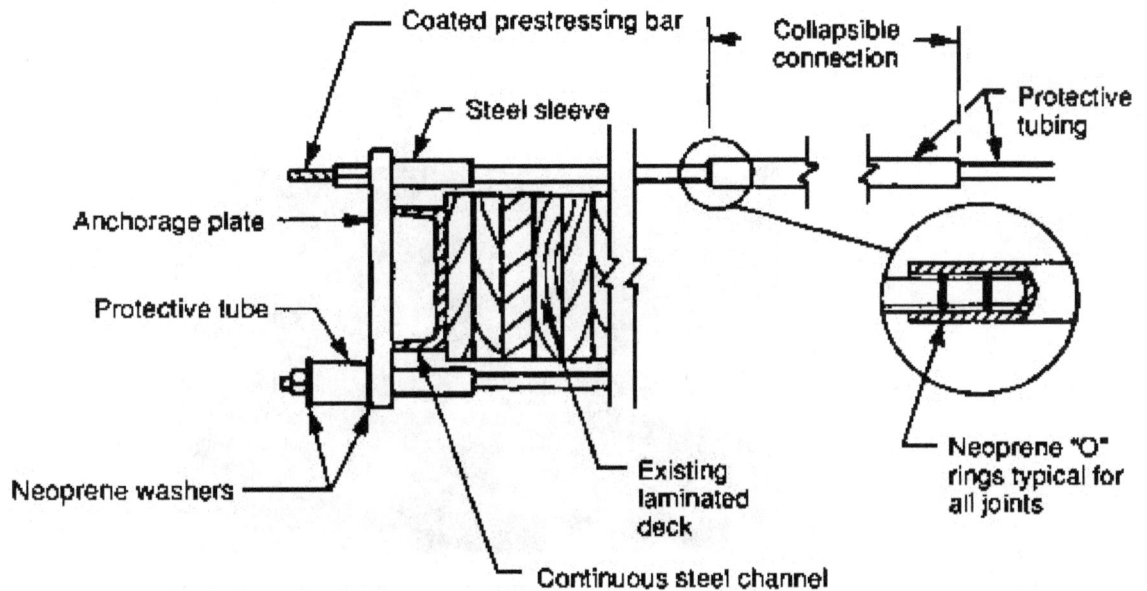

Figure 15-39. - Prestressing detail for rehabilitation of existing decks.

Figure 15-40 displays the anchorage detail used at the Pickerel River Bridge near Thunder Bay, Ontario. The bars were galvanized and enclosed in protective polyvinyl chloride (PVC) tubing to guard against deterioration. Several new wood laminations were added to offset the narrowing of the deck as it was squeezed transversely. Generally, about 1 to 2 percent of the initial bridge width is added depending on the extent of separation that has occurred in the existing deck.

Figure 15-40. - Prestressing anchorage used at the Pickerel River Bridge.

The joint between the PVC pipe and the steel sleeve that extends from the steel anchorage plate (Figure 15-40) is sealed with neoprene O-rings. The same O-ring joint is used at the collapsible connection shown in Figure 15-39. Figure 15-41 displays these O-rings, before assembly, as applied at the Pickerel River Bridge.

Figure 15-41. - PVC sleeve connection with neoprene O-rings.

The installation of the prestressing system is usually facilitated by the complete removal of the asphalt surface. However, in several bridge rehabilitation projects, the top bars were placed in transverse grooves cut in the asphalt surface (Figure 15-42). This was done to maintain traffic but was not considered to be successful because the asphalt cutting took considerable time and the final asphalt surface was badly deteriorated by traffic. To date, the best method of maintaining traffic during the stressing

operation has been the installation of a temporary plank surface
(Figure 15-43). The two-way plank system includes a bottom layer
parallel to the bars spaced so that the bars can be installed beneath the
running planks. Maintenance of the plank surface has not been a problem.

Figure 15-42. - Top prestressing bars installed in grooves cut in the asphalt wearing
surface of the North Pagwatchuan River Bridge near Terrace Bay, Ontario.

Figure 15-43. - Temporary plank surface placed over prestressing bars during construction
at the Pickerel River Bridge.

Figure 15-44. - Multijack hydraulic system used at the North Pagwatchuan River Bridge.

The actual stressing of the bars was originally performed using only a pair of hydraulic jacks so that only the two bars at one location could be stressed simultaneously. This did not prove to be efficient, as the initial stressings at each position had only local effects and the stressing of one pair of bars (station) would loosen the adjacent station. On the Hebert Creek Bridge in 1976, twelve passes along the bridge were required to reach an acceptable level of stress in all bars. Today, the stressing is performed using a multijack hydraulic system (Figure 15-44). This 24-jack system uses 530-kN (60-ton) capacity jacks with steel back-up plates that allow each jack to stress one pair of bars at the same time. These jacks are of hollow cylinder design, so they can also be used for single-bar construction as used in new decks.

Stressing an existing deck has not resulted in any visible distress at the bearing support of the laminations. The original toenailing of the laminations may have deteriorated or may simply provide little resistance to the transverse movement of the laminates. In any case, no repairs have been necessary other than tying down the final deck to the supports. Simple steel rods and plates are used to clamp the deck to the supporting member (Figure 15-45).

Figure 15-45. - Tie-down of the deck to supports at the Kabaigon River Bridge.

A number of bridges rehabilitated by transverse prestressing have been load tested. Typical test results are displayed in Figure 15-46, which compares midspan deflections on the Hebert Creek Bridge before and after transverse prestressing. These results demonstrate the benefits of rehabilitating old nail-laminated decks by this method; load distribution is improved, making the bridge stronger than its original design.

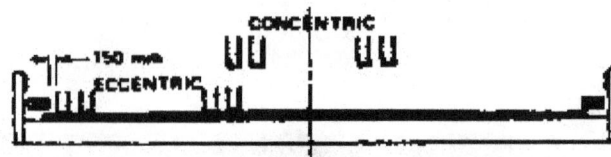

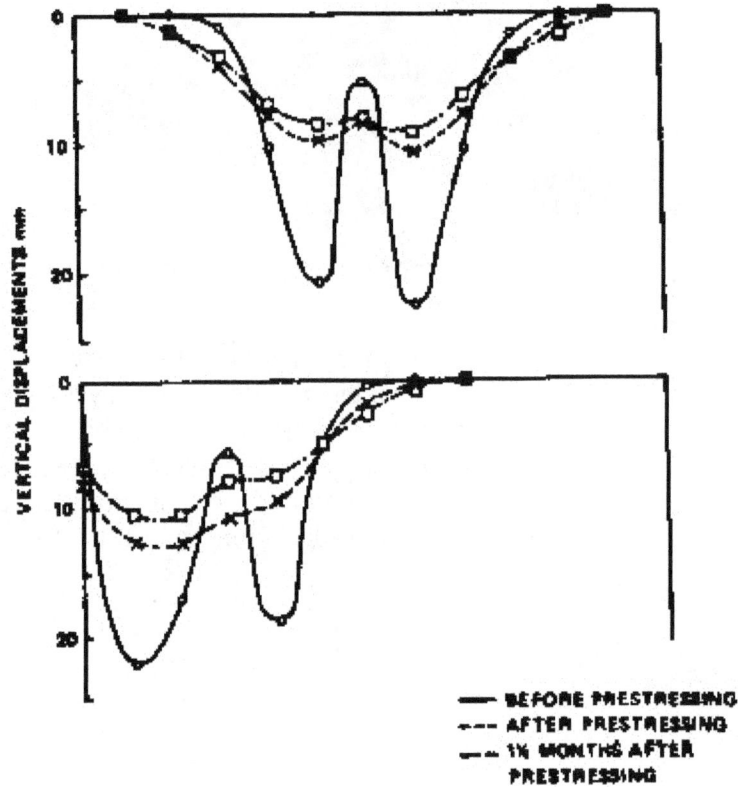

Figure 15-46. - Load test results at the Hebert Creek Bridge, before and after transverse prestressing was applied.

Contributed by Ken Johnson, Civil Engineer, Wheeler Consolidated, Inc.

Sauk County, a rural county in south-central Wisconsin, contains a portion of the prime recreation area around the Wisconsin Dells. This is one of the major tourist attractions in the upper Midwest. The major arterial route serving this area is County Trunk Highway A, which has the highest average daily traffic of all county trunks in Sauk County. The route is carried over the narrows of Mirror Lake on an overhead steel truss bridge. The 96-foot structure was originally constructed with a cast-in-place concrete deck.

Annual inspections of the structure revealed serious deterioration of the deck, which had been overlayed with a bituminous wearing course. The use of de-icing chemicals and the freeze-thaw cycles accelerated the rate of deterioration. The bridge was posted for restricted loads because of the condition of the deck and the additional dead load from the bituminous overlay. Sauk County was faced with the problem of eliminating this hazard. Their options were to replace the entire structure; replace the existing concrete deck, in kind; or replace the existing deck with some other type of construction.

Selection of the best option was easy for the county once each option was fully evaluated. Replacement of the entire structure, the most costly option, was not selected because the main structural components of the existing bridge were still adequate. The main truss, the floor beams, and the steel stringers were in excellent condition. The structure was hydraulically adequate and had sufficient roadway width. The choice narrowed to replacement of the deck.

The choice of replacing the concrete deck with one constructed of treated timber was made based on cost, ease of construction, time required for construction, weather restraints for construction, and deck weight. The estimated cost for the treated timber deck was considerably less than the estimated cost for replacement with a concrete deck. The treated timber deck could be placed by Sauk County's highway crew using existing county equipment and could proceed during wet or freezing weather (the highway could not be closed during the summer). The entire project could be completed within 1 week, which was an important consideration because there was no reasonable detour. In addition, the dead load of the timber deck would be only one-third that of concrete, which would substantially increase the bridge live load capacity. All factors indicated that replacement of the concrete deck with one constructed of treated timber was the most prudent decision for Sauk County.

The county provided the timber vendor with an as-built plan of the structure. The vendor provided detailed construction plans as part of the material purchase price. A transverse laminated lumber deck was pre-fabricated into panels that were matched to the lifting capacity of the county equipment.

Work on the project was started on Monday, November 15, 1982, and the bridge was completed and open for traffic on Friday, November 19, 1982. The concrete deck was removed in sections approximately 5-feet square. Once a portion of the old deck was removed, the new timber deck panels were placed. The panels were prefabricated so as to be placed without removing any of the bridge railing. The transverse panels were attached to the steel stringers using a 5/8-inch-diameter domehead bolt with an offset shoe that gripped the flange of the stringer. A compression spring was placed between the nut and the offset shoe. This type of hardware compensates for changes in deck thickness from moisture changes. The uncompressed length of the compression spring is 1-1/2 inches and the compressed length is 7/8 inch.

The completed timber deck was surfaced with a bituminous wearing course. The entire project was completed in 5 working days and the total cost to Sauk County was $31,184.44, which is $12.49/ft^2 of deck. The following is a breakdown of the total project cost:

Material delivered to jobsite	$21,621.37
Labor	4,396.29
Equipment	2,942.26
Labor overhead costs	2,224.52
TOTAL PROJECT COST	$31,184.44

Figures 15-47 through 15-54 present a sequence of photos describing the project.

Figure 15-47. - The existing bridge was a steel truss, 96 feet long, with a 26-foot roadway width. The noncomposite concrete deck was supported by steel stringers (WF16 x 50) spaced 4 feet 6 inches on center. The stringers were supported by steel floor beams (WF30 x 116).

Figure 15-48. - The concrete deck was removed in approximately 5-foot-square sections by the Sauk County Highway maintenance crew. The concrete was broken using jackhammers, and reinforcing bars were flame cut. The sections were lifted and loaded on the trucks with a hydraulic excavator.

Figure 15-49. - Exposed steel floorbeams and stringers as deck removal proceeds.

Figure 15-50. - New, prefabricated treated-timber deck panels were placed with a hydraulic excavator. The panels were designed to fit transversely on the stringers and were placed without removing the existing railing.

Figure 15-51. - Deck replacement proceeded in stages along the bridge length. After a portion of the concrete was removed, timber deck sections were placed. Steel straps were welded to the ends of the stringers (across floorbeams) to provide continuous support for the timber deck panels.

Figure 15-52. - Timber deck panels were provided with four eye-bolts and were banded with straps to facilitate shipping and handling. Panels were fabricated from 3-inch-thick, 6-inch-wide (S1S) creosote-treated Douglas-fir lumber, and were 6 feet 3 inches wide. The individual lumber laminations were laminated together with 5/16-inch-diameter, 8-inch-long ring-shank spikes. Ship-lap-type joints were used between panels, with 1/2-inch-diameter by 5-inch-long drive spikes placed vertically through the joints.

Figure 15-53. - Panels were attached to the stringers with 3/4-inch-diameter by 9-inch-long machine bolts, using a 3/4-inch offset shoe and spring. The offset shoe grips the flange of the stringer, and the compression spring compensates for changes in deck thickness from moisture changes in the timber.

Figure 15-54. - The completed treated-timber deck, before placement of an asphalt-pavement wearing surface.

CASE HISTORY 15.7-
BRUNEAU RIVER BRIDGE REHABILITATION

Contributed by Steve Bunnell, Regional Bridge Engineer, USDA Forest Service, Intermountain Region

The Bruneau River Bridge is located in north-central Nevada. Although located in a remote area, it is a vital link in the back-country traffic patterns in northern Nevada. The bridge, constructed in 1953, was originally 74 feet long and consisted of two 14-foot side spans of longitudinal nail-laminated lumber decking and a 46-foot steel-beam center span with a transverse nail-laminated lumber deck. The abutments were timber cap and bulkhead-type, and the piers were reinforced concrete contained by corrugated metal pipe on spread footings.

During the spring of 1984, flooding occurred throughout northern Nevada because of a record snowfall the previous winter. A small stream, Meadow Creek, has its confluence with the Bruneau River just upstream from the bridge site. Extreme high flow in Meadow Creek forced the main flow in the Bruneau River against the north abutment, breaching the north approaches and the abutment-supporting foundation material, and the north approach span fell into the stream. No other damage to the bridge occurred from this incident. However, previous technical inspections of this bridge identified that the transverse nail-laminated lumber decking was reaching the end of its expected life, and that there were some structural problems with the south bulkhead abutment.

Immediate temporary repair measures consisted of replacing the approach embankment and raising the existing span back in place to restore traffic on the bridge. Analysis in preparation for a permanent repair to the bridge determined that the existing river piers and the steel portions of the super-structure were structurally adequate and had sufficient life remaining for all alternatives considered. Hence, a decision was made to rehabilitate the bridge rather than replace it.

The rehabilitation proposal consisted of replacing the south approach span with four longitudinal glulam deck panels, 8-3/4 inches thick by 3 feet 10-1/2 inches wide by 18 feet 8 inches long, and lengthening the span from 14 feet to 19 feet. The longitudinal panel interfaces were doweled with 1-1/8-inch-diameter steel dowels. The north approach span was lengthened from 14 feet to 50 feet by using two steel beams (and bracing) salvaged from another dismantled bridge and secured by the contractor as government-furnished material. The two new steel beams were connected to the existing steel beams by a field-pinned connection because there was not sufficient room on the existing cap for an additional bearing. The existing bearing assembly was adequate to carry the additional reaction. The existing nail-laminated lumber deck on the steel girder span was removed, and 24 new treated-timber transverse glulam deck panels, 6-3/4 inches thick by 4 feet wide by 15 feet 6 inches long, were placed on the remaining portion of the bridge, including the new steel girder span. Each panel interface was doweled with fifteen 1-1/4-inch-diameter steel dowels. Both abutments were replaced with treated-timber bearing caps and bulkhead-type abutments. This new design resulted in an increase in traveled-way width from 12 feet to 14 feet 4 inches. Treated-timber glulam curb/wheel guards and a bituminous wearing surface were included in the contract proposal.

In September 1985, a contract was awarded to Thorton Construction of Burley, Idaho, for $65,103.80, which included the cleaning and painting of the steel work. The contract was completed on schedule in August 1986. The new facility provides an additional hydraulic opening of 36 feet on the north side, which may preclude damage from incidents similar to the 1984 flooding; it also provides an additional 2 feet 4 inches of traveled way on the deck. Photos of the project are shown in Figures 15-55 through 15-65.

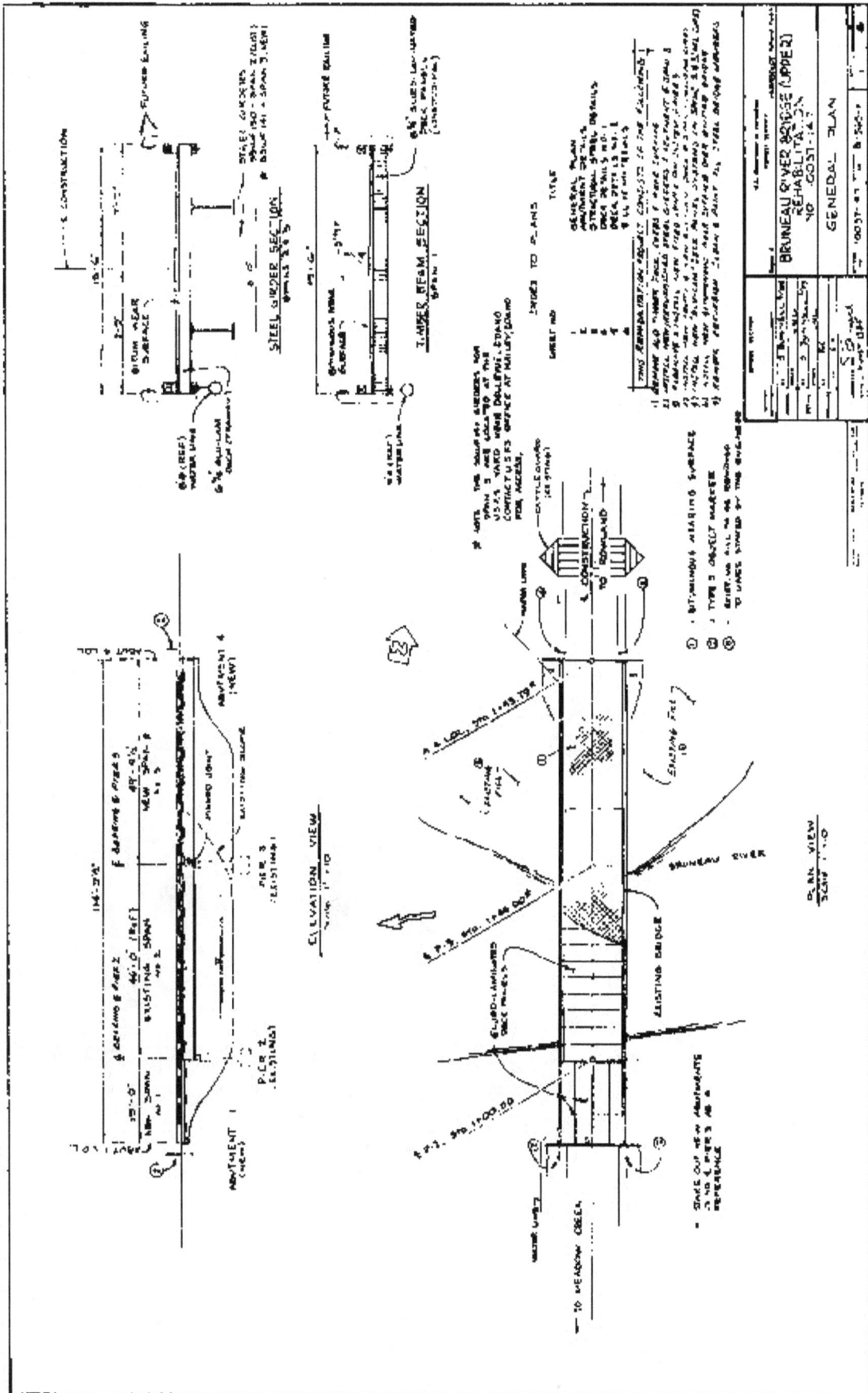

Figure 15-55.—General plan for the Bruneau River Bridge rehabilitation.

Figure 15-56. - Collapsed span and breached approaches of the existing bridge.

Figure 15-57. - Close-up of the collapsed longitudinal nail-laminated lumber approach span.

Figure 15-58. – The temporary repair made to the bridge consisted of replacing the approach embankment and lifting the nail-laminated approach span back into position.

Figure 15-59. – A view of the bridge as rehabilitation began, with the deck, curbs, and timber abutments removed.

Figure 15-60. – Blast cleaning of the existing steel work in preparation for painting, while a backhoe removes material that was placed for the temporary repair.

Figure 15-61. – Steel beam extensions in place and pinned to the existing beams.

Figure 15-62. - A portion of the final paint coat has been applied to the steel and the new treated-timber abutments are in-place. The backhoe is positioned to lift the longitudinal glulam panels for the far side approach span.

Figure 15-63. - Transverse glulam deck panels are lifted into place.

Figure 15-64. - Backhoe pushes panels together after dowels have been aligned. A large timber block was placed across the panel edge to prevent damage from the backhoe bucket.

Figure 15-65. - The completed bridge with glulam curb/wheel guards and asphalt wearing surface in place.

CASE HISTORY 15.8-
UINTA CANYON CANAL BRIDGE REPLACEMENT

Contributed by Steve Bunnell, Regional Bridge Engineer, USDA Forest Service, Intermountain Region

The Uinta Canyon Canal Bridge is located on the Ashley National Forest near Roosevelt, Utah. It serves mostly recreational traffic on the south slope of the Uinta Mountains. A recent timber sale necessitated an increase in the load capacity of the Uinta Canyon Canal Bridge.

The existing bridge consisted of a single-lane longitudinal nail-laminated lumber deck bridge, 26 feet long with a 27-degree skew, on treated timber abutments. The existing deck was severely delaminated, with openings of 1 to 2 inches between laminations. Previous technical inspection reports documented the delaminating condition, but revealed that the treated timber abutments had remaining life and were structurally adequate to support the timber-sale loads. Hence, a decision was made to replace just the deck and curb portions of the bridge as a measure to rehabilitate the bridge for the heavier loads.

The contract proposal was to replace the deck with longitudinal glulam deck panels and wheel guard/curbs. Design requirements showed that depths of single laminates normally used (10-3/4 inches, 8-3/4 inches, 6-3/4 inches, etc.) were not adequate for the existing 26-foot span. A check with a glulam supplier revealed that using edge-glued laminations consisting of two pieces, a 2 by 6 and a 2 by 8 (surfaced), placed alternately top and bottom to build up 14-inch-deep panel laminations (surfaced to 12-1/4 inches), would provide adequate structure depth to span the 26 feet.

Four pressure-treated glulam deck panels, measuring 12-1/4 inches deep, 4 feet wide and 26 feet long, were proposed, with glulam curbs measuring 6-3/4 inches by 12 inches. Panel interfaces were connected with 1-1/2-inch-diameter dowels. In September 1986, a contract was awarded to Niedermeyer-Martin Company of Portland, Oregon, for $9,430 to supply the treated-timber members. The bridge construction was performed by Ashley National Forest maintenance crews. Because of the skew and the interference of the existing abutment backwalls, the panels could not be jacked together in place because the dowels were to be placed normal to panel interfaces. The four deck panels were assembled together on two well casings in a staging area adjacent to the bridge and the entire deck unit was jacked together. The assembled deck was erected as one complete unit, 16 feet wide and 26 feet long. Sling connections were attached to the well casings and used to erect the unit with a crane. The new deck fit perfectly on the old abutments. The panels were then connected to the existing caps, the wheel guard/curbs were installed, and a temporary surface was placed on the deck.

The timber members were delivered to the site on Monday, November 24. All the construction was completed and traffic was restored across the bridge at 5 p.m. on Tuesday, November 25. When moderate weather permits, the temporary aggregate wearing surface will be removed and replaced with a permanent bituminous wearing surface.

The use of the glulam panels allowed salvaging the existing abutments and increasing the load capacity of the bridge. In addition, the traveled way was increased from 12 feet to 14 feet 4 inches. The new deck is free from any fasteners that penetrate the treated wood, which ensures improved protection against decay and increased life over nail-laminated construction. Assembly was made quick by simple, efficient, state-of-the-art equipment.

The project sequence is presented in Figures 15-66 through 15-75.

Figure 15-66. - The longitudinal nail-laminated lumber-deck bridge before replacement. The bridge was 26 feet long with a 27-degree skew.

Figure 15-67. – The existing deck was carefully removed in sections to avoid damage to the treated-timber abutment caps.

Figure 15-68. – A view of the abutments after the bridge was removed.

Figure 15-69. - The first two panels of the new bridge are interconnected with steel dowels and pulled together with a come-a-long. Because the crossing was skewed, the doweled panels could not be assembled on the abutments. Assembly was completed adjacent to the site on two well casings to facilitate assembly and lifting.

Figure 15-70. - Another view of the first two panels being pulled together. Note the corner protectors under the come-a-long chains and the can of roofing cement on the deck. All panel joints were sealed with the roofing cement before being pulled together.

Figure 15-71. - Steel dowels are aligned before joining the fourth panel. The dowels are 1-1/2 inches in diameter, 19-1/2 inches long, and are spaced 12 inches on center.

Figure 15-72. - The completed deck, resting on the well casings, is lifted into place on the abutments. Note that the curbs have been preassembled adjacent to site.

Figure 15-73. - The replacement deck in place.

Figure 15-74. - Curbs are attached to the deck.

Figure 15-75. - The completed structure with a temporary aggregate wearing surface. The deck was later provided with an asphalt pavement wearing surface.

CASE HISTORY 15.9
COOK COUNTY BRIDGE REPLACEMENT

Contributed by Michael G. Oliva, Associate Professor of Structural Engineering, University of Wisconsin, Madison

The Gunflint Trail is a historic route used by early fur traders to transport their goods from the northwestern territories to the shores of Lake Superior. It is currently a paved county road in Cook County, Minnesota, and is used primarily for access to the Boundary Waters Canoe Area and the Superior National Forest. A bridge located over the Cross River on a gravel-surfaced Gunflint Trail access road consisted of a two-span timber beam superstructure supported by timber cribbing constructed of logs. The superstructure consisted of sawn lumber beams that were judged to be untreated with preservatives, and a transverse plank deck spiked to the beams (Figure 15-76). The wearing surface was provided by lumber running planks. The bridge was structurally and functionally deficient and lacked curbs and railing (Figure 15-77). The increased recreational traffic and existing deficiencies required full replacement of both the abutments and superstructure.

Figure 15-76. - The existing single-lane bridge was a simply supported, two-span, untreated-timber beam bridge supported by log cribbing (photo courtesy of Wheeler Consolidated, Inc.).

Figure 15-77. - The existing bridge was structurally deficient and load-posted. In addition, it was too narrow and lacked curbs and railing (photo courtesy of Wheeler Consolidated, Inc.).

The bridge was located on the sole access route to a series of recreational and sporting locations on Gunflint Lake and was being used for fishing in the Cross River. Cook County's primary concern in planning a replacement bridge was economy, accompanied by a desire to increase load capacity and to widen the roadway to accommodate fishing activities. Wheeler Consolidated, Inc., a Midwest timber bridge supplier, used the opportunity to attempt construction of one of the first longitudinal stress-laminated lumber bridges in this country and provided cost incentives to Cook County for erection of this prototype structure.

The new bridge design consisted of a single-span longitudinal stress-laminated lumber deck measuring 44 feet long, 18 feet wide, and 16 inches thick. The deck was fabricated of nominal 4-inch wide by 16-inch deep Douglas Fir-Larch lumber laminations, visually graded No. 1 or better. The laminations were pressure treated with creosote and were 4 feet to 20 feet in length. Butt jointing of the laminations was achieved through the stress-laminating process, using 1-inch-diameter high-strength steel prestressing rods placed transversely through the center of the laminations at 4-foot intervals. The replacement bridge was supported on two new creosote-treated timber crib abutments. The bridge was still intended for one lane of traffic, but was widened to accommodate pedestrians, and included curbs and railing. The hydraulic configuration was unchanged although the center crib was no longer needed.

The entire replacement project was completed by the Cook County road crew of seven men, along with a hired crane and loader with operators. The timber cribbing materials for the abutments were precut and treated (with markings to indicate location of each piece) and were shipped to the Cook County shop. The county started work on March 11 during a lull in normal snow-removal activities. The old superstructure was removed using two backhoes and incinerated at the site. Explosives were used to remove the old cribbing abutments from the frozen ground, but the center support could not be removed because it was frozen in the lake. The new cribs were constructed in place using the precut treated Douglas Fir-Larch (Figure 15-78). Each of the cribs was backfilled with rock, granular material, and soil.

Figure 15-78.-The new crib abutment under construction using precut and creosote-treated Douglas-fir timbers. The only field fabrication involved drilling holes in the timbers for drive spikes used for connections. Such penetration of the protective envelope should have been avoided by drilling before pressure treatment (photo courtesy of Wheeler Consolidated, Inc.).

The county had arranged delivery of the replacement bridge to occur at the start of construction. The deck was prefabricated and prestressed in two stress-laminated panels, each 9 feet wide and 44 feet long with 2 inches of upward camber. The two panels were placed in adjacent positions over the span after the cribbing was completed (Figure 15-79). The two separate panels had to be connected to achieve the desired longitudinal deck action. A special coupling process was used to stress-laminate the panels together using the existing stressing hardware in the separate panels.

Since previous research at the University of Wisconsin had shown that little stiffness loss would occur if the stress in a stress-laminated deck was reduced to one-half the design value, every other rod in either panel could be released without losing the integrity of the panel. Alternate rods were released in each panel, with a released rod in one panel being opposite to the unreleased rod in the other panel. The released rods were attached to the unreleased rods in the adjacent panel using special high-strength couplers (Figure 15-80). After all rods were coupled, the released rods were restressed and the two panels were pulled and laminated together by the prestress. Two separate stressing passes along the rods, restressing each individually, had to be completed to obtain relatively uniform forces in all the rods. The two prefabricated panels then formed a single integral bridge deck (Figure 15-81).

Figure 15-79. - One of the prefabricated, prestressed, bridge deck panels is lifted into position on the treated-timber crib abutments.

Figure 15-80. - Prestressing rods protruding from the two prefabricated deck panels before connection. The rod opposite each coupler was released and threaded into the coupler. After all couplings were made, the released rods were restressed, causing the deck panels to be stress-laminated together as an integral bridge deck.

Figure 15-81. - The completed longitudinal stress-laminated deck bridge replaced the existing two-span structure with a single span. The new deck increased load capacity to AASHTO HS 20-44 and included curbs and railing. Remains of the old center support are visible below the replacement structure (photo courtesy of Wheeler Consolidated, Inc.).

The entire sequence of cribbing erection, superstructure placement, and stressing was accomplished by the Cook County crew in 13 hours over a 2-day period. The roadway was then reopened for traffic, although curb and rail installation was not completed until the third day. The connection between the prefabricated panels would have been much easier if the alignment of the rods in each of the panels had been more carefully monitored during the plant fabrication. Misalignment in some of the bars resulted in difficulty during coupling. One bar that was not sufficiently threaded into the coupler pulled out during the restressing. Sufficient insertion into couplers, and locking bars in position in couplers, could have been achieved by placing lock nuts on the bars at correct locations before threading into the coupler. In addition, the exterior laminations had large natural defects such as splits and knots. These defects usually induced further splitting of the exterior laminations during the prestressing operation because of the concentrated compression forces transferred into the wood. The splitting did not detract from the structural performance but the aesthetic appearance would have been improved if better-quality lumber without defects had been used for exterior laminations. One of the exterior laminations at the end of the deck was only 4 feet long with a single prestressing rod inserted through its face and the lamination rotated slightly during the restressing. The minimum length of laminations should be such that each lamination has at least two rods through its face.

The erection of this prototype bridge proved that the longitudinal stress-laminated bridge could be prefabricated in panels and connected at the site to form an integral deck. The erection of such a bridge can proceed very rapidly, in this case in less than two full days. The bridge was covered with a gravel wearing surface after completion of the superstructure erection and attachment of curbs and rails. Observation of the behavior of the bridge during the 2 years since it was erected has shown that the original camber was insufficient to balance the dead-load deflections and time-related creep deflection has resulted in a permanent sag of the deck. The deflections indicate that the span was probably longer than should normally be used with decks of 16-inch thickness.

CHAPTER 17

GLOSSARY OF TERMS

Absorption, gross. Total amounts of preservative indicated in the wood at the termination of the pressure period.

Abutment. A substructure supporting the end of a single span or the extreme end of a multispan superstructure.

Aggregate. Sand, gravel, broken or crushed stone, or combinations thereof.

Air-dried. Wood dried by exposure to the atmosphere without artificial heat.

All Heart. Wood that is heartwood throughout, i.e., free of sapwood.

Allowable stress. The maximum allowable material stress used for the design of timber members. Allowable stress equals the tabulated stress adjusted by all applicable modification factors.

Anchor bolt. A bolt or boltlike piece of metal commonly threaded and fitted with a nut, used to secure the superstructure to the substructure.

Anisotropic. Not isotropic; that is, not having the same properties in all directions.

Annual growth ring. The layer of wood growth put on a tree during a single growing season.

Arch. In general, any structure having a curved shape, either actual or approximated, and producing at its supports reactions having both horizontal and vertical components.

> **Two-hinge arch.** An arch that is supported by a pinned connection at each support.

> **Three-hinge arch.** An arch with end supports pinned and a third hinge (or pin) located somewhere near midspan.

Assay. Determination, by appropriate physical and chemical means, of the amount of preservative or fire retardant in a sample of treated wood.

Axial combinations. Glulam members manufactured primarily for axial loads or bending loads applied parallel to the wide faces of the laminations.

Axle load. The total load transferred by one axle of a traffic vehicle.

Backfill. Material (soil or rock) placed behind and within the abutment and wingwalls to fill the unoccupied portion of the foundation excavation.

Backwall. The topmost portion of an abutment above the elevation of the bearings, functioning primarily as a retaining wall.

Bark pocket. A natural opening between annual growth rings that contains bark.

Batter. The inclination of a surface in relation to a horizontal or vertical plane.

Batter pile. A pile driven in an inclined position to resist forces acting in a direction other than vertical.

Beam. A structural member supporting a load applied transversely to it. Beams used in bridge construction include stringers, girders, and floorbeams.

Bearing. The assembly or connection between the superstructure and the substructure at the superstructure reactions.

> **Fixed bearing.** A type of bearing that does not allow longitudinal movement of the superstructure.

> **Expansion bearing.** A type of bearing that allows small longitudinal movements of the superstructure, generally those resulting from thermal expansion and contraction.

Bearing pad. A thin layer of material, generally elastomeric rubber, placed between the superstructure and the substructure to provide an even surface at the reaction and allow for longitudinal and rotational movement of the superstructure.

Bearing plate. A steel plate placed at the reaction of a structural member (beam, column, etc.) to distribute and transmit loads to supporting members.

Bending combinations. Glulam members manufactured primarily for bending loads applied perpendicular to the wide face of the laminations.

Bent. A type of pier consisting of two or more column or columnlike members connected at their top ends by a cap, strut, or other member holding them in their correct positions.

> **Pile bent.** A type of bent using timber piles as the column members.

> **Frame bent.** A type of bent using a timber frame as the column.

Bleeding. The secretion of liquid preservative from treated wood. The secreted preservative may evaporate, remain liquid, or harden into a semisolid or solid state.

Board foot. A unit of measurement of lumber represented by a board 1 foot long, 1 foot wide, and 1 inch thick, or its cubic equivalent. In practice, the board foot calculation for lumber 1 inch or more thick is based on its nominal thickness, width, and length. Lumber with a nominal thickness of less than 1 inch is calculated as 1 inch.

Boards. Lumber that is nominally less than 2 inches thick and 2 inches or more wide.

Bole. The main stem of a tree of substantial diameter capable of yielding sawtimber, veneer logs, or large poles.

Bound water. Water (or moisture) contained in the cell walls of wood.

Bow. The distortion of lumber in which there is a deviation, in a direction perpendicular to the flat face, from a straight line from end to end of the piece.

Boxed heart. The term used when the pith falls entirely within the four faces of a piece of wood anywhere in its length. Also called boxed pith.

Bracing. A system of tension and/or compression members that provides strength, support, or stability to beam, truss, or frame structures.

Branding. Permanent marking on a treated wood product to identify the supplier, date of treatment, and other information as specified.

Bridging. A carpentry term applied to wood cross bracing fastened between lumber beams.

Brown rot. In wood, any decay in which the attack concentrates on the cellulose and associated carbohydrates rather than on the lignin, producing a light to dark brown friable residue.

Brush curb. A narrow curb, 9 inches or less in width, that prevents a vehicle from brushing against the traffic railing.

Buckle. To fail by an inelastic change in alignment, usually a result of compressive stress.

Bulkhead. A retaining wall-like structure commonly composed of driven piles supporting a wall or barrier of wooden timbers functioning as a constraining structure resisting the thrust of earth or other material bearing against the assemblage.

Burl. A distortion of grain, usually caused by abnormal growth from injury of the tree.

Camber. A slight amount of convex curvature provided in a single span or in a multiple-span structure to compensate for dead load deflection and to secure a more substantial and aesthetic appearance than is obtained when uniformly straight lines are produced.

Cambium. A thin layer of tissue between the bark and wood that repeatedly subdivides to form new wood and bark cells.

Cap. A sawn lumber or glulam member placed horizontally on an abutment or pier to distribute the live load and dead load of the superstructure. Also a metal, wood, or mastic cover to protect exposed wood end grain from wetting.

Cellulose. The carbohydrate that is the principal constituent of wood and that forms the framework of the wood cells.

Charge. All the wood treated together in one cylinder or treating tank at one time.

Check. A lengthwise separation of the wood that usually extends across the rings of annual growth and commonly results from stress set up in wood during seasoning.

Chord. In a truss, the upper and lower longitudinal members, extending the full length and carrying the tensile and compressive forces that form the internal resisting moment.

Clear. Free or practically free of all blemishes and strength-reducing characteristics.

Clear span. The unobstructed space or distance between the substructure elements measured between faces of abutments and/or piers.

Column. A general term applying to a member resisting compressive stress and having, in general, a considerable length in comparison with its transverse dimensions.

Combination symbol. A designation used for glulam to indicate the combination of laminations used to manufacture the member.

Compression failure. Deformation of the wood fibers from excessive compression along the grain either in direct end compression or in bending. It may develop in standing trees or result from stresses imposed after the tree is cut. In surfaced lumber, compression failures may appear as fine wrinkles across the face of the piece.

Compression wood. Wood formed on the lower side of branches and inclined trunks of softwood trees.

Conditioning. The removal of moisture from unseasoned or partially seasoned wood.

Connector, timber. Metal rings, plates, or grids that are embedded in the wood of adjacent members to increase the strength of the joint.

Continuous spans. A beam or truss-type superstructure designed to extend continuously over one or more intermediate support.

Creep. An inelastic deformation that increases with time while the stress is constant.

Creosote. A wood preservative that is a distillate of coal tar produced by high-temperature carbonization of bituminous coal.

Crib. A structure consisting of a foundation grillage combined with a superimposed framework providing compartments that are filled with gravel, stones, or other material satisfactory for supporting the structure to be placed thereon.

Crook. The distortion of lumber in which there is a deviation, in a direction perpendicular to the edge, from a straight line from end to end of the piece.

Cross frames. Transverse bracing between two main longitudinal beams or other structural members.

Cross section. The surface obtained when cutting a log perpendicular to its long axis or a piece of wood perpendicular to the longitudinal direction.

Cup. A distortion of a board in which there is a deviation flatwise from a straight line across the width of the board.

Curb. A barrier paralleling the side limit of the roadway to guide the movement of vehicle wheels and protect railings or other elements outside the roadway limit.

Dead load. The static load imposed by the weight of the materials that make up the structure.

Decay. A disintegration of the wood substance from action of wood-destroying fungi.

> **Advanced (or typical) decay.** The older stage of decay in which the destruction is readily recognized because the wood has become punky, soft, spongy, stringy, pitted, or crumbly. Decided discoloration or bleaching of the rotted wood is often apparent.

> **Incipient decay.** The early stage of decay that has not proceeded far enough to soften or otherwise perceptibly impair the hardness of the wood. It is usually accompanied by a slight discoloration or bleaching of the wood.

Deck. That portion of a bridge which provides direct support for vehicular and pedestrian traffic. While normally distributing loads to a system of beams and stringers, a deck may also be the main supporting element of a bridge, as with a longitudinally laminated timber bridge or a stressed-deck system.

Deformation, elastic. Deformation from applied loads; when the loads are removed, the material will return to its original shape.

Deformation, inelastic. Deformation from applied loads; when the loads are removed, the material will not return to its original shape.

Delamination. The separation of layers in a laminate through failure within the adhesive or at the mechanical bond between laminae.

Density. As usually applied to wood of normal cellular form, density is the mass of wood substance enclosed within the boundary surfaces of a wood-plus-voids complex having unit volume.

Design load. The loading comprising magnitudes and distributions of all loads used in the determination of the stresses, stress distributions, and ultimately the cross-sectional areas and compositions of the various portions of a bridge structure.

Design stress. The stress produced in a member by the design loading.

Diaphragm. Blocking between two main longitudinal beams consisting of solid lumber or glued-laminated timber.

Dimensional stability. Resistance of wood to swelling (or shrinking) upon adsorption or loss of water.

Distribution factor. The fractional portion of the forces produced by one wheel line of a design vehiclethat is distributed to a component of the structure.

Dowel. A short length of round metal bar used to interconnect or attach two members and prevent movement and displacement.

Drift bolt. A drift pin with a head formed or welded at one end for driving.

Drift pin. A length of metal bar, either round or square, used to connect and hold in position timber members placed in contact. Drift pins are commonly driven in holes having a diameter slightly less than the pins.

Dry. As applied to wood, having a relatively low moisture content, by definition 19 percent for sawn lumber and 16 percent for glued laminated timber.

Dry rot. A term loosely applied to any dry, crumbly rot but especially to that which, when in an advanced stage, permits the wood to be crushed easily to a dry powder. The term is actually a misnomer for any decay, since all fungi require considerable moisture for growth.

Dual treatment. Treatment of wood to be used under severe conditions of exposure with two dissimilar preservatives (usually creosote and an inorganic arsenical) in two separate treating cycles.

Durability. A general term for permanence or resistance to deterioration. As applied to wood, its lasting qualities or permanence in service, with reference to its resistance to decay and other forms of deterioration.

Duration of load factor. A factor expressing the dependence of wood strength on the duration of the loading.

Earlywood. The portion of the annual growth ring that is formed during the early part of the growing season.

Edge. The narrow face of rectangular-shaped pieces of lumber. Eased edges mean slightly rounded corners. Lumber 4 inches or less in thickness is frequently shipped with eased edges unless otherwise specified.

Elastomeric. Having elastic, rubberlike properties.

Empty-cell process. Any process for impregnating wood with preservatives or chemicals in which air, trapped in the wood under pressure, is released to drive out part of the injected preservative or chemical. The aim is to obtain good preservative distribution in the wood and leave the cell cavities only partially filled, thus minimizing future bleeding.

Equilibrium. In statics, the condition in which the forces acting upon a body are such that no external effect (or movement) is produced.

Equilibrium moisture content (EMC). The moisture content at which wood neither gains nor loses moisture when surrounded by air at a given relative humidity and temperature.

Equivalent uniform load. A load having a constant intensity per unit length producing an effect equal or practically equal to that of one or more concentrated loads.

Factor of safety. A factor or allowance predicated by common engineering practice upon the failure stress or stresses assumed to exist in a structure or a member or part thereof. Its purpose is to provide a margin in the strength, rigidity, deformation, and endurance of a structure or its component parts compensating for irregularities existing in structural materials and workmanship or other unevaluated conditions.

Fatigue. The decrease in member strength when subjected to cyclical loading as compared to static loading.

Fiber saturation point (FSP). The stage in the drying or wetting of wood at which the cell walls are saturated and the cell cavities free from water. It applies to an individual cell or group of cells, not to whole boards. It is usually taken as approximately 30 percent moisture content, based on ovendry weight.

Fibril. A threadlike component of cell walls, visible under a light microscope.

Flashing. Metal sheets placed over the top of timber piles or posts to protect them from water. Also, metal sheets placed on the top of glulam beams to protect them from draining water at the joint between glulam deck panels.

Floorbeam. A beam located transverse to the bridge alignment that supports the deck or other components of the floor system.

Footing. The enlarged lower portion of a substructure that distributes the structure loads either to the earth or to supporting piles.

Foundation. The supporting material upon which the substructure portion of a bridge is placed.

Frame. A structure having its parts or members so arranged and secured that the entire assemblage may not be distorted when supporting the loads, forces, and physical pressures considered in its design.

Frass. Insect droppings.

Free water. Water (or moisture) contained in the cell cavities of wood.

Frost heave. The upward movement of soil from alternate freezing and thawing of retained moisture.

Full-cell process. Any process for impregnating wood with preservative chemicals in which a vacuum is drawn to remove air from the wood before admitting the preservative. This favors heavy adsorption and retention of preservatives in the treated portions.

Galleries. Tunnels made in wood by insects.

Girder. A flexural member that is the main or primary support for the structure. In general, a girder is any large beam, especially if built up.

Glue line. The layer of adhesive that attaches two adherents.

Glued-laminated timber (glulam). An engineered, stress-rated product of a timber laminating plant comprising assemblies of specially selected and prepared wood laminations securely bonded together with adhesives.

Grade. The designation of the quality of a manufactured piece of wood.

Grade mark. Identification of lumber with symbols or lettering to certify its quality or grade.

Grain. The direction, size, arrangement, appearance, or quality of the fibers in wood or lumber. To have a specific meaning the term must be qualified.

> **Close-grained.** Wood with narrow, inconspicuous annual rings.
>
> **Coarse-grained.** Wood with wide, conspicuous annual rings in which there is considerable difference between springwood and summerwood.
>
> **Cross-grained.** Wood in which the fibers deviate from a line parallel to the sides of the piece.
>
> **Edge-grained.** Lumber that has been sawn so that the wide surfaces extend approximately at right angles to the annual growth rings. Lumber is considered edge-grained when the rings form an angle of 45 to 90 degrees with the wide surface of the piece.
>
> **End-grained.** The grain as seen on a cut made at right angles to the direction of the fibers.
>
> **Flat-grained.** Lumber that has been sawn parallel to the pith and approximately tangent to the growth rings. Lumber is considered flat-grained when the annual growth rings make an angle of less than 45 degrees with the wide surface of the piece.
>
> **Straight-grained.** Wood in which the fibers run parallel to the axis of a piece.

Green. Freshly sawed or undried wood. Wood that has become completely wet after immersion in water would not be considered green, but may be said to be in the green condition.

Grillage. A platformlike construction or assemblage used to ensure distribution of loads upon unconsolidated soil material.

Gross vehicle weight (GVW). The maximum total weight of a traffic vehicle.

Gusset. A plate serving to connect the members of a joint and hold them in correct alignment and position.

Hanger. A tension element or member serving to suspend or support a portion of the floor system of a truss, arch, or suspension span.

Hardness. A property of wood that enables it to resist indentation.

Hardwood. Generally, one of the botanical groups of trees that have broad leaves, in contrast to the conifers or softwoods. The term has no reference to the actual hardness of the wood.

Heartwood. The wood extending from the pith to the sapwood, the cells of which no longer participate in the life processes of the tree.

Hydroplaning. Loss of contact between a tire and the deck surface when the tire planes or glides on a film of water covering the deck.

Impact. As applied to bridge design, a dynamic increment of stress equivalent in magnitude to the difference between the stresses produced by a static load and those produced by the same loads applied dynamically.

Incising. The practice of puncturing the lateral surfaces of wood as an aid in securing more uniform penetration of preservative.

Increment borer. An augerlike instrument with a hollow bit and equipped with an extractor used to sample wood internally without destroying the piece.

Inventory rating. The load capacity rating for a bridge that represents the vehicle load level that can safely utilize an existing structure for an indefinite period of time.

Isotropic. The quality of having properties that are independent of the direction in which they are measured; properties are equal in all directions.

Joint. The junction of two pieces of wood or veneer.

> **Butt joint.** An end joint formed by abutting the square ends of two pieces.
>
> **Edge joint.** The place where two pieces of wood are joined together edge to edge.
>
> **End joint.** The place where two pieces of wood are joined together end to end, commonly by scarf or finger jointing.
>
> **Face joint.** The joint occurring between the wide faces of lamination.
>
> **Finger joint.** An end joint made up of several meshing wedges or fingers of wood bonded together with an adhesive.
>
> **Lap joint.** A joint made by placing one member partly over another and bonding the overlapped portions.
>
> **Scarf joint.** An end joint formed by joining with glue the ends of two pieces that have been tapered or beveled to form sloping plane surfaces.
>
> **Starved joint.** A glue joint that is poorly bonded because an insufficient quantity of glue remained in the joint.

Joint efficiency. The strength of a joint expressed as a percentage of the strength of clear straight-grained material.

Juvenile wood. The wood formed adjacent to the pith.

Kiln. A chamber having controlled air flow, temperature, and relative humidity, for drying lumber, veneer, and other wood products.

Kiln-dried. Dried in a kiln with the use of artificial heat.

Kneebrace. A member engaging at its ends two other members which are joined at right angles, or approximately right angles. It serves to strengthen the joint and make it more rigid.

Knot. That portion of a branch or limb that has been surrounded by subsequent growth of the stem.

> **Encased knot.** A knot whose rings of annual growth are not intergrown with those of the surrounding wood.

> **Intergrown knot.** A knot whose rings of annual growth are completely intergrown with those of the surrounding wood.

Laminate. A product made by bonding together two or more layers (laminations) of material or materials.

Laminated veneer lumber (LVL). Lumber made by laminating veneers in which the grain of all the veneers is essentially parallel to the longitudinal axis of the piece (as opposed to plywood).

Laminated wood. An assembly made by bonding layers of veneer or lumber with an adhesive so that the grain of all laminations is essentially parallel.

> **Horizontally laminated.** Laminated wood in which the laminations are arranged with their wider dimension approximately perpendicular to the direction of load.

> **Vertically laminated.** Laminated wood in which the laminations are arranged with their wider dimension approximately parallel to the direction of load.

Laminating. The process of bonding laminations together with adhesive, including the preparation of the laminations, preparation and spreading of adhesive, assembly of laminations in packages, application of pressure, and curing.

Lamination. A full-width and full-length layer contained in a member bonded together with adhesive. It may be composed of one or several wood pieces in width or length.

Lateral bracing. The bracing assemblage engaging the chords and inclined end posts of a truss, or the longitudinal beams of a beam superstructure, in the horizontal or inclined plane of the members to function in resisting the transverse forces tending to produce lateral movement and deformation.

Latewood. The portion of the annual growth ring that is formed late in the growing season, after the earlywood formation has ceased.

Lignin. The thin cementing layer between wood cells. It is the second most abundant constituent of wood and is located principally in the secondary wall and the middle lamella of the cells.

Live load. A dynamic load that is applied to a structure suddenly or that is accompanied by vibration, oscillation, or other physical condition affecting its intensity.

Longitudinal. For bridges, the direction parallel to the bridge span. For wood, parallel to the direction of the wood fibers.

Lumber. The product of the saw and planing mill not further manufactured than by sawing, resawing, passing lengthwise through a standard planing machine, crosscutting to length, and matching.

> **Dimension lumber.** Lumber with a nominal thickness of from 2 up to but not including 5 inches and a nominal width of 2 inches or more.
>
> **Dressed lumber.** Lumber that has been surfaced by a planing machine on one or more sides or edges.
>
> **Factory and shop lumber.** Lumber intended to be cut up for use in further manufacture, not for structural engineered uses.
>
> **Machine stress rated (MSR) lumber.** A grade of structural lumber determined by measuring the stiffness of each piece by a grading machine.
>
> **Matched lumber.** Lumber that is edge dressed and shaped to make a close tongued-and-grooved joint at the edges or ends when laid edge to edge or end to end.
>
> **Rough lumber.** Lumber that has not been dressed (surfaced) but that has been sawed, edged, and trimmed.
>
> **Structural lumber.** Lumber that is intended for use where allowable properties are required. The grading of structural lumber is based on the strength of the piece as related to anticipated uses.
>
> **Visual stress grade lumber.** A grade of structural lumber determined by estimating the influence of strength-reducing characteristics by visual examination of the surfaces.
>
> **Yard lumber.** A little-used term for lumber of all sizes and patterns that is intended for general property requirements.

Lumen. In wood anatomy, the cell cavity.

Manufacturing defects. Includes all defects or blemishes that are produced in manufacturing.

Marine borers. Marine organisms that attack wood in the submerged portions of structures located in salt or brackish waters.

Modification factor. A multiplicative factor applied to tabulated stress for lumber and glulam to compensate for various design and/or use conditions.

Modulus of rupture (MOR). Maximum stress at the extreme fiber in bending, calculated from the maximum bending moment on the basis of an assumed stress distribution. In clear wood the value of the modulus of rupture is intermediate between the tensile and compressive strengths.

Moisture content (MC). The amount of water contained in the wood, usually expressed as a percentage of the weight of the ovendry wood.

Moisture meter. An electrical instrument used to indicate the moisture content of wood.

Mud sill. A single piece of timber or a unit composed of two or more timbers placed upon a soil foundation as a support for a column, framed bent, or other similar member of a structure.

Neutral axis. The axis of a member in bending along which the strain is zero. On one side of the neutral axis the fibers are in tension, on the other side they are in compression.

Nominal size. As applied to timber or lumber, the size by which it is known and sold in the market; often differs from the actual size.

Nondestructive evaluation (NDE). The measurement of mechanical properties using test procedures that do not destroy the tested material.

Nondestructive testing (NDT). See nondestructive evaluation.

Occasional pieces. In lumber shipments, not more than 10 percent of the pieces in a parcel or shipment.

Old growth. Timber in or from a mature, naturally established forest.

Operating rating. The load capacity rating for a bridge that represents the absolute maximum vehicle load level to which the structure may be subjected.

Orthotropic. Having unique and independent properties in three mutually orthogonal (perpendicular) planes of symmetry. A special case of anisotropy.

Overload. In general, any load that is in excess of the design load.

Peck. Pockets or areas of disintegrated wood caused by advanced stages of localized decay in the living tree. It is usually associated with cypress and incense cedar.

Penetrant. A liquid used as a carrier for a soluble wood preservative.

Penetration. The depth to which preservative enters the wood.

Pentachlorophenol (penta). A chlorinated phenol used as a wood preservative, usually in petroleum oil.

Pier. A substructure built to support the ends of the spans of a multiple-span superstructure at intermediate points between the abutments.

Pile. A shaftlike linear member driven into the earth through weak material to provide a secure foundation for structures built on soft, wet, or submerged sites.

> **Bearing pile.** A pile that receives its support in bearing through the tip or lower end.
>
> **Friction pile.** A pile that receives its support through friction resistance along its lateral surface.

Pile cap. A lumber or glulam member attached to the tops of several piles to provide support and a point of attachment for the superstructure or other structural components.

Pile shoe. A metal piece fixed to the penetration end of a pile to protect it from damage in driving and to facilitate penetration in very dense material.

Pitch. An accumulation of resinous material in wood.

Pitch pocket. A natural opening extending parallel to the annual growth rings that contains, or has contained, pitch, either solid or liquid.

Pitch streaks. A well-defined accumulation of pitch in a more or less regular streak in the wood of certain conifers.

Pith. The small, soft core occurring near the center of a tree trunk, branch, twig, or log.

Plank. A broad board, usually more than 1 inch thick, laid with its wide dimension horizontal and used as a bearing surface.

Pocket rot. Advanced decay in wood that appears in the form of a hole or pocket, usually surrounded by apparently sound wood.

Preservative. Any substance that, for a reasonable length of time, is effective in preventing the development and action of wood-rotting fungi, borers of various kinds, and harmful insects that deteriorate wood.

> **Preservative, oil-borne.** A wood preservative that is introduced into wood in the form of a solution in oil.
>
> **Preservative, oil-type.** Preservatives such as creosote, creosote/coal-tar solutions, creosote-petroleum solutions and oil-borne preservatives, or other preservatives strictly of an oily nature that are generally insoluble in water.
>
> **Preservative, waterborne.** A wood preservative that is introduced into wood in the form of a solution in water.

Press-lam. A type of laminated veneer lumber developed at the FPL.

Pressure process. Any process of treating wood in a closed container whereby the preservative or fire retardant is forced into the wood under pressures greater than atmospheric pressure. The American Wood Preservers' Association usually denotes pressure as greater than 50 lb/in^2.

Radial. A direction in wood that is coincident with a radius from the axis of the tree or log to the circumference. A radial section is a lengthwise section in a plane that passes through the centerline of the tree trunk.

Rays, wood. Strips of cells extending radially within a tree and varying in height from a few cells in some species to 4 inches or more in others. The rays serve primarily to store food and transport it horizontally in the tree.

Reaction wood. Wood with more or less distinctive anatomical characters, formed typically in parts of leaning or crooked stems and in branches. In hardwoods this consists of tension wood and in softwoods of compression wood.

Refractory. Very difficult to penetrate with wood preservatives.

Refusal point. The point beyond which the rate of absorption of preservatives in wood at the maximum permitted pressure and temperature is too slow to be significant.

Resin. Inflammable, water-soluble, vegetable substances, secreted by certain plants or trees, and characterizing the wood of many coniferous species. The term is also applied to synthetic organic products related to the natural resins.

Retention. The amount of preservative, in lb/ft^3, remaining in the wood immediately after completion of the treating operation.

Retort. A steel tank, commonly horizontal, in which wood is placed for pressure treatment.

Roadway. The portion of the bridge deck intended for use by vehicular and pedestrian traffic.

Sapwood. The wood of pale color near the outside of the log. Sapwood generally has no natural resistance to decay.

Saw kerf. (1) Grooves or notches made in cutting with a saw; (2) that portion of a log, timber, or otherpiece of wood removed by the saw in parting the material into two pieces; (3) an artificial, predetermined split of limited length made by sawing through and parallel to the axis of a piece, thus preventing the uncontrolled location and direction of a possible natural split or check.

Scupper. An opening in the bridge deck, commonly located adjacent to the curb or wheel guard, provided to drain water from the roadway.

Seasoning. Removing moisture from green wood to improve its serviceability.

Second growth. Timber that has grown after the removal, whether by cutting, fire, wind, or other agency, of all or a large part of the previous stand.

Service load. The vehicle live load used for design which represents the maximum load level that can use the structure on a continual basis.

Shake. A separation along the grain, the greater part of which occurs between the rings of annual growth. Usually considered to have occurred in the standing tree or during felling.

Simple span. A superstructure span having, at each end, a single pinned, roller, or hinged support designed to be unaffected by load transmission to or from an adjacent span or structure.

Skewed bridge. A bridge with a superstructure forming an angle other than 90 degrees with the direction of the stream channel or the substructure.

Slenderness ratio. Measure of stiffness of a member, expressed as the length of the member divided by its radius of gyration.

Soft rot. A special type of decay developing under very wet conditions in the outer wood layers, caused by cellulose-destroying microfungi.

Softwoods. Generally, one of the botanical groups of trees that in most cases have needlelike or scalelike leaves; the conifers, also the wood produced by such trees. The term has no reference to the actual hardness of the wood.

Span. When applied to the design of beam, girder, truss, or arch superstructures, the distance center to center of the end bearings or the distance between the lines of action of the reactions.

Specific gravity. Ratio of the density of a material to the density of water. In wood it is the ratio of the weight of wood to the weight of an equal volume of water, the volume being gross volume of the wood and not of the wood substance itself.

Split. A separation of the wood from the tearing apart of the wood cells.

Stain. A discoloration in wood that may be caused by such diverse agencies as micro-organisms, metal, or chemicals. The term also applies to materials used to impart color to wood.

Stiffness. Resistance to deformation by loads that cause bending stress.

Stirrup. A U-shaped rod, bar, or angle providing a stirruplike support for a member.

Strain. The distortion of a body produced by the application of one or more external forces.

Strength. The ability of a member to sustain stress without failure.

Strength ratio. The hypothetical ratio of the strength of a structural member to that which it would have if it contained no strength-reducing characteristics (knots, cross-grain, shake, and so forth).

Stress. The intensity of forces distributed over a given section measured as force per unit area.

Stress grades. Lumber grades having assigned working stress and modulus of elasticity values in accordance with accepted basic principles of strength grading.

Stringer. A longitudinal beam supporting the bridge deck.

Structural composite lumber. A structural reconstituted lumber-type product of uniform cross section comprised of parallel-to-the-grain veneer strands, strips, or sheets predominantly bonded together parallel to each other using exterior grade adhesive.

Substructure. The abutments, piers, bents, or other constructions built to support the superstructure and transmit loads to the foundation.

Superelevation. The transverse inclination of the roadway surface within a horizontal curve. The purpose of superelevation is to provide a means of resisting or overcoming the centrifugal forces from moving vehicles.

Superstructure. The entire portion of a bridge structure that primarily receives and supports highway, pedestrian, or other traffic loads and transfers the applied loads to the bridge substructure.

Surface-hardened. A condition of the surface of timbers that appears to be from improper seasoning and may result in resistance to penetration of preservatives. Sometimes incorrectly called case-hardened.

Tabulated stress. The permissible material stress tabulated in appropriate design specifications. Tabulated stresses must be adjusted by all applicable modification factors to arrive at the allowable stress used for design.

Tangential. The direction in wood coincident with a tangent at the circumference of a tree or the annual growth rings. A tangential section is a longitudinal section through a tree perpendicular to a radius.

Tension wood. A form of wood found in leaning trees of some hardwood species and characterized by the presence of gelatinous fibers and excessive longitudinal shrinkage.

Threshold. The minimum amount of wood preservative that is effective in preventing significant decay by a particular fungus.

Timbers. Lumber that is nominally 5 inches or more in least dimension.

Toughness. A quality of wood that permits the material to absorb a relatively large amount of energy, to withstand repeated shocks, and to undergo considerable deformation before breaking.

Tracheid. The elongated cells that constitute the greater part of the structure of the softwoods (frequently referred to as fibers).

Track width. The transverse center-to-center spacing between the wheels of a traffic vehicle, equal to the distance between two wheel lines.

Transverse. For bridges, the direction perpendicular to the bridge span. For wood, the direction at right angles to the wood fibers, including the radial and tangential directions.

Transverse bracing. The bracing assemblage between beams or columns that serves to resist and distribute lateral loads and provides support for stability of the members.

Trestle. A bridge structure consisting of beam or truss spans supported upon bents.

Truss. A jointed structure having an open web construction so arranged that the frame is divided into a series of triangles with members primarily stressed axially only.

Twist. A distortion caused by the turning or winding of the edges of a board so that the four corners of any face are no longer in the same plane.

Unseasoned. Wood that is freshly sawn from green logs, specifically wood not dried to 19 percent or lower moisture content.

Uplift. A negative reaction or a force tending to lift a beam, truss, pile, or any other bridge element upwards.

Veneer. A thin layer or sheet of wood.

Waler (or wale). A horizontal member used for bracing the sheeting of a trench, cofferdam, retaining wall, bulkhead, or similar structure.

Wane. Bark or lack of wood from any cause on edge or corner of a piece.

Warp. Any variation from a true or plane surface. Warp includes bow, crook, cup, and twist, or any combination thereof.

Wearing surface. A topmost layer or course of material applied upon a roadway to receive the traffic service loads and to resist the abrading, crushing, or other disintegrating action resulting therefrom.

> **Full wearing surface.** A wearing surface that covers the entire bridge deck.

> **Partial wearing surface.** A wearing surface that covers only the portion of the bridge deck intended for vehicle tracking.

Weathering. The mechanical or chemical disintegration and discoloration of the surface of wood caused by exposure to light, the action of dust and sand carried by winds, and the alternate shrinking and swelling of the surface fibers with the continual variation in moisture content brought by changes in the weather. Weathering does not include decay.

Web. The portion of a beam or truss, located between and connected to the flanges or the chords. It serves mainly to resist shear stress.

Wet-use. Use conditions where the moisture content of the wood in service exceeds 16 percent for glulam and 19 percent for sawn lumber.

Wheel guard. A timber member placed longitudinally along the side limit of the roadway to guide the movement of vehicle wheels and protect railings or other elements outside the roadway limit.

Wheel line. The series of wheel loads measured along the length of a traffic vehicle. The total weight of one wheel line is one-half the gross vehicle weight.

Wheel load. The total load transferred by one wheel of a design vehicle.

White rot. Any wood decay or rot attacking both the cellulose and the lignin, producing a generally whitish residue.

Wing wall. The retaining wall extension of an abutment intended to restrain and hold in place the side slope of an approach roadway embankment.

Working stress. The unit stress in a member under design load.

CHAPTER 18

TIMBER BRIDGE BIBLIOGRAPHY

American Association of State Highway and Transportation Officials. 1983. Manual for maintenance inspection of bridges. Washington, DC: American Association of State Highway and Transportation Officials. 50 p.

American Association of State Highway and Transportation Officials. 1983. Standard specifications for highway bridges. 13th ed. Washington, DC: American Association of State Highway and Transportation Officials. 394 p.

American Association of State Highway and Transportation Officials. 1976. AASHTO manual for bridge maintenance. Washington, DC: American Association of State Highway and Transportation Officials. 251 p.

American Institute of Timber Construction. 1988. Glulam bridge systems, a manual to assist in the design of glued laminated timber bridges. Vancouver, WA: American Institute of Timber Construction. 33 p.

American Institute of Timber Construction. 1985. Timber construction manual. 3d ed. New York: John Wiley and Sons, Inc. 836 p.

American Institute of Timber Construction. 1975. Glulam report, bridging a problem. Englewood, CO: American Institute of Timber Construction. 4 p.

American Institute of Timber Construction. 1973. Modem timber highway bridges, a state of the art report. Englewood, CO: American Institute of Timber Construction. 79 p.

American Institute of Timber Construction in conjunction with the Virginia Highway Research Council. [1973]. Typical timber bridge design and details. Englewood, CO: American Institute of Timber Construction. 10 p.

American Railway Engineering Association. 1966. Wood bridges and trestles. In: Manual of recommended practice. Chicago, IL: American Railway Engineering Association: Chapter 7.

American Society of Civil Engineers. 1986. Evaluation and upgrading of wood structures: case studies. New York: American Society of Civil Engineers. 111 p.

American Society of Civil Engineers. 1982. Evaluation, maintenance, and upgrading of wood structures. Freas, A., ed. New York: American Society of Civil Engineers. 428 p.

American Society of Civil Engineers. 1980. A guide for the field testing of bridges. ASCE Working Committee on Safety of Bridges. New York: American Society of Civil Engineers. 72 p.

American Society of Civil Engineers. 1976. American wooden bridges. ASCE Historical Pub. No. 4. New York: American Society of Civil Engineers. 176 p.

American Society of Civil Engineers. 1975. Wood structures, a design guide and commentary. New York: American Society of Civil Engineers. 416 p.

American Wood-Preservers' Association. 1941. Timber-concrete composite decks. Chicago: American Wood Preservers' Association. 28 p.

Archibald, R. 1952. A survey of timber highway bridges in the United States. Civil Engineering. September: 171-176.

Avent, R.R. 1986. Repair of timber bridge piling by posting and epoxy grouting. In: Trans. Res. Rec. 1053. Washington, DC: Transportation Research Board, National Research Council: 70-79.

Barnhart, J.E. 1986. Ohio's experiences with treated timber for bridge construction. In: Trans. Res. Rec. 1053. Washington, DC: Transportation Research Board, National Research Council: 56-58.

Bell, L.C.; Yoo, C.H. 1984. Seminar on fundamentals of timber bridge construction. Course notes. Auburn, AL: Auburn University. [150 p.]

Berger, R.H. 1978. Extending the service life of existing structures. In: Bridge Engineering Volume 1. Transportation Research Record 664. Washington, DC: Transportation Research Board, National Academy of Sciences: 47-55.

Better Roads. 1976. Glulam helping to solve America's bridge problem. Better Roads 46(5): 36-37.

Better Roads. 1976. World's longest laminated-timber bridge. Better Roads 46(5): 10.

Blew, J.O., Jr. 1961. What can be expected from treated wood in highway construction. Rep. No. 2235. Madison, WI: U.S. Department of Agriculture, Forest Service, Forest Products Laboratory. 16 p.

Bohannan, B. 1972. FPL timber bridge deck research. Journal of the Structural Division, American Society of Civil Engineers 98(ST3): 729-740.

Bohannan, B. 1972. Glued-laminated timber bridges reality or fantasy. Paper presented at the annual meeting of the American Institute of Timber Construction; 1972 March 13-16; Scottsdale, AZ. Madison, WI: U.S. Department of Agriculture, Forest Service, Forest Products Laboratory. 12 p.

Boomsliter, G.P.; Cather, C.H.; Worrell, D.T. 1951. Distribution of wheel loads on a timber bridge floor. Res. Bull. 24. Morgantown, WV: West Virginia University, Engineering Experiment Station. 31 p.

British Columbia Logging News. 1977. Log bridges. British Columbia Logging News. February: 774-777.

British Columbia Logging News. 1976. Changing art of bridge building. British Columbia Logging News 7(10): [p. 25].

British Columbia Lumberman. 1962. Timber bridges. British Columbia Lumberman 46(6): 10-14, 72-75.

Bruesch, L.D. 1982. Forest service timber bridge specifications. Journal of the Structural Division, American Society of Civil Engineers 108(ST12): 2737-2746.

Bruesch, L.D. 1977. Timber bridge systems. Paper presented at the 1977 FCP review conference on new bridge design concepts; October 3-7; Atlanta, GA. 7 p.

Canadian Institute of Timber Construction. 1970. Modern timber bridges, some standards and details. 3d ed. Ottawa, Can.: Canadian Institute of Timber Construction. 48 p.

Carsen, E.W.; Rankenburg, B. 1978. Nomograph for load rating log stringer bridges. U.S. Department of Agriculture, Forest Service. Engineering Field Notes 10(6): 15-18.

Civil Engineering. 1971. Who says wooden bridges are dead? Civil Engineering June: 53.

Clark, J.W.; Eslyn, W.E. 1977. Decay in wood bridges: inspection and preventive & remedial maintenance. Madison, WI: U.S. Department of Agriculture, Forest Service, Forest Products Laboratory. 51 p.

Commonwealth of Pennsylvania, Department of Transportation. 1984. Standard plans for low cost bridges. Series BLC-540, timber spans. Pub. No. 130. [Pittsburgh, PA]: Commonwealth of Pennsylvania, Department of Transportation. 28 p.

Construction Digest. 1976. New breed of glulam structures bringing back the 'old wooden bridge'. Construction Digest. February 5: 55-58.

Csagoly, P.F.; Taylor, R.J. 1979. A development program for wood highway bridges. 79-SRR-7. Downsview, ON, Canada: Ministry of Transportation and Communications. 57 p.

Culmann, K. 1968. Remington's wood bridges. Steinhaus, M., trans. Civil Engineering 38(3): 60-61.

Culmann, K. 1966. Brown's timber railroad bridges. Steinhaus, M., trans. Civil Engineering 36(11): 72-74.

Dimakis, A.G. 1966. New ideas for timber bridges. Thesis proposal. Madison, WI: University of Wisconsin-Madison, Department of Civil and Environmental Engineering. 58 p.

Doyle, D.V.; Wilkinson, T.L. 1969. Evaluating Appalachian woods for highway posts. Res. Pap. FPL 111. Madison, WI: U.S. Department of Agriculture, Forest Service, Forest Products Laboratory. 20 p.

Eby, R.E. 1986. Timber & glulam as structural materials, general history. Paper presented at the Engineered Timber Workshop; 1986 March 17; Portland, OR. 15 p.

Engineering [Can.] 1976. Timber! Engineering [Can.] March: 8.

Erickson, E.C.O.; Romstad, K.M. 1965. Distribution of wheel loads on timber bridges. Res. Pap. FPL 44. Madison, WI: U.S. Department of Agriculture, Forest Service, Forest Products Laboratory. 62 p.

Eslyn, W.E.; Clark, J.W. 1979. Wood bridges-decay inspection and control. Agric. Handb. 557. Washington, DC: U.S. Department of Agriculture, Forest Service. 32 p.

Forest Industries. 1976. Glulam, computerized engineering making B.C. log bridges obsolete. Forest Industries 103(2): 46-47.

Forest Industries [Can.]. 1975. CanCel bridge opens new timber country. Forest Industries July: [68].

Forest Products Journal. 1968. Wood structures...successful through many decades. Forest Products Journal 18(7): 13.

Freas, A.D. 1952. Laminated timber permits flexibility of design. Civil Engineering 22(9): 173-175.

Gand, W.W. 1965. Timber in highway bridges, opportunities and challenges. American Society of Civil Engineers Specialty Conference on Wood; 1965 June 9-11; Chicago, IL. U.S. Department of Agriculture, Forest Service, Pacific Northwest Region. 11 p.

Gower, E. 1984. The logging bridge standards dilemma. Logging and Sawmilling Journal [Can.] 15(2): 16-19.

Gower, L.E. 1986. Remaining glulam bridges should be inspected carefully. Logging and Sawmilling Journal [Can.] 17(8): 42-43.

Gower, L.E. 1979. Maintenance and inspection of logging bridges. White Rock, BC, Canada: Big Wheel Publications Ltd. 46 p.

Gower, L.E. 1977. Bridge location. B.C. [Can.] Logging News (1): 700-701. Gower, L.E. 1977. Calculating stress distribution. B.C. [Can.] Logging News (3): 808-811.

Gower, L.E. 1977. Log bridges of the future? B.C. [Can.] Logging News (4): 948-950.

Gromala, D.S.; Moody, R.C.; Sprinkel, M.M. 1985. Performance of a press-lam bridge a 5-year load testing and monitoring program. Res. Note FPL-0251. Madison, WI: U.S. Department of Agriculture, Forest Service, Forest Products Laboratory. 7 p.

Gurfinkel, G. 1981. Wood engineering. 2d ed. Dubuque, IA: Kendall/Hunt Publishing Co. 552 p.

McCutcheon, W.J.; Tuomi, R.L. 1973. Procedure for design of glued-laminated orthotropic bridge decks. Res. Pap. FPL 210. Madison, WI: U.S. Department of Agriculture, Forest Service, Forest Products Laboratory. 42 p.

McDonald, R.H.; Anderson, G.R. 1978. Island Lake Creek timber culvert. U.S. Department of Agriculture, Forest Service. Field Notes 10(5): 6-8.

McGee, D. 1975. The timber bridge inspection program in Washington State. Portland, OR: U.S. Department of Transportation, Federal Highway Administration, Region 10. 52 p.

McGee, W.D. 1975. Timber bridge inspection. Northwest Bridge Engineers Seminar; 1975 September 15-18; Boise, ID. Idaho Falls, ID: Argonne National Laboratory. 13 p.

Mielke, K.F. 1977. Experimental project for glued-laminated timber deck panels on highway bridges. Juneau, AK: State of Alaska, Department of Highways. [50 p.].

Millbank, P. 1974. Timber bridges. Civil Engineering. May: 37.

Ministry of Transportation and Communications. 1983. Ontario highway bridge design code. Downsview, ON, Canada: Ministry of Transportation and Communications. 357 p.

Ministry of Transportation and Communications. 1983. Ontario highway bridge design code commentary. Downsview, ON, Canada: Ministry of Transportation and Communications. 279 p.

Moody, R.C.; Tuomi, R.L.; Eslyn, W.E. [and others]. 1979. Strength of log bridge stringers after several year's use in southeast Alaska. Res. Pap. FPL 346. Madison, WI: U.S. Department of Agriculture, Forest Service, Forest Products Laboratory. 17 p.

Muchmore, F.W. 1986. Designing timber bridges for long life. In: Trans. Res. Rec. 1053. Washington, DC: Transportation Research Board, National Res. Council: 12-17.

Muchmore, F.W. 1984. Techniques to bring new life to timber bridges. Journal of Structural Engineering 110(8): 1832-1846.

Muchmore, F.W. 1983. Timber bridge maintenance, rehabilitation, and replacement. GPO 693-015. Missoula, MT: U.S. Department of Agriculture, Forest Service, Northern Region. 31 p.

Muchmore, F.W. 1976. Analysis and load rating of native log stringer bridges. U.S. Department of Agriculture, Forest Service. Field Notes 8(8): 26-30.

Muchmore, F.W. 1976. Design guide for native log stringer bridges. U.S. Department of Agriculture, Forest Service. Field Notes 8(8): 7-25.

Nagy, M.M.; Trebett, J.T.; Wellburn, G.V. 1980. Log bridge construction handbook. Vancouver, Canada: Forest Engineering Research Institute of Canada. 421 p.

Neilson, G. 1971. Rubbing shoulders with the past. DuPont Magazine 65(6): 10-13.

Nowak, A.S.; Taylor, R.J. 1986. Ultimate strength of timber deck bridges. In: Trans. Res. Rec. 1053. Washington, DC: Transportation Research Board, National Research Council: 26-30.

Oliva, M.G.; Dimakis, A. 1986. Behavior of a prestressed timber highway bridge. Madison, WI: University of Wisconsin, Department of Civil and Environmental Engineering. 32 p.

Oliva, M.G.; Dimakis, A.G.; Tuomi, R.L. 1985. Interim report: behavior of stressed-wood deck bridges. Report 85-1/A. Madison, WI: University of Wisconsin, College of Engineering, Structures and Materials Test Laboratory. 40 p.

Oliva, M.G.; Tuomi, R.L.; Dimakis, A.G. 1986. New ideas for timber bridges. In: Trans. Res. Rec. 1053. Washington, DC: Transportation Research Board, National Research Council: 59-64.

Ou, F.L. 1986. An overview of timber bridges. In: Trans. Res. Rec. 1053. Washington, DC: Transportation Research Board, National Research Council: 1-12.

Ou, F.L. 1985. The state of the art of timber bridges: a review of the literature. Washington, DC: U.S. Department of Agriculture, Forest Service. [30 p.].

Park, S.H. 1989. Bridge rehabilitation and replacement. Trenton, NJ: S.H. Park. 818 p.

Parry, J.D. 1986. A prefabricated modular timber bridge. In: Trans. Res. Rec. 1053. Washington, DC: Transportation Research Board, National Research Council: 49-55.

Parry, J.D. 1981. The Kenyan low cost modular timber bridge. TRRL Rep. 970. Crowthorne, Berkshire, England: Transport and Road Research Laboratory. 35 p.

Quimby, A.W. 1974. The Comish-Windsor covered bridge. The Plain Facts 2(1): 1-2.

Rear, G.W. 1935. Experience of Southern Pacific with treated timber in bridge construction. Wood Preserving News 13(4): 49-51, 56-57.

Sackowski, A.S. 1963. Reconstructing a covered timber bridge. Civil Engineering October: 36-39.

St. Regis Paper Company, Wheeler Division. 1979. Load test of wood bridge. Lab. Rep. No. 4-0997. St. Louis Park, MN: St. Regis Paper Company, Wheeler Division. 16 p.

Sanders, W.W. 1984. Distribution of wheel loads on highway bridges. National Cooperative Highway Research Program Synthesis of Highway Practice. No. 3. Washington, DC: National Academy of Sciences, Transportation Research Board. 22 p.

Sanders, W.W., Jr. 1980. Load distribution in glulam timber highway bridges. Report ISU-ERI-AMES-80124. Ames, IA: Iowa State University, Engineering Research Institute. 21 p.

Sanders, W.W., Jr.; Elleby, H.A. 1970. Distribution of wheel loads on highway bridges. Cooperative Highway Research Program. Rep. 83. Washington, DC: Highway Research Board, National Academy of Sciences. 56 p.

Sanders, W.W., Jr; Klaiber, F.W.; Wipf, T.J. 1985. Load distribution in glued laminated longitudinal timber deck highway bridges. Ames, IA: Iowa State University, Engineering Research Institute. 47 p.

Sanders, W.W.; Laboube, R.A.; Woodworth, J.R. 1978. Distribution of wheel loads on Alaska native log stringer bridges. Final report ISU-ERI-Ames-78185. Ames, IA: Iowa State University, Engineering Research Institute. 85 p.

Sanders, W.W.; Muchmore, F.W. 1978. Behavior of Alaskan native log stringer bridges. In: Bridge engineering. Trans. Res. Rec. 665. Washington, DC: Transportation Research Board, National Research Council: 228-235. Vol. 2.

Scales, W.H. 1959. Standard treated timber bridges. In: Standardization of highway bridges. Bull. No. 244. Washington, DC: American Road Builders' Associaton: 22-26.

Scarisbrick, R.G. 1976. Laminated timber logging bridges in British Columbia. Journal of the Structural Division, American Society of Civil Engineers. 102(ST1). [10 p.].

Schaffer, E.L.; Jokerst, R.W.; Moody, R.C. [and others]. 1977. Press-Lam: progress in technical development of laminated veneer structural products. Res. Pap. FPL 279. Madison, WI: U.S. Department of Agriculture, Forest Service, Forest Products Laboratory. 27 p.

Scholten, J.A. 1944. Structural timbers for bridge construction in Central America. Madison, WI: U.S. Department of Agriculture, Forest Service, Forest Products Laboratory. 24 p.

Schuessler, R. 1972. America's antique bridges. Passages, Northwest Orient's Inflight Magazine 3(1): 16-19,

Seiler, J.F. 1935. Timber bridge structures their economy, safety, and utility. Bull. 43-A. Washington, DC: American Road Builders Association. 16 p.

Selbo, M.L. 1966. Laminated bridge decking (progress report). Madison, WI: U.S. Department of Agriculture, Forest Service, Forest Products Laboratory. 32 p.

Selbo, M.L.; Knauss, A.C.; Worth, H.E. 1966. 20 years of service prove durability of pressure-treated glulam bridge members. Wood Preserving News 44(3): 5-8, 16.

Smith, A.K. 1945. Timber connectors in highway structures. British Columbia [Can.] Lumberman 29(4): 40-41, 104.

Sprinkel, M.M. 1985. Prefabricated bridge elements and systems. National Cooperative Highway Research Program, Synthesis of Highway Practice 119. Washington, DC: National Research Council, Transportation Research Board. 75 p.

Sprinkel, M.M. 1982. Final report of evaluation of the performance of a press-lam timber bridge. Bridge performance and load test after 5 years. VHTRC 82-R56. Charlottesville, VA: Virginia Highway and Transportation Research Council. 21 p.

Sprinkel, M.M. 1978. Evaluation of the performance of a press-lam timber highway bridge. Interim rep. 2. Charlottesville, VA: Virginia Highway and Transportation Research Council. 13 p.

Sprinkel, M.M. 1978. Glulam timber deck bridges. VHTRC 79-R26, Charlottesville, VA: Virginia Highway and Transportation Research Council. 33 p.

Stacey, W.A. 1949. Longitudinal laminated decks giving good service. Wood Preserving News 27(9): 109-112.

Stacey, W.A. 1935. The design of laminated timber bridge floors. Wood Preserving News 13(4): 44-46, 55-56.

Stone, M.F. [1975]. New concepts for short span panelized bridge design of glulam timber. Tacoma, WA: Weyerhaeuser Co. 8 p.

Sunset Foundry Company, Inc. [1970]. Deck brackets for treated timber bridges. AIA File 17F. Kent, WA: Sunset Foundry. 4 p.

Suprenant, B.A.; Videon, F.; Ehlert, R.E.; Jackson, A. 1986. Lateral stability considerations of timber beams in old bridges. In: Trans. Res. Rec. 1053. Washington, DC: Transportation Research Board, National Research Council: 18-25.

Taylor, R.J. 1984. Prestressed wood applications. Paper presented at Western Area Association of State Highway and Transportation Officials meeting; Rapid City, SD. 12 p.

Taylor, R.J. 1984. Design of wood bridges using the Ontario highway bridge design code. SRR-83-02 revised. Downsview, ON, Canada: Ministry Of Transportation and Communications. 24 p.

Taylor, R.J. 1983. Appendix to SRR-83-02. Downsview, ON, Canada: Ministry of Transportation and Communications. 108 p.

Taylor, R.J. 1983. Design of prestressed wood bridges using the Ontario highway bridge design code. SRR-83-03. Downsview, ON, Canada: Ministry of Transportation and Communications. 30 p.

Taylor, R.J. 1983. Wood bridge calibration study for the Ontario highway bridge design code. SRR-83-04. Downsview, ON, Canada: Ministry of Transportation and Communications. 37 p.

Taylor, R.J.; Batchelor, B.; Van Dalen, K. 1983. Prestressed wood bridges. SRR-83-01. Downsview, ON, Canada: Ministry of Transportation and Communications. 15 p.

Taylor, R.J.; Csagoly, P.F. 1979. Transverse post-tensioning of longitudinally laminated timber bridge decks. Downsview, ON, Canada: Ministry of Transportation and Communications. 16 p.

Taylor, R.J.; Walsh, H. 1984. A prototype prestressed wood bridge. SRR-83-07. Downsview, ON, Canada: Ministry of Transportation and Communications. 75 p.

Timber Structures, Inc. [1955]. Permanent timber bridges. Portland, OR: Timber Structures, Inc. 4 p.

Tuomi, R.L. 1980. Full-scale testing of wood structures. In: W.R. Schriever, ed. Full scale load testing of structures. ASTM STP 702. Washington, DC: American Society for Testing and Materials: 44-22.

Tuomi, R.L. 1976. Erection procedure for glued-laminated timber bridge decks with dowel connectors. Res. Pap. FPL 263. Madison, WI: U.S. Department of Agriculture, Forest Service, Forest Products Laboratory. 15 p.

Tuomi, R.L. 1972. Advancements in timber bridges through research and engineering. In: Proceedings, 13th annual Colorado State University bridge engineering conference; 1972; Ft. Collins, CO. Colorado State University: 34-61.

Tuomi, R.; Bohannan, B. 1971. Timber bridges go mod. Wood Preserving. 49(12): 4-9.

Tuomi, R.L.; McCutcheon, W.J. 1973. Design procedure for glued laminated bridge decks. Forest Products Journal 23(6): 36-42.

Tuomi, R.L.; Wolfe, R.W.; Moody, R.C.; Muchmore, F.W. 1979. Bending strength of large Alaskan sitka spruce and western hemlock log bridge stringers. Res. Pap. FPL 341. Madison, WI: U.S. Department of Agriculture, Forest Service, Forest Products Laboratory. 16 p.

U.S. Department of Agriculture, Forest Service. 1985. Forest Service specifications for construction of bridges & other major drainage structures. EM-7720-100B. Washington, DC: U.S. Department of Agriculture, Forest Service. 241 p.

U.S. Department of Agriculture, Forest Service, Northern Region. 1985. Bridge design manual. Missoula, MT: U.S. Department of Agriculture, Forest Service, Northern Region. 299 p.

U.S. Department of Transportation, Federal Highway Administration. 1979. Standard plans for highway bridges. Timber bridges. Washington, DC: U.S. Department of Transportation, Federal Highway Administration. 19 p. Vol. 3.

Victor, R.F. 1978. Orthotropic bridge saves old covered bridge. In: Bridge engineering. Trans. Res. Rec. 664. Washington, DC: Transportation Research Board, National Academy of Sciences: 80-85. Vol. 1.

Walford, G.B. 1975. What is happening to timber bridge decking? What's New in Forest Research 26 (June). Rotorua, New Zealand: Forest Research Institute of New Zealand. 4 p.

West Coast Lumbermen's Association. 1952. Highway structures of Douglas fir. Portland, OR: West Coast Lumbermen's Association. 55 p.

Western Builder. 1976. Bridges are big business. Western Builder. April 15. 2 p.

Weyerhaeuser Company. 1980. Weyerhaeuser glulam wood bridge systems. Tacoma, WA: Weyerhaeuser Co. 114 p.

Weyerhaeuser Company. [1980]. Digest of five research studies made on longitudinal glued laminated wood bridge decks. St. Paul, MN: Weyerhaeuser Co. 19 p.

Weyerhaeuser Company. 1976. Weyerhaeuser bridge deck bracket. SL-495. Tacoma, WA: Weyerhaeuser Co. 4 p.

Weyerhaeuser Company. 1975. Weyerhaeuser panelized bridge system for secondary roadways, highways & footbridges. SL-1318. Tacoma, WA: Weyerhaeuser Co. 4 p.

Weyerhaeuser Company. 1974. Weyerhaeuser panelized bridge system. SL-1318. Tacoma, WA: Weyerhaeuser Co. 4 p.

Weyerhaeuser News. 1944. Log bridge carries 62-ton loads. Weyerhaeuser News 8(3): 3.

Wheeler Consolidated, Inc. [1985]. Timber bridge design. St. Louis Park, MN: Wheeler Consolidated, Inc. 42 p.

White, K.R.; Minor, J.; Derocher, K.N.; Heins, C.P., Jr. 1981. Bridge maintenance inspection and evaluation. New York: Marcel Dekker, Inc. 257 p.

Wilson, J. 1951. Truck road bridges. Weyerhaeuser Magazine. September: 1-2.

Wipf, T.J.; Klaiber, F.W.; Sanders, W.W. 1986. Load distribution criteria for glued-laminated longitudinal timber deck highway bridges. In: Trans. Res. Rec. 1053. Washington, DC: Transportation Research Board, National Research Council: 31-40.

Wood [Eng.]. 1968. Bridge at Titchfield: an 82 ft. span prefabricated structure. Wood. January: 28-30.

Wood Construction and Building Materialist. 1969. Is there a market for covered bridges? Wood Construction and Building Materialist. March: 10-11.

Wood Preserving. 1964. Wilderness bridge. Wood Preserving 50(2): 5-7.

Wood Preserving News. 1969. Pressure-treated wood bridges win civil engineering award. Wood Preserving News 47(4): 12-22.

Wood Preserving News. 1964. A sturdy all timber bridge built for logging operations. Wood Preserving News 42(9): 5.

Wood Preserving News. 1964. Wood bridge features economy and appearance. Wood Preserving News 42(7): 9-10, 21.

Wood Preserving News. 1960. Typical low cost timber bridges. Wood Preserving News 38(1): 10-11.

Wood Preserving News. 1958. Why glulam timber bridges are popular. Wood Preserving News August: [18-22].

Woodworking Industry [Eng.]. 1975. Progress on timber bridges could open up new business. Woodworking Industry 32(9): 8-9.

Youngquist, J.A.; Gromala, D.S. 1978. Press-lam timbers for exposed structures. ASCE spring convention; 1978 April 24-28; Pittsburgh, PA. Madison, WI: U.S. Department of Agriculture, Forest Service, Forest Products Laboratory. 19 p.

Youngquist, J.A.; Gromala, D.S.; Jokerst, R.W. [and others]. 1979. Design, fabrication, testing, and installation of a press-lam bridge. Res. Pap. FPL 332. Madison, WI: U.S. Department of Agriculture, Forest Service, Forest Products Laboratory. 19 p.

U.S. GOVERNMENT PRINTING OFFICE: 1990-723-248/20300